T0282522

Deep Learning in Personalized Healthcare and Decision Support

Deep Learning in Personalized Healthcare and Decision Support

Edited by

Harish Garg
School of Mathematics, Thapar Institute of Engineering & Technology,
Patiala, Punjab, India

Jyotir Moy Chatterjee
Department of Information Technology, Lord Buddha Education Foundation,
Kathmandu, Nepal

Academic Press is an imprint of Elsevier
125 London Wall, London EC2Y 5AS, United Kingdom
525 B Street, Suite 1650, San Diego, CA 92101, United States
50 Hampshire Street, 5th Floor, Cambridge, MA 02139, United States
The Boulevard, Langford Lane, Kidlington, Oxford OX5 1GB, United Kingdom

Notices
Knowledge and best practice in this field are constantly changing. As new research and experience broaden our understanding, changes in research methods, professional practices, or medical treatment may become necessary.

Practitioners and researchers must always rely on their own experience and knowledge in evaluating and using any information, methods, compounds, or experiments described herein. In using such information or methods they should be mindful of their own safety and the safety of others, including parties for whom they have a professional responsibility.

To the fullest extent of the law, neither the Publisher nor the authors, contributors, or editors, assume any liability for any injury and/or damage to persons or property as a matter of products liability, negligence or otherwise, or from any use or operation of any methods, products, instructions, or ideas contained in the material herein.

ISBN: 978-0-443-19413-9

For information on all Academic Press publications visit our website at
https://www.elsevier.com/books-and-journals

Publisher: Stacy Masucci
Acquisitions Editor: Linda Versteeg-Buschman
Editorial Project Manager: Tracy I. Tufaga
Production Project Manager: Sajana Devasi P K
Cover Designer: Greg Harris

Typeset by TNQ Technologies

Contents

Contributors xi
Preface xv
Acknowledgments xvii

1. The future of health diagnosis and treatment: an exploration of deep learning frameworks and innovative applications

Imran Zafar, Syed Umair Ahmad, Mohd Ashraf Rather, Azmat Ali Khan, Qurat Ul Ain, Arfa Safder and Sheikh Arslan Sehgal

1. Introduction 1
2. Computational deep learning frameworks for health monitoring 3
3. Advanced architectures and core concepts of deep learning in smart health 5
4. Comparative analysis of deep learning frameworks for different disease detection 6
5. Advantages of deep learning in smart medical healthcare analytics 9
6. Deep leering applications for disease prediction 10
7. Deep learning in research and development 14
8. Future challenges of deep learning in smart health diagnosis and treatment 15
9. Limitations of deep learning frameworks 15
10. Conclusion and future scope 15

References 16

2. Fermatean fuzzy approach of diseases diagnosis based on new correlation coefficient operators

Paul Augustine Ejegwa and Arun Sarkar

1. Introduction 23
2. Fermatean fuzzy sets and their correlation operators 24
3. New Fermatean fuzzy correlation operators 27
4. Application example of medical diagnosis 32
5. Conclusion 36

References 36

3. Application of Deep-Q learning in personalized health care Internet of Things ecosystem

Yamuna Mundru, Manas Kumar Yogi and Jyotir Moy Chatterjee

1. Introduction 39
2. Related work 40
3. Proposed mechanism 42
4. Experimental results 44
5. Future directions 45
6. Conclusion 46

References 46

4. Dia-Glass: a calorie-calculating spectacles for diabetic patients using augmented reality and faster R-CNN

Natasha Tanzila Monalisa, Shinthi Tasnim Himi, Nusrat Sultana, Musfika Jahan, Nayeema Ferdous, Md. Ezharul Islam and Mohammad Shorif Uddin

1. Introduction 49
2. Related works 50
3. Diabetes: categories, concerns and prevalence 51
4. System methodology 52
5. Result and analysis 59
6. Conclusion 65

References 65

5. Synthetic medical image augmentation: a GAN-based approach for melanoma skin lesion classification with deep learning

Nirmala Veeramani and Premaladha J.

1. Introduction 69
2. Skin lesion classification methodology 71
3. Generation of synthetic skin lesions 73
4. Experimental results 76

5. Performance comparison 77
6. Conclusion and future work 79
7. Conflict of interest 79
References 79

6. Artificial intelligence representation model for drug–target interaction with contemporary knowledge and development

M. Arvindhan, A. Daniel, N. Partheeban and Balamurugan Balusamy

1. Introduction 81
2. AI privacy and security challenges 82
3. Ensuring transparency, explain ability, and intelligibility 83
4. Drug discovery and precision medicine with deep learning 84
5. Clinical decision support and predictive analytics 85
6. Natural language processing in drug 89
7. Predictive analytics has a wide range of practical applications, including the following 91
8. Conclusion 93
References 93

7. Review of fog and edge computing–based smart health care system using deep learning approaches

Mamata Rath, Subhranshu Sekhar Tripathy, Niva Tripathy, Chhabi Rani Panigrahi and Bibudhendu Pati

1. Introduction 95
2. Literature review 96
3. Healthcare using artificial intelligence 97
4. Efficient health care system with improved performance 99
5. Conclusion 104
References 104

8. Deep learning in healthcare: opportunities, threats, and challenges in a green smart environment solution for smart buildings and green cities—Towards combating COVID-19

Palak Maurya, Anurag Verma and Ranitesh Gupta

1. Introduction 107
2. Green infrastructure measures in the legislature 108
3. Employment creation as part of the sustainable recovery 109
4. Carbon power 110
5. Smart buildings 112
6. Climate change disclosure laws 112
7. The action of biodiversity 113
8. Case study 114
9. Future pandemic preparedness 114
10. Recent literature 115
11. Conclusions 115
References 116
Further reading 117

9. Hybrid and automated segmentation algorithm for malignant melanoma using chain codes and active contours

Nirmala Veeramani, Premaladha J., Raghunathan Krishankumar and Kattur Soundarapandian Ravichandran

1. Introduction 119
2. Materials and methods 120
3. Proposed methodology 122
4. Results and discussions 125
5. Conclusion and future scope 126
References 128

10. Development of a predictive model for classifying colorectal cancer using principal component analysis

Micheal Olaolu Arowolo, Happiness Eric Aigbogun, Eniola Precious Michael, Marion Olubunmi Adebiyi and Amit Kumar Tyagi

1. Introduction 131
2. Related works 132
3. Methodology 133
4. Results and discussions 136
5. Conclusion 142
References 142

11. Using deep learning via long-short-term memory model prediction of COVID-19 situation in India

Saroja Kumar Rout, Bibhuprasad Sahu, Amar Kumar Das, Sachi Nandan Mohanty and Ashish K. Sharma

1. Introduction 143
2. Literature review 144
3. How to protect yourself from COVID-19 147
4. Facts about the vaccine against the COVID-19 149

5. Materials and methods 152
6. Results discussion 155
7. Conclusion 159
References 161

12. Post-COVID-19 Indian healthcare system: Challenges and solutions

Sanjeev Kumar Mathur, Akash Saxena, Ali Wagdy Mohamed, Karam M. Sallam and Shivani Mathur

1. Creation of robust healthcare system—A nationwide priority and its emergence 163
2. Pandemonium scenes 163
3. Efforts for healthcare system development 164
4. Providing treatment to all amidst difficulties 165
5. Corona warriors and their woes 165
6. Transformation of Indian healthcare sector post COVID-19 165
7. Healthcare component 1—Hospitals 166
8. Healthcare component 2—Pharmaceutical industry 167
9. Healthcare component 3—Medical devices and equipment 167
10. Healthcare component 4—Diagnostics 168
11. Healthcare component 5—Telemedicine 168
12. Healthcare component 6—medical insurance 168
13. Healthcare component 7—Medical tourism 169
14. No need for complacency 169
15. Strength weakness threat opportunity (SWOT) analysis of Indian healthcare industry in the post-COVID-19 era 171
16. Future of Indian healthcare system 172
17. Conclusion 172
References 172

13. SWOT Perspective of the Internet of Healthcare Things

M.S. Sadiq, I.P. Singh and M.M. Ahmad

1. Introduction 175
2. The IoT in healthcare systems 176
3. Key merits of the IoT during the COVID-19 pandemic 177
4. IoT processes for combatting COVID-19 178
5. The IoT's overall impacts in relation to COVID-19 concerns 178
6. Global technology developments to quickly treat COVID-19 cases 179
7. IoT applications to fight COVID-19 179
8. IoT challenges in the aftermath of COVID-19 181
9. Strengths, weaknesses, opportunities, and threats (SWOT) analysis 182
10. An answer to the problems of the COVID-19 battle 183
11. Conclusion 185
References 185
Further reading 186

14. Deep learning for clinical decision-making and improved healthcare outcome

Russell Kabir, Haniya Zehra Syed, Divya Vinnakota, Madhini Sivasubramanian, Geeta Hitch, Sharon Akinyi Okello, Sharon-Shivuli-Isigi, Amal Thomas Pulikkottil, Ilias Mahmud, Leila Dehghani and Ali Davod Parsa

1. Introduction 187
2. Application of deep learning machine (DLM) in clinical decision-making in the diagnosis of different diseases 188
3. Application of DLM in patient care plan in chronic disease management and rehabilitation 189
4. Application in cardiovascular medicine 190
5. Application in Pulmonary Disease 190
6. Role of DLM in population level future disease prediction 194
7. Effectiveness of DLM in screening and referral of clinical cases in remote and poor access areas 195
8. Conclusion 195
References 196
Further reading 201

15. Development of a no-regret deep learning framework for efficient clinical decision-making

Yamuna Mundru, Manas Kumar Yogi, Jyotir Moy Chatterjee, Madhur Meduri and Ketha Dhana Veera Chaitanya

1. Introduction 203
2. Related works 207
3. Proposed mechanism 208
4. Experimental results 211
5. Future directions 212
6. Conclusion 212
References 213

16. Symptom-based diagnosis of diseases for primary health check-ups using biomedical text mining

B. Rakesh and Sukanta Nayak

1. Introduction 215
2. Natural language processing 216
3. System architecture of primary diagnosis of diseases 217
4. Case study 221
5. Results and discussion 222

References 222

17. "Deep learning" for healthcare: Opportunities, threats, and challenges

Russell Kabir, Madhini Sivasubramanian, Geeta Hitch, Saira Hakkim, John Kainesie, Divya Vinnakota, Ilias Mahmud, Ehsanul Hoque Apu, Haniya Zehra Syed and Ali Davod Parsa

1. Introduction 225
2. Machine learning and deep learning 225
3. Deep learning and its opportunities in healthcare 226
4. Data analysis and statistical modules 227
5. Deep learning and immunization management 228
6. Role in "COVID-19" vaccination 228
7. Role in future disease prediction 229
8. Public health misinformation 229
9. Contact tracing and pandemic management 230
10. Remote healthcare 230
11. Telemedicine 230
12. Virtual healthcare platforms 231
13. Online healthcare education: opportunities and challenges 231
14. Hospital, clinical research, and healthcare facilities 231
15. Electronic health records (EHRs) 232
16. Clinical imaging 234
17. Genomics 236
18. Drug discovery and precision medicine 236
19. Biobank 237
20. Conclusion 238

References 239

Further reading 244

18 Deep learning IoT in medical and healthcare

Ashwani Sharma, Anjali Sharma, Reshu Virmani, Girish Kumar, Tarun Virmani and Nitin Chitranshi

1. Introduction 245
2. Review of literature 246
3. Deep learning: Algorithms and its application 247
4. IoT and its applications 252
5. Conclusion 253

References 257

19. Deep learning in drug discovery

Meenu Bhati, Tarun Virmani, Girish Kumar, Ashwani Sharma and Nitin Chitranshi

1. Introduction 263
2. Literature review 265
3. Approaches of machine and deep learning for drug target 267
4. Practical applications of QSAR-based virtual screening 270
5. Conclusion and future scope 272

References 273

20. Avant-garde techniques in machine for detecting financial fraud in healthcare

S. Geetha, G. Soniya Priyatharsini, N. Ethiraj and G. Victo Sudha George

1. Introduction 277
2. Survey of frequently used algorithms for machine learning 279
3. Background study 280
4. Materials and methods 280
5. Conclusion 283
6. Future work 283

References 283

21. Predicting mental health using social media: A roadmap for future development

Ramin Safa, S.A. Edalatpanah and Ali Sorourkhah

1. Introduction 285
2. Mental disorders and big social data 286
3. Assessment strategies 289
4. Social data configuration 291

5. Prediction algorithms 293
6. Evaluation 297
7. Status, challenges, and future direction 298
8. Conclusion 299
References 299

22. Applied picture fuzzy sets with its picture fuzzy database for identification of patients in a hospital

Van Hai Pham, Quoc Hung Nguyen, Kim Phung Thai and Le Phuc Thinh Tran

1. Introduction 305
2. Research background 306
3. The proposed model 308
4. Experimental results 310
5. Conclusions 312
Acknowledgments 312
References 312

23. A deep learning framework for surgery action detection

Prabu Selvam and Joseph Abraham Sundar K

1. Introduction 315
2. Literature review 315
3. Proposed approach 317
4. YOLOv5 317
5. EfficientDet 318
6. EfficientNet 319
7. Proposed ensemble network 319
8. Experiment 320
9. Result and discussion 320
10. Conclusion 327
References 327
Further reading 328

24. Understanding of healthcare problems and solutions using deep learning

Rajesh Kumar Shrivastava, Simar Preet Singh, Simranjit Singh and Mohit Sajwan

1. Introduction 329
2. Related work 330
3. Methods 331
4. Standard methods for learning 333
5. Conclusion and future directions 340
References 340

25. Deep convolution classification model-based COVID-19 chest CT image classification

R. Sujatha and Jyotir Moy Chatterjee

1. Introduction 343
2. Related works 345
3. Proposed architecture and workflow 345
4. Experiment 349
5. Experimental results and discussions 350
6. Conclusions and future work 354
References 355

26. Internet of Medical Things in curbing pandemics

M.S. Sadiq, I.P. Singh and M.M. Ahmad

1. Introduction 357
2. Cognitive Internet of Things (CIoT) 358
3. Internet of Medical Things 361
4. Smart e-healthcare 363
5. IoMTS emerging technologies 364
6. POC biosensing for infectious illnesses with IoMT assistance 365
7. IoT's primary benefits for the COVID-19 pandemic 366
8. Significant applications of IoT for COVID-19 pandemic 366
9. Technologies of IoT for the healthcare during COVID-19 pandemic 366
10. IoT-enabled healthcare helpful during COVID-19 pandemic 368
11. Conclusion 369
References 369

Index 373

Contributors

Marion Olubunmi Adebiyi, Department of Computer Science, Landmark University, Omu-Aran, Kwara, Nigeria

Syed Umair Ahmad, Department of Bioinformatics, Hazara University Mansehra, Pakistan

M.M. Ahmad, Department of Agricultural Economics and Extension, BUK, Kano, Nigeria

Happiness Eric Aigbogun, Department of Computer Science, Landmark University, Omu-Aran, Kwara, Nigeria

Micheal Olaolu Arowolo, Department of Electrical Engineering and Computer Science, University of Missouri, Columbia, SC, United States; Department of Computer Science, Landmark University, Omu-Aran, Kwara, Nigeria

M. Arvindhan, Galgotias University, Greater Noida, Uttar Pradesh, India

Balamurugan Balusamy, Shiv Nadar University, Greater Noida, Uttar Pradesh, India

Meenu Bhati, School of Pharmaceutical Sciences, MVN University, Palwal, Haryana, India

Ketha Dhana Veera Chaitanya, Pragati Engineering College, Surampalem, Andhra Pradesh, India

Jyotir Moy Chatterjee, Department of Information Technology, Lord Buddha Education Foundation (Asia Pacific University), Kathmandu, Nepal

Nitin Chitranshi, Faculty of Medicines, Health and Human Sciences, Macquarie University, Sydney, NSW, Australia

A. Daniel, ASET- CSE, Amity University, Gwalior, Madhya Pradesh, India

Amar Kumar Das, Department of Mechanical Engineering, Gandhi Institute for Technology, Bhubaneswar, Odisha, India

Leila Dehghani, CRN Eastern (NIHR), Cambridge University Hospitals NHS Foundation Trust, Cambridge, United Kingdom

S.A. Edalatpanah, Department of Applied Mathematics, Ayandegan Institute of Higher Education, Tonekabon, Mazandaran, Iran

Paul Augustine Ejegwa, Department of Mathematics, University of Agriculture, Makurdi, Nigeria

N. Ethiraj, Department of Mechanical Engineering, Dr.M.G.R Educational and Research Institute of Technology, Madhuravoyal, Chennai, Tamil Nadu, India

Nayeema Ferdous, Department of Computer Science and Engineering Jahangirnagar University, Savar, Dhaka, Bangladesh

S. Geetha, Department of Computer Science and Engineering, Dr.M.G.R Educational and Research Institute of Technology, Madhuravoyal, Chennai, Tamil Nadu, India

Ranitesh Gupta, Electrical Engineering Department, Institute of Engineering and Technology, Lucknow, Uttar Pradesh, India

Saira Hakkim, Faculty of Health Sciences and Wellbeing, University of Sunderland, London, United Kingdom

Shinthi Tasnim Himi, Department of Computer Science and Engineering Jahangirnagar University, Savar, Dhaka, Bangladesh

Geeta Hitch, Faculty of Health Sciences and Wellbeing, University of Sunderland, London, United Kingdom

Ehsanul Hoque Apu, Department of Biomedical Engineering, Institute of Quantitative Health Science and Engineering, Michigan State University, East Lansing, MI, United States; Division of Hematology and Oncology, Department of Internal Medicine, Michigan Medicine, University of Michigan, Ann Arbor, MI, United States

Md. Ezharul Islam, Department of Computer Science and Engineering Jahangirnagar University, Savar, Dhaka, Bangladesh

Premaladha J., School of Computing, SASTRA Deemed to be University, Thanjavur, Tamil Nadu, India

Musfika Jahan, Department of Computer Science and Engineering Jahangirnagar University, Savar, Dhaka, Bangladesh

Joseph Abraham Sundar K, SASTRA Deemed University, Thanjavur, India

Russell Kabir, School of Allied Health, Faculty of Health, Education, Medicine and Social Care, Anglia Ruskin University, Essex, United Kingdom

John Kainesie, Faculty of Health Sciences and Wellbeing, University of Sunderland, London, United Kingdom

Azmat Ali Khan, Pharmaceutical Biotechnology Laboratory, Department of Pharmaceutical Chemistry, College of Pharmacy, King Saud University, Riyadh, Saudi Arabia

Raghunathan Krishankumar, Department of Computer Science and Engineering, Amrita School of Computing, Coimbatore, Amrita Vishwa Vidyapeetham, India

Girish Kumar, School of Pharmaceutical Sciences, MVN University, Palwal, Haryana, India

Ilias Mahmud, College of Public Health and Health Informatics, Qassim University, Buraydah, Saudi Arabia; Department of Public Health, College of Public Health and Health Informatics, Qassim University, Al Bukairiyah, Saudi Arabia

Sanjeev Kumar Mathur, Faculty of Management & Commerce, Poornima University, Jaipur, Rajasthan, India

Shivani Mathur, Human Development and Family Studies, University of Rajasthan, Jaipur, Rajasthan, India

Palak Maurya, Electrical Engineering Department, Institute of Engineering and Technology, Lucknow, Uttar Pradesh, India

Madhur Meduri, Pragati Engineering College, Surampalem, Andhra Pradesh, India

Eniola Precious Michael, Department of Computer Science, Landmark University, Omu-Aran, Kwara, Nigeria

Ali Wagdy Mohamed, Operations Research Department, Faculty of Graduate Studies for Statistical Research, Cairo University, Giza, Egypt; Department of Mathematics and Actuarial Science, School of Sciences Engineering, The American University, Cairo, Egypt

Sachi Nandan Mohanty, School of Computer Science & Engineering (SCOPE), VIT-AP University, Amaravati, Andhra Pradesh, India

Natasha Tanzila Monalisa, Department of Computer Science and Engineering Jahangirnagar University, Savar, Dhaka, Bangladesh

Yamuna Mundru, CSE Department, Pragati Engineering College, Surampalem, Andhra Pradesh, India

Sukanta Nayak, School of Advanced Sciences, VIT-AP University, Amaravati, Andhra Pradesh, India; Center of Excellence, AI and Robotics, VIT-AP University, Amaravati, Andhra Pradesh, India

Quoc Hung Nguyen, University of Economics, Ho Chi Minh City (UEH), Ho Chi Minh City, Viet Nam

Sharon Akinyi Okello, School of Allied Health, Faculty of Health, Education, Medicine and Social Care, Anglia Ruskin University, Essex, United Kingdom

Chhabi Rani Panigrahi, Department of Computer Science, Rama Devi Women's University, Bhubanewar, Odisha, India

Ali Davod Parsa, School of Allied Health, Faculty of Health, Education, Medicine and Social Care, Anglia Ruskin University, Essex, United Kingdom

N. Partheeban, Galgotias University, Greater Noida, Uttar Pradesh, India

Bibudhendu Pati, Department of Computer Science, Rama Devi Women's University, Bhubanewar, Odisha, India

Van Hai Pham, Hanoi University of Science and Technology, Hanoi, Viet Nam

G. Soniya Priyatharsini, Department of Computer Science and Engineering, Dr.M.G.R Educational and Research Institute of Technology, Madhuravoyal, Chennai, Tamil Nadu, India

Amal Thomas Pulikkottil, School of Allied Health, Faculty of Health, Education, Medicine and Social Care, Anglia Ruskin University, Essex, United Kingdom

B. Rakesh, School of Advanced Sciences, VIT-AP University, Amaravati, Andhra Pradesh, India

Mamata Rath, Department of Computer Science and Engineering, DRIEMS (Autonomous), Cuttack, Odisha, India

Mohd Ashraf Rather, Division of Fish Genetics and Biotechnology, Faculty of Fisheries Rangil Ganderbal, Sher-e-Kashmir University of Agricultural Science and Technology, Jammu and Kashmir, India

Kattur Soundarapandian Ravichandran, Department of Mathematics, Amrita School of Physical Sciences, Coimbatore, Amrita Vishwa Vidyapeetham, India

Saroja Kumar Rout, Department of Information Technology, Vardhaman College of Engineering (Autonomous), Hyderabad, Telangana, India

M.S. Sadiq, Department of Agricultural Economics and Extension, FUD, Dutse, Jigawa, Nigeria

Ramin Safa, Department of Computer Engineering, Ayandegan Institute of Higher Education, Tonekabon, Mazandaran, Iran

Arfa Safder, Institute of Molecular Biology and Biotechnology, The University of Lahore Sialkot, Punjab, Pakistan

Bibhuprasad Sahu, Department of Artificial Intelligence & Data Science, Vardhaman College of Engineering (Autonomous), Hyderabad, Telangana, India

Mohit Sajwan, Department of Information Technology, Netaji Subhash University of Technology, Dwarka, Delhi, India

Karam M. Sallam, School of IT and Systems, University of Canberra, Canberra, ACT, Australia

Arun Sarkar, Department of Mathematics, Heramba Chandra College, Kolkata, India

Akash Saxena, School of Engineering and Technology, Central University of Haryana, Mahendergarh, India

Sheikh Arslan Sehgal, Department of Bioinformatics, The Islamia University of Bahawalpur, Bahawalpur, Pakistan; Department of Bioinformatics, University of Okara, Okara, Pakistan

Prabu Selvam, SASTRA Deemed University, Thanjavur, India

Ashwani Sharma, School of Pharmaceutical Sciences, MVN University, Palwal, Haryana, India

Anjali Sharma, School of Pharmaceutical Sciences, MVN University, Palwal, Haryana, India

Ashish K. Sharma, Department of Computer Science & Engineering, Bajaj Institute of Technology (BIT), Wardha, Maharashtra, India

Sharon-Shivuli-Isigi, School of Allied Health, Faculty of Health, Education, Medicine and Social Care, Anglia Ruskin University, Essex, United Kingdom

Rajesh Kumar Shrivastava, School of Computer Science Engineering and Technology (SCSET), Bennett University, Greater Noida, Uttar Pradesh, India

I.P. Singh, Department of Agricultural Economics, SKRAU, Bikaner, Rajasthan, India

Simar Preet Singh, School of Computer Science Engineering and Technology (SCSET), Bennett University, Greater Noida, Uttar Pradesh, India

Simranjit Singh, School of Computer Science Engineering and Technology (SCSET), Bennett University, Greater Noida, Uttar Pradesh, India

Madhini Sivasubramanian, Faculty of Health Sciences and Wellbeing, University of Sunderland, London, United Kingdom

Ali Sorourkhah, Department of Management, Ayandegan Institute of Higher Education, Tonekabon, Mazandaran, Iran

R. Sujatha, School of Information Technology and Engineering, Vellore Institute of Technology, Vellore, Tamil Nadu, India

Nusrat Sultana, Department of Computer Science and Engineering Jahangirnagar University, Savar, Dhaka, Bangladesh

Haniya Zehra Syed, School of Allied Health, Faculty of Health, Education, Medicine and Social Care, Anglia Ruskin University, Essex, United Kingdom

Kim Phung Thai, University of Economics, Ho Chi Minh City (UEH), Ho Chi Minh City, Viet Nam

Le Phuc Thinh Tran, University of Economics, Ho Chi Minh City (UEH), Ho Chi Minh City, Viet Nam

Subhranshu Sekhar Tripathy, Department of Computer Science and Engineering, DRIEMS (Autonomous), Cuttack, Odisha, India

Niva Tripathy, Department of Computer Science and Engineering, DRIEMS (Autonomous), Cuttack, Odisha, India

Amit Kumar Tyagi, Department of Computer Science, Vellore Institute of Technology, Chennai, Tamil Nadu, India; Department of Fashion Technology, National Institute of Fashion Technology, New Delhi, India

Mohammad Shorif Uddin, Department of Computer Science and Engineering Jahangirnagar University, Savar, Dhaka, Bangladesh

Qurat Ul Ain, Department of Chemistry, Government College Women University Faisalabad (GCWUF), Punjab, Pakistan

Nirmala Veeramani, School of Computing, SASTRA Deemed to be University, Thanjavur, Tamil Nadu, India

Anurag Verma, Electrical Engineering Department, Institute of Engineering and Technology, Lucknow, Uttar Pradesh, India

G. Victo Sudha George, Department of Computer Science and Engineering, Dr.M.G.R Educational and Research Institute of Technology, Madhuravoyal, Chennai, Tamil Nadu, India

Divya Vinnakota, Faculty of Health Sciences and Wellbeing, University of Sunderland, London, United Kingdom

Tarun Virmani, School of Pharmaceutical Sciences, MVN University, Palwal, Haryana, India

Reshu Virmani, School of Pharmaceutical Sciences, MVN University, Palwal, Haryana, India

Manas Kumar Yogi, CSE Department, Pragati Engineering College, Surampalem, Andhra Pradesh, India

Imran Zafar, Virtual University of Pakistan, Bioinformatics and Computational Biology, Virtual University of Pakistan, Lahore, Punjab, Pakistan

Preface

Healthcare today is known to suffer from siloed and fragmented data, delayed clinical communications, and disparate workflow tools due to the lack of interoperability caused by vendor-locked healthcare systems, lack of trust relationships among data holders, and security/privacy concerns regarding data sharing. The present generation and scenario are a time for big leaps and bounds in terms of growth and advancement of the health information industry. This book is an attempt to unveil the hidden potential of the vast health information and technology. Through this book, we attempt to combine numerous compelling views, guidelines, and frameworks on enabling personalized healthcare service options through the successful application of deep learning (DL) frameworks. The progress of the healthcare sector shall be incremental as it learns from associations between data over time through the application of suitable artificial intelligence (AI), deep net frameworks, and patterns. The major challenge healthcare is facing is effective and accurate learning of unstructured clinical data through the application of precise algorithms. Incorrect input data leading to erroneous outputs with false positives shall be intolerable in healthcare as patients' lives are at stake. This book is being formulated with the intent to uncover the stakes and possibilities involved in realizing personalized healthcare services through efficient and effective DL algorithms. The specific focus of this book will be on the application of DL in any area of healthcare, including clinical trials, telemedicine, health records management, etc. For this book, we have only considered the articles that focus on the intersection of DL, healthcare, and computer engineering (CE) approaches. The features of the book will be the following:

- Machine learning: better billing/coding error detection (leading to reduced claims denials); optimization of the supply chain for pharmaceuticals; and enhanced data mining, thus guiding effective diagnosis.
- NLP: Text-to-speech and vice versa, document and data conversions, patient notes, processing of unstructured data, and query support systems.
- Deep learning and cognitive computing tools: Enable processing of very large data sets, help with a precise and comprehensive forecast of risks, and deliver recommended actions that improve outcomes for consumers.
- It is a novel application domain of deep learning that is of prime importance to human civilization. It has been predicted as the next big thing in personal health monitoring by the government as well as Forbes.
- Engineers gain insights into the real-world usage of the products they designed with all the benefits that bring.

Chapter 1 offered a cutting-edge perspective on current developments in DL and how they are being implemented in healthcare systems to achieve various objectives in medicine. In Chapter 2, two new correlation coefficient operators that reliably measure the correlation between any two arbitrary FFSs are developed for disease diagnosis. Chapter 3 features the clinical utilizations of Deep Q-learning procedures for patient therapy choice and other suitable medication aspects. Chapter 4 proposed a calorie-estimation system named as Dia-Glass which consists of Augmented Reality (AR)—based spectacles that can capture pictures of food in real-time to detect food type and estimate calorie through a deep convolutional neural network technique. Chapter 5 presents a deep neural network (DNN) that scales up from efficiency and is based on CNN with VGG-16 architecture to create high-quality targeted skin lesions from dermoscopy images. Chapter 6 examined AI applications from the past, present, and future, with a focus on the prediction of drug synergy in hematological malignancies using DNNs. Chapter 7 gives a thorough analysis of the device knowledge and subterranean erudition techniques utilized to address various tribulations in submissions connected to fog and edge computing.

The major findings of Chapter 8 are that it provides alternative solutions for modern-day's problems related to energy like renewables being the new alternative to coal, smart buildings being the future for clean energy, Saudi Arabia's Vision 2030, and the NEOM project. Chapter 9 presents a hybrid and automated segmentation algorithm to precisely separate the affected region from a melanoma skin lesion. Chapter 10 gives a review of colon-rectal cancer prediction techniques using machine learning classifiers. Chapter 11 forecasts COVID-19 infections using recurrent neural networks, including LSTM and encoder-decoder LSTM models. Chapter 12 presents the exertions to build a robust healthcare system in India and a Strength Weakness Opportunity Threat (SWOT) analysis of private hospitals to work in tandem with the government-initiated multilayer healthcare system. Chapter 13 presents the role of IoT-based technologies during the pandemic, IoT-based solutions in containing the pandemic, the state-of-the-art architecture, applications, platforms, and the SWOT of IoHT. Chapter 14 presents how DL can be used for clinical decision making and improved healthcare outcomes.

Chapter 15 introduces a novel mechanism of no-regret learning in deep architecture to minimize error while making clinical decisions. Chapter 16 includes a natural language processing approach to classify certain types of diseases based on the symptoms reported. Chapter 17 discusses "big data" and how it helps to manage healthcare issues and investigates the challenges and threats in using DL in healthcare practices. Chapter 18 mainly focuses on the various approaches enabled by AI and DL and their significant applications in the field of medicine and healthcare. Chapter 19 presents an outline of these expanding topics related to drug discovery, the key concepts of prevalent deep learning algorithms, and the motivation to investigate these techniques for their potential applications in computer-assisted drug discovery and design. Chapter 20 presents how Avant-Garde Techniques in Machines can be used for detecting financial fraud in healthcare. Chapter 21 research offers a roadmap for analysis, where mental state detection can be based on machine learning techniques. In Chapter 22, a picture fuzzy set is an extension of the fuzzy and intuitionistic fuzzy set combined with the criminal database for reasoning to improve human recognition and features under uncertain environments. Chapter 23 proposes a novel DL-based ensemble framework to detect surgery actions in the operating room. Chapter 24 aims to consider DL methodologies for healthcare systems by reviewing recent trends, cutting-edge network topologies, applications, and industry developments. Chapter 25 proposed a deep convolution neural network (DCNN) model named DCCM (Deep Convolution Classification Model), an 11-layered architectural model for recognizing COVID-19 cases from Chest CT images. Chapter 26 explains in detail the role of the Internet of Medical Things in curbing pandemics.

We thank all the authors for their valuable contribution which makes this book possible. Among those who have influenced this project are our family and friends, who have sacrificed a lot of their time and attention to ensure that we remained motivated throughout the time devoted to the completion of this crucial book.

Harish Garg, India
Jyotir Moy Chatterjee, Nepal

Acknowledgments

I would like to acknowledge the most important people in my life, that is, my Grandfather late Shri. Gopal Chatterjee, Grandmother late Smt. Subhankori Chatterjee, Father Shri. Aloke Moy Chatterjee, Mother late Ms. Nomita Chatterjee, and Uncle Shri. Moni Moy Chatterjee. This book has been my long-cherished dream which would not have been turned into reality without the support and love of these amazing people. They have continuously encouraged me despite my failure to give them proper time and attention. I am also grateful to my friends, who have encouraged and blessed this work with their unconditional love and patience.

Jyotir Moy Chatterjee
Department of IT
Lord Buddha Education Foundation
Kathmandu, Nepal-44600

Chapter 1

The future of health diagnosis and treatment: an exploration of deep learning frameworks and innovative applications

Imran Zafar[1], Syed Umair Ahmad[2], Mohd Ashraf Rather[3], Azmat Ali Khan[4], Qurat Ul Ain[5], Arfa Safder[6] and Sheikh Arslan Sehgal[7,8]

[1]*Virtual University of Pakistan, Bioinformatics and Computational Biology, Virtual University of Pakistan, Lahore, Punjab, Pakistan;* [2]*Department of Bioinformatics, Hazara University Mansehra, Pakistan;* [3]*Division of Fish Genetics and Biotechnology, Faculty of Fisheries Rangil Ganderbal, Sher-e-Kashmir University of Agricultural Science and Technology, Jammu and Kashmir, India;* [4]*Pharmaceutical Biotechnology Laboratory, Department of Pharmaceutical Chemistry, College of Pharmacy, King Saud University, Riyadh, Saudi Arabia;* [5]*Department of Chemistry, Government College Women University Faisalabad (GCWUF), Punjab, Pakistan;* [6]*Institute of Molecular Biology and Biotechnology, The University of Lahore Sialkot, Punjab, Pakistan;* [7]*Department of Bioinformatics, The Islamia University of Bahawalpur, Bahawalpur, Pakistan;* [8]*Department of Bioinformatics, University of Okara, Okara, Pakistan*

1. Introduction

The era of big data in medicine and health has started due to the rapid development of computational and therapeutic research and the exponential growth of genomic information. An interdependent shift in scientific and technological advancement has forced the human generations to co-evolve with them in synergy [1]. The incredible transformation of information and communication technology has paved the foundation for innovative solutions in diverse industry domains, including healthcare [2], agriculture [3], transportation [4], logistics, and many others [5]. Emerging deep learning (DL) sectors involving automation alongside decentralized intelligence are way ahead on the road of innovation [6]. The constantly evolving DL technologies touch every dimension of life while imitating a living entity [7]. A significant revolution in medical science is observed since the information technology installed in contemporary healthcare applications started remotely obtaining, tracking, and controlling the status of a patient's condition [8]. Hence, a neoteric quantum leap in healthcare leading to its revolution is DL derived, mainly based on data acquired physically through wearable devices and sensor networking approaches from patients [9].

For efficient decision-making of quantitative and qualitative data analysis, a systematic analysis of the information has to be carried out [10]. Whereas predictive analytics originates from advanced analytics intending to engender the prognosis of upcoming events employing the data present [11]. Healthcare analytics is a direct channel to promote crucial tasks like early risk assessment, distant health monitoring, and clinical decision-making [12]. Contemplate reduction of associated risk factors based on the present and past history of the patient under study is salient a feature of medical science [13]. The inclusion of corresponding information from diverse sources embracing computerized health records, medical imaging, screening outcomes, and administrative information validating expeditious settlements is efficiently handled by healthcare analytics [9,14]. Clinicians often encounter situations manifesting a high degree of uncertainty; however, as a result of advancements in predictive analytics healthcare, they will be able to provide more informed decisions. Predictive analytics is a cutting-edge approach that can easily identify discrepancies beforehand, thus avoiding complicated risks, ameliorating chronic illnesses, evading hospital readmission, sustaining healthcare-related research aid, and cutting down overhanging costs [15].

Deep Learning in Personalized Healthcare and Decision Support. https://doi.org/10.1016/B978-0-443-19413-9.00002-3

Extensive research techniques by predictive analytics are taken into consideration for reforming algorithms of machine learning (ML) and artificial intelligence (AI) from their conventional linear modules [16]. DL, a subfield of the broader ML family, is recognized as a reliably expeditious self-regulated way to memorize and tackle complex data that, in addition to delivering actionable insights, also provide resolutions to complicated situations [17]. Its integration into diverse healthcare-related applications transcends the outcomes of standard modules. Distinctively, the recurrent neural networks (RNNs), adept at maintaining long-term dependencies of input data [18], become eminent in temporal observations regarding time-sequential applications as mentioned in (Fig. 1.1).

Intelligent health integrates medical care planning and delivery using digital technology [19]. It involves using, authenticating, and transmitting data for comprehensive healthcare research that uses multilayer DL algorithms to evaluate complex data. To improve the delivery of professional healthcare to customers, Google, for instance, has integrated the "DeepMind" health mobile application [20]. ML, which has gained popularity and been adopted by all industries, has provided advantages that have increased output and assisted us in resolving complex problems. DL, a subfield of AI, was developed to mimic cognitive functions [21] closely. It enables a computer to complete jobs that people take care of automatically. It is a method that is regularly used to organize and spot patterns in unstructured or unmanaged data. DL has had a considerable impact on the field of medical science because of its applications in healthcare pharmaceutical research, medical image processing, genome assembly, sickness diagnoses, and many other fields. The approaches and data types used in these fields have significantly increased DL growth.

The advancement of computational healthcare is a result of the development of medical innovation and the emergence of the significant data era, which is empowered and facilitated by artificial intelligence algorithms [22]. Individuals need to separate the helpful information to further the development of molecular diagnostics. In the past, pattern recognition techniques were used to detect traits in the data and extract biological information. These methods require a lot of time and money since they rely on the pattern design and specialized knowledge of experts. In contrast to traditional approaches, transfer learning, a cutting-edge field of computer vision, offers the benefit of dynamically discovering robust, sophisticated structures from raw data without requiring feature engineering [23]. We looked at how DL is used in computational medicine including genomics, drug discovery, and therapeutic neuroimaging. It is possible to guide clinicians better and raise the standard of medical care by using deep convolutional neural networks (CNN) to handle large-scale biological data. This book chapter also highlights the issues and difficulties in incorporating DL to digital healthcare. It offers a

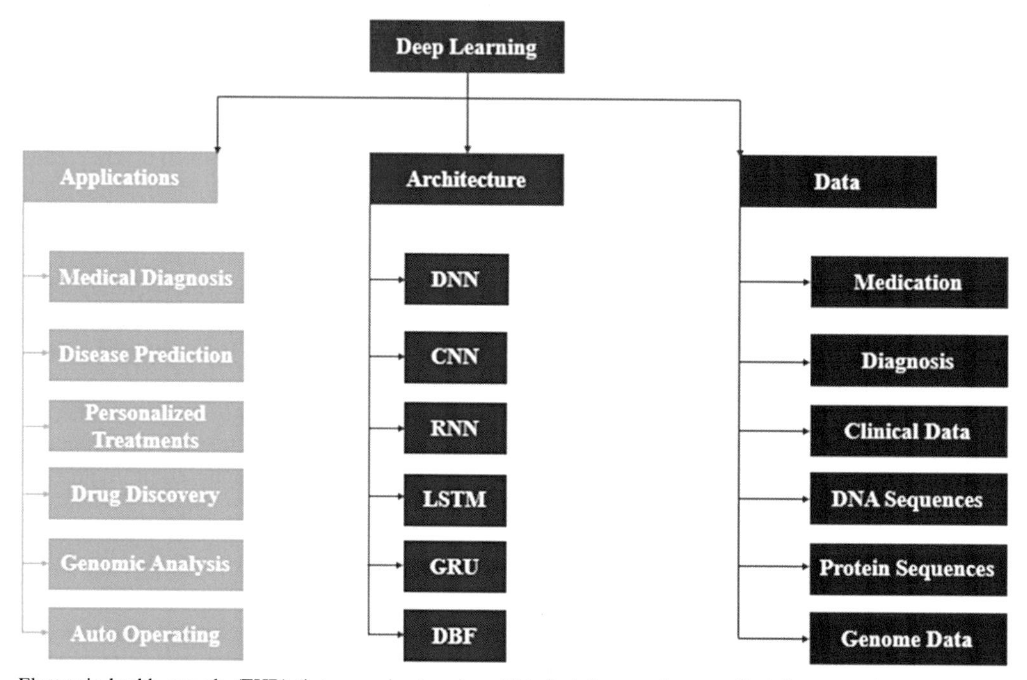

FIGURE 1.1 Electronic health records (EHR) that comprise based on historical data, such as medical documentation, lab results, diagnoses, and prescription information, may be diagnosed using deep learning neural network (DLN) at a phenomenal rate and with the highest levels of precision. Deep learning algorithms streamline complicated data processing, improving the accuracy with which irregularities are identified and given more priority. Convolutional neural networks (CNNs) offer insights that aid medical personnel in individuals' health problems earlier and more precisely.

resource and a means of enhancing computational intelligence applications in the future for the medical and health care industries.

Telemedicine makes the associated healthcare system advanced, conversational, and accessible through DL implementation tools for determining the natural causes of diseases [24]. The extensive use of neural network tools has fueled the expansion of numerous medical subject areas, reducing data complexity and enhancing 3-4D overlap images using coordinate value label data sensing devices that support knowledge strengthening, un-semi-supervised acquiring knowledge, multimodality, and the effect was discovered design. Health research has recently focused on machine learning to provide healthcare, treat illnesses, encourage healthy lifestyles, and make diagnoses [25].

2. Computational deep learning frameworks for health monitoring

The computational proficiency of DL models has empowered fastidious and systematic operations in healthcare [26]. DL networks play a fundamental role in transforming patient care within clinical practices [27]. Some of healthcare's most commonly used DL techniques are reinforced learning, lingual processing, and computerized vision. Rose et al. [28] executed hierarchical cluster analysis, an unsupervised clustering algorithm to perceive mammographic images to better understand malignancy. The following image proposes an end-to-end DL approach to classify life science applications accommodating pervasive sensors worn or implanted in the body for capturing ambient inertia motion to pursue physical activity such as smartwatches, ECG patches, EEG, and prosthetics [29]. These ambient sensing devices strictly check overall well-being, precisely evaluate food intake as per body requirement, access minute details of energy expenditures, inform beforehand about noncommunicable illnesses, and obesity and take care of patients with deformities [29]. Assisting devices are frequently implanted in patients requiring critical care or rehabilitation to detect their vitals [30]. Quickly escalated health issues such as cardiovascular disorders, diabetes, and obesity are successfully controlled by a physical activity recognition device that can supervise average calorie intake to sustain a balanced diet [31]. Alternatively, CNN adopt cloud computing, size calibration, and distant estimation strategies to accelerate innovation and improve the efficiency of accurate food intake monitoring [20]. Gasmi [20] consolidated DL algorithms with hierarchical visual representation incorporating double-layer 3D convolution and mx pooling inputs to acknowledge day-to-day activities of humans.

The Researcher Camara et al. [32] took advantage of RGB-D video sequencing to visualize human behavior while focusing surveillance on children and elders. The alarming feature of CNN facilitates caregivers to detect infants fall and crawling. The RBMs work seamlessly on watches and smartphones, with apps for aural and tactile feedback specifically recognizing vision impairment. AI-based DL algorithms of CNN encoded in several assistive devices promote successful recognition of sign language modalities and static hand gestures and authorize a contactless user-friendly human–robot interaction (HRI) to meet patients' unique needs. Lu et al. [33] designed a deep neural network (DNN) with depth perception capabilities specifically for sign language recognition that can synchronize finger joint data as input directly in real time without handcraft features to enhance communication between disabled and normal persons. Data-driven machine health monitoring is achieved through the softmax layer installed as the output layer of DNNs to reduce human effort and time [33].

An intelligent pattern recognition model focusing on a single layer auto encoder with minimized dimensionality for accurate induction of motor fault diagnosis and damage repair. In contrast, Lu et al. [33] primarily worked on fault diagnosis of rotating machines incorporating stacked triple-layered denoising auto encoders encoder, which has become a popular approach to identify specific health sates. Patient monitoring devices referred to as vital sign monitors, for example, wearable BCG sensors, ECG, or EKG, have DL applications integrated infrastructure to enhance diagnostic efficiency.

Extensive studies on accelerometer bracelets specialized for tracing pathological gait patterns, including stumbling, falling, and inpatient seizure onset, claimed that longer duration of a relatively inactive lifestyle can deteriorate a person's health, thus compelling clinicians to suggest the use of wearable health monitoring devices along with exercises to boost their stamina [34]. Wu and Luo [35], included wearable sensors in their research to detect lingual pattern recognition between a mother and her child, affecting their physiological and psychological development. The researchers [20] preferred employing unobtrusive smart sensors featuring SC/CT technology patients for automatic stress monitoring through electro dermal activity (EDA), heart activity, and temperature measurement in both adult and younger patients.

According to research [36], EDA sensors assist in measuring skin conductance level and skin conductance response as their expressions vary due to endocrine secretions in stress feedback. Lin et al. [37] monitored cardiac arrhythmia by implementing fuzzy-based clinical detection method to represent inferential signaling in AI instruments. Portable point-to-care diagnostic devices equipped with wireless commination in healthcare have transformed clinical patient management through tools like EMS also knowns as emergency medical service, activated in case of prehospital treatment. Inan et al. [38] found valuable wearable devices that can monitor heart rate, record ECG, and 3D represent sternal seism cardiogram (SCG) as mentioned in (Table 1.1).

TABLE 1.1 Exploring disease prediction applications: evaluating publication categories and key characteristics.

Data	Model	Application	References
Clinical big data analytics and health information system	Stacked sparse autoencoder	Early stages Alzheimer's disease diagnosis from brain MRIs	[39]
	Restricted Boltzmann machine	Identification of Alzheimer's disease mode of variation using manifold of brain MRIs	[40]
	Conventional Neural network	Prediction of the risk of osteoarthritis using MRIs	[41]
	Restricted Boltzmann machine	Multiple sclerosis lesions separation using 3D MRIs	[42]
	Stacked sparse autoencoder	Using ultrasound images analysis Detect the breast nodules	[43]
	Conventional Neural network	Identification of the most common disease using machine photographs in the context of diabetic retinopathy	[44]
	Conventional Neural network	Skin cancer classification	[45]
EHRs	Conventional Neural network	Analysis and prediction of heart functions and explore failure site	[46]
	LSTM RNN	Classification and identification of clinical measurements unit care	[47]
	LSTM RNN	Prediction of medicine based on patient history	[48]
	Stacked sparse autoencoder	Prediction of future clinical events	[49]
	Stacked sparse autoencoder	Based on HER detection diseases in patient	[11]
	Restricted Boltzmann machine	Robotically check patients to assign diagnoses from their clinical status	[50]
	Restricted Boltzmann machine	The chance of suicide in patients with mental health disorders may be predicted using a limited approximation of the disease categories in the EHRs	[48]
	Stacked sparse autoencoder	Identifying and detecting physiology's distinctive characteristics in the therapeutic data set	[51]
	Stacked sparse autoencoder	To identify various demographic subsets and to differentiate between the uric-acid petitions of diseases and symptomatic leukemia, prototype spatial sequence data of uric acid serum measurements	[52]
	Gated recurrent unit—recurrent neural network	Employ medical histories to forecast diagnoses and treatments	[53]
	Conventional Neural network	Deeper: An online platform for predicting unscheduled readmissions in outpatient therapy	[54]
	Gated recurrent unit—recurrent neural network	Disease occurrence detection from continuous laboratory tests	[55]
	Gated recurrent unit—recurrent neural network	Individual medical records that are anonymous	[56]

TABLE 1.1 Exploring disease prediction applications: evaluating publication categories and key characteristics.—cont'd

Data	Model	Application	References
Whole genome identification and analysis	Conventional Neural network	Given nucleotide sequence, infer nucleosome scores	[57]
	Conventional Neural network	Basset: Online system to estimate the impact of SNVs on nucleosome permeability and anticipate DNase I responsiveness throughout several cellular processes	[58]
	Conventional Neural network	DeepBind: Estimate the specific characteristics of proteins that adhere to RNA and DNA	[59]
	Conventional Neural network	Determine single-cell bisulfite genotyping experiments' acetylation levels	[60]
	Conventional Neural network	Applying these strategies to various nucleosome markers	[61]
	Stacked sparse AE	Cancer identification using the profile of gene expression	[62]
	Stacked sparse AE	Cancer identification using the profile of gene expression	[63]
m-Health information and record	Conventional Neural network	HAR to identify gait halting in PD patients	[64]
	Conventional Neural network	Measurement of EE with biosensors	[65]
	Conventional Neural network	Telemonitoring using photoplethysmography signals recognition	[66]
	Conventional Neural network	Analyzing information obtained from local field possibilities and electroencephalograms	[67]
	Conventional Neural network	Predicting the sleeping quality using information from wearables used for aerobic exercise throughout the day	[68]

3. Advanced architectures and core concepts of deep learning in smart health

DL networks are structural frameworks for classifying or regressing data containing hidden or visible output layers [69]. Some have more than two hidden layers that permit the expression of complex or nonlinear hypotheses. DNNs are successfully used in bioinformatics but lack training authentication due to their backpropagated layers and steady learning process, talk about a deep auto-encoder designed to extract features and later dimensional reduction with its input, hidden, and output layers.

Closely resemble input and output nodes of autoencoder recreates vector input specialize in handling unsupervised learning [70]. The potential DL-based auto encoder architecture is prerequisite for robust representation of labeled data. However, rigorous pretraining is required before its implementation in the end-to-end communication systems [71]. Restricted Boltzmann machine (RBM) is referred to as a bare block of deep belief networking, a sub-class of the DNN containing two stacked layers of hidden and visible nodes to empower unsupervised training of the neural network, which in turn dictate networking commands necessary for maintaining sampling and training operations.

Hinton [72] emphasizes the essential variance among deep belief networking and differences between restricted Boltzmann machine. Independent dual layers in Boltzmann make it an undirected stochastic neural network in which each neuron behaves randomly upon activation resulting in robust inference representation even with ambiguous inputs. It is also capable of handling time inference, a crucial factor that assists machines in bridging the gap between pretrained and ambiguous real-time data better than DL models along other significant database parameters. Progressive neural networks give rise to the recurrent neural network, a subpart of an artificial neural network having infinite impulse response that can significantly empower extensive data stream analysis, enable model time dependency, process sequential data, and natural language recognition.

Exploding gradients are one of the significant challenges RNN networks face as they can turn model unstable and incapacitate learning during training. A complex DNN is preferably employed in infectious disease modeling due to its 2D compatibility, which can be conveniently transformed into the rotational 3D dataset. CNN-supported biological neuron models reconstruct and use neuron follow connection with other applications like clarifa and Google net to perform visual cortex. The primary obstacle in dealing with CNN is the hierarchical feature exploited to input enormously labeled information. The GPU parallel acceleration provides the hardware capacity needed to evaluate DNN on multicore microprocessors and the cloud. The RNN can examine multiple datasets, Shi et al. [73] investigate an RNN variant known as long-short-term memory unit (LSTM). LSTM utilizes highly integrated input sequences to alleviate gradient disappearance issue.

LSTM uses information that has been accurately saved, read, and written without interruption at the time of training. It supports natural language processing like speech recognition, mapping, and image characterization graded equal to RNN. Fischer and Igel [74] emphasized using the Gibbs sampler, an RMB variant through the Gaussian method. With the help of Gibbs sampling, researchers can modify weights, reduce errors, and predict incongruous probabilities relation. Wang et al. [75], explained the conditional independence among directed acyclic graphs and variables in stochastic units of the probabilistic graphic model.

The earlier researcher [75] illustrated how contrastive divergence (CDA), an alternative training approach along RMB, can tackle unstructured learning. Both positive and negative phases are included in this method where the former stimulates network configuration, the latter one the former stimulates network configuration, and the latter replicates network configuration. The neurobiological model of the visual cortex backed by CNN exhibits consistently correlated data integrated with multidimensional inputs significant for back-propagation, an essential practice required for fine-tuning neural net weight. Local receptive field maps in CNN convey anterior granularity input to a sophisticated sub-sampled output using minimal filters.

4. Comparative analysis of deep learning frameworks for different disease detection

In addition to minimal data processing requirement, DL networks automatically supervise multiple filtration and normalization tasks of ML that demand programmer assistance. ML approaches limit the biological preprocessing of raw data-driven DL systems involve in network representation directly filter information through layers necessary for predicting or classifying datasets to escalate the extraction of actionable insights that have never been extensively curated. Extraordinarily complex mathematical equations incorporated in DL models [76]with multiple variations lift other sub-strategies in the domain. Rapidly evolving DL models backed by some highly advanced computational approaches [77] are adding significant value to the user expertise by dominating almost every field and enhancing competitive decision making for health industry to solve big data analytics problems.

Table 1.2 shows that AI techniques have proven beneficial for detecting diseases with upgraded results. AI uses DL/ML models that are functional upon training and testing data sets so that the system can detect and diagnose the disease early. In the AI-based model, we initially need to train humans to memorize the data and provide accurate outcomes. However, it also deals with problematic situations. Suppose the training data produced the incorrect analysis of disease due to insufficient information, which DL cannot comprehend. As a result, it will become a nightmare for the patients as DL cannot assure us whether the prediction regarding disease detection is absolute.

The precision of DL algorithms made them a competent option for disease detection. In addition to scalp disease diagnosis, DL models can successfully identify Alzheimer, thyroid disorder, heart arrhythmia, diabetes, and skin problems with 95%–99% accuracy.

TABLE 1.2 Deep learning models' efficiency values through applying on different diseases.

Disease	DL models	Results	References
Skin disease	MLP/ANN	Precision: 76.67%	[78]
Liver disease	CBR, BPNN, LRC	Precision: 95% Sensitivity: 98% Sensitivity: 94%	[79]
Liver disease	ANN	Precision: 99%	[80]
Urology disease	CRML, NN, DSS	Precision: 71.8%	[81,82]
Arrhythmia disease	DGEC, ECG	Sensitivity: 94.62% Precision: 99.37% Sensitivity: 99.66%	[83]
Kidney disease	ANN, KC, LQS	Precision: 99.61%	[84]
Gastrointestinal disease	RN, LSTM	Area under Curve: 97.057%	[85]
Gastrointestinal cancer	GRAIDS, CPM	Precision: 95%	[86]
Gastrointestinal disease	VGG 16, ANN, DL	Precision: 98.4%	[87]
COVID-19 disease	AI	Sensitivity: 90.9% Sensitivity: 87.5%	[88]
COVID-19 disease	DLM, VGG16, DenseNet121, ResNet50	A precision 98.8%	[89]
Covid-19 disease	CNN, ResNet 18, ResNet 50, Squeeze Net, DenseNet121	Sensitivity: 97% Sensitivity: 90%	[90]
Hypertension disease	Vess-net method, AI, semantic segmentation	Sensitivity: 80.22% Sensitivity: 98.1% Precision: 96.55%	[91]
Hypertension disease	XGBoost, ensemble, logistic regression	AUC of XGBoost: 0.877 Ensemble: 0.881 Logistic regression: 0.859	[92]
Pulmonary arterial hypertension	GBTA	Sensitivity: 99.99%	[93]
Diabetic disease	SVM, RBF, KNN, ANN, MDR	Precision of SVM: 0.89 KNN: 0.88 ANN: 0.86 MDR: 0.83	[94]
Diabetic disease	FSVM, SVM	Precision: 89.02%	[95]
Diabetic disease	SVM, CNN, LSTM	Precision: 95.7%	[96]
Tuberculosis	ANN, RF	Precision: 88.67% Sensitivity: 80% Sensitivity: 90.4%	[97]
Tuberculosis	DL, ResNet	Precision: 85.29%	[98]
Tuberculosis	CNN, IP	Recall: 97.13% Precision: 78.4% F-score: 86.76%	[99]
Diabetic retinopathy retina disease	AI software	Sensitivity: 95% Sensitivity: 80.2%	[99]
Retinal fluid detection	AI	Precision: 0.805 Sensitivity: 0.468 Sensitivity: 0.970	[100]
Retinopathy detection	IAS, AI	Sensitivity: 90.8% Sensitivity: 75.3%	[101]

Continued

TABLE 1.2 Deep learning models' efficiency values through applying on different diseases. cont'd

Disease	DL models	Results	References
Bladder tumor detection	CystoNet, DL	Sensitivity: 90.9% Sensitivity: 98.6%	[102]
Tumor detection	MP, ANN	Precision: 76.67%	[78]
Alzheimer's disease detection	LSTM, RNN, DLM	AUC: 0.98–0.99	[103]
Alzheimer's disease	ML, PPR	Precision: 86.84%	[87]
Alzheimer's disease	SVM, KNN, DT	Accuracy: 73.46%	[104]
Cardiac arrest	ML, KNN, IoT	Precision: 96%	[105]
Heart disease	KNN, NB, DT, ECG	Precision: 80% Sensitivity: 60%	[106]
Cardiovascular disease	DMML, SVM, WEKA	Precision: 97.53% Sensitivity: 94.94% Sensitivity: 97.50%	[107]
Chronic obstructive pulmonary disease	AI	AUC: 0.886	[108]
Chronic disease	SVM, LR	Precision: 73.1%–91.6%	[109]
Chronic disease	SVM, KNN, NB, RF,	Precision: 80.55% Sensitivity: 80.14% Sensitivity: 80.14% Precision: 90% F-score 84.78%	[110]
Skin lesion	CNN, VGG Net, KNN, SVM, RF	Accuracy: 96.805%	[111]
Liver cancer	GMM, DNNC	Precision: 99.38%	[112]
Breast cancer	SVM, ML	Accuracy: 99% Sensitivity: 98% Sensitivity: 99%	[57]
Oral cancer	ANN, FL	Precision: 78.89%	[113]
Vertical root fracture	CNN, DetectNet	Precision: 0.93 Recall: 0.75 F measure: 0.83	[113]
Oral cancer	KNN, DT SVM, LR, PCA	Precision: 70.59% Sensitivity: 41.98% Sensitivity: 84.12%	[114]
Large artery occlusion detection stroke	LVO-AAI	Sensitivity: 92% Sensitivity: 90%	[115]
Cerebrovascular large vessel detection	LVOA, AI	Sensitivity: 94% Sensitivity: 82%	[116]
Diabetic retinopathy detection	CBIR	Precision: 99.6% Precision: 0.991 Recall: 0.9932 AUC: 0.995	[117]
Chronic disease detection	RF, NB, KNN,	Precision: 93%	[118]
Thyroid disease	ANN, RF, MR	Precision: 98.22%	[108]
Thyroid disease	ANN	Precision: 99%	[119]
Alzheimer's disease	CNN	Precision: 86.60%	[120]
Covid 19 disease	DL, HMA	Sensitivity: 60%–70%	[121]

TABLE 1.2 Deep learning models' efficiency values through applying on different diseases.—cont'd

Disease	DL models	Results	References
Thyroid disease	DT, RF, RT	Precision of DT: 98% RF: 99%	[122]
Hypertension	FLMLP, SVM, DT	Sensitivity: 90.48% Sensitivity: 71.79% Predictively: 81.48%	[123]
Pulmonary disease	DL, DRN	AUC: 0.886	[124]
Alzheimer's disease	RNN, CNN	Precision: 96.0%	[125]
Squamous cell carcinoma	ANN	Precision: 91.7% Sensitivity: 97.6% Sensitivity: 85.7%	[126]
Scalp disease	DL, CNN, MLP	Precision: 80%	[127]
Scalp disease	DL, RNN	Precision: 97.41%–99.09%	[128]

5. Advantages of deep learning in smart medical healthcare analytics

Current industrial DL models are running as pilot projects in their extensive precommercialize stage to ensure the validity of all required resources. Despite all challenges, DL science is progressing at a high pace toward the development of high-end real-time applications that can bring revolution to medical science. Patient-facing systems are an exceptional example of how DL improves patient interaction within healthcare by proving Internet-based services to boost user experience [129].

Innovation in DL and ML science is impacting almost every field of life by introducing revolutionary changes that are capable enough to shift traditional systems to modern practices. It is convenient to assume that most of the major changes are observed in healthcare due to online and offline services [130]. A significant role of automatic cancer cell detection by DL models is widely known, for human assistance to accomplish various tasks differentiates ML models from DL systems that: The demand for human assistance to accomplish various tasks differentiates ML models from DL systems that can decode entire program independently. DL proves to be more efficient in massive data analysis for multiple disorders, DL is more beneficial for massive analysis of multiple diseases [131], such as heart problems in younger children and coma patients. Different advance DL approaches are represented in Fig. 1.2.

Deep learning is a promising source of problem recognition leading data-driven performance as it makes strategic decisions firmly based on critical analysis, thus enabling organizations to handle complicated situations efficiently [33]. Effortless learning algorithms involved in ML are very easy to learn and perform, making them readily available for manipulation in versatile applications. ML is nonlinear, whereas traditional DL models influence ML for language acquisition to figure out accurate data for patients' help. With the help of DL networks, visualization of patient data, including text and images, can be made possible. The latest DL applications for health monitoring [132] require sophisticated tools and much computational resources for data interpretation. Using these high-end neural networks requires a longer duration to train and evaluate data. The level of precision required enhances with network sophistication and forecast duration. Real-time discoveries are vital in healthcare and other fields; hence, this has been a significant problem.

DL automation competently delivers potential control systems to regulate the behavior of other devices. These networks examine the dynamic systems to bring change in the surroundings by identifying various forms of data. With the assistance of DL big data analytics [131] and genome editing applications [133], which provide a comprehensive way for

FIGURE 1.2 Deep learning advantages in intelligent healthcare system.

detecting genetic anomalies via exploring genomic regions, disorders like hemophilia, Turner syndrome, and anemia are feasibly studied, enabling physicians to recommend best possible treatments and future medication. DL can automatically distinguish between normal and aberrant data allowing clinicians to know the patients' status.

6. Deep leering applications for disease prediction

Drug development and precision medicine are also on the agenda of DL developers [134]. To discover unidentified connections between genes, medication, and physical environment, both tasks necessitate the constant processing of massive volumes of clinical, genomic, and population-level data [135]. Given that most precision medicine researchers are unaware of their particular interests, DL seems to become the best method for pharmaceutical stakeholders who wish to highlight novel trends in their largely new data sets [136]. Remarkably unexpected discoveries in the field of genomic medicine due to being so young have provided an exciting testing ground for creative ideas capable of transferring lives.

A mutual strategy of predictive analytics and molecular remodeling will provide a certain understanding of cancer spread in a particular patient [137]. DL techniques speed up the data analysis by cutting the processing time for crucial elements from days or weeks to limited hours. The private sector is equally dedicated to proving it is equally dedicated to proving DL approaches' strong impact on precision medicine.

Several difficulties are associated with user-graded cardiac rate-detecting sensor employment in medical settings, as routine activities can bewilder basic heuristics. Sensors employed in outpatient care error-prone have short-span battery life and can modify the measurement rate. An experiment for semi-supervised learning outcomes was stationed. A Deep heart sequence autoencoder [138] initially referred to as the starting point of the second supervised phase with autoencoder was set up. This mindset contributes significantly to high cholesterol, blood pressure, and sleep apnea.

Brain—computer interface (BCI), a computer-based technology specific for capturing and processing brain signals, has the potential to forecast epileptic seizures to regulate epilepsy [139]. An automatic BCI system gathers and processes data from computer to implement requisite action in a highly systemized way through three main stages [140]—an array of electrodes placed throughout the scalp recorded ECG outcomes displaying the brains electrical activity spontaneously. A practical BCI system needs computational skills for signal analysis, DL, and electroencephalography prognosis acquired from users in real time. Calculations made through cloud open an easy door to access the databases for calculating or processing incoming data in real time. Creating a patient localization and seizure prediction system from an endless unsupervised database to enable learning is the central concept of BCI [141]. The detail of major healthcare domain objective and diease prediction applications are mentioned in (Tables 1.3 and 1.4).

Real-time condition identification of the patient is achieved via DL, a significant aspect of ML or AI. Based on either raw data obtained or the patients' prior information, DL can recreate the real-time state of the patients. This capability subjects it to manipulation in various domains, including healthcare and the protection of the cybercrime business. Taking into account the data, DL is necessary for breakthroughs.

Cardiovascular diseases (CD) are considered one of the prime reasons for one of the prime reasons for death in modern times. As wearable devices irritate newborns and the elderly, DL approaches are employed in such circumstances to gather data, whereas these devices can still find patients' conditions. The likelihood of acute diseases like leukemia, Alzheimer, diabetes, respiratory issues, and other genetic diseases can be reduced by integrating smartness in healthcare systems with user-friendly computational approaches such as SHM or IOT along with the assistance of qualified physicians who can detect patient's condition in real-time.

A recent medical assessment found that cardiovascular problems affect about 30% of the world's population. Given that the intelligent health monitoring system combines artificial intelligence, IoT, and health monitoring devices, we may assume that robotic learning is applied. The SHM is fundamental to today's healthcare system. AI is used in healthcare systems to leverage machine learning software and algorithms. The data utilized in machine learning determines how AI performs assigned tasks. Since CNN is a type of ANN utilized in the visualization of data such as MRI, ultrasound, or CT scans, their operations are connected to DL systems.

Computational learning is employed in healthcare when the internet of things and other intelligent monitoring equipment of healthcare becomes unable to identify patient's information. The behavioral data these algorithms use to build their models can be recognized. ML is a technique that combines ANN and CNN to create an artificial neural network (AI) capable of functioning in two separate ways: Algorithms are literal, and black boxes are used to apply DL. DL uses linear approaches rather than the more conventional, nonlinear approach, which requires patients to make unwelcome hospital visits but does not allow for computational communication with the doctor.

Numerous devices, which are the fundamental components of IoTs, are used for various types of disease monitoring, including pacemakers, cardiovascular defibrillators, accelerometers for breathing rate detection, and accelerometers for

TABLE 1.3 Deep learning in healthcare domains and their major objectives.

Healthcare domains	Major objective	References
Disease exploration and identification	Recognizing and examining diseases that are thought to be challenging to diagnose represents one of the most important applications of ML and DL in medical care.	[142]
Drug development	The early stages of the drug identification process are a separate area where ML and DL may considerably develop. To detect patterns within data without making any predictions, solo AI is beneficial.	[142]
Customized medicine	When given to just issues related to health, medications are at their most effective. Currently, a doctor's assessment of the patient's safety might err toward a lack of determination or an unspecified threat depending on the person's medical y and the available facts.	[116]
Digital health records	They are trying to keep up with the costly and lengthy cycle of preserving crucial healthcare information. They now play a crucial role in promoting the database access strategy.	[86]
Medical trials	It is based on ML and DL, which use expository analysis to identify major therapeutic initial applications and allows scientists to narrow their pool from a wide variety of data.	[143]
Information crowdsourcing	The medical community has officially supported the method, and today's experts use it to access many data that people share.	[115]
Outbreak prediction	ML and DL-based procedures are utilized to screen and expect flare-ups about the world to anticipate the scourge	[81,82]
Medical imaging diagnostics	Computational intellectual ability techniques become more expansive and effective in gathering data from a wider selection of medical images.	[78]

TABLE 1.4 Classification of recent studies in disease prediction applications.
The categories of the publications mentioned above are shown in the following table and their critical characteristics for evaluating illness prediction applications. The three primary BCI phases in the illness prediction are based on the DL network, which includes an LSTM neural network for forecasting air quality index (AQI) in smart cities and seizure prediction. The supervised learning process is utilized to identify cardiovascular risk factors.

Research	Main context	Strengths	Weakness	New finding
[144]	DL for monitoring forecast heart disease risk	Significant enhancement	Difficulty of deployment	Architecture
		High cholesterol	The necessity of interpretability	Architecture
		High blood pressure	N/A	Architecture
		Sleep apnea	N/A	Architecture
[145]	Seizure prediction BCI	Intelligence facilitated by massive, unsupervised data	N/A	Architecture
		Effective approach for real-time seizure prediction and localization system that relies on patients	N/A	Architecture

blood pressure and body temperature measurement. When the entire dataset is gathered utilizing microphone sensor arrays that utilize a two-dimensional spectrum CNN, DL effectively detect routine activities (DCNN). Numerous nonwearable and wearable gadgets and smart houses that are fully digitalized may monitor patients' everyday activities and behavior by installing sensors in places like toilet seats, smart beds, and other objects.

Behavioral activities are further subdivided into two small-term and large-term categories. The repetitive behavior of routine activities is an indication of Alzheimer's. Short-term behaviors last from few hours to days, such as frequently going to the bathroom, which an upset stomach may bring on. DL science can play a fundamental role in identifying certain illnesses. DL comprises of two major stages the labeling of enormous data sets and the identification of the feature are compromised by 1. training and 2. inferring training. Memorization is required for inference, which bases its conclusions on previously collected data (Table 1.5).

Healthcare management systems continue to significantly benefit from digitalization and medical computerization due to high unpredictability and complexity of healthcare operations. DL can enhance human cognition to govern healthcare tasks leading to significantly more efficient patient care. The author has discovered a research gap in the public domain for a comprehensive analysis of ML, AI, and DL roles in healthcare. This factor has motivated the author to provide a deep insight into how the current integration of ML, AI, and DL technologies to raise the standard of patient care and well-being is bringing a revolution into the healthcare framework. In the quick-paced world of technological innovation where we have survived today, disease outbreaks have become a significant problem. Regarding disease prevalence and control measures, technology has become significantly essential for practitioners working in healthcare sectors. Due to tight work schedules, it has become quite challenging to maintain a generally excellent and healthy lifestyle. The answer to the aforementioned problems is an innovative health monitoring system. Smart, inexpensive sensors have been created due to the recent 5.0 and 5G revolutions, which can help with real-time public health monitoring.

Current SHM technology has made timely, dependable and economical delivery of monitoring services from distant localities possible, which was not attainable in the past due to conventional healthcare setup. The author in this chapter has offered an up-the-mark evaluation on some of the widely used applications specific for disease detection, treatment, data management and information security in the healthcare sector. Similar to how they have transformed practically every other business, machine learning and artificial intelligence have transformed the healthcare sector. From the potential integration of neural networks and DL in pharmacological research to more complex applications of intellectual diagnostic testing and surgical humanoids, AI is transforming how we build automobiles, maximize our energy consumption, and manage our finances. At the same time, it is presenting new opportunities and threats for managing human health. The capacity of AI to merge vast amounts of data sets specifically concentrating on all facets of patient to sharpen decision-making is considered the main player of health sector from pharmaceutical inventors to healthcare practitioners, which ties it all together (Table 1.6).

TABLE 1.5 Categories of recent research conducted in medical diagnosis and differentiation applications.

Main context	Strength	Weakness	New finding	References
An easy-to-use fog technology and efficient data administration can instantly identify cardiac problems.	Low latency	In real-time fog situations, it is challenging to optimize Quality of Service (QoS) characteristics due to latency and reaction time.	Framework	[146,147]
Optimization of QoS parameters in real-time fog environments	Energy-efficient solutions to process data			
Automatically identifying and diagnosing EEG illness advances	Enhanced EEG decoding	The majority of EEGs are tiny and inappropriate for DL models.	A cloud-based framework	[148–151]
Recognizing unusual medical conditions				
Prompt diagnosis can tackle CVA		The proposed structure cannot be utilized in additional medical imaging.	An IoT framework	[152,153]
MRI and CT imaging are key tools for stroke diagnosis	Less human-dependent area			
Through CT scans and an IoT framework, stroke may be classified	Fewer human errors			
The DL system built on LSTM was tested using labeled HR signals	Less constrained than machine learning methods	Training during the job is not addressed	A creation of DL model to detect AF using HR	[154]
	A larger dataset may generalize information from a smaller training dataset.			
The cornerstone of Chinese medicine is differentiating syndromes	Reduce over-fitting	Cannot distinguish between multiple fever cases that are infectious and instances that were seen in clinics	An adaptive DL model	[155]
Differentiating infectious fever syndromes with the aid of computers	Enhance the classification accuracy			
Computer-assisted lung cancer detection and treatment using DL reinforcement learning models	We need to find a solution for the lung cancer localization issue.	Q-value shall be updated in every action	Several models to represent DL	[156]
Melanoma is a serious skin cancer	Accessible usage in different regions	Requires a good connection	Using transfer learning and DL in IoT systems	[157]
Skin lesions are categorized using an IoT-based approach	Easy to handle manner			
The offered techniques for displaying the learned characteristics	ConvNets suit end-to-end learning	They may show false predictions		[158]
Illustrates the ConvNets demodulation for task-related data	They are sufficiently scalable for large datasets.	A vast amount of data is required for training and testing purposes.		

TABLE 1.6 A Classification of newly conducted researches in home-based and personal healthcare applications.

Research	Main context	Case study	Advantage	Weakness	New finding
[159]	Establish an intelligent environment for multi morbidity patients with chronic diseases to receive home health.	Multimorbidity chronic patients	A new wave of care-giving amenities	There are no statistics to prove the effectiveness	Algorithm
			Controlling costs		Framework
[160]	Establish a link using deep convolutional neural network among sensor data from devices and personal health.	Personal health assistance	Simple structure Low computation load High performance	More genres of sensor data should be researched.	Architecture Framework
[161]	The smart dental health IoT system focuses on intelligent technology, transfer learning, and enabled mobility.	Smart dental Health-IoT system	Its coverage varies from 1 to 6.5 cm, is 5.5 mm broad and 4 mm thick, and its light color may be altered to match the environment. The affordable price of hardware	Incomplete coverage of larger teeth	– Prototype – Implementation
[162]	A DNN was presented to detect the signal features in the sensor system to comprehend the patients' physiological conditions.	Patient monitoring system	High accuracy Low cost	Require a large amount of data	Algorithm Architecture
[163]	Algorithms to detect falls are created using traditional (Support Vector Machine and Naive Bayes) and non-traditional (DL) methods.	Fall detection	Learn formats that can help in generalization	The DL model is not very flawless on ADLs	Algorithm Architecture
[164]	Intelligent chips or patches that use IoT sensors to track individual medical conditions	The multi-access physical monitoring system	Auspicious upshots based on: Precision, efficiency Mean residual error-Delay Using energy	Advanced multimedia methods should be suggested cutting expenses and privacy	A novel optimized neural network
[165]	A label, fully autonomous nutritional assessment system (smart-log)	Nutrition monitoring	Cost-efficient high accuracy	Another method should be suggested for more accurate diet prediction	Algorithm
[166]	Implementations of IoT to reduce sports injuries	Sports injury	Working better in packet loss rate and accuracy	The small size of system experiment feedback population	Framework

7. Deep learning in research and development

With the technical advancement, contributions of DL models toward potential therapeutic entities and disease prediction have increased. Medical researchers employ DL tools to analyze recent trends in vast oceans of data. Upgraded interpretability of DL models leads to detailed comprehension of biological information. CNNs are commonly employed, empowering researchers to extract attributes of window-size DNA sequences. DL approaches proactively employ DL approaches to improvize therapeutic application in mental health treatment. According to scientists, trained DL models may perform better than conventional ML models specific contexts. For instance, DL algorithms can identify specific CNS biomarkers on their own.

8. Future challenges of deep learning in smart health diagnosis and treatment

The main reasons for the development of the medical sector today are DL and ML. There are significant restrictions, though, such as the need for a sizable data collection for recognizing and producing results, which is occasionally unattainable. These computing approaches are more expensive since heavy data might be complicated and lead to complex data models. Since this is not relevant to everyone, they cannot adequately benefit from this strategy. Due to the volume of data, DL is more vulnerable to mistakes that happen during testing and training data. Continuous updating can occasionally cause data to become inconsistent.

DL science undoubtedly performs an essential role in all disciplines, but it also has several certain drawbacks. Initially, it is difficult to read, requires a lot of data, occasionally struggles with dealing with direct symbols and decision-making, and interacts with natural languages. Because the DL approaches require such a large quantity of data, managing it is significantly more complex, time-consuming, and space-intensive. Only practitioners or healthcare professionals with a thorough understanding of language grammar and patient health information are capable of understanding the study of DL methods.

9. Limitations of deep learning frameworks

The main DL algorithm constraints are privacy concerns or security issues. Vulnerabilities in data gathering occur when DL are applied to huge datasets, which also requires human time and labor. Diagnostic errors may result from erroneous or changed datasets. The instrumental noise from intelligent machines because of unnecessary commotion, particularly with multishot MRI, high sensitivity modal-motion, which raises the probability of misdiagnosis brought about by errors in artifacts, which is hazardous to the overall well-being of humans.

Doctors who are unfamiliar with data analytics might make severe mistakes in medical diagnosis. The researcher [167] addresses the challenges faced during the labeling and annotation of data while highlighting the ambiguous classification of medical imaging that leads to misunderstanding and conflict among clinicians. The earlier researcher, while explaining methodical approaches of data labeling to reduce shortcomings [73], warned that incorrect annotation of computational algorithms in healthcare could cost life. Scattered data can result into undefined or repetitive samples, which can ultimately be the unfortunate death reason of hundred and thousands of people. The researcher Liu et al. [168] considered that model training vulnerabilities might cause model poisoning or incomplete privacy violations due to incorrect training. Data poisoning, also known as the corruption of already-collected data, necessitates protection, particularly during digital forensics and biometrics. In the event of a breach during DL implementation, distribution adjustments in real-world healthcare settings may occur, making test data partially vulnerable.

10. Conclusion and future scope

Innovative health is a progressing and exceedingly critical research field with a probability of noteworthy results in the conventional healthcare industry. The recent research demonstrates how DL technology plays its role in chronic illness detection by introducing novel applications and outbreak eradication from different parts of the world. This work also elaborates an overview of the challenges, pipeline research, and innovative techniques of intelligent health. Traditional healthcare systems are assisted by a rational data processing pipeline encompassing data dissemination, data security, networking, and computational technologies.

Recent advances in DL neural network have been impressively exhibited by the exceptional performance of intelligent health networks and services. Additionally, DL represent a strong record of achievements in clinical decision-making under strict supervision. Despite several data analysis methodologies in healthcare explained in this chapter, multiple significant attributes such as quality and security in patient care still need to be investigated with utmost scrutiny.

In the future, DL algorithms will be applied to detect anomalies in the data of intensive care patients. DL models can extract meaningful information from complicated data sets produced by healthcare systems in the form of medical imaging and E-health records from wearable sensors that are specialize in real-time monitoring. The current effort will assist in formulating naïve concepts of DL deployment based on innovation in multiple healthcare sectors. Meanwhile, the present study does not include much about the function of DL in the healthcare security framework as this is a broad and complex field requiring vast amounts and vast amounts of continuous research. The author sees the review under study as a plan to investigate intelligent healthcare monitoring diagnosis and treatment in the context of DL frameworks.

References

[1] P.A. Corning, Synergy and self-organization in the evolution of complex systems, Systems Research 12 (1995) 89–121.
[2] F. Bukachi, N. Pakenham-Walsh, Information technology for health in developing countries, Chest 132 (2007) 1624–1630.
[3] P.K. Reddy, R. Ankaiah, A framework of information technology-based agriculture information dissemination system to improve crop productivity, Current Science 88 (2005) 1905–1913.
[4] J.-C. Thill, Geographic information systems for transportation in perspective, Transportation Research Part C: Emerging Technologies 8 (2000) 3–12.
[5] S.S. Intille, Emerging Technology for Studying Daily Life, 2012.
[6] D. Ali, S. Frimpong, Artificial intelligence, machine learning and process automation: existing knowledge frontier and way forward for mining sector, Artificial Intelligence Review 53 (2020) 6025–6042.
[7] R. Balasubramanian, A. Libarikian, D. McElhaney, Insurance 2030—The Impact of AI on the Future of Insurance, McKinsey & Company, 2018.
[8] S. Negash, P. Musa, D. Vogel, S. Sahay, Healthcare Information Technology for Development: Improvements in People's Lives through Innovations in the Uses of Technologies, vol 24, Taylor & Francis, 2018, pp. 189–197.
[9] Z. Ahmed, K. Mohamed, S. Zeeshan, X. Dong, Artificial intelligence with multi-functional machine learning platform development for better healthcare and precision medicine, Database 2020 (2020).
[10] R. Morton, A. Tong, K. Howard, P. Snelling, A. Webster, The views of patients and carers in treatment decision making for chronic kidney disease: systematic review and thematic synthesis of qualitative studies, BMJ 340 (2010).
[11] R. Miotto, L. Li, J.T. Dudley, Deep learning to predict patient future diseases from the electronic health records, in: European Conference on Information Retrieval, 2016, pp. 768–774.
[12] O.S. Albahri, A. Zaidan, B. Zaidan, M. Hashim, A.S. Albahri, M. Alsalem, Real-time remote health-monitoring Systems in a Medical Centre: a review of the provision of healthcare services-based body sensor information, open challenges and methodological aspects, Journal of Medical Systems 42 (2018) 1–47.
[13] P. Greenland, M.D. Knoll, J. Stamler, J.D. Neaton, A.R. Dyer, D.B. Garside, et al., Major risk factors as antecedents of fatal and nonfatal coronary heart disease events, JAMA 290 (2003) 891–897.
[14] M. Herland, T.M. Khoshgoftaar, R. Wald, A review of data mining using big data in health informatics, Journal of Big data 1 (2014) 1–35.
[15] D. Mechanic, The Truth About Health Care, Rutgers University Press, 2006.
[16] F. Galbusera, G. Casaroli, T. Bassani, Artificial intelligence and machine learning in spine research, JOR spine 2 (2019) e1044.
[17] G. Nguyen, S. Dlugolinsky, M. Bobák, V. Tran, Á. López García, I. Heredia, et al., Machine learning and deep learning frameworks and libraries for large-scale data mining: a survey, Artificial Intelligence Review 52 (2019) 77–124.
[18] F. Abid, M. Alam, M. Yasir, C. Li, Sentiment analysis through recurrent variants latterly on convolutional neural network of Twitter, Future Generation Computer Systems 95 (2019) 292–308.
[19] S.B. Zaman, N. Hossain, S. Ahammed, Z. Ahmed, Contexts and opportunities of e-health technology in medical care, Journal of Medical Research and Innovation 1 (2017) AV1–AV4.
[20] A. Gasmi, Deep Learning and Health Informatics for Smart Monitoring and Diagnosis, 2022 arXiv preprint arXiv:2208.03143.
[21] M. Aghbashlo, W. Peng, M. Tabatabaei, S.A. Kalogirou, S. Soltanian, H. Hosseinzadeh-Bandbafha, et al., Machine learning technology in biodiesel research: a review, Progress in Energy and Combustion Science 85 (2021) 100904.
[22] A. Bohr, K. Memarzadeh, The rise of artificial intelligence in healthcare applications, in: Artificial Intelligence in Healthcare, Elsevier, 2020, pp. 25–60.
[23] D. George, H. Shen, E. Huerta, Deep Transfer Learning: A New Deep Learning Glitch Classification Method for Advanced LIGO, arXiv preprint arXiv:1706.07446, 2017.
[24] M. Chan, D. Estève, J.-Y. Fourniols, C. Escriba, E. Campo, Smart wearable systems: current status and future challenges, Artificial Intelligence in Medicine 56 (2012) 137–156.
[25] C. Yu, J. Liu, S. Nemati, G. Yin, Reinforcement learning in healthcare: a survey, ACM Computing Surveys 55 (2021) 1–36.
[26] S. Benzekry, Artificial intelligence and mechanistic modeling for clinical decision making in oncology, Clinical Pharmacology & Therapeutics 108 (2020) 471–486.
[27] T. Chen, E. Keravnou-Papailiou, G. Antoniou, Medical analytics for healthcare intelligence–Recent advances and future directions, Artificial Intelligence in Medicine 112 (2021) 1–5.
[28] D.C. Rose, I. Arel, T.P. Karnowski, V.C. Paquit, Applying deep-layered clustering to mammography image analytics, in: 2010 Biomedical Sciences and Engineering Conference, 2010, pp. 1–4.
[29] E. Thomaz, I. Essa, G.D. Abowd, A practical approach for recognizing eating moments with wrist-mounted inertial sensing, Proceedings of the 2015 ACM international joint conference on pervasive and ubiquitous computing (2015) 1029–1040.
[30] C.L. Wells, Physical therapist management of patients with ventricular assist devices: key considerations for the acute care physical therapist, Physical Therapy 93 (2013) 266–278.
[31] K.E. Sleeman, M. De Brito, S. Etkind, K. Nkhoma, P. Guo, I.J. Higginson, et al., The escalating global burden of serious health-related suffering: projections to 2060 by world regions, age groups, and health conditions, Lancet Global Health 7 (2019) e883–e892.
[32] F. Camara, N. Bellotto, S. Cosar, D. Nathanael, M. Althoff, J. Wu, et al., Pedestrian models for autonomous driving Part I: low-level models, from sensing to tracking, IEEE Transactions on Intelligent Transportation Systems 22 (2020) 6131–6151.

[33] C. Lu, Z. Wang, B. Zhou, Intelligent fault diagnosis of rolling bearing using hierarchical convolutional network based health state classification, Advanced Engineering Informatics 32 (2017) 139–151.
[34] I. Domingues, G. Pereira, P. Martins, H. Duarte, J. Santos, P.H. Abreu, Using deep learning techniques in medical imaging: a systematic review of applications on CT and PET, Artificial Intelligence Review 53 (2020) 4093–4160.
[35] M. Wu, J. Luo, Wearable technology applications in healthcare: a literature review, Online Journal of Nursing Informatics 23 (2019).
[36] A. Sanchez-Comas, K. Synnes, D. Molina-Estren, A. Troncoso-Palacio, Z. Comas-González, Correlation analysis of different measurement places of galvanic skin response in test groups facing pleasant and unpleasant stimuli, Sensors 21 (2021) 4210.
[37] S.-S. Lin, C.-W. Lan, H.-Y. Hsu, S.-T. Chen, Data analytics of a wearable device for heat stroke detection, Sensors 18 (2018) 4347.
[38] O.T. Inan, P.-F. Migeotte, K.-S. Park, M. Etemadi, K. Tavakolian, R. Casanella, et al., Ballistocardiography and seismocardiography: a review of recent advances, IEEE Journal of Biomedical and Health Informatics 19 (2014) 1414–1427.
[39] S. Liu, S. Liu, W. Cai, S. Pujol, R. Kikinis, D. Feng, Early diagnosis of Alzheimer's disease with deep learning, in: 2014 IEEE 11th International Symposium on Biomedical Imaging (ISBI), 2014, pp. 1015–1018.
[40] T. Brosch, R. Tam, A. S. D. N. Initiative, Manifold learning of brain MRIs by deep learning, in: International Conference on Medical Image Computing and Computer-Assisted Intervention, 2013, pp. 633–640.
[41] A. Prasoon, K. Petersen, C. Igel, F. Lauze, E. Dam, M. Nielsen, Deep feature learning for knee cartilage segmentation using a triplanar convolutional neural network, in: International Conference on Medical Image Computing and Computer-Assisted Intervention, 2013, pp. 246–253.
[42] Y. Yoo, T. Brosch, A. Traboulsee, D.K. Li, R. Tam, Deep learning of image features from unlabeled data for multiple sclerosis lesion segmentation, in: International Workshop on Machine Learning in Medical Imaging, 2014, pp. 117–124.
[43] J.-Z. Cheng, D. Ni, Y.-H. Chou, J. Qin, C.-M. Tiu, Y.-C. Chang, et al., Computer-aided diagnosis with deep learning architecture: applications to breast lesions in US images and pulmonary nodules in CT scans, Scientific Reports 6 (2016) 1–13.
[44] V. Gulshan, L. Peng, M. Coram, M.C. Stumpe, D. Wu, A. Narayanaswamy, et al., Development and validation of a deep learning algorithm for detection of diabetic retinopathy in retinal fundus photographs, JAMA 316 (2016) 2402–2410.
[45] A. Esteva, B. Kuprel, R.A. Novoa, J. Ko, S.M. Swetter, H.M. Blau, et al., Dermatologist-level classification of skin cancer with deep neural networks, Nature 542 (2017) 115–118.
[46] Y. Cheng, F. Wang, P. Zhang, J. Hu, Risk prediction with electronic health records: a deep learning approach, in: Proceedings of the 2016 SIAM International Conference on Data Mining, 2016, pp. 432–440.
[47] Z.C. Lipton, D.C. Kale, C. Elkan, R. Wetzel, Learning to Diagnose with LSTM Recurrent Neural Networks, arXiv preprint arXiv:1511.03677, 2015.
[48] T. Pham, T. Tran, D. Phung, S. Venkatesh, Deepcare: a deep dynamic memory model for predictive medicine, in: Pacific-asia Conference on Knowledge Discovery and Data Mining, 2016, pp. 30–41.
[49] R. Miotto, L. Li, B.A. Kidd, J.T. Dudley, Deep patient: an unsupervised representation to predict the future of patients from the electronic health records, Scientific Reports 6 (2016) 1–10.
[50] Z. Sicong, X. Xiaoyao, X. Yang, Intrusion detection method based on a deep convolutional neural network, Journal of Tsinghua University 59 (2019) 44–52.
[51] Z. Che, D. Kale, W. Li, M.T. Bahadori, Y. Liu, Deep computational phenotyping, in: Proceedings of the 21th ACM SIGKDD International Conference on Knowledge Discovery and Data Mining, 2015, pp. 507–516.
[52] T.A. Lasko, J.C. Denny, M.A. Levy, Computational phenotype discovery using unsupervised feature learning over noisy, sparse, and irregular clinical data, PLoS One 8 (2013) e66341.
[53] E. Choi, M.T. Bahadori, A. Schuetz, W.F. Stewart, J. Sun, Doctor ai: predicting clinical events via recurrent neural networks, in: Machine Learning for Healthcare Conference, 2016, pp. 301–318.
[54] N. Wickramasinghe, A Convolutional Net for Medical Records, 2017.
[55] N. Razavian, J. Marcus, D. Sontag, Multi-task prediction of disease onsets from longitudinal laboratory tests, in: Machine Learning for Healthcare Conference, 2016, pp. 73–100.
[56] F. Dernoncourt, J.Y. Lee, O. Uzuner, P. Szolovits, De-identification of patient notes with recurrent neural networks, Journal of the American Medical Informatics Association 24 (2017) 596–606.
[57] M.H. Memon, J.P. Li, A.U. Haq, M.H. Memon, W. Zhou, Breast cancer detection in the IOT health environment using modified recursive feature selection, Wireless Communications and Mobile Computing 2019 (2019).
[58] D.R. Kelley, J. Snoek, J.L. Rinn, Basset: learning the regulatory code of the accessible genome with deep convolutional neural networks, Genome Research 26 (2016) 990–999.
[59] B. Alipanahi, A. Delong, M.T. Weirauch, B.J. Frey, Predicting the sequence specificities of DNA-and RNA-binding proteins by deep learning, Nature Biotechnology 33 (2015) 831–838.
[60] C. Angermueller, H.J. Lee, W. Reik, O. Stegle, DeepCpG: accurate prediction of single-cell DNA methylation states using deep learning, Genome Biology 18 (2017) 1–13.
[61] P.W. Koh, E. Pierson, A. Kundaje, Denoising genome-wide histone ChIP-seq with convolutional neural networks, Bioinformatics 33 (2017) i225–i233.
[62] R. Fakoor, F. Ladhak, A. Nazi, M. Huber, Using deep learning to enhance cancer diagnosis and classification, in: Proceedings of the International Conference on Machine Learning, 2013, pp. 3937–3949.

[63] J. Lyons, A. Dehzangi, R. Heffernan, A. Sharma, K. Paliwal, A. Sattar, et al., Predicting backbone Cα angles and dihedrals from protein sequences by stacked sparse auto-encoder deep neural network, Journal of Computational Chemistry 35 (2014) 2040–2046.
[64] N.Y. Hammerla, S. Halloran, T. Plötz, Deep, Convolutional, and Recurrent Models for Human Activity Recognition Using Wearables, arXiv preprint arXiv:1604.08880, 2016.
[65] J. Zhu, A. Pande, P. Mohapatra, J.J. Han, Using deep learning for energy expenditure estimation with wearable sensors, in: 2015 17th International Conference on E-health Networking, Application & Services (HealthCom), 2015, pp. 501–506.
[66] V. Jindal, J. Birjandtalab, M.B. Pouyan, M. Nourani, An adaptive deep learning approach for PPG-based identification, in: 2016 38th Annual International Conference of the IEEE Engineering in Medicine and Biology Society (EMBC), 2016, pp. 6401–6404.
[67] E. Nurse, B.S. Mashford, A.J. Yepes, I. Kiral-Kornek, S. Harrer, D.R. Freestone, Decoding EEG and LFP signals using deep learning: heading TrueNorth, in: Proceedings of the ACM International Conference on Computing Frontiers, 2016, pp. 259–266.
[68] A. Sathyanarayana, S. Joty, L. Fernandez-Luque, F. Ofli, J. Srivastava, A. Elmagarmid, et al., Sleep quality prediction from wearable data using deep learning, JMIR mHealth and uHealth 4 (2016) e6562.
[69] L. Zhang, L. Zhang, B. Du, Deep learning for remote sensing data: a technical tutorial on the state of the art, IEEE Geoscience and Remote Sensing Magazine 4 (2016) 22–40.
[70] S. Min, B. Lee, S. Yoon, Deep learning in bioinformatics, Briefings in Bioinformatics 18 (2017) 851–869.
[71] D. Griffiths, J. Boehm, A review on deep learning techniques for 3D sensed data classification, Remote Sensing 11 (2019) 1499.
[72] G.E. Hinton, Deep belief networks, Scholarpedia 4 (2009) 5947.
[73] Z. Shi, C. Miao, U.J. Schoepf, R.H. Savage, D.M. Dargis, C. Pan, et al., A clinically applicable deep-learning model for detecting intracranial aneurysm in computed tomography angiography images, Nature Communications 11 (2020) 1–11.
[74] A. Fischer, C. Igel, An introduction to restricted Boltzmann machines, in: Iberoamerican Congress on Pattern Recognition, 2012, pp. 14–36.
[75] X. Wang, H. Wang, Z. Wang, S. Lu, Y. Fan, Risk spillover network structure learning for correlated financial assets: a directed acyclic graph approach, Information Sciences 580 (2021) 152–173.
[76] A.M. Saxe, J.L. McClelland, S. Ganguli, A mathematical theory of semantic development in deep neural networks, Proceedings of the National Academy of Sciences 116 (2019) 11537–11546.
[77] Y.-C. Wu, J.-W. Feng, Development and application of artificial neural network, Wireless Personal Communications 102 (2018) 1645–1656.
[78] I.M. Nasser, S.S. Abu-Naser, Predicting Tumor Category Using Artificial Neural Networks, 2019.
[79] C.-L. Chuang, Case-based reasoning support for liver disease diagnosis, Artificial Intelligence in Medicine 53 (2011) 15–23.
[80] M.M. Musleh, E. Alajrami, A.J. Khalil, B.S. Abu-Nasser, A.M. Barhoom, S.A. Naser, Predicting liver patients using artificial neural network, International Journal of Academic Information Systems Research (IJAISR) 3 (2019).
[81] J. Chen, D. Remulla, J.H. Nguyen, Y. Liu, P. Dasgupta, A.J. Hung, Current status of artificial intelligence applications in urology and their potential to influence clinical practice, BJU International 124 (2019) 567–577.
[82] P.-H.C. Chen, K. Gadepalli, R. MacDonald, Y. Liu, S. Kadowaki, K. Nagpal, et al., An augmented reality microscope with real-time artificial intelligence integration for cancer diagnosis, Nature Medicine 25 (2019) 1453–1457.
[83] Ö. Yõldõrõm, P. Pławiak, R.-S. Tan, U.R. Acharya, Arrhythmia detection using deep convolutional neural network with long duration ECG signals, Computers in Biology and Medicine 102 (2018) 411–420.
[84] A. Nithya, A. Appathurai, N. Venkatadri, D. Ramji, C.A. Palagan, Kidney disease detection and segmentation using artificial neural network and multi-kernel k-means clustering for ultrasound images, Measurement 149 (2020) 106952.
[85] M. Owais, M. Arsalan, J. Choi, T. Mahmood, K.R. Park, Artificial intelligence-based classification of multiple gastrointestinal diseases using endoscopy videos for clinical diagnosis, Journal of Clinical Medicine 8 (2019) 986.
[86] H. Luo, G. Xu, C. Li, L. He, L. Luo, Z. Wang, et al., Real-time artificial intelligence for detection of upper gastrointestinal cancer by endoscopy: a multicentre, case-control, diagnostic study, The Lancet Oncology 20 (2019) 1645–1654.
[87] M.A. Khan, M.A. Khan, F. Ahmed, M. Mittal, L.M. Goyal, D.J. Hemanth, et al., Gastrointestinal diseases segmentation and classification based on duo-deep architectures, Pattern Recognition Letters 131 (2020) 193–204.
[88] W. Gouda, R. Yasin, COVID-19 disease: CT pneumonia analysis prototype by using artificial intelligence, predicting the disease severity, Egyptian Journal of Radiology and Nuclear Medicine 51 (2020) 1–11.
[89] S. Vasal, S. Jain, A. Verma, COVID-AI: an artificial intelligence system to diagnose COVID 19 disease, Journal of Engineering Research and Technology 9 (2020) 1–6.
[90] S. Minaee, R. Kafieh, M. Sonka, S. Yazdani, G.J. Soufi, Deep-COVID: predicting COVID-19 from chest X-ray images using deep transfer learning, Medical Image Analysis 65 (2020) 101794.
[91] M. Arsalan, M. Owais, T. Mahmood, S.W. Cho, K.R. Park, Aiding the diagnosis of diabetic and hypertensive retinopathy using artificial intelligence-based semantic segmentation, Journal of Clinical Medicine 8 (2019) 1446.
[92] H. Kanegae, K. Suzuki, K. Fukatani, T. Ito, N. Harada, K. Kario, Highly precise risk prediction model for new-onset hypertension using artificial intelligence techniques, The Journal of Clinical Hypertension 22 (2020) 445–450.
[93] D.G. Kiely, O. Doyle, E. Drage, H. Jenner, V. Salvatelli, F.A. Daniels, et al., Utilising artificial intelligence to determine patients at risk of a rare disease: idiopathic pulmonary arterial hypertension, Pulmonary Circulation 9 (2019), 2045894019890549.
[94] H. Kaur, V. Kumari, Predictive modelling and analytics for diabetes using a machine learning approach, Applied Computing and Informatics (2020).

[95] R.B. Lukmanto, E. Irwansyah, The early detection of diabetes mellitus (DM) using fuzzy hierarchical model, Procedia Computer Science 59 (2015) 312–319.
[96] G. Swapna, R. Vinayakumar, K. Soman, Diabetes detection using deep learning algorithms, ICT Express 4 (2018) 243–246.
[97] N.-H. Lai, W.-C. Shen, C.-N. Lee, J.-C. Chang, M.-C. Hsu, L.-N. Kuo, et al., Comparison of the predictive outcomes for anti-tuberculosis drug-induced hepatotoxicity by different machine learning techniques, Computer Methods and Programs in Biomedicine 188 (2020) 105307.
[98] X.W. Gao, C. James-Reynolds, E. Currie, Analysis of tuberculosis severity levels from CT pulmonary images based on enhanced residual deep learning architecture, Neurocomputing 392 (2020) 233–244.
[99] R. Rajalakshmi, R. Subashini, R.M. Anjana, V. Mohan, Automated diabetic retinopathy detection in smartphone-based fundus photography using artificial intelligence, Eye 32 (2018) 1138–1144.
[100] T.D. Keenan, T.E. Clemons, A. Domalpally, M.J. Elman, M. Havilio, E. Agrón, et al., Retinal specialist versus artificial intelligence detection of retinal fluid from OCT: age-related eye disease study 2: 10-year follow-on study, Ophthalmology 128 (2021) 100–109.
[101] V. Sarao, D. Veritti, P. Lanzetta, Automated diabetic retinopathy detection with two different retinal imaging devices using artificial intelligence: a comparison study, Graefe's Archive for Clinical and Experimental Ophthalmology 258 (2020) 2647–2654.
[102] E. Shkolyar, X. Jia, T.C. Chang, D. Trivedi, K.E. Mach, M.Q.-H. Meng, et al., Augmented bladder tumor detection using deep learning, European Urology 76 (2019) 714–718.
[103] B. Ljubic, S. Roychoudhury, X.H. Cao, M. Pavlovski, S. Obradovic, R. Nair, et al., Influence of medical domain knowledge on deep learning for Alzheimer's disease prediction, Computer Methods and Programs in Biomedicine 197 (2020) 105765.
[104] R. Janghel, Y. Rathore, Deep convolution neural network based system for early diagnosis of Alzheimer's disease, Irbm 42 (2021) 258–267.
[105] F. Ahmed, An Internet of Things (IoT) application for predicting the quantity of future heart attack patients, International Journal of Computer Applications 164 (2017) 36–40.
[106] D.P. Isravel, S. Silas, Improved heart disease diagnostic IoT model using machine learning techniques, Neuroscience 9 (2020) 4442–4446.
[107] S. Nashif, M.R. Raihan, M.R. Islam, M.H. Imam, Heart disease detection by using machine learning algorithms and a real-time cardiovascular health monitoring system, World Journal of Engineering and Technology 6 (2018) 854–873.
[108] J.-E. Bibault, L. Xing, Screening for chronic obstructive pulmonary disease with artificial intelligence, The Lancet Digital Health 2 (2020) e216–e217.
[109] G. Battineni, G.G. Sagaro, N. Chinatalapudi, F. Amenta, Applications of machine learning predictive models in the chronic disease diagnosis, Journal of Personalized Medicine 10 (2020) 21.
[110] T.H. Aldhyani, A.S. Alshebami, M.Y. Alzahrani, Soft clustering for enhancing the diagnosis of chronic diseases over machine learning algorithms, Journal of Healthcare Engineering 2020 (2020).
[111] D.d.A. Rodrigues, R.F. Ivo, S.C. Satapathy, S. Wang, J. Hemanth, P.P. Reboucas Filho, A new approach for classification skin lesion based on transfer learning, deep learning, and IoT system, Pattern Recognition Letters 136 (2020) 8–15.
[112] A. Das, U.R. Acharya, S.S. Panda, S. Sabut, Deep learning based liver cancer detection using watershed transform and Gaussian mixture model techniques, Cognitive Systems Research 54 (2019) 165–175.
[113] M. Fukuda, K. Inamoto, N. Shibata, Y. Ariji, Y. Yanashita, S. Kutsuna, et al., Evaluation of an artificial intelligence system for detecting vertical root fracture on panoramic radiography, Oral Radiology 36 (2020) 337–343.
[114] C.S. Chu, N.P. Lee, J. Adeoye, P. Thomson, S.W. Choi, Machine learning and treatment outcome prediction for oral cancer, Journal of Oral Pathology & Medicine 49 (2020) 977–985.
[115] J.C. Rodrigues, A.M. Amadu, A.G. Dastidar, G.V. Szantho, S.M. Lyen, C. Godsave, et al., Comprehensive characterisation of hypertensive heart disease left ventricular phenotypes, Heart 102 (2016) 1671–1679.
[116] N.S. Parikh, A. Chatterjee, I. Díaz, A. Pandya, A.E. Merkler, G. Gialdini, et al., Modeling the impact of interhospital transfer network design on stroke outcomes in a large city, Stroke 49 (2018) 370–376.
[117] T. Nazir, A. Irtaza, Z. Shabbir, A. Javed, U. Akram, M.T. Mahmood, Diabetic retinopathy detection through novel tetragonal local octa patterns and extreme learning machines, Artificial Intelligence in Medicine 99 (2019) 101695.
[118] S. Ansari, I. Shafi, A. Ansari, J. Ahmad, S.I. Shah, Diagnosis of liver disease induced by hepatitis virus using artificial neural networks, in: 2011 IEEE 14th International Multitopic Conference, 2011, pp. 8–12.
[119] M. Hosseinzadeh, O.H. Ahmed, M.Y. Ghafour, F. Safara, S. Ali, B. Vo, et al., A multiple multilayer perceptron neural network with an adaptive learning algorithm for thyroid disease diagnosis in the internet of medical things, The Journal of Supercomputing 77 (2021) 3616–3637.
[120] K. Oh, Y.-C. Chung, K.W. Kim, W.-S. Kim, I.-S. Oh, Classification and visualization of Alzheimer's disease using volumetric convolutional neural network and transfer learning, Scientific Reports 9 (2019) 1–16.
[121] A. Ostovar, E. Ehsani-Chimeh, Z. Fakoorfard, The diagnostic value of CT scans in the process of diagnosing COVID 19 in medical centers, Health Technology Assessment in Action (2020).
[122] D.C. Yadav, S. Pal, Prediction of thyroid disease using decision tree ensemble method, Human-Intelligent Systems Integration 2 (2020) 89–95.
[123] D. Tegunov, P. Cramer, Real-time cryo-electron microscopy data preprocessing with Warp, Nature Methods 16 (2019) 1146–1152.
[124] L.Y. Tang, H.O. Coxson, S. Lam, J. Leipsic, R.C. Tam, D.D. Sin, Towards large-scale case-finding: training and validation of residual networks for detection of chronic obstructive pulmonary disease using low-dose CT, The Lancet Digital Health 2 (2020) e259–e267.
[125] T. Jo, K. Nho, A.J. Saykin, Deep learning in Alzheimer's disease: diagnostic classification and prognostic prediction using neuroimaging data, Frontiers in Aging Neuroscience 11 (2019) 220.

[126] G. Damiani, E. Grossi, E. Berti, R. Conic, U. Radhakrishna, A. Pacifico, et al., Artificial neural networks allow response prediction in squamous cell carcinoma of the scalp treated with radiotherapy, Journal of the European Academy of Dermatology and Venereology 34 (2020) 1369–1373.
[127] F.C. Morabito, M. Campolo, C. Ieracitano, J.M. Ebadi, L. Bonanno, A. Bramanti, et al., Deep convolutional neural networks for classification of mild cognitive impaired and Alzheimer's disease patients from scalp EEG recordings, in: 2016 IEEE 2nd International Forum on Research and Technologies for Society and Industry Leveraging a Better Tomorrow (RTSI), 2016, pp. 1–6.
[128] W.-C. Wang, L.-B. Chen, W.-J. Chang, Development and experimental evaluation of machine-learning techniques for an intelligent hairy scalp detection system, Applied Sciences 8 (2018) 853.
[129] M.K. Sherwani, A. Aziz, F. Calimeri, Role of deep learning for smart health care, in: Computational Intelligence Techniques for Green Smart Cities, Springer, 2022, pp. 169–186.
[130] R.A. Rayan, I. Zafar, C. Tsagkaris, Artificial intelligence and big data solutions for COVID-19, in: Intelligent Data Analysis for COVID-19 Pandemic, Springer, 2021, pp. 115–127.
[131] R.A. Rayan, C. Tsagkaris, I. Zafar, D.V. Moysidis, A.S. Papazoglou, Big data analytics for health: a comprehensive review of techniques and applications, in: Big Data Analytics for Healthcare, 2022, pp. 83–92.
[132] R.A. Rayan, I. Zafar, Monitoring technologies for precision health, in: The Smart Cyber Ecosystem for Sustainable Development, 2021, pp. 251–260.
[133] I. Zafar, A. Rafique, J. Fazal, M. Manzoor, Q.U. Ain, R.A. Rayan, Genome and gene editing by artificial intelligence programs, in: Advanced AI Techniques and Applications in Bioinformatics, CRC Press, 2021, pp. 165–188.
[134] D.M. Roden, J.M. Pulley, M.A. Basford, G.R. Bernard, E.W. Clayton, J.R. Balser, et al., Development of a large-scale de-identified DNA biobank to enable personalized medicine, Clinical Pharmacology & Therapeutics 84 (2008) 362–369.
[135] I. Degtiar, A review of international coverage and pricing strategies for personalized medicine and orphan drugs, Health Policy 121 (2017) 1240–1248.
[136] G.S. Ginsburg, K.A. Phillips, Precision medicine: from science to value, Health Affairs 37 (2018) 694–701.
[137] M.A. Hamburg, F.S. Collins, The path to personalized medicine, New England Journal of Medicine 363 (2010) 301–304.
[138] H. Bolhasani, M. Mohseni, A.M. Rahmani, Deep learning applications for IoT in health care: a systematic review, Informatics in Medicine Unlocked 23 (2021) 100550.
[139] R. Alkawadri, Brain–computer interface (BCI) applications in mapping of epileptic brain networks based on intracranial-EEG: an update, Frontiers in Neuroscience 13 (2019) 191.
[140] L. Huang, G. van Luijtelaar, Brain computer interface for epilepsy treatment, in: Brain-Computer Interface Systems-Recent Progress and Future Prospects, 2013.
[141] M.P. Hosseini, Brain-computer Interface for Analyzing Epileptic Big Data, Rutgers University-School of Graduate Studies, 2018.
[142] M. Momin, N.S. Bhagwat, S. Chavhate, A.V. Dhiwar, N. Devekar, Smart body monitoring system using IoT and machine learning, International Journal of Advanced Research in Electrical, Electronics and Instrumentation Engineering 8 (2019) 1501–1506.
[143] L.W. Braun, M.A. Martins, J. Romanini, P.V. Rados, M.D. Martins, V.C. Carrard, Continuing education activities improve dentists' self-efficacy to manage oral mucosal lesions and oral cancer, European Journal of Dental Education 25 (2021) 28–34.
[144] B. Ballinger, J. Hsieh, A. Singh, N. Sohoni, J. Wang, G.H. Tison, et al., DeepHeart: semi-supervised sequence learning for cardiovascular risk prediction, in: Thirty-Second AAAI Conference on Artificial Intelligence, 2018.
[145] W. He, Y. Zhao, H. Tang, C. Sun, W. Fu, A wireless BCI and BMI system for wearable robots, IEEE Transactions on Systems, Man, and Cybernetics: Systems 46 (2015) 936–946.
[146] S. Tuli, N. Basumatary, S.S. Gill, M. Kahani, R.C. Arya, G.S. Wander, et al., HealthFog: an ensemble deep learning based smart healthcare system for automatic diagnosis of heart diseases in integrated IoT and fog computing environments, Future Generation Computer Systems 104 (2020) 187–200.
[147] N. Mäkitalo, A. Ometov, J. Kannisto, S. Andreev, Y. Koucheryavy, T. Mikkonen, Safe, secure executions at the network edge: coordinating cloud, edge, and fog computing, IEEE Software 35 (2017) 30–37.
[148] K. Kamnitsas, C. Ledig, V.F. Newcombe, J.P. Simpson, A.D. Kane, D.K. Menon, et al., Efficient multi-scale 3D CNN with fully connected CRF for accurate brain lesion segmentation, Medical Image Analysis 36 (2017) 61–78.
[149] B. Albert, J. Zhang, A. Noyvirt, R. Setchi, H. Sjaaheim, S. Velikova, et al., Automatic EEG processing for the early diagnosis of traumatic brain injury, Procedia Computer Science 96 (2016) 703–712.
[150] U.R. Acharya, S.L. Oh, Y. Hagiwara, J.H. Tan, H. Adeli, D.P. Subha, Automated EEG-based screening of depression using deep convolutional neural network, Computer Methods and Programs in Biomedicine 161 (2018) 103–113.
[151] M.S. Hossain, S.U. Amin, M. Alsulaiman, G. Muhammad, Applying deep learning for epilepsy seizure detection and brain mapping visualization, ACM Transactions on Multimedia Computing, Communications, and Applications 15 (2019) 1–17.
[152] P.P. Rebouças Filho, R.M. Sarmento, G.B. Holanda, D. de Alencar Lima, New approach to detect and classify stroke in skull CT images via analysis of brain tissue densities, Computer Methods and Programs in Biomedicine 148 (2017) 27–43.
[153] H. Masoumi, A. Behrad, M.A. Pourmina, A. Roosta, Automatic liver segmentation in MRI images using an iterative watershed algorithm and artificial neural network, Biomedical Signal Processing and Control 7 (2012) 429–437.
[154] O. Faust, A. Shenfield, M. Kareem, T.R. San, H. Fujita, U.R. Acharya, Automated detection of atrial fibrillation using long short-term memory network with RR interval signals, Computers in Biology and Medicine 102 (2018) 327–335.

[155] M. Jiang, C. Lu, C. Zhang, J. Yang, Y. Tan, A. Lu, et al., Syndrome differentiation in modern research of traditional Chinese medicine, Journal of Ethnopharmacology 140 (2012) 634−642.

[156] F. Bray, J. Ferlay, I. Soerjomataram, R.L. Siegel, L.A. Torre, A. Jemal, Global cancer statistics 2018: GLOBOCAN estimates of incidence and mortality worldwide for 36 cancers in 185 countries, CA: A Cancer Journal for Clinicians 68 (2018) 394−424.

[157] Z. Ma, J.M.R. Tavares, A novel approach to segment skin lesions in dermoscopic images based on a deformable model, IEEE Journal of Biomedical and Health Informatics 20 (2015) 615−623.

[158] R.T. Schirrmeister, J.T. Springenberg, L.D.J. Fiederer, M. Glasstetter, K. Eggensperger, M. Tangermann, et al., Deep learning with convolutional neural networks for EEG decoding and visualization, Human Brain Mapping 38 (2017) 5391−5420.

[159] D. Mendes, M. Lopes, P. Parreira, C. Fonseca, Deep Learning and IoT to Assist Multimorbidity Home Based Healthcare, 2017.

[160] G.M. Sandstrom, N. Lathia, C. Mascolo, P.J. Rentfrow, Opportunities for smartphones in clinical care: the future of mobile mood monitoring, The Journal of Clinical Psychiatry 77 (2016) 13476.

[161] L. Liu, J. Xu, Y. Huan, Z. Zou, S.-C. Yeh, L.-R. Zheng, A smart dental health-IoT platform based on intelligent hardware, deep learning, and mobile terminal, IEEE Journal of Biomedical and Health Informatics 24 (2019) 898−906.

[162] S. Sharma, K. Chen, A. Sheth, Toward practical privacy-preserving analytics for IoT and cloud-based healthcare systems, IEEE Internet Computing 22 (2018) 42−51.

[163] J. Klenk, L. Schwickert, L. Palmerini, S. Mellone, A. Bourke, E.A. Ihlen, et al., The FARSEEING real-world fall repository: a large-scale collaborative database to collect and share sensor signals from real-world falls, European Review of Aging and Physical Activity 13 (2016) 1−7.

[164] L.P. Malasinghe, N. Ramzan, K. Dahal, Remote patient monitoring: a comprehensive study, Journal of Ambient Intelligence and Humanized Computing 10 (2019) 57−76.

[165] J. Wei, A.D. Cheok, Foodie: play with your food promote interaction and fun with edible interface, IEEE Transactions on Consumer Electronics 58 (2012) 178−183.

[166] T. Kinnison, S.A. May, Evidence-based healthcare: the importance of effective interprofessional working for high quality veterinary services, a UK example, Veterinary Evidence 1 (2016).

[167] R.S. Tubbs, J. Malefant, M. Loukas, W. Jerry Oakes, R.J. Oskouian, F.N. Fries, Enigmatic human tails: a review of their history, embryology, classification, and clinical manifestations, Clinical Anatomy 29 (2016) 430−438.

[168] Q. Liu, P. Li, W. Zhao, W. Cai, S. Yu, V.C. Leung, A survey on security threats and defensive techniques of machine learning: a data driven view, IEEE Access 6 (2018) 12103−12117.

Chapter 2

Fermatean fuzzy approach of diseases diagnosis based on new correlation coefficient operators

Paul Augustine Ejegwa[1] **and Arun Sarkar**[2]

[1]Department of Mathematics, University of Agriculture, Makurdi, Nigeria; [2]Department of Mathematics, Heramba Chandra College, Kolkata, India

1. Introduction

Medical diagnosis procedure is often complex because of the presence of vagueness and incomplete data in most cases and as such, a mechanism for check-mating incomplete data needed to be incorporated in diagnostic process. Albeit, with the institutionalization of fuzzy set [1], the process of medical-diagnostic decision-making has been made a lots easy, though with some limitation in the construct of fuzzy sets. The surety of the setback of fuzzy set is because it only recognizes the membership grade of the decision parameters. Subject to this, Atanassov [2] initiated the conception of intuitionistic fuzzy set (IFS), which recognized both grades of membership and nonmembership, which are defined within $[0, 1]$) with the possibility of hesitation margin, and in so doing gives a trustworthy assessment unlike the fuzzy scenario. In a search for an enhanced approach, Yager [3] initiated the conception of Pythagorean fuzzy set (PFS), which has an extensive scope and an improved decision output due to its flexibility. Be that as it may, PFS cannot be applied in a number of decision-making problems when the aggregate of the square of the grades of nonmembership and membership exceeds one. And so, the idea of Fermatean fuzzy set (FFS) [4,5] was conceived to cater for such cases.

In terms of application, IFS has been used in many areas of applications [6–8]. Similarly, solutions of many real-world problems have been proffered using the idea of PFSs [9–12] based on distance measures, similarity measures, aggregate operators, etc. Quite a lot of the applications of FFSs have been investigated by means of some information measures. Liu et al. [13] considered Fermatean fuzzy linguistic set (FFLS) and deliberated on the application in decision-making. Liu et al. [14] discussed Fermatean fuzzy distance measure (FFDM) by means of linguistic scale function to demonstrate TODIM and TOPSIS methods. Senapati and Yager [15] deliberated on the application of FFSs using weighted averaging/geometric operators, and some operations on FFSs were studied and used in decision-making [16]. Mehdi et al. [17] discussed the process of green construction supplier evaluation by means of an innovative Fermatean fuzzy decision-making approach. Garg et al. [18] discussed analysis of decision-making using aggregation operators under FFSs with application to COVID-19 testing facility. Ejegwa et al. [19] presented Fermatean fuzzy composite relation and discussed its application in diagnostic analysis, and a number of score functions on FFSs were discussed and applied in the process of bride selection [20]. Aydin [21] deliberated on fuzzy multicriteria decision-making method using Fermatean fuzzy theories. The concepts of SAW, VIKOR, and ARAS were extended to FFS and applied to the selection of COVID-19 testing laboratory [22], and the ideas of FFSs were studied from continuous perspective, and Fermatean fuzzy differential calculus was presented in Ref. [23]. By using Hamacher interactive geometric operators, Shahzadi et al. [24] discussed the application of FFSs in multiple-attribute decision-making. A case of admission process into a higher institution was resolved by means of Fermatean fuzzy data based on similarity operators [25]. A new-fangled Fermatean fuzzy composite relation which enriched the approach in Ref. [19] was investigated and applied in medical-diagnostic decision-making [26] and pattern recognition [27], respectively. Zeng et al. [28] presented a contemporary application of FFSs in the evaluation of online teaching quality based on analytical hierarchy processes using a Fermatean fuzzy aggregation operator. Most recently, Akram et al. [29] studied a hybrid model of FFSs with complex fuzzy sets and soft sets and discussed its application.

Deep Learning in Personalized Healthcare and Decision Support. https://doi.org/10.1016/B978-0-443-19413-9.00021-7

Amid several applicable information measures in computational intelligence, the notion of correlation coefficient is quite prominent in machine learning, data analysis, pattern recognition, etc. Correlation coefficient is a tool that evaluates the strength of association between two data. In fact, not a few researchers have deployed correlation coefficient to investigate possible relations between two data. Albeit with the advent of fuzzy data, it became sufficient to investigate fuzzy correlation coefficient. The concept of correlation coefficient in intuitionistic fuzzy setting was first studied in Ref. [30] in consonant with classical correlation coefficient. Quite a lot of approaches for the computation of intuitionistic fuzzy correlation coefficient have been investigated with applications in diverse areas [31−34]. In the same way, the idea of Pythagorean fuzzy correlation coefficient was first established in the work of Garg [35]. Sundry authors have examined correlation coefficient under Pythagorean fuzzy setting with applications in numerous of areas [36−40]. Since FFS is more reliable compare to PFS, it is appropriate to as well study correlation coefficient under FFSs, and so the concept has been initiated in Fermatean fuzzy setting and applied to medical-diagnosis decision-making by the means of hypothetical data [41].

To be more specific, Kirisci [41] introduced two approaches of Fermatean fuzzy correlation coefficient (FFCC). Although the two approaches of calculating correlation coefficient in Ref. [41] are novelty in Fermatean fuzzy set theory, we observed they lack reliability due to the models formulation. The motivations for this research include;

(i) The both approaches in Ref. [41] will yield misleading result whenever anyone of the grades of membership, nonmembership, and margin of hesitation is zero.
(ii) In addition to (i), the second approach in Ref. [41] only considers the maximum information energy, which will definitely leads to error of exclusion.

Owing to these setbacks, this chapter introduces two new approaches of computing FFCC to correct the setbacks in Ref. [41]. The contributions of this article include the following, which are to;

(i) assess the FFCC approaches in Ref. [41] for possible enhancement,
(ii) develop two new-fangled approaches of FFCC, which rectifies the limitations in the existing FFCC approaches.
(iii) discuss the properties of the new FFCC approaches to validate their concord with the classical correlation coefficient's properties.
(iv) apply the new FFCC approaches to discuss diagnostic analysis using Fermatean fuzzy simulated medical data.
(v) present the significance of the new FFCC approaches over to the existing FFCC approaches based on comparative analysis.

The residue of the chapter is outlined thus: The fundamental of FFSs and some existing approaches of FFCC are presented in Section 2; Section 3 presents the two new approaches of FFCC and discusses their properties with numerical computations; Section 4 applies the new approaches to discuss the process of medical diagnostic analysis and presents comparative analysis to display the relevance of the work; and Section 5 sum-ups the chapter with recommendations for upcoming research.

2. Fermatean fuzzy sets and their correlation operators

2.1 Fermatean fuzzy sets

Some basic definitions of FFSs and operations on them are examined in this section in order to construct their correlation operator approaches. By the inclusion of a nonmembership degree, the idea of fuzzy sets was stretched to intuitionistic fuzzy sets. We designate $\mathscr{U}$ as nonempty set throughout the chapter.

Definition 2.1 [2] An IFS $\widetilde{N}$ in $\mathscr{U}$ is given by

$$\widetilde{N} = \left\{ \langle u, \mu_{\widetilde{N}}(u), \nu_{\widetilde{N}}(u) \rangle \,\middle|\, u \in \mathscr{U} \right\}, \tag{2.1}$$

where $\mu_{\widetilde{N}}, \nu_{\widetilde{N}} : \mathscr{U} \to [0, 1]$ denote the grade of membership and nonmembership for $u \in \mathscr{U}$ to the set $\widetilde{N}$, such that

$$\mu_{\widetilde{N}}(u), \nu_{\widetilde{N}}(u) \in [0, 1], \text{and } 0 \leq \mu_{\widetilde{N}}(u) + \nu_{\widetilde{N}}(u) \leq 1. \tag{2.2}$$

Indeterminacy degree for IFS is presented by $\pi_{\widetilde{N}}(u) = 1 - \mu_{\widetilde{N}}(u) - \nu_{\widetilde{N}}(u)$. For usefulness, $\left(\mu_{\widetilde{N}}(u), \nu_{\widetilde{N}}(u) \right)$ is taken as intuitionistic fuzzy number (IFN) and is denoted by $\widetilde{N} = (\mu, \nu)$.

By extending IFS, Yager [3] presented a novel set called PFS as defined in the following manner.

Definition 2.2 [3] A PFS, $\widetilde{P}$ in $\mathscr{U}$ is presented by

$$\widetilde{P} = \left\{ \langle u, \mu_{\widetilde{P}}(u), \nu_{\widetilde{P}}(u) \rangle \,|\, u \in \mathscr{U} \right\}, \tag{2.3}$$

where $\mu_{\widetilde{P}}(u), \nu_{\widetilde{P}}(u) \in [0, 1]$ denote the grades of membership and nonmembership for $u \in \mathscr{U}$ to the set $\widetilde{P}$, such that

$$0 \le \left(\mu_{\widetilde{P}}(u)\right)^2 + \left(\nu_{\widetilde{P}}(u)\right)^2 \le 1. \tag{2.4}$$

Indeterminacy degree for PFS is given by $\pi_{\widetilde{P}}(u) = \sqrt{1 - \left(\mu_{\widetilde{P}}(u)\right)^2 - \left(\nu_{\widetilde{P}}(u)\right)^2}$. With the introduction of PFS, it is clearly realized that the space of the membership and nonmembership values have been enlarged to model many real-world problems.

Definition 2.3 [5] An FFS, $\widetilde{\mathscr{F}}$ on $\mathscr{U}$ is represented by

$$\widetilde{\mathscr{F}} = \left\{ \langle u, \xi_{\widetilde{\mathscr{F}}}(u), \eta_{\widetilde{\mathscr{F}}}(u) \rangle \,|\, u \in \mathscr{U} \right\}, \tag{2.5}$$

where $\xi_{\widetilde{\mathscr{F}}}(u) \in [0, 1]$ and $\eta_{\widetilde{\mathscr{F}}}(u) \in [0, 1]$ denote the grades of membership and nonmembership, respectively, of the element $u \in \mathscr{U}$ to the set $\widetilde{\mathscr{F}}$ satisfying the condition that

$$0 \le \left(\xi_{\widetilde{\mathscr{F}}}(u)\right)^3 + \left(\eta_{\widetilde{\mathscr{F}}}(u)\right)^3 \le, \forall u \in \mathscr{U}. \tag{2.6}$$

For any FFS $\widetilde{\mathscr{F}}$ and $u \in \mathscr{U}$, the grade of indeterminacy/hesitancy, $\pi_{\widetilde{\mathscr{F}}}(u)$ of u in $\widetilde{\mathscr{F}}$ is given as

$$\pi_{\widetilde{\mathscr{F}}}(u) = \left[1 - \left(\xi_{\widetilde{\mathscr{F}}}(u)\right)^3 - \left(\eta_{\widetilde{\mathscr{F}}}(u)\right)^3\right]^{\frac{1}{3}}, \tag{2.7}$$

For simplicity, $\left(\xi_{\widetilde{\mathscr{F}}}(u), \eta_{\widetilde{\mathscr{F}}}(u)\right)$ is called a Fermatean fuzzy number (FFN) and is symbolized by $\widetilde{F} = (\xi, \eta)$.

Fig. 2.1 differentiates IFS and PFS from FFS.

Here, a definition that explains the idea of correlation coefficient under FFSs is provided.

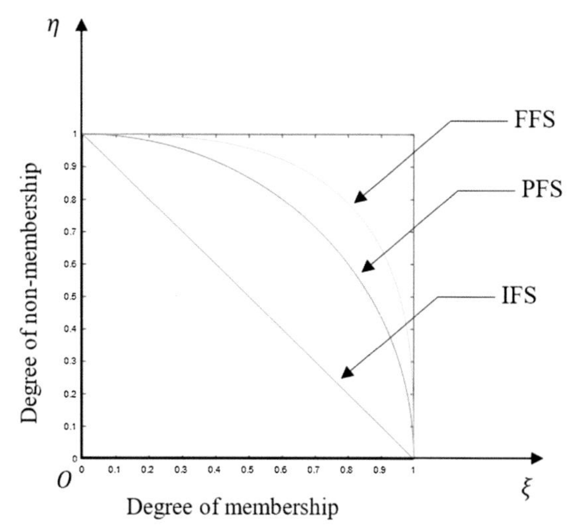

FIGURE 2.1 Graphical representation of Fermatean fuzzy set (FFS).

Definition 2.4 [41] Suppose $\widetilde{\mathscr{F}}_1$ and $\widetilde{\mathscr{F}}_2$ are FFSs in $\mathscr{U} = \{u_1, u_2, \ldots, u_n\}$, then the correlation coefficient under FFSs $\mathscr{F}_1$ and $\mathscr{F}_2$ denoted by $\varrho\left(\widetilde{\mathscr{F}}_1, \widetilde{\mathscr{F}}_2\right)$ is a function, $\varrho : \widetilde{\mathscr{F}}_1 \times \widetilde{\mathscr{F}}_2 \rightarrow [0, 1]$ which satisfies

i. $\varrho\left(\widetilde{\mathscr{F}}_1, \widetilde{\mathscr{F}}_2\right) \in [0, 1]$,

ii. $\varrho\left(\widetilde{\mathscr{F}}_1, \widetilde{\mathscr{F}}_2\right) = \varrho\left(\widetilde{\mathscr{F}}_2, \widetilde{\mathscr{F}}_1\right)$,

iii. $\varrho\left(\widetilde{\mathscr{F}}_1, \widetilde{\mathscr{F}}_2\right) = 1$ if and only if $\widetilde{\mathscr{F}}_1 = \widetilde{\mathscr{F}}_2$.

As $\varrho\left(\widetilde{\mathscr{F}}_1, \widetilde{\mathscr{F}}_2\right)$ moves closer to 1, it shows that the correlation between $\widetilde{\mathscr{F}}_1$ and $\widetilde{\mathscr{F}}_2$ is strong. On the other hands, as $\varrho\left(\widetilde{\mathscr{F}}_1, \widetilde{\mathscr{F}}_2\right)$ moves closer to 0, it shows that the correlation between $\widetilde{\mathscr{F}}_1$ and $\widetilde{\mathscr{F}}_2$ is very weak. Whereas, $\varrho\left(\widetilde{\mathscr{F}}_1, \widetilde{\mathscr{F}}_2\right) = 1$ and $\varrho\left(\widetilde{\mathscr{F}}_1, \widetilde{\mathscr{F}}_2\right) = 0$ indicate a perfect correlation and no correlation exist between $\widetilde{\mathscr{F}}_1$ and $\widetilde{\mathscr{F}}_2$, respectively.

2.2 Existing correlation operators for Fermatean fuzzy sets

The only work on correlation operators under Fermatean fuzzy environment was carried out by Kirisci [41]. We recall the Fermatean fuzzy operators developed in Ref. [41] as follow:

Let $\widetilde{\mathscr{F}} = \left\{\langle u_i, \xi_{\widetilde{\mathscr{F}}}(u_i), \eta_{\widetilde{\mathscr{F}}}(u_i)\rangle \,|\, u_i \in \mathscr{U}\right\}$ be an FFS, where $\xi_{\widetilde{\mathscr{F}}}(u_i), \eta_{\widetilde{\mathscr{F}}}(u_i) \in [0, 1]$ and $0 \leq \left(\xi_{\widetilde{\mathscr{F}}}(u_i)\right)^3 + \left(\eta_{\widetilde{\mathscr{F}}}(u_i)\right)^3 \leq 1$ for each $u_i \in \mathscr{U}$. The informational energy of $\widetilde{\mathscr{F}}$ is:

$$IE\left(\widetilde{\mathscr{F}}\right) = \sum_{i=1}^{n}\left(\left(\xi_{\widetilde{\mathscr{F}}}(u_i)\right)^6 + \left(\eta_{\widetilde{\mathscr{F}}}(u)\right)^6 + \left(\pi_{\widetilde{\mathscr{F}}}(u_i)\right)^6\right). \tag{2.8}$$

Suppose there are FFSs

$$\widetilde{\mathscr{F}}_1 = \left\{\langle u_i, \xi_{\widetilde{\mathscr{F}}_1}(u_i), \eta_{\widetilde{\mathscr{F}}_1}(u_i)\rangle \,|\, u_i \in \mathscr{U}\right\} \textit{ and}$$

$$\widetilde{\mathscr{F}}_2 = \left\{\langle u_i, \xi_{\widetilde{\mathscr{F}}_2}(u_i), \eta_{\widetilde{\mathscr{F}}_2}(u_i)\rangle \,|\, u_i \in \mathscr{U}\right\},$$

then the correlation of $\widetilde{\mathscr{F}}_1$ and $\widetilde{\mathscr{F}}_2$ is defined by:

$$C\left(\widetilde{\mathscr{F}}_1, \widetilde{\mathscr{F}}_2\right) = \sum_{i=1}^{n}\left(\xi^3_{\widetilde{\mathscr{F}}_1}(u_i)\xi^3_{\widetilde{\mathscr{F}}_2}(u_i) + \eta^3_{\widetilde{\mathscr{F}}_1}(u)\eta^3_{\widetilde{\mathscr{F}}_2}(u_i) + \pi^3_{\widetilde{\mathscr{F}}_1}(u_i)\pi^3_{\widetilde{\mathscr{F}}_2}(u_i)\right). \tag{2.9}$$

Choose any two FFSs $\widetilde{\mathscr{F}}_1$ and $\widetilde{\mathscr{F}}_2$ in $\mathscr{U}$, the correlation coefficients between $\widetilde{\mathscr{F}}_1$ and $\widetilde{\mathscr{F}}_2$ are:

$$\varrho_1\left(\widetilde{\mathscr{F}}_1, \widetilde{\mathscr{F}}_2\right) = \frac{C\left(\widetilde{\mathscr{F}}_1, \widetilde{\mathscr{F}}_2\right)}{\sqrt{IE\left(\widetilde{\mathscr{F}}_1\right).IE\left(\widetilde{\mathscr{F}}_2\right)}}, \tag{2.10}$$

$$\varrho_2\left(\widetilde{\mathscr{F}}_1,\widetilde{\mathscr{F}}_2\right)=\frac{C\left(\widetilde{\mathscr{F}}_1,\widetilde{\mathscr{F}}_2\right)}{\max\left[IE\left(\widetilde{\mathscr{F}}_1\right),IE\left(\widetilde{\mathscr{F}}_2\right)\right]},\tag{2.11}$$

where $C\left(\widetilde{\mathscr{F}}_1,\widetilde{\mathscr{F}}_2\right)$, $IE\left(\widetilde{\mathscr{F}}_1\right)$ and $IE\left(\widetilde{\mathscr{F}}_2\right)$ are given in Eqs. (2.8) and (2.9), respectively.

Now, we show the limitation of the approaches in Ref. [41] with the following example:

Suppose $\widetilde{\mathscr{F}}_1$ and $\widetilde{\mathscr{F}}_2$ are FFSs defined in $\mathscr{U}=\{u_1,u_2,u_3\}$ as follow:

$$\widetilde{\mathscr{F}}_1=\{\langle u_1,0.6,0.2\rangle,\langle u_2,0.8,0\rangle,\langle u_3,1,0\rangle\}\ and\ \widetilde{\mathscr{F}}_2=\{\langle u_1,0.5,0.3\rangle,\langle u_2,0.6,0.2\rangle,\langle u_3,0.8,0.1\rangle\}.$$

After calculating the hesitation margins for the FFSs, we apply Eqs. (2.10) and (2.11), respectively and get $\varrho_1\left(\widetilde{\mathscr{F}}_1,\widetilde{\mathscr{F}}_2\right)=\varrho_2\left(\widetilde{\mathscr{F}}_1,\widetilde{\mathscr{F}}_2\right)=1.5875$. In fact, whenever $IE\left(\widetilde{\mathscr{F}}_1\right)$ equals $IE\left(\widetilde{\mathscr{F}}_2\right)$, then $\varrho_1\left(\widetilde{\mathscr{F}}_1,\widetilde{\mathscr{F}}_2\right)=\varrho_2\left(\widetilde{\mathscr{F}}_1,\widetilde{\mathscr{F}}_2\right)$ as we have seen in this example. Furthermore, we observe that $\varrho_1\left(\widetilde{\mathscr{F}}_1,\widetilde{\mathscr{F}}_2\right)=\varrho_2\left(\widetilde{\mathscr{F}}_1,\widetilde{\mathscr{F}}_2\right)=1.5875\notin[0,1]$. Hence, we infer that the Fermatean fuzzy correlation coefficients in Ref. [35] are not consistent correlation operators (Table 2.1).

3. New Fermatean fuzzy correlation operators

Here, we introduce two efficient approaches of calculating correlation coefficient for FFSs. The first one is obtained by modifying the approach in Ref. [32] by way of including the attributes of FFSs and incorporating the hesitation margin of FFSs. The second one is obtained by modifying the approach in Ref. [41].

For FFSs $\widetilde{\mathscr{F}}_1$ and $\widetilde{\mathscr{F}}_2$ in $\mathscr{U}=\{u_1,u_2,\ldots,u_n\}$ where $n<\infty$, the correlation coefficients between the FFSs can be measured by:

$$\overline{\varrho}\left(\widetilde{\mathscr{F}}_1,\widetilde{\mathscr{F}}_2\right)=\frac{1}{3n}\sum_{i=1}^{n}\left(\mu_i(1-\Delta\xi_i)+\nu_i(1-\Delta\eta_i)+\phi_i(1-\Delta\pi_i)\right),\tag{2.12}$$

where $\mu_i=\frac{3-\Delta\xi_i-\Delta\xi_{\max}}{3-\Delta\xi_{\min}-\Delta\xi_{\max}}$, $\nu_i=\frac{3-\Delta\eta_i-\Delta\eta_{\max}}{3-\Delta\eta_{\min}-\Delta\eta_{\max}}$, $\phi_i=\frac{3-\Delta\pi_i-\Delta\pi_{\max}}{3-\Delta\pi_{\min}-\Delta\pi_{\max}}$,

$$\Delta\xi_{\min}=\min_i\left\{\left|\xi^3_{\widetilde{\mathscr{F}}_1}(u_i)-\xi^3_{\widetilde{\mathscr{F}}_2}(u_i)\right|\right\},\Delta\eta_{\min}=\min_i\left\{\left|\eta^3_{\widetilde{\mathscr{F}}_1}(u_i)-\eta^3_{\widetilde{\mathscr{F}}_2}(u_i)\right|\right\},$$

$$\Delta\pi_{\min}=\min_i\left\{\left|\pi^3_{\widetilde{\mathscr{F}}_1}(u_i)-\pi^3_{\widetilde{\mathscr{F}}_2}(u_i)\right|\right\},$$

TABLE 2.1 Computational values.

X	$\Delta\xi_i$	$\Delta\eta_i$	$\Delta\pi_i$
x_1	0	0.19	0.019
x_2	0	0.007	0.0071
x_3	0	0.001	0.0009

$$\Delta\xi_{\max} - \max_i\left\{\left|\xi^3_{\widetilde{\mathscr{F}}_1}(u_i) - \xi^3_{\widetilde{\mathscr{F}}_2}(u_i)\right|\right\}, \Delta\eta_{\max} = \max_i\left\{\left|\eta^3_{\widetilde{\mathscr{F}}_1}(u_i) - \eta^3_{\widetilde{\mathscr{F}}_2}(u_i)\right|\right\},$$

$$\Delta\pi_{\max} = \max_i\left\{\left|\pi^3_{\widetilde{\mathscr{F}}_1}(u_i) - \pi^3_{\widetilde{\mathscr{F}}_2}(u_i)\right|\right\},$$

$$\Delta\xi_i = \left|\xi^3_{\widetilde{\mathscr{F}}_1}(u_i) - \xi^3_{\widetilde{\mathscr{F}}_2}(u_i)\right|, \Delta\eta_i = \left|\eta^3_{\widetilde{\mathscr{F}}_1}(u_i) - \eta^3_{\widetilde{\mathscr{F}}_2}(u_i)\right|,$$

$$\Delta\pi_i = \left|\pi^3_{\widetilde{\mathscr{F}}_1}(u_i) - \pi^3_{\widetilde{\mathscr{F}}_2}(u_i)\right|.$$

The second approach is given as Eq. (2.13):

$$\widetilde{\varrho}\left(\widetilde{\mathscr{F}}_1, \widetilde{\mathscr{F}}_2\right) = \frac{\left(\sum_{i=1}^n\left(\xi^3_{\widetilde{\mathscr{F}}_1}(u_i)\xi^3_{\widetilde{\mathscr{F}}_2}(u_i) + \eta^3_{\widetilde{\mathscr{F}}_1}(u_i)\eta^3_{\widetilde{\mathscr{F}}_2}(u_i) + \pi^3_{\widetilde{\mathscr{F}}_1}(u_i)\pi^3_{\widetilde{\mathscr{F}}_2}(u_i)\right)\right)^{\frac{1}{3}}}{\sqrt{\left(\sum_{i=1}^n\left(\xi^6_{\widetilde{\mathscr{F}}_1}(u_i) + \eta^6_{\widetilde{\mathscr{F}}_1}(u_i) + \pi^6_{\widetilde{\mathscr{F}}}(u_i)\right)\right)^{\frac{1}{3}}\left(\sum_{i=1}^n\left(\xi^6_{\widetilde{\mathscr{F}}_2}(u_i) + \eta^6_{\widetilde{\mathscr{F}}_2}(u_i) + \pi^6_{\widetilde{\mathscr{F}}_2}(u_i)\right)\right)^{\frac{1}{3}}}}, \tag{2.13}$$

which can be rewritten as:

$$\widetilde{\varrho}\left(\widetilde{\mathscr{F}}_1, \widetilde{\mathscr{F}}_2\right) = \left(\frac{\left(\sum_{i=1}^n\left(\xi^3_{\widetilde{\mathscr{F}}_1}(u_i)\xi^3_{\widetilde{\mathscr{F}}_2}(u_i) + \eta^3_{\widetilde{\mathscr{F}}_1}(u_i)\eta^3_{\widetilde{\mathscr{F}}_2}(u_i) + \pi^3_{\widetilde{\mathscr{F}}_1}(u_i)\pi^3_{\widetilde{\mathscr{F}}_2}(u_i)\right)\right)^2}{\sum_{i=1}^n\left(\xi^6_{\widetilde{\mathscr{F}}_1}(u_i) + \eta^6_{\widetilde{\mathscr{F}}_1}(u_i) + \pi^6_{\widetilde{\mathscr{F}}}(u_i)\right)\sum_{i=1}^n\left(\xi^6_{\widetilde{\mathscr{F}}_2}(u_i) + \eta^6_{\widetilde{\mathscr{F}}_2}(u_i) + \pi^6_{\widetilde{\mathscr{F}}_2}(u_i)\right)}\right)^{\frac{1}{6}} \tag{2.14}$$

Proposition 4.1. For FFSs $\widetilde{\mathscr{F}}_1$ and $\widetilde{\mathscr{F}}_2$ in $\mathscr{U}$, $\overline{\rho}\left(\widetilde{\mathscr{F}}_1, \widetilde{\mathscr{F}}_2\right)$ satisfies

(i) $\overline{\varrho}\left(\widetilde{\mathscr{F}}_1, \widetilde{\mathscr{F}}_2\right) = \overline{\varrho}\left(\widetilde{\mathscr{F}}_2, \widetilde{\mathscr{F}}_1\right)$,

(ii) $\overline{\varrho}\left(\widetilde{\mathscr{F}}_1, \widetilde{\mathscr{F}}_2\right) = 1$ iff $\widetilde{\mathscr{F}}_1 = \widetilde{\mathscr{F}}_2$.

Proof. Firstly, we establish (i). Recall that

$$\overline{\varrho}\left(\widetilde{\mathscr{F}}_1, \widetilde{\mathscr{F}}_2\right) = \frac{1}{3n}\sum_{i=1}^n(\mu_i(1 - \Delta\xi_i) + \nu_i(1 - \Delta\eta_i) + \phi_i(1 - \Delta\pi_i)),$$

and then

$$\overline{\varrho}\left(\widetilde{\mathscr{F}}_1, \widetilde{\mathscr{F}}_2\right) = \frac{1}{3n}\sum_{i=1}^n\left(\mu_i\left(1 - \left|\xi^3_{\widetilde{\mathscr{F}}_1}(u_i) - \xi^3_{\widetilde{\mathscr{F}}_2}(u_i)\right|\right) + \nu_i\left(1 - \left|\eta^3_{\widetilde{\mathscr{F}}_1}(u_i) - \eta^3_{\widetilde{\mathscr{F}}_2}(u_i)\right|\right) + \phi_i\left(1 - \left|\pi^3_{\widetilde{\mathscr{F}}_1}(u_i) - \pi^3_{\widetilde{\mathscr{F}}_2}(u_i)\right|\right)\right)$$

$$= \frac{1}{3n}\sum_{i=1}^n\left(\mu_i\left(1 - \left|\xi^3_{\widetilde{\mathscr{F}}_2}(u_i) - \xi^3_{\widetilde{\mathscr{F}}_1}(u_i)\right|\right) + \nu_i\left(1 - \left|\eta^3_{\widetilde{\mathscr{F}}_2}(u_i) - \eta^3_{\widetilde{\mathscr{F}}_1}(u_i)\right|\right) + \phi_i\left(1 - \left|\pi^3_{\widetilde{\mathscr{F}}_2}(u_i) - \pi^3_{\widetilde{\mathscr{F}}_1}(u_i)\right|\right)\right)$$

$$= \overline{\varrho}\left(\widetilde{\mathscr{F}}_2, \widetilde{\mathscr{F}}_1\right)$$

which prove (i).

Again, suppose $\widetilde{\mathscr{F}}_1 = \widetilde{\mathscr{F}}_2$. Then

$$\left|\xi^3_{\widetilde{\mathscr{F}}_1}(u_i) - \xi^3_{\widetilde{\mathscr{F}}_2}(u_i)\right| = 0, \left|\eta^3_{\widetilde{\mathscr{F}}_1}(u_i) - \eta^3_{\widetilde{\mathscr{F}}_2}(u_i)\right| = 0,$$

$$\left|\pi^3_{\widetilde{\mathscr{F}}_1}(u_i) - \pi^3_{\widetilde{\mathscr{F}}_2}(u_i)\right| = 0.$$

Consequently, $\Delta\xi_i = \Delta\eta_i = \Delta\pi_i = 0$, $\Delta\xi_{\min} = \Delta\eta_{\min} = \Delta\pi_{\min} = 0$, and $\Delta\xi_{\max} = \Delta\eta_{\max} = \Delta\pi_{\max} = 0$. Thus $\mu_i = \nu_i = \phi_i = 1$, and so $\overline{\varrho}\left(\widetilde{\mathscr{F}}_1, \widetilde{\mathscr{F}}_2\right) = 1$ then $\widetilde{\mathscr{F}}_1$ and $\widetilde{\mathscr{F}}_2$ have perfect relation.

Conversely, suppose $\overline{\varrho}\left(\widetilde{\mathscr{F}}_1, \widetilde{\mathscr{F}}_2\right) = 1$. Then it is straightforward that $\widetilde{\mathscr{F}}_1 = \widetilde{\mathscr{F}}_2$. Hence, (ii) holds.

Proposition 4.2. For FFSs $\widetilde{\mathscr{F}}_1$ and $\widetilde{\mathscr{F}}_2$ in $\mathscr{U}$, $\widetilde{\varrho}\left(\widetilde{\mathscr{F}}_1, \widetilde{\mathscr{F}}_2\right)$ satisfies

(i) $\widetilde{\varrho}\left(\widetilde{\mathscr{F}}_1, \widetilde{\mathscr{F}}_2\right) = \widetilde{\varrho}\left(\widetilde{\mathscr{F}}_2, \widetilde{\mathscr{F}}_1\right)$,

(ii) $\widetilde{\varrho}\left(\widetilde{\mathscr{F}}_1, \widetilde{\mathscr{F}}_2\right) = 1$ *iff* $\widetilde{\mathscr{F}}_1 = \widetilde{\mathscr{F}}_2$.

Proof. We first prove (i) as follows:

$$\widetilde{\varrho}\left(\widetilde{\mathscr{F}}_1, \widetilde{\mathscr{F}}_2\right) = \left(\frac{\left(\sum_{i=1}^n\left(\xi^3_{\widetilde{\mathscr{F}}_1}(u_i)\xi^3_{\widetilde{\mathscr{F}}_2}(u_i) + \eta^3_{\widetilde{\mathscr{F}}_1}(u_i)\eta^3_{\widetilde{\mathscr{F}}_2}(u_i) + \pi^3_{\widetilde{\mathscr{F}}_1}(u_i)\pi^3_{\widetilde{\mathscr{F}}_2}(u_i)\right)\right)^2}{\sum_{i=1}^n\left(\xi^6_{\widetilde{\mathscr{F}}_1}(u_i) + \eta^6_{\widetilde{\mathscr{F}}_1}(u_i) + \pi^6_{\widetilde{\mathscr{F}}}(u_i)\right)\sum_{i=1}^n\left(\xi^6_{\widetilde{\mathscr{F}}_2}(u_i) + \eta^6_{\widetilde{\mathscr{F}}_2}(u_i) + \pi^6_{\widetilde{\mathscr{F}}_2}(u_i)\right)}\right)^{\frac{1}{6}}$$

$$= \left(\frac{\left(\sum_{i=1}^n\left(\xi^3_{\widetilde{\mathscr{F}}_2}(u_i)\xi^3_{\widetilde{\mathscr{F}}_1}(u_i) + \eta^3_{\widetilde{\mathscr{F}}_2}(u_i)\eta^3_{\widetilde{\mathscr{F}}_1}(u_i) + \pi^3_{\widetilde{\mathscr{F}}_2}(u_i)\pi^3_{\widetilde{\mathscr{F}}_1}(u_i)\right)\right)^2}{\sum_{i=1}^n\left(\xi^6_{\widetilde{\mathscr{F}}_2}(u_i) + \eta^6_{\widetilde{\mathscr{F}}_2}(u_i) + \pi^6_{\widetilde{\mathscr{F}}_2}(u_i)\right)\sum_{i=1}^n\left(\xi^6_{\widetilde{\mathscr{F}}_1}(u_i) + \eta^6_{\widetilde{\mathscr{F}}_1}(u_i) + \pi^6_{\widetilde{\mathscr{F}}}(u_i)\right)}\right)^{\frac{1}{6}}$$

$$= \widetilde{\varrho}\left(\widetilde{\mathscr{F}}_1, \widetilde{\mathscr{F}}_2\right).$$

Now, we prove (ii). Suppose $\widetilde{\mathscr{F}}_1 = \widetilde{\mathscr{F}}_2$, then

$$\widetilde{\varrho}\left(\widetilde{\mathscr{F}}_1, \widetilde{\mathscr{F}}_2\right) = \left(\frac{\left(\sum_{i=1}^n\left(\xi^6_{\widetilde{\mathscr{F}}_2}(u_i) + \eta^6_{\widetilde{\mathscr{F}}_2}(u_i) + \pi^6_{\widetilde{\mathscr{F}}_2}(u_i)\right)\right)^2}{\left(\sum_{i=1}^n\left(\xi^6_{\widetilde{\mathscr{F}}_2}(u_i) + \eta^6_{\widetilde{\mathscr{F}}_2}(u_i) + \pi^6_{\widetilde{\mathscr{F}}_2}(u_i)\right)\right)^2}\right)^{\frac{1}{6}}$$

$$= \frac{\left(\sum_{i=1}^{n}\left(\xi^{6}_{\widetilde{\mathscr{F}}_2}(u_i) + \eta^{6}_{\widetilde{\mathscr{F}}_2}(u_i) + \pi^{6}_{\widetilde{\mathscr{F}}_2}(u_i)\right)\right)^{\frac{1}{3}}}{\left(\sum_{i=1}^{n}\left(\xi^{6}_{\widetilde{\mathscr{F}}_2}(u_i) + \eta^{6}_{\widetilde{\mathscr{F}}_2}(u_i) + \pi^{6}_{\widetilde{\mathscr{F}}_2}(u_i)\right)\right)^{\frac{1}{3}}},$$

$$= 1.$$

Conversely, assume $\widetilde{\varrho}\left(\widetilde{\mathscr{F}}_1, \widetilde{\mathscr{F}}_2\right) = 1$, then it is easy to see that $\widetilde{\mathscr{F}}_1 = \widetilde{\mathscr{F}}_2$.

Theorem 4.3. If $\overline{\varrho}\left(\widetilde{\mathscr{F}}_1, \widetilde{\mathscr{F}}_2\right)$ and $\widetilde{\varrho}\left(\widetilde{\mathscr{F}}_1, \widetilde{\mathscr{F}}_2\right)$ are correlation coefficients between FFSs $\widetilde{\mathscr{F}}_1$ and $\widetilde{\mathscr{F}}_2$ in $\mathscr{U}$, then $\overline{\varrho}\left(\widetilde{\mathscr{F}}_1, \widetilde{\mathscr{F}}_2\right), \widetilde{\varrho}\left(\widetilde{\mathscr{F}}_1, \widetilde{\mathscr{F}}_2\right) \in [0, 1]$.

Proof. In this case, we need to first and foremost prove that $0 \leq \overline{\varrho}\left(\widetilde{\mathscr{F}}_1, \widetilde{\mathscr{F}}_2\right) \leq 1$, that is, $\overline{\varrho}\left(\widetilde{\mathscr{F}}_1, \widetilde{\mathscr{F}}_2\right) \geq 0$ and $\overline{\varrho}\left(\widetilde{\mathscr{F}}_1, \widetilde{\mathscr{F}}_2\right) \leq 1$. The fact that $\overline{\varrho}\left(\widetilde{\mathscr{F}}_1, \widetilde{\mathscr{F}}_2\right) \geq 0$ is clear. So, it suffices to show that $\overline{\varrho}\left(\widetilde{\mathscr{F}}_1, \widetilde{\mathscr{F}}_2\right) \leq 1$. To establish this, let us assume that

$$\sum_{i=1}^{n}\mu_i(1 - \Delta\xi_i) = \Gamma, \sum_{i=1}^{n}\nu_i(1 - \Delta\eta_i) = \Psi, \sum_{i=1}^{n}\phi_i(1 - \Delta\pi_i) = \Omega.$$

We then apply the principle of Cauchy-Schwarz inequality, and so

$$\overline{\varrho}\left(\widetilde{\mathscr{F}}_1, \widetilde{\mathscr{F}}_2\right) = \frac{1}{3n}\sum_{i=1}^{n}\left(\mu_i(1 - \Delta\xi_i) + \nu_i(1 - \Delta\eta_i) + \phi_i(1 - \Delta\pi_i)\right)$$

$$\leq \frac{\sum_{i=1}^{n}\mu_i(1 - \Delta\xi_i) + \sum_{i=1}^{n}\nu_i(1 - \Delta\eta_i) + \sum_{i=1}^{n}\phi_i(1 - \Delta\pi_i)}{3n}$$

$$= \frac{\Gamma + \Psi + \Omega}{3\text{n}}.$$

Thus,

$$\overline{\varrho}\left(\widetilde{\mathscr{F}}_1, \widetilde{\mathscr{F}}_2\right) - 1 = \frac{\Gamma + \Psi + \Omega}{3n} - 1$$

$$= \frac{\Gamma + \Psi + \Omega - 3n}{3n}$$

$$= -\frac{(3n - \Gamma - \Psi - \Omega)}{3n} \leq 0,$$

which implies that $\overline{\varrho}\left(\widetilde{\mathscr{F}}_1, \widetilde{\mathscr{F}}_2\right) \leq 1$. Hence, $\overline{\varrho}\left(\widetilde{\mathscr{F}}_1, \widetilde{\mathscr{F}}_2\right) \in [0, 1]$.

Now, we show that $\widetilde{\varrho}\left(\widetilde{\mathscr{F}}_1, \widetilde{\mathscr{F}}_2\right) \in [0, 1]$. We need to show that $\widetilde{\varrho}\left(\widetilde{\mathscr{F}}_1, \widetilde{\mathscr{F}}_2\right) \leq 1$ since it is certain that $\widetilde{\varrho}\left(\widetilde{\mathscr{F}}_1, \widetilde{\mathscr{F}}_2\right) \geq 0$. Let us assume that

$$\sum_{i=1}^{n}\xi_{\widetilde{\wp}_1}^3(u_i)=\Gamma_1,\sum_{i=1}^{n}\xi_{\widetilde{\wp}_2}^3(u_i)=\Gamma_2,$$

$$\sum_{i=1}^{n}\eta_{\widetilde{\wp}_1}^3(u_i)=\Psi_1,\sum_{i=1}^{n}\eta_{\widetilde{\wp}_2}^3(u_i)=\Psi_2,$$

$$\sum_{i=1}^{n}\pi_{\widetilde{\wp}_1}^3(u_i)=\Omega_1,\sum_{i=1}^{n}\pi_{\widetilde{\wp}_2}^3(u_i)=\Omega_1.$$

Then

$$\widetilde{\varrho}\left(\widetilde{\mathscr{F}}_1,\widetilde{\mathscr{F}}_2\right)=\frac{\left(\sum_{i=1}^{n}\left(\xi_{\widetilde{\mathscr{F}}_1}^3(u_i)\xi_{\widetilde{\mathscr{F}}_2}^3(u_i)+\eta_{\widetilde{\mathscr{F}}_1}^3(u_i)\eta_{\widetilde{\mathscr{F}}_2}^3(u_i)+\pi_{\widetilde{\mathscr{F}}}^3(u_i)\pi_{\widetilde{\mathscr{F}}_2}^3(u_i)\right)\right)^{\frac{1}{3}}}{\left(\sum_{i=1}^{n}\left(\xi_{\widetilde{\mathscr{F}}_1}^6(u_i)+\eta_{\widetilde{\mathscr{F}}_1}^6(u_i)+\pi_{\widetilde{\mathscr{F}}_1}^6(u_i)\right)\sum_{i=1}^{n}\left(\xi_{\widetilde{\mathscr{F}}_2}^6(u_i)+\eta_{\widetilde{\mathscr{F}}_2}^6(u_i)+\pi_{\widetilde{\mathscr{F}}_2}^6(u_i)\right)\right)^{\frac{1}{6}}}$$

$$=\left(\frac{\sum_{i=1}^{n}\left(\xi_{\widetilde{\mathscr{F}}_1}^3(u_i)\xi_{\widetilde{\mathscr{F}}_2}^3(u_i)+\eta_{\widetilde{\mathscr{F}}_1}^3(u_i)\eta_{\widetilde{\mathscr{F}}_2}^3(u_i)+\pi_{\widetilde{\mathscr{F}}}^3(u_i)\pi_{\widetilde{\mathscr{F}}_2}^3(u_i)\right)}{\left(\sum_{i=1}^{n}\left(\xi_{\widetilde{\mathscr{F}}_1}^6(u_i)+\eta_{\widetilde{\mathscr{F}}_1}^6(u_i)+\pi_{\widetilde{\mathscr{F}}_1}^6(u_i)\right)\sum_{i=1}^{n}\left(\xi_{\widetilde{\mathscr{F}}_2}^6(u_i)+\eta_{\widetilde{\mathscr{F}}_2}^6(u_i)+\pi_{\widetilde{\mathscr{F}}_2}^6(u_i)\right)\right)^{\frac{1}{2}}}\right)^{\frac{1}{3}}$$

$$\leq\left(\frac{\sum_{i=1}^{n}\left(\xi_{\widetilde{\mathscr{F}}_1}^3(u_i)\xi_{\widetilde{\mathscr{F}}_2}^3(u_i)\right)+\sum_{i=1}^{n}\left(\eta_{\widetilde{\mathscr{F}}_1}^3(u_i)\eta_{\widetilde{\mathscr{F}}_2}^3(u_i)\right)+\sum_{i=1}^{n}\left(\pi_{\widetilde{\mathscr{F}}_1}^3(u_i)\pi_{\widetilde{\mathscr{F}}_2}^3(u_i)\right)}{\left(\sum_{i=1}^{n}\left(\xi_{\widetilde{\mathscr{F}}_1}^{3+2}(u_i)+\eta_{\widetilde{\mathscr{F}}_1}^{3+2}(u_i)+\pi_{\widetilde{\mathscr{F}}_1}^{3+2}(u_i)\right)\sum_{i=1}^{n}\left(\xi_{\widetilde{\mathscr{F}}_2}^{3+2}(u_i)+\eta_{\widetilde{\mathscr{F}}_2}^{3+2}(u_i)+\pi_{\widetilde{\mathscr{F}}_2}^{3+2}(u_i)\right)\right)^{\frac{1}{2}}}\right)^{\frac{1}{3}}$$

$$=\left(\frac{\Gamma_1\Gamma_2+\Psi_1\Psi_2+\Omega_1\Omega_2}{\left(\left(\Gamma_1^2+\Psi_1^2+\Omega_1^2\right)\left(\Gamma_2^2+\Psi_2^2+\Omega_2^2\right)\right)^{\frac{1}{2}}}\right)^{\frac{1}{3}}.$$

Thus $\widetilde{\varrho}^3\left(\widetilde{\mathscr{F}}_1,\widetilde{\mathscr{F}}_2\right)\leq\frac{\Gamma_1\Gamma_2+\Psi_1\Psi_2+\Omega_1\Omega_2}{\left(\left(\Gamma_1^2+\Psi_1^2+\Omega_1^2\right)\left(\Gamma_2^2+\Psi_2^2+\Omega_2^2\right)\right)^{\frac{1}{2}}}$, and consequently we have $\widetilde{\varrho}^6\left(\widetilde{\mathscr{F}}_1,\widetilde{\mathscr{F}}_2\right)\leq\frac{\left(\Gamma_1\Gamma_2+\Psi_1\Psi_2+\Omega_1\Omega_2\right)^2}{\left(\Gamma_1^2+\Psi_1^2+\Omega_1^2\right)\left(\Gamma_2^2+\Psi_2^2+\Omega_2^2\right)}$ and so,

$$\widetilde{\varrho}^6\left(\widetilde{\mathscr{F}}_1,\widetilde{\mathscr{F}}_2\right)-1=\frac{\left(\Gamma_1\Gamma_2+\Psi_1\Psi_2+\Omega_1\Omega_2\right)^2}{\left(\Gamma_1^2+\Psi_1^2+\Omega_1^2\right)\left(\Gamma_2^2+\Psi_2^2+\Omega_2^2\right)}-1$$

$$=\frac{\left(\Gamma_1\Gamma_2+\Psi_1\Psi_2+\Omega_1\Omega_2\right)^2-\left(\Gamma_1^2+\Psi_1^2+\Omega_1^2\right)\left(\Gamma_2^2+\Psi_2^2+\Omega_2^2\right)}{\left(\Gamma_1^2+\Psi_1^2+\Omega_1^2\right)\left(\Gamma_2^2+\Psi_2^2+\Omega_2^2\right)}$$

$$=-\frac{\left(\left(\Gamma_1^2+\Psi_1^2+\Omega_1^2\right)\left(\Gamma_2^2+\Psi_2^2+\Omega_2^2\right)-\left(\Gamma_1\Gamma_2+\Psi_1\Psi_2+\Omega_1\Omega_2\right)^2\right)}{\left(\Gamma_1^2+\Psi_1^2+\Omega_1^2\right)\left(\Gamma_2^2+\Psi_2^2+\Omega_2^2\right)}$$

$$\leq 0.$$

Hence, $\widetilde{\varrho}^6\left(\widetilde{\mathscr{F}}_1,\widetilde{\mathscr{F}}_2\right)\leq 1$. Thus, $\widetilde{\varrho}\left(\widetilde{\mathscr{F}}_1,\widetilde{\mathscr{F}}_2\right)\leq 1$.

3.1 Computational example

Some numerical examples of FFSs are provided to verify the innovative approaches of estimating correlation coefficient in the Fermatean fuzzy domain. The FFSs in both cases are very similar, and it is suggested that their correlation coefficient must be almost one.

Example 1. Assume $\widetilde{\mathscr{F}}_1$ and $\widetilde{\mathscr{F}}_2$ are FFSs in $\mathscr{U} = \{u_1, u_2, u_3\}$ defined by:

$$\widetilde{\mathscr{F}}_1 = \{\langle u_1, 0.1, 0.2\rangle, \langle u_1, 0.2, 0.1\rangle, \langle u_1, 0.29, 0.0\rangle\},$$

$$\widetilde{\mathscr{F}}_2 = \{\langle u_1, 0.1, 0.3\rangle, \langle u_1, 0.2, 0.2\rangle, \langle u_1, 0.29, 0.1\rangle\}.$$

Firstly, we use Eq. (2.12) to find the correlation coefficient of $\widetilde{\mathscr{F}}_1$ and $\widetilde{\mathscr{F}}_2$ and get Table 2.1.

So, $\Delta\xi_{\min} = \Delta\xi_{\max} = 0$, $\Delta\eta_{\min} = 0.01$, $\Delta\eta_{\max} = 0.019$, $\Delta\pi_{\min} = 0.0009$, $\Delta\pi_{\max} = 0.019$. Thus, $\mu_1 = 1$, $\nu_1 = 0.994$, $\phi_1 = 0.9939$, $\mu_2 = 1$, $\nu_2 = 0.998$, $\phi_2 = 0.9979$, $\mu_3 = \nu_3 = \phi_3 = 1$. Hence, $\overline{\varrho}\left(\widetilde{\mathscr{F}}_1, \widetilde{\mathscr{F}}_2\right) = 0.9922$.

Using Eq. (2.13), we have $\widetilde{\varrho}\left(\widetilde{\mathscr{F}}_1, \widetilde{\mathscr{F}}_2\right) = \frac{\sqrt[3]{0.0007+0.0002+2.8893}}{\sqrt{\sqrt[3]{2.9168\times 2.8645}}} = 0.99996 \approx 1.$

From Eqs. (2.12) and (2.13), we see that $\widetilde{\mathscr{F}}_1$ and $\widetilde{\mathscr{F}}_2$ are closely related, notwithstanding, Eq. (2.13) gives more accurate result compare to Eq. (2.12).

4. Application example of medical diagnosis

Under this section, we present a prototype of diagnostic process using simulated dataset structured based on basic medical knowledge. Assume that there is a Fermatean fuzzy data of some diseases viz Dx_1 (malaria fever), Dx_2 (viral fever), Dx_3 (typhoid fever), Dx_4 (stomach problem), and Dx_5 (chest problem) in Table 2.2, simulated from medical knowledge in the space of some related symptoms $S = \{S_1, S_2, S_3, S_4, S_5\}$, where S_1 stands for temperature, S_2 stands for headache, S_3 stands for stomachache, S_4 stands for cough, and S_5 stands for chest ache. Suppose five sick folks represented by FFSs $P = \{P_1, P_2, P_3, P_4\}$, approach a medical center to determine their medical status. The patients are showing certain symptoms of headache, cough, stomachache, high temperature, and chest ache. During the medical examinations, the health information of the patients is also represented in Fermatean data as seen in Table 2.3.

The diseases and symptoms' relation is represented by $K_1 : Dx_i \rightarrow S$, and the symptoms and patients' relation is defined by $K_2 : S \rightarrow P_j$, for $i = 1, 2, 3, 4, 5$ and $j = 1, 2, 3, 4$.

The diagnosis of the sick fellows is ascertained by calculating the correlation coefficients between each patient and each ailment, respectively. Then, the diagnosis is determined by spotting the disease that has the greatest correlation coefficient. By deploying Eqs. (2.10)–(2.13) between (P_j, Dx_i), we get Tables 2.4–2.7, respectively.

The results from Tables 2.4–2.7 are plotted as Figs. 2.2–2.5 to graphically show the edge of the new FFCCOs over the Kirisci's FFCCOs [41].

Judging by the greatest correlation coefficient value from Table 2.4 and Fig. 2.2, we glean that patient P_1 should be treated for malaria fever. It is interesting that both the operators in Ref. [41] and the new FFCCOs give the same diagnosis.

TABLE 2.2 Fermatean fuzzy data of five diseases.

K_1	Dx_1	Dx_2	Dx_3	Dx_4	Dx_5
S_1	$\langle 0.7, 0.2\rangle$	$\langle 0.6, 0.0\rangle$	$\langle 0.5, 0.3\rangle$	$\langle 0.1, 0.8\rangle$	$\langle 0.1, 0.7\rangle$
S_2	$\langle 0.7, 0.1\rangle$	$\langle 0.5, 0.3\rangle$	$\langle 0.6, 0.2\rangle$	$\langle 0.3, 0.5\rangle$	$\langle 0.1, 0.8\rangle$
S_3	$\langle 0.1, 0.8\rangle$	$\langle 0.2, 0.7\rangle$	$\langle 0.2, 0.7\rangle$	$\langle 0.9, 0.0\rangle$	$\langle 0.2, 0.7\rangle$
S_4	$\langle 0.8, 0.1\rangle$	$\langle 0.5, 0.3\rangle$	$\langle 0.3, 0.6\rangle$	$\langle 0.1, 0.7\rangle$	$\langle 0.6, 0.3\rangle$
S_5	$\langle 0.1, 0.8\rangle$	$\langle 0.2, 0.7\rangle$	$\langle 0.1, 0.8\rangle$	$\langle 0.2, 0.7\rangle$	$\langle 0.9, 0.0\rangle$

TABLE 2.3 Fermatean fuzzy medical information of patients.

K_1	S_1	S_2	S_3	S_4	S_5
P_1	$\langle 0.8, 0.1\rangle$	$\langle 0.7, 0.2\rangle$	$\langle 0.1, 0.8\rangle$	$\langle 0.7, 0.2\rangle$	$\langle 0.2, 0.7\rangle$
P_2	$\langle 0.0, 0.9\rangle$	$\langle 0.5, 0.4\rangle$	$\langle 0.7, 0.1\rangle$	$\langle 0.2, 0.7\rangle$	$\langle 0.2, 0.8\rangle$
P_3	$\langle 0.9, 0.0\rangle$	$\langle 0.8, 0.1\rangle$	$\langle 0.0, 0.7\rangle$	$\langle 0.2, 0.6\rangle$	$\langle 0.1, 0.5\rangle$
P_4	$\langle 0.6, 0.1\rangle$	$\langle 0.5, 0.3\rangle$	$\langle 0.4, 0.5\rangle$	$\langle 0.7, 0.1\rangle$	$\langle 0.4, 0.5\rangle$

TABLE 2.4 Correlation coefficient of (P_1, Dx_i).

Methods	(P_1, Dx_1)	(P_1, Dx_2)	(P_1, Dx_3)	(P_1, Dx_4)	(P_1, Dx_5)
Kirisci's method 1	0.9684	0.9353	0.9005	0.6865	0.6932
Kirisci's method 2	0.4842	0.4652	0.4490	0.3427	0.3464
Our method 1	0.9112	0.8454	0.8119	0.6908	0.6770
Our method 2	0.9893	0.9779	0.9657	0.8822	0.8850

TABLE 2.5 Correlation coefficient of (P_2, Dx_i).

Methods	(P_2, Dx_1)	(P_2, Dx_2)	(P_2, Dx_3)	(P_2, Dx_4)	(P_2, Dx_5)
Kirisci's method 1	0.6962	0.7899	0.8121	0.9201	0.6778
Kirisci's method 2	0.3476	0.3943	0.4059	0.4600	0.3389
Our method 1	0.6784	0.7395	0.7933	0.8514	0.6598
Our method 2	0.8863	0.9244	0.9330	0.9726	0.8784

TABLE 2.6 Correlation coefficient of (P_3, Dx_i).

Methods	(P_3, Dx_1)	(P_3, Dx_2)	(P_3, Dx_3)	(P_3, Dx_4)	(P_3, Dx_5)
Kirisci's method 1	0.8108	0.8476	0.8090	0.6430	0.6092
Kirisci's method 2	0.4038	0.4237	0.4045	0.3214	0.3043
Our method 1	0.7441	0.7858	0.7810	0.6511	0.6517
Our method 2	0.9325	0.9464	0.9318	0.8631	0.8477

TABLE 2.7 Correlation coefficient of (P_4, Dx_i).

Methods	(P_4, Dx_1)	(P_4, Dx_2)	(P_4, Dx_3)	(P_4, Dx_4)	(P_4, Dx_5)
Kirisci's method 1	0.8860	0.9634	0.9189	0.7645	0.7711
Kirisci's method 2	0.4410	0.4816	0.4591	0.3816	0.3843
Our method 1	0.7992	0.8871	0.8252	0.7098	0.7112
Our method 2	0.9614	0.9876	0.9722	0.9144	0.9170

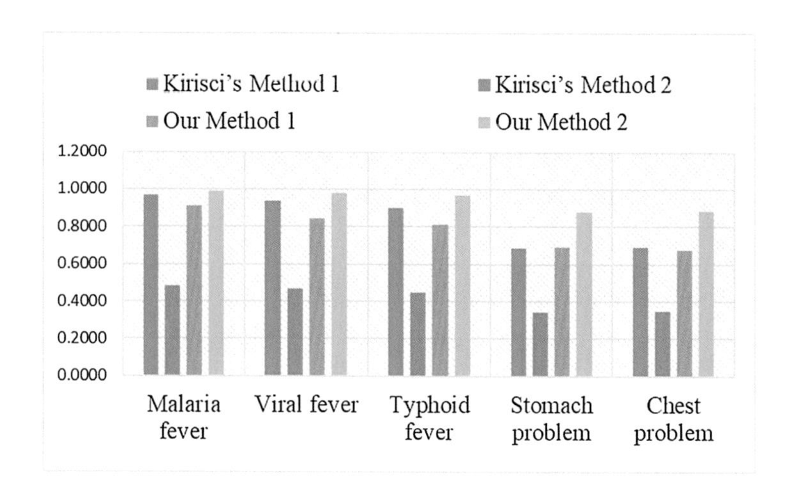

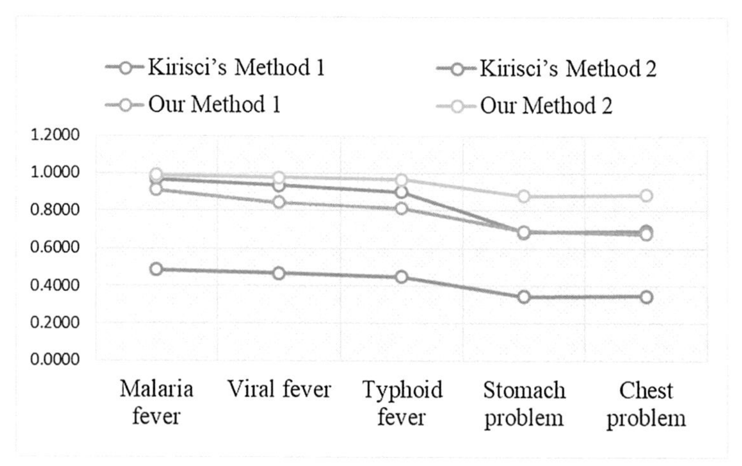

FIGURE 2.2 Patient P_1 versus diseases.

From Table 2.5 and Fig. 2.3, we see that patient P_2 should be treated for stomach problem since the correlation of the disease and the patient is the greatest. Similarly, both the operators in Ref. [41] and the new FFCCOs give the same diagnosis.

Judging by the greatest correlation coefficient values from Table 2.6 and Fig. 2.4, it follows that patient P_3 should be treated for viral fever. Likewise, both the operators in Ref. [41] and the new FFCCOs give the same diagnosis.

From Table 2.7 and Fig. 2.5, we see that patient P_4 should be treated for viral fever since the correlation of the disease, and the patient is the greatest. Similarly, both the operators in Ref. [41] and the new FFCCOs give the same diagnosis.

By means of comparison, we can say that our approaches are the most accurate and reliable FFCCOs compare to FFCCOs in Ref. [41] because the values of their correlation coefficient are the greatest. The limitations of the FFCCOs in Ref. [41] are as follows.

(i) The approaches will yield misleading result whenever anyone of the membership grade, nonmembership grade, and hesitation margin is zero.
(ii) In addition to (i), the second method in Ref. [41] only considers the maximum information energy, which will certainly leads to error of exclusion.

Going by the model formulation, our first method seems to be more reliable because our second method shares the first limitation observed in Ref. [41].

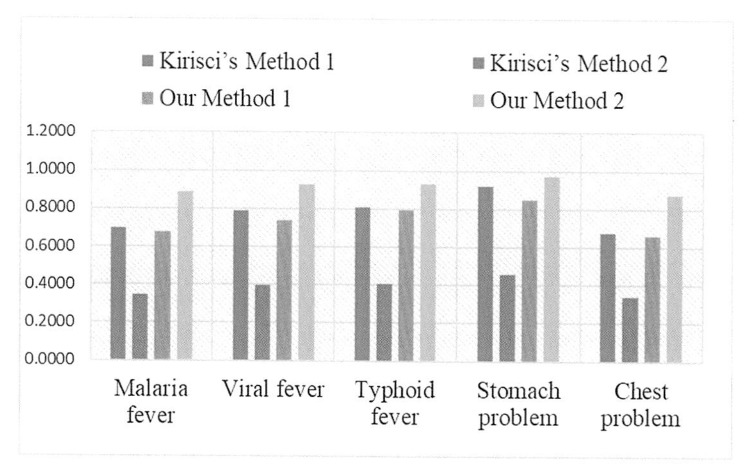

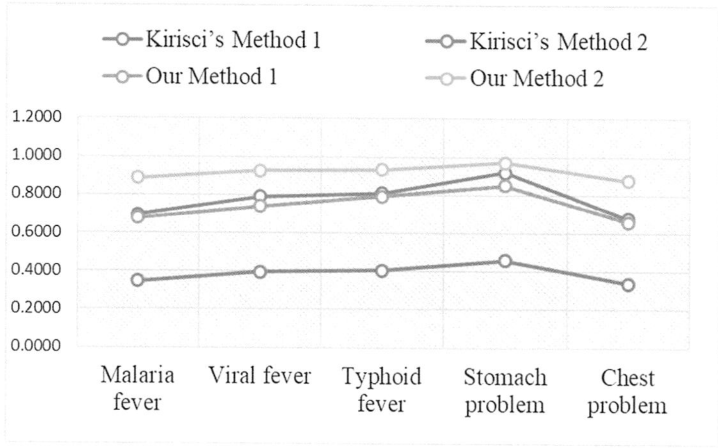

FIGURE 2.3 Patient P_2 versus diseases.

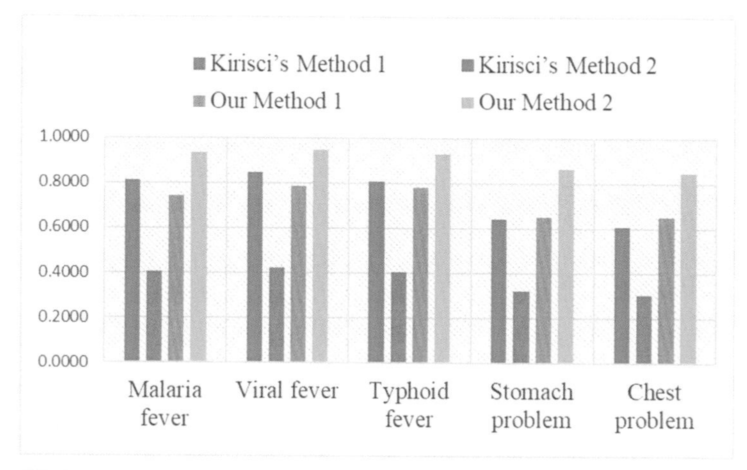

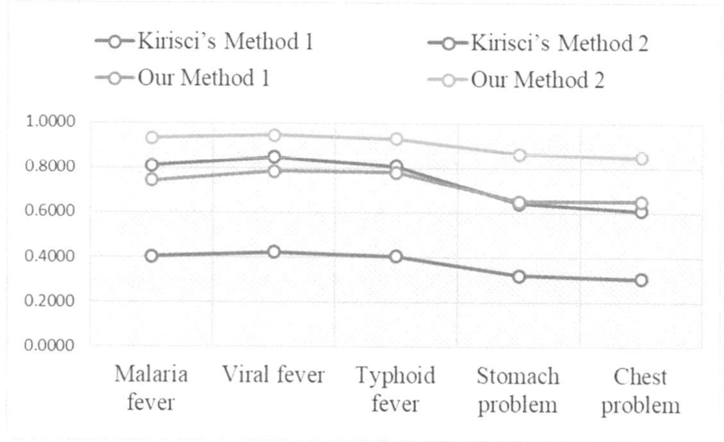

FIGURE 2.4 Patient P_3 versus diseases.

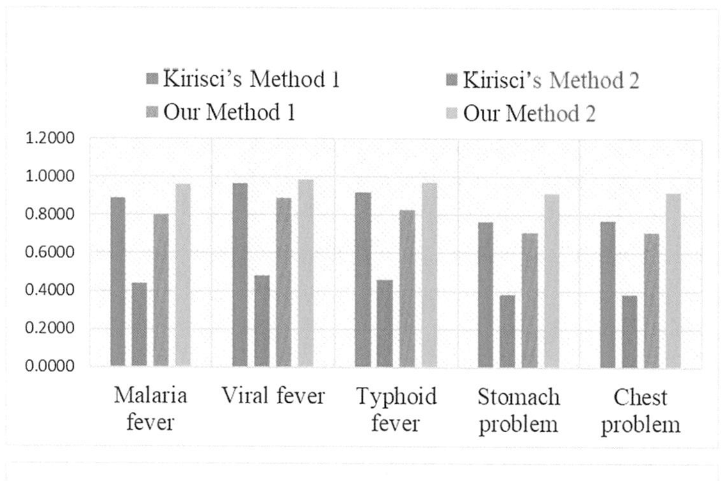

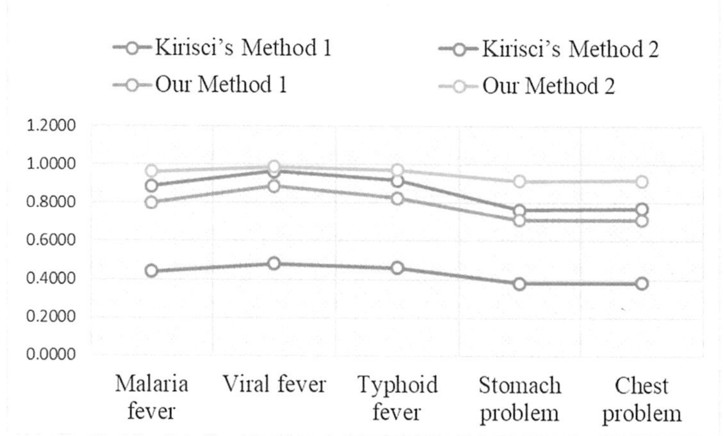

FIGURE 2.5 Patient P_4 versus diseases.

5. Conclusion

An exploration of FFS and its application in medical diagnosis has been discussed in details. Due to the relevance of correlation coefficient measure in computational intelligence, the idea has been introduced under Fermatean fuzzy environment in Ref. [41]. From the review of the two approaches of computing Fermatean fuzzy correlation coefficient in Ref. [41], we observed that the approaches will yield misleading result whenever anyone of the grades of membership, nonmembership, and margin of hesitation is zero. In addition, the second method in Ref. [41] only considers the maximum information energy, which leads to error of exclusion. Consequence upon these setbacks, we have developed two new Fermatean fuzzy correlation coefficient approaches with better model formulations. An elaborate discussion of the properties of the new Fermatean fuzzy correlation coefficient approaches were carried out in consonant with the properties of classical correlation coefficient. In addition, the process of medical diagnosis was carried out using Fermatean fuzzy medical information based on the new Fermatean fuzzy correlation coefficient approaches. To buttress the relevance of the new Fermatean fuzzy correlation coefficient approaches, the approaches in Ref. [41] were also applied on the Fermatean fuzzy medical information, and it is found that our approaches yield better results. With some minimal modifications, these Fermatean fuzzy correlation coefficient approaches could be used to model certain complex decision-making problems under picture fuzzy sets, complex fuzzy sets, spherical fuzzy sets, and other fuzzy variants in the future.

References

[1] L.A. Zadeh, Fuzzy sets, Information and Control 8 (1965) 338–353.
[2] K.T. Atanassov, Intuitionistic fuzzy sets, Fuzzy Sets and Systems 20 (1986) 87–96.
[3] R.R. Yager, Pythagorean Membership Grades in Multicriteria Decision Making, Technical Report MII-3301, Machine Intelligence Institute, Iona College, New Rochelle, 2013.

[4] S.S. Begum, R. Srinivasan, Some properties on intuitionistic fuzzy sets of third type, Ann Fuzzy Math Inform 10 (5) (2015) 799−804.
[5] T. Senapati, R.R. Yager, Fermatean fuzzy sets, Journal of Ambient Intelligence and Humanized Computing 11 (2020) 663−674.
[6] S.K. De, R. Biswas, A.R. Roy, Some operations on intuitionistic fuzzy sets, Fuzzy Sets and Systems 114 (3) (2000) 477−484.
[7] P.A. Ejegwa, Novel correlation coefficient for intuitionistic fuzzy sets and its application to multi-criteria decision-making problems, International Journal of Fuzzy System Applications 10 (2) (2021) 39−58.
[8] P.A. Ejegwa, I.C. Onyeke, A novel intuitionistic fuzzy correlation algorithm and its applications in pattern recognition and student admission process, International Journal of Fuzzy System Applications 11 (1) (2022). Article ID: 84, https://doi.org/10.4018/IJFSA.285984.
[9] R.R. Yager, Pythagorean membership grades in multicriteria decision making, IEEE Transactions on Fuzzy Systems 22 (2014) 958−965.
[10] X.L. Zhang, Z.S. Xu, Extension of TOPSIS to multiple criteria decision making with Pythagorean fuzzy sets, International Journal of Intelligent Systems 29 (2014) 1061−1078.
[11] H. Garg, A new generalized Pythagorean fuzzy information aggregation using Einstein operations and its application to decision making, International Journal of Intelligent Systems 31 (9) (2016) 886−920.
[12] P.A. Ejegwa, S. Wen, Y. Feng, W. Zhang, J. Liu, A three-way Pythagorean fuzzy correlation coefficient approach and its applications in deciding some real-life problems, Applied Intelligence 53 (1) (2022) 226−237, https://doi.org/10.1007/s10489-022-03415-5.
[13] D. Liu, Y. Liu, X. Chen, Fermatean fuzzy linguistic set and its application in multicriteria decision making, International Journal of Intelligent Systems 34 (5) (2019) 878−894.
[14] D. Liu, Y. Liu, L. Wang, Distance measure for Fermatean fuzzy linguistic term sets based on linguistic scale function: an illustration of the TODIM and TOPSIS methods, Journal of Intelligence Systems 34 (11) (2019) 2807−2834.
[15] T. Senapati, R.R. Yager, Fermatean fuzzy weighted averaging/geometric operators and its application in multi-criteria decision-making methods, Engineering Applications of Artificial Intelligence 85 (2019) 112−121.
[16] T. Senapati, R.R. Yager, Some new operations over Fermatean fuzzy numbers and application of Fermatean fuzzy WPM in multiple criteria decision making, Informatica 30 (2) (2019) 391−412.
[17] K.G. Mehdi, A. Maghsoud, H.T. Mohammad, K.Z. Edmundas, K. Arturas, A new decision-making approach based on Fermatean fuzzy sets and WASPAS for green construction supplier evaluation, Mathematics 8 (12) (2020). Article ID: 2202, https://doi.org/10.3390/math8122202.
[18] H. Garg, G. Shahzadi, M. Akram, Decision-making analysis based on Fermatean fuzzy Yager aggregation operators with application in COVID-19 testing facility, Mathematical Problems in Engineering 2020 (2020). Article ID: 7279027, https://doi.org/10.1155/2020/7279027.
[19] P.A. Ejegwa, K.N. Nwankwo, M. Ahmad, T.M. Ghazal, M.A. Khan, Composite relation under Fermatean fuzzy context and its application in disease diagnosis, Informatica 32 (10) (2021) 87−101.
[20] L. Sahoo, Some score functions on Fermatean fuzzy sets and its application to bride selection based on TOPSIS method, International Journal of Fuzzy System Applications 10 (3) (2022). Article ID: 2, https://doi.org/10.4018/IJFSA.2021070102.
[21] S. Aydin, A fuzzy MCDM method based on new Fermatean fuzzy theories, International Journal of Information Technology and Decision Making 20 (3) (2021) 881−902.
[22] S. Gül, Fermatean fuzzy set extensions of SAW, ARAS, and VIKOR with applications in COVID-19 testing laboratory selection problem, Expert Systems 38 (8) (2021) e12769, https://doi.org/10.1111/exsy.12769.
[23] Z. Yang, H. Garg, X. Li, Differential calculus of Fermatean fuzzy functions: continuities, derivatives, and differentials, International Journal of Computational Intelligence Systems 14 (1) (2021) 282−294.
[24] G. Shahzadi, F. Zafar, M.A. Alghamdi, Multiple-attribute decision-making using Fermatean fuzzy Hamacher interactive geometric operators, Mathematical Problems in Engineering 2021 (2021). Article ID: 5150933, https://doi.org/10.1155/2021/5150933.
[25] P.A. Ejegwa, I.C. Onyeke, Fermatean fuzzy similarity measure algorithm and its application in students' admission process, International Journal of Fuzzy Computation and Modelling 4 (1) (2022) 34−50.
[26] P.A. Ejegwa, G. Muhiuddin, E.A. Algehyne, J.M. Agbetayo, D. Al-Kadi, An enhanced Fermatean fuzzy composition relation based on a maximum-average approach and its application in diagnostic analysis, Journal of Mathematics 2022 (2022). Article ID: 1786221, https://doi.org/10.1155/2022/1786221.
[27] P.A. Ejegwa, D. Zuakwagh, Fermatean fuzzy modified composite relation and its application in pattern recognition, Journal of Fuzzy Extension and Application 3 (2) (2022) 140−151.
[28] S. Zeng, Y. Pan, H. Jin, Online teaching quality evaluation of business statistics course utilizing Fermatean fuzzy analytical hierarchy process with aggregation operator, Office Systems 10 (3) (2022). Article ID: 63, https://doi.org/10.3390/systems10030063.
[29] M. Akram, U. Amjad, J.C.R. Alcantud, G. Santos-García, Complex Fermatean fuzzy N-soft sets: a new hybrid model with applications, Journal of Ambient Intelligence and Humanized Computing (2022), https://doi.org/10.1007/s12652-021-03629-4.
[30] T. Gerstenkorn, J. Manko, Correlation of intuitionistic fuzzy sets, Fuzzy Sets and Systems 44 (1) (1991) 39−43.
[31] Z.S. Xu, On correlation measures of intuitionistic fuzzy sets, in: E. Corchado, et al. (Eds.), IDEAL 2006, LNCS 4224, Springer-Verlag, Berlin Heidelberg, 2006, pp. 16−24.
[32] H.L. Huang, Y. Guo, An improved correlation coefficient of intuitionistic fuzzy sets, Journal of Intelligence Systems 28 (2) (2019) 231−243.
[33] N.X. Thao, A new correlation coefficient of the intuitionistic fuzzy sets and its application, Journal of Intelligent and Fuzzy Systems 35 (2) (2018) 1959−1968.
[34] P.A. Ejegwa, I.C. Onyeke, Intuitionistic fuzzy statistical correlation algorithm with applications to multi-criteria based decision-making processes, International Journal of Intelligent Systems 36 (3) (2021) 1386−1407.

[35] H. Garg, A novel correlation coefficients between Pythagorean fuzzy sets and its applications to decision-making processes, International Journal of Intelligent Systems 31 (12) (2016) 1234–1252.
[36] N.X. Thao, A new correlation coefficient of the Pythagorean fuzzy sets and its applications, Soft Computing 24 (2020) 9467–9478.
[37] P.A. Ejegwa, S. Wen, Y. Feng, W. Zhang, Determination of pattern recognition problems based on a Pythagorean fuzzy correlation measure from statistical viewpoint, in: Proceedings of the 13th International Conference of Advanced Computational Intelligence, Wanzhou, China, 2021, pp. 132–139.
[38] M. Lin, C. Huang, R. Chen, H. Fujita, X. Wang, Directional correlation coefficient measures for Pythagorean fuzzy sets: their applications to medical diagnosis and cluster analysis, Complex and Intelligent Systems 7 (2021) 1025–1043.
[39] S. Singh, A.H. Ganie, On some correlation coefficients in Pythagorean fuzzy environment with applications, International Journal of Intelligent Systems 35 (2020) 682–717.
[40] P.A. Ejegwa, Y. Feng, W. Zhang, Pattern recognition based on an improved Szmidt and Kacprzyk's correlation coefficient in Pythagorean fuzzy environment, in: H. Min, Q. Sitian, Z. Nian (Eds.), Advances in Neural Networks, Lecture Notes in Computer Science (LNCS) 12557, Springer Nature, Switzerland, 2020, pp. 190–206.
[41] M. Kirisci, Correlation coefficients of Fermatean fuzzy sets with a medical application, Journal of Mathematical Science and Modelling 5 (1) (2022) 16–23.

Chapter 3

Application of Deep-Q learning in personalized health care Internet of Things ecosystem

Yamuna Mundru[1], Manas Kumar Yogi[1] and Jyotir Moy Chatterjee[2]

[1]*CSE Department, Pragati Engineering College, Surampalem, Andhra Pradesh, India;* [2]*Department of Information Technology, Lord Buddha Education Foundation (Asia Pacific University), Kathmandu, Nepal*

1. Introduction

Contemporary global social and economic advances are intended to enhance the future for people by addressing their wellness and prosperity. Certain technical headways in have assisted in accomplishing this plan. To this end, different committed specialized drives have been presented. One such drive is Health Care 5.0, which was created because of the rise in digital health and digital standards for health care administrations [1]. Ongoing improvements in innovation have empowered remote and programmed observations in health care administration through medical devices to screen for the different health states of a patient. These devices are committed to explicit health conditions and work autonomously. For example, circulatory devices screen only for heart-related health conditions, and insulin pumps manage only the health of diabetic patients by maintaining the right amount of blood insulin. Previous to Health Care 5.0, Health Care 4.0 arose from Industry 4.0, which changed the health care sector into a digital format. X-rays and magnetic resonance imaging have developed into computer tomography and ultrasound scans using electronic medical records [2]. These devices are client-driven and are used by medical practitioners to screen and treat patients' conditions for prevention and financial purposes. With later advances (eg, industrial Internet of Things [IoT], industrial digital actual frameworks), the use of IoT devices and applications has developed dramatically. According to a review, by the end of 2021, there were expected to be 212 billion IoT devices and applications health care ($\sim$41%) [3]. Moreover, as announced by reports, the health care IoT market is expected to be worth US $534.3 billion by 2025. The rule objective of innovation-driven applications in health care is to help health care tasks by automatically controlling different health states of patients through constant and remote observations of circumstances. According to the rule, Health Care 5.0 plans to help different objectives [4] (i.e., continuous, solid, strong, and personalized health care [PH] administration). In this chapter, we discuss the characteristics of PH. PH benefits commonly work in the strictest mode by supporting the customization of a particular health condition under unambiguous circumstances. This does not work in practice, because patients with chronic health conditions frequently have various health problems. Thus, we understand the personalization of health care administration as the capacity of the help that results in determinant-based (e.g., hereditary qualities, conduct, ecological and actual impacts, medical consideration, and social factors) improvement of various health states of a patient. Streamlining expands the patient's future by decreasing the symptoms of different health conditions. In light of our scientific categorization, we add to Health Care 5.0 to PH benefits intended to improve the different related clinical health states of a patient [5]. As observed, ways to deal with lay out unwavering quality, versatility, and personalization of health care administrations neglect to help persistent checking of health conditions in Health Care 5.0 principally in light of the fact that they are either health condition-explicit, or climate explicit, or specialist explicit. Also, the developments of new IoT medical devices manufactured by various merchants and their documentation through various conventions have made the task of laying out these necessities difficult [6]. Consequently, in this overview, we deliberately study and examine different methodologies that examine key prerequisites of the administrations and recognize gaps to be filled in assistance to explore Health Care 5.0 administrations in practice.

Deep Learning in Personalized Healthcare and Decision Support. https://doi.org/10.1016/B978-0-443-19413-9.00024-2

2. Related work

Personalized electronic health care is not safe to analyze and has downsides. It has the central problems of IoT and machine learning (ML). This section portrays a situation in which an adult uses a sensor-based PH framework [7]. The sensor gathers data such as the pulse, electroencephalogram, glucose, and pressure, and sends it to a dataset. The data are put there for later use. The dataset also uses ML calculations to investigate the gathered data to decide a patient's risk and improvements in health and to propose further activities in view of those. Even in this basic situation, there are explicit problems and difficulties [8]. The transmission of data and problem of group inferencing drawbacks, security, and confirmation by tactile devices should be addressed. Generally, the financial prosperity regarding the innovation is not determined by the value to the client. A Body Area Network (BAN) utilizes sensors to gather body related well-being data from clients. To administer PH, BAN should communicate and change detected peculiarities into important data, and it needs to guarantee that it will meet other framework necessities, such as energy productivity [9]. Moreover, its capacity specifically process and conveys data at constant levels and rates appropriate to the data's objective, whether to a runner asking about her pulse or a doctor needing a patient's electrocardiogram [10]. These application requirements demand various levels of data and incorporate BAN frameworks into the data innovation foundation. To get the full advantage of the BAN for PH, guaranteeing good data respectability and energy productivity of the sensors is required, as well as the ability to coordinate with the existing infrastructure. The BAN architecture is robust with state-of-art capabilities [11]. ML is firmly connected with factual similarity, decision-making from existing data, and synthesis from past experience. When checking a patient, the ML-based technique looks at what is happening according to the prepared dataset. The prepared dataset has a pivotal job in effectively foreseeing the future pattern of a given problem. Feature engineering is considered when preparing a dataset for efficient model training and validation [12]. This dataset might be one-sided and might not be different enough to cover numerous situations. Data which sometimes include noise as well as missing and inadequate data could prompt a lower likelihood proportion to distinguish and foresee a health-related conclusion and warning notification are also facilitated [13]. A total dataset of all contextual investigations may not be accessible to track examples, which could prompt unacceptable assessments in PH. When using IoT and ML to drive PH, the framework might have to be chosen for analysis, and the client might have to be cautioned regarding expectations. There are a few occurrences in which an ML-based choice could be off-base, and is difficult to determine why a specific choice was made. This choice should also have a fair degree of trust with respect to the privacy of the user [14]. For instance, in the case of driverless vehicles, not many mishaps arose because of wrong choices made by the vehicle itself. The basic question was how to understand the choice made by an artificial intelligence (AI) machine while driverless learning was employed. This prompted a moral inquiry regarding who would be responsible in the event of a false prediction, and how to distinguish or amend that defect in the dynamic cycle and understand how a driverless machine might work. These downsides can restrict the purposes of ML in PH for a delicate application such as personalized medicine. Personalized medicine is becoming a challenge in modern e-health care systems owing to its diversity and difficulty in obtaining curated data. Ethical issues during disease identification and false-positive cases compound the difficulty for researchers working in this direction [15]. While using ML-based PH administration, predictive investigation can help the emergency clinic identify patients who ought to be readmitted. Predictive analysis is useful for creating a risk delineation model in which certain patients with higher risk are managed with extra effort. Such effort includes, but is not limited to, providing extra monitoring devices and continuous development. Generally, such models are created based on cases and previous historical data. The powerful PH system, which aid in readmission preadmission efforts, should also use dynamic patient data and to predict future prospects and initiate an activity to mitigate possible complications. Studies have shown the way that incorporating clinical variables and relevant vital signs can significantly predict readmissions. Predictive models for readmission are being developed that can help minimize costs associated with readmitting patients. The major challenges of developing these predictive models are the heterogeneity of diseases and huge amount of patient data, which is not deductive in nature but discrete in appearance [16]. The PH framework can be used for this situation. However, problems of data transmission loss, noise in data, and incomplete data need to be addressed. Traditional ML methods have demonstrated efficiency in simple and well-constrained classification and regression problems. However, most of them have constraints in addressing convoluted problems because of their shallow models, which lead to a single and often direct transition from input data to problem-specific features. Consequently, important feature engineering and domain knowledge are required to extract, select features, and develop appropriate representations from raw data sources. The current application of shallow learning models has resulted in 60% predictive ability, which is not desired in the medical domain. Thus, the next step is to deploy deep reinforcement learning approaches that provide a high degree of predictive value, even with fewer patient data [17]. In any case, deep learning (DL) is a type of representation learning strategy that permits the framework to discover features required for a particular task from raw data automatically by building numerous processing layers with nonlinear tasks to learn representations at different levels. DL excels at finding complex designs in high-dimensional data. For instance, convolutional neural networks are suited to handle visual and other two-dimensional data such as images and speech. Recurrent neural networks (RNNs) have shown superior outcomes in sequential data, such as in the use of natural language processing. Deep belief networks, commonly composed of restricted Boltzmann machines (BMs), are generative deep models particularly useful for improving modeling and faster convergence of the training stage, even with limited training data. They can easily approximate the target function by virtue of efficient feature representation. Deep BMs are even more efficient because they work according to human behavior by analyzing from higher to lower levels of abstraction. In this process, they collect and store every element of the structural property of a model [18]. DL has been applied in precision medicine. The availability of biomedical data, such as clinical imaging, electronic health records (EHRs), genomics, laboratory tests, patient histories, sensors, and wearable devices presents increasing opportunities to obtain more accurate and in-depth insights for patients. However, challenges also arise. These biomedical data are normally large-scale, noisy, sparse, inadequate, unpredictable, heterogeneous, high-dimensional, mostly unstructured, and poorly annotated. A main advantage of DL models is their power to link a patient's complete health record to a low-dimensional format. In most research investigations, the patient representations used to obtain visual clarity tend to form groups, but the representations from end-to-end depictions often develop into a continuum. However, cases of emergency admission and heart failure are clustered in distinct areas. Also, during profiling of the concerned clusters, it helped in identifying and assessing common characteristics of patients. This led to improved performance of the models and aided in future clinical research [19]. In addition, the lack of spatial data hinders the adoption of traditional ML, which is limited by its ability to handle raw data and therefore strongly depends on manual engineering to create compelling and robust elements. As a part of the goals of this work to provide guidance to work with DL for EHR, we recognized that the central concerns are connected with the choice of consideration measures and preprocessing of the data. The consideration models normally rely on the clinical question and result of interest. This is a critical piece of a good study design. Moreover, it is of great importance to divide the data into the standard training, validation, and test sets, because this would help to avoid overfitting and appropriately assess the DL models. In terms of data preprocessing, we highlighted the importance of using clinical clusters to diminish the cardinality of the two diagnoses and medications. This could be considered a type of dimensionality reduction that prevents overly long and sparse data vectors that would hamper the training process of the DL models [20]. DL provides a more successful worldview than traditional methods to deal with complex data:

1) Appropriate elements can be advanced by employing a universally useful learning system. By stacking layers progressively, deep neural networks can reveal hidden patient characteristics from various layers.
2) Deep neural networks give a more important level of semantic construction with which to understand relationships among input data. They are useful for detecting patterns from irregular data.
3) The mechanism of deploying DL outputs more exact expectations by pretraining hidden layers individually to find fundamental connections, and then fine-tunes the whole organization with backpropagation of characteristics for accuracy.

An study using DL incorporated a variational autoencoder and its semidirected type to determine reactions to drugs by examining the passive condition of genetic representation during the use of such drugs. Many years earlier, discoveries and improvements in drugs were limited to researcher chemists working in a laboratory with colossal measurement testing, approval, and engineered systems, all of which added to an impressive amount of time and work to get one medicine out into the clinics. Advances in computational procedures and an explosion in multiomics data prompted improvements in bioinformatics devices that have assisted in accelerating improvements in the way medicine is handled. Yet, with the advent of emerging technologies like AI, ML and DL, the regular medicine revelation process has been additionally think. Broad organic data are present in different datasets across the globe. They are propagated as not model fitting entities for ML/DL-based approaches and assist in precise distinguishing pieces of proof of examples and models which with canning be utilized to recognize restoratively dynamic atoms with many less ventures on time, labor force and abundance [21]. An essential quality of this exploration is the reception of a semimanaged learning technique that uses the outcomes from several predeveloped autoencoders as well as new data to predict reactions. It has been observed that machine learning articulations are dispersed non uniformly in the search space thus eliminating search bias and representing the benefit of using a deep neural organization. Because of the continuously expanding volume of data in botany and the power of DL applications in numerous different domains of science, DL strategies have also been assessed regarding predictive ability in genetic science. Frequently, outcomes are blended with the possibly overstated assumptions for datasets with moderately small samples of people. Here, we survey DL applications for genome structure (GS) to provide a metaimage of their results predicted versus actual results compared with traditional genomic expectation models. We incorporate a prolog to DL basics and its prerequisites regarding the data size, training process, data, kind of data, computational assets, and so on, to apply DL appropriately. We also break down the advantages and disadvantages of this procedure compared with customary genomic forecast models, as well as future patterns using this strategy [22]. Other research centers on matching comparative drugs using similarities between drugs and medical conditions. Matches between novel drugs and infections are related by positioning collective proof using strategic regression. Known drug–sickness associations are combined with drug similitude measures to predict symptoms in light of the rule that comparable drugs will probably model comparative illnesses. Existing techniques created to work with the manufactured course of novel medicine candidates include a consistent circle of self-learning in which at first, it exploits current information about known bioactive mixtures and is supposed to deliver appropriately restorative frameworks. With regard to certain successive ML tasks such as speech recognition, RNNs continually arrive at levels of predictive accuracy that no other calculation can coordinate. Most artificial neural networks, such as feedforward neural networks, have no memory of the information they received only 1 second previously. RNNs recall what they have recently experienced and at a strikingly complex level, especially RNNs integrating long short-term memory (LSTM). Medical researchers are investigating how much an RNN with LSTM cells can sort out reasonable substance manages and produce artificially achievable atoms subsequent to being trained on existing mixtures [23]. The research is further supported by supplanting infection names with genetic marks to assess sickness illness likeness by looking at genetic cosmetics. Drug comparability is assessed by hashed chemical fingerprints, text mining–separated aftereffects, drug-related quality succession arrangement and closeness in Privacy Preserving Identity. Most e-health care systems try to achieve certain privacy objectives within the context of collaborative neural-network training: to save the privacy of data used during training, and to help users control the learning objective so they can choose which data they wish to remain hidden. Many implementations are developed that are able to achieve these privacy objectives at a considerably lower cost of performance compared with cryptographic approaches such as secure multiparty computation and homomorphic encryption. They can be used for deployment in big data–based DL systems [24]. Similarities of medical conditions are assessed from phenotypic and genetic points of view. Another likely area of use is the prediction of adverse drug reactions. The World Health Organization dataset containing adverse drug response contains 2,000,000 reports of adverse drug reactions and can be used for prediction. Researchers have created a Bayesian certainty spread neural organization to recognize signs of adverse drug reactions in the dataset. The benefits of implementing a neural organization are the chance of equal and successive calculation, and the power of managing missing data, as well as reproducibility and straightforwardness. Moreover, a strategy consolidating the neighborhood-based technique and the limited Boltzmann machine technique was created to lessen the risk for poor results for a particular task and dataset; enhancements from a conventional ML strategy were also noticed. Limited BMs can predict using even latent factors, so when handling patient situations with the need for privacy, the data remain hidden and the model can predict efficiently. Even in an IoT-based ecosystem, in which the sensor measures and stores patient data, owing to the sensitivity of the data, the learning rate of the model may suffer but this proposed Q-learning approach will handle such situations effectively [25]. In this manner, to predict adverse drug reactions, DL has also shown guarantees.

3. Proposed mechanism

Here, we prepare a novel mechanism that applies principles present in the Deep-Q learning architecture (Fig. 3.1). For a limited number of states and actions Q-learning performs well, but when we consider a relatively complex system such as an IoT system, the number of states increases and consequently, the number of actions to be taken also increases. The Deep-Q learning function approximator works efficiently in such cases.

In PH based on an IoT system, factors to be considered are:

1. Patient history
2. Current observation state
3. Action to be taken based on the current state
4. If alternative action is taken, what the future state will be

The agent guided by the function approximator, which considers the current state reward obtained after taking an action the learning rate, discount factor which decrease the current value the rewards over time when more immediate rewards get more value. To obtain this, the Bellman equation is used:

$$\mathrm{Q}(S_t, A_t) = (1 - \mathrm{LR})\ \mathrm{Q}(S_t, A_t) + \mathrm{LR} * \left(\mathrm{R}W_t + D_f * \max_a \mathrm{Q}(S_{t+1,}\mathrm{LR})\right) \quad (3.1)$$

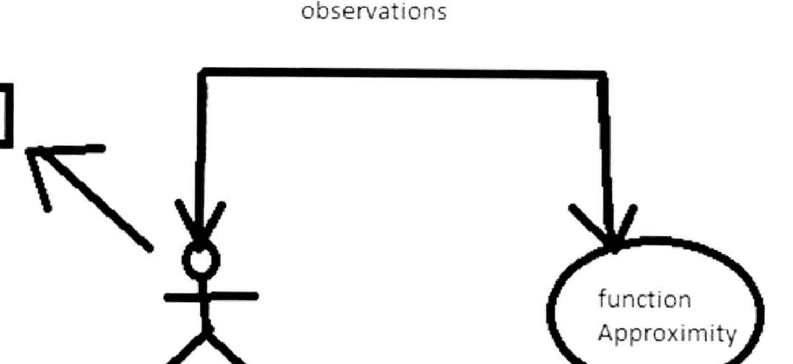

FIGURE 3.1 Proposed architecture.

where S = state/observation; A = agents' action; RW = reward obtained from action; T = time step; LR = learning; and D_f = discount factor. The overall Deep-Q learning architecture works based on steps of the Deep-Q network algorithm:

Step 1: Initialize the source and destination neural networks.

Step 2: Choose an appropriate action using the Epsilon-greedy exploration technique.

Step 3: Apply the Bellman equation to update the network weights (rewards).

Explanation of these steps:

Step 1: Initialization of source and destination neural networks:

The source neural network will contain the patient history and will be regarded as the states. These states are dynamic in nature (i.e., the write time changes). The most challenging aspect is predicting the destination neural network. The output states existing in the destination neural network will indicate the states that represent the healthy state of a patient. This is made possible by applying the Deep-Q learning technique. Each state in the destination network inherits some optimal reward value based on the action taken by the agent.

Step 2: Use the Epsilon-greedy exploration technique to select the best action with the highest reward. The agent picks a random action with epsilon probability and applies the best known action with 1-epsilon probability. Both the source and destination network map input states to output actions that denote the model's predicted Q-value. The action that corresponds to the highest predicted Q-value is the best known action at the current state.

Step 3: Update the network weights using the Bellman equation.

Deep-Q learning agents have the benefit of experience replay to gain knowledge about their environment and update the source and destination networks.

Leverage provided by experience replay:

This is the act of storing and replaying game states. Main advantage of this mechanism is in off-policy technique that can update the parameters of the source and destination networks using the saved history. In PH, the history of the patient is important for taking current and future actions. Another benefit is that nonskewed data can be generated to provide better results. In our novel approach, we deploy experience replay to train on small batches in terms of five steps. This results the quick implementation of our approach.

Bellman Equation is used to obtain new temporal difference after the weights are updated.

In the Fig. 3.2, the old Q-value is replaced with new Q-value 9. Then, proceed to retrain the network with the new Q-values.

The temporal difference is the factor:

$$\left(RW_t + D_f * \max.\ Q(S_{t+1,} LR)\right) \tag{3.2}$$

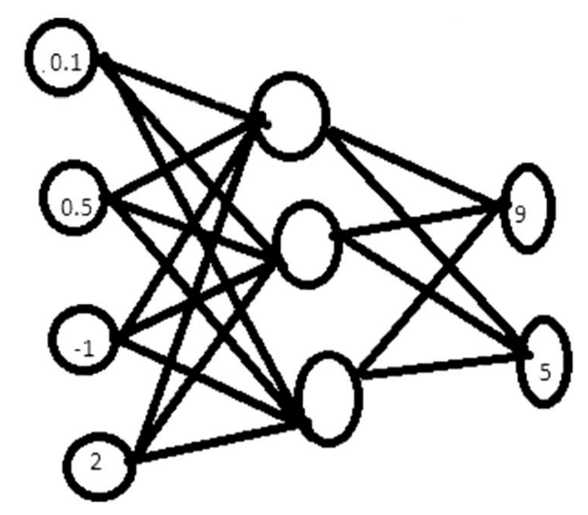

FIGURE 3.2 Old Q-values are replaced with new Q-values.

TABLE 3.1 Input values with linked feature set.

Input state values (weights)	Feature in health care domain
0.1	Prone to disease
0.5	Recovered
−1	Very sick
2	Healthy

TABLE 3.2 Output Q-values with action proposed for agent.

Output Q values	Action
9	Report as healthy
5	Report as recovered

For our proposed mechanism for PH, we model the input state values (weights) and corresponding actions through Tables 3.1 and 3.2.

We consider values less than 2 to be in unhealthy states. Thus, we adjust the weights to obtain output Q values 6, 7, 8, 9 are considered to be optimal values for our proposed approach. If our treatment results into action with output Q values of more than 5, we advocate the same action to treat the patient in future courses. However, if Q values of 2, 3, and 4 are obtained, we need to change the treatment policy so that it undergoes training with input state values. The training cycle should not be too fast or slow.

We have to follow these practices before the agent in the IoT environment starts to learn to formulate an optimal treatment policy for patients who are under observation:

1. The frequency of updating the model parameter should not be too often.
2. Deploying the Huber loss function helps the agent to learn better than does using mean squared error loss function.
3. The correct frequency to copy weights from the source network to the destination network should not be too low or high.

4. Experimental results

The algorithm is implemented to execute on TensorFlow with the Python program, and the system configuration is a CPU with an Intel Core i6 CPU with 32 GB RAM. To evaluate the performance of the proposed novel algorithms and compare them against performance, we tested three treatment policies, T1, T2, and T3 with assumptions that the personalized reaction for four types of health conditions. For unambiguity, each simulation is executed 100 times. We observe that each simulation converges in 3000 steps and takes less than 400 s. For indicative inferences, Figs. 3.3 and 3.4 show the daily rewards obtained and the clustering of treatment policies during the 30 days of observation. Our Deep-Q learning–based treatment policy can adapt to patients' current health conditions.

Fig. 3.5 shows that the comparative performance of the Deep-Q learning–based approach outperforms the two popular techniques of LSTM and multilayer perceptron. The average running time with our proposed mechanism is reduced to nearly 2%, and the average number of converging steps is also reduced by 11.9% compared with multilayer perceptron. These cumulative studies suggest that our technique is robust enough to apply to a PH scenario without questioning performance problems.

5. Future directions

A few important difficulties are related to improving DL calculations, mostly because of the design and nature of data or because of engineering of the deep neural networks.

Data source: Medical data are stored various databases, some for drug properties and others for research facility test results. Furthermore, assuming patients switch specialists, medical records are retrieved from various databases. Due to uncertainty in drug effect results are exposed to irregular and methodical blunders across various test climates. Data from electronic medical records are commonly noisy, sparse, and sporadically coordinated, which makes them difficult to prepare using DL.

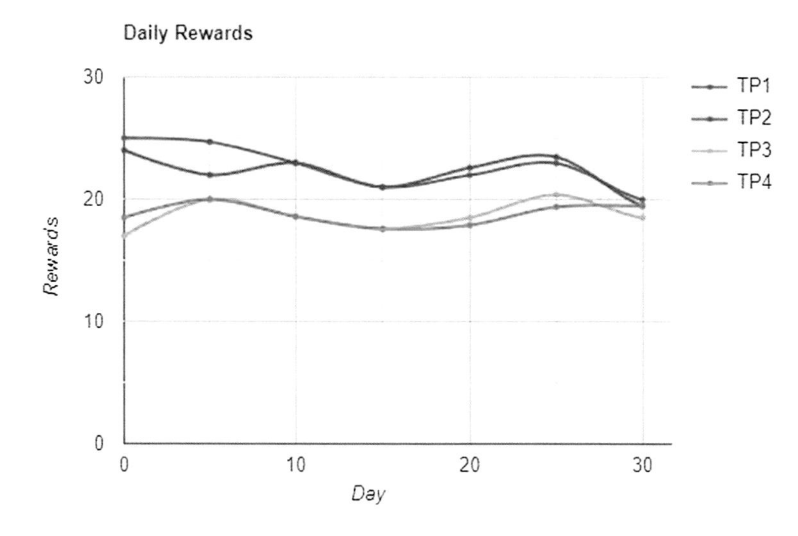

FIGURE 3.3 Daily rewards for treatment policies TP1 to TP4 after training phase.

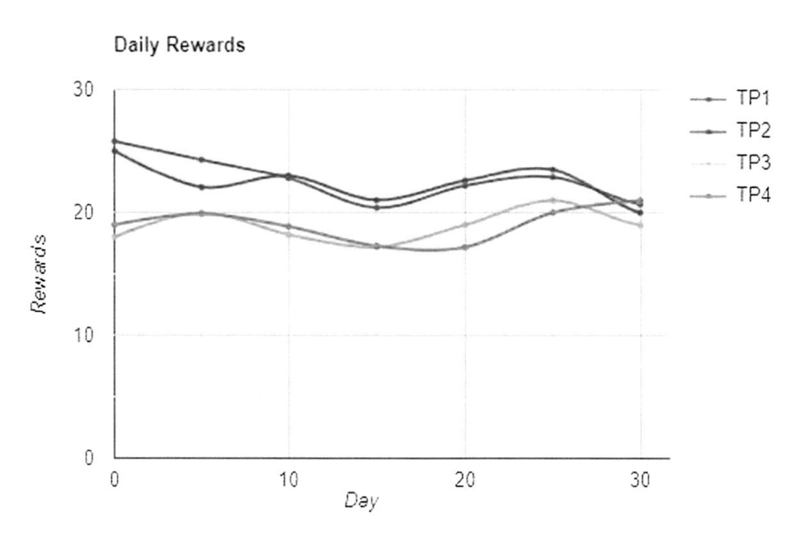

FIGURE 3.4 Daily rewards for treatment policies TP1 to TP4 after testing phase.

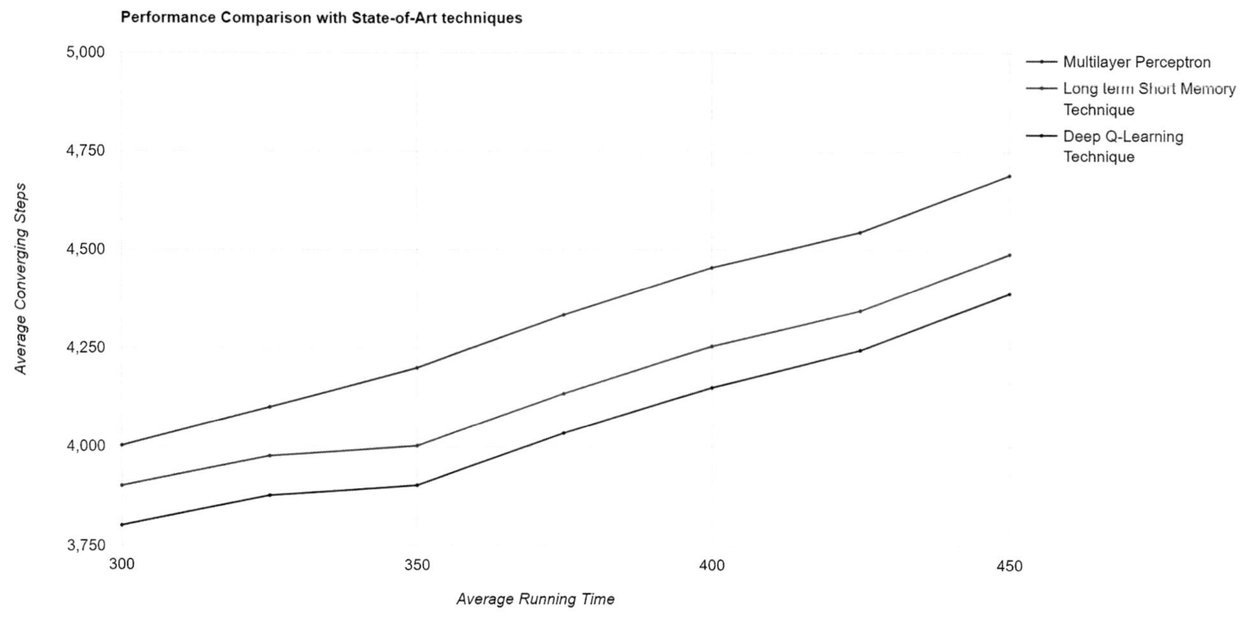

FIGURE 3.5 Performance comparisons with current popular methods.

Data size: With the huge use of e-health records frameworks, an ever increasing amount of data is gathered and requires consolidation into the learning framework. However, the data size is considered small for preparing a deep brain network, which requires a great many data per class to accomplish reasonable accuracy. For better execution, a huge number of estimations are required.

Reproducibility: The preparation of a DL network requires calibrating with a certain degree of adaptability. Different preparation series might lead to various predictive results that can be inconvenient for medical communications. This tests the use of DL in medical rounds.

Overfitting: Data control including hereditary qualities ordinarily incorporates data with enormous dimensionality and hence frequently meets the problem of overfitting even with appropriate regulations.

Data quirkiness: The adequacy of calculation training is profoundly subject to the independence of the data. Many examinations that look at the adequacy of calculations depend entirely on the preparation dataset and might be exposed to data-explicit varieties. This may also make sense for the fluctuated presentation of various computations thought about in contrast to one another in customized medicine applications.

6. Conclusion

The results of our novel approach show us that efficiency leads us to advocate the application of Deep-Q learning in PH. Based on the daily rewards and cumulative rewards over time, the treatment policy of a patient can be personalized so that the cost and time of treatment will be reduced. Because PH solutions are increasingly evolving in the technological market, our solution is worth the promise. This chapter will serve as a baseline research document for future researchers who do not want to compromise on the quality of service in the PH domain.

References

[1] J.R.A. Solares, et al., Deep learning for electronic health records: a comparative review of multiple deep neural architectures, Journal of Biomedical Informatics 101 (2020) 103337.

[2] M. Cherrington, et al., Deep learning decision support for sustainable asset management, in: Advances in Asset Management and Condition Monitoring, Springer, Cham, 2020, pp. 537–547.

[3] R. Gatta, et al., Integrating radiomics into holomics for personalised oncology: from algorithms to bedside, European radiology experimental 4 (1) (2020) 1–9.

[4] V.P. Gurupur, et al., Analysing the power of deep learning techniques over the traditional methods using medicare utilisation and provider data, Journal of Experimental and Theoretical Artificial Intelligence 31 (1) (2019) 99–115.

[5] Q. Zhu, Z. Chen, Y.C. Soh, A novel semisupervised deep learning method for human activity recognition, IEEE Transactions on Industrial Informatics 15 (7) (2018) 3821−3830.

[6] T. Panch, et al., Artificial intelligence: opportunities and risks for public health, The Lancet Digital Health 1 (1) (2019) e13−e14.

[7] S. Zhang, et al., Learning for personalized medicine: a comprehensive review from a deep learning perspective, IEEE reviews in biomedical engineering 12 (2018) 194−208.

[8] B.P. Nguyen, et al., Predicting the onset of type 2 diabetes using wide and deep learning with electronic health records, Computer Methods and Programs in Biomedicine 182 (2019) 105055.

[9] A.K. Sahoo, C. Pradhan, H. Das, Performance evaluation of different machine learning methods and deep-learning based convolutional neural network for health decision making, in: Nature Inspired Computing for Data Science, Springer, Cham, 2020, pp. 201−212.

[10] V. Jain, et al. (Eds.), Deep Learning for Personalised Healthcare Services, De Gruyter, 2021.

[11] D. Kaul, H. Raju, B.K. Tripathy, Deep learning in healthcare, in: Deep Learning in Data Analytics, Springer, Cham, 2022, pp. 97−115.

[12] A. Motwani, Piyush Kumar Shukla, and Mahesh Pawar. "Smart predictive healthcare framework for remote patient monitoring and recommendation using deep learning with novel cost optimization, in: International Conference on Information and Communication Technology for Intelligent Systems, Springer, Singapore, 2020.

[13] J. Fang, V. Lee, H. Wang, Dynamic Physical Activity Recommendation on Personalised Mobile Health Information Service: A Deep Reinforcement Learning Approach, arXiv, 2022 preprint arXiv:2204.00961.

[14] O. Costilla-Reyes, et al., Deep learning in gait analysis for security and healthcare, in: Deep Learning: Algorithms and Applications, Springer, Cham, 2020, pp. 299−334.

[15] C. Fei, et al., Machine and deep learning algorithms for wearable health monitoring, in: Computational Intelligence in Healthcare, Springer, Cham, 2021, pp. 105−160.

[16] M. Hammad, et al., Deep learning models for arrhythmia detection in IoT healthcare applications, Computers & Electrical Engineering 100 (2022) 108011.

[17] M. Gomes, A. Santos, et al., Machine learning applied to healthcare: a conceptual review, Journal of Medical Engineering and Technology (2022) 1−9.

[18] G. Sood, N. Raheja, Recent advancements on recommendation systems in healthcare-assisted system, Emergent Converging Technologies and Biomedical Systems (2022) 587−605.

[19] Banegas-Luna, A. Jesús, H. Pérez-Sánchez, SIBILA: High-Performance Computing and Interpretable Machine Learning Join Efforts toward Personalised Medicine in a Novel Decision-Making Tool, arXiv, 2022 preprint arXiv:2205.06234.

[20] A. Merkin, R. Krishnamurthi, O.N. Medvedev, Machine learning, artificial intelligence and the prediction of dementia, Current Opinion in Psychiatry 35 (2) (2022) 123−129.

[21] R.M. Devi, H. Gunasekaran, A.K. Prabavathy, K. Ramalakshmi, S. Akila, K.V.G. Kanth, Design of intelligent IoT for smart healthcare monitoring system using optimal neural network (IONN) model, International Journal of Health Sciences 6 (S2) (2022) 13422−13434, https://doi.org/10.53730/ijhs.v6nS2.8535.

[22] R. Jeyaraj, Pandia, E.R. Samuel Nadar, Smart-monitor: patient monitoring system for IoT-based healthcare system using deep learning, IETE Journal of Research 68 (2) (2022) 1435−1442.

[23] M. Javaid, et al., Significance of machine learning in healthcare: features, pillars and applications, International Journal of Intelligent Networks 3 (2022) 58−73, https://doi.org/10.1016/j.ijin.2022.05.002.

[24] P.-H. Huang, K.-hun Kim, M. Schermer, Mapping the ethical issues of digital twins for personalised healthcare service, Journal of Medical Internet Research 24 (2022) 1.

[25] E.A. Mantey, et al., Maintaining privacy for a recommender system diagnosis using blockchain and deep learning, Human-centric computing and information sciences (2022).

Chapter 4

Dia-Glass: a calorie-calculating spectacles for diabetic patients using augmented reality and faster R-CNN

Natasha Tanzila Monalisa, Shinthi Tasnim Himi, Nusrat Sultana, Musfika Jahan, Nayeema Ferdous, Md. Ezharul Islam and Mohammad Shorif Uddin

Department of Computer Science and Engineering Jahangirnagar University, Savar, Dhaka, Bangladesh

1. Introduction

Food is necessary for body growth as it helps to get nutrition and energy. Good food or a healthy diet helps protect our bodies against malnutrition and other noncommunicable diseases such as diabetes, cancer, heart disease, and stroke [1]. On the contrary, overeating harms the human body. Excess food consumption causes stomach enlargement, excessive secretion of hormones and enzymes, and storage of extra consumed food as fat [2]. For decades, people thought that fat was a passive site in the body for energy storage. Views about fat have changed dramatically in the past 10 years of research. It is now classified as an endocrine organ. Before modern humanization, when there were abundant foods, the body released specific molecules to decrease glucose utilization in the form of insulin resistance and induce fat formation. Though insulin resistance was meant to be a temporary state to help the species survive when food was in abundance, it has now become the permanent state of type 2 diabetes [3].

Eating food containing saturated fat, high amounts of salt, and added sugar causes the lifelong disease called "diabetes" [4]. According to International Diabetes Federation (IDF), an individual dies from diabetes or its complications every 7 s. Almost 50% (4 million in total per year) of those deaths are under the age of 60 [5]. Among all diabetes types, type 2 is the most common, and excess body fat is the major reason behind this. The good news is that people can reduce risk and even reverse type 2 diabetes by losing as little as 10% of their body weight. The most effective and safest way to shed excess body weight is to exercise and reduce calorie intake. Reducing calorie consumption has different methods to work for different people. The most precise one of those is adding up the number of calories per serving to all the foods one eats and then taking measures by excluding some foods or decreasing portion sizes and reducing 250–500 cal/day [6]. To follow this method, diabetes patients should know the calorie intake for every portion of food being taken.

The total number of people with type 2 diabetes is ever on increase. People having obesity and high-fat distribution are at more risk of developing it. So, efficient dietary management system is a must for managing a prediabetic individual or having diabetes mellitus type 2. This paper describes a calorie consumption management system for diabetic patients using augmented reality through Faster R-CNN-based food recognition mechanism by monitoring and showing the calorie intake of users per serving. It will also recommend healthy foods and warn users about excessive calories. These will ultimately create a cognitive change in the users for avoiding excessive foods. The fundamental contributions of this research are as follows.

- Design of a food recognition model based on augmented reality-based spectacles using Faster R-CNN.
- Estimation of calorie intake to avoid excessive calories.
- Providing a comparative analysis of this method with the existing state-of-the-art methods from diverse perspectives.

Deep Learning in Personalized Healthcare and Decision Support. https://doi.org/10.1016/B978-0-443-19413-9.00015-1

We have organized the remaining parts of this paper as follows: section II manifests some related works, section III concisely discusses about diabetes categories, its concerns and prevalence, section IV demonstrates system methodology, section V shows result and analysis, and section VI concludes our paper.

2. Related works

Many researchers are working on the improvement of calorie measurement, food management, object recognition, and their applications. Their works include the use of augmented reality, mixed reality, and other modern technologies.

Shen et al. [7] proposed a prototype CNN model that works for the improvement in the accuracy of the food classification in specific categories. The drawback of this model is insufficient datasets and parameter settings that cause inability to recognize various processed, mixed, and cooked foods. As a result, it is not possible to detect the proper food proportion through volume calculation. Another related research by Parisa et al. [8] proposed a system for food detection and calorie estimation, which focused on fighting against diet-related health conditions. They claim to achieve food recognition accuracy of around 99% by using deep convolutional neural networks. In spite of all their claims, this system still lacks features that are necessary to support users in this modern automated world. Takumi and Keiji [9] showed calorie estimation by detecting each dish in a multiple-dish food image. They performed the detection with Faster R-CNN by using school lunch images as datasets. This system offers very common usage of food calorie measurement and does not focus on any special problem. In today's fast-paced world, with advances in technology and virtual reality, this device is no longer a viable option. Similar work has been carried out by Shaikh et al. [10], but they worked with a system that calculates calorie by measuring the volume and mass of a single food image instead of a multidish food image. It also did not work with cooked dishes and fast food items. Kasyap et al. [11] proposed a deep learning—based algorithm to estimate food volume and calorie. However, they worked with only fresh fruits not fast and cooked foods. The majority of these strategies are based on the relative or geographical relationships between features. However, these technologies have a high computational cost when used on a big scale. They demonstrated how the model responds to hyperparameters tuning by altering factors such as the learning rate and amount of neurons in each hidden layer. This assessment is based on the pictures of food, such as fruits and vegetables. Their methodology only works on simple food images and gives a poor performance on complicated food variations such as soup, sandwiches, and so on, which are very insufficient compared to the whole range of foods a person consumes in day-to-day life. A bit different approach has been followed by Sirichai et al. [12]. In their paper, they have proposed an ingredient-based food calorie calculation technique by using nutrition knowledge along with thermal and brightness information. The system at first detects the food and the ingredients and then fetches nutritional information about the ingredients from their database. For implementation, they used fuzzy logic. Though their results were found good, it needs complex hardware and software systems for computation, which may not be cost-efficient. A food calorie estimator using wearable sensor has been proposed by Ghulam et al. [13] where they embedded a piezoelectric sensor in necklace. The system attempts to measure calorie intake by sensing swallow count. Although they asserted that making healthy meal choices lowers the risk of obesity, calculating calories after food has been consumed seems to act contrary to their claims. Additionally, it could be uncomfortable and inconvenient to wear such a necklace. Shobana et al. [14] presented a nutritional food intake tracking system that employs convolutional neural network to determine the number of nutrients from a food image. The fact that the system only evaluates data every 7 days, which could be lengthy enough to result in any health problems, serves as evidence of its inefficiency. Apart from these framework for food nutritional value, estimation based on social media rich food photos has also been proposed [15]. All these systems leverage one or more machine learning approaches to calculate food nutritional values but does a little help in managing the daily dietary of a diabetic patient.

To make the tasks like food detection and calorie estimation more convenient, smart, and suitable for the present automated world, the incorporation of certain level of automation, augmented reality (AR), and mixed reality (MR) is found effective. Han et al. [16] introduce the Automatic Ingestion Monitor (AIM), a food intake sensor that collects dining scene photos for dietary assessment. AIM uses a hands-free automated way to gather photos of the dining situation and calculate the chew count. While AIM has various advantages, the photographs it captures are occasionally obscure. As a result, food image analysis, such as food recognition, can be significantly harmed by blurry photos. However, while correcting the blurry photographs, they completely overlooked people's privacy concerns. AIM does not just snap pictures of the food but also takes pictures of the entire dining environment, which is an infringement of people's privacy. Muhammad et al. [17] developed an AR prototype application that detects and tracks food objects then shows nutrition information visually in the form of a gauge meter with some color-coding. But their system recognizes all food elements as one food item, and food calories are shown. The system cannot deal with mixed food portions, and their insufficient database is also a problem to recognize food items. However, our Dia-Glass system can detect each food element as individuals and shows food value; database is also not an issue here. Klaus et al. [18] proposed a model using mixed reality (MR) that uses computer vision for detecting dietary activities and shows results continuously in real time. As a result, front-of-package labels influence consumers to give up unhealthy beverages or foods and take up selected healthy ones. However, it is a vision not an actual prediction of healthy or unhealthy foods. They focused on choosing food by the preferences of one's taste rather than food calories. Their chosen foods through MR headset is still questionable whether those foods are healthy or not. Moreover, wearing such a big headset is not so convenient and could be troublesome.

The above papers with their great contributions have some crucial limitations, such as some of them lack sufficient datasets and high accuracy, some use a quite troublesome method like capturing images through a mobile camera, and some use complex hardware, while some are less handy. All these limitations have mostly been overcome in our proposed Dia-Glass system. Compared to the above-related works, Dia-Glass provides a more feasible as well as flexible solution. Firstly, the light-weight spectacles are handy to use all day long. Also, its easy-to-access features make it effortless and trouble-free to use for diabetic patients of every age group. Secondly, its robust framework does food detection with great efficiency using faster R-CNN. Thirdly, the use of a large volume data helps the system to recognize varieties of cooked meals and mixed foods. Fourthly, efficient estimation of calories of detected foods and showing of the augmented calories via the spectacles in real time.

3. Diabetes: categories, concerns and prevalence

Increased blood glucose (or blood sugar) levels are a long-term metabolic complication of diabetes, which can seriously harm the heart, blood vessels, eyes, kidneys, and nerves. Around 463 million people worldwide have diabetes, the majority of whom reside in low- and middle-income nations. Diabetes also causes 1.6 million fatalities annually [19].

When you consume a carbohydrate, your body turns it into the sugar glucose and transfers it into your circulation. Your pancreas secretes the hormone insulin, which helps transfer glucose from your bloodstream into your cells, where it is used as fuel. If you have diabetes and do not receive treatment, your body does not utilize insulin as effectively as it should. When an excess of glucose accumulates in your blood, a condition known as high blood sugar results. Health problems that could be serious or even fatal could emerge from this.

3.1 Categories

1) *Type 1 diabetes:* Insulin-dependent diabetes is another name for type 1 diabetes. Because it typically develops in infancy, it was previously known as juvenile-onset diabetes. Diabetes type 1 is an autoimmune disease. It happens when your body attacks your pancreas with antibodies. The organ is no longer releasing insulin since it has been damaged. This form of diabetes might be caused by your genes.
2) *Type 2 diabetes:* Non-insulin-dependent diabetes and adult-onset diabetes were the prior names for type 2 diabetes. However, due to an increase in the proportion of overweighted or obese young people over the past 20 years, it has become more common among kids and teenagers. When you have type 2 diabetes, your pancreas often generates some insulin. But either it is not enough or your body does not use it appropriately. Your cells do not react to insulin when you have insulin resistance. It primarily affects muscle, liver, and fat cells. Typically, type 2 diabetes is less severe than type 1 diabetes. Obesity increases a person's risk of developing type 2 diabetes and its consequences, especially in the tiny blood vessels that run through their kidneys, nerves, and eyes. Keeping a healthy weight, eating well, and exercising are all part of type 2 diabetes treatments. Some folks require medicine as well [20].
3) *Gestational diabetes:* Gestational diabetes is a form of diabetes that develops during pregnancy. Similar to the other two forms of diabetes, it affects the use of glucose in the body. Insulin production in the mother's body is reduced due to the secretion of certain hormones by the placenta, which aids in the growth of the fetus. This causes excess blood glucose to be accumulated as body fat and gradually develops health issues for the fetus. Preventing gestational diabetes can be controlled by maintaining a healthy weight and exercising before becoming pregnant.

3.2 Concerns

Each of the several types of diabetes is connected with a number of issues. Type 2 diabetes, which is the most common of the three, has a number of risk factors. As a result, proper diagnosis and treatment are unavoidable if you want to live longer. In the studies of medicine, diabetes and its type can be diagnosed by assessing a set of symptoms learned through running some tests on suspected person and asking a set of questions [21], which are shown in Table 4.1.

TABLE 4.1 Comparative assessment for clinical symptoms of type 1 and type 2 diabetes.

	Type 1	Type 2
Typical age at onset	<40 yrs	>50 yrs
Duration of symptoms	Weeks	Months to years
Body weight	Normal or low	Obese
Rapid death without treatment with insulin	Yes	No
Ketonuria	Yes	No
Autoantibodies	+ve in 80%–90%	No
Diabetic complications at diagnosis	No	25%
Family history of diabetes	Uncommon	Common
Other autoimmune disease	Common	Uncommon

TABLE 4.2 Daily proportion (quantity and calorie) of individual food category for type 2 diabetic patients of different body frames for a healthy life.

Daily meal plans (cal)	1200	1500	1800	2000	2500
Carbohydrate (Starch) oz.	5	7	8	9	11
Protein (Meat) oz.	4	4	6	6	8
Vegetables oz.	2	2	3	4	5
Fat oz.	3	4	4	5	6
Fruits oz.	3	3	4	4	6
Milk oz.	2	2	3	3	3

Controlling diabetes is not just resisting the temptation on some special occasions, it is a daily routine for lifetime to lead a healthier life. So, people need a dietary management for daily life, which has already been determined by the dietitian for different body frames. For example, type-2 diabetic patients with different body frames can consume 1500–1800 cal/day [22]. Table 4.2 shows daily consumable proportion of each food nutrients for different levels of per day calorie intake.

3.3 Diabetes prevalence

Diabetes is considered to trigger 463 million people worldwide in 2019, accounting for 9.3% of the global adult population (20–79 years) [23]. This number is anticipated to rise to 578 million (10.2%) in 2030 and 700 million (10.9%) in 2045. Diabetes is expected to affect 9.0% of females and 9.6% of the men in 2019. Fig. 4.1 shows a detail picture of prevalence of diabetes over a several age groups where we can see that diabetes prevalence rises with age and becomes 19.9% (111.2 million) among persons aged 65–79 years.

4. System methodology

This section gives a brief overview of our model with related figures for a better illustration of the system. Fig. 4.2 presents the overall system architecture using schematic diagram where the user first needs to register into the system with some basic information including his/her diabetic type. The system at the same time is wirelessly connected with the smart spectacles "Dia-Glass." When the user wears the spectacles and powers them on by tapping its switch, the spectacles start capturing the food image in front of it. The captured image is then sent for detection and calorie estimation processes, which are encapsulated in a mobile-phone system remotely connected to a cloud server. After accomplishing these major

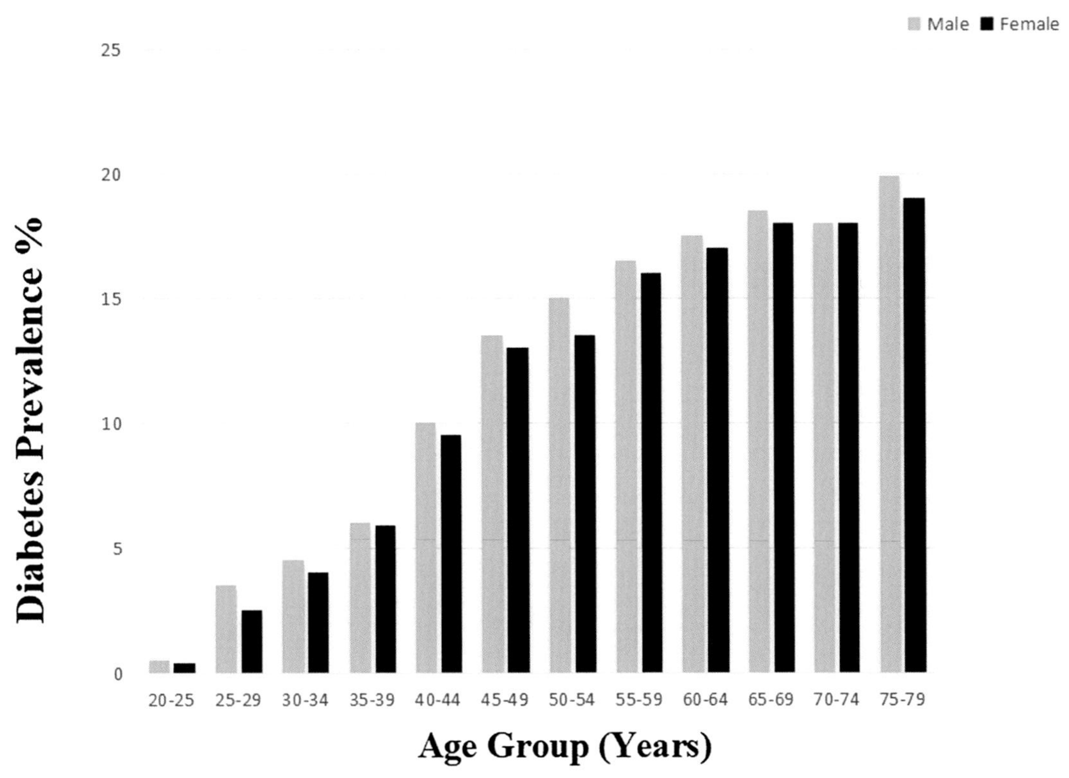

FIGURE 4.1 Prediction of diabetes prevalence over several age groups [23].

tasks, augmentation is performed on the detected food with the obtained calorie information and is shown using the Dia-Glass. Calorie intake per serving and total calorie intake for the day are repeatedly calculated, and if the threshold limit is exceeded for a day, notification alert is immediately sent to the user.

Fig. 4.3 is an elaboration of Fig. 4.2 focusing on two main processes performed by the system: object detection and calorie estimation. For the training purpose of the convolutional neural network (CNN), we need a large volume of data that can be collected data from Kaggle [24], ImageNet [25], and Open Images [26]. In addition to the dataset for CNN training, the calorie measurement system must communicate with a dedicated database server for collecting calorie information for each of the food ingredients and for storing calculated calories per serving. All these required system components have been sketched in Fig. 4.3. Fig. 4.4 depicts a visual representation of different steps followed by the system in developing a prototype.

In this section, we devote to present a detailed discussion on the overall working procedure, technologies to be used, and services that will be provided by our system. We have divided our system into six basic parts, which are discussed as follows.

4.1 User information insertion module

Dia-Glass model is dedicated to controlling all days calorie intake of diabetes patients. Following the goal, the system must acknowledge the type of diabetes, it is about to deal with. Because there are variations in calorie intake restrictions for each type. For certain reasons, some information is to be fed to the system before use. Firstly, the (name) field is required so that the system can distinguish between the persons for whom it is going to operate. Then, if someone inputs the diabetes type, the system can work as it is supposed to be. But the system is also capable of handling patients who do not know about their specific diabetes type. The system is loaded with a set of answers for some specific questions that are presented in Table 4.1. Fig. 4.5 shows the user registration page of the Dia-Glass. Consulting a doctor and inserting the correct diabetic type is always appreciated as the system will work most efficiently in that case. Secondly, as soon as the system will know the type or height, weight, and age of the user, it knows how many calories to intake as shown in Table 4.2 and will be ready to operate. Lastly, it will wait for getting connected to the smart spectacles.

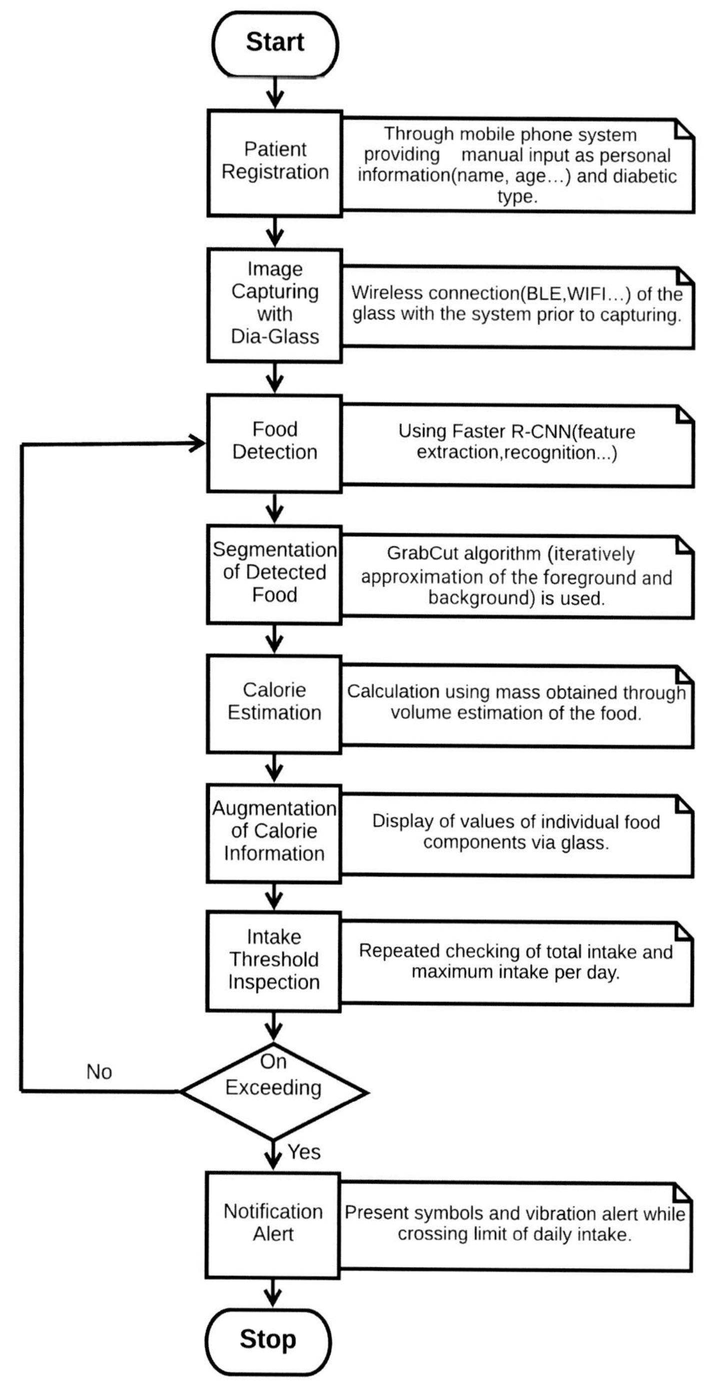

FIGURE 4.2 Schematic framework of Dia-Glass.

4.2 Data acquisition through spectacles

To perform several data transfer operations, like sending captured food images for processing or getting any notifications from the system the spectacles need to be connected to the system. Wireless connectivity such as Bluetooth and Wi-Fi are widely used to serve this purpose. The Dia-Glass is also no exception, it can connect to the mobile-based system with Bluetooth low energy (BLE) and Wi-Fi at the same time. BLE technology will enable the spectacles to operate in low power and will ensure reliable data transmission at short distances [27]. Wi-Fi will ensure a high transfer rate and increase mobility.

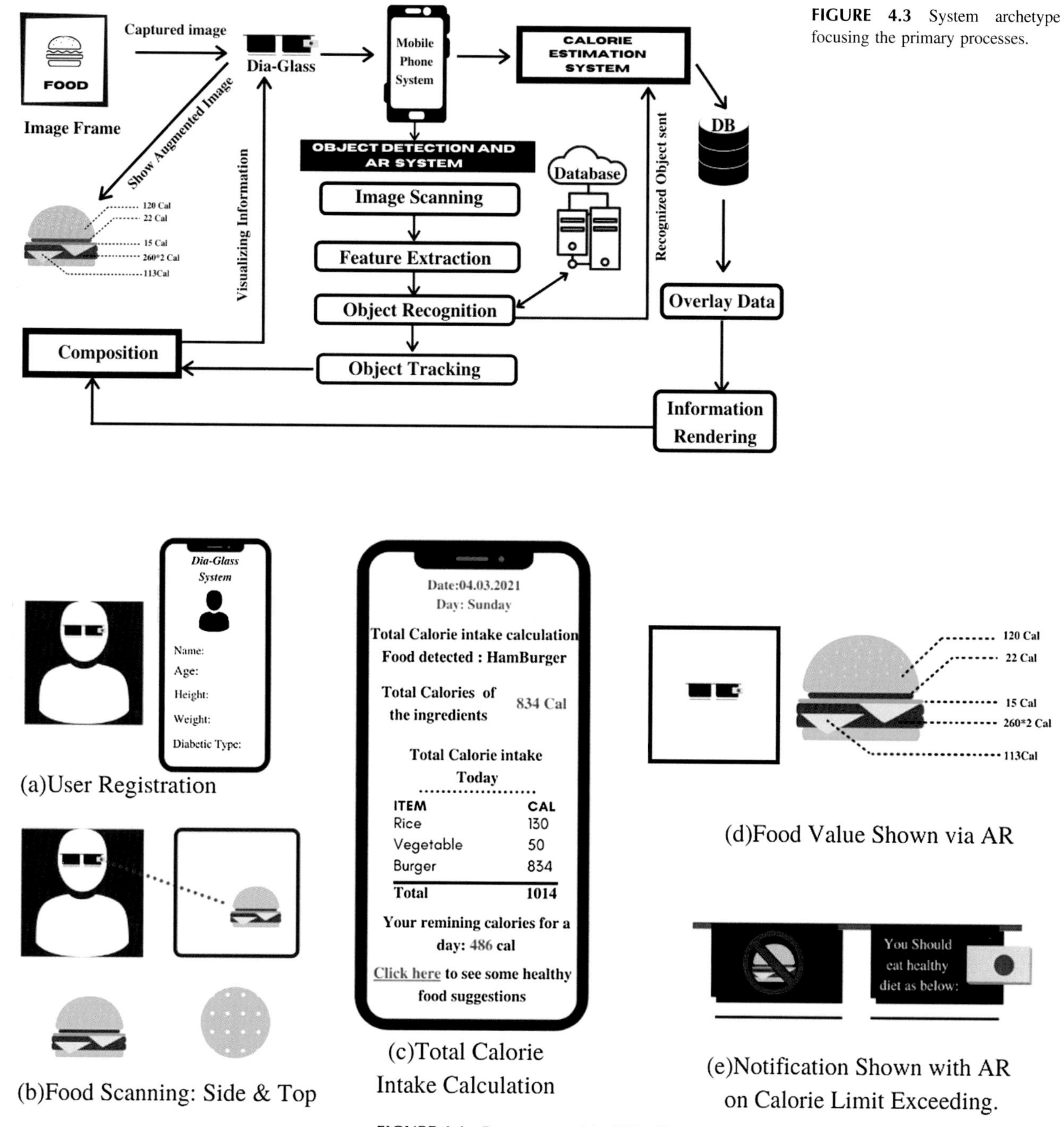

FIGURE 4.3 System archetype focusing the primary processes.

FIGURE 4.4 Prototype model of Dia-Glass.

Wi-Fi has become a ubiquitous technology to be used for connectivity in smart wearable devices. It can transfer high volume of data and is flexible, secure, and interoperable [28].

4.3 Food recognition using faster R-CNN

Like most other object detection systems, the Dia-Glass starts with capturing an image of the food and then sends it to the system for processing. The system has two major works to perform: (1) Food recognition and (2) calorie estimation. Object recognition, a common and popular computer vision problem, can be solved with several techniques of deep learning. For the proposed system, we have chosen Faster R-CNN [21] to perform this primary task. Faster R-CNN has been proven to

FIGURE 4.5 Mobile application interface of Dia-Glass for user registration.

be the fastest and most efficient among all the other predecessor networks of CNN. Faster R-CNN utilizes the mechanism of Fast R-CNN along with the use of Region Proposal Network (RPN) other than selective search, which improves the performance and efficiency of recognition.

ALGORITHM 4.1 Backbone VGG Net

Input: Image [416 × 416]
Output: Feature Map
Process.

i. Input image preprocessing [Convert into a fixed size of 60 × 40]
ii. Generate anchors
iii. Apply max-pool operation using [2 × 2] filter

---

1) *Image processing in RPN:* The recognition process starts with feeding the captured image with Dia-Glass into a backbone convolutional neural network. The output produces an image of a comparatively smaller size than the input image called the output feature map. The network learns about the presence of each point in the output feature map by placing "Anchors" to the corresponding point in the input image. These set of points are then checked to see if they contain the object. The region of interest (ROI) is found by providing a bounding box with the refined coordinates of anchors. The output feature map passes through the 3 × 3 convolution layer of 512 units, which gives out 512 features for each location followed by two 1 × 1 convolution layers of 18 units to assist in classification and regression. The class of the object and the bounding area can be determined by the following equations:

$$L[(p_u),(t_u)] = \frac{1}{N_{\text{class}}}\sum_i L_{\text{class}}(p_u, p_u^*) + \lambda\frac{1}{N_{\text{reg}}}\sum_i p_u^* L_{reg}(t_u, t_u^*) \quad (4.1)$$

$$L_{\text{reg}}\left[t_u, t_u^*\right] = \sum_{i\varepsilon(x,y,w,h)} \text{smooth}_{L1}\left[t_{ui} - t_{ui}^*\right] \quad (4.2)$$

where

$$\text{smooth}_{\text{L1}} = (\text{x}) \rightarrow \text{piecewise}\left\{\text{abs}(\text{x}) < 1, 0.5\text{x}^2, \text{abs}(\text{x}) - 0.5\right\} \quad (4.3)$$

Here, u denotes the index of the anchor. $L_{\text{class}}(p_u, p_u^*)$ is the classification loss determined as the log loss over two classes.

p_u denotes the resultant score from classification unit and $P^*{}_u$ denotes the ground truth label.

$L_{\text{reg}}(t_u, t_u^*)$ represents the regression loss, which is activated only in cases of an anchor containing any object. The term x, y, w, h denotes the top left (x,y) position on the box and the logarithmic width and height, respectively.

Here, smooth_{L1} loss is used, which is said to be less susceptible to outliers.

ALGORITHM 4.2 Faster R-CNN Region Proposal Network (RPN)

Input: Feature map from Backbone VGG Net

Output: Classification and regression bounding box proposal Process.

i. Apply single convolution on the obtained feature map rpn = slim.conv2d(net, 512 [3], ...)

ii. Obtain classification score

rpn cls score = slim.conv2d(rpn, 9*2 [1], ...)

iii. Obtain bounding box regressor score

rpn bbox pred = slim.conv2d(rpn, 9*4 [1], ...)

---

2) *Recognition mechanism with Faster R-CNN:* The regression bounding box obtained from the RPN unit is then used for feature extraction by the "ROI Pooling" layer. The ROI Pooling layer takes the regions corresponding to each proposal in the feature map, divides it into several sub-windows, and performs max-pooling in each of the sub-windows to produce fixed size output. These output proposals are then passed through the fully connected layers and features are fed into the regression and classification units as shown in Fig. 4.6. The classification unit determines the probability to which a proposal belongs to a class and the regression unit provides the coefficient for improving the bounding boxes.

ALGORITHM 4.3 Faster R-CNN Classification

Input: Classification score from RPN

Output: Classified object on the basis of highest score Process.

i. Flatten output from pooling layer and feed it to the first fully connected layer [pool5, scope = 'flatten']

ii. Feed the second fully connected layer with the output of the first layer

iii. Input the result obtained from second fully connected layer to the classifier

iv. Apply softmax to convert feature vector obtained from classifier into probabilities

---

As shown in Fig. 4.2, after the food has been recognized, the system performs image segmentation. The GrabCut [29] algorithm has been used here for image segmentation. This algorithm has a better performance record and efficiently extracts foreground with the least user interaction.

4.4 Calorie estimation

After segmentation, we obtain the image portion containing food that needs to be analyzed to calculate calories. The calorie measurement starts with volume estimation and then the mass calculation.

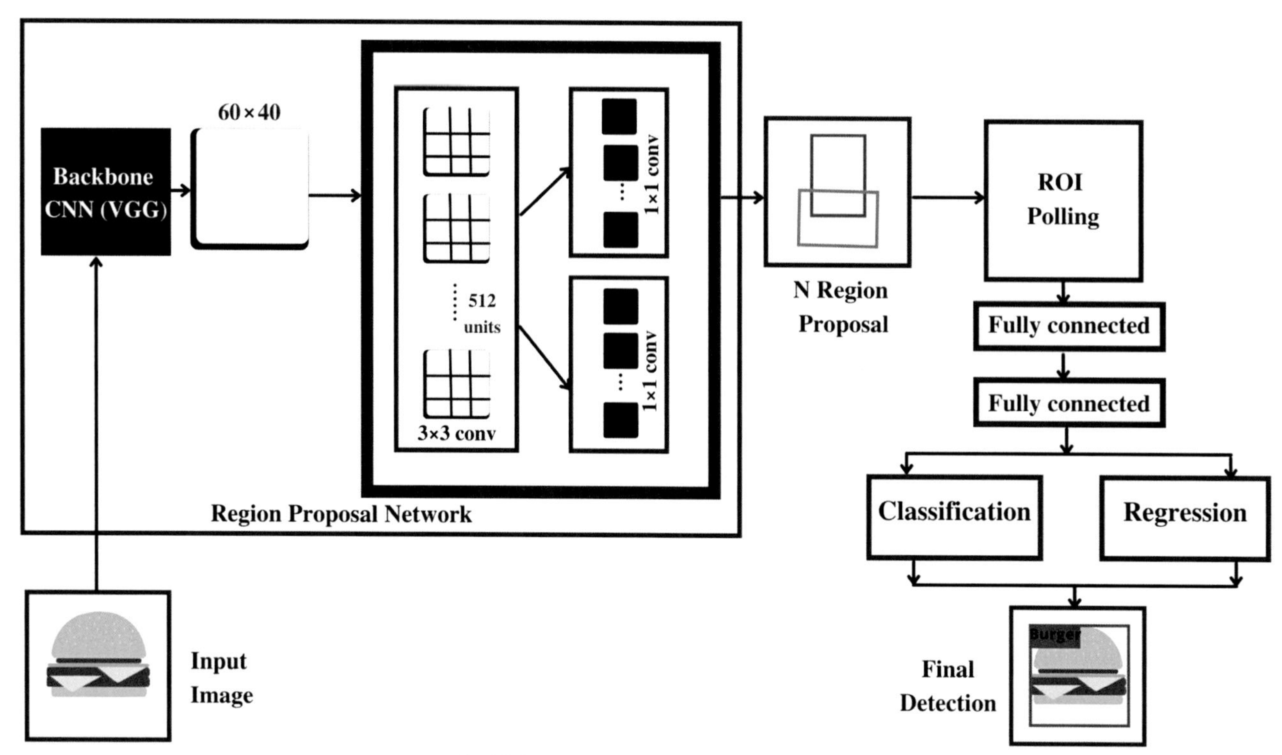

FIGURE 4.6 Architecture of faster R-CNN

1) *Volume estimation:* The food image for calorie measurement is captured from both top and side-wise. These two obtained parameters are then used to calculate the volume of the recognized food.
2) *Mass calculation:* Mass is calculated from the volume by using Eq. (4.4).

$$\text{Mass(m)} = \text{Density}(\rho) * \text{Volume(V)} \tag{4.4}$$

As the food is recognized and both its mass and volume are known, calories can be easily calculated by finding the number of calories present per unit of that food from the database.

4.5 Information rendering and user notification

This step begins with tracking the recognized food to find a pattern. The tracking is performed constantly for the position of the pattern in the real-world space. This is a way to position the overlay data at the accurate position of the food and make the augmentation more realistic. The information obtained from the calorie calculation step is used as overlay data, which is combined with the tracked object (food). It gives out a visual representation of food through the smart spectacles with the calorie information of each of the ingredients present in the food item. Fig. 4.4(d) illustrates the augmentation performed on food. As the calorie of the food ingredients will be shown to the user with augmented reality, the information will also be sent to the system in their mobile device for further processing. This step contains a daily report of total intake, where the total calorie calculation of the current intake as well as the total calorie intake for the whole day is shown. The report also shows healthy food suggestions to the user, as presented in Fig. 4.7. Since it has been proven that diabetes is incurable for the time being, but it can go into remission, the body will not show any symptoms although the disease is technically still present [30]. As a result, we must understand and adhere to the proper diet at all times. If the calorie intake exceeds a particular limit for a day (the limit defined for the diabetic type of the user), the system sends out an immediate alert to the Dia-Glass, restricting any further high-calorie food intake. Fig. 4.4 presents for a better understanding of the above discourse.

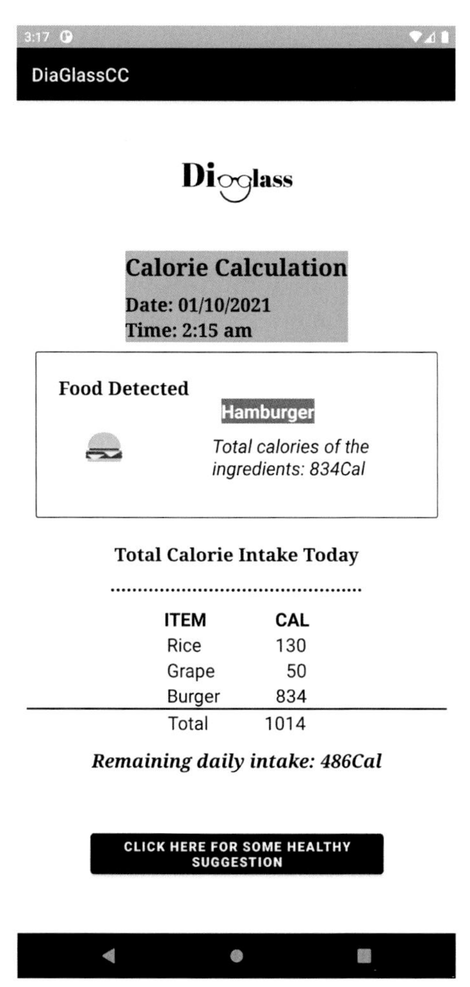

FIGURE 4.7 Mobile application interface of Dia-Glass presenting daily calorie report.

5. Result and analysis

5.1 Dataset

To evaluate the proposed system, at first we have described the dataset used and then we have shown the results obtained by applying the Faster R-CNN algorithm on the dataset. We have considered 17 different common food items in Bangladesh food perspectives for our experimental dataset. These are stuffed tomato, pudding, apple, macaron, burger, coffee, donut, noodles, pasta, rice, pizza, ice-cream, chicken wings, chocolate cake, spring roll, samosa, and juice. Our initial training dataset contains 4703 images. The majority of the images are collected from the Kaggle Food Images (Food-101) dataset [31]. This Food-101 dataset contains 101 categories of food items with 1000 images for each. From this dataset, we used eight common food classes which are chocolate cake, spring roll, pizza, ice-cream, samosa, chicken wings, donut, and macaron. In our experimentation, for each class, we have taken first 300 images. For another class "Apple," we have used the Kaggle Fruits fresh and rotten dataset [32]. This dataset includes six classes with 13,600 images. However, we have taken only fresh apple class images of 263 images. We have prepared the images for the rest eight common of the food classes (rice, juice, burger, noodles, coffee, pasta, pudding, and stuffed tomato) by downloading from the Internet sources. Details of our used datasets are given in Table 4.3. For availability and convenience to other

TABLE 4.3 Details of our used dataset.

Class	Number of images	Data source
Apple	263	Kaggle [32]
Juice	255	Internet sources
Coffee	255	Internet sources
Burger	255	Internet sources
Noodles	255	Internet sources
Pizza	300	Kaggle [31]
Pasta	255	Internet sources
Macaron	300	Kaggle [31]
Rice	255	Internet sources
Chocolate cake	300	Kaggle [31]
Ice cream	300	Kaggle [31]
Samosa	300	Kaggle [31]
Spring roll	300	Kaggle [31]
Donut	300	Kaggle [31]
Pudding	255	Internet sources
Stuffed tomato	255	Internet sources
Chicken wings	300	Kaggle [31]

researchers, we have kept our all experimental images (Kaggle + internet sources) at: https://github.com/nusratshimu/Diaglass-Dataset.git.

All of the images are 416×416 pixels in size in RGB format. The dataset has been divided into three subsets for experimental purposes: 70% of the images are used as a training set, 20% are used as a validation set, and rest 10% are used as the testing set [33]. The selected food item names with sample images are shown in Fig. 4.8, and their corresponding calories/unit has been provided in Fig. 4.9 [34].

5.2 Performance analysis

The percentage (%) of total correctly identified samples by the classifier is known as accuracy. Precision is the percentage of total anticipated positive samples that were actually positives as determined by the classifier. The recall is the percentage of all positive samples that the classifier correctly identified as positive. The harmonic mean of precision and recall represents the F1-score [35]. Mathematical representations of these metrics are shown in Eqs. (4.5)–(4.8).

$$\text{Accuracy}(\%) = \frac{\text{TruePositive(TP)} + \text{TrueNegative(TN)}}{\text{TruePositive(TP)} = \text{TrueNegative(TN)} + \text{FalsePositive(FP)} + \text{FalsePositive(FP)}} \times 100 \tag{4.5}$$

$$\text{Precision}(\%) = \frac{\text{TruePositive(TP)}}{\text{TruePositive(TP)} = \text{FalsePositive(FP)}} \times 100 \tag{4.6}$$

$$\text{Recall}(\%) = \frac{\text{TruePositive(TP)} + \text{TrueNegative(TN)}}{\text{TruePositive(TP)} + \text{FalseNegative(FN)}} \times 100 \tag{4.7}$$

$$F1 - \text{score}(\%) = 2 \times \frac{\text{Precision} \times \text{Recall}}{\text{Precision} + \text{Recall}} \times 100 \tag{4.8}$$

FIGURE 4.8 Sample images of the 17 food items.

CALORIE SHEET		
Food Item	**Quantity**	**Caloric Value**
Pudding	100g	113
Stuffed_tomato	1 cup	196
Apple	1	95
Ice-cream	100g	207
Burger	100g	308
Pasta	1 cup	221
Macaron	1	97
Noodles	1 cup	219
Coffee	1 cup	33
Rice	1 cup	206
Pizza	1 slice	285
Donut	1	269
Chocolate_cake	100g	367
Juice	1 cup	136
Spring_roll	1	148
Samosa	1	262
Chicken_wings	1	88

FIGURE 4.9 Calories per unit mass of the food items. *Collected from www.nutritionix.com.*

where,

True positive (TP)—the model predicts the correct compound character as correct. True negative (TN)—the model predicts the incorrect compound character as incorrect. False positive (FP)—the model predicts the incorrect compound character as correct. False negative (FN)—the model predicts the correct compound character as incorrect.

Class-wise TP, TN, FP, and FN are calculated by using Eqs. (4.9)–(4.12) [35].

$$TP_i = C_{ii} \tag{4.9}$$

$$TN_i = \sum_{k=1,k\neq i}^{n} \sum_{j=1,j\neq i}^{n} C_{jk} \tag{4.10}$$

$$\mathrm{FP_i} = \sum_{\mathrm{j=1,j\neq i}}^{\mathrm{n}} \mathrm{C_{jk}} \tag{4.11}$$

$$\mathrm{FN_i} = \sum_{j=1,j\neq i}^{n} C_{ij} \tag{4.12}$$

where i represents the class and n represents the total number of compound character classes. Here, C_{jk} is known as the jk-th component of the confusion matrix, j and k are the confusion matrix's row and column [35].

For food recognition, our suggested approach constructs a classifier based on the retrieved features of faster R-CNN. The recognition as well as classification success rate was visualized using a confusion matrix, as shown in Fig. 4.10. A confusion matrix is a contingency table that contains information about the actual and expected classifications of a classification system [36]. Fig. 4.11 presents some of the food images where our system succeeded to perform correct classification. Due to the presence of similar characteristics in the food items, some of the resultant classification values are found to be overlapped. Fig. 4.12 depicts some of the food images where our system failed to perform correct classification. This is due to the interclass similarities among the food items, low resolution of the photos, visibility of small portion of the food items, etc. More specifically, in Fig. 4.12(a), the output image depicts this is stuffed tomato, but the sample image contains pizza. This error is occurred due to its similar color like stuffed tomato. In Fig. 4.12(b), the output

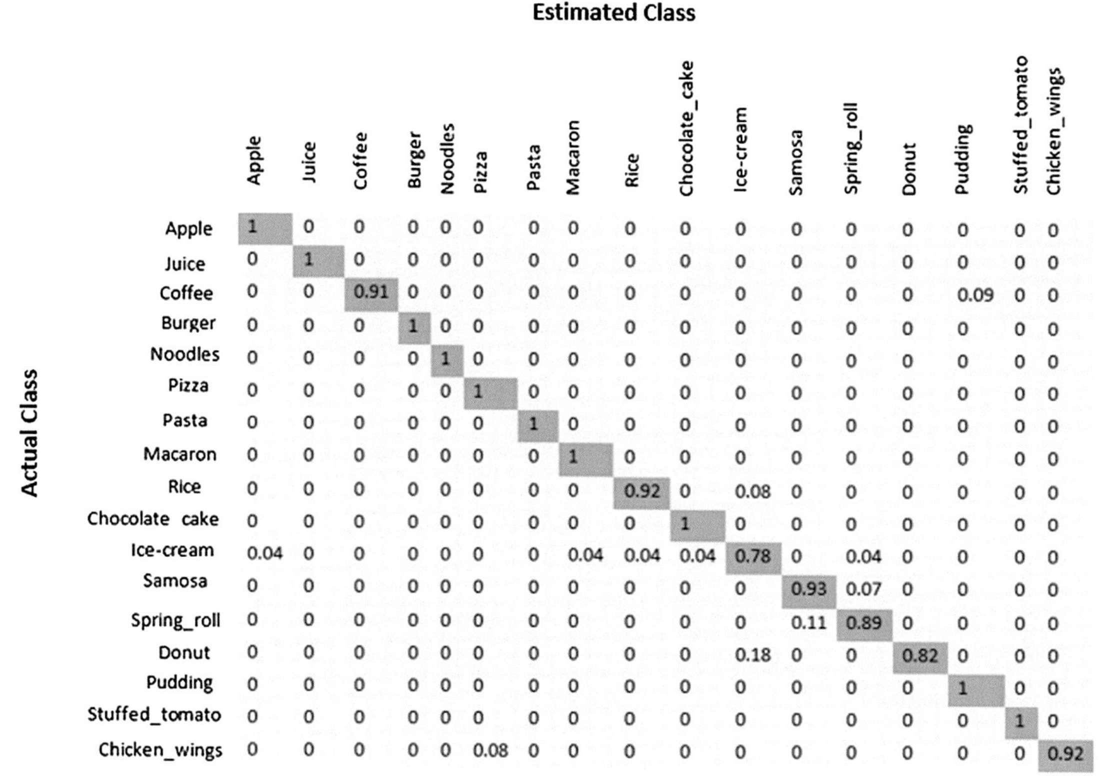

	Apple	Juice	Coffee	Burger	Noodles	Pizza	Pasta	Macaron	Rice	Chocolate_cake	Ice-cream	Samosa	Spring_roll	Donut	Pudding	Stuffed_tomato	Chicken_wings
Apple	1	0	0	0	0	0	0	0	0	0	0	0	0	0	0	0	0
Juice	0	1	0	0	0	0	0	0	0	0	0	0	0	0	0	0	0
Coffee	0	0	0.91	0	0	0	0	0	0	0	0	0	0	0	0.09	0	0
Burger	0	0	0	1	0	0	0	0	0	0	0	0	0	0	0	0	0
Noodles	0	0	0	0	1	0	0	0	0	0	0	0	0	0	0	0	0
Pizza	0	0	0	0	0	1	0	0	0	0	0	0	0	0	0	0	0
Pasta	0	0	0	0	0	0	1	0	0	0	0	0	0	0	0	0	0
Macaron	0	0	0	0	0	0	0	1	0	0	0	0	0	0	0	0	0
Rice	0	0	0	0	0	0	0	0	0.92	0	0.08	0	0	0	0	0	0
Chocolate cake	0	0	0	0	0	0	0	0	0	1	0	0	0	0	0	0	0
Ice-cream	0.04	0	0	0	0	0	0	0.04	0.04	0.04	0.78	0	0.04	0	0	0	0
Samosa	0	0	0	0	0	0	0	0	0	0	0	0.93	0.07	0	0	0	0
Spring_roll	0	0	0	0	0	0	0	0	0	0	0	0.11	0.89	0	0	0	0
Donut	0	0	0	0	0	0	0	0	0	0	0.18	0	0	0.82	0	0	0
Pudding	0	0	0	0	0	0	0	0	0	0	0	0	0	0	1	0	0
Stuffed_tomato	0	0	0	0	0	0	0	0	0	0	0	0	0	0	0	1	0
Chicken_wings	0	0	0	0	0	0.08	0	0	0	0	0	0	0	0	0	0	0.92

FIGURE 4.10 Normalized confusion matrix.

FIGURE 4.11 Samples of food images that are correctly classified.

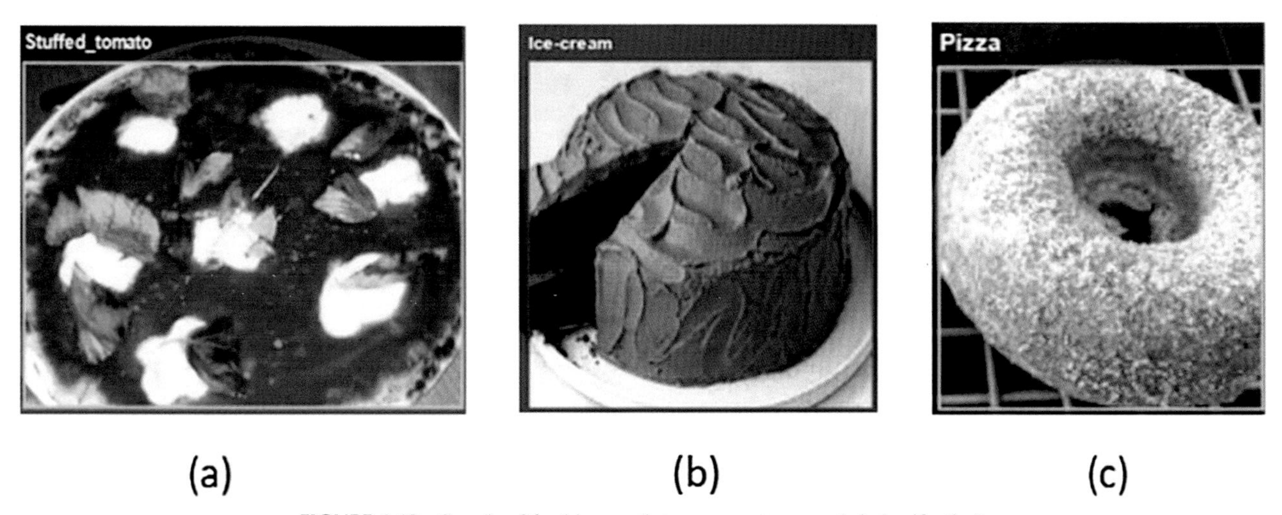

FIGURE 4.12 Sample of food images that are prone to errors (misclassification).

image depicts this is ice cream, but the sample image contains chocolate cake. This error is occurred due to its similar texture like ice-cream. In Fig. 4.12 (c), the output image depicts this is pizza but the sample image contains donut. This error is occurred due to its similar shape like pizza.

There are some concerns that may arise while evaluating the Dia-Glass are mentioned below.

- Categories of food for detection
- Dataset management and efficiency
- Cost-effectiveness
- User data security
- Privacy concern
- Power consumption and optimality

Our system is able to detect any kind of food with each of its ingredients. It is able to detect ranges from regular organic food, fast food, packaged food to cooked meals. This whole range of food detection requires very enriched datasets, which is a bit challenging. For this challenging task, we relied on a cloud server for efficiency, scalability, and cost-effectiveness (Table 4.4) [37]. Focusing on user privacy and data security, we employed a dedicated cloud database to handle user data that follows data encryption standards [38]. To protect the privacy concern completely, the spectacles only turns on when the user taps its switch. And it shows a red light as an indication of the system capturing food images for detection. Thus, the privacy of people will not be hampered in public places, and no embarrassing situation will be captured. Segmentation helps to store and compare only the necessary food part of the image leaving other background objects excluded. It is fully

TABLE 4.4 Class-wise precision, recall, and F1-score of the testing food dataset.

Class	Precision	Recall	F1 score
Apple	0.95	1.00	0.97
Juice	0.95	1.00	0.97
Coffee	0.98	0.96	0.97
Burger	1.00	1.00	1.00
Noodles	1.00	1.00	1.00
Pizza	0.92	1.00	0.96
Pasta	1.00	1.00	1.00
Macaron	0.95	1.00	0.97
Rice	0.93	0.95	0.94
Chocolate cake	0.97	0.90	0.93
Ice cream	0.75	0.82	0.78
Samosa	0.96	0.94	0.95
Spring roll	0.90	0.92	0.91
Donut	1.00	0.82	0.90
Pudding	0.97	1.00	1.00
Stuffed tomato	1.00	1.00	1.00
Chicken wings	1.00	0.96	0.98
Average	**95.97%**	**96.67%**	**95.48%**

TABLE 4.5 Performance comparison of food items recognition with existing state-of-the-art methods [35].

Method	Applied technique	Accuracy(%)
Ghazanfar et al., 2017 [39]	CNN	92
Ege and Yanai, 2017 [9]	CNN	96.93
Haotian et al., 2018 [40]	RIS and text mining	87.9
Ryosuke et al., 2018 [41]	CNN (inception-v3)	76.9
Takumi et al., 2019 [42]	Computer vision	91.67
YUNUS et al., 2019 [43]	CNN (inception-v4)	90.63
LANDU et al., 2020 [44]	DCNN	93.1
Kasyap et al., 2021 [11]	CNN (TensorFlow)	97
Han et al., 2021 [16]	Faster R-CNN	96
Rakib et al., 2022 [45]	CNN	86
Akhil et al., 2022 [46]	CNN	95.5
Proposed Model	Faster R-CNN and GrabCut	98.33

power-efficient as the system stays in the idle state until it is turned on by the user. The performance results of the Dia-Glass are shown in Table 4.5, which clearly depicts that the proposed Dia-glass performs better than any other state-of-the-art systems. Though, our Dia-Glass can effectively help a diabetic patient's dietary management. However, doctors' recommendations should be followed first, and then the system's assistance should be sought.

6. Conclusion

Considering the fact that there is no permanent cure for diabetes, so management through controlling calorie intake is must for a diabetic patient. As diabetes is a lifelong condition once diagnosed, so our Dia-Glass attempts to make the life of a diabetic patient much easier through an AR-based spectacles along with Faster R-CNN modules for food recognition and calorie management. It could be a next-generation gadget that will be with every diabetic patient as a day-to-day accessory due to its portability. People can visually experience calorie control as it is enhanced with AR in the spectacles as if a doctor is always with them. Implementing the model with an enriched database makes the system stronger and faster R-CNN along with GrabCut makes it more speedy than other existing systems. The initial training and testing with 17 common categories of foods yielded an average efficiency of 98.33%. Dia-Glass is also highly secured because there is a button in the spectacles which controls the camera and optimizes energy. This paper concludes that all of the technologies used in the model are efficient and will allow diabetic patients to choose any food anywhere with proper guidance, providing them with a healthy life. Thus, the Dia-Glass acts as a full-time assistant in helping diabetic patients to achieve their dietary goals avoiding excessive calories for maintaining a healthy life. In future, we will work with more enhanced dataset that will conform to even better real-time performance. The system may also incorporate MR with AR to make it more suitable.

Author statement

Natasha Tanzila Monalisa: Conceptualization, Software, Draft preparation, Abstract, Introduction, Diabetes: Categories, Concerns and Prevalence, Related works, System Methodology, Conclusion. Shinthi Tasnim Himi: Software, Draft preparation, Abstract, Introduction, Diabetes: Categories, Concerns and Prevalence, Related works, System Methodology, Conclusion. Nusrat Sultana: Software, Formal analysis, Data Curation, Abstract, Result and Analysis. Musfika Jahan: Software, Formal analysis, Data Curation, Abstract, Result and Analysis. Nayeema Ferdous: Abstract, Background Investigation, Diabetes: Categories, Concerns, and Prevalence, Related works. Md. Ezharul Islam: Visualization, Review and Editing. Mohammad Shorif Uddin: Supervision, Validation, Review, and Editing.

Conflict of interest: The authors declare that the research was conducted in the absence of any commercial or financial or personal relationships that could have appeared to influence the work reported in this paper.

Data availability statement: The datasets used in the paper are publicly available secondary data. Proper acknowledgments with citation guidelines are maintained for the use of these datasets.

Human and animal rights: This article does not contain any studies with animals and human performed by any of the authors.

References

[1] Healthy Diet, Who.int, 2022. Available: https://www.who.int/news-room/fact-sheets/detail/healthy-diet. (Accessed 18 April 2022).

[2] K. Blackburn, What Happens When You Overeat? MD Anderson Cancer Center, 2021. Available: https://www.mdanderson.org/publications/focused-on-health/What-happens-when-you-overeat.h23Z1592202.html. (Accessed 17 March 2021).

[3] H. Information, D. Overview, Diabetes Diet, Diabetes Diet, N. Health, Diabetes Diet, Eating, and Physical Activity—NIDDK, National Institute of Diabetes and Digestive and Kidney Diseases, 2021. Available: https://www.niddk.nih.gov/health-information/diabetes/overview/diet-eating-physical-activity. (Accessed 17 March 2021).

[4] L. Stehno-Bittel, Intricacies of fat, Physical Therapy 88 (11) (2008) 1265−1278.

[5] Escardioorg, Global Statistics on Diabetes, 2021. Available at: https://www.escardio.org/Education/Diabetes-and-CVD/Recommended-Reading/global-statistics-on-diabetes. (Accessed 17 March 2021).

[6] Cutting calories to control diabetes-Harvard Health. (online) Harvard Health. Available at: https://www.health.harvard.edu/healthbeat/cutting-calories-to-control-diabetes (Accessed March 17, 2021).

[7] Z. Shen, A. Shehzad, S. Chen, H. Sun, J. Liu, Machine learning based approach on food recognition and nutrition estimation, Procedia Computer Science 174 (2020) 448−453, https://doi.org/10.1016/j.procs.2020.06.113. Available:. (Accessed 28 March 2021).

[8] P. Pouladzadeh, A. Yassine, S. Shirmohammadi, FooDD: food detection dataset for calorie measurement using food images, New Trends in Image Analysis and Processing−ICIAP 2015 Workshops (2015) 441−448, https://doi.org/10.1007/978-3-319-23222-554. Available:. (Accessed 28 March 2021).

[9] T. Ege, K. Yanai, Estimating food calories for multiple-dish food photos, in: 2017 4th IAPR Asian Conference on Pattern Recognition (ACPR), 2017, https://doi.org/10.1109/acpr.2017.145. Available:. (Accessed 28 March 2021).

[10] S. Wasif, S. Thakery, A. Nagauri, S. Pereira, Food calorie estimation using machine learning and image processing, International Journal of Advance Research, Ideas and Innovations in Technology 5 (2) (2019) 1627–1630.

[11] V. Kasyap, N. Jayapandian, Food calorie estimation using convolutional neural network, in: 2021 3rd International Conference on Signal Processing and Communication, ICPSC), 2021, https://doi.org/10.1109/icspc51351.2021.9451812. Available:. (Accessed 16 September 2021).

[12] S. Turmchokkasam, K. Chamnongthai, The Design and implementation of an ingredient-based food calorie estimation system using nutrition knowledge and fusion of brightness and heat information, IEEE Access 6 (2018) 46863–46876, https://doi.org/10.1109/access.2018.2837046. Available:. (Accessed 24 March 2021).

[13] G. Hussain, et al., Smart piezoelectric-based wearable system for calorie intake estimation using Machine Learning, Applied Sciences 12 (12) (2022) 6135, https://doi.org/10.3390/app12126135. Available at:.

[14] V.D. Shobana Kumar, Deep learning-based implementation of food nutritional intake tracking system using convolution neural network algorithm, International Journal of Discoveries and Innovations in Applied Sciences 2 (4) (2022) 1–18. Retrieved from, https://www.openaccessjournals.eu/index.php/ijdias/article/view/1160.

[15] C. Ilias, S. Georgios, Machine learning for all: a more robust Federated Learning Framework, in: Proceedings of the 5th International Conference on Information Systems Security and Privacy, 2019, https://doi.org/10.5220/0007571705440551 [Preprint]. Available at:.

[16] Y. Han, S. Yarlagadda, T. Ghosh, F. Zhu, E. Sazonov, E. Delp, Improving food detection for images from a wearable egocentric camera, Electronic Imaging (8) (2021) 286, https://doi.org/10.2352/issn.2470-1173.2021.8.imawm-286, 1-286-7, 2021. Available:. (Accessed 16 September 2021).

[17] M. Bayu, H. Arshad, N. Ali, Nutritional information visualization using mobile augmented reality technology, Procedia Technology 11 (2013) 396–402, https://doi.org/10.1016/j.protcy.2013.12.208. Available:. (Accessed 26 March 2021).

[18] K. Fuchs, M. Haldimann, T. Grundmann, E. Fleisch, Supporting food choices in the Internet of People: Automatic detection of diet-related activities and display of real-time interventions via mixed reality headsets, Future Generation Computer Systems 113 (2020) 343–362, https://doi.org/10.1016/j.future.2020.07.014. Available:. (Accessed 27 March 2021).

[19] Who.int., Diabetes, 2021 [online] Available at: https://www.who.int/westernpacific/health-topics/diabetes. (Accessed 12 August 2021).

[20] WebMD, Types of Diabetes Mellitus, 2021 [online] Available at: https://www.webmd.com/diabetes/guide/types-of-diabetes-mellitus. (Accessed 12 August 2021).

[21] J. Innes, Davidson's Essentials of Medicine, Elsevier Health Sciences UK, 2015. Paperswithcode.com. 2021. Papers with Code - Faster R-CNN Explained. [online] Available at: https://paperswithcode.com/method/faster-r-cnn. (Accessed 17 March 2021).

[22] F. James Norman, Treatment of Diabetes: The Diabetic Diet, EndocrineWeb, 2021. Available at: https:/www.endocrineweb.com/conditions/diabetes/treatment-diabetes. (Accessed 18 March 2021).

[23] P. Saeedi, et al., Global and regional diabetes prevalence estimates for 2019 and projections for 2030 and 2045: results from the international diabetes federation diabetes atlas, 9th edition, in: Diabetes Research and Clinical Practice, 157, 2019, p. 107843. Available: https://www.diabetesresearchclinicalpractice.com/action/showPdf?pii=S0168-8227%2819%2931230-6. (Accessed 12 August 2021).

[24] Kagglecom, Find Open Datasets and Machine Learning Projects — Kaggle, 2021. Available at: https://www.kaggle.com/datasets. (Accessed 20 September 2021).

[25] J. Deng, W. Dong, R. Socher, L. Li, K. Li, L. Fei-Fei, ImageNet: a large-scale hierarchical image database, in: IEEE Conference on Computer Vision and Pattern Recognition, Miami, FL, USA, 2009, pp. 248–255, https://doi.org/10.1109/CVPR.2009.5206848, 2009.

[26] A. Kuznetsova, et al., The open images dataset V4, International Journal of Computer Vision 128 (7) (2020) 1956–1981, https://doi.org/10.1007/s11263-020-01316-z. Available:. (Accessed 26 March 2021).

[27] M. Ghamari, H. Arora, R. Sherratt, W. Harwin, Comparison of low-power wireless communication technologies for wearable health-monitoring applications, in: 2015 International Conference on Computer, Communications, and Control Technology (I4CT), 2015.

[28] Wi-fiorg, The Benefit of Wi-Fi® Connectivity in Wearable Devices Wi-Fi Alliance, 2021. Available at: https://www.wi-fi.org/ko/beacon/jay-white/the-benefit-of-wi-fi-connectivity -in-wearable-devices. (Accessed 18 March 2021).

[29] C. Rother, V. Kolmogorov, A. Blake, GrabCut", ACM Transactions on Graphics 23 (3) (2004) 309–314, https://doi.org/10.1145/1015706.1015720. Available:. (Accessed 29 March 2021).

[30] Medicalnewstodaycom, Can Diabetes Be Cured? A Review of Therapies and Lifestyle Changes, 2021. Available at: https://www.medicalnewstoday.com/articles/317074#Managing-type-2-diabetes. (Accessed 19 March 2021).

[31] "Food Images (Food-101)", Kaggle.Com, 2021. Available: https://www.kaggle.com/kmader/food41?select=images. (Accessed 7 October 2021).

[32] "Fruits Fresh and Rotten for Classification", Kaggle.Com, 2021. Available: https://www.kaggle.com/sriramr/fruits-fresh-and-rotten-for-classification. (Accessed 7 October 2021).

[33] R. Hafiz, M. Haque, A. Rakshit, M. Uddin, Image-based soft drink type classification and dietary assessment system using deep convolutional neural network with transfer learning, Journal of King Saud University - Computer and Information Sciences (2020), https://doi.org/10.1016/j.jksuci.2020.08.015. Available:. (Accessed 14 August 2021).

[34] S. Park, A. Palvanov, C. Lee, N. Jeong, Y. Cho, H. Lee, The development of food image detection and recognition model of Korean food for mobile dietary management, Nutrition Research and Practice 13 (6) (2019) 521, https://doi.org/10.4162/nrp.2019.13.6.521. Available:. (Accessed 14 August 2021).

[35] M. Khan, M. Uddin, M. Parvez, L. Nahar, A Squeeze and Excitation ResNeXt-Based Deep Learning Model for Bangla Handwritten Compound Character Recognition", Journal of King Saud University-Computer and Information Sciences, 2021 https://doi.org/10.1016/j.jksuci.2021.01.021. Available:. (Accessed 14 August 2021).

[36] F. Akter, T. Khatun, M.S. Uddin, Recognition and classification of fast food images, Global Journal of Computer Science and Technology 18 (2018).

[37] E. Subramanian, L. Tamilselvan, A focus on future cloud: machine learning-based cloud security, Service Oriented Computing and Applications 13 (3) (2019) 237–249.

[38] W. Ahmed, S. Garg, A cloud computing-based advanced encryption standard, in: Conference: 2019 Third International Conference on I-SMAC (IoT in Social, Mobile, Analytics and Cloud) (I-SMAC), ResearchGate, 2019, pp. 205–210. Available at: https://www.researchgate.net/publication/339906204 A Cloud computing-based Advanced Encryption Standard. (Accessed 19 March 2021).

[39] G. Latif, B. Alsalem, W. Mubarky, N. Mohammad, J. Alghazo, Automatic fruits calories estimation through convolutional neural networks, in: Proceedings of the 2020 6th International Conference on Computer and Technology Applications, 2020, https://doi.org/10.1145/3397125.3397154. Available:. (Accessed 14 August 2021).

[40] H. Jiang, J. Starkman, M. Liu, M. Huang, Food nutrition visualization on google glass: Design tradeoff and field evaluation, IEEE Consumer Electronics Magazine 7 (3) (2018) 21–31, https://doi.org/10.1109/mce.2018.2797740. Available:. (Accessed 14 August 2021).

[41] T. Ege, Y. Ando, R. Tanno, W. Shimoda, K. Yanai, Image-based estimation of real food size for accurate food calorie estimation, in: 2019 IEEE Conference on Multimedia Information Processing and Retrieval (MIPR), 2019, https://doi.org/10.1109/mipr.2019.00056. Available:. (Accessed 14 August 2021).

[42] R. Yunus, et al., A framework to estimate the nutritional value of food in real time using deep learning techniques, IEEE Access 7 (2019) 2643–2652, https://doi.org/10.1109/access.2018.2879117. Available:. (Accessed 14 August 2021).

[43] L. Jiang, B. Qiu, X. Liu, C. Huang, K. Lin, DeepFood: food image analysis and dietary assessment via deep model, IEEE Access 8 (2020) 47477–47489, https://doi.org/10.1109/access.2020.2973625. Available:. (Accessed 14 August 2021).

[44] R. Tanno, T. Ege, K. Yanai, AR DeepCalorieCam: an iOS app for food calorie estimation with augmented reality, MultiMedia Modeling (2018) 352–356, https://doi.org/10.1007/978-3-319-73600-631. Available:. (Accessed 14 August 2021).

[45] R.U. Haque, R.H. Khan, A.S. Shihavuddin, M.M. Syeed, M.F. Uddin, Lightweight and parameter-optimized real-time food calorie estimation from images using CNN-based approach, Applied Sciences 12 (19) (2022) 9733.

[46] A. Kumar Singh, A. Srivastava, A. Mani Tripathi, D.C. Dalela, Real time nutrition detection for raw food using CNN, SSRN Electronic Journal (2022).

Chapter 5

Synthetic medical image augmentation: a GAN-based approach for melanoma skin lesion classification with deep learning

Nirmala Veeramani and Premaladha J.
School of Computing, SASTRA Deemed to be University, Thanjavur, Tamil Nadu, India

1. Introduction

The dataset imbalance problem is the most challenging task in the health care environment, especially, in the medical imaging domain. The collection of datasets is a rigorous task, and only limited samples of real-time images are available. For researchers, it is crucial to work on the benchmark resources; at the same time, they need a larger training sample with annotations when deploying the machine learning model for medical image diagnosis. Though the availability of challenge datasets is publicized, it is limited to certain organs such as the liver, lungs, and brain. Also, the collection of 2D slices and 3D volumes is complex which is to be communicated with collaborative research consultants and radiologists. Researchers use data augmentation to try to solve this problem. The most typical data augmentation procedures are simple manipulations of dataset images, such as image translation, image transformation, image flip-flop, and size scaling. In computer vision tasks, using traditional data augmentation to accelerate the training process of networks is a routine technique [1]. Small changes to the photos, on the other hand, yield little new information (e.g., translating the image into a few pixels of the quadrants either to the right or left). Synthetic data augmentation of high-quality instances is a novel and advanced type of data augmentation. Synthetic data samples learned using a generative model add richness and expand the dataset, letting the system training be optimized further.

GANs are the combinatory module of generators and discriminators which effectively contributes to training a model that synthesizes images based on game theory. The model is composed of two networks that have been trained in an adversarial process in which one network generates fake images, while the second network constantly distinguishes between real and fake images. GANs have gained a lot of traction in the computer vision area, and various varieties of GANs have lately been proposed for creating high-quality realistic natural images. GAN has several fascinating applications, including the ability to generate images of one new style of a lesion from another (i.e., the process of image-to-image translation) and image inpainting.

1.1 Background and motivation

On top of all the other diseases, skin cancers are one of the most deadly diseases in which increasing rapidly due to the depletion of the ozone layer and high incident of ultraviolet rays toward the earth's surface which provokes the skin cells, and DNA inside is annihilated and thus causes unusual symptoms and damage to the human skin later diagnosed as cancers.

The incidence rate of melanoma is in increasing pace, and people often regret the timely skin check routines. It is reported that cases of melanoma skin cancer have a global occurrence of about 132,000 cases each year [2]. Melanoma diagnosis is quite difficult to diagnose, and its classification in terms of its early stages differentiates from its malignancy and benignness. According to the research findings, it is predicted that an approximate rate of 5.4 million cases of skin cancer in the United States is to be diagnosed per year [3]. The most powerful type of skin cancer is malignant melanoma,

Deep Learning in Personalized Healthcare and Decision Support. https://doi.org/10.1016/B978-0-443-19413-9.00026-6

which starts from the cell called melanocytes which are located in the second layer known as the epidermis of the skin. This kind of malignancy may become fatal for the patients if it is unattended or delayed in the diagnosis in the process of differentiating it from the other benign ones. It has to be encompassing rapidly, and on the other hand, it is also harder to treat if it has metastasis. As this motivation, early detection of skin cancer can undoubtedly lead to proper computer-aided diagnosis and treatment, resulting in a lower death rate of human life.

1.2 Contributions

In this research article, we found that most of the authors worked with different algorithms to identify the disease, that is, the malignant lesions, but the works are found limited with the utilization of the datasets. And most of the researchers carried out the pre-enhanced images for their works in the process of identifying the skin lesion classes as benign or malignant. This work is focused mainly on the generation of inputs from the archive dataset to improve generalization, and the classification results achieved are efficient with the use of CNN-based model which can help in the computer-based diagnosis of melanoma.

We applied deep learning to achieve the goal of skin lesion classification in the proposed study. CNN combined with deep learning has emerged as a valuable method to be cherished in computer vision. Around the world, medical imaging, and disease diagnosis from image studies in recent years has used CNNs and demonstrated enhanced performance for a beneficiary range of medical activities [4].

Our suggested CNN for skin lesion classification is combined with GAN-based synthetic skin lesion synthesis. The following are the contributions of this work.

1) Using GANs to create high-quality targeted skin lesions from dermoscopy images.
2) Development of a CNN-based skin lesion classification solution that outperforms state-of-the-art approaches.
3) Using the generated synthetic data, augment the CNN training set for improved classification results.

1.3 Related works

Melanoma is a major type of skin cancer which takes a high probability to evolve as a malignant cancerous tumor on the skin. This skin cancer can be detected and diagnosed by using dermatological photos called dermoscopic images. For the computer-based efficient diagnosis, the machine learning algorithms are practiced based on the notable performance with the images inputs used for the detection of skin cancer that resulted in good efficiency in identification [5]. However, the model accuracy has to be increased by the process of extraction concerning more features and also resulted in more deviation in sensitivity. Here, in this article [6], the study offered a traditional state-of-the-art employing image processing procedures that lend hands to improve skin cancer diagnosis accuracy. However, they are not succinct enough to describe a feature of the model that can identify cancer efficiently. In another work, the author proposed an architecture-driven solution that combines a deep learning algorithm to detect skin cancer.

The model-driven architecture of DL-based algorithms can be created quickly enough that the precise model can anticipate the output as soon as humanly possible. It performed well in detecting skin cancer [7]. Furthermore, the system is dependent on real-time interfacing with clinically directed medical imaging to improve classification accuracy. The author of this study [8] presented a CNN-based skin cancer diagnosis using characteristics extracted from dermoscopic dataset images. About the features techniques, they achieved 89.5% accuracy detection in the phase test datasets. On the other hand, the metrics of the accuracy were not evaluated sufficiently and badly need improvement. Another issue was the occurrence of overfitting between their test and train split datasets phases, which was a relatively minor imbalance discovered in that study. The technique proposed by the author in this work [9] is a lesion indexing network (LIN) technique that is exclusively based on DL to identify and categorize skin cancer. As a result, the obtained result with DL-based LIN by extracting more characteristics with desired quality. However, the problem persists in the segmentation performance where it must be needed to be considered and can be increased for further improvement of the result. Applying the dermoscopic dataset images, the investigator performed the task of using CNN algorithms to detect skin cancer and distinguish it from pigmented melanocytic lesions of the human skin in this study [10].

However, screening for nonmelanocytic and nonpigmented skin cancer was difficult. It also got decreased detection accuracy. In this paper [11], a deep CNN incorporates three major steps, which perform intentionally to detect skin lesions where first, it focuses on The color transformation is mostly used to improve the image's contrast scenario; proceeding with the second step of the CNN approach, the author proposed is to extract lesion boundaries, which are referred as the most region of the interest. Finally, deep features are extracted via transfer learning. Though the technique produced good results in certain datasets, the outcomes differ depending on the sample.

Another CNN-based model [12] improved melanoma skin cancer detection where they employed the preprocessing techniques for noise removal and postprocessing techniques for the image as the phase of image enhancement carried out before and after the process of segmentation, respectively. The method marks the borderline of the lesion regions by integrating local and global contextual information, resulting in good results with a contrast time delay in the part of execution that might not be mentioned, which can enhance the value of the results proposed as a transfer learning model known as ResNet50 with better accuracy without preprocessing stage or feature handcraft selection [13].

A few of the experts who performed the classification and segmentation used MATLAB software to analyze and scrutinize the optimal formats for doing the analysis. This study used preprocessing techniques such as image denoising, image scaling, shadowing effect, vignette effect, and black border cropping [14]. Popular techniques including Otsu's Thresholding [15] and other color space transformations, the Watershed technique, and C-Means algorithms were utilized for lesion segmentation [16]. The multisupport vector machine classifiers were included in this research work as an introduction to image-based screening techniques to discern similar ailments in the field of medical image classification [17].

2. Skin lesion classification methodology

We begin this section by describing the data and its features. After that, we explain the CNN architecture for differentiating the malignant lesions from the noncancerous skin lesion called classification. The fundamental difficulty is the limited quantity of data available for CNN training. The methods for artificially enlarging data are described in the next section.

2.1 Dataset

The dataset required to train this model is taken from the ISIC Archive dataset [18] Fig. 5.1 for melanoma skin lesions with an image dimension of 1022 X 767. The dataset is utilized with handcrafted experiments and trials for analyzing the effectiveness of the classification model. Notably, no preprocessing is done for the dataset before it is fed into the adversarial network.

The ISIC Archive dataset is been used with the handcrafted images also the balanced test sets are taken and fed into the deep neural network. The classes are three, namely, the class benign, class premalignant, and class malignant. The use of imbalanced data sets may have a greater impact on the decision of the results classification where the premalignant lesions are identified from the suspicious benign lesions. To avoid such circumstances carefully, the datasets are adopted for this research work as shown in the below chart in Fig. 5.2.

2.2 CNN architecture with modified VGG16

CNN with VGG-16 is used for the classification of malignant and benign lesions. It takes data from a resulting image saved from the GAN model.

The CNN algorithms had been the primary use to teach the model. CNN is an artificial neural network (ANN). This is analogous to the human nervous system, that is made of multiple layers upon layer of interconnected neurons. CNN makes use of deep mastering algorithms to perceive images by assigning weights to them. Because of its efficiency in identifying photos, face recognition, digit processing, and other domains, it is widely employed. Fig. 5.3 depicts the structure of the proposed skin lesion classification system. In computer vision, CNNs are commonly employed to solve image

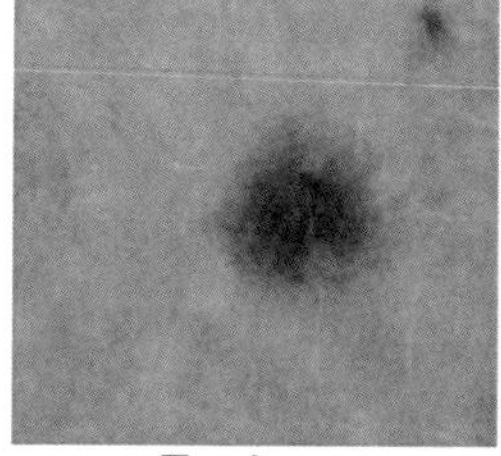
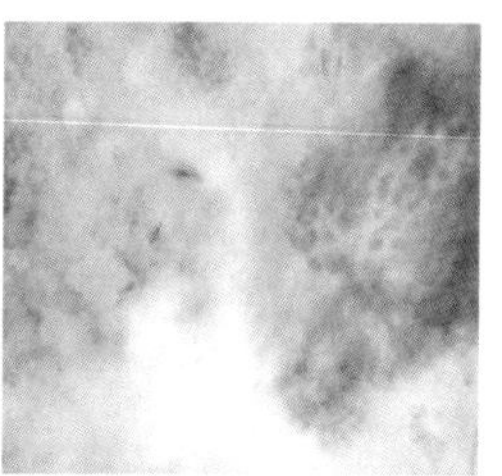

Benign Malignant

FIGURE 5.1 Sample dataset images.

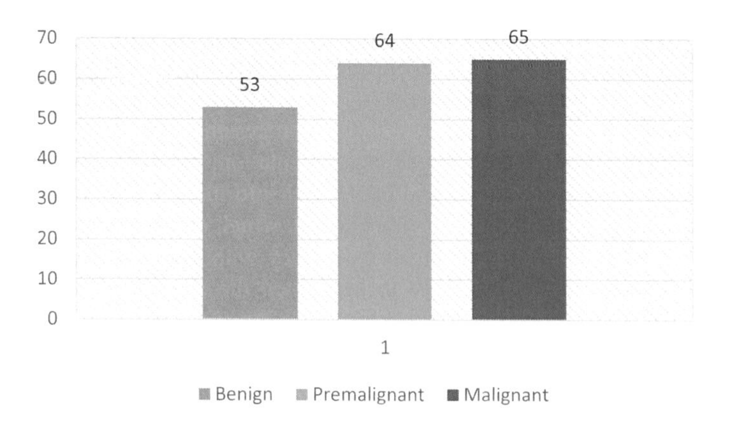

FIGURE 5.2 Visualization of handpicked dataset.

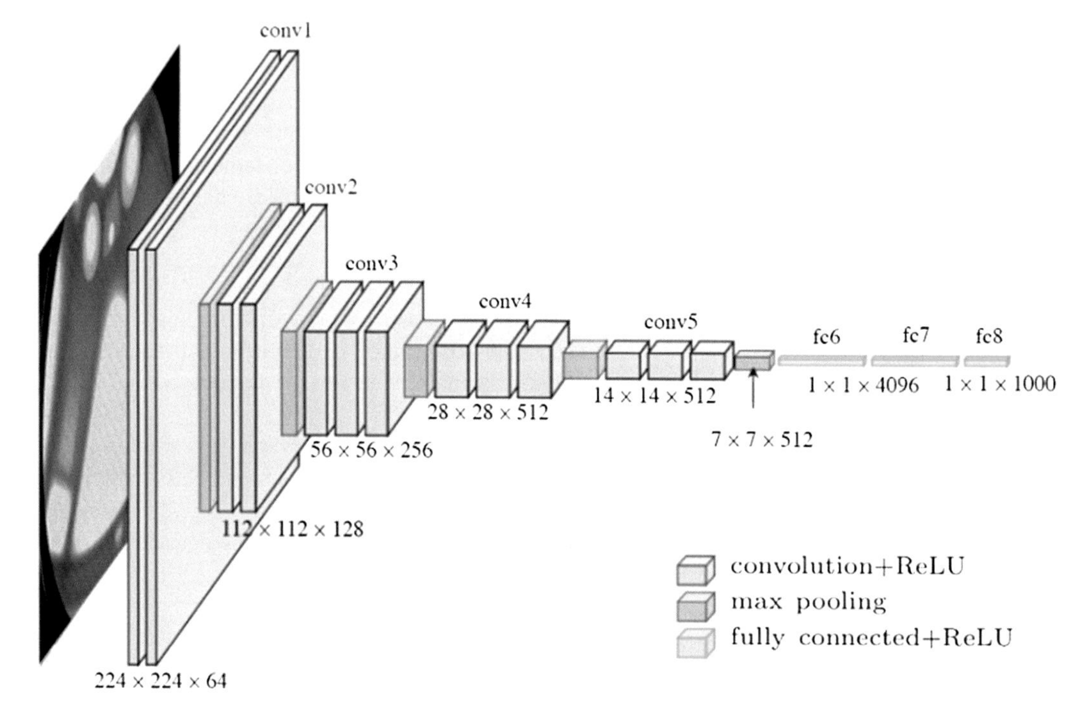

FIGURE 5.3 VGG-16 architecture.

classification tasks [19]. Deep learning architectures for medical imaging have been proposed [20,21] due to the restricted datasets and input size. These architectures often have fewer convolutional layers. Our classification CNN handles input with the modified layers incorporating the VGG16 on the ROIs with a fixed size of 64x64 and an intensity range rescaled, and the handcrafted data excluding the artifacts problems are utilized for the proposed work.

To classify lesions into three groups, the framework consists of convolutional layers which are referred out in the triple pairs, each preceded by a combination of one max-pooling layer, and two dense fully connected layers, ending in a softmax layer to generate network predictions. As an activation function, we employ ReLU. There were roughly 1.3 million parameters in the network. There were roughly 1.3 million parameters in the network. A dropout layer [22] with a value of 0.5 probabilities is used during our training session of the module to reduce error significantly.

Procedure for training: From each image entered into the CNN, it subtracts the mean value of the training images. We trained for 100 epochs with a 64 batch size and 0.001 is used as a learning rate. Gradient descent optimization with the help of Nesterov momentum updates [23] was utilized for the optimal solution thereby employing stochastic, which analyzed the gradient at the "look ahead" regions rather than the actual state.

3. Generation of synthetic skin lesions

3.1 Traditional data augmentation

Thousands of parameters must be taught in even a tiny CNN. There is a risk of overfitting when employing deep networks with numerous layers or when working with a small number of training images. Data augmentation, which artificially expands the dataset, is the conventional approach for reducing overfitting [24]. On gray-scale pictures, the most common augmentation techniques are affine transforms such as translation for image quadrants, angle rotation, size scaling, mirroring, flipping, and shearing. Fig. 5.4 shows the structure of the skin lesion classification system. In CNN, we avoided changes that would induce shape distortion to retain the skin lesion's features (such as shearing). Furthermore, the ROI was concentrated on the lesion.

The input skin lesion images were rotated several numbers called N_{rot} times at random angles ranging between 0 and 180°. Following that, each of these rotated images was flipped N_{flip} times, i.e., (top to bottom, left to right), and translated several times denoted as N_{trans}, in which we collected shuffled pairings of pixel values (x, y) coordinate lies between (p, p).

$p = Min(4, 0.1x\ d)$. Finally, the N_{scale} number of times from a stochastic scale range $s = [0.1x\ d, 0.4\ x\ d]$ was used to scale the ROI. The scale was created by altering the number of cancerous lesions or the affected skin region surrounding the lesion. The number of image augmentations was the total $N = N_{rot}\ (1 + N_{flip} + N_{trans} + N_{scale})$ as a consequence of the augmentation procedure. Fig. 5.5 depicts an example lesion and the augmentations that go with it. The bicubic interpolation was adopted to perform the scaling operation on all of the ROIs to a consistent size of 64x64 pixels.

Where each notation is explained with its terminologies as follows,

N denotes the number of augmentations of an image
N_{rot} indicates the image augmentation with rotation to angle θ
N_{flip} refers to the flipping of the image on the horizontal and vertical axis
N_{trans} refers to the translation of the images from the point coordinates
N_{scale} indicated the change in the size of the original images

3.2 Generative adversarial networks for skin lesion synthesis

GAN generates synthetic images from real images. GAN is widely used for text classification and this technique is adopted in our work for melanoma skin lesion classification which can also play a vital role in the field of medical image classification.

In general, the GAN's workflow is divided into two parts.

1. A generator is a program that learns to generate convincing facts. It creates images that the discriminator can use as negative training examples.
2. Discriminator learns to perceive the distinguishability between fake and real data generated by the generator. Generators can be penalized by discriminators if they produce improbable results.

The GAN is primarily concerned with the random input vectors directed toward the generative model and thus automatically generates the fake images of the fed inputs. Whereas, the discriminator identifies and differentiates the real and fake generated images in the training with the given epochs and updates the model with the learned parameters.

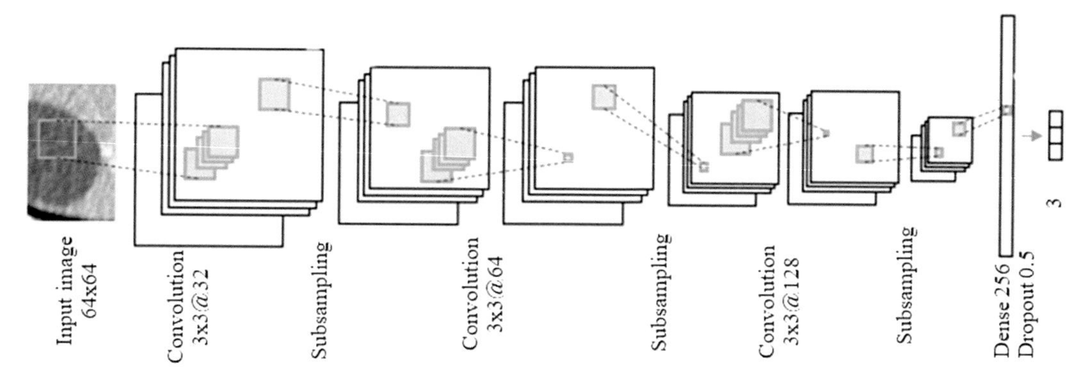

FIGURE 5.4 The convolutional neural network (CNN) classification architecture for the skin lesion.

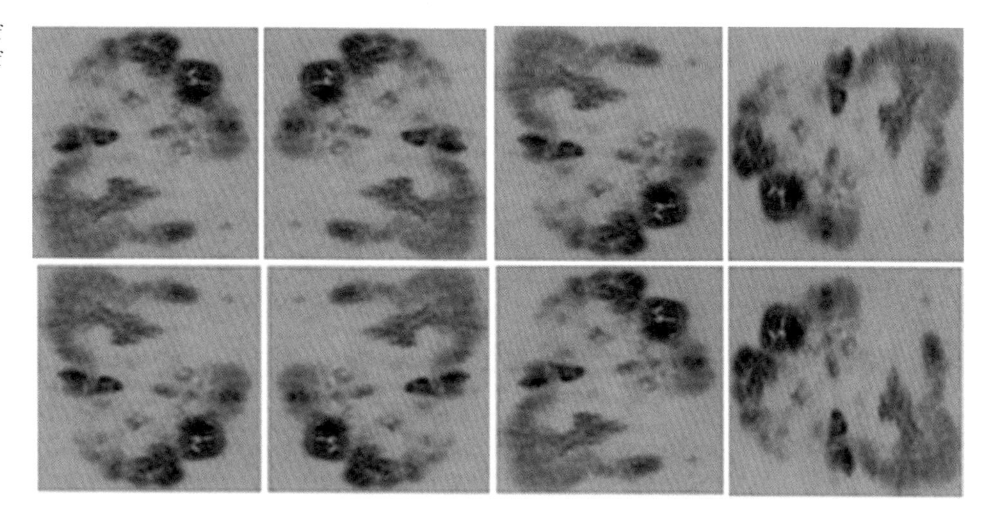

FIGURE 5.5 Lesion region of interest (ROI) and augmentation of skin lesion samples.

Fig. 5.6 shows the general workflow of the deep GAN, and the datasets are multiplied, which, in turn, helps us work with huge data for training the CNN model to get the generalized computer-based diagnosis method for the melanoma classification. The loss in the generator and discriminator model is observed, and those are reduced necessarily for the perfect training in the ISIC archive dataset.

Generator: The generative model generates a 64 x 64 x one skin lesion image from a vector of 100 random integers chosen from a uniform distribution, as seen in Fig. 5.7. The network architecture is made up of a fully connected layer reshaped to size 4 x 4 ×1024 and four fractionally stridden convolutional layers to up-sample the picture with a 5 x 5 kernel size [25]. A fractionally stridden convolution (also known as "deconvolution") is a method of enlarging pixels by interpolating zeros across them. Convolution over the expanded image will result in a larger output image. All layer of the network is batch-normalized, besides the output layer. The GAN learning process is stabilized by normalizing answers to have a zero mean value and a unit variance value over the whole mini-batch, which prevents the generator from compressing every sample to a single point [26]. All layers employ ReLU activation functions excluding the output layer, which uses a tanh for an activation function.

Discriminator: The discriminator employs a typical CNN architecture to take a 64 x 64 x 1 (lesion ROI) input image and make a single decision stating that this lesion is a real one or a fake. The network is made up of a completely linked layer and four convolution layers with a kernel size of 5 x 5. Instead of applying pooling layers, five stridden convolutions are employed for every convolution layer that works on minimizing the spatial dimensionality. Each layer of the network is

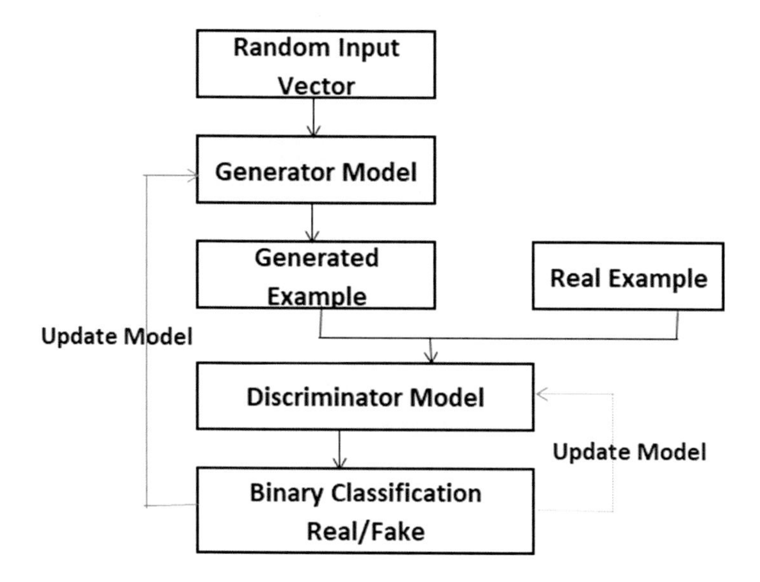

FIGURE 5.6 Block diagram.

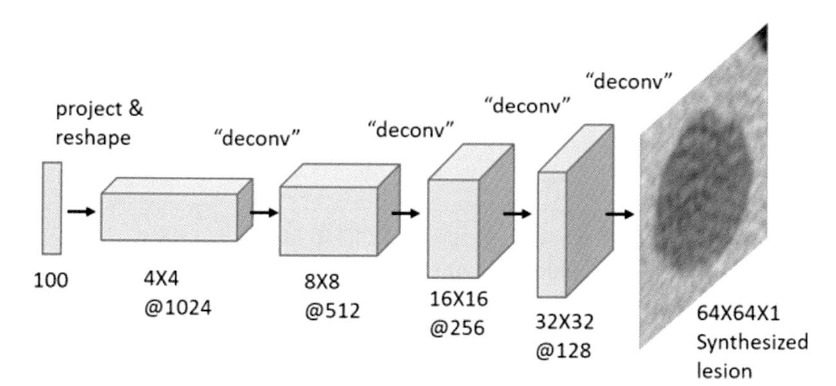

FIGURE 5.7 Generative adversarial network (GAN) generator architecture of DCNN.

batch-normalized, except for the exception of the hidden layer and output layer. Except for the output layer, which employs the leaky rectified linear functions f(x) = max, all layers use the sigmoid function for the likelihood (0, 1) probability score of the image (x, leak x).

Procedure for Training: We trained the deep convolutional GAN (DCGAN) to create ROIs for each type of skin lesion separately. For both the generator and the discriminator, the training procedure was done iteratively. For each lesion type l (benign, premalignant, malignant), we utilized mini-batches of size 64 for the skin lesion ROI with different instances of x(1)l, …, x(m)l and m = 64 for the depiction of random noise samples z(1), …, z(m) taken from a uniform distribution between [1]. Scaling the training images to the range of the tanh activation function (1, 1) was the sole preprocessing step employed. The default value is set to 0.2 for the slope of the leak in the Leaky ReLU. The weights were set to a zero-centered normal distribution with a standard deviation of 0.02. We used stochastic gradient descent with the Adam optimizer [38], an adaptive moment method of estimation in which that takes into account the gradient's first and second moments, which are regulated by parameters for the first gradient is set to 1 = 0.5 and the second gradient parameter is set to 0.999. For the iterations of 100 epochs, we utilized a learning rate of 0.0002.

3.3 Conditional skin lesion synthesis

The GAN version with the auxiliary classifier is the second GAN version (ACGAN). Conditional GANs are a type of GAN model that allows your model to be conditionally on external input to increase the size in terms of the number of samples produced [27,28] suggested GAN designs that produce labeled samples using class labels. Instead of providing side information to the discriminator, Odena et al. [29] advocated that perhaps the discriminator be tasked with recreating it. This is accomplished by adding extra decoder architecture to the discriminator that outputs both the class label and the true or false conclusion.

To synthesize the labeled lesions of all three types, we employed the design supplied with just minor alterations. The ACGAN discriminator modifies the DCGAN by using size 33kernels; also it strides convolutions only on the odd layer, and a dropout of 0.5 in all layers saves the final. After optimizing our little dataset, we employ the ACGAN discriminator without these adjustments. The three types of lesions were categorized using the discriminator auxiliary decoder.

Procedure for training: Except for a learning rate of 0.0001 for 60 epochs, the settings for the training were comparable to those given in III-B. Our training dataset contained skin lesion ROIs and their corresponding labels for all lesion types l (benign, premalignant, and malignant), as well as noise samples of the image given as z(1), …, z(m) chosen from a uniformly distributed values between [1].

To integrate the label information, the loss function has to be updated. Let us simplify the fundamental GAN discriminator process of maximization on equation over the log-likelihood depicted as Eq. (5.1) as follows:

$$L = E[\log P(S = Real|XReal)] + E[\log(S = Fake|XFake)] \tag{5.1}$$

where P(S|X) is equal to D(X) and Xfake = G(z). The generator module is trained enough to minimize that objective. In DCGAN, the discriminator reveals the desired outputs such as P(S|X) and P(C|X) are said to be equal to D(X), and similarly the Xfake is equal to the G(c, z) where C is referred to class labels. The loss has two parts:

$$Ls = E[\log P(S = real|Xreal)] + E[\log P(S = fake|Xfake)] \tag{5.2}$$

$$Lc = E[\log P(C = c|Xreal)] + E[\log P(C - c|Xfake)] \tag{5.3}$$

The discriminator module is aimed to increase the added value of both Ls and Lc, and on the other hand, this generator module is intended to train to aim to maximize the difference value of Lc − Ls.

4. Experimental results

The following experiments and results are presented. We used the CNN architecture of VGG16 outlined in Section II-B to put the categorization findings to the test. The results of data augmentation utilizing fake skin lesions were then compared to traditional data augmentation techniques. We used the couple of synthetic lesion production methods described in Sections III-B and III-C. We discovered that the deep convolutional GAN (DCGAN) technique performs better in our experiments. As a result, the results reported below concentrate on that strategy. In Section IV-E, the outcomes of the CNN traditional AUG and the VGG16+GAN synthesized lesion augmentation will be compared.

4.1 Dataset evaluation and performance metrics

The dataset is divided into three, namely, train, test, and validate images. Each folder in turn consists of 53 benign, 64 premalignant, and 65 malignant lesion images. The input datasets are split into three separate folders for the training, testing, and validation 80%, 10%, and 10%, respectively. Our state of the artwork shows better accuracy in these best splits.

The performance evaluation metric for our proposed work is accuracy. It is generally defined as the ratio of the number of correct class predictions to the sum of all input samples involved in training [30]. Extending to that, we computed confusion matrices for sensitivity and specificity measures to all lesion class labels. All the available parametric measures of our research work are presented in the following equations:

$$\text{Total accuracy} = \frac{\sum \text{True } Positive}{\text{Amount of Skin lesions}} \tag{5.4}$$

$$\text{Sensitivity} = \frac{\text{True Positive}}{\text{True Positive} + \text{False Negative}} \tag{5.5}$$

$$\text{Specificity} = \frac{\text{True Negative}}{\text{True Positive} + \text{False Positive}} \tag{5.6}$$

True positives (TP): A skin lesion image in which the expected class matched the actual class.

True negatives (TN): An instance in which we predicted a false result, and the actual result was also false.

False positives (FP): We expect something to be true, but it turns out to be false.

False Negatives (FN): The occurrence in which we expected something to be false, but it turned out to be true.

In the field of healthcare, the FN should be given more care because the intermediate lesion like actinic keratosis can be turned into a cancerous melanoma if not diagnosed well before in hand.

4.2 Implementation specifications

Skin lesion classification implementation required the following system configurations to work with DCNN; we used the framework named Keras [31]. We have also adopted the tensor flow framework [32] for implementing the architectures of the GAN. For training the model, the following GPU was used an NVIDIA GeForce GTX 970 Ti.

The generator produces duplicate data during the training process, and the discriminator quickly classifies and tells the decision that it is fake. If generator module training results well, the generator generates images closer to real data and fools the discriminator. The discriminator fails to distinguish between real and fake data. Thus, its accuracy decreases. Through the backpropagation process, the discriminator classifies and provides a signal that the generator used to update its weights.

GAN generates. The final CNN layer of the generator network has an activation tanh. Discriminator uses Leaky ReLU is more balanced than ReLU and may learn faster. Leaky ReLU can have a slope for negative values, instead of altogether zero. Both networks use Adam optimizer. The GAN model is trained for 350 epochs. After each epoch produces one image that contains four rows and seven columns of a set of data that are taken as input from the base dataset as shown in Fig. 5.8. There is no preprocessed image, simply the raw input synthesized image is loaded to the model, and basic traditional layers

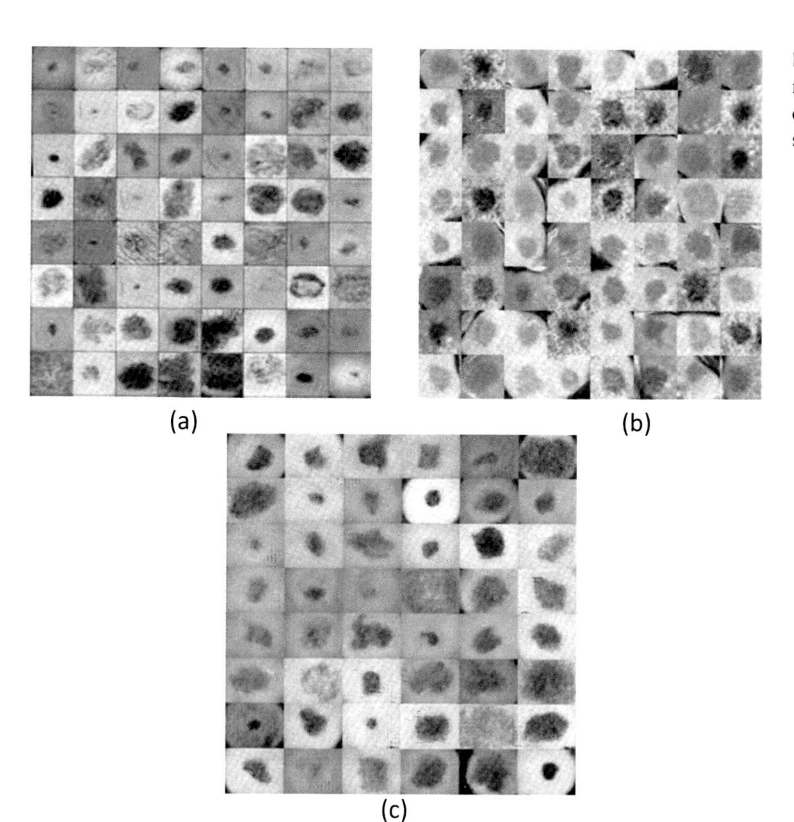

FIGURE 5.8 Deep convolutional generative adversarial network (DCGAN)-generated synthetic skin lesions of classes: (A) Benign, (B) premalignant, and (C) malignant samples.

are inclusive in the process. Max pooling layer, convolution layers, and dense layers are employed in the proposed model for designated services for the images that are to be trained. The fully connected layers lead to the vanishing gradient points, that is, simply known as the disappearance of gradients. The optimal dropout is used in our work 0.25. The additional VGG16 layer has been intentionally added to this model, and the resulting output is shown in Fig. 5.8. The images are taken as one for malignant and 0 for benign data. It is then built with the capability to find more relevant features, and hence it guarantees the best performance as well. The learning rate for the model is 0.001 and trained for five epochs. Using VGG along with CNN gives an accuracy of 96.33%.

4.3 Training and test data

When working with CNN VGG-16, the result so much depends on the criteria of how well the VGG-16 is being trained along the model of the CNN for the GAN-generated synthetic image inputs. So, the increased number of training images VGG-16 gives us the trained results with higher accuracy. We tried training the VGG-16 with different handcrafted image sets of dermoscopic images as input, and we achieved an accuracy of 96.33% as its accuracy, and loss graph is shown in Fig. 5.9. The early stopping mechanism was employed when working beyond the optimal epochs were generalization. Because, many epcohs doesnot have any impact on the accuracy were it already attained its maximum value. And this early stopping is ensured to eliminate the problem of overfitting.

5. Performance comparison

When collated with the other existing state-of-the-art methods [33], our proposed model of deep generative CNN with VGG16 gives us decent results of accuracy without having to work with the preprocessing steps for the input image datasets. The synthesized lesion images are used for the classification which excels the state-of-the-art methods and shows an increased accuracy with that of the traditional augmentation techniques. The values of the method and detail are shown in below Tables 5.1–5.3.

When there is no augmentation carried out, the results were obtained for low values, and this may be due to the problem of overfitting. The results were improvized as the number of training samples increased.

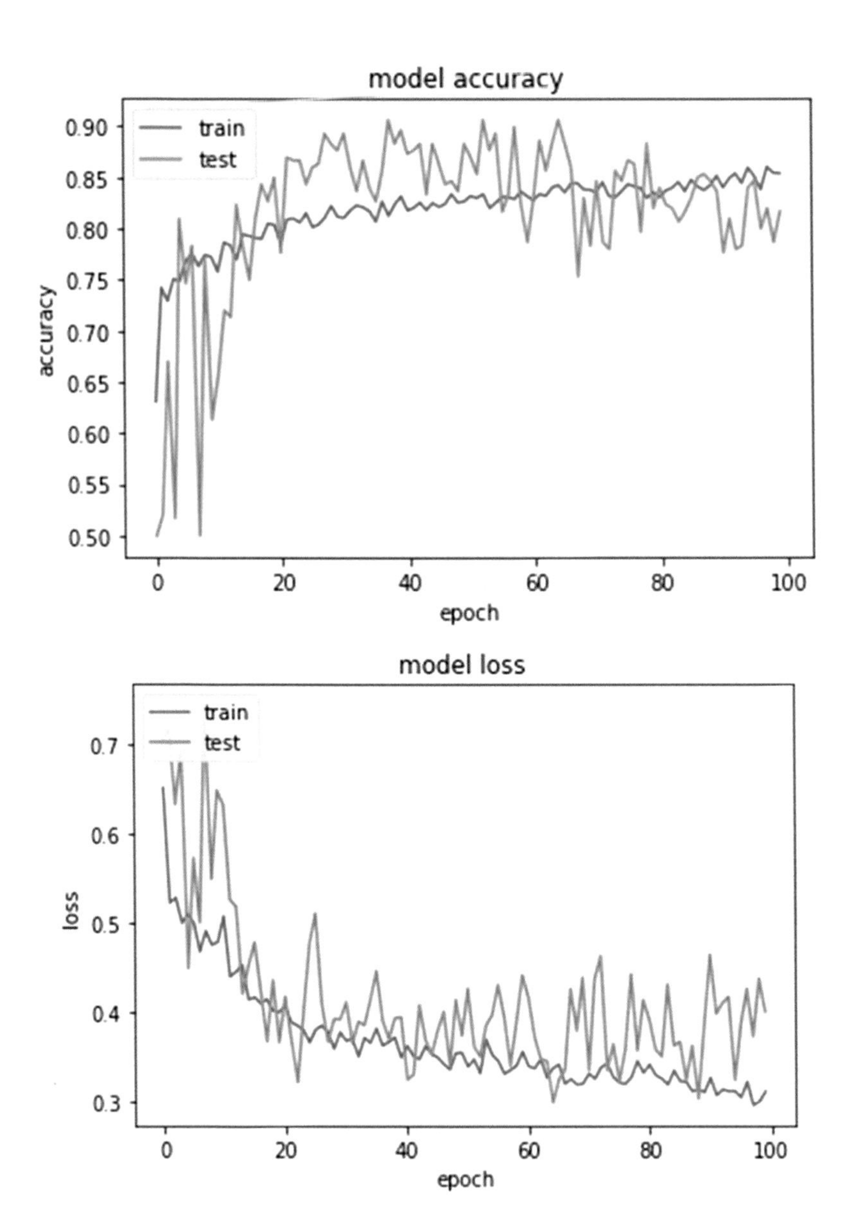

FIGURE 5.9 Loss and accuracy.

TABLE 5.1 Performance comparison.

S. No	Model	AUC-ROC score	Pre-processing
1.	VGG16	89.2%	Yes
2.	Efficient-B5	91.9%	Yes
3.	ResNet50	91.0%	Yes
4.	GAN + VGG16	96.33%	**No**

TABLE 5.2 Traditional augmentation of skin lesions grouped as CNN-AUG.

True/Auto	Benign	Premalignant	Malignant	Sensitivity
Benign	50	1	1	96.1%
Premalignant	2	44	20	68.7%
Malignant	1	17	47	72.3%
Specificity	94.34%	80.9%	84.6%	

TABLE 5.3 Synthetic augmentation of skin lesion grouped as CNN-VGG16+GAN.

True/Auto	Benign	Premalignant	Malignant	Sensitivity
Benign	52	0	1	99.4%
Premalignant	2	52	10	81.2%
Malignant	1	13	51	78.5%
Specificity	97.9%	89%	91.4%	

6. Conclusion and future work

In this article, we present a deep neural network that scales up from efficiency and is based on CNN with VGG-16 architecture. In comparison to previous popular CNN models, the suggested model is capable of evaluating the skin lesion classification of their corresponding classes in a reasonable amount of time and with higher accuracy without any need for preprocessing. Extensive experiments were conducted on the large publicly available dataset ISIC Archive Challenge Dataset with the handcrafted dataset, demonstrating the power of our model, which outperforms the classic VGG16 [34] and ResNet5 [35] in melanoma classification, using the demonstration of optimal runs in the python environment. In the future, we intend to investigate the differences between skin cancer and melanoma to a greater extent, to produce high-quality results that can be used for clinical purposes in melanoma detection and classification at an early stage of diagnosis [36], not limited to other types of skin cancer [37].

7. Conflict of interest

The authors declare that they have no known competing financial interests or personal relationships that could have appeared to influence the work reported in this paper.

References

[1] H.R. Roth, L. Lu, J. Liu, J. Yao, A. Seff, K. Cherry, L. Kim, R.M. Summers, Improving computer-aided detection using convolutional neural networks and random view aggregation, IEEE Transactions on Medical Imaging 35 (5) (May 2016) 1170–1181.

[2] Skin cancers, World Health Organization, http://www.who.int/uv/faq/skincancer/en/index1.html Accessed December, 2018.

[3] H.W. Rogers, M.A. Weinstock, S.R. Feldman, B.M. Coldiron, Incidence estimate of nonmelanoma skin cancer (keratinocyte carcinomas) in the US population, 2012, JAMA Dermatology 151 (10) (2015) 1081–1086.

[4] H. Greenspan, B. van Ginneken, R.M. Summers, Guest editorial deep learning in medical imaging: overview and future promise of an exciting new technique, IEEE Transactions on Medical Imaging 35 (5) (May 2016) 1153–1159.

[5] V. Srividhya, K. Sujatha, R. Ponmagal, G. Durgadevi, L. Madheshwaran, et al., Vision-based detection and categorization of skin lesions using deep learning neural networks, Procedia Computer Science 171 (2020) 1726–1735.

[6] A.N. Hoshyar, A. Al-Jumaily, A.N. Hoshyar, The beneficial techniques in preprocessing step of skin cancer detection system comparing, Procedia Computer Science 42 (2014) 25–31.

[7] M.A. Kadampur, S. Al Riyaee, Skin cancer detection: applying a deep learning-based model-driven architecture in the cloud for classifying dermal cell images, Informatics in Medicine Unlocked 18 (2020). Article 100282.

[8] M. Hasan, S.D. Barman, S. Islam, A.W. Reza, Skin cancer detection using convolutional neural network, in: Proceedings of the 2019 5th International Conference on Computing and Artificial Intelligence, 2019, pp. 254–258.
[9] Y. Li, L. Shen, Skin lesion analysis towards melanoma detection using deep learning network, Sensors 18 (2) (2018) 556.
[10] P. Tschandl, C. Rosendahl, B.N. Akay, G. Argenziano, A. Blum, R.P. Braun, et al., Expert-level diagnosis of non-pigmented skin cancer by combined convolutional neural networks, JAMA Dermatology 155 (1) (2019) 58–65.
[11] T. Saba, M.A. Khan, A. Rehman, S.L. Marie-Sainte, Region extraction and classification of skin cancer: a heterogeneous framework of deep CNN features fusion and reduction, Journal of Medical Systems 43 (9) (2019) 289.
[12] M.H. Jafari, N. Karimi, E. Nasr-Esfahani, S. Samavi, S.M.R. Soroushmehr, K. Ward, et al., Skin lesion segmentation in clinical images using deep learning, in: 2016 23rd International Conference on Pattern Recognition (ICPR), IEEE, 2016, pp. 337–342.
[13] D.N. Le, H.X. Le, L.T. Ngo, H.T. Ngo, Transfer Learning with Class Weighted and Focal Loss Function for Automatic Skin Cancer Classification, 2020.
[14] S. Mustafa, A.B. Dauda, M. Dauda, Image processing and SVM classification for melanoma detection, in: 2017 International Conference on Computing Networking and Informatics (ICCNI), IEEE, 2017.
[15] H.J. Vala, B. Astha, A review on Otsu image segmentation algorithm, International Journal of Advanced Research in Computer Engineering & Technology (IJARCET) (2013) 387–389, 2.2.
[16] M.S. Manerkar, et al., Automated Skin Disease Segmentation and Classification Using Multi-Class SVM Classifier, 2016.
[17] M.S. Manerkar, et al., Classification of skin disease using multi SVM classifier, in: International Conference on Electrical, Electronics, Engineering Trends, Communication, Optimization and Sciences, 2016.
[18] International Skin Imaging Collaboration, SIIM-ISIC 2020 Challenge Dataset, International Skin Imaging Collaboration, 2020, https://doi.org/10.34970/2020-DS01.
[19] I.S. Krizhevsky, G.E. Hinton, Imagenet classification with deep convolutional neural networks, in: Advances in Neural Information Processing Systems, 2012, pp. 1097–1105.
[20] H.C. Shin, H.R. Roth, M. Gao, L. Lu, Z. Xu, I. Nogues, J. Yao, Mollura, R.M. Summers, Deep convolutional neural networks for computer-aided detection: Cnn architectures, dataset characteristics and transfer learning, IEEE Transactions on Medical Imaging 35 (5) (May 2016) 1285–1298.
[21] A.A. Setio, F. Ciompi, G. Litjens, P. Gerke, C. Jacobs, S.J. van Riel, M.M.W. Wille, M. Naqibullah, C.I. Snchez, B. van Ginneken, Pulmonary nodule detection in ct images: false positive reduction using multi-view convolutional networks, IEEE Transactions on Medical Imaging 35 (5) (May 2016) 1160–1169.
[22] N. Srivastava, G.E. Hinton, A. Krizhevsky, I. Sutskever, R. Salakhutdinov, Dropout: a simple way to prevent neural networks from overfitting, Journal of Machine Learning Research 15 (1) (2014) 1929–1958.
[23] Y. Nesterov, A method for unconstrained convex minimization problem with the rate of convergence o (1/k2), Doklady an SSSR 269 (3) (1983) 543–547.
[24] R. Yeh, C. Chen, T.Y. Lim, M. Hasegawa-Johnson, M.N. Do, Semantic Image Inpainting with Perceptual and Contextual Losses, 2016 arXiv preprint arXiv:1607.07539.
[25] L.M. Radford, S. Chintala, Unsupervised Representation Learning with Deep Convolutional Generative Adversarial Networks, 2015 arXiv preprint arXiv:1511.06434.
[26] S. Ioffe, C. Szegedy, Batch normalization: Accelerating deep network training by reducing internal covariate shift, in: International Conference on Machine Learning, 2015, pp. 448–456.
[27] M. Mirza, S. Osindero, Conditional Generative Adversarial Nets, 2014 arXiv preprint arXiv:1411.1784.
[28] T. Salimans, I. Goodfellow, W. Zaremba, V. Cheung, A. Radford, X. Chen, Improved techniques for training gans, in: Advances in Neural Information Processing Systems, 2016, pp. 2234–2242.
[29] C.O. Odena, J. Shlens, Conditional Image Synthesis with Auxiliary Classifier Gans, 2016 arXiv preprint arXiv:1610.09585.
[30] J.-A. Almaraz-Damian, Melanoma and Nevus skin lesion classification using handcraft and deep learning feature fusion via Mutual information measures, Entropy (2020), https://doi.org/10.3390/e22040484.
[31] Chollet, et al., Keras. [Online]. Available: https://github.com/fchollet/keras, 2015.
[32] M. Abadi, A. Agarwal, P. Barham, E. Brevdo, Z. Chen, C. Citro, G.S. Corrado, A. Davis, J. Dean, M. Devin, et al., Tensorflow: Large-Scale Machine Learning on Heterogeneous Distributed Systems, 2016 arXiv preprint arXiv:1603.04467.
[33] jiahao1 Wang, X.2 JIN, Deep neural network for melanoma classification in dermoscopic images, IEEE International Conference on Consumer Electronics and Computer Engineering (ICCECE 2021) (2021) 666–669.
[34] Z. Li, Y. Zhao, J. Ma, J. Ai, Y. Dong, Fault Detection and Classification of Aerospace Sensors Using a VGG16-Based Deep Neural Network, 2022 arXiv preprint arXiv:2207.13267.
[35] T. Castilla, M.S. Martínez, M. Leguía, I. Larrabide, J.I. Orlando, A ResNet Is All You Need? Modeling A Strong Baseline for Detecting Referable Diabetic Retinopathy in Fundus Images, 2022 arXiv preprint arXiv:2210.03180.
[36] M. Janda, C.M. Olsen, J. Mar, A.E. Cust, Early detection of skin cancer in Australia - current approaches and new opportunities, Public Health Res Pract 32 (1) (March 10, 2022) 3212204, https://doi.org/10.17061/phrp3212204. PMID: 35290997.
[37] D. Mukherkjee, P. Saha, D. Kaplun, et al., Brain tumor image generation using an aggregation of GAN models with style transfer, Scientific Reports 12 (2022) 9141, https://doi.org/10.1038/s41598-022-12646-y.

Chapter 6

Artificial intelligence representation model for drug–target interaction with contemporary knowledge and development

M. Arvindhan[1], A. Daniel[2], N. Partheeban[1] and Balamurugan Balusamy[3]

[1]Galgotias University, Greater Noida, Uttar Pradesh, India; [2]ASET- CSE, Amity University, Gwalior, Madhya Pradesh, India; [3]Shiv Nadar University, Greater Noida, Uttar Pradesh, India

1. Introduction

According to academic studies and relevant formalities, the phrase "artificial intelligence," or "AI," is widely used in societal structure; nevertheless, its precise meaning has been brought into question. Rather than insisting on a single definition, we would prefer to focus on a few subgroups: Machine learning (ML), a select group of AI, has been the most common strategy to existing AI healthcare applications in recent years because it enables computational methods to learn from the data and increase their performance without even being pattern recognition. This really is due to the fact that ML was among the most common AI technologies [1]. Deep learning is a branch of ML that uses artificial neural networks to multiple layers of processing capabilities to identify patterns in some very massive data. It outlines 10 principles for organizations to keep in mind when developing strategies for AI technologies. People's confidence in AI is the first of these principles, followed by public participation, followed by a thorough risk assessment and project management, followed by a consideration of both the advantages and disadvantages of AI Fig. 6.1. A more difficult question is whether using AI to support in patient outcomes conflicts with consent forms guidelines in any way. Despite the reality that gaining permission from the participants is among the most important concerns in integrating AI into clinical practice, the ethical debate has not given one such problem nearly as much attention. We need to look into whether or not informed consent can be implemented in clinical AI in order to better understand how it works. "Black-box" algorithms are used by AI to perform tasks that clinicians cannot fully understand because they may be the result of difficult-to-understand ML techniques [2].

These AI wellness apps as well as chatbots can be used for everywhere from dietary guidelines and physical examinations to improving medication compliance and monitoring the information recorded by wearable sensors. Agreement is an issue that medical ethicists were also left to ponder when it comes to these apps' terms of agreement. In comparison to the conventional process of explicit consent, a user agreement is a legal document that one party signs without either party getting a face-to-face discussion [3]. It is common knowledge that most people do not bother to read the fine print of user agreements and thus frequently break them. In additament, the software is updated on a regular basis, making it even harder for users to comply with the requirements of service those who have consented to.

1.1 AI will challenge the status Quo in healthcare

In the field of healthcare, AI faces one of its greatest challenges in terms of safety. AI algorithms are used by IBM Watson for Oncology to analyses data taken from the medical records of patients in order to provide physicians with assistance in determining the best cancer treatment options for their patients. This application has received significant media attention. On the other hand, it has recently been the target of criticism for allegedly providing recommendations for cancer

Deep Learning in Personalized Healthcare and Decision Support. https://doi.org/10.1016/B978-0-443-19413-9.00005-9

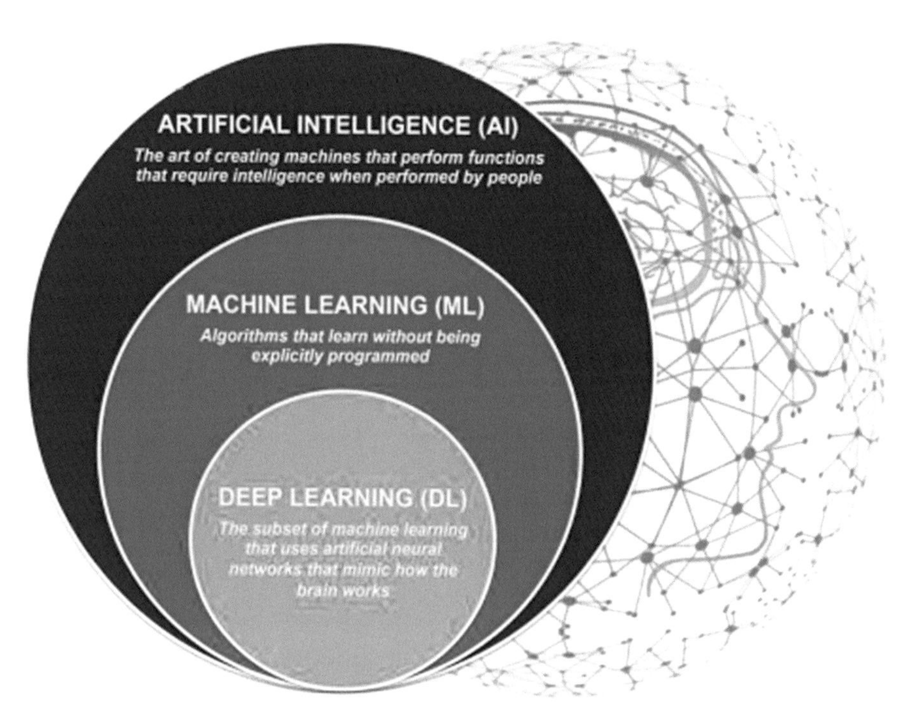

FIGURE 6.1 Basic structure of an artificial intelligence in neural network.

treatments that are "unsafe and incorrect." This one unfortunate incident has cast a negative light on the entire industry. This demonstrates that it is of the utmost importance for AI to be both secure and efficient. However, how can we make sure that AIs live up to their commitments? Both the reliability and validity of the datasets as well as transparency need to be ensured before the full potential of AI can be tapped. This is especially true for those involved in the development of AI. To begin, the datasets that were utilized have to be reliable and valid. In the field of imitation intelligence, the adage "garbage in, garbage out" holds true. The more accurate the training data (also known as labeled data) is, the more accurate the AI will be. In addition, the algorithms almost always require additional tweaking so that they can produce accurate results. Another significant challenge is the exchange of data: In situations in which the AI must have an extremely high level of confidence (for example, self-driving cars), massive amounts of data and, as a result, increased data sharing will be required. Second, in the interest of preserving both the safety of patients and their confidence in the healthcare system, a certain amount of transparency is required to be maintained. Even though in a perfect world all data and algorithms would be available for inspection by the general public, there may be some valid concerns regarding the protection of investments and intellectual property, as well as the prevention of an increase in the risk of cybersecurity breaches. Particular concerns have been raised in response to the recommendations of additional "black-box" systems [4]. To determine the means by which transparency can be achieved in this setting will be a difficult task. Even if the model could be simplified into a mathematical relationship that links symptoms and diagnosis, the process might still have complex transformations that are beyond the capabilities of clinicians (and especially patients) to comprehend. On the other hand, it is possible that there is no requirement to look inside the "black box": It is possible that positive results from randomized trials or other forms of testing will serve as sufficient demonstration of the safety and effectiveness of AI, at least in some cases. This would be the case at least in some cases.

2. AI privacy and security challenges

Even for the remotest parts of the world can benefit from AI's potential, as it can democratize expertise, "globalize" healthcare, and bring it even to the most impoverished regions. Data used to prepare any ML system or algorithm, on the other hand, are what ultimately determine the system's level of trustworthiness, effectiveness, and fairness. The possibilities for stigmatization in AI is also present. AI developers must be aware of this risk and work hard to reduce the impact of implicit weaknesses throughout the product development process. As a starting point, they should consider the possibility of biases when making decisions about. Age and a disability are just two examples of other factors that could lead to discrimination. It is possible that there are multiple facets to each of these biases. The datasets taken by individual (which are not representative) can be the cause of these inaccuracies, as well as the selection and analysis of data-by-data scientists and ML systems and the context in which AI is used. Biased AI might, for instance, lead to false diagnosis and treatment and treatment ineffectiveness for some population groups in the medical field, trying to put their people's lives at risk. Phenotypic expression and occasionally genetic markers details are implicated. Suppose an AI-powered clinical decision support (CDS) tool could assist physicians in determining the best course of action to take in treating patients who have skin cancer.

2.1 Ensuring transparency, explain ability, and intelligibility

2.1.1 Algorithmic fairness and biases

There is a possibility that some of these biases will be eliminated as a result of an increase in the amount of data that is available, as well as efforts to better gather information from minority populations and to more precisely specify for which populations the algorithm should and should not be used. However, one issue that persists is the fact that numerous algorithms are complex and opaque. This presents a challenge. In furthermore, like we have seen in the frame of reference of police enforcement, some businesses that create software are resistant to disclose certain information and claim that their work is protected by trade secrecy. Consequently, the task of collecting the data and demonstrating the biases may very well be delegated to nongovernmental organizations.

2.1.2 Data availability

There is a possibility that any of these preconceptions will be eliminated as a result of an increase in the amount of data that is available, as well as efforts to effectively gather information from minority populations and to better specify about which compared to participants the algorithm should not be used appropriately. Nevertheless, one issue that persists is the fact that numerous algorithms are complex and opaque. This presents a challenge. In addition, as we have seen in the frame of reference of policing, some businesses that develop software are resistant to disclosure as well as claim that there own work is protected by trade secrecy. Consequently, the task of collecting the data and demonstrating the biases may very well be delegated to nongovernmental organizations [5].

2.1.3 Privacy concerns

The issue of privacy also presents its own unique set of dangers. Because large amounts of data are required, developers have an opportunity to collect this kind of information to as many patient populations as possible. There is a possibility that certain patients will be concerned this accumulation might very well violate there own privacy, and there have been cases in which litigation have already been filed because of the sharing of information among both large healthcare systems and AI designers. Some other manner that AI might violate patients' privacy is that it might be able to guess personal details about patients even if the algorithm has never seen those details before. (In point of fact, this is frequently the objective of AI research in the medical field.) An AI system could, for instance, determine that a person has Parkinson's disease predicated on the quivering of a computer mouse, even though the individual never had disclosed that relevant data to anyone else. Service users could view this as an invasion of their privacy, particularly if the implication made by the Information system were accessible to third parties like financial institutions or insurance carriers.

3. Ensuring transparency, explain ability, and intelligibility

There could still be issues although if Ai technologies learn from correct and accurate data if the information reflects underlying biases and inequalities in the healthcare system. AI systems that are responsible for resource distribution to make disparity worse by allotting limited resources to service users who health systems deem to be less beneficial or less financially viable for a wide assortment of troublesome purposes.

There is a possibility that some of these biases will be eliminated as a result of an increase in the amount of data that is available, as well as efforts to better collect data from minority populations and to effectively specify for which compared to participants the algorithm should or should not be used appropriately Fig. 6.2. However, one issue that persists is the fact that numerous algorithms are complex and opaque. This presents a challenge. In addition, as we have seen in the context of policing, some businesses that develop software are resistant to disclosure and claim that their work is protected by trade secrecy. Therefore, the task of collecting the data and demonstrating the biases may very well be delegated to nongovernmental organizations [6].

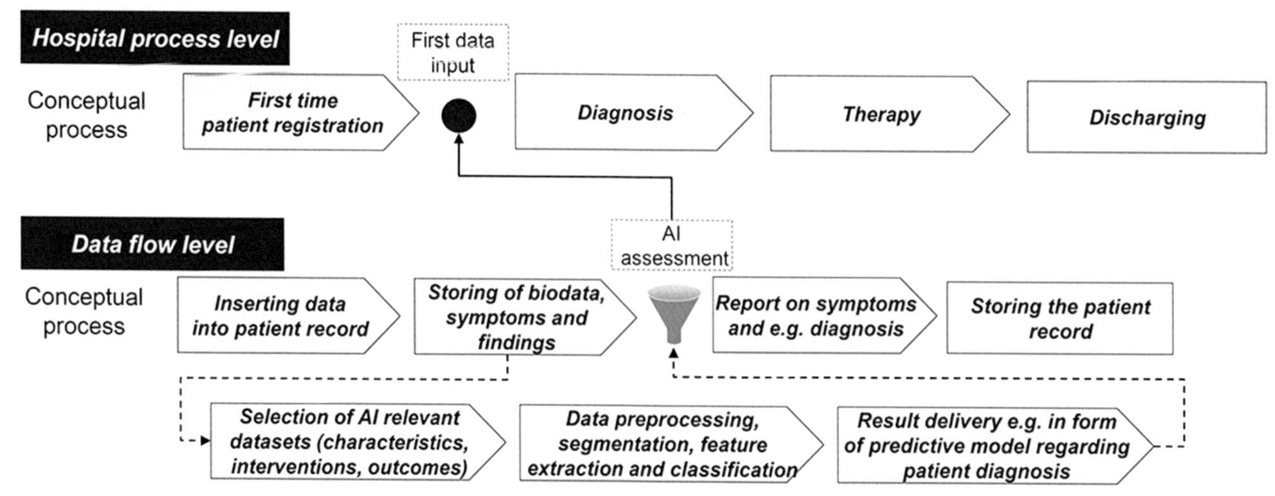

FIGURE 6.2 Concerning fair treatment and preconceptions in algorithms.

3.1 Data availability

In order to train AI systems, large amounts of data must be collected from a variety of sources, including health records, pharmacy records, insurance claims records, or information generated by consumers themselves, such as fitness trackers or purchasing history. However, data on health are frequently unreliable. In most cases, data are scattered across a large number of different systems. Even apart from the variety that was just mentioned, patients generally see a variety of providers and switch insurance providers, which results in data that is split across different systems and multiple formats. Because of this fragmentation, the risk of error is higher, the comprehensiveness of datasets is lower, and the cost of gathering data is higher. This also restricts the types of entities that are able to develop efficient AI for the healthcare industry.

3.2 Concerns regarding privacy

The issue of privacy also presents its own unique set of dangers. Because large datasets are required, there is an incentive for developers to gather this kind of information from as many patients as possible. There is a possibility that some patients will be concerned that this collection may violate their privacy, and there have been cases in which lawsuits have been filed because of the sharing of data between large health systems and AI developers. Another way that AI could compromise patients' privacy is that it could guess personal details about patients even if the algorithm had never been given that information An AI system could, for instance, determine that a person has Parkinson's disease based on the trembling of a computer mouse, even if the person had never disclosed that information to anyone else. This would be possible even if the mouse was connected to a computer (or did not know). Patients could view this as an invasion of their privacy, particularly if the inference made by the AI system were accessible to third parties like financial institutions or life insurance companies.

4. Drug discovery and precision medicine with deep learning

Artificial technologies learn from precise statistical information; there still may be problems if the knowledge takes into account fundamental biases and social inequality in the health system. This is because artificial technology systems are still learning. AI systems that are willing to take responsibility for allocation of resources to make disparity become worse allotting limited options to patient populations who health system design deem and become less desirable or less financially viable for a variety of troublesome causes. This could happen for a number of separate purposes [7].

4.1 Discrimination and unequal treatment

There is a potential for bias and inequality to arise from the use of AI in healthcare. AI systems can take on the biases of the data on which they are proficient because they learn from the data and use it to train themselves. For instance, if the majority of the data that can be used for AI are collected in academic medical centers, the AI systems that are developed as a result will have less information about patients will check for their academic report in the center in the academic, and as a result, they will treat these patients less effectively. In a similar vein, if speech-recognition AI systems are used to transcribe encounter notes, it is possible that these AI systems will perform less well when the provider is of a race or gender that is underrepresented in the training data.

4.2 The production of data and its availability

Because it is difficult to compile high-quality data in a way that is consistent with protecting patient privacy, there is a greater potential for a number of risks. The establishment of infrastructural resources for data by the government is the basis for one set of potential solutions. These solutions range from creating standards for electronic health records to directly providing technical support for high-quality data-gathering pains in health systems that would not then have access to such resources [8].

4.3 The supervision of quality

Monitoring the quality of AI systems will be helpful in reducing the risk of patients being injured. The Food and Drug Administration (FDA) is in charge of monitoring the safety of certain AI (AI) health care products that are sold on the market. The agency has already given the go-ahead for several products to be sold on the market, and it is currently pondering novel approaches to the supervision of AI systems used in the medical field. On the other hand, the FDA will not be responsible for regulating the majority of the AI systems that are used in the healthcare industry. Considering this fact that these AI system will not perform any function in medical back-ends and in business allocations developments while in deploying the health system process. These artificially intelligent health care systems fall into something of a regulatory oversight void. If the FDA chooses not to exercise its regulatory authority over a given system, it may be necessary for other entities, such as health systems and hospitals, professional organizations like the American College of Radiology and the American Medical Association, or insurers, to increase the amount of oversight they perform on such systems.

5. Clinical decision support and predictive analytics

Moreover, medicine is being fundamentally transformed as a result, along with virtually every other facet of the pharmaceutical industry, improving diagnostics, automating administrative tasks, increasing the pace of research, and being able to diagnose and avoid getting sick in the first place. It is more important than ever to undergo digital transformation in light of increasing costs as well as an increasing number of doctors, nurses, and administrative staff who are burning out. When a powerful AI data infrastructure is in place, there is an almost infinite number of ways in which costs can be cut, staff can be assisted, and patient care can be improved. There is a possibility that AI could automate 40% of the tasks performed by healthcare support staff and 33% of the tasks performed by healthcare practitioners. Increasing the efficiency of these tasks by automating them with solutions powered by AI improves efficiency and frees up staff to do more high-value work. It has been shown that the use of remote monitoring devices can reduce the likelihood of hospitalization by 76% in patients who suffer from chronic diseases Fig. 6.3. This results in a significant savings in healthcare costs for both the providers and the patients.

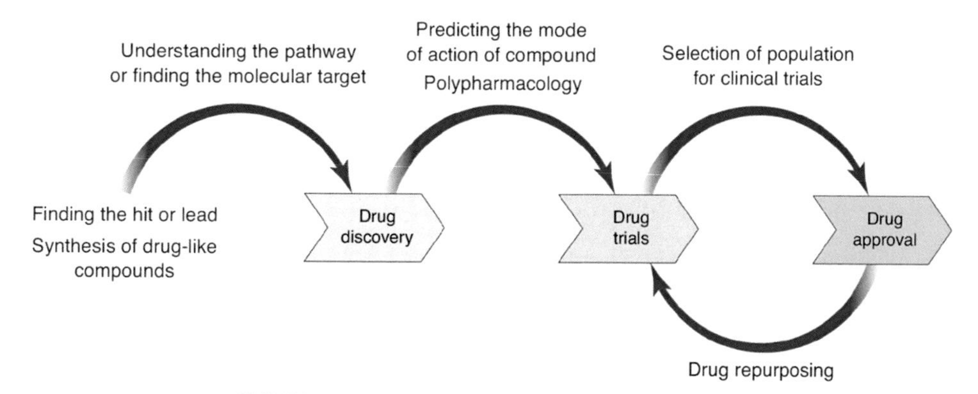

FIGURE 6.3 Drug cycle in approval of pathway discovery.

(1) The main policy for framework with capability, accountability, and liability.
(2) Confirming with the deign infrastructure some of the practically applicable privacy compliant.
(3) In routine healthcare basis some of the impact in AI have the significant on QALY.

Private companies can use the success factor analysis already today to build and scale AI services, such as through high-quality economic measurements and comprehensive technological planning regarding data processing and privacy-by-design structures. While some of the success factors require input from public institutions, private companies can use the analysis to build and scale AI services.

Many fields of study rely heavily on chemical entities (MDs). Physicochemical information can be expressed numerically in terms of the molecule's molecular characterizations. However, only a portion of a molecule's information can be gleaned through experimentation. An big manufacturing in past few decades has indeed been placed upon understanding the molecular framework so that one or even more numbers can be used to develop mathematical relation among both properties and structure, bioactivities, as well as other experimental characteristics. In this way, MDs have become such a helpful tool for searching molecular directories because they can find particles with comparable physicochemical characteristics based on the calculated descriptors. Descriptors for properties derived from two-as well as three-dimensional (2D and 3D) constructions, which describe extra particular characteristics but require more computation effort, have been defined since the commencement of their application, resulting in thousands of single-molecule descriptions [9].

These descriptions encapsulate molecules in various ways, giving a generic summary of the entire molecule (1 D descriptors). Many scientists believe that the atomic and molecular descriptors that have been developed so far provide a sufficient toolkit for future drug discovery efforts. However, the essence of the model or the incorrect collection of structural signifiers may be to blame for the lack of adaptation in the models. The latter could be the result of a flaw in the selection process or a model's inability to adequately capture the phenomenon. New identifiers that could be used in QSAR-based prototypical studies are needed because of all of this. There are two major groups of characteristic parameters. In the methods of physical–chemical properties and theoretical molecular descriptors derived from a powerful symbol of the chemical compound and can be much farther classified according the various types of molecular representations that can be used to perform the experiment. On the other hand, theories are divided into three categories:

1. Constitutional: reflects molecular humanity's greatest basic characteristics.
2. Graph theory is used to perform the arithmetic of its topological properties
3. Based on empirical schemes, they encode this same particle's rights to engage in various interactions.
4. Physicochemicals: describe how a molecule responds to reactions occurring outside of it.

The basic types emerge when we take into account the measurements of the theoretical molecular descriptors' representations of physicochemical parameters.

Only when drugs bind to specific proteins in the body will they have any effect. Stickiness is an important consideration in drug discovery and screening. Some promising new results coming from chemistry and AI research are on the way. It calculates the affinity of drug candidates very quickly to their targets with the new technique, called. "In comparison to the previous state-of-the-art methodologies, in the method of DeepBAR that yield more precise calculation done with short amount of time could indeed day speed up drug discovery and protein engineering, according to the researchers. Our technique is substantially faster than before." Journal of Physical Chemistry Letters published the findings today. When it comes to a drug's affinity for a specific protein, the binding free energy is the yardstick used to gauge how tightly the two molecules are bound. One way to predict a drug's potential efficacy is to determine its binding free energy. However, determining an exact number is difficult. There are two broad categories of methods for calculating binding free energy, both of which have their own drawbacks. Each classification requires a lot of time and computational power to calculate the exact quantity. However, it only provides an approximation of the binding free energy in the second category, which is less computationally costly. In order to get the best of both worlds, Zhang and Ding devised a strategy. precise and time-saving [10].

5.1 Natural language processing could translate EHR jargon for patients

However, compared to previous methodologies, DeepBAR only necessitates a tiny fraction of the computation time. Using a combination of ML and traditional chemistry calculations, the new method improves on previous approaches. Binding affinity can be precisely calculated using the decades-old "Bennett recognition ratio" algorithm, which is the "BAR" in DeepBAR. As a result, using the Bennet acceptance ratio typically necessitates knowledge of two "endpoint" states (e.g., drug molecules bound to proteins and those that have been completely dissociated), as well as many intermediate states (e.g., varying levels of partial binding). DeepBAR uses the Bennett acceptance ratio in machine-learning frameworks known as deep generative models to eliminate the in-between states Fig. 6.4. Using these models, Zhang explains, "the bound and the unbound states are established for each endpoint." The Bennett acceptance ratio can be applied directly to these two different states without the need for costly intermediate steps. The researchers drew inspiration from computer vision by employing deep generative models. When it comes to computer image syntheses, Zhang says, "it's basically the same model." "We're treating each molecular structure as an image that the model can pick up on," says Dr. Wang. In other words, this project is a continuation of the ML effort. There have been some difficulties in adapting the computer vision approach used in DeepBAR's key innovation to chemistry. Ding explains that these models were originally designed for 2D images. This is a 3D structure, but here we have proteins and molecules [11].

5.2 Faster drug screening in the future

DeepBAR calculated binding unlimited electricity 50 s faster than conventional techniques in tests involving small nutrient molecules. "Humans can start to think regarding using this for pharmaceutical analysis, particularly in the context of Covid," Zhang says regarding effectiveness. Faster than the gold standard, DeepBAR is as accurate as the gold standard. Furthermore, the experts claim that, in addition to drug screening, DeepBAR can be used to model the interactions among both multiple proteins. He claims that DeepBAR would need to be tested against a wide range of complex data. "Which will undoubtedly add complexity and necessitate the inclusion of additional approximations." Deep BAR's ability to perform calculations on large proteins will be enhanced in the future thanks to recent advancement of computer science. This investigation is an example of integrating traditional quantum chemistry methods, which have been around for decades, with the most recent developments in the field of ML.As a result, we've accomplished something that couldn't have been accomplished prior to now. The National Institutes of Health contributed to the cost of this study Fig. 6.5.

In just about any research field, the implementation of an ML research methods must be transversal, even though the experiment is the same in all fields. There are four distinct steps with in ML methodology used for drug development: As a starting point, the data are gathered; the mathematical descriptors are generated; the best subset of variables is sought, and the models are trained and tested; finally, the models are validated.

5.3 Machine learning predictions rely on input data

The training of the model relies heavily on the ability of the descriptive words to capture the characteristics as well as structural characteristics of the particles. Descriptors ranging from simple molecular structures to sophisticated three-dimensional as well as complex single-molecule fingerprint formulations have indeed been described in the literature. These include vectors of hundreds and thousands of elements.

5.4 Connection between quantifiable construction and function

The quantitative connection between framework and exercise (QSAR) models, which quantitatively connect the inorganic compounds of the particles with their bioactivity, enable, through arithmetical systems, to anticipate the physicochemical and living fate characteristics that a chemical discovery will get from the known characteristics of the substance Fig. 6.6.

5.4.1 Support for clinical decision-making and predictive analytics are included in this section

Symptomatic illness: When we believe of predictive, persistent illness typically comes immediately in to the mind since it represents the greatest opening for cost avoidance. With the help of data, we hope to identify patients who are headed in the wrong direction and intervene before costs gain momentum into something unmanageable. A large and immediate return on investment is possible with this type of strategic prevention if it succeeds (ROI) [12].

5.4.2 In order to make prescriptive modeling useful, they must be put into practice

Predictive analytics cannot be put into practice without real-time access to integrated data. In order for it to be a great achievement, providers must take action to achieve the desired result. The future of CDS will be in virtual assistants that was made in processing of ML to analyze data and make real-time recommendations. Patients will benefit from more individualized care, but providers will also benefit from a significant reduction in workload Fig. 6.7.

Relevant information will be the most important factor in their success They must use both discrete and nondiscrete data, such as free text from a physician's note in the patient's record, to accomplish this. It is possible to obtain a more

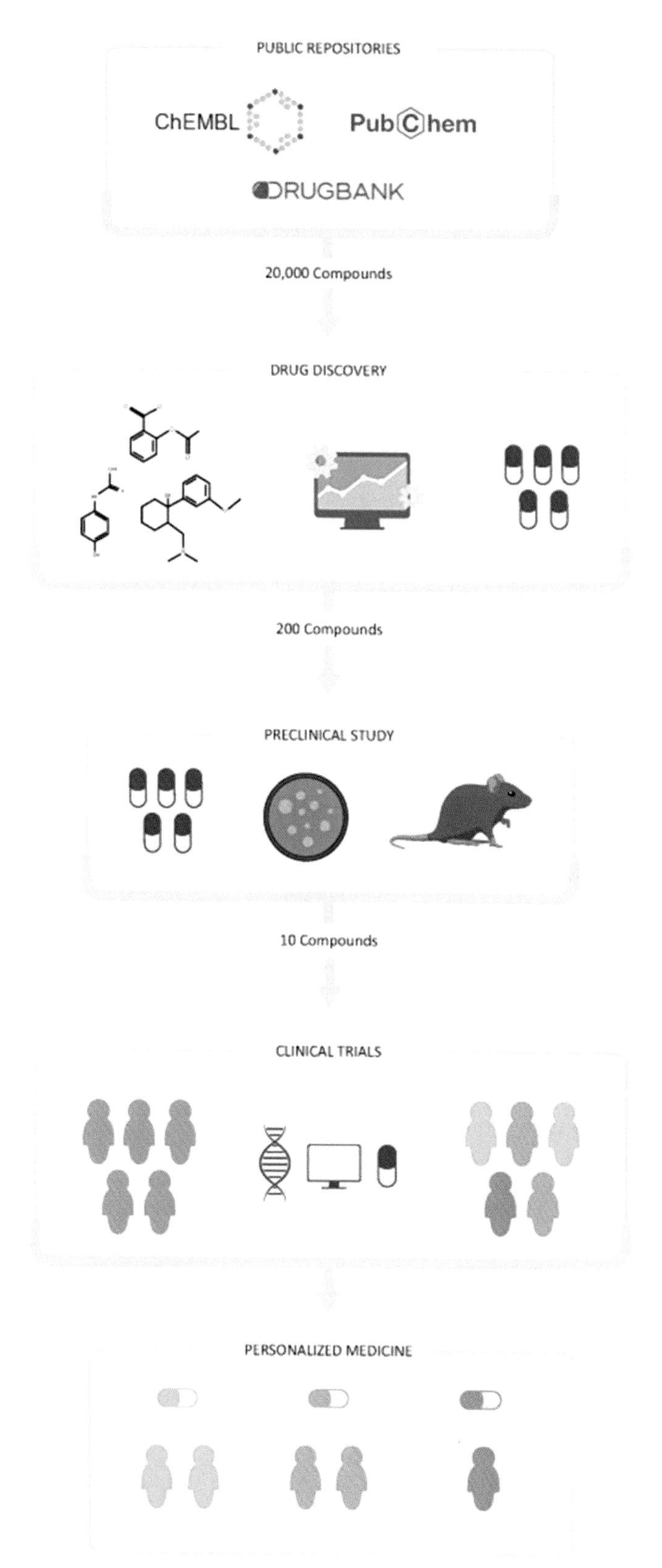

FIGURE 6.4 3D structural view of clinical trails compounds in drug discovery.

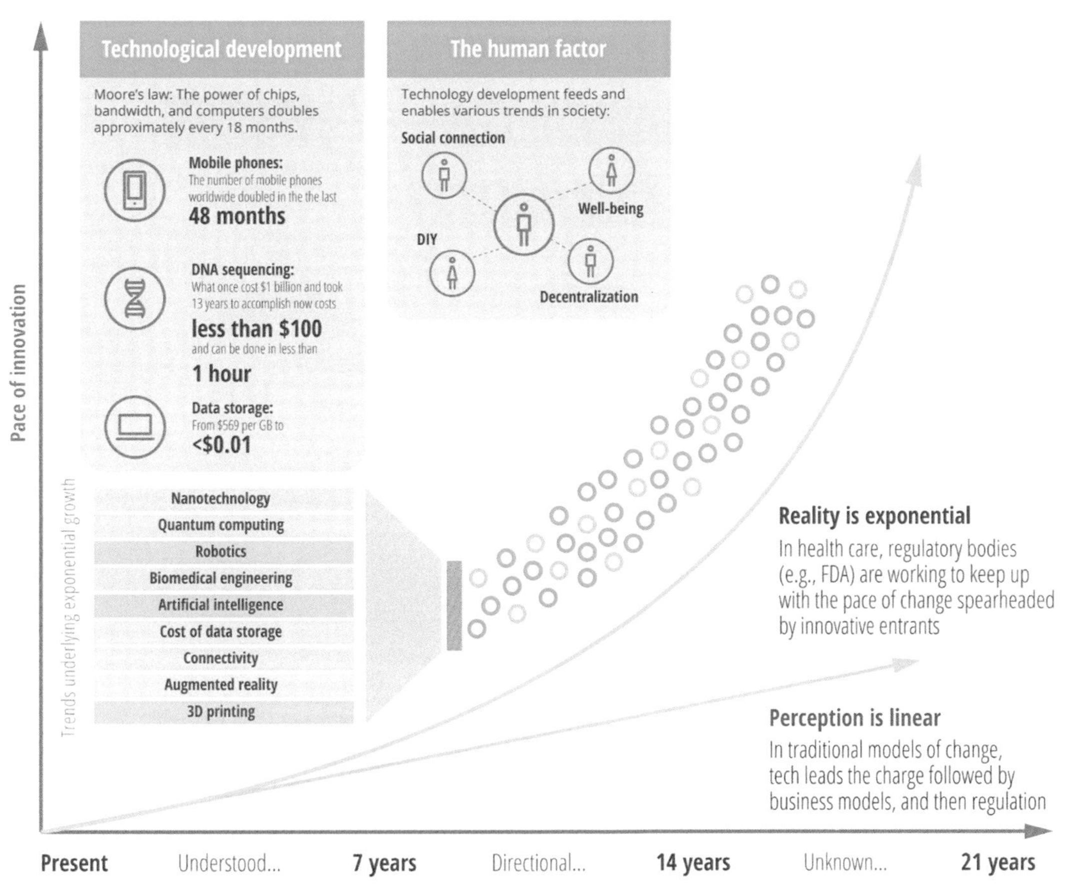

FIGURE 6.5 Technology development in human factors in pace of innovation.

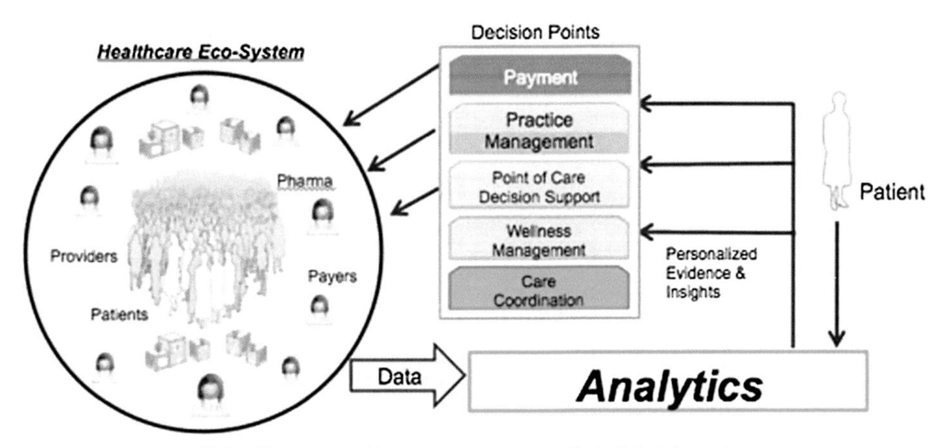

FIGURE 6.6 Healthcare eco-system in analytical decision point.

detailed clinical picture by combining the two types of data. Amazon Alexa and Google Home are being used in exam rooms to listen to provider-patient conversations, interpret the information, and provide guidance.

6. Natural language processing in drug

To begin, there is a significant underutilization of natural language processing (NLP) in the fields of allergy, asthma, and immune disease. Allergic response and breathing problems have seen some NLP research, but dermatitis and allergic rhinitis have seen much less NLP research. NLP algorithms for difficult ideas such as predefined allergy disorder

FIGURE 6.7 Prescriptive modeling for drug patients representation.

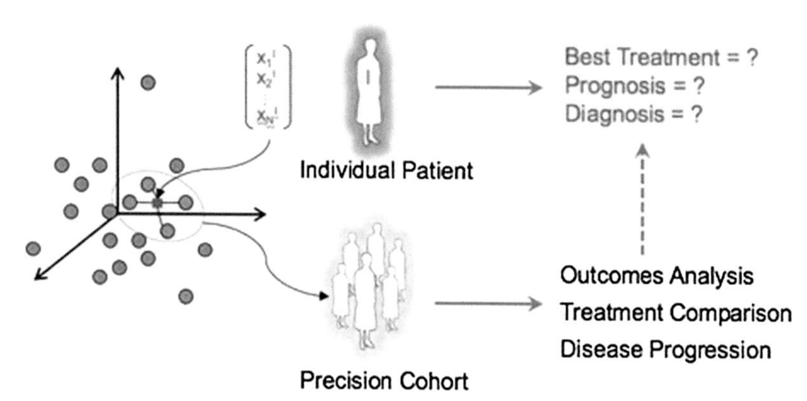

FIGURE 6.8 Natural language processing methodologies conditional feature for training model in asthma.

requirements are extremely diminished, despite the fact that valuable data requested for complicated predetermined allergy disorder criteria exists in an EHR free-text format, aside from those reported by our group. It is possible to use NLP to recognize a group of patients with distinctive clinical character traits in a condition like asthma that is highly heterogeneous Fig. 6.8. Using NLP as an example, researchers can change the current NLP methodologies for a complaint (such as asthma) in order to determine a subset of sick people with distinguishable patient features that might not be possible with indicators for asthma.

Quality problems, privacy concerns, algorithmic bias, and dearth of interoperability are just some of the issues to be addressed in the future for NLP-based investigation, as well as health information technology (HIT)-related clinician exhaustion and workflow concerns. AI (AI) may face challenges related to trust, comprehensibility, functionality, transparency and fairness when it comes to its use in health care. Even though studies published for asthma or allergy are regularly conducted, as discussed in our literature review for the existing literature, investigation ability to leverage free texts throughout EHRs via NLP is seriously restricted at the moment. EHRs are biased, but NLP could help us identify algorithmic biases stemming from EHRs' systematic biases. Notwithstanding the these drawbacks, FDA has recently identified EHR-based investigations as a valuable supplement to conventional RCT-based evidence. This information systems' tool is to be accepted as well as incorporated into repetitive workflow.

Deep learning has the potential to enhance radiology medical diagnostic tasks, such as image classification, picture quality improved performance, image classification, having found identification, and investigation prioritization based on immediate diagnoses. In order to train deep learning algorithms, a large amount of data are needed. Deep learning algorithms in radiology are hindered by the difficulty of getting sufficient data. Using generative adversarial networks (GANs), we can get around this problem. GAN-generated augmented data have been used in several studies to train deep learning algorithms, as demonstrated in this review. The achievement of CNN algorithms was greatly enhanced by the addition of generated images to the data. With GANs, the quantity of clinical data required for training can be reduced. As a result, the success of radiology's automated image analysis could be impacted by the growing interest in GANs. There has been a significant amount of progress made with GANs since they were first introduced. However, they have not yet been tested in clinical settings. Several factors could be at play as to why this innovation is still in the prototype stage. For starters, successfully developing and training GANs is a difficult task. For example, if the generator and discriminator are out of balance, the training will fail. In most cases, it is the discriminator who gains the upper hand. These images can be easily recognized by the discriminator when this occurs. At some point, it stops generating useful output for the generator, and as a result, the images it generates stop getting better. Complete or partial modes may also collapse at any time. GANs in medical imagination have shown probable. In the meantime, it is too early to tell if GANs will have a significant impact on computer-assisted radiology (CAR). Small data sets and brand-new research are used in the peer-reviewed articles in every area of medical imaging. A longer period of time is required for follow-up studies involving larger numbers of participants. Furthermore, it is difficult to evaluate and compare the performance of GANs. That is due to the variability in image processing and analysis initiatives between studies. Objective numerical metrics of various kinds are used in some studies, but they vary from one study to the next. Some examples of these metrics are the mean absolute error (MAE), the peak signal-to-noise ratio (PSNR), the structural similarity index (SSIM), and the inception score (IS). Due to the lack of standardization, some studies use open to interpretation physicians' evaluations of image quality. Everyone else assess down-stream tasks. Examining, for example, whether the images obtained are better used for algorithm training [13].

1) The Health Care Benefits of Predictive Analytics
2) Predictive Analytics in Healthcare: Use Cases

7. Predictive analytics has a wide range of practical applications, including the following

Enhanced results for patients. Healthcare organizations can detect warning signs of serious medical events and prevent their occurrence by integrating patient records with other health data. Holistic health care support. Rather than focusing solely on outcomes, many patient-centric models are emphasizing the individual as a whole. It is now possible to gather and integrate lifestyle, symptom, and treatment data to create holistic treatment plans with predictive tools. Improvements to the way things are done. A reduction in overall healthcare costs can be achieved by incorporating predictive tools into internal processes such as equipment procurement and staffing requirements Fig. 6.9. Dedicated attention to the needs of the customer. As the pandemic threat grows, the need for patient-centered care is becoming increasingly important. Predictive tools allow for truly personalized treatment plans that are tailored to the specific needs of each patient.

The use of both strategy and technology is required on the part of healthcare providers in order to realize the potential offered by predictive modeling. According to Phan, a statement of the problem is the first step in a successful implementation. "First, come up with some theories about how the issue might be solved, and then build out now to one or more algorithms that you're going to put into action." This is related to the second pillar of data analytics, which consists of advanced ML and AI technologies. It is possible to create reliable and able to respond models by providing these tools with verified healthcare data. The above features are able to analyses incoming data to determine potential patient concerns, improve the current operational processes, and predict recent developments [14].

7.1 Efforts made to lessen potential dangers to healthcare organizations' security

Additionally, companies have a responsibility to be aware of any potential risks. For instance, the Deloitte paper notes that regulatory guidance is still emerging around predictive analysis in healthcare, particularly as applied to machine-driven

FIGURE 6.9 Benefits of predictive modeling in healthcare.

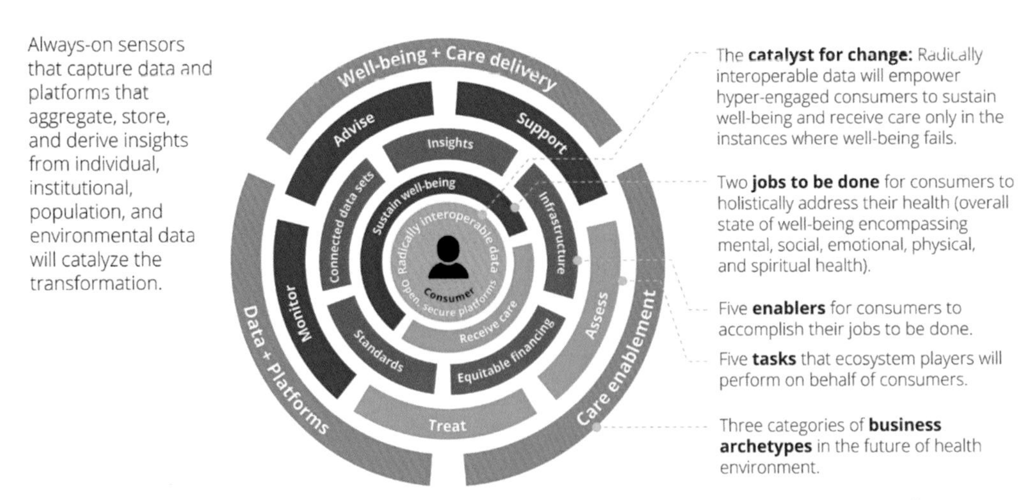

FIGURE 6.10 Machine learning–driven application in healthcare model for organization security.

interventions in patient-centered care Fig. 6.10. This raises the question of who is responsible for errors made by predictive models and whether or not these models should be used at all [15].

7.2 Future of deep learning in healthcare

The incorporation of always-on biosensors and software into medical devices that are capable of data generation, collection, and sharing has already begun at a number of MedTech companies. It may be possible to develop more advanced cognitive technologies that could analyze a knowingly large set of limitations and produce personalized insights into the health of a consumer. The obtainability of data and adapted AI can make it possible to enable exactness well-being and real-time microinvolvements, which enables us to get forward of illness and far gaining of appalling disease. It is possible that customers, armed with such extremely specific personal material about their own health, will claim that their health information be portable. Regulars have become habituated to the changes that have taken place in other industries, such as e-commerce and agility. These changes have also had an effect on the consumers. These customers will election with their feet and their cases to show that they want health to follow the same trajectory as other aspects of their lives and become an integrated component of their lives. Consumers do have the ability to control their thermostats, set alarms, and turn on their lights from a remote location. Turn the clock forward to a house that has telemonitoring biosensors installed in it. This may include a highly connected restroom, in which the mirror as well as other tech-enabled fixtures process, detect, and analyses data pertaining to the user's health. A highly sensitive sensor could be embedded in a bathroom mirror, for instance, to monitor a person's body temperature and blood pressure. The sensor could then identify abnormalities by comparing these vitals to the individual's previous biometric data. Perhaps, this smart device even plays a skin care tutorial and reminds the user to wear sunscreen predicated on the user's plans for the day as well as the weather forecast. This would be done by taking into account both of these factors. Analyses that are carried out by a toilet that is equipped with technology may be able to identify biomarkers that direct a potential change in health position well in advance of the manifestation of symptoms [16]. Environmental sensors placed outside of a home can evaluate factors such as changes in air pressure, pollen concentrations, and UV levels. Consumers could benefit from such information by remaining more aware of their own health and being better able to recognize red flags that may signal the beginning stages of an illness or disease. When necessary, personalized therapies determined by a person's genomic information could be delivered by drone instead of having to pick them up at the pharmacy like a regular prescription.

Designers of health products (also known as automakers, software vendors, inventors/innovators, and others) design the products in various costs based on the utilization. The economic model of these organizations is determined by their capacity to facilitate the delivery of care and well-being to their constituents. It is possible that pharmaceuticals and medical devices would not be the only types of medical products available in the future. In addition to that, they might include software, applications, wellness products, or even foods that are focused on health. In the home bathrooms of the future, for instance, there may be a "smart" toilet that has sensors that are always on and can perform tests for nitrites, glucose, protein, and pH. These tests can detect infections, diseases, and even pregnancy. It is possible that a smart mirror with facemask acknowledgment technology could tell the difference between a mole and melanoma. A smart toothbrush equipped with breath biome sensors may be able to detect genetic changes that point to the beginning stages of disease. It is possible to modify foods so that they contain bacteria that kill cancer cells and that add to the consumer's microbiome.

8. Conclusion

Accurate binding capacity can be difficult to predict quantitatively by contrast with the detection of a communication. Despite data-driven models have a number of advantages, but they also have a significant drawback. reliability in forecasting Incorporating this could be done to solve the problem. For constructing and running tests legitimate models including using elevated datasets, domain knowledge all through feature extraction notwithstanding, experimental results are unavailable or limited for some target areas. There are a few available data points on models that can be relied on in some cases. Due to a major lack of high-quality datasets, it is impossible to build. Reading comprehension is one ML technique that could be in situations where only a limited set of information is accessible, as a data strategy create a model that are both dependable and accurate. As a result, the discovery and development of new medicines can be expedited. However, scalable deep learning models and so they are adaptable, they have not been fully integrated into lead optimization strategies yet. the difficulty of developing new medicines because of a lack of data However, the use of deep-learning models has been attempted recently. By using machine-learning and deep-learning techniques to explore the vast chemical space, we are making a lot of progress and by speeding up the drug discovery process by automating the early stages of the drug discovery research stages.

References

[1] G. Liang, W. Fan, H. Luo, X. Zhu, The emerging roles of artificial intelligence in cancer drug development and precision therapy, Biomedicine & Pharmacotherapy 128 (August 2020) 110255, https://doi.org/10.1016/J.BIOPHA.2020.110255.

[2] M.H. Jarrahi, D. Askay, A. Eshraghi, P. Smith, Artificial intelligence and knowledge management: a partnership between human and AI, Business Horizons (March 2022), https://doi.org/10.1016/J.BUSHOR.2022.03.002.

[3] K.K. Mak, M.R. Pichika, Artificial intelligence in drug development: present status and future prospects, Drug Discovery Today 24 (3) (March 2019) 773–780, https://doi.org/10.1016/J.DRUDIS.2018.11.014.

[4] M. Sahu, R. Gupta, R.K. Ambasta, P. Kumar, Artificial intelligence and machine learning in precision medicine: a paradigm shift in big data analysis, Progress in Molecular Biology and Translational Science (April 2022), https://doi.org/10.1016/BS.PMBTS.2022.03.002.

[5] M. Schmidt, et al., Learning experience design of an mHealth intervention for parents of children with epilepsy, International Journal of Medical Informatics 160 (April 2022) 104671, https://doi.org/10.1016/J.IJMEDINF.2021.104671.

[6] P. Rani, K. Dutta, V. Kumar, Artificial intelligence techniques for prediction of drug synergy in malignant diseases: past, present, and future, Computers in Biology and Medicine 144 (May 2022) 105334, https://doi.org/10.1016/J.COMPBIOMED.2022.105334.

[7] S. He, L.G. Leanse, Y. Feng, Artificial intelligence and machine learning assisted drug delivery for effective treatment of infectious diseases, Advanced Drug Delivery Reviews 178 (November 2021) 113922, https://doi.org/10.1016/J.ADDR.2021.113922.

[8] M. Hervey, Harnessing AI in drug discovery without losing patent protection, Drug Discovery Today 25 (6) (June 2020) 949–950, https://doi.org/10.1016/J.DRUDIS.2020.03.007.

[9] C. Wong, AI tool could be used to make tests to spot new drugs fast, New Scientist 252 (3361) (November 2021) 13, https://doi.org/10.1016/S0262-4079(21)02051-0.

[10] R. Wieder, N. Adam, Drug repositioning for cancer in the era of AI, big omics, and real-world data, Critical Reviews in Oncology 175 (July 2022) 103730, https://doi.org/10.1016/J.CRITREVONC.2022.103730.

[11] J.D. Piette, et al., Artificial Intelligence (AI) to improve chronic pain care: evidence of AI learning, Intelligence-Based Medicine 6 (January 2022) 100064, https://doi.org/10.1016/J.IBMED.2022.100064.

[12] A. Čartolovni, A. Tomičić, E. Lazić Mosler, Ethical, legal, and social considerations of AI-based medical decision-support tools: a scoping review, International Journal of Medical Informatics 161 (May 2022) 104738, https://doi.org/10.1016/J.IJMEDINF.2022.104738.

[13] H.M. Tsai, A mechanistic approach to the diagnosis and management of atypical hemolytic uremic syndrome, Transfusion Medicine Reviews 28 (4) (October 2014) 187–197, https://doi.org/10.1016/J.TMRV.2014.08.004.

[14] E. González-Esteban, P. Calvo, Ethically governing artificial intelligence in the field of scientific research and innovation, Heliyon 8 (2) (February. 2022) e08946, https://doi.org/10.1016/J.HELIYON.2022.E08946.

[15] Z. Latinovic, S.C. Chatterjee, Achieving the promise of AI and ML in delivering economic and relational customer value in B2B, Journal of Business Research 144 (May 2022) 966–974, https://doi.org/10.1016/J.JBUSRES.2022.01.052.

[16] K. Chakravarty, V. Antontsev, Y. Bundey, J. Varshney, Driving success in personalized medicine through AI-enabled computational modeling, Drug Discovery Today 26 (6) (June 2021) 1459–1465, https://doi.org/10.1016/J.DRUDIS.2021.02.007.

Chapter 7

Review of fog and edge computing–based smart health care system using deep learning approaches

Mamata Rath[1], Subhranshu Sekhar Tripathy[1], Niva Tripathy[1], Chhabi Rani Panigrahi[2] and Bibudhendu Pati[2]

[1]*Department of Computer Science and Engineering, DRIEMS (Autonomous), Cuttack, Odisha, India;* [2]*Department of Computer Science, Rama Devi Women's University, Bhubanewar, Odisha, India*

1. Introduction

Fog-based technology and the currently well-demanding Internet of things (IoT) are excellent technologies for smart health care applications and security models in the present wireless network communication environment. Applications utilizing deep learning and artificial intelligence (AI) have also received a lot of praise in recent years. The situation changed even further with the start of the COVID-19 pandemic. People saw a rapid digital transition throughout the crisis in both rural areas and smart cities, as well as the adoption of disruptive technology across several industries.

One of the major sectors that benefited greatly from the preface of disruptive expertise was healthcare. Simulated aptitude, engine knowledge, and profound knowledge techniques are now essential components of life. Deep learning has a significant impact on the healthcare industry and has made it possible to improve patient monitoring and diagnosis. The IoT digitized devices in many ways, particularly in business and healthcare, and they continued to exchange a wealth of crucial data for the efficient implementation of healthcare systems. Instead of using local servers to handle data packages, cloud computing relies on sharing computer resources. Through the use of cloud computing, we have come to realize that all data-related processing and calculations are carried out in the cloud. Previously, we used our computer server or another server for the computing work, but today, the same computing can be done in the cloud. In a typical architecture for a health care application, Fig. 7.1 shows the functioning of the body sensing stratum, the fog stratum, and the cloud stratum. Every company, endeavor, and endeavor is transitioning to cloud computing with the added benefit of fog computing [1].

A software uses machine learning (ML) algorithms to carry out prediction task and make choices from the given data. Big data approaches and ML collaborate closely to harvest vast amounts of unstructured data for precise and insightful insights [2]. ML has made a significant contribution to epidemiologists' understanding of the dangers of infectious illnesses. Electronic health systems have been altered by ML applications as well [3]. Choosing the best ML model to handle the diagnostics is a significant challenge for the healthcare sector [4,5]. Data from electronic health record apps are a topic of discussion. In order to anticipate the diseases of patients, ML aids in the creation of a trained model [6]. By utilizing ML algorithms, one may forecast a patient's prognosis and identify patterns in large amounts of data [7]. For the early detection and diagnosis of diseases, ML approaches employ data-mining algorithms. A paradigm for a secure health information system has been put forth by researchers [8] that uses ML and security procedures to safeguard patient data. The effectiveness of algorithms for forecasting diseases has been examined, and DM approaches have been investigated to build projecting models using unremitting renal data sets [9]. Researchers have used the Naive Bayes classification techniques to forecast diseases. The nave Bayes method is appropriate for using big data, or enormous data sets, in experimental settings [10]. Expert systems created with AI and ML, according to these authors [11], are useful for diagnosing patient diseases. A disease prediction system based on fuzzy-based methodologies has been put forth by others [12]. Sentiment analysis is carried through using unsupervised ML algorithms on social media messages.

Deep Learning in Personalized Healthcare and Decision Support. https://doi.org/10.1016/B978-0-443-19413-9.00012-6

Authors	Used Technique	Proposal / Approach
H. Harutyunyan,etal.[12]	MLclassification	Theauthorsproposedclinicalpredictionbenchmarks.Thetaskincludes(1)modelingrisk ofmortality,(2)detectingphysiologicdecline,and(3)phenotypeclassification
		Informationanalysisofthedatafromelectronichealthrecordsintotheappropriatedatafo rmatisdonetoimproveclinicalmachinelearningtasks
W.WengandP.Szolovits [13]	Machine Learning	Theseauthorsperformedclassificationofcellimagestodeterminetheadvancementands everityofbreastcancerusingaartificialneuralnnetwork.Inaddition,theauthorsstudieda nddemonstrateduseoftheBayesianMLtechniquefordiagnosingAlzheimerdisease
S.Das[14]	Artificial neural network	Anexperimentalapproachisfollowedbytheauthorstoanalyze the role of big data in the healthcare industry.They proposed a novel design of smart and secure HISusingML.handlingbigdatafromthemedicalindustry Authors have proposed the convergence of blockchainand machine learning for achieving accurate results inthehealthcaresectoranddecisionmaking.
		AnMLalgorithmisimplementedfortheeffectivepredictionofachronicdiseaseoutbrea k.TheauthorsproposedanewCNN-basedmodelfordiseaseriskpredictionalgorithmsusingstructuredandunstructureddata
P.Kaur,etal.[7]	Smarthealthcareinforma tion systemusingML	MLalgorithmsforprocessingbigdatainhealthcarearediscussedbasedonthestudyofthe already-publishedbigdataarchitectures

FIGURE 7.1 Tabular view of research carried out using machine learning techniques.

2. Literature review

According to a study [13], the IoT explosion is making smart fitness (s-fitness) an attractive new paradigm. It can provide accurate disease predictions and make healthcare more enjoyable. Information sanctuary and user privacy concerns, however, continue to be issues that need to be resolved. Ciphertext coverage feature-based encryption (CP-ABE) approaches increase challenges, including hefty transparency and quality seclusion of the give up consumers [14]. They are a high capability and prospective method to steady IoT-oriented s-fitness programs. In order to effectively change the admission policy and consumer attribute set into corresponding vectors of lower duration at the same time as other methods result in duplicated data, an optimized vector transformation technique is first provided to address such limitations. In-depth analysis and examination of the fitness IoT structure and related implementation technologies are presented in research work [15], with a focus on both theoretical and practical aspects. Laptop simulations are used to evaluate the viability and performance of the QoS architecture provided in [15].

The ability to track and control physical activity and healthcare is unexpectedly evolving with the IoT and system mastering-based structures [16]. It focused on the idea of a real-time, fog-centric, advanced wearable and IoT platform for omnipresent health and health analysis in a cutting-edge gym setting. Based on body vitals, frame motion, and fitness-related statistics, the suggested framework aims to resource the health and fitness business. The framework is intended to assist doctors, athletes, and running shoes in the translation of many biological signals and lift messages in the event of any health risks. Researchers put forth a method for compiling and analyzing exercise-specific data that can be utilized to gauge the impact of exercise on an athlete's fitness and work as a machine that makes recommendations to aspiring athletes. Researchers tested the validity of the proposed framework by providing a 6-week exercise schedule that included 6 days a week of training focused on all muscular tissues along with a time frame for recovering. Researchers measured the athlete's mobility using a 3-D accelerometer and recorded the athlete's ECG, coronary heart rate, heart rate variability, and breath rate. Two modules make use of the study's records that were gathered. A health quarter module that uses frame vitals data to classify athletes' fitness levels into several groups.

Hzone module is responsible for identifying and alerting to health concerns. Amazingly, the Hzone module can identify an athlete's physical condition with 97% accuracy. A fitness center activity recognition (GAR) module is used to identify training activity in real-time using body movements and frame vital information. The GAR module's goal is to compile and evaluate exercise-specific data. The athlete impartial version of the GAR module, which is based on muscle arrangement, produced accuracy levels above 89%.

Research in [17] asserts that the greatest possible usage of the IoT, particularly cutting-edge wearable, will be essential for enhancing medical care, delivering comfort to patients, and raising hospital control levels. However, because of the limitations of communication protocols, there is currently no single architecture that can link all intelligent objects in smart hospitals, but the narrowband IoT is allowing this to happen (NB-IoT). In light of this, researchers suggest introducing area computing to address the need for delay in clinical technique and recommending an architecture to connect advanced objects in advanced hospitals based entirely on NB-IoT. As a case take a look at, researchers increase an infusion tracking device to display the real-time drop price and the quantity of closing drug in the course of the intravenous infusion. Finally, researchers talk the demanding situations and destiny directions for building a smart health facility through connecting intelligent things.

In urbanized areas [18], one of the most crucial aspects of health control is hospital control, which is engaged. Accurate patient fitness information analysis is required to deliver outstanding scientific care for ill patients. Data collection is therefore crucial for tracking patient fitness. Large-scale IoT devices, which are well-known to be resource-constrained, are used to achieve this. Light-weight index generation is required from this. Additionally, when professional medical technologies advance, the hospital administrator must select a wide variety of exceptional professional medical doctors. To do an accurate analysis and provide each patient with a green healing timetable, they require the shared patient fitness information. The impacted person's fitness information does contain a lot of confidential information, and the privacy of facts is maintained. Researchers promote a new gadget that can examine traceable patient health records for medical institution control in developed towns in this newsletter. In this technology, the system administrator transmits the encrypted patient fitness data to special doctors seated at hospital beds. Each doctor, as it should be, may identify a patient with a certain condition from the patient health tracking records. The functions of unauthorized search question blocking off and internal malicious consumer tracing are created to stop the leakage of patient health records. The performance assessment demonstrates that its equipment is useful for portable data collection tools.

3. Healthcare using artificial intelligence

AI is utilized in the healthcare industry to solve problems by employing ML techniques. AI has various uses in the healthcare industry. To increase productivity and precision, simulated aptitude involves the discovery of hereditary regulations and the use of medical robots. Partial medical and formless data arrangement are examples of human errors that AI is effectively minimizing and supporting in the diagnosis process [19]. A different area of AI someplace investigates and expansion be accelerating the identification, and creation of new medications is drug design and development. Neuroscience and the chief region of AI in medication development are useful. Additionally, clinical trial patients' features are tracked using AI, and deep learning is employed to retrieve data pertaining to medical sciences.AI is a data science that integrates pre-existing healthcare applications through a cloud service [20].

Telemedicine, which enables in-person doctor consultations, is another application of AI. For patient diagnosis, experimental judgment maintain, extrapolative analytics, etc., AI leverages ML approaches. For precise patient diagnosis, Google's subterranean intellect and Watson systems leverage ML and data mining methods [21,22]. Robotic surgical assistance has been made possible by the implementation of AI-enabled virtual reality approaches. Fig. 7.1 provides an overview of the work in table form that various authors have done over the past 5 years utilizing ML approaches in the healthcare sector. Fig. 7.2 provides an overview of the work in table form that various authors have done in the field of ML, Cloud, and Swarm Intelligence.

The development of wearables that give data-driven mental health therapy [23] is facilitated by AI techniques. Sensors in the device monitor alter in sympathy speed [24], warmth, and progress. The fields of AI and psychology are partnering on investigate and progress [25] to provide personalized healthcare and accurate patient care [26]. The sentiment gathered by the digital sensing devices antenna is detected and studied with the aid of AI algorithms and psychological data.

Fig. 7.3 depicts the systematic functionality of the one chatbot application. The following can be accomplished with the aid of AI techniques: (1) illness prophecy, (2) diagnosis, (3) therapy, (4) prediction, and (5) diagnosis estimate. AI techniques are frequently utilized for studies linked to stroke. NLP is a method for obtaining valuable information from unstructured data that closely collaborates with ML. The information obtained is useful for structured medical data that are already readily available. Semantic web is also produced by NLP techniques in conjunction with ML [27].

In addition to real-time clinical support and decision optimization, AI and ML techniques help enhance and streamline healthcare operations. Clinical and important trends are found using ML algorithms. Screening data, image categorization, classification, data production, attribute collection, and clinical data set prediction are among the tasks carried out utilizing

Authors	Used Technique	Proposal / Approach
A.Abdelaziz,etal.[18]	Cloud-enabled environment	AnMLalgorithmisimplementedfortheeffectivepredictionofachronicdiseaseoutbreak.Theauthorsproposeda newCNN-basedmodelfordiseaseriskpredictionalgorithmsusingstructuredandunstructureddata
		MLalgorithmsforprocessingbigdatainhealthcarearediscussedbasedonthestudyofthealready-publishedbigdataarchitectures
J.Wiens and E.S.Shenoy [2]	ML tools and techniques	A new modelisproposedfor healthcareservicesbasedonacloudenvironmentusingparallelparticleswarmoptimizationtooptimizethevirtua lmachineselection.Inaddition,anewmodelforchronickidneydiseasediagnosisandpredictionisproposed
	Swarm optimization	TheauthorsreviewedMLtechniques,andtheirapplicationstotransformpatientriskstratificationinthemedicalfi eld. epidemiologists and to reduce the spreadofhealthcare-associatedpathogens
A.Qayyumetal.[19]		TheauthorsusedMLtechniquesforthepredictionofcardiacarrestfrom1-Dheartsignalswithcomputer-aideddiagnosisusingmultidimensionalmedicalimages
		MLmodelprinciplesarediscussedforclinicalapplicationssuchasimagesegmentationforradiationtherapyplan ningandmeasuringcardiacparametersfromechocardiography
P.H.C.Chen,etal.[20]	MLmodelforimagesegm entation	MLtechniquesarereviewed,especiallyfor healthcare

FIGURE 7.2 Tabular view of research carried out using ML, Cloud, and swarm.

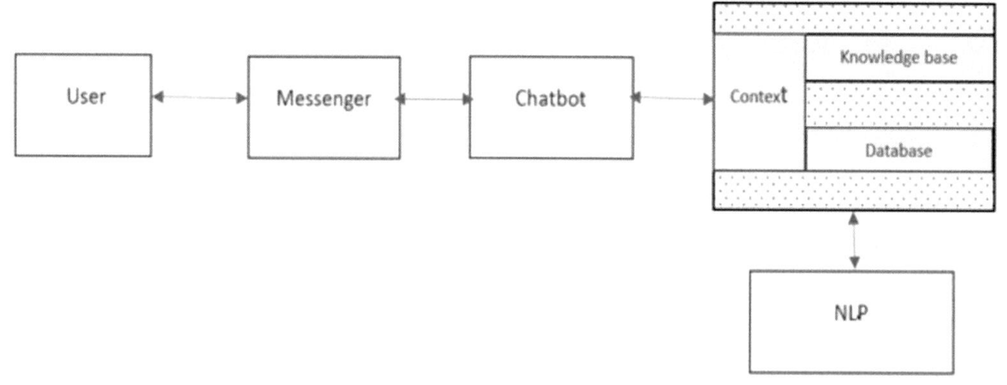

FIGURE 7.3 Artificial intelligence–based systematic functionality of chat bot design in healthcare.

ML and deep learning. Given that ML approaches can handle enormous datasets, the availability of electronic health records presents a significant opportunity for the development of new drugs and the enhancement of medical treatment [28]. In order to safeguard patients from life-threatening hazards and to ensure their safety and fitness, wearable technology, including smartphones, has been developed [29].

Apps based on artificial intelligence are being created and used in wearable devices like watches. The following are accomplished for effective clinical decision-making and preventing in accurate diagnosis with the aid of AI apps: Users receive notifications for abnormal heartbeats, diabetic retinopathy picture analysis, and real-time MRI image analysis utilizing cutting-edge ML techniques, among other things [30]. (4) Radiologists and doctors have access to medical diagnostic techniques that are effective predictors of heart failure, brain stroke, etc. For the diagnosis of eye illnesses, ML algorithms have been created, as well as speech therapy employing AI and NLP techniques [31]. Another application of AI that benefits patients' comfort is telemedicine. Patients receive advice via cellphones, schedule appointments online, and consent to blood testing [23,24].

4. Efficient health care system with improved performance

For intelligent healthcare, it is desirable to use IoT, AI, and factual discipline junction technologies. The healthcare industry has benefited greatly from ML. Personalized healthcare, EHR, sickness calculation, diagnostic drawing, and hazard assessment are some of the uses of ML in the healthcare industry. Supervised learning and unsupervised learning are the two main learning methods used in ML [32]. Classification and regression algorithms are examples of supervised learning methods. The healthcare framework is depicted in next section.

At the registration counter, the patient is initially asked about their symptoms. The IoT enables proactive patient communication and remote health monitoring with alarms [27]. The IoT is useful in devices that automate statistics torrent analysis and enable isolated tools pattern. The projected system uses connected sensors to gather patient data from wearable devices. For figures storeroom and systematic dispensation, the gathered data are kept on the cloud. The cloud can be used to get patient health records, which can then be given to a doctor for early disease identification and diagnosis. The patient data can be analyzed using ML classification and clustering techniques to produce the trained model for future predictions [33]. NLP and online ontology are used in ML, an additional component of AI [34] (Fig. 7.4).

Fig. 7.5 demonstrates predictive analysis and ML workflow. Both supervised and unsupervised learning methods are used in ML. Precategories data are used in supervised learning techniques. Regression and classification techniques make up the majority of this. The supervised learning method's goal is to make prophecy and build extrapolative sculpt [35]. Regression, decision trees, and random forests are the supervised ML algorithms for continuous data [36,37].

4.1 Dataset

The statistics deposit is initially gathered from the communal sphere, namely the "UCI Machine Learning Repository." The experiment's dataset relates to the risk calculation for initial phase of diabetes. The statistics collection is topical and includes patients who are recently diagnosed with diabetes or who are at risk of developing the disease. The data collection consists of 520 cases and has 17 attributes. The characteristics of the dataset are multivariate in nature. Table 1.2 displays a sample dataset.

Data preprocessing—Preprocessing the data is the main challenge after data gathering. Outliers must be eliminated in data preprocessing since the acquired data contain missing values, which makes it necessary to do so in order to improve classification outcomes. To improve the performance of the ML algorithms, extreme data are found and absent principles are eliminated [38] (Fig. 7.6).

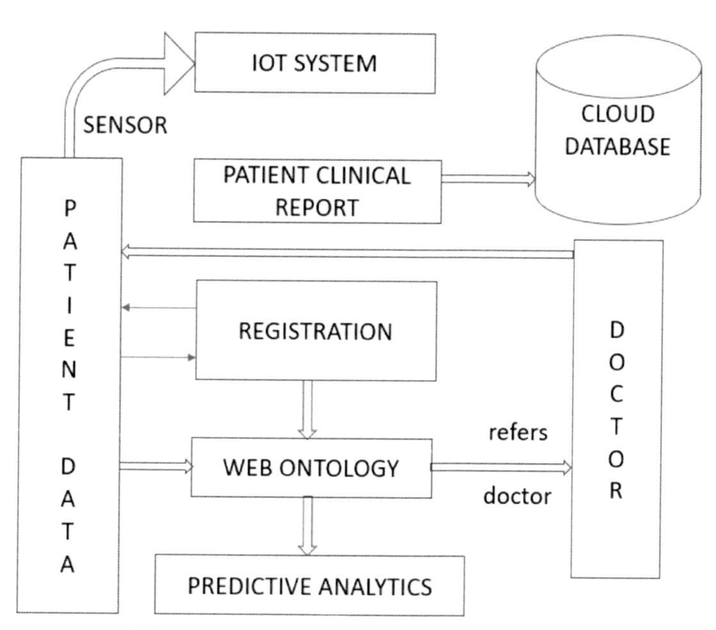

FIGURE 7.4 Suggested approach in smart healthcare.

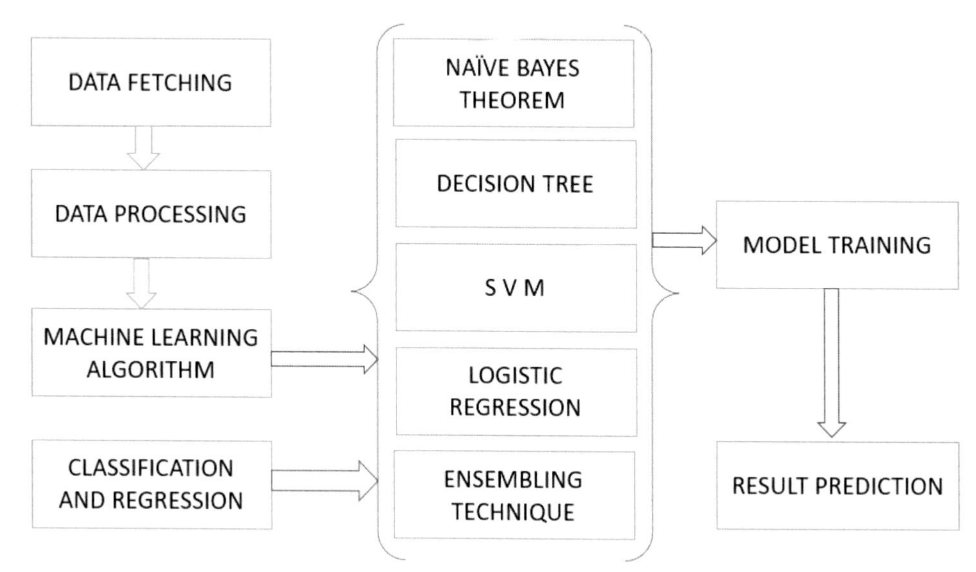

FIGURE 7.5 Predictive analysis and machine learning workflow.

Age	Gender	Polyuria	Polydipsia	Suddenweightloss	Weakness	Polyphagia	Genitalthrush	Visualblurring
40	Male	No	Yes	No	Yes	No	No	No
58	Male	No	No	No	Yes	No	No	Yes
41	Male	Yes	No	No	Yes	Yes	No	No
45	Male	No	No	Yes	Yes	Yes	Yes	No
60	Male	Yes	Yes	Yes	Yes	Yes	No	Yes
55	Male	Yes	Yes	No	Yes	Yes	No	Yes

Age	Gender	Itching	Irritability	Delayedhealing	Partialparesis	Musclestiffness	Alopecia	Obesity
40	Male	Yes	No	Yes	No	Yes	Yes	Yes
58	Male	No	No	No	Yes	No	Yes	No
41	Male	Yes	No	Yes	No	Yes	Yes	No
45	Male	Yes	No	Yes	No	No	No	No
60	Male	Yes	Yes	Yes	Yes	Yes	Yes	Yes
55	Male	Yes	No	Yes	No	Yes	Yes	Yes

FIGURE 7.6 Snapshot of the information set.

Fig. 7.7 portrays comparison of accuracy among different ML classifiers. A supervised ML approach for data classification and regression is the decision tree. The data collection is subjected to if-then rules to create a decision tree, which has leaf nodes and decision nodes. The input features and target class in a decision tree are used to calculate the likelihood of an event [39]. To split the tree further, get the optimal outcome, and make predictions using the feature with the largest information gain, each node's information gain is determined. The decision tree's construction requires consideration of the entropy calculation.

Fig. 7.8 illustrates comparison of training time among different ML classifiers. To reduce the likelihood of outcomes that are either over appropriate or under decent, the data position is crack into guidance and taxing phases in this stage. Fig. 7.9 displays confusion matrix evaluation of various classifiers (Figs. 7.10 and 7.11).

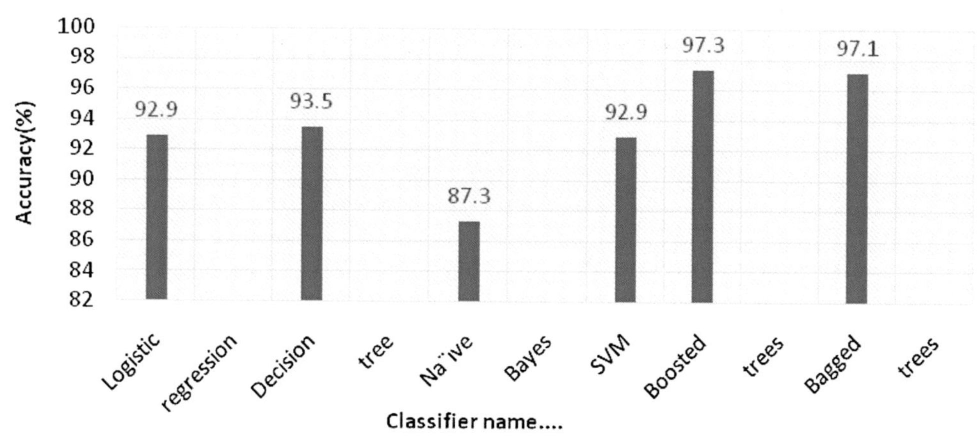

FIGURE 7.7 Comparison of accuracy among different machine learning classifiers.

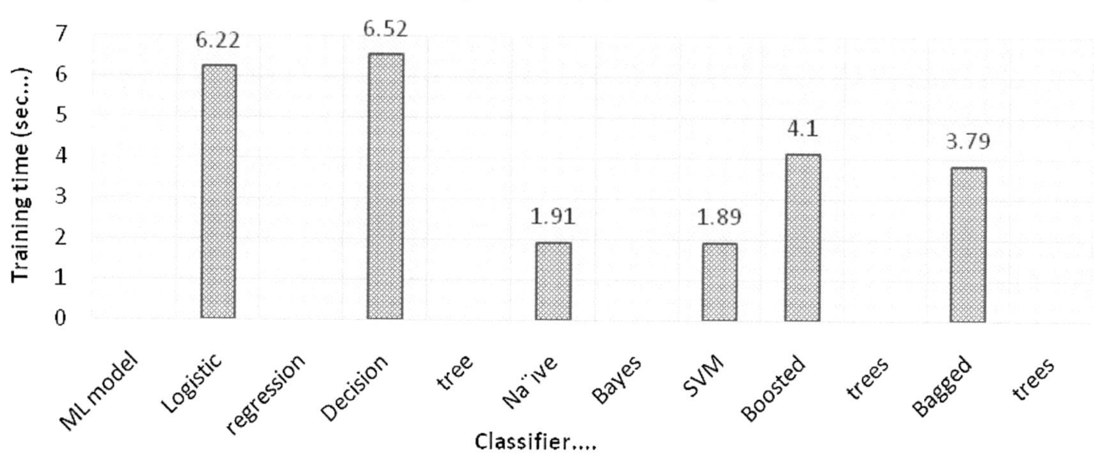

FIGURE 7.8 Comparison of training time among different machine learning classifiers.

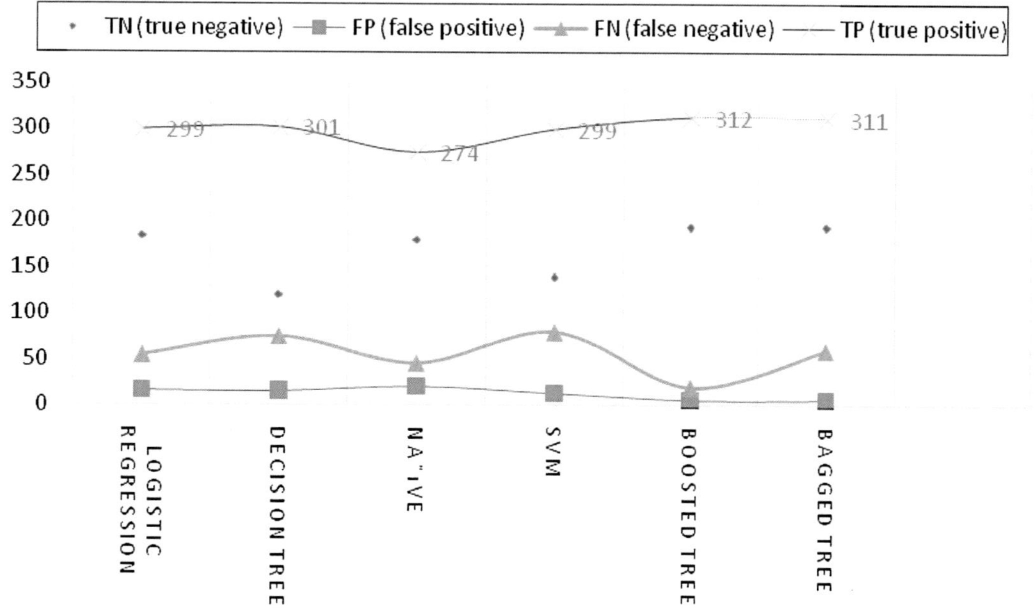

FIGURE 7.9 Confusion matrix evaluation of classifiers.

ML model	TPR (true positive rate)	FPR (falsepositiverate)	NPV (negativepredictive value)	TNR (truenegativerate)	PPV (positivepredictivevalue)	FNR (falsenegativerate)	F-measure
Logistic regression	93.4	8	89.7	92	94.9	6.56	94.14
Decision tree	98	7.5	90.6	92.5	95.2	5.93	94.63
Na¨ıve Bayes	76	10	79.6	91	93.1	14.37	89.19
SVM	90	6.5	88.6	93.5	95.8	7.43	94.12
Boosted trees	97.5	3	96.03	97	98.1	2.5	97.8
Bagged trees	97.1	3	95.5	97	98.1	2.8	97.6

FIGURE 7.10 Confusion matrix values of multiple machine learning classifiers.

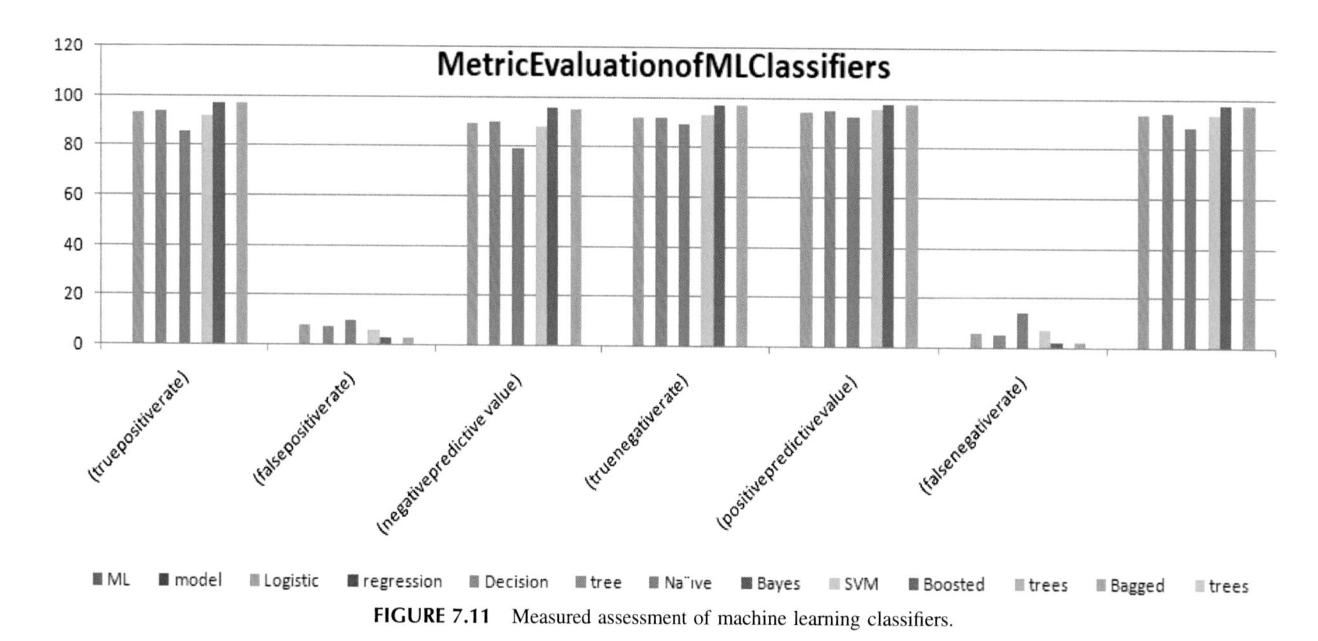

FIGURE 7.11 Measured assessment of machine learning classifiers.

Fig. 7.12 demonstrates proportional examination of training time among three approaches: ML model proposed in [8], ANN model planned in [21], and blockchain-based ML in [22].

Fig. 7.13 demonstrates comparative analysis of accuracy among three approaches: ML model proposed in [8], ANN model planned in [21], and blockchain-based ML in [22].The use of the clustering technique is another illustration of an ML method that may be capable of being applied in the medical industry. This is an unsupervised learning technique, and unlike supervised learning techniques, it does not offer the input label or the end values. Clustering examines statistics substance devoid of a notorious group sticker, in contrast to organization and forecast, which evaluate group brand statistics substance.

The training information does not have class labels, but clustering can be used to create them. The substance within a group contain an elevated degree of resemblance to one another, but they differ greatly from objects in other clusters due to the manner the clusters are formed. The iterative, data-partitioning technique k-means clusters use the squared Euclidean distance method. One method that can be used to group diabetic patients, whether they have diabetes or not, is the k-means clustering technique.

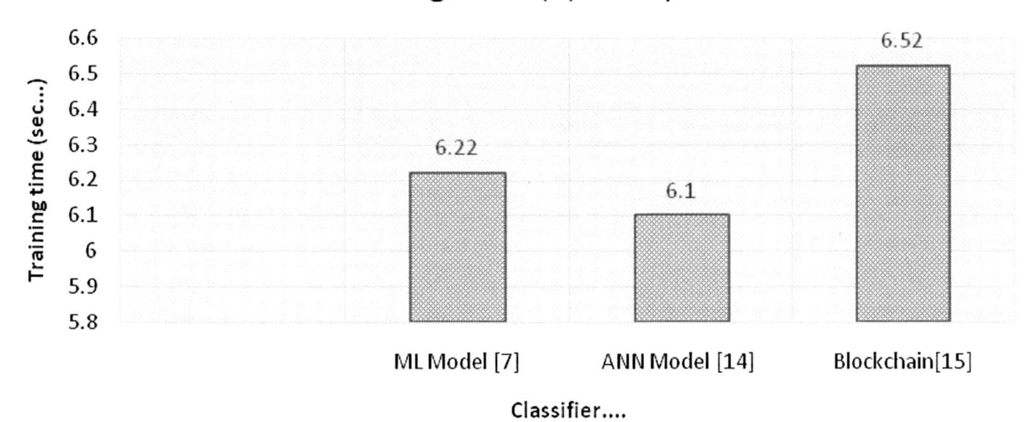

FIGURE 7.12 Proportional examination of training time.

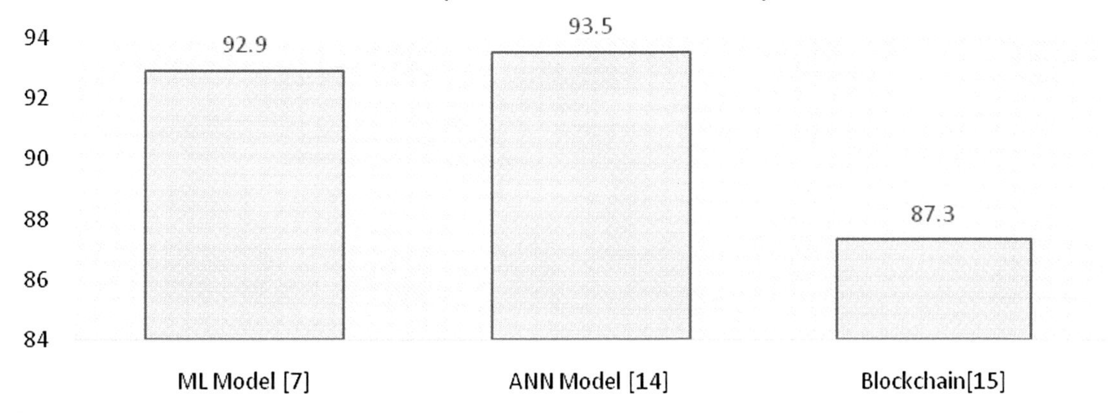

FIGURE 7.13 Comparison of accuracy.

FN	Falsenegative
FNR	Falsenegativerate
FP	Falsepositive
FPR	Falsepositiverate
IoT	InternetofThings
KNN	K-nearestneighbor
LR	Logisticregression
ML	Machinelearning
NLP	Naturallanguageprocessing
NPV	Negativepredictive value
PCA	Principalcomponentanalysis

FIGURE 7.14 Abbreviations used in this chapter.

5. Conclusion

Technologies utilizing AI and ML closely collaborate with one another. While ML algorithms allow the machine to understand complicated correlations and build patterns for successful decision-making, AI aids in the extraction of relevant information from massive amounts of unstructured data. Current technologies as illustrated here are already in use in several industries, but their entry into the healthcare industry is helping to preserve electronic health data and provide clinical support for patients.

Fig. 7.14 shows the abbreviations used in this chapter. The writers of this chapter have suggested a framework for smart healthcare leveraging emerging technology. It is explained how the junction of IoT, cloud-computing, AI, and ML will enable an elegant medical diagnosis project to deliver experimental assistance at patients' doors. For the purpose of early diabetes patient detection, ML categorization methodologies are executed, and presentation parameters are assessed. Additionally, ML approaches can be applied to similar patient datasets to create trained models that can make precise predictions, provide concern to diseased people, and transform conventional healthcare into well turned-out healthcare projects.

References

[1] S.S. Tripathy, R.K. Barik, D.S. Roy, Secure-M2FBalancer: a secure mist to fog computing-based distributed load balancing framework for smart city application, in: S. Dhar, S.C. Mukhopadhyay, S.N. Sur, C.M. Liu (Eds.), Advances in Communication, Devices and Networking, Lecture Notes in Electrical Engineering, vol 776, Springer, Singapore, 2022, https://doi.org/10.1007/978-981-16-2911-2_30.

[2] A.L. Beam, I.S. Kohane, Big data and machine learning in health care, JAMA 319 (13) (2018) 1317–1318.

[3] J. Wiens, E.S. Shenoy, Machine learning for healthcare: on the verge of a major shift in health care epidemiology, Clinical Infectious Diseases 66 (1) (2018) 149–153.

[4] M.A. Ahmad, C. Eckert, A. Teredesai, Interpretable machine learning in healthcare, in: Proceedings of The 2018 ACM International Conference on Bioinformatics, Computational Biology, and Health Informatics, 2018, pp. 559–560.

[5] A. Esteva, A. Robicquet, B. Ramsundar, V. Kuleshov, M. DePristo, K. Chou, J. Dean, A guide to deep learning in healthcare, Nature Medicine 25 (1) (2019) 24–29.

[6] P. Chowriappa, S. Dua, Y. Todorov, Introduction to machine learning in healthcare informatics, in: Machine Learning in Healthcare Informatics, Springer, Berlin,Heidelberg, 2014, pp. 1–23.

[7] A. Dhillon, A. Singh, Machine learning in healthcare data analysis: a survey, Journal of Biology and Today's World 8 (6) (2019) 1–10.

[8] P. Kaur, M. Sharma, M. Mittal, Big data and machine learning-based secure healthcare framework, Procedia Computer Science 132 (2018) 1049–1059.

[9] A. Charleonnan, T. Fufaung, T. Niyomwong, W. Chokchueypattanakit, S. Suwannawach, N. Ninchawee, Predictive analytics for chronic kidney disease using machine learning techniques, in: 2016 Management and Innovation Technology International Conference(MITicon), IEEE, October 2016, p. MITe80.

[10] R. Venkatesh, C. Balasubramanian, M. Kaliappan, Development of big data predictive analytics model for disease prediction using machine learning technique, Journal of Medical Systems 43 (8) (2019) 1–8.

[11] I. Kononenko, Machine learning for medical diagnosis: history, state of the art, and perspective, Artificial Intelligence in Medicine 23 (1) (2001) 89–109.

[12] M. Nilashi, O. Binbrahim, H. Ahmadi, L. Shahmoradi, Ananalytical method for diseases prediction using machine learning techniques, Computers & Chemical Engineering 106 (2017) 212–223.

[13] J. Sun, H. Xiong, X. Liu, Y. Zhang, X. Nie, R.H. Deng, Lightweight and privacy-aware fine-grained access control for IoT-oriented smart health, IEEE Internet of Things Journal 7 (2020) 6566–6575, 2327-4662.

[14] M. Rath, B. Pattanayak, Technological improvement in modern health care applications using Internet of Things (IoT) and proposal of novel health care approach, International Journal of Human Rights in Healthcare (2018) 2056–4902, https://doi.org/10.1108/IJHRH-01-2018-0007 (Scopus Indexed).

[15] H. Yu, Z. Zhou, Optimization of IoT-based artificial intelligence assisted telemedicine health analysis system, IEEE Access 9 (2021) 85034–85048, 2169-3536.

[16] A. Hussain, K. Zafar, A.R. Baig, Fog-centric IoT based framework for healthcare monitoring, management and early warning system, IEEE Access 9 (2021) 74168–74179, 2169-3536.

[17] H. Zhang, J. Li, B. Wen, Y. Xun, J. Liu, Connecting intelligent things in smart hospitals using NB-IoT, IEEE Internet of Things Journal 5 (2018) 1550–1560, 2327-4662.

[18] R. Zhou, X. Zhang, X. Wang, G. Yang, N. Guizani, X. Du, Efficient and traceable patient health data search system for hospital management in smart cities, IEEE Internet of Things Journal 8 (2021) 6425–6436, 2327-4662.

[19] H. Harutyunyan, H. Khachatrian, D.C. Kale, A. Galstyan, Multi task learning and benchmarking with clinical time series data, Scientific Data 6 (2019).

[20] W. Weng, P. Szolovits, Representation Learning for Electronic Health Records, ArXiv, 2019 abs/1909.09248.

[21] N.G. Maity, S. Das, Machine learning for improved diagnosis and prognosis in healthcare, in: 2017 IEEE Aerospace Conference, IEEE, 2017, pp. 1–9.
[22] S. Vyas, M. Gupta, R. Yadav, Converging blockchain and machine learning for healthcare, in: 2019 Amity International Conference on Artificial Intelligence(AICAI), IEEE, 2019, pp. 709–711.
[23] R.J. Gillies, P.E. Kinahan, H. Hricak, Radiomics: images are more than pictures, they are data, Radiology 278 (2) (2016) 563–577.
[24] G. Karakulah, O. Dicle, O. Kosaner, A. Suner, C.¸.C. Birant, T. Berber, S. Canbek, Computer based extraction of phenotypic features of human congenital anomalies from the digital literature with natural language processing techniques, in: MIE, 2014, pp. 570–574.
[25] A.M. Darcy, A.K. Louie, L.W. Roberts, Machine learning and the profession of medicine, JAMA 315 (6) (2016) 551–552.
[26] M. Chen, Y. Hao, K. Hwang, L. Wang, L. Wang, Disease prediction by machine learning over big data from healthcare communities, IEEE Access 5 (2017) 8869–8879.
[27] F. Jiang, Y. Jiang, H. Zhi, Y. Dong, H. Li, S. Ma, Y. Wang, Artificial intelligence in healthcare: past, present and future, Stroke and Vascular Neurology 2 (4) (2017) 230–243.
[28] G. Manogaran, D. Lopez, A survey of big data architectures and machine learning algorithms in healthcare, International Journal of Biomedical Engineering and Technology 25 (2e4) (2017) 182–211.
[29] A. Abdelaziz, M. Elhoseny, A.S. Salama, A.M. Riad, A machine learning model for improving healthcare services on cloud computing environment, Measurement 119 (2018) 117–128.
[30] A. Qayyum, J. Qadir, M. Bilal, A. Al-Fuqaha, Secure and Robust Machine Learning for Healthcare: A Survey, arXiv, 2020. PreprintarXiv:2001.08103.
[31] P.H.C. Chen, Y. Liu, L. Peng, How to develop machine learning models for healthcare, Nature Materials 18 (5) (2019) 410.
[32] Rani Panigrahi C, Pati B, Rath M, Buyya R. "Computational Modeling and Data Analysis in COVID-19 Research (Emerging Trends in Biomedical Technologies and Health Informatics)". CRC Press, ISBN-978-0-367-68036-7.
[33] S.P. Somashekhar, R. Kumarc, A. Rauthan, K.R. Arun, P. Patil, Y.E. Ramya, Abstract S6-07: double blinded validation study to assess performance of IBM artificial intelligence platform, Watson for Oncology in Comparison with Manipal Multidisciplinary Tumour Boarde First Study of 638 Breast Cancer Cases S6-07 (2017).
[34] A. Esteva, B. Kuprel, R.A. Novoa, J. Ko, S.M. Swetter, H.M. Blau, S. Thrun, Dermatologist-level classification of skin cancer with deep neural networks, Nature (London) (7639) (2017) 115–118.
[35] C.E. Bouton, A. Shaikhouni, N.V. Annetta, M.A. Bockbrader, D.A. Friedenberg, D.M. Nielson, A.G. Morgan, Restoring cortical control of functional movement in a human with quadriplegia, Nature 533 (7602) (2016) 247–250.
[36] S.S. Tripathy, K. Mishra, R.K. Barik, D.S. Roy, A novel task offloading and resource allocation scheme for mist-assisted cloud computing environment, in: S.K. Udgata, S. Sethi, X.Z. Gao (Eds.), Intelligent Systems, Lecture Notes in Networks and Systems, vol 431, Springer, Singapore, 2022, https://doi.org/10.1007/978-981-19-0901-6_10.
[37] S.S. Tripathy, D.S. Roy, R.K. Barik, M2FBalancer: a mist-assisted fog computing-based load balancing strategy for smart cities, 2021, pp. 219–233, https://doi.org/10.3233/AIS-210598.
[38] S.E. Dilsizian, E.L. Siegel, Artificial intelligence in medicine and cardiac imaging: harnessing big data and advanced computing to provide personalized medical diagnosis and treatment, Current Cardiology Reports 16 (1) (2014) 441.
[39] G. Battinenni, N. Chintalapudi, F. Amenta, AI chatbot design during an epidemic like the novel coronavirus, MDPI, Healthcare 8 (2) (2020).

Chapter 8

Deep learning in healthcare: opportunities, threats, and challenges in a green smart environment solution for smart buildings and green cities—Towards combating COVID-19

Palak Maurya, Anurag Verma and Ranitesh Gupta
Electrical Engineering Department, Institute of Engineering and Technology, Lucknow, Uttar Pradesh, India

1. Introduction

1.1 Major findings and motivation

Global environmental challenges, including the loss of biodiversity, climate variability, atmosphere and water contamination, and the management of waste are all intertwined with the pandemic. Governments face problems in terms of environmental sustainability and well-being as a result of the current crisis. In the end though, the recovery is a chance to make the world a better place.

To control and preserve forests in a sustainable manner, governments and organizations and agencies must (1) support sustainable forest management best practices, including public engagement; (2) support a single health system that brings both experts as well as policy makers in forest, natural deposit, cultivation, livestock farming, and health service and nourishment to create sufficient well-being resolutions; and (3) support opportunities for forest communities to earn a living.

> *The coronavirus has a rapid and devastating effect. However, there is another major issue on the horizon: the planet's looming environmental disaster. Biodiversity is rapidly dwindling. Climate disruption is rapidly approaching a tipping point ... The current crisis is a wake-up call like no other. We must transform the current economic rebound to be a genuine opportunity to do the right thing in the future.*
>
> United Nations Secretary-General Antonio Guterres [1].

1.2 The coal crisis creates the need for alternatives [2]

The current problem is also the world's opportunity to break free from coal dependence and push for renewable energy for the long term. India is the second largest importer, consumer, and producer of coal and yet, on October 6, 2021 it had enough coal stock to generate electricity for about 4 days. On the 10th of October, 2021 this number was revised to a 1-day total of coal storage, and as coal produced about 70% of India's electricity, there were widespread fears of massive power outages.

Deep Learning in Personalized Healthcare and Decision Support. https://doi.org/10.1016/B978-0-443-19413-9.00001-1

This shortage threatens India's economic stability after the epidemic and the well-being of its citizens. The current shortage of coal is a long-standing issue of service inequality. With the recent economic growth, electricity consumption has risen by about 17% in the last 2 months. On the supply side, heavy rains have caused disruption and overcrowding of mines has disrupted domestic coal production. It is difficult to address this shortage of imports as global coal prices are rising by 50% and coal exports in India are down over the past 2 years.

Adding to these woes, countries such as China purchase more coal from the world's coal mines, which adds to the pressure on global prices.

1.3 With minor delays caused by COVID-19, renewable volumes at auction continue to break records [3]

The highest ever quantity of extra sustainable capacity, to be connected online between 2021 and 2024, has been produced by thirteen countries in the first half of 2020, totaling nearly 50 GW. In June 2020, China's domestic solar photovoltaic (PV) sales reached 25 GW, reflecting a global trend.

India has produced 11.3 GW of solar electricity and more than 1 GW of air volume at major and regional markets33333, despite the sharp fall in construction activities. This is a return to declining practise that started in the second half of 2019.

2. Green infrastructure measures in the legislature [4]

Emphasize laws and regulations that call for public investment in green infrastructure and encourage private investment in it, especially in urban and metropolitan areas, which are responsible for the great majority of greenhouse gas emissions worldwide. Pass legislation that encourages the creation of jobs in green infrastructure and increases the potential for green infrastructure-related businesses. Offer legislative oversight of the government's efforts to fulfil SDGs like providing global access to safe, complete, and readily available greenery and populated areas.

2.1 Approaches within the direction of a green economy

The green economy promotes economic activities that conserve biodiversity by reducing, reusing, and recycling commodities.

2.1.1 An account of the ongoing power situation in Iceland: a global paradigm [5]

Today, renewable energy makes up almost all electricity used in this small country of 330,000 people. In addition, geothermal energy is used to heat nine out of 10 homes. The story of Iceland's departure from the use of fossil energy could be used as a model for other nations cautious about increasing their share of sustainable energy.

2.1.2 Saudi Arabia's Vision 2030 [6,7]

Saudi Vision 2030 is a strategic plan for Saudi Arabia to diminish its oil dependency, vary its economies, as well as expand sectors of government services, for example, in health, education, construction, food, and tourist areas. Strengthening economic activity as well as investment, expanding oil-free foreign trade, and creating a soft, public image of the state are key objectives. It includes increasing government spending on the armed forces, as well as the production of related equipment and weapons. Prince Mohammed bin Salman revealed the first information on this on April 25, 2016. The board of Ministers charged the National Board for Financial Development and Social Growth with establishing and enforcing key policies and practices to achieve Saudi Arabia Vision 2030.

Oil-based fuel makes up 30%–40% of Saudi Arabia's gross domestic product, without including the share of the economy that is also dependent on the oil supply. Since the 1970s, another agenda of the government has been diminishing reliance on oil.

2.1.3 NEOM [8]

NEOM is the abbreviation for "New Enterprise Operating Model" and "New Future." NEOM is a unique perspective of what the future might look like. It is an endeavor at something unprecedented, as well as there coming a point when the world needs new ideas and solutions. NEOM will be a home to people who have great dreams as well as desire to be a partner in building homes that are new example of self-sufficiency. In the future, $500 billion will be used to cover 10,230 square kilometers northeast of Saudi Arabia, on the border of the Gulf of Aqaba. The venue highlights the importance of NEOM's strategy in Mohammed bin Salman's foreign policy. The goal of this corporation, which is completely controlled by a Public Investment Fund, a sovereign wealth fund, is to develop the NEOM economic zone. The project will be entirely powered by renewable energy sources.

2.1.4 Saudi Arabia, Line [9,10]

The city of Line, stretching for over 170 miles, is completely carbon neutral and sustains 95% of its environment within the NEOM. The town is a part of Saudi Vision 2030, and it is projected to create 380,000 jobs as well as adding $48 billion to the nation's GDP. NEOM, a city costing $500 billion, as a revolution in urban life, is proposed as a pattern of how people can live in peace with the world, where citizens are have free access to a healthy environment.

- maximum travel time of 201 min—a short travel time between communities;
- A city connected through the environment for improving people's health as well as prosperity and ability to live a happy life—refreshing our connection with the environment to improve health and well-being and live a happy life;
- Empowered clean energy—empowered plentiful renewable energy for a sustainable future;
- Empowered high technology that is constantly learning and developing to make life easier for everyone.

3. Employment creation as part of the sustainable recovery

As a major concern, green recovery should address job creation. Green growth will contribute to overall employment. Various "green" industries and jobs provide great opportunities through the creation of sustainable energy, especially solar PV, which utilizes many people per unit as well as power than the output of older fuels [11].

If the international community uses its sustainable energy sources, the International Renewable Energy Agency estimates that sustainable energy could employ over 40 million people by 2050, and the number of jobs in the power sector could reach 100 million by 2050, from −58 million currently. Energy efficiency also offers many benefits.

3.1 Organic agriculture has the ability to create jobs [12]

Several studies have shown that the demand for labor per hectare on organic crops is higher than on conventional farms due to the production process that requires more workers. Organic horticulture farms require far more workers than conventional horticulture farms, although organic grain farms as well as dairy farming may need no more work than their non-organic equivalents.

Organic farming requires more acreage than conventional agriculture, and greenhouse gas (GHG) discharges per unit of manufacture can be greater. In addition, yields per hectare are low on organic farming, and the constant cost per unit of production is often high.

While the energy transition is expected to have a net beneficial effect on business, many fossil-fuel-based employees will need to find new careers. Policies for a just transition can also help with the process of upgrading the at-risk fossil-fuel workforce [13].

Support is needed for Ministry of Micro, Small and Medium Enterprises (MSMEs) in providing energy-effective products as well as in providing renewable energy technology and apparatuses.

Spending in power-efficient and sustainable energy has a beneficial impact on GDP and services, improved human well-being, as well as general well-being, and decreases pollution and GHG emissions. There is a 0.3%-1.3% positive effect on GDP from energy savings due to EU activity. If renewable energy sources made up half of the world's energy consumption by 2030, global GDP would rise by 1.1%, or the equivalent of US $1.3 trillion. Direct as well as indirect employment within the renewable power sectors can achieved 24.4 million people by 2030 (7.7 million 1n 2014). By meeting energy requirements as well as energy resources, the value of the potential business by 2030 is set to be more than US $4.3 trillion at present price.

3.1.1 Brief explanation

The trends mentioned above refer to present business possibilities in MSMEs, which are considered to be well-established allocation, establishment, operation, and maintenance of technology and apparatus. According to the Atmosphere and Growth Awareness Network, the credit gap to MSME that supplies weather change technology is anticipated to be $4-5 billion in the rising nation. Access to climate finance is a challenge for MSME due to a lack of understanding, a lack of suitable climate, and limited resources accessibility to economical products for horticulture activity and climate technology. That the shift to clean energy is so crucial is reflected in the 164 countries that have set their own renewable energy targets. The government can take a variety of actions to assist and encourage MSME to supply energy-efficient products as well as renewable energy technology. They may apply the law and policy to pure as well as green energy; use economic compensations to reduce power tax or set tax rates that encourage investment; encourage (native) banks to provide loans, low interest rate or interest rate loan, or else green credit line for MSME operating with less carbon technology as well as sustainable energy field promote the creation to organisations that assist members in finding new vendor or else customer; making it easier to obtain funding with global growth agencies, for example Climate Change Development Fund (CDF), Green Climate Fund (GCF), and Renewable Energy and Energy Efficiency.

Partnership for Renewable Energy and Energy Efficiency, or associates with foreign enterprises such as be able to invest within MSME get access to new markets.

4. Carbon power

Progress in green technology can be measured using trends in carbon dioxide (CO_2) emission and GDP, as CO_2 discharges are a major cause of climate variability (Fig. 8.1).

4.1 There are many rehabilitation strategies that have a favorable impact on the environment

Color blurring depicts the total number of actions that have a positive impact on the environment that are being implemented in all OECD and major partner nations by August 2020. European Commission (2020), European Green Deal Website [14] (Table 8.1).

Improving city transport with better city planning , can be done by switching to sustainable transportation, better fuel effectiveness, and electrical installation.

Despite improvements in vehicle efficiency, the transportation industry contributed 23% of global GHG emissions in 2010 and was another fast-growing source.

Transportation-related environment mitigation action as well as laws for urban areas could save 2.8 Gt of GHG emissions per year by the end of 2050 as well as having significant financial, societal, and health-related benefits. Investment within growing public transportation as well as enhancing vehicle efficiency could produce a further three million

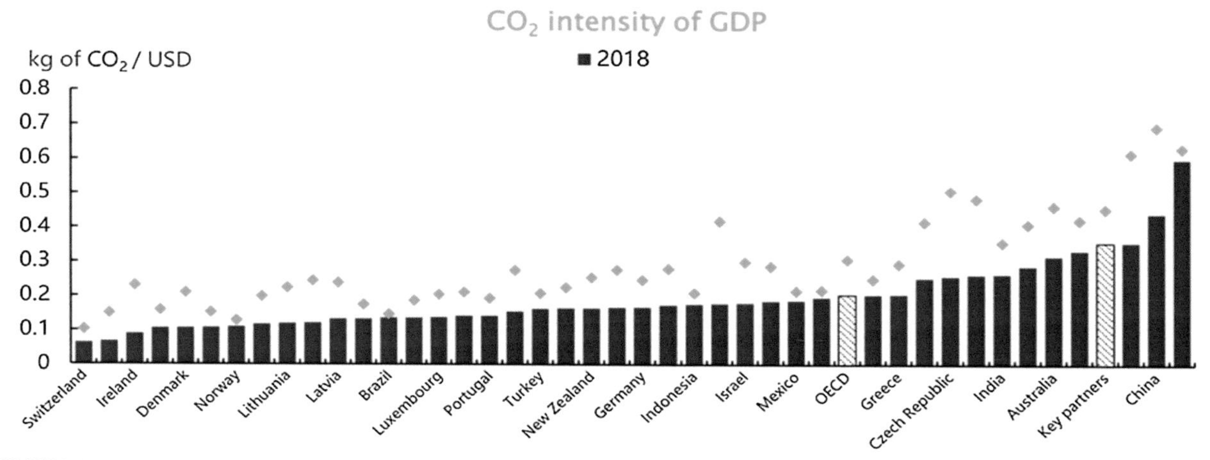

FIGURE 8.1 World Energy Statistics and Balances. *Source: IEA (2020), "Extended world energy balances" IEA World Energy Statistics and Balances (database).*

TABLE 8.1 Environmentally beneficial recovery methods by sector and its type.

	Power	Aviation	Ground Transportation	Maritime transportation	Hard Industrial	Properties	Agricultural	Forests	Waste managers	Other
Tax cuts / other subsidies										
Grant / Loan (including interest-free loan)										
R&D support										
Control change										
Skill training										
Other										

Source-EC (2020), European green deal website [14].

jobs per year within OECD cities, as well as between 3 million and 23 million net jobs per year in non OECD cities, up to 2050. Unprotected populations, which are more likely to survive and work in polluted areas, would gain unevenly from air quality improvements.

4.2 Benefits

There exist a broad range of options available to lawmakers for promoting low-carbon transport within cities, however the choice should be based on the particular city conditions. The following steps can be integrated into these areas:

(1) Integrated city planning as well as travel and reducing the need for people to travel, results in short journeys to works, schools, and other facilities, and as a result reduces the requirement for public transport; while producing a lower carbon footprint, it must be properly incorporated with the city development plans;
(2) Change the routes for private cars to more credible modes of transportation, which include walking, cycling, and increasing clean public transport. This will involve construction of the needed framework as well as knowledge, and also provides other attractive substitutes (for example shared urban bikes or electric bikes), as travel outlooks and practice are often very deep-seated and can be difficult to change. Action taken must also be on the basis of a knowledge of gender inequality within the decision-making as well as transportation decisions considering that women use public transportation more than men, with many non-work trips, and they are also much more likely to have a shorter distance to work, which they mean less efficient hours and selective flexibility;
(3) To improve fuel economy, passenger transport electrification through sustainable transportation technology (for example fuel cell buses, as well as the use of biofuel) and the promotion of electrification of passenger vehicles building the required infrastructure;
(4) Development of the delivery of goods (for example, from freight train, delivery patterns are very efficient, delivery over the weekend), as well as the efficiency of freight vehicles and electrical installation. In supplement, several nations present national as well as domestic financial and constant measure, similar to petrol tax, fuel tax based on the fuel effectiveness, and/or CO_2 emissions, fuel-efficient management, road user charging, parking prices, entry restriction, and registration restriction.

5. Smart buildings [15]

Homeowners are starting to look beyond their four walls to examine the impact of their construction on the electric grid, and the associated goals, as well as the climate. To achieve such goals, the building must include more than just systems that provide comfort, light, and security. Future structures must integrate multiple elements in an integrated, flexible, and efficient way. This is a vision of an institution that achieves its goal while saving energy costs, maintaining a strong electricity grid, and limiting its environmental impact.

Intelligent buildings provide essential construction aids that benefit residents (for instance, light, heat exchange, ambient air quality, material protection, hygiene, etc.) with low prices as well as a reduced impact on the environment throughout the construction life span. Until such a goal is achieved, ingenuity should begin from the commencement of the design stage to the completion of the construction work. Throughout the work, intelligent structures use information technology to connect a range of subsystems that often work separately so that information is shared to improve the overall performance of the building. Intelligent buildings consider features other than the building material within their four walls.

Advanced materials and nanotechnology to enhance energy efficiency and environmental performance provide a comprehensive examination of the most recent advances in building material and design of the building that are used to enhance building efficiency in both new and existing buildings.

6. Climate change disclosure laws [16]

New Zealand was the first country in the world to enact climate change disclosure laws for financial firms. According to officials, New Zealand was the first country to enact legislation that obliges banks, insurance corporations, and investment fund managers to account for the impact of climate change on their operations.

About 200 New Zealand's major financial institutions, including banks with more than NZ $1 billion in assets ($ 718.90 million), key insurance agents, shareholders, and credit providers on the national stock list, will be required to provide this information.

Several international companies that meet the regulatory requirements of NZ $1 billion will be subject to the law, including four major Australian banks: the Bank of Australia, Australia and New Zealand Banking Company, Westpac Corp, and domestic Australian banks.

New Zealand is a world leader in the field, as well as having the possibility to set international standards for compulsory climate change, the Minister for Climate Change, James Shaw, said in a report:

> *The fresh law will force economic institutions to define in what way they intend to address environment-associated hazard as well as possibilities, and the exposure conditions will be founded on the principles established by the New Zealand Private Center, the External Reporting Board. These requirements will be on the basis of the Task Force on Climate-related Financial Disclosures, and disclosures will be required from the 2023 financial year.*

The New Zealand government has done a lot of work in this area.

7. The action of biodiversity [17]

Biodiversity into COVID-19 rehabilitation projects reduces epidemic risk, strengthens economic stability, and protects people and animals.

Loss of biodiversity is a major cause of rising contagious disorders, and also has other risks to companies, communities, and the global economy. Spending on biodiversity during the course of tackling the epidemic can help reduce risk, while providing jobs and boosting the economy. According to study, spending $260 billion annually for more than ten years on initiatives such as desertification prevention, improved land administration, animal welfare, and disease control in wildlife and domesticated animals could reduce the danger of another epidemic. This cost is roughly 2% of the estimated COVID-19 pandemic cost [13]. According to a research of 163 industrial departments and their supply chains, US $44 trillion in land was added (more than half of the world's GDP) is dependent on its environment and resources.

About 1.2 billion jobs worldwide are directly dependent on ecosystem services [14]. Biodiversity protection measures should also aim at avoiding violations of domestic and international laws, as well as extreme and unnecessary suffering to animals.

7.1 Smart houses and smart buildings: weather [17]

7.1.1 Measures

The energy efficiency of buildings should be enhanced, including colleges and clinics, through renovating present properties, and the replacement of boilers, light, and domestic equipment with energy-saving equivalents.

7.1.2 Benefits

Tasks to improve the energy effectiveness of buildings are cost-efficient, with a payback time of 15–20 year to restore a full residence or, for a model, only 3–4 years to deploy spray froth on a new site property. Spatial distribution can encourage the growth of common, small, and medium firms since jobs require more workers. They help to reduce greenhouse emissions when fossil fuels are used to heat or produce electricity, and they also offer opportunities to use the generated electricity for other purposes. The pattern is well defined, and both the results and the performance can be measured. Better living conditions will reduce some of the most unpleasant and stressful aspects of women's everyday lives and expand the options for advancement available to women, their families, and society as a whole.

7.1.3 Concise description

Compulsory power productivity demands, widely used in European countries as well parts of Asia, offer an excellent opportunity to embrace and enforce power-efficiency standards and to promote the use of power-efficient technology in the construction industry, using sustainable and supportable construction methods. In pursuit of such progress, countries, depending on their requirements, can take steps to focus on various areas, such as the following. (1) The legal and administrative framework for power efficiency in the construction area, which includes the progress of institutional skills as well as better communication among organs of state, local authorities, and the private sector. (2) The efficient operation of buildings using suitable equipment, for example power research, domestic administration businesses, efficient real estate certification schemes, and access to financing to improve energy efficiency. (3) Capable construction and behavioral modifications in practice as well as education programs for manufacture experts, repair technicians, residents of domestic buildings, and users of people buildings. (4) Technical achievements that include intelligent as well as available technologies including water as well as polishing, space heating, solar collection systems, to intelligent and advanced methods (for example, intelligent measurement, detector, IoT, new building materials as well as new, stable construction; smart structures as suggested by ITU). (5) Economic systems, such as effective borrowing or green recognition by local or foreign banks, subsidies, credit or charge modifications held by individual investors as well as other human beings (ESCO, power-saving producers as well as providers). Enhanced power efficiency goes hand in hand with sustainable energy solutions, specifically in key infrastructure, for example schools and hospitals.

8. Case study

8.1 Acceptable approved cases of integrating biodiversity in response to COVID-19 and rehabilitation programs [18]

Finland provided a EUR 5.5 billion package for improvement action comprising EUR 53 million (US$ 62 million) for schemes including fresh fields, water resources, as well as forestry preservation; and EUR 13.1 million to protect an rehabilitate natural reserves and to promote eco-tourism. The UK provided GBP 40 million (US $ 51 million) for a "fresh healing challenge trust", to finance 2000 projects and promote 3000 unique small-term and long-term projects to plant trees, restore habitats, and create green spaces. Austria committed to EUR 350 million for research to improve climate change adaptation, which includes the protection and development of biodiversity and the expansion of protected areas of natural forests.

8.2 Securing Georgia's forest by space [19,20]

Administrators used Georgia's fresh forestry and ground use atlas, an online tool that allows them to create better plans, perform, and monitor forest activities, including unlawful or unreasonable exploitation. Atlas partners global satellite data on forestry National forestry information for Georgia is based on global satellite data. It is also widely used by scholars, nongovernment organizations, researchers, visitors, and the general public who can find details on the Internet.

8.3 Kazakhstan needs waste management and community well-being technologies [21]

In August 2020, UNDP gave Phthisiopulmonology an autoclave and secure waste collection containers at the Nur-Sultan Town Center. Phthisiopulmonology now operates a secure collection and elimination centre for contaminated medical waste that was previously renovated to assist clients and COVID. This program is being applied as part of a joint UNDP and Asian scheme. The hospital waste management system is the focus of the growing store, which strives to improve preparation for the domestic people's health emergency.

This is a shared project that will assist in redesigning typical working processed as well as a program for teaching health staff to improve their ability to correctly gather, organize, transport, and monitor medical waste in hospitals. The scheme will additionally grow an efficient digital health service waste disposal website for all parts of Kazakhstan. The new system is anticipated to create the basis for a sustained medicinal waste disposal managing system not only for the novel coronavirus illness circumstances, but also further ahead.

9. Future pandemic preparedness [22]

As the world struggles with the damage caused by the COVID-19 disruption, controversies over how to prevent future zoonotic outbreaks have intensified. According to the IPBES Pandemics report, surviving the epidemic is a possibility, but it will require a change of strategy from response to prevention. Although COVID-19 is derived from animal-borne pathogens, its emergence, like previous epidemics, is largely driven by human actions.

9.1 We should be concerned about the legal wildlife trade in order to prevent the next pandemic [23]

According to a former wildlife inspector, millions of living creatures enter the United States every year without infection testing, putting the country at risk of further outbreaks.

10. Recent literature

10.1 A comparative analysis of data-driven based optimization models for energy-efficient buildings [24]

For energy-efficient buildings to use energy resources effectively and maintain optimum comfort, an efficient energy management system is necessary. This study compares data-driven optimization strategies to concurrently optimize the comfort index and reduce energy usage. To maintain interior thermal comfort, heating—cooling systems are utilized, which uses energy based on the difference between inside and outside temperatures. Particle swarm optimization (PSO), genetic algorithm (GA), bat, neural network algorithm (NNA), and artificial bee colony are thus used to optimize the ambient temperature parameter. The major goal of this optimization is to reduce the discrepancy between the user-specified temperature and the ambient temperature. After that, the constructed machine learning (ML)-based controller receives the difference between the optimal and actual temperatures. The output of the controller is also sent to the coordinator agents, who then provide the actuators with power as necessary. The performance of the ML method is enhanced by the suggested model's linearity. An ML controller was created using the dataset provided from the fuzzy temperature controller.

10.2 Machine learning forecasting model for the COVID-19 pandemic in India [25]

The year 2019 resulted in the global COVID-19 pandemic, and several studies are being conducted using various numerical models to predict the likely progression of this pestilence. These mathematical models depend on various factors. Investigations are based on potential inclination, whereas numerical models depend on other variables. In this work, the researchers proposed a COVID-2019 spread prediction model. They used the COVID-19 Kaggle data to carry out linear regression, multilayer perceptron, and vector auto regression in order to predict the epidemiological example of the disease and the rate of spread of COVID-19 cases in India based on information obtained from Kaggle, and predicted the possible COVID-19 trends in India. The ability to predict and estimate into the not too distant future is made possible by using the common data on confirmed, deceased, and recovered cases across India for an extended period of time.

10.3 AI-based building management and information system with multi-agent topology for an energy-efficient building: toward occupant comfort

A multi-agent topology building management and information system powered by artificial intelligence (AI) is suggested for energy-efficient buildings. The multi-agent topology building management and information system is based on lowering the error between the real environmental parameters and the planned environmental parameters in order to minimize energy consumption and maximize comfort. To provide a collection of optimum solutions, a restricted nonlinear optimization technique is used in the initial optimization, followed by an artificial intelligence-based optimization that incorporates deep learning concept training and validation. These solutions include temperature, lighting, and CO_2 concentration settings for the highest degree of thermal, visual, and air quality while using the lowest amount of energy possible. The created system uses less energy and retains a high degree of comfort [26].

11. Conclusions

Nowadays, we are seeing that climate change is having a great impact on people's lives. We are feeling the impact of the climate crisis, with its urgent demand of the time that people live in harmony with the environment. This has resulted in a greater attraction for smart cities and buildings.

Smart buildings will support health and safety measurement and will also be energy efficient. Since COVID-19 it has become even more important that public health should be given maximum importance in the workplace.

11.1 Advantages

- Social benefit: for a healthy ecosystem green buildings are very useful.
- Environmental benefits: Green buildings are designed for best use of natural sources.
- Economical benefit: Green buildings use less resources like water and energy, so the overall cost is less than in typical buildings.

11.2 Limitations

- Green economy is a time-consuming process.
- The initial cost for establishment is high.
- High Yielding Variety Seeds (HYV seeds) need plenty of water and also chemical fertilizers.

References

[1] UN secretary-general's remarks at launch of report on the socio-economic impacts of Covid-19 in Philippines. In: United Nations. https://philippines.un.org/en/42362-un-secretary-generals-remarks-launch-report-socio-economic-impacts-covid-19. Accessed 13 Nov 2021.

[2] India Today, Why the Coal Crisis Is a Wake-Up Call for Both Coal India and the Centre—India Today Insight News, October 25, 2021. https://www.indiatoday.in/india-today-insight/story/why-the-coal-crisis-is-a-wake-up-call-for-both-coal-india-and-the-centre-1869244-2021-10-25.

[3] https://www.iea.org/reports/renewables-2020/covid-19-and-the-resilience-of-renewables 1https://www.oecd.org/coronavirus/en/themes/green-recovery.

[4] Why Does Green Economy Matter? | UNEP—UN Environment Programme. (n.d.). UNEP - UN Environment Programme. https://www.unep.org/explore-topics/green-economy/why-does-green-economy-matter.

[5] Nations, U. (n.d.). Iceland's Sustainable Energy Story: A Model for the World? | United Nations. United Nations. https://www.un.org/en/chronicle/article/icelands-sustainable-energy-story-model-world.

[6] Homepage. In: Vision 2030. https://www.vision2030.gov.sa/. Accessed 13 Nov 2021.

[7] Wikipedia, Saudi Vision 2030 - Wikipedia, April 25, 2016. https://en.wikipedia.org/wiki/Saudi_Vision_2030.

[8] S. Said, R.J. Stephen Kalin, B., Saudi Crown Prince's Vision for Neom, a Desert City-State, Tests His Builders - WSJ, WSJ, May 1, 2021. https://www.wsj.com/articles/saudi-crown-princes-vision-for-neom-a-desert-city-state-tests-his-builders-11619870401.

[9] The Line. In: NEOM. https://www.neom.com/en-us/whatistheline. Accessed 13 Nov 2021.

[10] 2021) The Line, Saudi Arabia. In: Wikipedia. https://en.wikipedia.org/wiki/The_Line,_Saudi_Arabia. Accessed on 13 Nov 2021.

[11] OECD, Investing in Climate, Investing in Growth, OECD Publishing, Paris, 2017, https://doi.org/10.1787/9789264273528-en.

[12] OECD, Farm Management Practices to Foster Green Growth, OECD Green Growth Studies, OECD, Publishing, Paris, 2016, https://doi.org/10.1787/9789264238657-en.

[13] OECD (2020), OECD Employment Outlook 2020: Worker Security and the COVID-19 Crisis, OECD Publishing, Paris, https://doi.org/10.1787/1686c758-en; IRENA (2020), Global Renewables Outlook: Energy transformation 2050, www.irena.org/publications.

[14] Making the Green Recovery Work for Jobs, Income and Growth, OECD, October 6, 2020. https://www.oecd.org/coronavirus/policy-responses/making-the-green-recovery-work-for-jobs-income-and-growth-a505f3e7/#component-d1e248.

[15] Building Efficiency Initiative (April 5, 2011). What Is a Smart Building? WRI Ross Centre for Sustainable Cities. https://buildingefficiencyinitiative.org/articles/what-smart-building.

[16] New Zealand Passes Climate Change Disclosure Laws for Financial Firms In World First World News. (October 21, 2021). The Indian Express. https://indianexpress.com/article/world/climate-change/new-zealand-passes-climate-change-disclosure-laws-for-financial-firms-in-world-first-7583510/.

[17] UNECE. (n.d.). Compendium of Best Practices on Standards and Technologies for Energy Efficiency in Buildings in the UNECE Region. Retrieved November 13, 2021, from http://www.unece.org/fileadmin/DAM/energy/se/pp/eneff/9th__Forum_Kiev_Nov.18/13_Novembe_2018/EE_Buildings/05_Vitaly_Bekker.pdf.

[18] Environment, U. N. (n.d.). Why does Nature Action Matter? UNEP. Retrieved November 21, 2021, from https://www.unep.org/explore-topics/ecosystems-and-biodiversity/why-does-nature-action-matter.

[19] Inspiring Fruit Tree Cultivation and use In Central Asia. (n.d.). Retrieved November 21, 2021, from https://www.bioversityinternational.org/ar2015/inspiring-fruit-tree-cultivation-and-use-in-central-asia/.

[20] Protecting Georgian forests from Space UNEP. (n.d.). Retrieved November 30, 2021, from https://www.unep.org/news-and-stories/story/protecting-georgian-forests-space.

[21] Covid-19 И Лкологизация Лкономики Стран Восточной Европы, Кавказа Центральной Азии. OECD. (n.d.). Retrieved November 21, 2021 from https://www.oecd.org/coronavirus/policy-responses/covid-19-and-greening-the-economies-of-eastern-europe-the-caucasus-and-central-asia-37dc59cf/.

[22] Geneva Environment Network, COVID-19 and the Environment— Geneva Environment Network, July 21, 2021. https://www.genevaenvironmentnetwork.org/resources/updates/updates-on-covid-19-and-the-environment/.

[23] J. Kolby, To Prevent the Next Pandemic, It's the Legal Wildlife Trade We Should Worry About, National Geographic, May 7, 2020. https://apinationalgeographiccom.cdn.ampproject.org/c/s/api.nationalgeographic.com/distribution/public/amp/animals/article/to-prevent-next-pandemic-focus-on-legal-wildlife-trade.
[24] A. Verma, S. Prakash, A. Kumar, A comparative analysis of data-driven based optimization models for energy-efficient buildings, IETE Journal of Research (2020) 1−17.
[25] R. Sujath, J.M. Chatterjee, A.E. Hassanien, A machine learning forecasting model for COVID-19 pandemic in India, Stochastic Environmental Research and Risk Assessment 34 (2020) 959−972, https://doi.org/10.1007/s00477-020-01827-8.
[26] A. Verma, S. Prakash, A. Kumar, AI-Based building management and information system with multi-agent topology for an energy-efficient building: towards occupants comfort, IETE Journal of Research (2020), https://doi.org/10.1080/03772063.2020.1847701.

Further reading

[1] IEA, Sustainable Recovery, IEA, Paris, 2020. https://www.iea.org/reports/sustainable-ecovery.

Chapter 9

Hybrid and automated segmentation algorithm for malignant melanoma using chain codes and active contours

Nirmala Veeramani[1], Premaladha J.[1], Raghunathan Krishankumar[2] and Kattur Soundarapandian Ravichandran[3]

[1]*School of Computing, SASTRA Deemed to be University, Thanjavur, Tamil Nadu, India;* [2]*Department of Computer Science and Engineering, Amrita School of Computing, Coimbatore, Amrita Vishwa Vidyapeetham, India;* [3]*Department of Mathematics, Amrita School of Physical Sciences, Coimbatore, Amrita Vishwa Vidyapeetham, India*

1. Introduction

Melanoma is widely known as the deadliest skin cancer in the world. This melanoma is an abnormal pigmentation in the human skin due to the high incidence of ultraviolet radiation which in turn later forms the normal skin lesion into the abnormal tumor cell. Melanocytes are solely responsible for this type of cancer. Melanoma is then classified into two major terminologies based on the characteristics of its normal and abnormal growth and the behavior of the cells called benign and malignant, respectively [1]. Thus, distinguishing a normal skin lesion from a melanocytic tumor lesion in the human body is a tedious and challenging task [2]. Dermoscopy lenses and digital cameras are the few common tools used for the diagnostics of malignant skin lesions, which facilitate the investigation of the characteristics of skin lesion pigments. Anyhow, these diagnostic tools have its way of advantages and disadvantages which may impact the results and can end up fatal for patients with malignant melanoma.

1.1 Motivation and contribution

The subjectivity of the specialist's choices is one of the fundamental issues in medical diagnosis. More precisely, in the realm of medical imaging interpretation, the specialist's experience might have a significant impact on the final diagnostic' result. Manual inquest of medical images is found to be difficult, lagging, and prone to interpretation mistakes in some circumstances. As a result, computer vision algorithms for medical diagnostics have grown in popularity. Deep learning's widespread success in computer vision has improved prediction accuracy, reigniting interest in computer-assisted medical diagnosis. Despite this, deep convolutional neural networks (CNNs) have a high number of trainable parameters and require a substantial amount of labeled data to train. Deep CNN [3] has been used in image classification problems when working with a large dataset. It can assimilate groundwork filters automatically and put them together in a hierarchy that enables the elucidation of the latent notions of pattern recognition. This might be a significant disadvantage in the medical imaging area, where the high costs of highly skilled specialists' annotation make the production of huge labeled datasets difficult.

In this work, we explore the segmentation of malignant skin lesions in melanoma of the human body based on the h-CEAC (hybrid—Chain code Euclidean Active Contour) with the EDRS-based prediction of the interested lesion region. The subject of this research work is listed in the following.

- Designing the CNN and Contrast Limited Adaptive Histogram Equalization (CLAHE) approach or the phases of image enhancement was used to construct and train a framework for medical imaging segmentation of melanoma skin cancer.
- The integrated hybrid approach h-CEAC is incorporated for effective medical image segmentation.
- The results can be used for the analysis of the characterization of malignant melanoma from benign and help the clinical experts to stratify the disease metastasis.

Deep Learning in Personalized Healthcare and Decision Support. https://doi.org/10.1016/B978-0-443-19413-9.00018-7

In this research article, the novel hybrid segmentation approach for melanoma skin lesion images with the use of benchmark datasets is tested. The optimization of this techniquc is collated to the prevail state-of-the-art with the selective parametric assessment. This work focuses only on the integration of segmentation approaches into a novel hybrid segmentation procedure that depicts the best boundary of the skin lesion for melanoma which can be utilized for the screening and diagnosis in a noninvasive method on skin cancer patients.

1.2 Related works

Segmentation is a word used often in machine vision systems to describe a variety of image decomposition and classification algorithms. Segmentation is exercised in the process of extraction of contours and patches of an image or scene [4], boundary detection [5], and voxel image analysis [6], among other applications. The initial step in image analysis is usually to segment the image in question. The term "segmentation" refers to the division of an image toward its integrant pieces or closed regions. The extent to which this sub-division is achieved by the problem being processed. Substantially, unsupervised segmentation (autonomous segmentation) [7] is one of the biggest difficulties in image processing jobs. This task is the procedure that decides if the analysis is beneficial or not. Effective segmentation, in fact, almost always leads to a good solution. As a result, considerable effort should be made to increase the likelihood of rugged segmentation. There are several applications for segmentation in diverse fields For example: In medical imaging, video surveillance, object recognition, medical imaging, etc. From the input images to segmentation and recognition, there are numerous methods and techniques there for segmentation that can be used at various phases. The chain code approach is one of the most intriguing segmentation strategies. In this work, researchers experiment with the initial method of encoding digital curves, with the utilization of chain codes. Traditional processing chain approaches are also discussed [8]. Since they maintain information and admit for significant data reduction, chain coding approaches are extensively employed to describe an object. A chain code approximates a curve by employing a square grid and a succession of directed vectors [9]. This article [10] proposes a simple recursive strategy for converting a chain code into a quadtree with a lookup table. However, the quantity of storage required increases as a result of this [11]. Devised a contour-based technique for binary form coding using a multiple grid chain code, which resulted in a significant computational cost. A synchronous chain coding scheme was suggested in the article [12] for image languages. In polynomial time, the procedure is used to determine if something is finite or infinite. Individual digits or outlines of objects were discovered via hierarchical search, which established rectangles or search lines within or along which they were detected. In image processing, this is the typical method for specifying the area within which image objects should be segregated (search regions, test windows, areas of interest). The success of feature-based inspection techniques is determined by the quality of feature detection. The most well-known two-dimensional feature detection challenges, edge detection, and area extraction are instances of inverse ill-posed problems.

There is no stable transformation function that can accurately describe a random observation. To tackle this problem, impose a priori knowledge of the problem region on the solution space to reduce the number of possible solutions. As a result, the detection procedure could be viewed as a breakdown of a series of sub-problems in medical imaging, particularly in the case of melanoma skin lesions.

1.3 Paper organization

The rest of this article describes as follows. Section 2 describes the materials and methods inculcated and the detailed information of the dataset used for our hybrid automated segmentation techniques and the enhancement phase of images carried out for our work and with the algorithms involved in the research work to improve the readability of the paper. Section 3 suggests the proposed methodologies including the explanation of the blocks and workflow. Section 4 depicts the results of the segmented malignant melanoma with our hybrid automated method and is discussed in detail in the aspect of the research goals. Section 5 concludes the various approaches to segmentation and the scope of our work with suggested future directions.

2. Materials and methods

2.1 Datasets

The benchmark PH2 datasets [13] with the handcrafted images were taken for the implementation work, which is specifically carried out on the segmentation of the melanoma skin lesions. To provide efficient results the dataset needs to be balanced hence, we included the dermoscopic images from the Med node [14] A dataset is also included in this implementation work. The PH2 database was established as a result of collaboration here between the Universidade do Porto, the Tecnico Lisboa, and the Hospital of Pedro Hispano's Dermatology service in Matosinhos in Portugal.

The dermoscopic images were procured using a Tuebinger Mole Analyzer system with a magnification ratio of 20 under indistinguishable conditions. These 8-bit color information graphics dimensions are 768 × 560-pixel resolution. There are 200 dermoscopic images in this collection, encompassing 80 common benign lesions, 80 uncommon nevi, and 40 malignant melanomas.

According to the categorization scale theory of Fitzpatrick skin type [15], the adopted skin types from the dermoscopic images are off from types II or III. As a result, the PH2 database's skin colors may range from white to creamy white. The images in the database were deliberately crafted, as shown in Fig. 9.1, based on their quality, resolution, and dermoscopic properties.

2.2 Image enhancement

CLAHE technique [16] is deployed here in our work for the elimination of foggy or unwanted blurred information from the images to improve the visibility level of the source image of the malignant skin lesions of melanoma. This method of image enhancement is CLAHE, which operates on the smaller regions of the images and focuses mainly on tile after tile rather than working on the entire image. This perspective of the CLAHE algorithm is applied to improve the contrast of images. The steps involved in this are as follows.

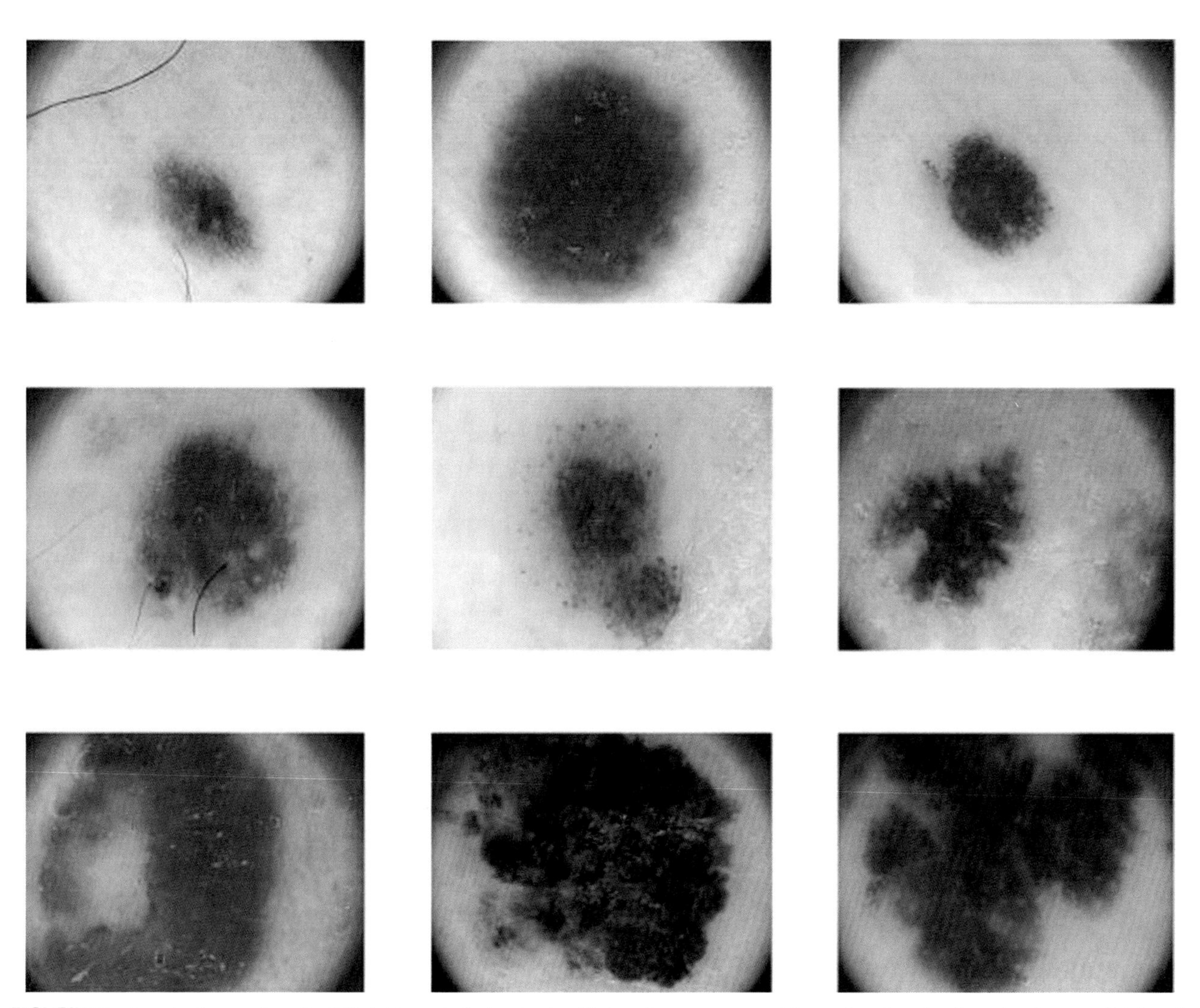

FIGURE 9.1 Samples images from the PH2 database, including benign (first row), uncommon nevi (second row), and malignant melanomas (third row).

Algorithm:	Steps involved in CLAHE
Step 1:	Make nonoverlapping contextual regions M X N out of the original intensity skin lesion image
Step 2:	Histogram calculation of each contextual region based on gray levels
Step 3:	Contrast limited histogram values are calculated Navg = (NrX × NrY)/Ngray
Step 4:	Redistribution of the remaining pixels until all leftover pixels have been distributed, i.e., N_{gray}/N_{remain}
Step 5:	Enhancing intensity values $y(i) = y_{min} + \sqrt{2\propto^2\left(\frac{1}{1-p_{input}(i)}\right)}$
Step 6:	Reducing abruptly changing effect $y(i) = \frac{x(i)-x_{min}}{x_{max}-x_{min}}$
Step 7:	New gray level assignment of pixels is calculated within a contextual region of the sub-matrix

3. Proposed methodology

The samples of malignant skin lesion dermoscopic images were carefully selected for the segmentation, which has a balanced sample, and then the initial steps for the preprocessing of the images are carried out, namely, the CLAHE method is employed after the contrast enhancement of the dermoscopic image to ensure the visibility of the disconnected borders and to brighten up the image, which matters a lot in the medical image segmentation. Thus, preprocessed images are passed to the phase of segmentation as shown in Fig. 9.2.

The second main focused work in this paper is to segment the accurate borders without any discontinuity for acquiring the characteristics of melanoma in future diagnosis without having many deviations to the ground truth and the real-time condition of the patient clinical condition. To ensure the most accuracy in terms of segmentation, the h-CEAC method is incorporated. Here, the combination of chain code with EDRS and the method of contours reveals precise results. The exact segmentation of the borders with effectiveness is obtained by the proposed hybrid approach.

3.1 Preprocessing phase

Preprocessing the image [17] is the most important step to be carried out after the step of image acquisition from the medical modality to improve the visualization of the image. The general preprocessing steps involved in our work include image noise removal, isolation of patterns of interest from the background, and image enhancement. In this work, the preprocessing consists of two stages: First, the conversion of the application of median and sharpening filter for noise removal. The median filter is applied to keep the borders of the skin lesion and its shape intact while minimizing the noise, especially for salt and pepper noise. The sharpening process is the application of an high-pass filter (HPF) to an image; this process is used to highlight the fine details and remove blurring from images. Image sharpening is to make the contour lines and image details clearer.

3.2 *h*-CEAC segmentation

3.2.1 *Feature extraction and chain codes*

Chain code [18] is a methodology used for representing an object in an image and also depicts the coordinates of any object's continuous border. A string of numbers can be used to symbolize the chain codes. Each number corresponds to a

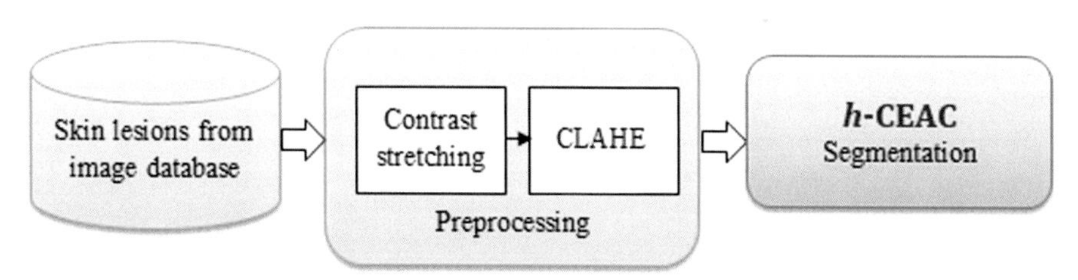

FIGURE 9.2 A workflow of our proposed methodology.

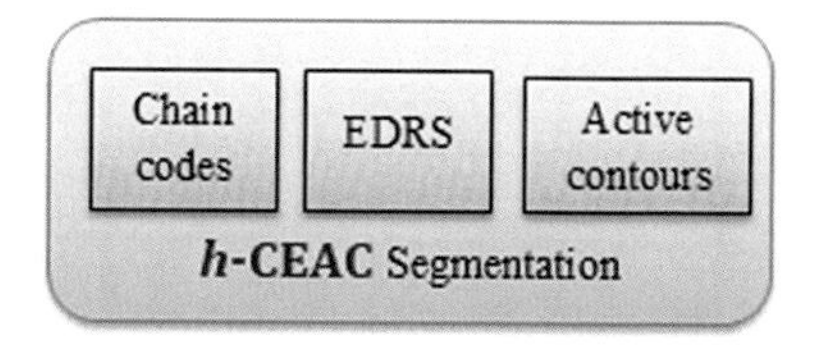

FIGURE 9.3 Segmentation process using the *h*-CEAC workflow is depicted which is carried out at the end phase of our work.

distinct direction to the next place on the connecting line that is present. The retrieved features from the input image under investigation are matched with those of the predefined models throughout the segmentation process. If the original input image to be segmented is noisy, the segmentation procedure becomes more complicated since attributes may appear at random places and orientations. In the medical image segmentation procedure, Fig. 9.3 displays a hybrid combination of the exact process workflow that is carried out meticulously to outperform this state-of-the-art technology.

The segmentation process using the h-CEAC can be availed by pixel gradients [19] in place of only pixel values, indicating the structure of the image instead of the texture. At the location (x, y), the gradient ∇f of an image is given as:

$$\nabla f = \begin{bmatrix} G_x \\ G_y \end{bmatrix} = \begin{bmatrix} \frac{\partial f}{\partial x} \\ \frac{\partial f}{\partial y} \end{bmatrix} \tag{9.1}$$

From vector analysis [20], it is generally known that the gradient vector points of the image lie in the direction of its maximum rate of change of the value of f at (x, y) coordinates. The normalization of the gradient function del f is then widely referred to simply as the gradient and defined as:

$$\|\nabla f\| = \sqrt{G_x^2 + G_y^2} \tag{9.2}$$

The inferences taken into account highly rely on particularly what features are to get a load which prevalently depends on lesion images that are to be identified. Features provide the necessary data. Whereas the reductional of the feature while conserving the knowledge from dominant features is required for medical image segmentation. According to the discrete objects of interest, an image has been segmented and fed into the next process. Numerous features can be used to describe necessary information that lies in the region of interest of an image. In particular, the procedures used for feature extraction necessitate the approaches of edge detection, line tracing, and as well as for the description of shape techniques.

3.2.2 Formulation of the chain code

The following three goals must be met by the coding line structure scheme in general. It must first maintain the information of the region of interest and then allow for compact storage and display, as well as any necessary processing. A boundary is represented by a chain code, which is a connected sequence of straight-line segments of a specific length and direction. This orientation is usually determined by the segment's four or eight connectivity. Let R stand for the full image area. The segmentation procedure can be thought of as dividing R into n sub-regions:

Here $R_1, R_2, \ldots, R_n$. are referred to as the sub-regions where the following axioms must be satisfied:
R_i is denoted for a connected region, $i = 1, 2, \ldots, n$,
$R_i \cap R_j = \acute{\varnothing}$ for all i and j, for every $i \neq j$.
$P(R_i) = \text{TRUE}$ for $i = 1, 2, \ldots, n$.
$P(R_i \cup R_j) = \text{FALSE}$ for every $i \neq j$.

Where P (R_i) is denoted for a logical predicate over the points in set R_i which signifies the characteristics of the relationship of the connected regions and ǿ is denoted for the null set.

Here, the axioms are represented as follows,
Axiom 1 stipulates that every pixel must be inside a region.
Axiom 2 needs that point in a region must be found connected.
Axiom 3 designates that the region must be disunited, that is, null intersection.
Axiom 4 if every pixel in R_i is said to have identical intensity.
Axiom 5 specifies the region of R_i and R_j are unlike in the sense of the predicate.

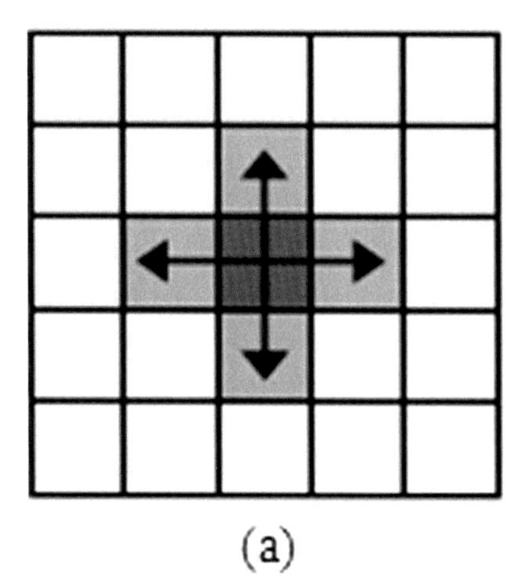

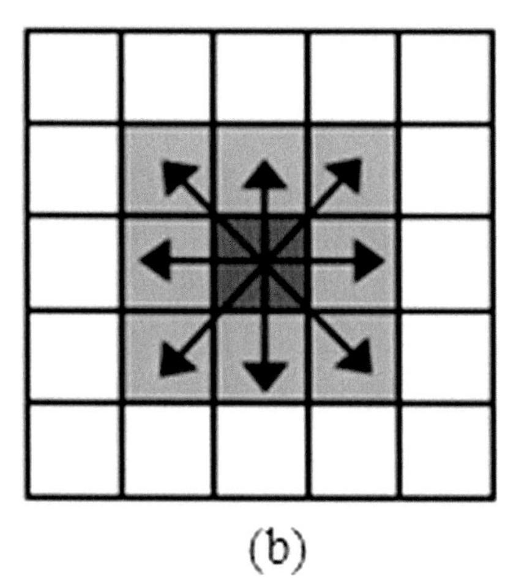

FIGURE 9.4 Definitions of pixel neighborhoods. (A) Four-Connectedness and (B) eight-connectedness. The image pixels' relationship with the neighborhood is depicted here with its actual working illustration.

3.2.3 Pixel interdependence

The image contouring process is in general explained as a closed line that runs along the border of the object, that is, consisting absolutely of the border pixels. There are two habitual methods to define the neighborhood operations for a given image pixel in digital image processing: They are four connectedness and eight connectedness [21], as depicted in Fig. 9.4.

The 4-connectedness permits only the horizontal and vertical movements between adjacent pixels, but on the other hand, the 8-connectedness method also utilizes diagonal connections. Chains are always represented as the boundary or simply said as contours of any discrete shapes that are composed of regular pixels. Here, the length measurement of every side of the pixel cell is one as default, and then these chains are represented in the form of closed boundaries, which results in all the chains being closed.

3.2.4 EDRS and active contours

The challenges of manual initialization to segment the human skin lesion are limited to capturing the range and have poor convergence to borders, and as well as it may end up in the issue of assigning energy functional parameters, these can be addressed by an automated image segmentation method. We train CNNs to forecast active contour weighting factors and build a ground truth mask for distance transformation and startup the circular extraction. The distance transform is used to construct a vector field also with the size of the Euclidean distance map, pointing from each pixel of the image to the nearest point on the boundaries.

All the MxN images can be articulated effortlessly in an MN-dimensional in a Euclidean space, which is typically called an image space [22]. It uses the $e_1, e_2, e_3, \ldots e_{MN}$ bases to establish the image space's coordinate system, with e_{kN+1} association to an ideal point source with unit intensity at a given location (k,l). $x=(x_1, x_2, x_3, \ldots x_{MN})$, where x^{kN+1} is the gray level at the $(k,l)^{th}$ pixel. The coordinate concerning e_{kN+1} is represented by x^{kN+1}. The picture space starts with a grayscale with negligible gray levels everywhere. The distance between comparable points in image space, known as the Euclidean distance between images, could not be determined until the basis metric coefficients were known.

The metric coefficients are denoted as c_{ij} $i,j = 1,2,3, \ldots$ MN in the equation as follows,

$$c_{ij} = \langle e_i, e_j \rangle = \sqrt{\langle e_i, e_i \rangle}\sqrt{\langle e_j, e_j \rangle} \cos \theta_{ij}. \tag{9.3}$$

where the angled brackets $\langle , \rangle$ is the represented form of scalar product, and the e_i and e_j have the in-between angle called θ_{ij}. Note that, $\langle e_i, e_i \rangle = \langle e_j, e_j \rangle$.

Then, c_{ij} the metric coefficients depends entirely on the θ_{ij} angle if at all the base vectors have the same length.

$$d(x, y) = \sum_{i,j=1}^{MN} c_{i,j}(x^i - y^i)(x^j - y^j) = (x - y)T\,G(x - y) \tag{9.4}$$

For every MN^{th} order symmetric and positive definite matrix G induces a Euclidean distance for pictures with fixed dimensions M by N. However, the vast majority of them are unsuitable for determining image distances. If any two base vectors e_i and e_j where ($i \neq j$) are mutually perpendicular, regardless of whose pixels they belong to, the basis then creates a Cartesian coordinate system [23]. As a result, G is the identity matrix and thus generates the standard Euclidean distance (G refers to a diagonal matrix if the base vectors' lengths differ). The sensitivity of classical metrics to deformation is driven by the fact that the two objects being compared are images, The orthogonality of the base vectors $e_1, e_2, \ldots,$ and e_{MN}, which correspond to pixels, causes this geometric flaw. All mutually perpendicular base vectors are often unable to reflect spatial information, such as the distances between pixels. Such data, on the other hand, is strongly mentioned in sensible image distance, as in the following sentence: The original image resembles a slightly distorted image. The term "slightly deformed" refers to the deformed image's pixels being close to the equivalent pixels in the original image.

This indicates that a decent Euclidean distance for images should include information about pixel distance. As a result, the metric coefficients would use to derive the Euclidean distance must be connected to pixel distances. The resulting Euclidean distance is then found to be completely resilient to modest deformation if the metric coefficients are appropriately dependent on the pixel distances. The reader should be aware that this article addresses two distances: the image distance in high-dimensional image space and the pixel distance. Let P_i, and P_j are the point coefficients, where the values of i, j = 1,2, ... MN all are represented in pixels. The pixel distance is the distance between P_i and P_j on the image lattice, denoted by $| P_i - P_j |$. The most important positive definite function, which we utilized in the construction of the metric coefficients, is the Gaussian function, which is denoted as,

$$g_{i,j} = f\left(|P_i - P_j|\right) = \frac{1}{2\pi\sigma^2} \exp\left\{-|P_i - P_j|2 \;/ 2\sigma^2\right\} \tag{9.5}$$

The width parameter σ is set to estimate the distance factor d, and the computation of the image distance for the segmentation algorithm with the enabled active contour iterations is given as,

$$d^2(x, y) = \frac{1}{2\pi} \sum_{i,j}^{MN} \exp\{-|P_i - P_j|2 \;/ 2\} \,(x_i - y_i)\left(x_j - y_j\right) \tag{9.6}$$

Two publicly accessible datasets, including PH2 datasets and med node datasets, are used to test our hybrid method, which also comprises the chain code mechanism as described. The active contour method of the segmentation approach has the main advantage of its ableness to delineate the image edges, especially in potentially noisy two-dimensional medical images [24]. In this method, the contour is known by a constrained spline, which is resulted in an effect by minimizing a cost function.

4. Results and discussions

The dice similarity coefficient (DSC) [25] is here returned as the scalar or vector numeric values range between 0 and 1. In general, between the two segmentations of X and Y target regions, it measures the spatial overlap. On the other hand, if the similarity is closer to 1 state, it is said to be completely similar and 0 depicts the nonsimilarity.

$$\text{DSC} = 2\,\frac{|X \cap Y|}{|X| + |Y|} \tag{9.7}$$

For the input arrays,

- The similarity is a scalar in terms of binary images
- The similarity is a vector in terms of label images

Where the first and second coefficient is the dice index for label 1 and label 2, respectively. For the categorical images, the similarity is considered as the vector, in which the first coefficient is the dice index to the first category and so on. The segmentation results are evaluated with the above-mentioned parametric equation of the dice similarity coefficient to measure the performance of our state of art method, and Table 9.1 depicts the results of our proposed segmented images on the benchmark image dataset.

TABLE 9.1 Segmented image results with its dice coefficient.

Melanoma dataset	Dice coefficient
IMD048.BMP	0.563456596
IMD049.BMP	0.767896423
IMD050.BMP	0.778523780
IMD051.BMP	0.236879747
IMD052.BMP	0.429297891
IMD053.BMP	0.891696371
IMD054.BMP	0.678934542
IMD055.BMP	0.872945754
IMD057.BMP	0.587987655
IMD058.BMP	0.693306654
IMD059.BMP	0.815453901
IMD060.BMP	0.645350247
IMD063.BMP	0.854353563
IMD065.BMP	0.846909078
IMD066.BMP	0.991176931
IMD067.BMP	0.832594442
IMD068.BMP	0.990893422
IMD079.BMP	0.544258675
IMD080.BMP	0.150479915
IMD081.BMP	0.385302118

The system has experimented with a set of challenging datasets. This approach facilitates optimizing the features of melanoma. It has succeeded to demonstrate robustness that is used for the decision analysis in relation and decisions of highly perplexed models and close to the performance of segmentation toward the ground truth images. The resulting image from our state-of-art-method with the h-CEAC-based automated segmentation of malignant melanoma produces a well-defined border from the human skin. It shows any shape of malignant melanoma in the human skin, and Fig. 9.5 shows the effectiveness of the said approach.

4.1 Comparison with existing algorithms

Our research outcome shows the feasibility of segmentation of melanoma that is compared with the existing algorithms as shown in Table 9.2. Here, the evaluation metrics of DSC are used to compare the proposed segmentation results. The ACC (accuracy) and SPEC (specificity) metrics [34] will be assessed for the classification of skin lesions with this hybrid segmentation results to be extended in our future work.

The proposed hybrid segmentation algorithm generates the mask for every lesion close enough to the ground truth; they are used to extract the features such as quantifying the color and texture information for quantifying the ABCD features [35] of the melanoma skin lesion.

5. Conclusion and future scope

This research work presents a new procedure for the segmentation of the malignant melanoma borders from the human skin lesion, which helps to identify the characteristics of the malignant melanoma in the diagnosis which facilities the treatment planning and surgery to be carried out on the patient. Here are the images that are taken by the dermoscopy clinical devices. This new method makes a difference in the quality of the processing and analysis of the dermoscopic

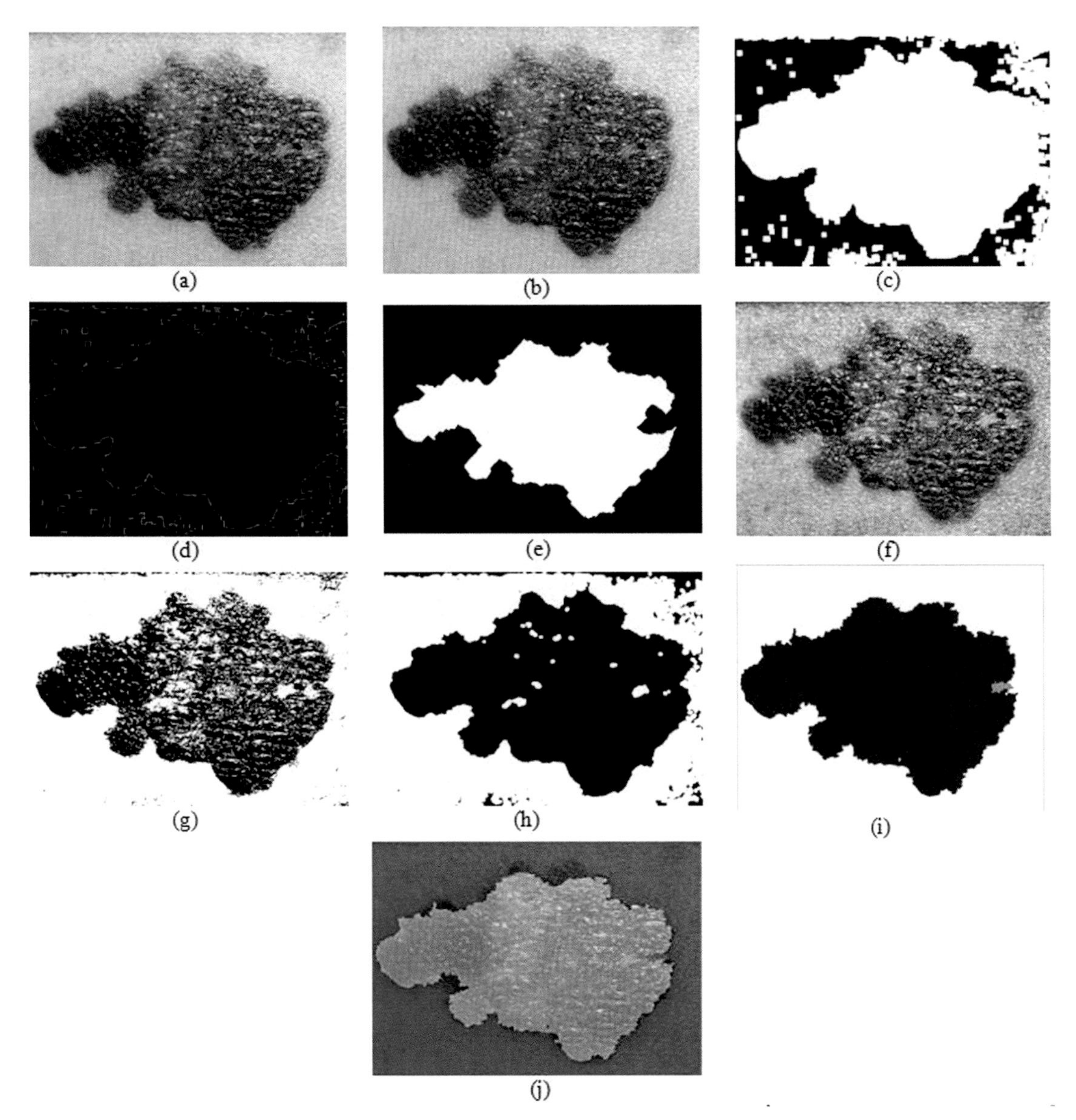

FIGURE 9.5 Automated segmentation results in malignant melanoma. (a) Input skin lesion from the PH2 dataset belongs to the class malignant (b) Gray scale conversion of the input color image, (c) thresholding for the region of interest, (d) mask creation for the image for subtraction, (e) morphological operation on the detected mask with erosion, (f) CLAHE image enhancements, (g and h) results of EDRS of the image, and (i and j) results of h-CEAC segmented image and the ground truth image.

image segmentation process. The active contour detection method with the EDRS is employed in our work to get the continuous contours along the boundaries of the malignant skin lesions of melanoma from the human body. This hybrid automated segmentation method can also be incorporated in character recognitions and numerical digits in vehicle number plates successfully. As a result, because a chain code provides a complete representation of an object or a curve, this technique can be used to compare distinct things. Hence, we infer that the referred chain code methodology for image segmentation as a lossless method outperforms generating connected lines of borders for successful image processing. Future work in this area can be utilized in upgrading this method effectual for images or videos with the illumination artifacts and will be incorporated in designing the ensemble model to test the effectiveness of this proposed hybrid segmentation approach.

TABLE 9.2 Comparative analysis of our proposed work with existing algorithms.

Methodology	DSC	ACC	SPEC
Mendonça et al. [26]	94.00	–	96.00
Pennisi et al. [27]	–	89.40	97.10
Maglogianisa et al. [28]	90.00	92.80	97.00
Ahn et al. [29]	91.50	–	–
Fan et al. [30]	89.30	93.60	–
Zamani et al. [31]	92.00	96.50	98.10
DRFI method [32]	95.20	97.90	98.90
GFAC method [33]	97.08	98.64	99.22
Proposed work	98.70	–	–

References

[1] M. Ramezani, et al., Automatic detection of malignant melanoma using macroscopic images, Journal of Medical Signal and Sensors 4 (4) (2014) 281–290. PMID: 25426432; PMCID: PMC4236807.

[2] Z. Kutlubay, et al., Current management of malignant melanoma: state of the art, Highlights in Skin cancer (2013), https://doi.org/10.5772/55304.

[3] J. Amin, A. Sharif, N. Gul, M.A. Anjum, M.W. Nisar, F. Azam, S.A.C. Bukhari, Integrated design of deep features fusion for localization and classification of skin cancer, Pattern Recognition Letters 131 (2020) 63–70.

[4] R. Falah, et al., A region edge cooperative approach to image segmentation, IEEE International Conference on Image Process 3 (1994) 470–474, https://doi.org/10.1109/ICIP.1994.413762.

[5] F. Ercal, et al., Detection of skin tumor boundaries in color images, Computer Journal of IEEE Transactions on Medical Imaging 12 (3) (1993) 624–625, https://doi.org/10.1109/42.241892.

[6] K. Vincken, et al., Probabilistic segmentation of partial volume voxels, Pattern Recognition Letters (1994), https://doi.org/10.1016/0167-8655(94)90139-2.

[7] M. Radu, et al., Autonomous image segmentation by competitive unsupervised GrowCut, in: 21st International Symposium on Symbolic and Numeric Algorithms for Scientific Computing (SYNASC), 2019.

[8] H. Freeman, On the encoding of arbitrary geometric configurations, in: Proceedings of IRE Translation Electron Computer, 1961.

[9] G. Wilson, Properties of contour codes, in: Proceedings of IEE Visual Image and Signal Processing, 1997.

[10] Z. Chen, et al., A simple recursive method for converting a chain code introduction a quadtree with a lookup table, Computer Journal of Image Vision and Computing (2001), https://doi.org/10.1016/S0262-8856(00)00080-9.

[11] P. Nunes, et al., A contour-based approach to binary shape coding using a multiple grid chain code, in: Proceedings of Image Communication Special Issue on Shape Coding, 2000.

[12] B. Truthe, On the finiteness of picture languages of synchronous deterministic chain code picture systems, Acta Cybernetica Archive 17 (1) (2005) 53–73.

[13] T. Mendonca, et al., PH2—a dermoscopic image database for research and benchmarking, in: Annual International Conference of the IEEE EMBS Osaka, 2013.

[14] Giotis, N. Molders, S. Land, M. Biehl, M.F. Jonkman, N. Petkov, MED-NODE: a computer-assisted melanoma diagnosis system using non-dermoscopic images, Expert Systems with Applications 42 (19) (2015) 6578–6585, https://doi.org/10.1016/j.eswa.2015.04.034.

[15] S. Sachdeva, Fitzpatrick skin typing: applications in dermatology, Indian Journal of Dermatology, Venereology and Leprology 75 (1) (2009) 93–96, https://doi.org/10.4103/0378-6323.45238, PMID: 19172048.

[16] Jinxiang, et al., Contrast Limited Adaptive Histogram Equalization Based Fusion for Underwater Image Enhancement, 2017.

[17] A.N. Hoshyar, A. Al-Jumailya, A.N. Hoshyar, The beneficial techniques in preprocessing step of skin cancer detection system comparing, in: International Conference on Robot PRIDE 2013-2014.

[18] W. Shahab, et al., A modified 2D chain code algorithm for object segmentation and contour tracing, The International Arab Journal of Information Technology 6 (3) (2009) 250–257.

[19] B.S. Min, et al., A novel method of determining parameters of CLAHE based on image entropy, International Journal of Soft Engineering and its Applications 7 (2013) 113–120, https://doi.org/10.14257/ijseia.2013.7.5.11.

[20] T.F. Chan, et al., Active contours without edges for vector-valued images, in: Journal of Visual Communication and Image Representation, 2000.

[21] S. El-Mashad, Y. Amani, A. Abdulwahab, E.H. Basem, Local features-based watermarking for image security in social media, Computers Materials and Methods 69 (3) (2021) 3857–3870, https://doi.org/10.32604/cmc.2021.018660.

[22] L. Wang, Y. Zhang, J. Feng, On the Euclidean distance of images, IEEE Transactions on Pattern Analysis and Machine Intelligence 27 (8) (2005) 1334–1339. Aug.

[23] J. Alakuijala, J. Oikarinen, Y. Louhisalmi, X. Ying and J. Koivukangas, Image transformation from polar to Cartesian coordinates simplifies the segmentation of brain images, in: 14th Annual International Conference of the IEEE Engineering in Medicine and Biology Society.

[24] S. Albahli, N. Nida, A. Irtaza, M.H. Yousaf, M.T. Mahmood, Melanoma lesion detection and segmentation using YOLOv4-DarkNet and active contour, in: IEEE Access, 2020.

[25] E. Tom, J. Bertels, M.B. Dirk, Optimization for medical image segmentation: theory and practice when evaluating with dice score or Jaccard index, IEEE Transactions on Medical Imaging (2020), https://doi.org/10.1109/TMI.2020.3002417.

[26] T. Mendonça, et al., PH2—a dermoscopic image database for research and benchmarking, in: 2013 35th Annual International Conference of the IEEE Engineering in Medicine and Biology Society, 2013.

[27] A. Pennisi, D.D. Bloisi, D. Nardi, A.R. Giampetruzzi, C. Mondino, A. Facchiano, Skin lesion image segmentation using Delaunay triangulation for melanoma detection, Computerized Medical Imaging and Graphics 52 (2016) 89–103.

[28] I. Maglogiannis, C.N. Doukas, S. Member, Overview of advanced computer vision systems for Skin Lesions Characterization, IEEE Transactions on Information Technology in Biomedicine 13 (September) (2009) 721–733.

[29] E. Ahn, J. Kim, L. Bi, A. Kumar, C. Li, M. Fulham, D.D. Feng, Saliency-based lesion segmentation via background detection in Dermoscopic images, IEEE Journal of Biomedical and Health Informatics 21 (6) (2017) 1685–1693, https://doi.org/10.1109/JBHI.2017.2653179. Accepted to be printed.

[30] H. Fan, F. Xie, Y. Li, Z. Jiang, J. Liu, Automatic segmentation of dermoscopy images using saliency combined with Otsu threshold, Computers in Biology and Medicine 85 (2017) 75–85.

[31] N. Zamani Tajeddin, B. Mohammad Zadeh Asl, A general algorithm for automatic lesion segmentation in dermoscopy images, in: 23rd Iranian Conference on Biomedical Engineering and 2016 1st International Iranian Conference on Biomedical Engineering (ICBME), 2016, pp. 134–139.

[32] M. Jahangir, N.Z. Tajeddin, A. Goodya, B.M. Asl, Segmentation of Lesions in Dermoscopy Images Using Saliency Map and Contour Propagation, 2017 arXiv preprint arXiv: 1703.00087.

[33] T. Sreelatha, M.V. Subramanyam, M.N.G. Prasad, Early detection of skin cancer using melanoma segmentation technique, Journal of Medical Systems 43 (7) (2019) 1–7.

[34] H. Wong, G.H. Lim, Measures of diagnostic accuracy: sensitivity, specificity, PPV, and NPV, Proceedings of Singapore Healthcare 20 (4) (2011) 316–318, https://doi.org/10.1177/201010581102000411.

[35] A. Murugan, S.A.H. Nair, K.P. Sanal Kumar, Detection of skin cancer using SVM, random forest and kNN classifiers, Journal of Medical Systems 43 (8) (2019) 1–9.

Chapter 10

Development of a predictive model for classifying colorectal cancer using principal component analysis

Micheal Olaolu Arowolo[1,2], Happiness Eric Aigbogun[2], Eniola Precious Michael[2], Marion Olubunmi Adebiyi[2] and Amit Kumar Tyagi[3,4]

[1]*Department of Electrical Engineering and Computer Science, University of Missouri, Columbia, SC, United States;* [2]*Department of Computer Science, Landmark University, Omu-Aran, Kwara, Nigeria;* [3]*Department of Computer Science, Vellore Institute of Technology, Chennai, Tamil Nadu, India;* [4]*Department of Fashion Technology, National Institute of Fashion Technology, New Delhi, India*

1. Introduction

The second-leading cause of mortality in the world and one of the most commonly diagnosed diseases is colorectal cancer (CRC). The prevalence and mortality rates of CRC in each state, about the report, are well over one million and 600,000 cases annually, respectively [1].

Recurrences occur in about 45% of cases during the first year of tumor excision, and in over 90% of cases within 4 years. Almost all cases of CRCs are caused by adenocarcinoma, cancer that grows in the inner lining of the colon rectum. Staging and pathological characteristics, which can aid to improve care choices, are one of the most important predictors of diagnosis in CRC. Only a few biomarkers can be utilized to predict CRC currently.

It is critical to explore additional biomarkers that are specific to staging. CRC is the leading second cause of cancer death cases in both genders in the United States. On the other hand, smoking, an unhealthy lifestyle, excessive alcohol use, physical inactivity, and excess body weight represent more than half of all cases and deaths, making them possibly avoidable. With effective screening and monitoring, CRC morbidity and death can be decreased [2].

Because early detection of CRC is a major challenge around the world, current treatment options must be delivered so late after tumor metastasis. As a result, the incidence and death rate of CRC may be lowered if tumors are found early enough and polyps are surgically removed. Many treatments are based on molecular investigations, which include extracting tumor tissue from paraffin blocks for sequencing, thanks to the development of targeted medications. By acting as a screening device, an automated system could potentially minimize pathologists' effort while also reducing diagnosis subjectivity [3].

MicroRNAs (miRNA) are small noncoding molecules that complement their targets to regulate gene expression and silence RNA. Experimenting with target prediction approaches can be time-consuming and expensive. As a result, using a computational approach to illuminate these issues through experimental research is suggested. However, in miRNA biology, an optimized strategy is still required ( ).

As a result, machine learning (ML) will bring in a new era of miRNA biology research focused on disease biomarkers. As a result, in the context of CRC, we examine the use of machine algorithms in miRNA identification and target prediction, as well as their functions and prospects. The application of a new era of computational techniques in this approach will lead to even more advanced levels of miRNA discoveries, lowering the fatality rate in CRC patients. Due to the significant limitations of molecular approaches to finding microRNAs for diagnostic and prognostic biomarkers for CRC, the in silico technique is critical (CRC) [1].

Deep Learning in Personalized Healthcare and Decision Support. https://doi.org/10.1016/B978-0-443-19413-9.00019-9

The status of the Kristen rat sarcoma viral oncogene homolog (KRAS) V-Ki-ras2 mutation has recently been discovered as a critical component in the treatment of CRC, with many studies demonstrating that KRAS mutation predicts a lack of responsiveness to epidermal growth factor receptor therapies (EGFR) [4].

Data mining techniques like classification and prediction have been adopted in solving problems related to health. Data mining techniques are used mostly in health care, which helps them in the early detection and prevention of cancerous diseases, such as CRC, among others. Late prediction and diagnosis of CRC is the major challenge encountered in cubing the fatality rate. By analyzing the past data, data mining can help the health sector and researchers predict patients that have high chances of developing colon cancer; thus, preventing high death rate and developing special means for reducing the chances of more people developing CRC [5].

Machine learning is a data analysis whereby analytical models are formed with the aid of artificial intelligence. It is a subset of artificial intelligence focused on the environment where data are learned, pattern recognized, and judgment made with little or no input from humans (Kumar et al.2019).

One of the major causes of high mortality or death rate is the problem associated with early detection or diagnosis of CRC worldwide. As a result, existing therapeutic techniques must be applied after tumor spread has occurred. CRC incidence and mortality rates may be lowered if tumors are detected early enough and polyps are surgically removed [1]. Preceding research of the relationship involving imaging characteristics of CRC and KRAS mutant status has relied heavily on particle emission tomography with 18F-fluorodeoxyglucose (18F-FDG PET). The outcome derived from studies, however, are inconclusive. Recently, radio genomics applying computed tomography (CT) texture analysis has advanced from the growing scientific field of deriving quantitative imaging features from medical pictures to project genetic status, degree of differentiation, and chemotherapy efficacy in diverse malignancies noninvasively. Validation of genes as indicators for clinical outcome prediction is a serious challenge in cancer research [1]. While some other studies had other limitations too, such as the data were related to one population of similar types. When data are gathered from a single location, there is a chance that some aspects are unique to Chang Gung Memorial Hospital in Taiwan, such as patient characteristics and healthcare professionals, which may not reflect the situation in other hospitals or settings [6].

In the realm of clinical research, the use of ML, which can be applied using supervised or unsupervised methodologies, has skyrocketed [6]. Neural network (NN), decision tree (DT), random forest (RF), linear model, support vector machine (SVM), extreme learning machines, model tree, logistic regression, multivariate adaptive regression splines, and bagged cart model are some ML methods used in predicting CRC. Problems encountered using the ML models include a lack of precision even when using the GPU, the models required a long time to train when the parameters were not initialized properly and hyperparameters were not stated correctly. In the case of incorrect initialization, the algorithms were unable to converge, resulting in poor performance [6].

This study shows how to combine SVM, RF, and PCA to create an ML method. This technique is used to predict CRC.

2. Related works

Yamashita et al., [7] microsatellite instability (MSI) detection in CRC is critical for clinical decision-making since it identifies patients with varying therapy responses and prognoses. MSI testing is suggested for all patients, yet many remain untested. There is a pressing need for widely available, cost-effective tools to help patients choose which tests to undergo. They look at the possibilities of a deep learning-based system for automated MSI prediction directly from whole-slide photos stained with hematoxylin and eosin (H&E) (WSIs). On holdout test set from the internal dataset, the MSINet model had an AUROC of 0931 (95% CI 0771−1000) and on the exterior dataset, it had an AUROC of 0779 (0720−0838). On the external dataset, using a sensitivity-weighted operating point, the model achieved an NPV of 93·7% (95% CI 90·3−96·2), the sensitivity of 76·0% (64·8−85·1), and specificity of 66·6% (61·8−71·2). The model had an AUROC of 0865 (95% CI 0735−0995) in the reader experiment (40 examples). The five pathologists' average AUROC performance was 0605 (95% CI 0453−0757).

Nirmalakumari et al., (2020), investigated the performance investigation of classifiers for colon cancer detection using dimensionality-reduced microarray gene data. PCA and fuzzy C-means clustering (FCM) approaches are used in ANOVA to pick the best genes inseparability and nonlinearity.

Specogna & Sinicrope (2020) used deep learning to define colon cancer biomarkers and accuracy of a deep learning-based method for estimating CRC-specific survival.

To investigate the diagnostic accuracy of artificial intelligence (AI) on histological prediction and detection of colorectal polyps, Lui et al. [8] conducted a meta-analysis of all published research. The histology prediction analysis comprised 7680 pictures of colorectal polyps from 18 studies. The accuracy of the AI (AUC) was 0.96 (95% confidence interval [CI], 0.95−0.98), with a corresponding pooled sensitivity of 92.3% (95% CI, 88.8%−94.9%) and specificity of

89.8% (95% CI, 85.3%–93.0%). AI with narrow-band imaging (NBI) had a considerably higher AUC than AI without NBI (0.98 vs. 0.84 P.01). In comparison to nonexpert endoscopists, AI performed better (0.97 vs.90, P.01). The pooled negative predictive value for characterization of tiny polyps using a deep learning model with nonmagnifying NBI was 95.1% (95% CI, 87.7%–98.1%). The pooled AUC for polyp identification was.90 (95% CI, 67–1.00), with a sensitivity of 95.0% (95% CI, 91.0%–97.0%) and a specificity of 88.0% (95% CI, 58.0%–99.0%).

In Skrede et al. [9], they enhanced prognostic indicators, which are needed to stratify patients with early-stage CRC and refine adjuvant therapy selection. The goal of this work was to use deep learning to construct a biomarker for patient outcomes following primary CRC resection by evaluating scanned conventional hematoxylin and eosin-stained sections. 828 patients from four cohorts were used as a training cohort to obtain clear ground truth. Tuning was done on 1645 patients who had a nondistinct outcome. In the primary analysis of the validation cohort, the biomarker had a hazard ratio of 384 (95% CI 272–543; *P*00001) for poor versus good prognosis, and a hazard ratio of 304 (207–447; *P*00001) after adjusting for established prognostic markers significant in univariable analyses of the same cohort, which were a pN stage, pT stage, lymphatic invasion, and venous.

Cardiol and Med [10] study used training, test, and validation cohorts to create and retrospectively deploy a deep learning-based algorithm for estimating CRC-specific survival. The training set included 828 individuals with stage I–III CRC (49% women; median age 69 years [IQR 61–75]) who were classified as having a favorable or poor disease outcome. These patients' data were used to train 10 convolutional neural networks (designed for categorizing diverse images), which were then combined into a predictive biomarker and applied to patients with varying disease outcomes. The marker was subsequently independently verified using a clinical trial cohort of patients with stage II and III CRC (QUASAR2; 1122 patients; 43% women; median age 65 years [IQR 59–71]) and another dataset (920 patients; 46% women; median age 71 years [IQR 64–78]).

Sánchez-Peralta et al. [11] conducted a thorough assessment of 35 studies published since 2015 that used deep learning algorithms to detect, localize, and segment polyps. We also looked at seven publicly available public databases of colonoscopy pictures, as well as the most frequent reporting parameters. The approaches were categorized into primary (end-to-end vs. hybrid methods) and secondary (feature extractor, classification, patch-based, bounding-box, and semantic segmentation) classifications depending on their approach. Despite the lack of a uniform dataset or framework for easy and direct method comparison, several trends, benefits, and drawbacks have been observed and explored. Finally, future issues and recommendations have been outlined.

In Takamatsu et al. [12], predicting lymph node metastasis (LNM) for early CRC is crucial for deciding treatment regimens following endoscopic resection. Although several histologic factors for predicting LNM have been developed, concerns such as evaluator error and interobserver disagreement remain unresolved. They present an ML-based LNM prediction system for submucosal invasive (T1) CRC. They studied 397 T1 CRCs in a retrospective single-institution investigation. Image J was used to derive several morphologic metrics from full slide images of cytokeratin immunohistochemistry. A RF technique was utilized to predict LNM for the test dataset (n = 120) using a training dataset (n = 277). The results were compared to traditional hematoxylin and eosin staining histology. On several datasets, ML outperformed the traditional method in terms of LNM prediction. There was no discernible difference between the approaches after cross-validation. In comparison to the traditional method, ML produced fewer false-negative cases. Conclusions: Using ML on entire slide images to determine treatment regimens for T1 CRC is a viable option.

Taguchi et al. [4] goal of the study was to see if an ML-based computed tomography (CT) texture analysis could predict V-Ki-ras2 Kirsten rat sarcoma viral oncogene homolog (KRAS) mutation status in CRC. When compared to the 18F-FDG-PET SUVmax, they discovered that a ML approach based on extensive CT texture features performed better for predicting the KRAS mutation status in CRC.

Kourou et al. (2015) ML applications in cancer prognosis and prediction were studied artificial neural networks (ANNs), Bayesian networks (BNs), SVMs, and decision trees were used to examine these methods on a picture dataset (DTs). They ran into problems with their investigation due to the tiny sample size of the dataset.

3. Methodology

The system in this study has developmental stages that include preprocessing and classification. The dataset is pre-processed using PCA. RF and SVM will be used to classify the data, and the results will be compared.

Principal component analysis is a method for reducing the dimensionality of such datasets while preserving information and improving interpretability. It achieves this by sequentially creating new uncorrelated variables, which optimize variance.

Support vector machine, K-Nearest Neighbor, and RF are the categorization algorithms used. The results of this technique are based on the classification performance. The methodologies employed in this study are broken down as follows.

i. Reduce the dimensionality of the picture collection using PCA.
ii. Then, for the classification strategy, utilize SVM K-nearest neighbor, and RF.
iii. Then compare the data that are been passed through the PCA and the one that is passed through the classification technique.
iv. Assess the accuracy, specificity, sensitivity, precision, and computational time of the results.

Fig. 10.1 depicts the proposed work's overall system design.

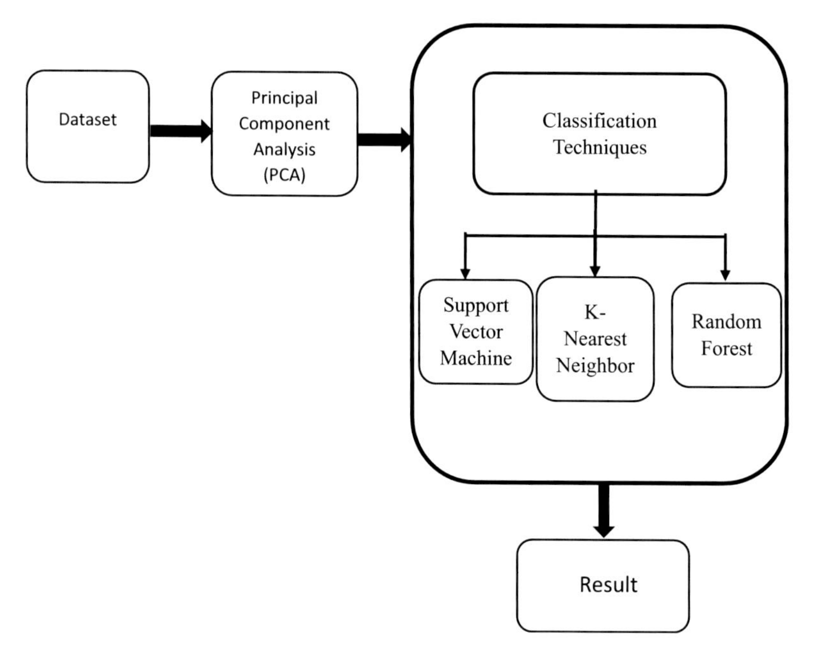

FIGURE 10.1 Proposed system design. Step 1: Get data set relating to the topic. The dataset was gotten from Kaggle (https://www.kaggle.com/kmader/colorectal-histology-mnist). Step 2: To minimize the dimensionality of the image dataset, PCA is utilized. Step 3: To carry out and simplify the performance of the dataset to predict, use SVM as a classification method. Step 4: K-nearest neighbor is used as a classification method to execute and simplify the dataset performance to predict. Step 5: Using Random Forest as a Classification method to execute and simplify the dataset's performance to predict. Step 6- Analyze the performance of the result acquired using PCA to the results obtained without it (PCA).

3.1 Experimental dataset

To achieve the desired goal, the proposed system uses a Colon Cancer Gene Expression dataset from Kaggle (https://www.kaggle.com/kmader/colorectal-histology-mnist).

3.2 Dimensionality reduction tool

Principal component analysis is a technique for reducing the dimensionality of such datasets, improving interpretability while minimizing loss of information. It achieves this by sequentially generating new uncorrelated variables that reduce variance.

Algorithms 1 PCA

i. Step 1: Get your data
ii. Step 2: Give your data a structure
iii. Step3: Standardize your data
iv. Step 4: Get Covariance of Z
v. Step 5: Calculate Eigen vectors and Eigen values
vi. Step 6: Sort the Eigen vectors
vii. Step 7: Calculate the new features
viii. Step 8: Drop unimportant features from the new set

3.3 Classification

Classification models heavily influence the ultimate performances. On any given data set, different classifiers perform differently.

3.3.1 Support vector machine

The SVM is an ML technique that can handle both small- and large-scale sample problems. SVM has been validated as one of the best and most accurate classifiers (Li & Zeng, 2007). The SVM polynomial kernel examines the combinations of input samples' provided features to identify their similarity, using regression analysis with a polynomial in the number of parameters to be learned. These features correspond to logical conjunctions of input features.

Support vector machine algorithms: SVM algorithms are a statistical classification tool that works well with high-dimensional datasets (Pang-ning et al., 2018).

SVM is designed by using support vectors, which are instances of training data, to create a decision border between classes. The grandest hyperplane isolation between the instances of the different classes is achieved by determining this. SVM has been better thought to discover world-optimal solutions than rules or neural network techniques (Pang-ning et al., 2018).

Kar et al. (2016) use the SVM classification and experiment in the SVM classification with various centrality metrics. Instead of text-based analysis, as is the case in that study, Choi et al., 2015 use an SVM-classifier.

Algorithms two support vector machine

- **i.** Step 1 candidate SV = {Contra class closest pair}
- **ii.** Step 2 while violations do occur
- **iii.** Step 3 Find a violator
- **iv.** Step 4 candidate SV = infringer candidate SV S
- **v.** Step 5 if any $\alpha p < 0$ due to addition of c to S then
- **vi.** Step 6 candidate SV = candidate SV\P
- **vii.** Step 7 Repeat until all these points have been cut
- **viii.** Step 8 ends if
- **ix.** Step 9 ends while

3.3.2 K-nearest neighbor

The K-nearest neighbor algorithm is a type of supervised learning technique that is used for classification and regression. A flexible approach may also be used to fill in missing values and resample datasets. K-nearest neighbor considers (data points) to predict the class or continuous value for a new data point, as the name suggests [13]. The algorithm learning is.

- **i.** Instance-based learning: Rather than learning weights from training data to predict output (as in model-based algorithms), we use full training instances to predict output for unseen data.
- **ii.** Lazy learning: The model is not learned using training data before the prediction is required on the new instance, and the learning process is postponed until the prediction is asked.
- **iii.** Nonparametric: In KNN, the mapping function has no specified form [13].

Algorithms 3 K-nearest neighbor

- **i.** Step 1 Load the training data.
- **ii.** Step 2 Prepare data by scaling, missing value treatment, and dimensionality reduction as required.
- **iii.** Step 3 Find the optimal value for K:
- **iv.** Step 4 Predict a class value for new data:
    - **a.** Calculate distance(X, Xi) from i = 1,2,3, ...,n. where X = new data point, Xi = training data, distance as per your chosen distance metric.
    - **b.** Sort these distances in increasing order with corresponding train data.
    - **c.** From this sorted list, select the top "K" rows.
    - **d.** Find the most frequent class from these chosen "K" rows. This will be your predicted class.

3.3.3 Random forest

The RF concept is used in the classification module. RF is a robust, easy-to-use machine-learning algorithm that, in the vast majority of cases, produces great results without hyperparameter adjustment. It is also one of the most extensively utilized algorithms due to its simplicity and versatility (it can be used for both classification and regression tasks). By fitting several decision tree classifiers on different sub-samples of the dataset, a meta estimator increases projected accuracy and controls overfitting. The sub-sample size is controlled by the max samples argument if bootstrap = True (default); otherwise, the whole dataset is used to generate each tree.

Algorithms four random forest

i. Step1: Randomly select **"k"** features from total **"m"**features.
ii. Step2: Where **k << m**
iii. Step3: Among the **"k"** features, calculate the node **"d"**using the best split point.
iv. Step4: Split the node into **daughter nodes** using the **best split.**
v. Step5: Repeat **1 to 3**rdeps until the "l" number of nodes has been reached.
vi. Step6: Build forest by repeating steps **1 to 4** for "n" number times to create **"n" number of trees.**
vii. Step7: end

3.4 Research tool

The study proposes to develop the implementation of this study using PYTHON. The system configuration used comprises of processor with Intel inside Corei7, 2.90 GHz speed, 8 GB RAM, 20 GB Hard disc, Windows 10 OS, with dataset https://www.kaggle.com/kmader/colorectal-histology-mnist.

3.5 Performance evaluation metrics

The performance evaluation metrics of the classifier are evaluated in terms of classification accuracy, time, sensitivity, specificity, and precision. The terms are defined below.

Sensitivity = TP/(TP + FN) %
Specificity = TN/(TN + FN) %
Accuracy = (TP + TN)/(TP + TN + FP + FN) %
Precision: TP/(TP + FP)

Where.

TP (True Positives) = correctly classified positive cases,
TN (True Negative) = correctly classified negative cases,
FP (False Positives) = incorrectly classified negative cases,
FN (False Negative) = incorrectly classified positive cases.

Sensitivity (true-positive fraction) is the probability that a diagnostic test is positive, given that the person has the disease.

Specificity (true-negative fraction) is the probability that a diagnostic test is negative, given that the person does not have the disease.

Accuracy is the probability that a diagnostic test is correctly performed.

4. Results and discussions

Classification algorithms were implemented using PYTHON on the Jupyter notebook 6.0.3 platform, thereafter, classification techniques were performed. Particularly, this section presents the results of studies for the proposed model. The application and comparison of these methods were performed using feature extraction (PCA). This study makes use of classification procedures such as SVM, KNN, and RF. This environment consists of eight functionality.

This is done by using Pandas; it is a package that makes importing and analyzing data much easier. To read the dataset used for this project we made use of: pd.*read_csv*, where pd stands for pandas and *.read_csv* signifies that only CSV files can be read into this program. The dataset read was named *colon data,* it was also stored in the download folder in the system and it contained 5000 instances and 785 attributes. The dataset was loaded using the following code *data* = pd.*read_csv('Downloads/colondata.csv')*

Data. While *data. head( )* is used to return the first five(5) rows of the dataset read. The return values can be manipulated to the required number of return rows need by adding a value to the bracket *( ).* This is represented as *data. shape* in the program written. It describes the shape attribute of the dataset; it stores the number of rows and columns as a tuple and displays it. Fig. 4.4 shows the data shape of the Colorectal Histology MNIST dataset.

Datasets are described in the program by using *data.describe( ),* this is a method used to calculate several statistical data like the mean, percentile, and the standard deviation (std) of the numerical values of the data frame. In this dataset, the count, mean, std, minimum, percentile of (25%, 50%,75%), and the maximum was calculated.

It is a technique used to convert the raw data into clean datasets. In this program, we made use of rescaling to optimize the algorithm and we used scikit-learn using the MinMaxScaler class and StandardScaler. Fig. 4.6 shows the scaled dataset.

This is a technique commonly used in ML to split data into a train, test, or validation set. Each algorithm divided the data into two subsets, training/testing. The training set was used to fit the model and testing for the evaluation. In this study, 80% was used for training and 20% is used for testing.

In this study, Principal Component Analysis (PCA) is used to reduce the dimensionality of the dataset to produce better performance. Fig. 10.2 shows the selected features of the dataset after using the principal component analysis.

Support vector machine is a supervised learning technique and it is applied with the PCA to produce a result for the confusion matrix. Fig. 10.3 shows the confusion matrix of the Support vector machine with Principal Component analysis.

K- Nearest Neighbor is a supervised learning technique and it is applied with the PCA to produce a result for the confusion matrix. Fig. 10.4 shows the Confusion Metrix of the SVM with PCA.

Random Forest is a supervised learning technique and it is applied with the PCA to produce a result for the confusion matrix. Fig. 10.5 shows the confusion matrix of Random Forest with Principal Component analysis.

Scatter plots are important in statistics because they can show the extent of correlation, if any, between the values of selected features or variables. Scatter plots are useful data visualization tools for illustrating a trend. Figs. 10.6 and 10.7 shows the scatter plot visualization before and after the application of PCA.

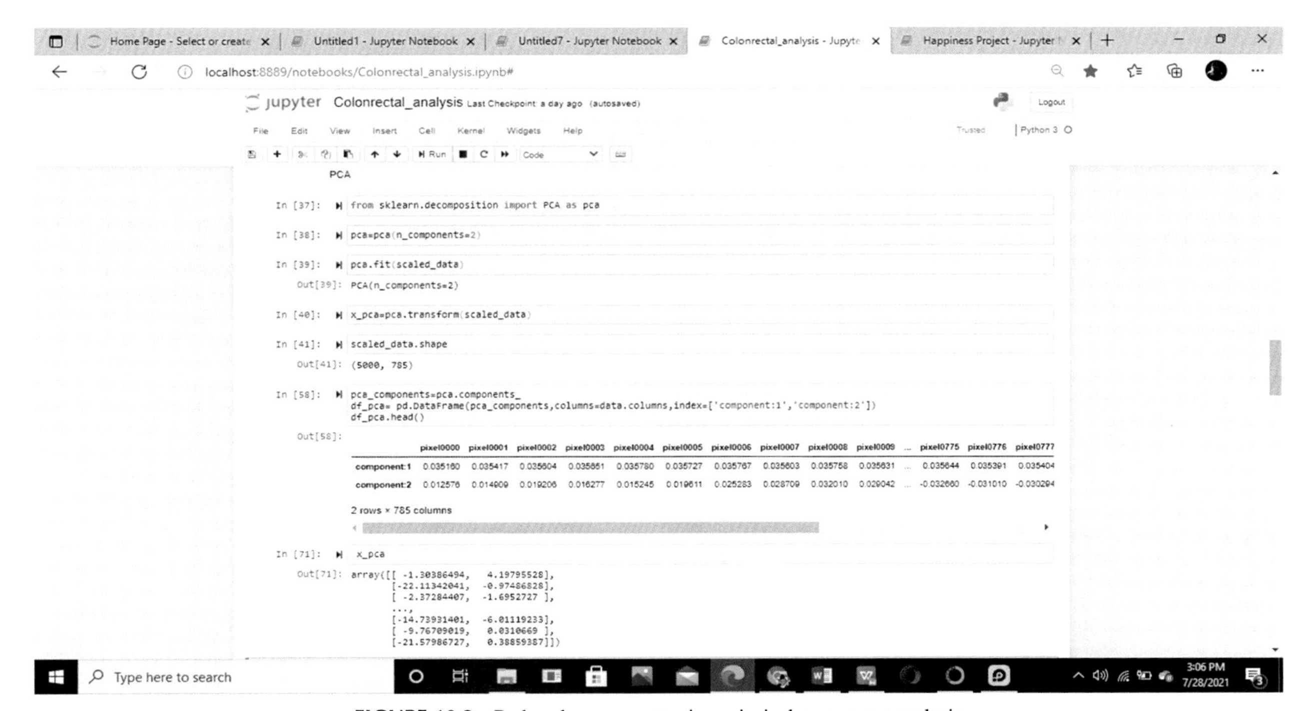

FIGURE 10.2 Reduced component using principal component analysis.

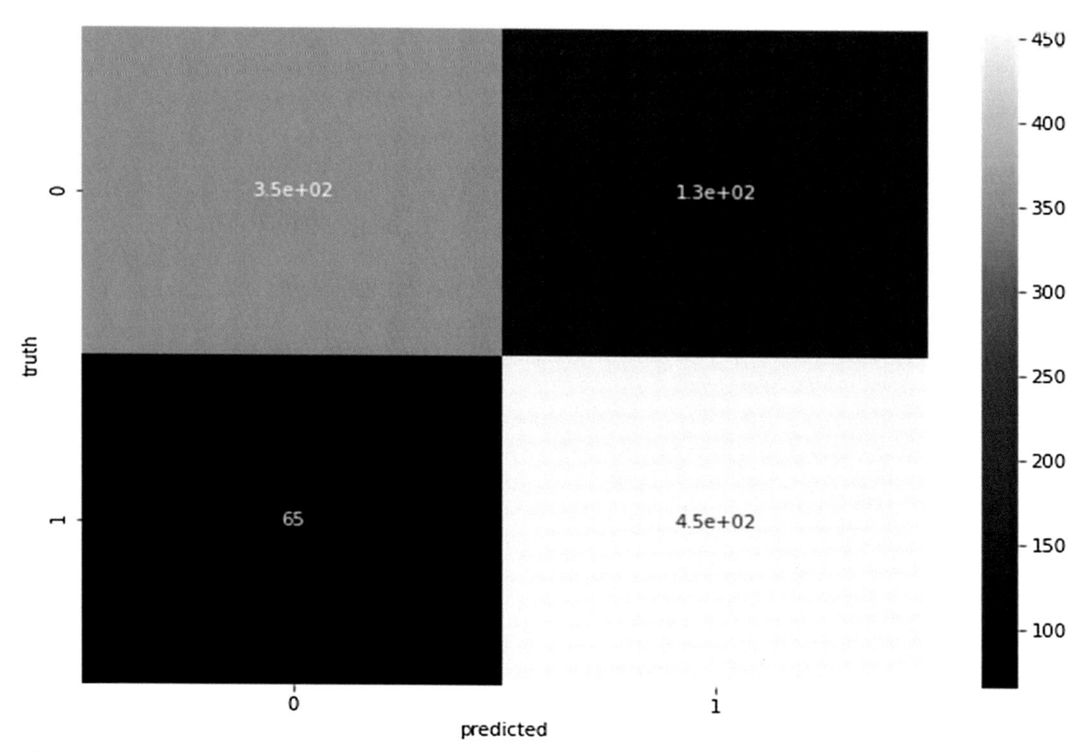

FIGURE 10.3 Confusion matrix for colon cancer using PCA with support vector machine (TP = 350; TN = 131; FP = 65; FN = 454).

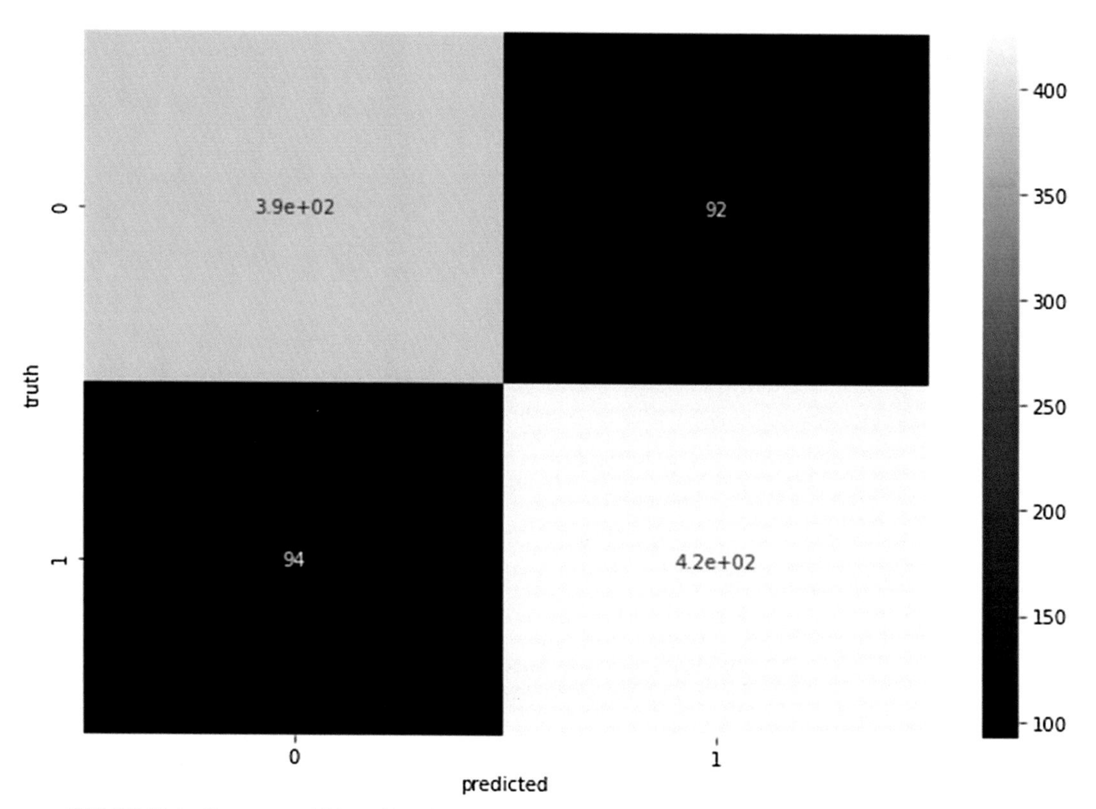

FIGURE 10.4 K-nearest neighbor with principal component analysis (TP = 389; TN = 92; FP = 94; FN = 425).

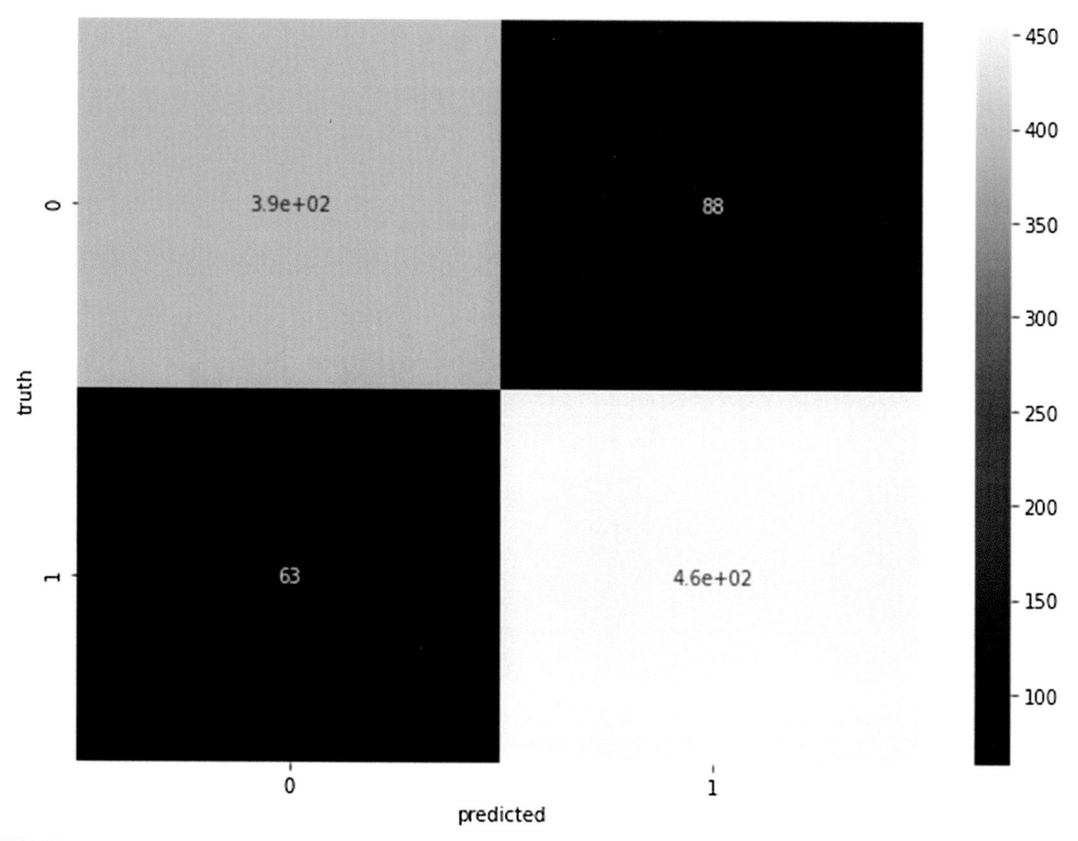

FIGURE 10.5 Confusion metrix of random forest with principal component analysis TP = 453; TN = 28; FP = 12; FN = 507).

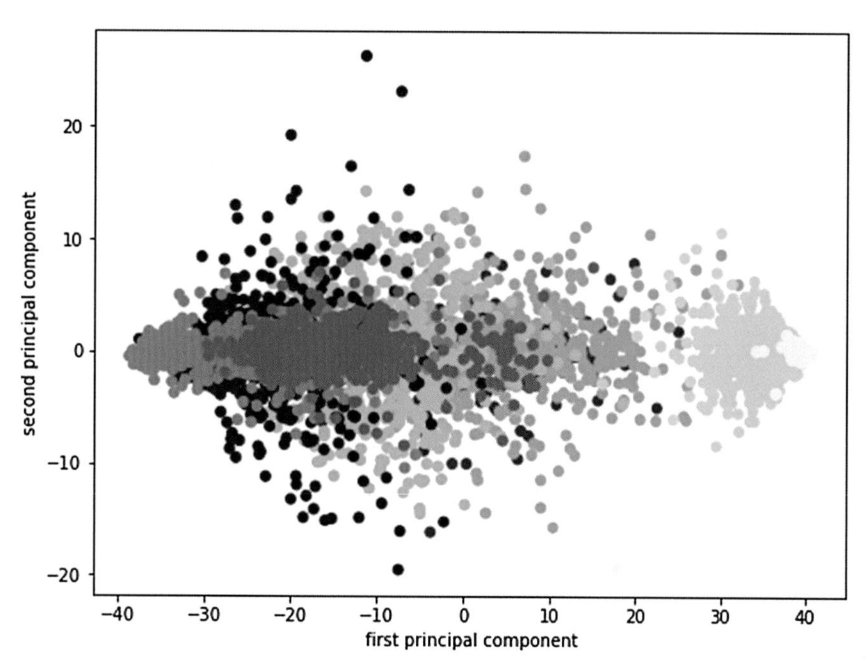

FIGURE 10.6 Scattered plot of colorectal histology MNIST before application of PCA.

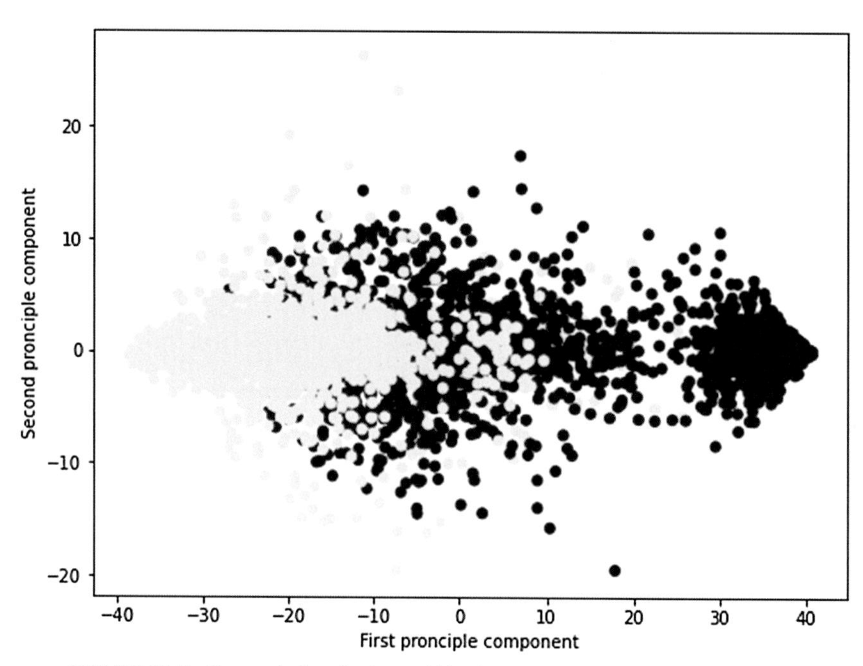

FIGURE 10.7 Scattered plot of colorectal histology MNIST after application of PCA.

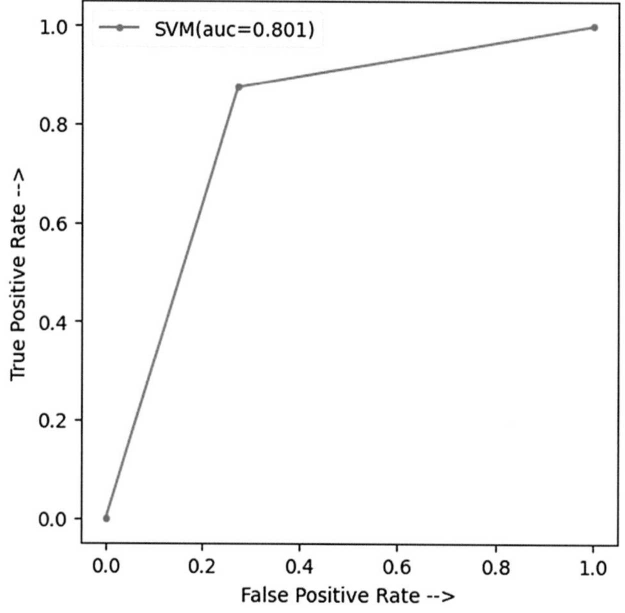

FIGURE 10.8 ROC curve for support vector machine.

The ROC curve shows the performance of the classification thresholds. The curve plots the True Positive and False Positive rates. Figs. 10.8, 10.9 shows the ROC curve for SVM, K-nearest neighbor classifiers, and RF.

The confusion matrices obtained are been evaluated using evaluation metrics such as sensitivity, specificity, precision, accuracy, and f1 score.

In this study, classification is performed on the data using SVM, K-NN, and random forest; the data are however passed in these stated classifiers; hence, data are passed through the PCA in each classifier used. Table 10.1 shows the performance measures of each of the classifiers with and without the PCA. Table 10.2 compares the obtained result with PCA.

In this study, several experiments were carried out and the table shows the evaluations, however, PCA + Random Forest outperformed with 96.00%accuracy. Table 10.3 shows the comparison of the results obtained with the state-of-art.

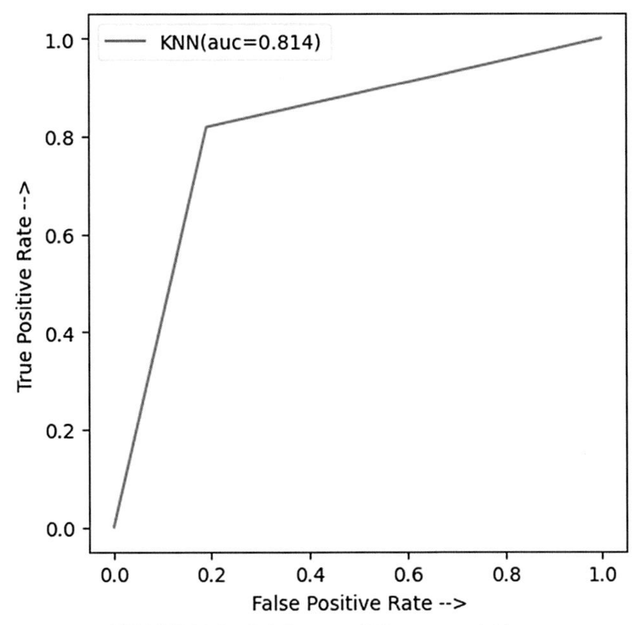

FIGURE 10.9 ROC curve of K-nearest neighbor.

TABLE 10.1 Calculation and comparison of performance before PCA.

Performance	Support vector machine	K-Nearest neighbor	Random forest	Derivation
Accuracy	77.75	81.00	84.00	ACC = (TP + TN)/(P + N)
Sensitivity	82.95	93.18	84.97	TPR = TP/(TP + FN)
Specificity	69.57	74.81	77.78	SPC = TN/(FP + TN)
Precision	81.08	65.28	96.08	PPV = TP/(TP + FP)
F1 score	82.00	76.77	90.18	F1 = 2 TP/(2 TP + FP + FN)

TABLE 10.2 Calculation and comparison of performance after PCA.

Performance	Support vector machine	K-Nearest neighbor	Random forest	Derivation
Accuracy	80.40	81.90	96.00	ACC = (TP + TN)/(P + N)
Sensitivity	84.34	94.12	97.42	TPR = TP/(TP + FN)
Specificity	77.61	75.61	94.77	SPC = TN/(FP + TN)
Precision	72.77	66.53	94.18	PPV = TP/(TP + FP)
F1 score	78.13	77.95	95.77	F1 = 2 TP/(2 TP + FP + FN)

TABLE 10.3

Authors	Methods	Results of accuracy
[4]	Univariate logistic regression Multivariate support vector machine	58% 82%
[8]	Artificial intelligence	96%
[9]	Deep learning	95%
[7]	Deep learning mode (msinet)	93.1%
[3]	Deep learning–based method in colorectal cancer detection and segmentation from digitized H&E-stained histology slides.	99.9% of normal slides and 94% of cancer slides compared to pathologist based diagnosis on H&E-stained slides digitized from clinical samples

5. Conclusion

This study made use of ML models such as SVM, K-nearest neighbor, and RF, in the development of a predictive model for CRC. This study made use of the Colon Cancer Gene Expression dataset gotten from Kaggle (https://www.kaggle.com/kmader/colorectal-histology-mnist). The dataset consists of 5000 instances and 785 attributes. The Colon Cancer Gene Expression dataset was preprocessed and split into the training and testing set. The training set (80%) was used to determine the optimal combinations of variables that will generate a good predictive model, and the testing set (20%) was used to provide an unbiased evaluation of a final model fit on the training dataset. The dataset was passed through the PCA first to fetch the relevant features that will provide improved performance before it is passed into SVM, K Nearest Neighbor, and Random Forest classifiers, and results were obtained with an accuracy of 80.40%, 81.90%, and 96.00% respectively, which was later compared to the results obtained in the related works. Further, to enhance this study, I recommend that future works should introduce more algorithms and hybridize them instead of using PCA or can make use of other feature extraction methods or techniques. In addition, other classification methods or algorithms can be included such as neural network, K-means, and logistic regression in other to improve the performance of the system, and then compared with.

References

[1] A.O. Fadaka, A. Klein, A. Pretorius, In silico identification of microRNAs as candidate colorectal cancer biomarkers, Tumor Biology 41 (11) (2019) 1–15, https://doi.org/10.1177/1010428319883721.

[2] R.L. Siegel, K.D. Miller, A. Goding Sauer, S.A. Fedewa, L.F. Butterly, J.C. Anderson, A. Cercek, R.A. Smith, A. Jemal, Colorectal cancer statistics, 2020, CA: A Cancer Journal for Clinicians 70 (3) (2020) 145–164, https://doi.org/10.3322/caac.21601.

[3] L. Xu, B. Walker, P. Liang, Y. Tong, C. Xu, Y.C. Su, A. Karsan, Colorectal Cancer Detection Based on Deep Learning, 2021, https://doi.org/10.4103/jpi.jpi, 1–13.

[4] N. Taguchi, S. Oda, Y. Yokota, S. Yamamura, M. Imuta, T. Tsuchigame, Y. Nagayama, M. Kidoh, T. Nakaura, S. Shiraishi, Y. Funama, S. Shinriki, Y. Miyamoto, H. Baba, Y. Yamashita, CT texture analysis for the prediction of KRAS mutation status in colorectal cancer via a machine learning approach, European Journal of Radiology 118 (March) (2019) 38–43, https://doi.org/10.1016/j.ejrad.2019.06.028.

[5] H. Liang, L. Yang, L. Tao, L. Shi, W. Yang, J. Bai, D. Zheng, N. Wang, Data mining-based model and risk prediction of colorectal cancer by using secondary health data, A systematic review 32 (2) (2021) 242–251, https://doi.org/10.21147/j.issn.1000-9604.2020.02.11.

[6] P. Gupta, S. Chiang, P.K. Sahoo, Prediction of colon cancer stages and survival, Cancers (2019) 1–16.

[7] R. Yamashita, J. Long, T. Longacre, L. Peng, G. Berry, B. Martin, J. Higgins, D.L. Rubin, J. Shen, Deep learning model for the prediction of microsatellite instability in colorectal cancer: a diagnostic study, The Lancet Oncology 22 (1) (2021) 132–141, https://doi.org/10.1016/S1470-2045(20)30535-0.

[8] T.K.L. Lui, C.G. Guo, W.K. Leung, Accuracy of artificial intelligence on histology prediction and detection of colorectal polyps: a systematic review and meta-analysis, Gastrointestinal Endoscopy 92 (1) (2020) 11–22, https://doi.org/10.1016/j.gie.2020.02.033, e6.

[9] O.J. Skrede, S. De Raedt, A. Kleppe, T.S. Hveem, K. Liestøl, J. Maddison, H.A. Askautrud, M. Pradhan, J.A. Nesheim, F. Albregtsen, I.N. Farstad, E. Domingo, D.N. Church, A. Nesbakken, N.A. Shepherd, I. Tomlinson, R. Kerr, M. Novelli, D.J. Kerr, H.E. Danielsen, Deep learning for prediction of colorectal cancer outcome: a discovery and validation study, The Lancet 395 (10221) (2020) 350–360, https://doi.org/10.1016/S0140-6736(19)32998-8.

[10] C.J. Cardiol, I.E. Med, Defining Colon Cancer Biomarkers by Using Deep Learning, 2020, https://doi.org/10.1016/S0140-6736(20)30034-9, 395, 314–316.

[11] L.F. Sánchez-Peralta, L. Bote-Curiel, A. Picón, F.M. Sánchez-Margallo, J.B. Pagador, Deep learning to find colorectal polyps in colonoscopy: a systematic literature review, Artificial Intelligence in Medicine 108 (March) (2020) 101923, https://doi.org/10.1016/j.artmed.2020.101923.

[12] M. Takamatsu, N. Yamamoto, H. Kawachi, A. Chino, S. Saito, Computer methods and programs in biomedicine prediction of early colorectal cancer metastasis by machine learning using digital slide images, Computer Methods and Programs in Biomedicine 178 (2019) 155–161, https://doi.org/10.1016/j.cmpb.2019.06.022.

[13] V. Book, Simple understanding and implementation of the KNN algorithm, AnalyticsVidhya 10000 (2021) 1–14. https://www.analyticsvidhya.com/.

Chapter 11

Using deep learning via long-short-term memory model prediction of COVID-19 situation in India

Saroja Kumar Rout[1], Bibhuprasad Sahu[2], Amar Kumar Das[3], Sachi Nandan Mohanty[4] and Ashish K. Sharma[5]

[1]*Department of Information Technology, Vardhaman College of Engineering (Autonomous), Hyderabad, Telangana, India;* [2]*Department of Artificial Intelligence & Data Science, Vardhaman College of Engineering (Autonomous), Hyderabad, Telangana, India;* [3]*Department of Mechanical Engineering, Gandhi Institute for Technology, Bhubaneswar, Odisha, India;* [4]*School of Computer Science & Engineering (SCOPE), VIT-AP University, Amaravati, Andhra Pradesh, India;* [5]*Department of Computer Science & Engineering, Bajaj Institute of Technology (BIT), Wardha, Maharashtra, India*

1. Introduction

What is COVID-19 and its analysis?

Some establishments found that the South China Seafood Region was the source of a large number of pneumonia cases during December 2019 [1]. South China Seafood City was promptly pursued by the Regional Health and Welfare Commission, which reported 27 cases, seven of which were real, but most were healthy and manageable. The Chinese government authoritatively outlined the COVID-19 outbreak on January 9, 2020, where the underlying cases in Wuhan City were reported. As of January 9, 2020, the China Center for Disease Control and Prevention announced that a novel coronavirus (SARS-CoV-2) has been identified as the cause of 15 of the 59 pneumonia cases. Regrettably, India upheld its first case on January 30, 2020. From that point on, we have seen a steady increase in the number of confirmed cases [2]. In India, the COVID-19 pandemic is part of the global coronavirus disease pandemic. Due to coronavirus disease in extreme respiratory conditions (SARS-CoV-2) (SARS-CoV-2 in India, as of December 29, 2020, a total of 10,224,303 confirmed cases have been registered in India (with 268,581 active cases representing 2.62% of total cases). 9,807,569 (95.92%) cases have recovered, while the fatality of cases is 1.45%, one of the lowest worldwide. Recorded the highest number of cases confirmed in India, Asia [3] and in the world, the third-largest number of instances reported [2,4]. As of December 17, 2020, India had registered over 9.9 million confirmed COVID-19 cases. Out of these, more than 9.5 million patients recovered, and 144,000 cases were fatal [5]. As of May 24, 2020, the only location that had not registered an incident is Lakshadweep [6]. Indian recoveries reached active events on June 10, eliminating for the first time, 49% of infections total. At the behest of Pm Narendra Modi, India faced a 14-h voluntary public curfew on March 22, prompted by compulsory lockdowns in COVID-19 zones and all major cities. On March 2 and April 4, 2020, respectively, containment plans to monitor clusters and major outbreaks have been issued by the Ministry of Health and Family Welfare, and these plans have been updated from time to time. The nation was coordinating regional assessments at the state and UT levels digitally, in recognition of the difficulties involved in extending the rollout during the COVID-19 pandemic. It is common for states/UTs to present and disseminate their best practices for ensuring the continuity of healthcare facilities to promote cross-learning. In this paper, we propose deep learning-based models for predicting COVID-19 confirmed and mortality cases in India and the United States, as well as a case comparison. We predicted the COVID-19 cases for both countries a month ahead of time. To create the proposed models, we used recurrent neural networks (RNNs) based on LSTM variations. The anticipated results are derived using stacked LSTM, bi-directional LSTM, and convolutional LSTM, which are more accurate than regular LSTM models in predicting future values. To our knowledge, no previous comparative case

Deep Learning in Personalized Healthcare and Decision Support. https://doi.org/10.1016/B978-0-443-19413-9.00010-2

study of the COVID-19 pandemic in India and the United States has been conducted. Our contribution to this comparative study will help both countries rebuild their COVID-19 preparation designs and demographics.

The Government of India, under the chairmanship of the Member (Health), NITI Aayog, established a National Expert Group on Vaccine Administration for COVID-19 on August 7, 2020, to coordinate COVID vaccine procurement and distribution. The use of telemedicine is already largely promoted to support the government in healthcare services in far areas both in COVID and non-COVID health issues. A comprehensive web-based telemedicine platform ("eSanjeevani") can be used (in 23 states) to broaden the reach of critical health care services to the masses across rural remote societies. More than 11 lakh teleconsultations were held on this digital platform as of December 29, 2020. The Ministry also launched a Clinical Center of Excellence (CoE) initiative to guide clinical management protocols in Delhi as the ultimate nodal agency and state level CoEs with AIIMS. AIIMS organizes weekly webinars to instruct these state-level CoEs on important medical problems doctors might have advice on case management for COVID-19. CoEs at the state level are expected to disseminate these in their districts as well.

There is now published public data on COVID-19 pandemic effects in India, similar to what is available for the rest of the world [7–9]. Despite being the world's second-most populous country with subpar healthcare services, India has been fortunate in having a low case-fatality rate till now [10,11]. The position of different transmission reduction measures is now better established [12–16]. The specific reasons for India's outcome may be multidimensional, but the different measures taken by policymakers to slow down the spread of the virus have also been important [17–19]. These have provided the scientific community an opportunity to better understand the dynamics of the vector host system, in addition to helping to flatten the curve [20,21]. These actions have also given time to schedule, enable, and reallocate resources for existing healthcare entities to better prepare for the surge. Within a limited period, a large number of patients needed to be taken care of. The time gained by several government programs also helped individuals and commercial partnerships achieve resource self-sufficiency, such as personal protective equipment, sanitizers, and hospital beds [22], during a time when the world was grappling with problems [23,24]. A mathematical model has been provided for predicting time series and analyzing the impact of a regional lockdown. Using RNN-based long short-term memory, Pathan et al. [24] created a model and predicted the mutation rate of COVID-19 (LSTM). Table 11.1 describes the statistics of COVID-19 data till (December 30, 2020). Table 11.2 also represents the statistics of COVID-19 data till (June 30, 2022).

1.1 Research gaps and motivation

In this way, we can compare the characteristics of COVID-19 articles found through ML with those related to non-SARS-CoV-2 coronaviruses to see how the focus of COVID-19 research differs from the research on other coronaviruses. Research gaps for COVID-19 may be identified based on these differences.

Traditional machine learning models can take a long time to develop with an acceptable level of accuracy, which could postpone forecasts. Third parties are responsible for preparing, cleaning, and reorganizing data so the model can be prepared; however, this may violate data privacy if the data are handled for this purpose. This paper presents a model for predicting COVID-19-confirmed cases based on LSTMs. We propose an LSTM-based prediction model based on the current COVID-19 infection status. The LSTM models form the core of this paper. To reduce costs, solve the cold start issue, and maintain data privacy, it is essential to develop an accurate machine learning methodology. Clients would also receive results immediately with this approach. LSTM models offer the advantage of effectively capturing long-term time dependencies. This model takes into account diagnoses, deaths, and whether the city is closed. It also includes features such as the number of newly diagnosed, deaths, and the growth rate of diagnoses and deaths. Infection rates are forecasted based on the final output.

2. Literature review

Literature evidence indicates that animals are the source of COVID-19 and that the virus has recently spread from humans to humans [25]. There is no reasonable evidence recently to clarify how easily pathogens are transmitted from person to person. In this case, people who sneeze, cough, or breathe out by respiratory droplets are effectively spreading the infection. It is currently estimated that COVID-19 broods for between 2 and 14 days after the initial infection. We understand at this stage that when the infected exhibits (influenza-like) symptoms, the infection can be transmitted.

The impact of COVID-19 on the health of individuals and populations in such settings has been modeled in several studies [26–29], but few primary studies document the dynamics of transmission and clinical outcomes in LMICs to validate the models and inform interventions.

TABLE 11.1 State-wise statistics for COVID-19 (till December 30, 2020).

State/UT	Confirmed	Active	Recovered	Deceased
Total	1,02,66,674	257,656	98,60,280	148,738
Maharashtra	19,28,603	54,206	18,24,934	49,463
Karnataka	918,544	11,629	894,834	12,081
Andhra Pradesh	881,948	3256	871,588	7104
Tamil Nadu	817,077	8615	796,353	12,109
Kerala	755,718	65,572	687,104	3042
Delhi	624,795	5838	608,434	10,523
Uttar Pradesh	584,966	14,155	562,459	8352
West Bengal	550,893	12,381	528,829	9683
Odisha	329,306	2332	325,103	1871
Rajasthan	307,554	9835	295,030	2689
Telangana	286,354	5974	278,839	1541
Chhattisgarh	278,540	11,939	263,251	3350
Haryana	262,054	3799	255,356	2899
Bihar	251,348	4799	245,156	1393
Gujarat	244,258	9979	229,977	4302
Madhya Pradesh	240,947	9387	227,965	3595
Assam	216,139	3258	211,838	1043
Punjab	166,239	3865	157,043	5331
Jammu and Kashmir	120,744	3034	115,830	1880
Jharkhand	114,873	1640	112,206	1027
Uttarakhand	90,616	4963	84,149	1504
Himachal Pradesh	55,114	2796	51,387	931
Goa	50,981	931	49,313	737
Puducherry	38,096	363	37,100	633
Tripura	33,264	128	32,751	385
Manipur	28,137	1182	26,601	354
Chandigarh	19,682	399	18,967	316
Arunachal Pradesh	16,711	106	16,549	56
Meghalaya	13,408	184	13,085	139
Nagaland	11,921	218	11,624	79
Ladakh	9447	188	9132	127

Reference: https://www.mygov.in/corona-data/covid19-tatewise-status/.

2.1 Symptoms of COVID-19

It can take 2–14 days for symptoms to develop after a person has contracted coronavirus. The normal duration for incubation tends to be approximately days 5–6. According to the World Health Organization (WHO), COVID-19 symptoms are mild and may develop slowly. According to Just the Lancet, this usually occurs from 7 days onward when hospital admission is required. COVID-19 has been associated with several symptoms, including shortness of breath and a dry cough, according to the Centers for Disease Control and Prevention (CDC). A fever is described as a temperature of 100.4°F or higher. Touching or checking a fevered person's chest will make them feel wet. Coughing that is dry, a dry

TABLE 11.2 State-wise data: India (June 30)-2022.

Sl no	State/UT	Confirmed cases	Active cases	Cured/discharged	Death
	India	**43,452,164**	**104,555**	**42,822,493**	**525,116**
1	Andaman and Nicobar Islands	10,157	42	9986	129
2	Andhra Pradesh	2,321,379	755	2,305,893	14,731
3	Arunachal Pradesh	64,518	4	64,218	296
4	Assam	724,788	395	716,405	7988
5	Bihar	832,581	934	819,388	12,259
6	Chandigarh	93,785	568	92,052	1165
7	Chhattisgarh	1,154,179	861	1,139,282	14,036
8	Dadra and Nagar Haveli and Daman and Diu	11,474	14	11,456	4
9	Delhi	1,934,009	4325	1,903,423	26,261
10	Goa	248,540	982	243,720	3838
11	Gujarat	1,231,483	2914	1,217,623	10,946
12	Haryana	1,015,501	2655	1,002,222	10,624
13	Himachal Pradesh	286,061	507	281,413	4141
14	Jammu and Kashmir	455,006	447	449,803	4756
15	Jharkhand	435,858	284	430,254	5320
16	Karnataka	3,968,365	5707	3,922,541	40,117
17	Kerala	6,634,722	28,860	6,535,869	69,993
18	Ladakh	28,411	78	28,105	228
19	Lakshadweep	11,408	3	11,353	52
20	Madhya Pradesh	1,044,243	490	1,033,012	10,741
21	Maharashtra	7,972,474	25,735	7,798,817	147,922
22	Manipur	137,266	18	135,128	2120
23	Meghalaya	93,947	65	92,288	1594
24	Mizoram	229,048	261	228,084	703
25	Nagaland	35,507	2	34,744	761
26	Odisha	1,289,602	627	1,279,849	9126
27	Puducherry	166,438	304	164,172	1962
28	Punjab	762,755	1079	743,903	17,773
29	Rajasthan	1,288,328	939	1,277,825	9564
30	Sikkim	39,224	26	38,744	454
31	Tamil Nadu	3,473,116	10,033	3,425,057	38,026
32	Telangana	800,476	4421	791,944	4111
33	Tripura	100,901	7	99,971	923
34	Uttar Pradesh	2,090,050	3541	2,062,971	23,538
35	Uttarakhand	438,663	787	430,180	7696
36	West Bengal	2,027,901	5885	2,000,798	21,218

cough produces no mucus. According to the National Health Service of the United Kingdom, if a person has been coughing for more than an hour or has three or more coughing episodes in a day, they might have coronavirus (NHS). Fatigue tiredness is a sense of exhaustion and a general energy shortage. An individual may feel exhausted, tired, or sluggish with fatigue. Shortness of breath subjective sensation of shortness of breath is many who experience breath shortage, however, can identify it as being choking or being unable to catch their breath.

Other symptoms of COVID-19 may include:

- Nose blocked
- Diarrhea infection
- Sputum, or mucus and saliva, coughed up
- Nausea
- Vomiting

Recent evidence shows that where the virus can be detected, the fundamental symptoms of COVID-19 are mild flu-like signs such as fever, nausea, difficulty breathing, muscle cramps, and weakness [25]. In more serious cases, extreme infection, acute renal failure, pneumonia, and septic shock are present, which could contribute to the patient's death. People with existing health problems tend to be more vulnerable to severe diseases. These symptoms are usually mild and begin progressively. Some people will get ill but still have very mild side effects.

Without seeking medical attention, most individuals recover from the disease (about 80%). Around one in five people with COVID-19 are critically ill and have trouble breathing. Seniors, but those with actual diseases like high blood pressure, complications with heart, diabetes, or cancer, are at increased risk for serious diseases. Medical treatment should be urgently sought by people from all backgrounds who develop a fever and/or cough in combination with difficulty breathing/breathlessness, chest pain/pressure, or loss of speech or motion. If practicable, it is recommended to contact the health services care provider or hospital first so that it becomes possible to guide the patient to the correct facility.

Arora et al. [28] used LSTM models to forecast COVID-19 instances for all states in India, and they forecasted the following day and 1-week COVID-19 cases with a 3% inaccuracy.

The research work presents LSTM-based models both with univariate and multivariate data creation methods. An early prediction of the possible number of infections can help to take sufficient precautions [30]. This paper proposes an LSTM-based RNN method for predicting active cases per day, confirmed cases per day, and cumulative confirmed cases per day for each Indian state [31], predicting health outcomes at the state level using deep learning in COVID-19. In Ref. [32], there was a spatial distribution found in the models used to predict COVID-19 mortality data. For relatively short forecast horizons, they use a vanilla LSTM model. As well as the susceptible-exposed-infectious-removed method, other conventional epidemiological methods have been employed to quantify viral dissemination [33–35]. Health systems like hospitals use predictive models to minimize the dangers of transmission by studying the influence of COVID-19 on outbreaks [36]. The result with a 92% accuracy rate and 98% AUC score can differentiate X-rays of COVID-19 patients, normal patients, and pneumonia patients. An X-ray-based COVID-19 detection pipeline based on vision transformer is described in Ref. [37]. Global supply chains have been strained, and the global economy has virtually crashed [38]. The purpose of this study is to devise a model for the diagnosis and classification of COVID-19. In this work, a fusion-based model of feature extraction, called FM-CNN, is applied to the diagnosis of COVID-19 using a convolutional neural network (CNN) [39]. The prediction models for COVID-19 have been developed using support vector regression (SVR) and deep neural networks [13]. The authors of this manuscript argue that AI-mediated methods are capable of predicting mortality rates. Healthcare professionals can handle this unpredictable situation better with an effective prediction model [40]. In comparison to conventional machine learning and deep learning algorithms, DeepSense provides improved accuracy during the prediction of the COVID-19 virus [41].

3. How to protect yourself from COVID-19

In the present situation, clinical professionals believe that the disease is transmitted from person to person, especially when you move from home to home wherever you go; you will eventually come into contact with others wearing a mask. Strategic distance from human interaction with individuals with hacking manifestations is necessary because of the way the infection is transmitted from one person to another, especially from the infected person to another. It is prudent to circumvent visiting areas where animals continue, keeping away from physical contact with any form of an insect, its discharges, or droppings, as the evidence indicates that the animals are the birthplace of the virus. Nonetheless, adherence to the general standards of hand cleanliness and dietary cleanliness and compliance with them is recommended. This means washing your hands with water and a cleanser before eating, after using the can, and after any contact with a creature by using a liquor-based disinfectant system. Fig. 11.1 described the COVID-19 process.

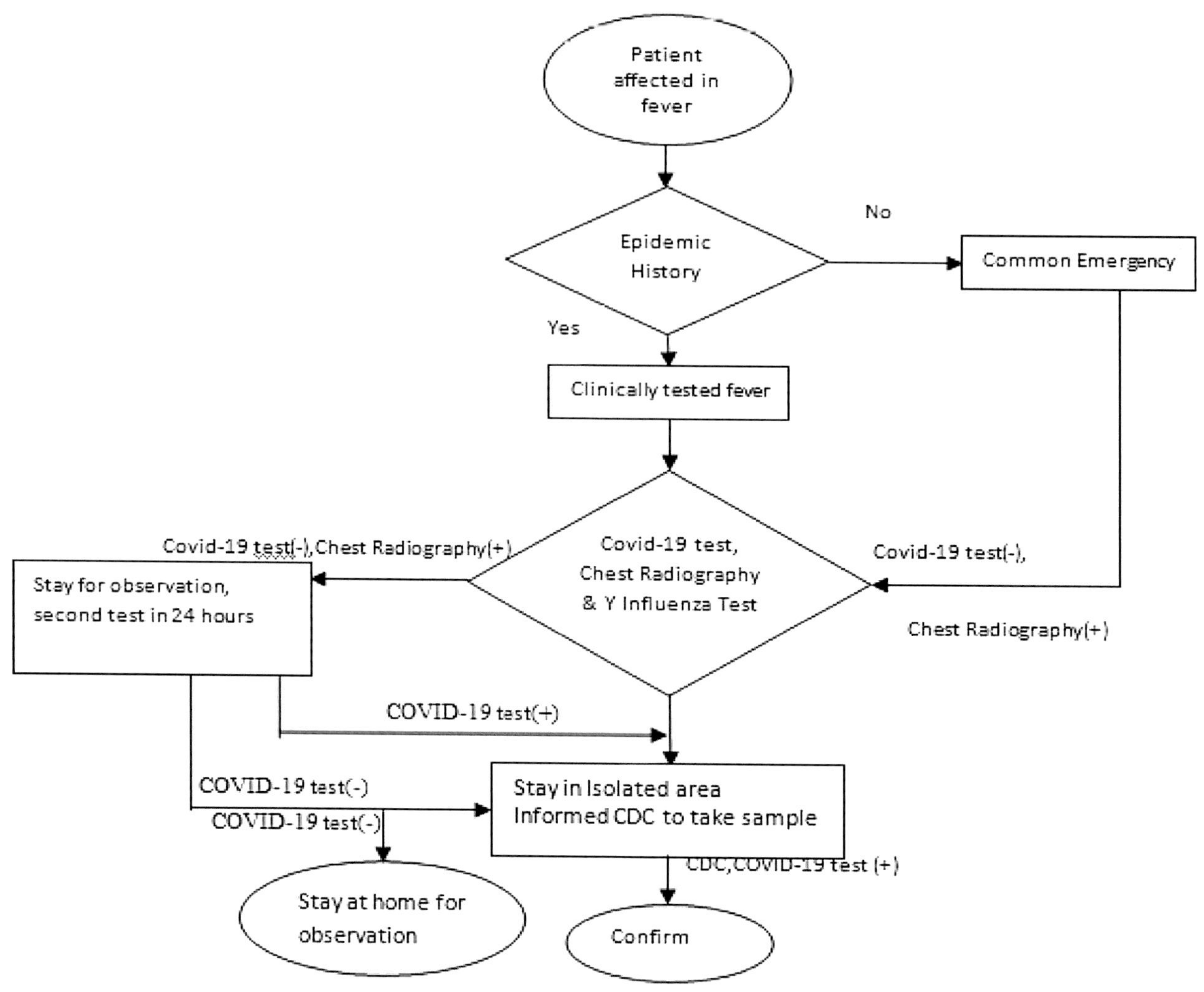

FIGURE 11.1 COVID-19 spreading process.

- If you have a fever, cough, sore throat, or other COVID-19 symptoms, you should not travel by airplane.
- At airports, train stations, trains, buses, and other locations, physical distance is practiced.
- To protect yourself and your fellow passengers, wear a medical face mask at airports, train stations, and on planes, trains, and buses.
- It is best to call your healthcare provider first if you become ill while flying. In the event of an illness while flying, notify the crew and seek immediate medical attention. You should use alcohol-based sanitizers when traveling or whenever you are dealing with sick people. Avoid touching your face with filthy hands or confronting sick people.
- It is advisable to seek medical attention within 14 days of returning home if you suffer from symptoms such as coughing, fever, sore throat, or the like.

To prevent the spread of COVID-19,

- Typically clean your face and hand by using soapy water or hand rubbing with an alcohol-based sanitizer.
- Keep a safe distance from someone sneezing or coughing.
- Do not touch your mouth, eyes, and nose at all.
- Instead of a hand, use a bent elbow to cover your nose and mouth, also use a mask and tissue paper while you cough or sneeze.
- Remain at home if you feel unwell.
- If you have issues with fever, cough, and breathing, seek medical treatment. Call in advance.
- Follow the local public health authority's instructions.
- Preventing unnecessary visits to medical facilities allows healthcare services to operate more effectively and protect you and others in this way.

3.1 High-risk groups

It applies only to people who are in the high-risk COVID-19 category (e.g., 60 years or older or suffering from underlying health issues). You should also presume that, in addition to the above preventive steps, avoid contact with people who are ill, particularly those who have a cough.

- In places where continuous group transmission exists, avoid physical meetings, festivals, and other social gatherings.
- Crowds should be avoided, particularly in enclosed areas with poor ventilation.
- Purchase and distribute medications, food, and other necessities at off-peak hours or enlist the help of family, friends, or volunteers.
- During rush hours, stop using public transportation.
- Outdoor practice instead of indoor settings.
- If you encounter symptoms associated with COVID-19, seek medical attention by phone fast.

4. Facts about the vaccine against the COVID-19

The production of vaccines is time-consuming. Several pharmaceutical firms work on candidates for vaccines (different kinds of vaccines). However, it will take months before any vaccine can be widely used because comprehensive studies need to be carried out to assess its safety and efficacy [25,42]. At least seven Indian pharmaceutical firms are attempting to create a coronavirus vaccine as part of global attempts to find a preventive way to prevent the spread of the deadly virus, which has infected more than 14 million people worldwide. Domestic pharmaceutical companies working on coronavirus vaccines in India include Bharat Biotech, ZydusCadila, Panacea Biotec, Indian Immunologicals, Mynvax, and Biological E, the Serum Institute. It is proven that the vaccine is safe and reliable. While it was produced in record time, it has undergone the same comprehensive process of Food and Drug Administration as any other vaccine, upholding all safety requirements. There were no steps missed. Instead, we should thank the unparalleled global cooperation and investment in the production of the vaccine itself for the shorter timeline.

The enormous exercise comes a day after India's regulatory Drugs Control Authority received approval for the Oxford COVID-19 vaccine made by the Serum Institute of India from a panel of government-appointed specialists. In all countries, a dry run to determine the optimal strategy to vaccinate people against COVID-19 and close logistical and preparation gaps have begun. This day-long drive will also evaluate the Cowin program's operational feasibility in a field environment. Cowin is a digital portal, short for COVID Vaccine Intelligence Network, to carry out and scale up the vaccination drive. The enormous exercise comes just a day after India's regulatory Drugs Control Authority received approval for it from a group of government-appointed specialists. Britain became the first country to approve the AstraZeneca vaccine this week, leapfrogging other western nations as it strives to contain a record-breaking outbreak of infections driven by a highly virulent strain of the virus that has recently appeared in India. The Serum Institute of India produces the COVID-19 vaccine. In all countries, a dry run to determine the optimal strategy to vaccinate people against COVID-19 and close logistical and preparation gaps have begun. A new coronavirus disease called COVID-19 is having scientists and researchers working on possible medications and vaccines.

- People who already have COVID-19 are being treated with antiviral drugs that have already been used against other diseases.
- Several firms are developing vaccines that could be used to control diseases.

Attempts to develop vaccines and drugs to delay the pandemic and mitigate the harm of the disease are being moved forward by scientists, with reported cases of COVID-19 reaching 6.4 million globally and continuing to rise. Medicines that are already approved for other diseases or have been tested on other viruses are likely to be some of the first therapies.

4.1 Vaccines in COVID-19

In vaccine platforms with the COVID-19 pandemic, there have never been so many vaccine candidates in different stages of development in the history of fighting infectious diseases. Technology systems of various types are used to increase vaccine production [43]. Depending on the source, there could be up to 163 vaccines being evaluated in preclinical studies and 52 vaccines being evaluated in clinical trials [44]. Broadly speaking, these vaccines can be divided into five major groups. A few of these include viruses-based vaccines (both replicating and nonreplicating), nucleic acids-based vaccines (DNA and RNA), recombinant proteins, peptide-based vaccines, and virus-like particles [45,46] (Table 11.3).

TABLE 11.3 Overview of soon-to-be-available COVID-19 vaccines.

Platform	Manufacturer/supplier	Route of administration	Doses	Cold chain requirement (°C)
Inactivated	Bharat Biotech (Whole virion) Sinova	IM	2	0–8
Viral vector adenovirus	Oxford/Astra Zeneca/SII Gamaleya Research Institute/ Dr. Reddy's Laboratory, India	IM	2	0–8
Protein subunit (recombinant)	Novavax GSK/Sanofi Biological E, India	IM	2	0–8
mRNA	Moderna Pfizer	IM	2	−20 at 2–8 × 30 days −70
Live attenuated virus	Codagenix/SII	Inhalation	1, 2	0–8
DNA	ZydusCadilla, India	Intradermal		0–8

IM, intramuscular; *SII*, Serum Institute of India.
Source: Ref. [47].

A vaccine is intended to avoid the exposure of individuals to a virus, such as SARS-CoV-2, the COVID-19-causing virus, The Trusted Source immune system is effectively trained by a vaccine to identify and attack the virus when it enters it. Vaccines protect both the person who is vaccinated and the community. Viruses do not kill people that are vaccinated, making sure that the infection will not be passed on to others by vaccinated individuals. This is known to be immunity from herds. Several Trusted Source organizations work together on potential SARS-CoV-2 vaccines, several of which are sponsored by the nonprofit Alliance for Excellence in Outbreak Preparedness (CEPI).

Around the world, there are more than 100 Trusted Source initiatives focused on creating a coronavirus vaccine. As of 11 May, eight candidate vaccines in individuals have been tested in clinical trials. In mid-May, an official from the National Institutes of Health said large-scale tests could start in July with a vaccine that could theoretically be available by January. Other experts suggest the summer or fall of 2021 is the most possible timeline.

4.2 Here's a peek at some of the initiatives

Bharat Biotech developed the vaccine in collaboration with the Indian Council of Medical Research (ICMR). Bharat Biotech has received clearance to perform phase I and II clinical trials for Covaxin, a candidate vaccine developed and manufactured at the firm's Hyderabad facility. Clinical experiments on humans began. India's Leading Vaccine Serum Institute announced its goal of achieving COVID-19 by the end of 2020. Today, on January 2, the center asked all states and UTs to perform dry-running vaccinations to ensure successful preparedness for the roll-out of the vaccine.

A high-level meeting with key Secretaries (Health) and other health administrators of all States/UTs was held by Union Health Secretary Rajesh Bhushan via video conference to review the preparedness at the COVID-19 vaccination sites. According to a senior government official, the COVID19 dry-run vaccine aims to assess the operational feasibility of using the Co-WIN application in the field setting, test the links between planning and implementation, identify the difficulties, and lead the way forward before actual implementation. Clinical trials for ZyCoV-D, a COVID-19 vaccine candidate, are expected to be completed by the end of 2020, according to ZydusCadila.

Panacea Biotec announced in June the launch a COVID-19 vaccination joint venture in Ireland with Refana Inc of the United States. The COVID-19 candidate vaccine was expected to be ready early next year, with Panacea Biotec collaborating with Refine to produce over 500 million doses. A COVID-19 vaccine is being developed by Indian Immunologicals, an affiliate of the National Dairy Development Board (NDDB). COVID-19 vaccines are also being improved by companies like Mynvax and Biological E.

4.3 Immunology and antigen detection for the COVID-19 vaccine

The novel virus is SARS-CoV-2. To confer immunity, the viral determinants that need to be attacked remain unknown. While other respiratory viruses, including coronaviruses, have favorable precedents [48]. In the COVID-19 vaccine, the

SARS-CoV-2 spike protein was selected as the antigen of choice, as it is a type 1 protein that should fold properly during use and storage. Infection or vaccination produces neutralizing antibodies that bind to the receptor which is responsible for converting angiotensin into nitric oxide and preventing the virus from entering. Regarding the vaccine, the initial suggestion that only receptor-binding protein should be used is revisited in light of the discovery that other areas beyond the receptor-binding domain may be targeted by antibodies [49].

4.4 Potential vaccine-related threats

A possible adverse effect of the COVID-19 vaccine, such as T-cell-mediated damage or antibody-related effects, has not been determined. In addition to influenza and measles [50], many other illnesses have been linked to these unintended and potentially dangerous side effects following vaccinations. In the case of the COVID-19 vaccine, these events can be detected through postmarketing surveillance. The long-term protection of vaccines should be demonstrated in credible data to encourage confidence in societies and their acceptance of vaccines (Table 11.4).

4.5 Vaccines have been approved in India

4.5.1 Covishield

In the first week of January 2021, India approved the Covishield vaccine, which was developed by Serum Institute of India (SII). Oxford University in the United Kingdom and AstraZeneca, a British-Swedish multinational pharmaceutical corporation, collaborated on the two-dose vaccine.

4.5.2 Covaxin

The DCGI gave the green light to Hyderabad-based Bharat Biotech's Covaxin and Covishield in early January. The vaccine Covaxin has been made inactive. It contains a dead virus that stimulates the immune system without infecting or harming the person.

4.5.3 Sputnik V

The COVID-19 vaccine is a preventative vaccine against the COVID virus. The Sputnik V (Gam-COVID-Vac) vaccine is a two-part adenoviral vaccine that protects against the coronavirus SARS-CoV-2. Sputnik V employs a weakened virus to deliver small amounts of pathogens and elicit an immune response. This will be the third COVID-19 vaccination available in India. The DCGI granted emergency usage authorization for two COVID-19 vaccines—Bharat Biotech's Covaxin and Oxford-Covishield, AstraZeneca's both manufactured by Serum Institute of India in Pune—in January.

TABLE 11.4 COVID-19 vaccine challenges.

Category	Challenge
Political	Immunization of vulnerable populations. Selection of vaccines based on efficacy and cost. Ethically procure and distribute vaccines.
Program	Reaching out to target populations by enhancing the health system
Management	Up to the last mile, there must be an adequate and functional cold chain
Technical	Monitor the efficacy and safety of the community phase and launch the Phase 4 safety surveillance
Operational	Timely procurement and efficient distribution. Managing cold chain. Enhancing capacity for efficient response. Data management. Community engagement including response to adverse reactions
Epidemiological	Adapting vaccination plans as pandemics progress
Research	Availability of better vaccines in the second generation.
Financial	Financing of the development effort through local or international sources.

Source: Ref. [51]

5. Materials and methods

This research study depicts the condition of utilizing and integrating accessible data, modeling that data using a variety of hyperparameter combinations in MLP, and then evaluating the outcomes in Python. Fig. 11.2 shows the overview of the computing process. It shows an example of how the most common model can be used to test future instances.

5.1 Artificial neural network (ANN)

Neural networks are effective at recognizing universal input/output mappings and solving complex nonlinear problems [29]. Consider a definition that is fully linked to acquiring a deeper understanding of the concept. Multilayer perceptron (MLP) networks consist of the input layer, hidden layer string, component layer, and output layer. Input, hidden layer string, component layer, and output layer are all part of the neural system. There are four layers to MLP networks: the input layer, the hidden layer string, the component layer, and the output layer. It comprises several neurons, and each sheet has been indexed with $l = [0, \ldots, L-1]$. Every example of input training x will be denoted as $I(x) = [I_1, I_2, I_3, \ldots, I_{n0}]$ and its output $(x) = [O_1, O_2, O_3, \ldots, O_{nL-1}]$.

An ANN is capable of modeling as well as learning nonlinear interactions. When training and testing, input parameters and outputs (0 or 1) were split 80/20 in proportion to the number of total COVID-19 cases versus non-COVID-19 cases. Learning can be classified in the following ways [52–58]. In this implementation of input-output mappings, reinforcement methods include the existence of a goal or desired outcome.

- An unsupervised system can only train itself if it is given high-quality, framework-independent calculation interpretations.
- Continually interacting with the environment can reduce the scalar performance of cognitive systems.

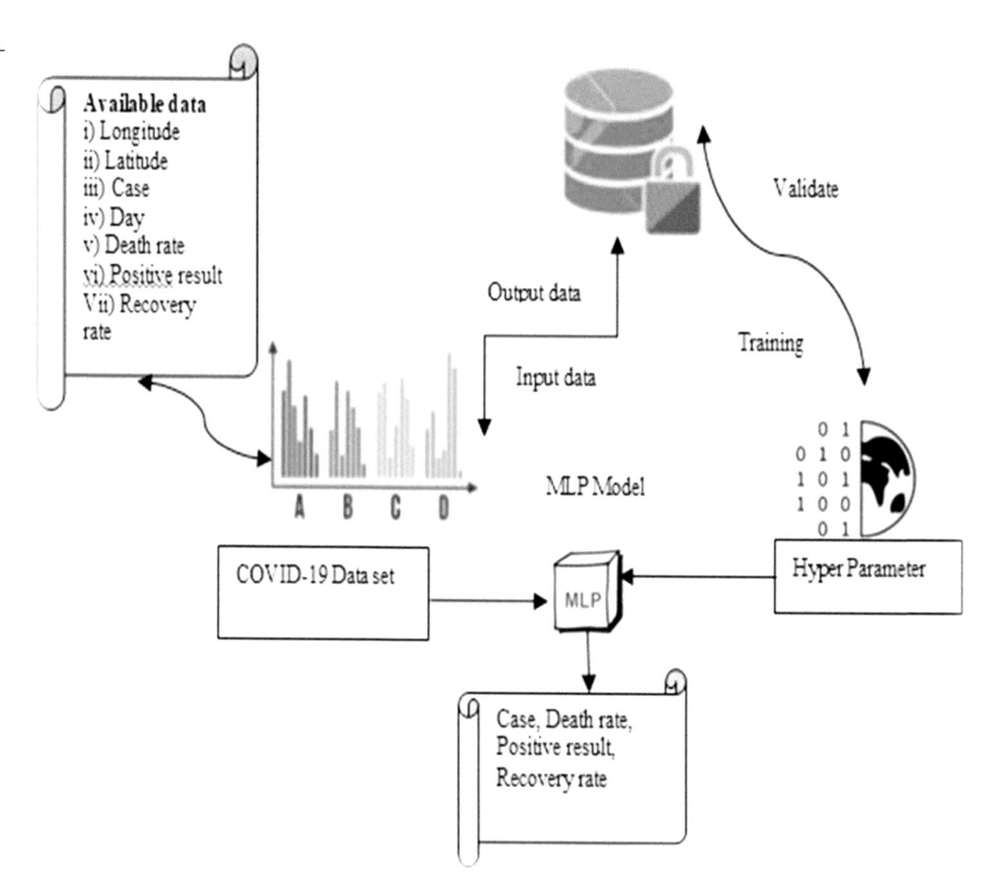

FIGURE 11.2 Phase for COVID-19 process in data analytics.

5.2 Model of a neuron

Neurons are typically designed with n input signals (accompanied by weights), an insect, an activation feature, and an output feature. In a neuron, an adder describes perfectly the weighted input control signal and bias. The activation function, shown in Fig. 11.3, sums together the weighted input signals and biases. This function might be linear or nonlinear. The transitional function is also known as the activation role. It transfers the output values to a restricted range depending on the function (between 0 and 1 or −1 to 1, etc.). The output could be used as a source of information for another neuron in the following layer.

5.2.1 Recurrent neural network (RNN)

Using prior data samples, the RNN is primarily used to forecast the future data sequence. The RNN is widely employed in the modeling of sequence data like speech or text. However, because these networks are difficult to train in such a way that they capture long-term dependencies, they have not been widely deployed [59]. RNN output is calculated by iterating the following equations from time $t = 1$ to $t = T$ (Fig. 11.4).

Fig. 11.3 depicts the essential design of an LSTM unit. The gates are represented by circles. An LSTM network has three gates: an input gate, a forget gate (f), and an output gate (o). Each gate is a neuron with a number from 0 to 1 as its inputs and outputs, with the past state vector (h_{t-1}) and current input (x_t) in equation representing the gating rate (11.1−11.4).

$$\mathbf{Z_t} = \boldsymbol{\sigma}(\mathbf{W_z} \cdot [\mathbf{h_{t-1,}x_t}] + \boldsymbol{\beta}_i) \tag{11.1}$$

$$r_t = \sigma\left(W_r \cdot [h_{t-1,}x_t] + \beta_f\right) \tag{11.2}$$

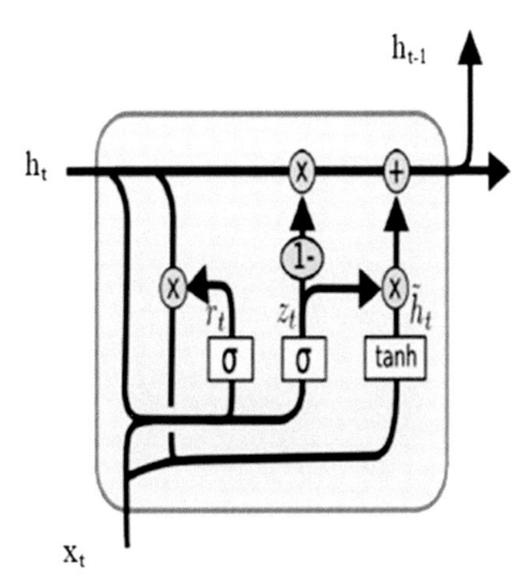

FIGURE 11.3 Long-short-term memory algorithm [60].

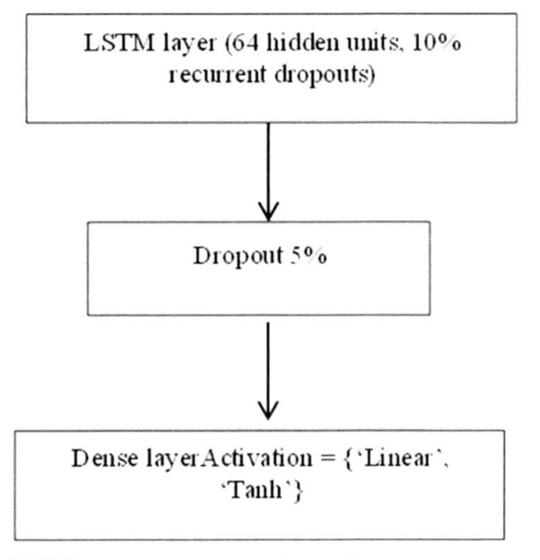

FIGURE 11.4 Flow chart of long-short-term memory.

$$\widetilde{h}_t = \tan h(W \cdot [x_t * h_{t-1}, x_t]) \tag{11.3}$$

$$h_t = [1 - Z_t] * \left(h_{t-1} + \left[z_t * \widetilde{h}_t \right] + \beta_o \right) \tag{11.4}$$

Several types of reduced LSTM architectures are created in Ref. [46]. These LSTM variants are created by deleting combinations of input signals, bias, and concealed input signals. As a result, these models have fewer parameters, which reduce the time required to prepare them. In this book, two of the reduced LSTM models are explored and analyzed. LSTM2 is one of the models we looked at in our research. Eq. (11.4) describes a model for LSTM2. The bias is removed, and the input is obtained solely from the prior output, h_{t-1}. The LSTM3 model was also used in our research. LSTM3 describes the gate functions as being in (11.1-11.4). Only the model's bias is present.

$$Z_t = \sigma(\beta_i) \tag{11.5}$$

$$f_t = \sigma(\beta_f) \tag{11.6}$$

$$f_t = \sigma(\beta_o) \tag{11.7}$$

LSTM3 is the other model utilized in our work. The gate functions are described by LSTM3 as being in Eqs. (11.5) −(11.7). There is only the bias value represented in Fig. 11.7 because it only shows the model's bias.

Following the identification and prediction of COVID-19 symptoms, AI can be used as the major tool for diagnosing this deadly condition. We developed several AI-based algorithms that may reliably detect and diagnose COVID-19 patients after conducting a thorough literature study. Fig. 11.5 shows how artificial intelligence became involved in the coronavirus outbreak. A study [51] found a relationship between the COVID-19, mortality and morbidity rates, and the strain on radiologists and healthcare facilities. The training of nurses and physicians during a pandemic is nearly impossible in large countries like India and China. Artificial intelligence can mimic human intelligence by creating intelligent machines.

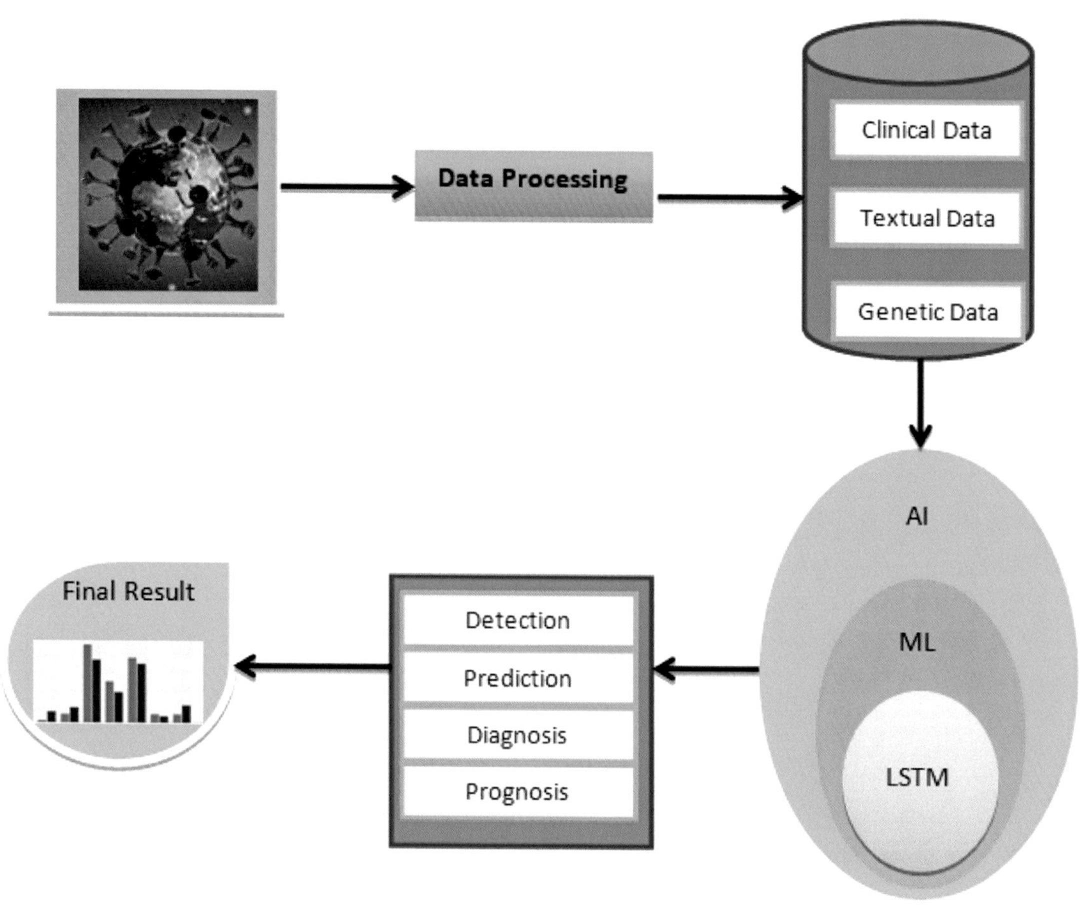

FIGURE 11.5 Artificial intelligence's vital role in COVID-19 prediction.

6. Results discussion

In this section, we present the results of the prediction of COVID-19 daily cases in India using prominent LSTM neural network models. Our investigations look at the prediction task from both a univariate and multivariate perspective. By following the methods below, we may provide a general overview of our studies. The results demonstrate that the created method beat both the machine learning and long-short-term memory (LSTM) models in terms of error rate and promising prognosticating outcomes for the current COVID-19 pandemic. The results of this study demonstrate how effectively the DL approach can forecast COVID-19 cases in the future. Overall, it can be said that the model's predictions are consistent with the virus's current state, which may help us understand and stop the virus's spread. In order to address the COVID-19 situation, it may be very helpful to act promptly and make informed judgments. In the future, to stop COVID-19 from spreading further, we suggest adopting a semi-supervised hybrid approach to identify social media platforms.

- Using the following two methods, determine the optimal strategy to separate training and testing datasets. From the commencement of COVID-19 until July 7, 2021, we use a static split of training samples, with the remainder being used for the test dataset. We employ data in random split to construct the train and test sets by randomly shuffling the dataset.
- Determine whether a univariate or multivariate prediction strategy yields better results.
- Show the findings for the full case of India, followed by the top 10 states with the highest number of COVID-19 infections.
- LSTM is used to assess the accuracy.

COVID-19, the coronavirus, which is spreading across the country, poses a serious threat and challenge to the nation, which the Indian government is taking every step to combat. Residents should be given the necessary information and be advised to follow the Ministry of Health and Family Welfare's guidance to avoid spreading the virus locally. When someone is determined to be positive for COVID-19, they are placed in a containment area. This region is deemed polluted and is cordoned off. Residents in the region have been asked not to leave their houses. Even a single instance of COVID-19, according to the health ministry, might be a hotspot for the government represented in Fig. 11.6.

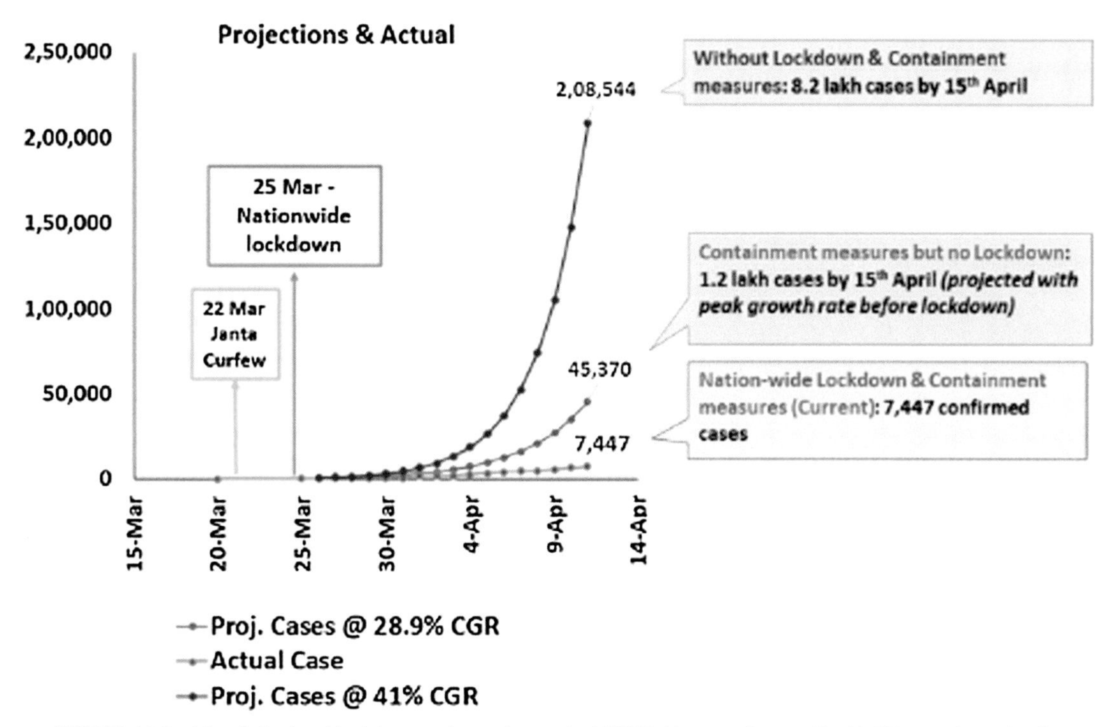

FIGURE 11.6 The vital role of lockdown and containment in COVID-19 cases. *Source: Health Ministry Govt of India.*

In Fig. 11.6, the trend lines follow the number of days associated with each month starting on the day, the first case of the virus was confirmed. As with the graph above, use the drop-down choices to visualize verified events.

Fig. 11.7 shows how the number of new COVID-19 cases has changed over time. As a result, the ANN mechanism captured a significant amount of the relationship between search behavior and COVID-19 occurrence frequencies, with a very small error term along the time axis.

Fig. 11.8 depicts the LSTM's mean square error (MSE). The MSE is another technique for measuring the predictive model's error information. LSTM models employ standardized data for each function they employ. This value is derived from the smallest MSE.

In this diagram, we are attempting to create a model that will predict the test data. As a result, we use the COVID-19 training data to match the model and testing details to test it and get the projected value. The developed models are intended to predict coronavirus outcomes, which are defined in Fig. 11.9 as the test set. Fig. 11.10 shows the number of new confirmed COVID-19 cases reported each month concerning the specified number. In terms of numbers, the technique is accurate in predicting confirmed instances month by month. Fig. 11.11 describes cumulative daily cases.

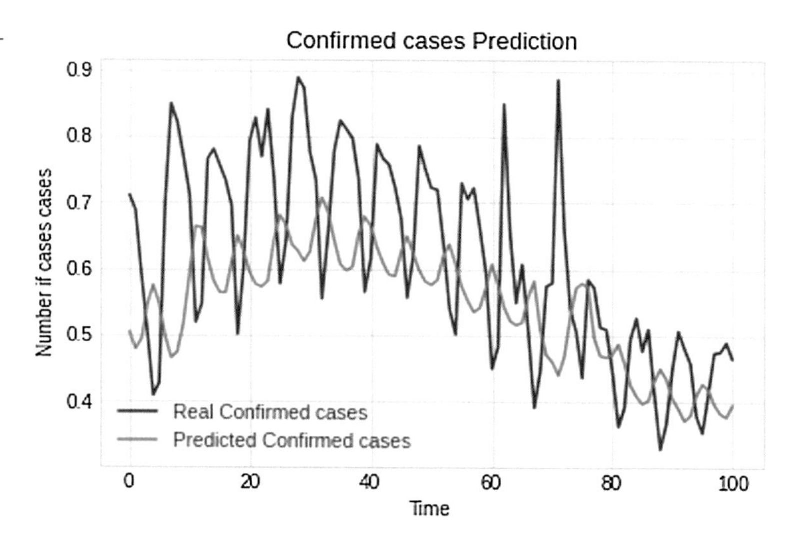

FIGURE 11.7 Prediction of the number of new COVID-19 cases.

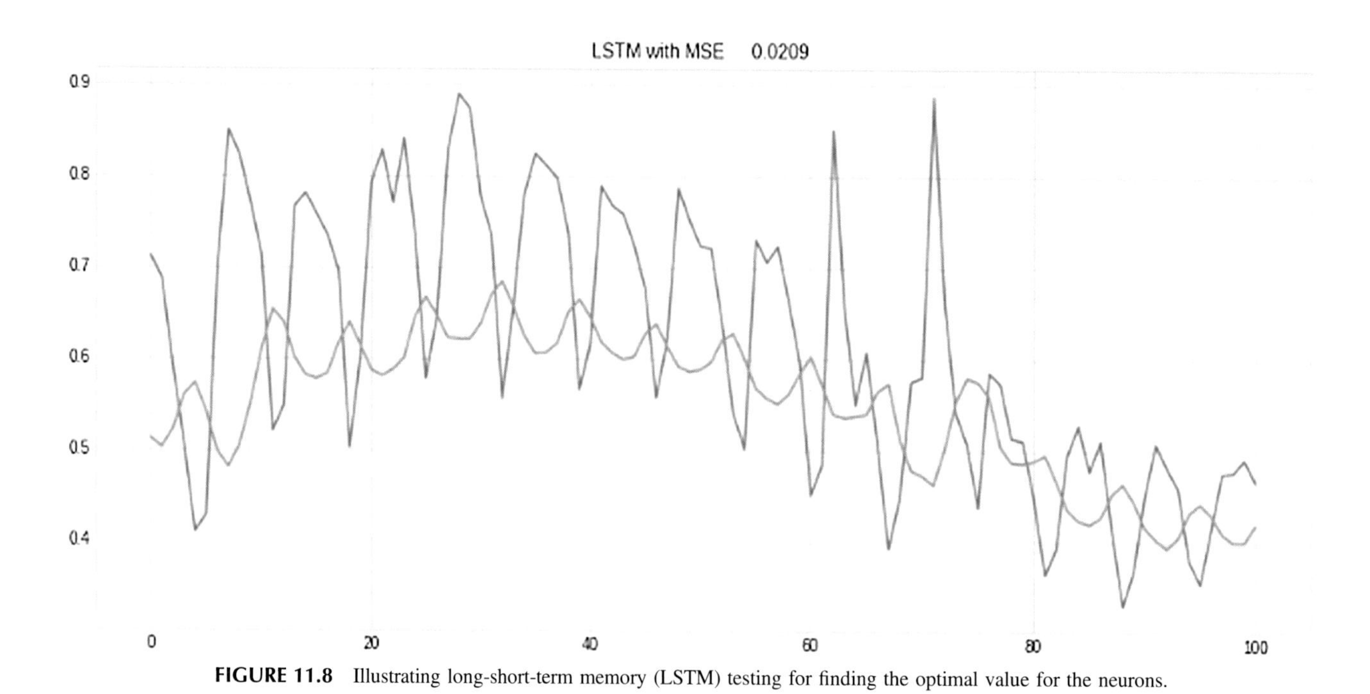

FIGURE 11.8 Illustrating long-short-term memory (LSTM) testing for finding the optimal value for the neurons.

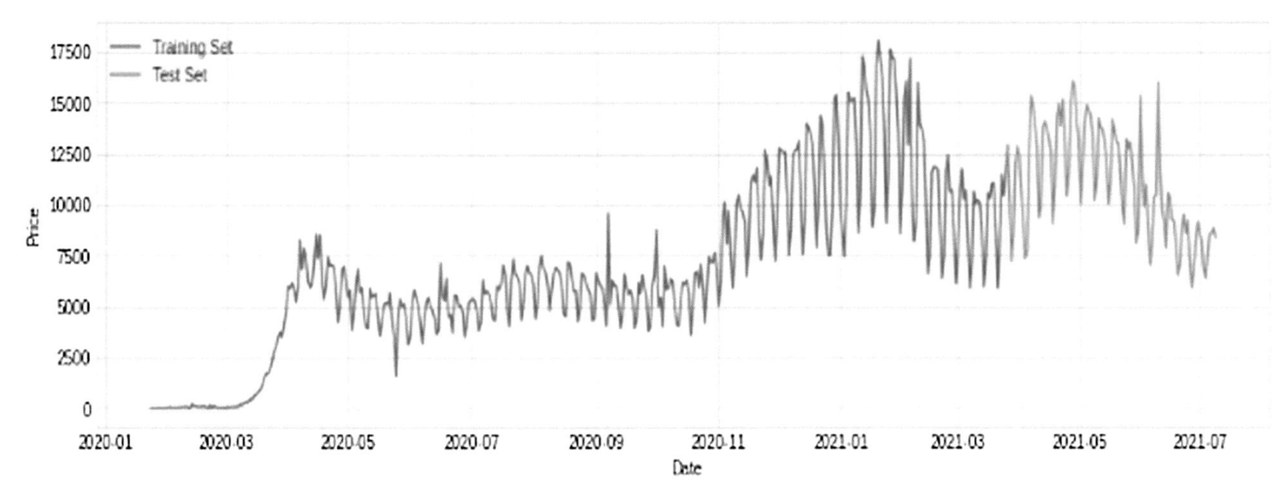

FIGURE 11.9 The model of training and test data.

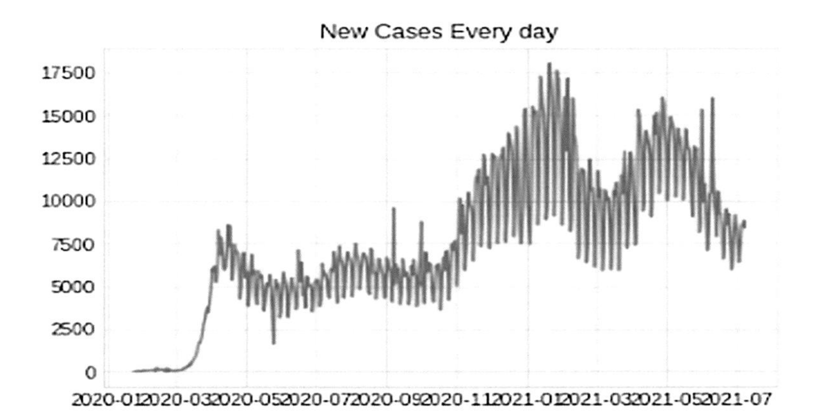

FIGURE 11.10 New confirm coronavirus case.

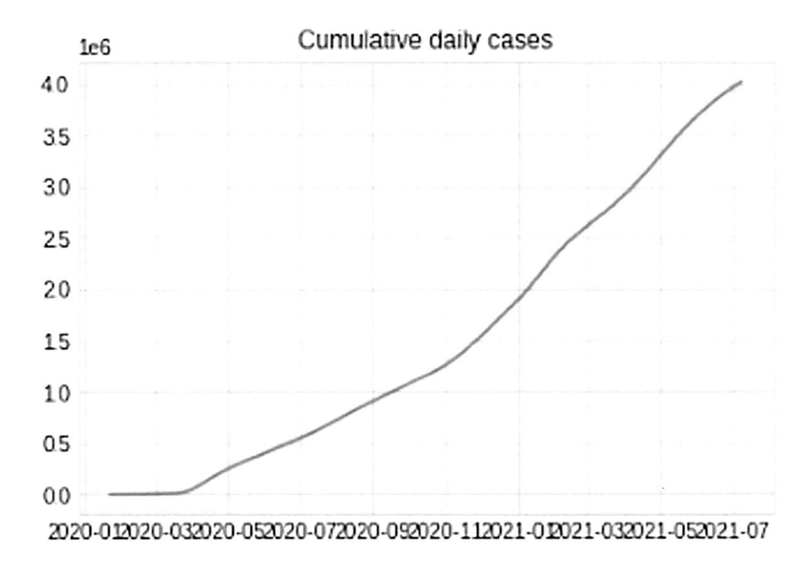

FIGURE 11.11 Cumulative daily cases.

Figs. 11.12 and 11.13 describes the number of deaths, cured, and confirmed COVID-19 cases in India concerning state-wise in diagnosed cases in India. Fig. 11.14: Described the outcome of the number of deaths from coronavirus cases concerning statewide diagnosed cases in India.

6.1 Top 10 states (confirmed cases and cured cases in Covid-19)

Fig. 11.15 described the outcome of newly confirmed COVID cases concerning States in diagnosed cases in India. Fig. 11.16 described the outcome of Cured cases in India in diagnosed cases state-wise.

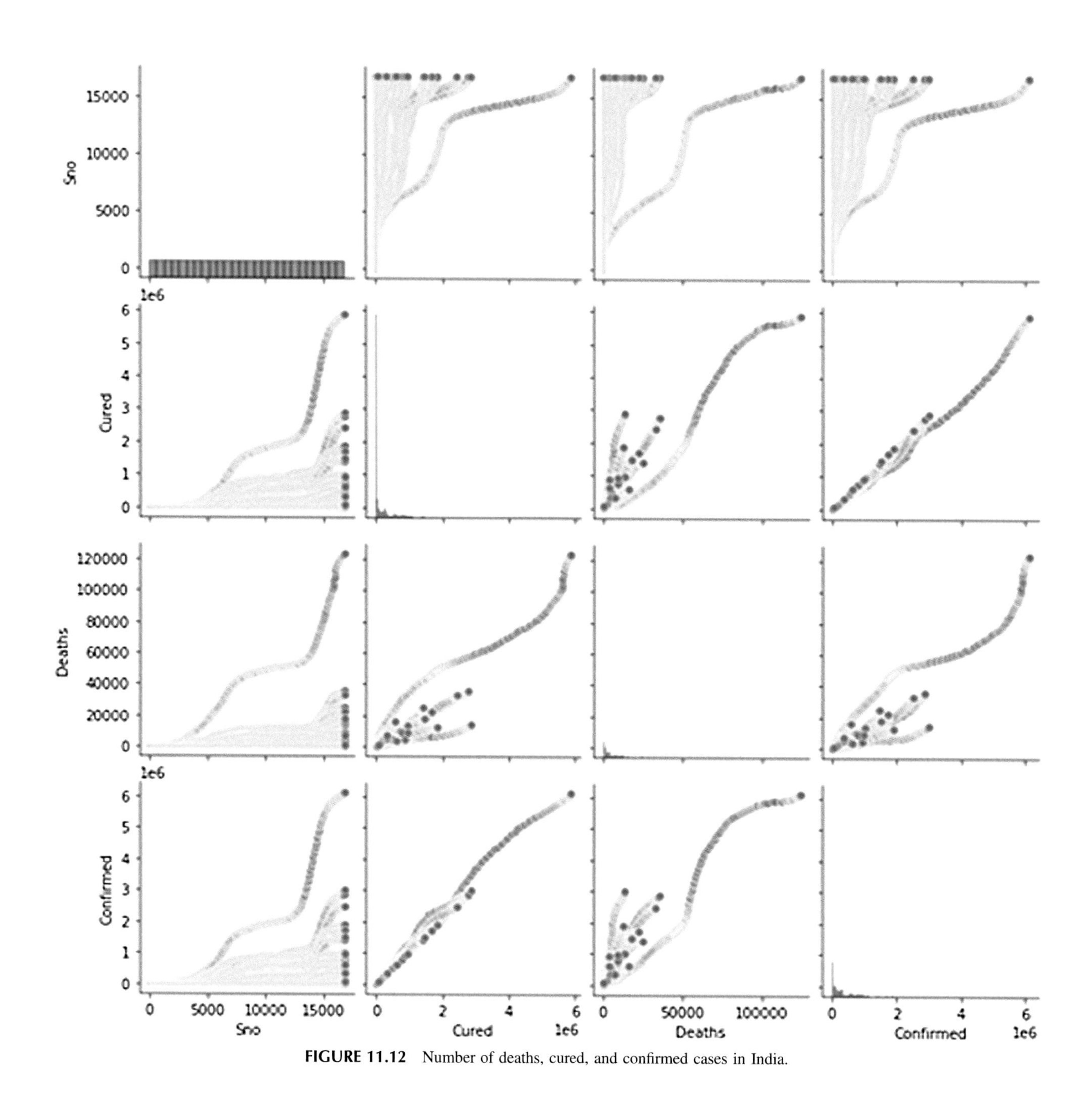

FIGURE 11.12 Number of deaths, cured, and confirmed cases in India.

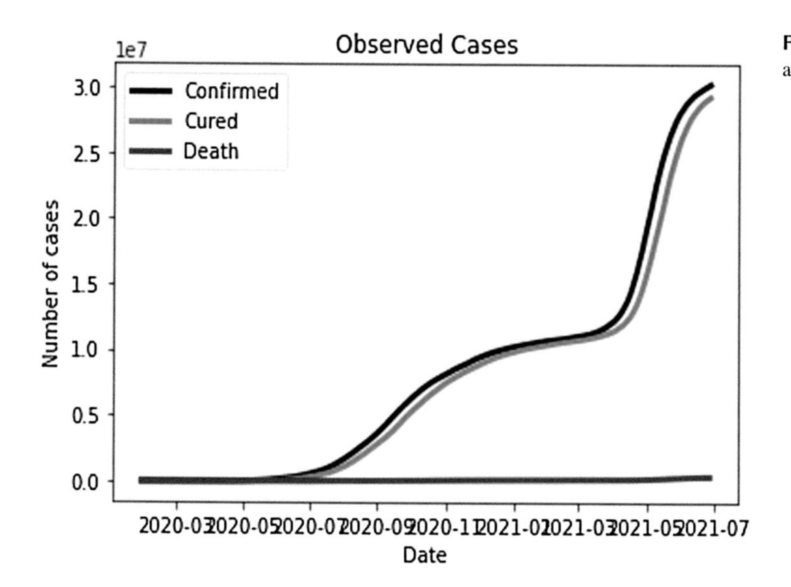

FIGURE 11.13 The outcome of the number of deaths, cured, and confirmed cases in India.

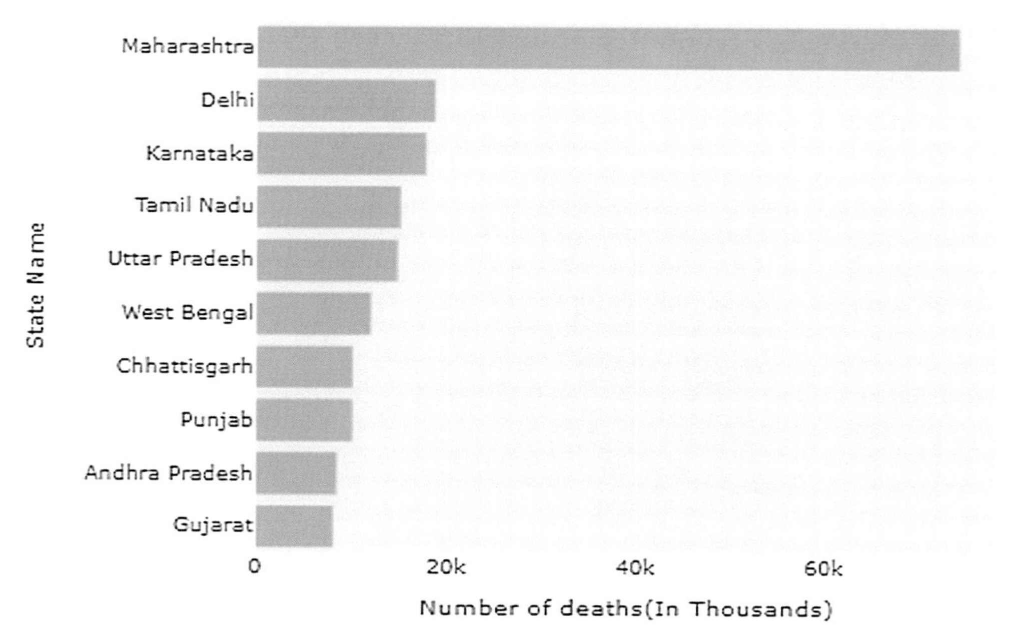

FIGURE 11.14 The outcome of the number of deaths cases in India.

7. Conclusion

The COVID-19 causes deaths around the world despite its unpredictability. The pace and extent of recovery can be assessed by a country's health, humanitarian, and socioeconomic policies. The current review was carried out to classify the current situation of COVID-19 in India, ranging from confirmed incidents, deaths, and recovered cases in India between 30 January and 8 June 2020. A steady number of recovered cases and few deaths have followed a rapid increase in COVID-19 cases in India. It emphasized the challenges associated with limited data and the spread of illnesses and used deep learning to combine some of the most cutting-edge forecasting technologies. It was difficult to pick a single model from the LSTM variants due to certain strengths and limitations in different scenarios. Overall, we found that LSTM

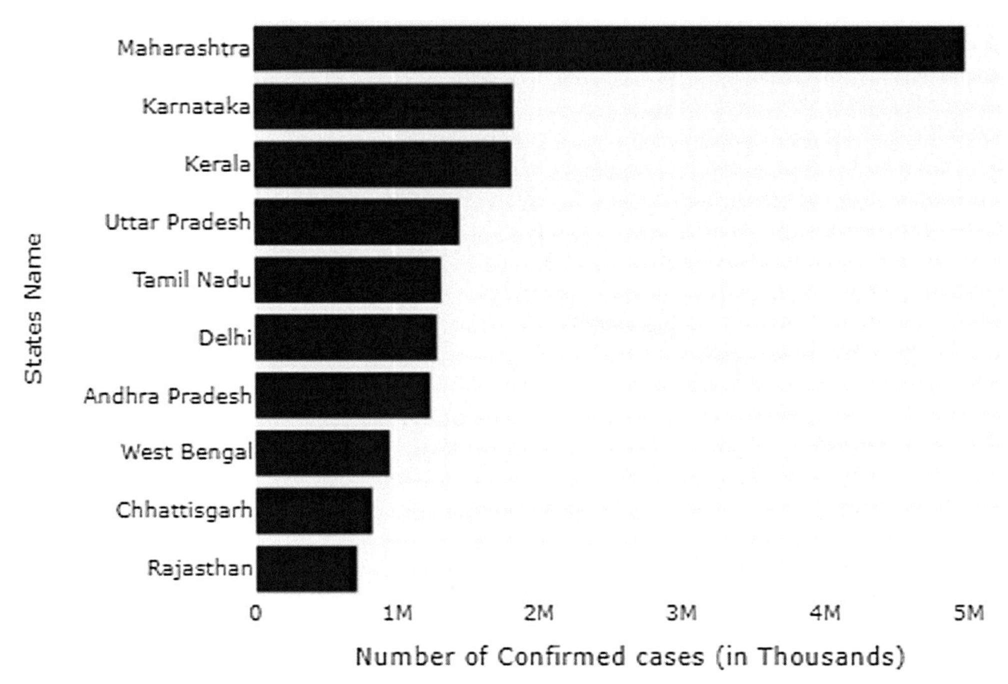

FIGURE 11.15 The outcome of the number of confirmed cases in India.

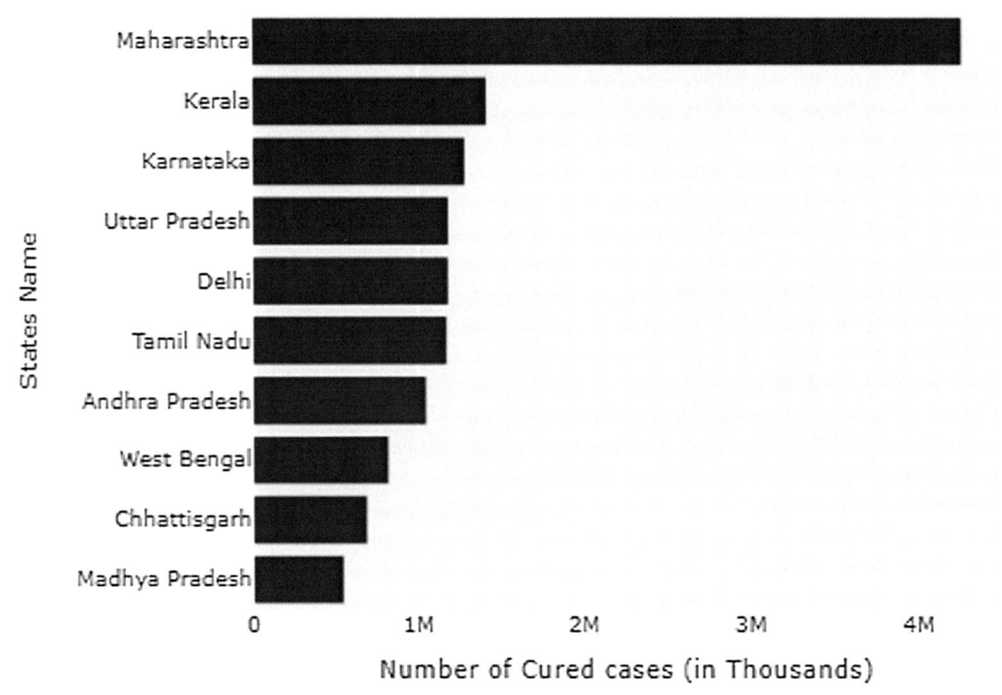

FIGURE 11.16 The outcome of cured cases in India.

models using random splits perform the best in comparison with multivariate LSTM models. Therefore, it was not possible to outperform the univariate model using the data from the adjacent states. COVID-19 recovered cases can be successfully fitted with our Deep Learning model and predictions can be made accurately based on that data. In general, our findings demonstrate that the LSTM model provides the greatest outcomes.

References

[1] Wuhan City Health Committee, Wuhan Municipal Health and Health Commission's Briefing on the Current Pneumonia Epidemic Situation in Our City 2019, 2019 [updated 31 December 2019−14 January 2020]. Available from: http://wjw.wuhan.gov.cn/front/web/showDetail/2019123108989.

[2] Home I Ministry of Health and Family Welfare I GOI. mohfw.gov. Retrieved 14 June 2020.

[3] India most infected by Covid-19 among Asian countries, leaves Turkey behind, Hindustan Times, May 29, 2020. Retrieved 30 May 2020.

[4] Jump up to: a b, India crosses UK to become fourth worst hit by coronavirus. NDTV.com. Retrieved 12 June 2020.

[5] S. Kumar, Covid-19: number of recoveries exceed active cases for first time, New Delhi: Hindustan Times (June 2020).

[6] Infections over 1 lakh, five cities with half the cases: India's coronavirus story so far, Metals Week. Retrieved 20 May 2020.

[7] COVID Data. Available from: https://covid19india.org/opendata/. (Accessed 16 October 2020).

[8] Ministry of Health & Family Welfare, Government of India. Available from: https://www.mohfw.gov.in. (Accessed 25 September 2020).

[9] World Meter. World/Counties/India. Available from: https://www.worldometers.info/coronavirus/country/india/. (Accessed 26 September 2020).

[10] R. Changotra, H. Rajput, P. Rajput, S. Gautam, A.S. Arora, Largest democracy in the world crippled by COVID-19: current perspective and experience from India, Environment, Development and Sustainability (2020) 1−19.

[11] D.M. Weinberger, T. Cohen, F.W. Crawford, F. Mostashari, D. Olson, V.E. Pitzer, et al., Estimating the early death toll of COVID-19 in the United States, bioRxiv (2020).

[12] P. Block, M. Hoffman, I.J. Raabe, J.B. Dowd, C. Rahal, R. Kashyap, et al., Social network-based distancing strategies to flatten the COVID-19 curve in a post-lockdown world, Nature Human Behaviour 4 (2020) 588−596.

[13] S. Dash, S. Chakravarty, S.N. Mohanty, C.R. Pattanaik, S. Jain, A deep learning method to forecast COVID-19 outbreak, New Generation Computing 39 (3) (2021) 515−539.

[14] T. Mitze, R. Kosfeld, J. Rode, K. Wälde, Face masks considerably reduce COVID-19 cases in Germany, medRxiv (2020), https://doi.org/10.1101/2020.06.21.20128181.

[15] S. Balachandar, S. Zaleski, A. Soldati, G. Ahmadi, L. Bourouiba, Host-to-host airborne transmission as a multiphase flow problem for science-based social distance guidelines, International Journal of Multiphase Flow 132 (2020) 103439.

[16] V.C. Cheng, S.C. Wong, V.W. Chan, S.Y. So, J.H. Chen, C.C. Yip, et al., Air and environmental sampling for SARS-CoV-2 around hospitalized patients with coronavirus disease 2019 (COVID-19), Infection Control & Hospital Epidemiology (2020) 1−8.

[17] Advisory Manual on Use of Homemade Protective Cover for Face & Mouth. Available from: https://www.mohfw.gov.in/pdf/Advisory&ManualonuseofHomemadeProtectiveCoverforFace&Mouth.pdf. (Accessed 20 September 2020).

[18] Indian Council of Medical Research, How India Ramped Up COVID Testing Capacity. Available from: https://main.icmr.nic.in/sites/default/files/press_realease_files/ICMR_Press_Release_India_testing_story_20052020.pdf. (Accessed 25 September 2020).

[19] Current Rules and Guidelines. Available from: https://covidindia.org/current-rules-and-regulations. (Accessed 25 September 2020).

[20] S.K. Rout, B. Sahu, D. Singh, Artificial neural network modeling for prediction of coronavirus (COVID-19), in: Advances in Distributed Computing and Machine Learning, Springer, Singapore, 2022, pp. 328−339.

[21] P. Mahajan, J. Kaushal, Epidemic trend of COVID-19 transmission in India during lockdown-1 phase, Journal of Community Health (2020) 1−10.

[22] National Portal of India, Building Atmanirbhar Bharat and Overcoming COVID-19. Available from: https://www.india.gov.in/spotlight/building-atmanirbhar-bharat-overcomingcovid-19. (Accessed 25 September 2020).

[23] D. Guan, D. Wang, S. Hallegatte, S.J. Davis, J. Huo, S. Li, et al., Global supply-chain effects of COVID-19 control measures, Nature Human Behaviour 4 (2020) 577−587.

[24] I.J. Bateman, A. Dannenberg, R. Elliott, M. Finus, P. Koundouri, K. Millock, et al., Economics of the environment in the shadow of coronavirus, Environmental and Resource Economics 76 (4) (2020) 519−523.

[25] T. Lupia, S. Scabini, S.M. Pinna, G. Di Perri, F.G. De Rosa, S. Corcione, 2019 novel coronavirus (2019-nCoV) outbreak: a new challenge, Journal of Global Antimicrobial Resistance 21 (2020) 22−27.

[26] M. Gilbert, G. Pullano, F. Pinotti, E. Valdano, C. Poletto, P.Y. Boëlle, et al., Preparedness and vulnerability of African countries against importations of COVID-19: a modelling study, The Lancet 395 (10227) (2020) 871−877.

[27] N.G. Davies, P. Klepac, Y. Liu, K. Prem, M. Jit, R.M. Eggo, Age-dependent effects in the transmission and control of COVID-19 epidemics, Nature Medicine 26 (8) (2020) 1205−1211.

[28] P. Arora, H. Kumar, B.K. Panigrahi, Prediction and analysis of COVID-19 positive cases using deep learning models: a descriptive case study of India, Chaos, Solitons & Fractals (2020), https://doi.org/10.1016/j.chaos.2020.110017 [PMC free article] [PubMed].

[29] B. Jacob, S. Kligys, B. Chen, M. Zhu, M. Tang, A. Howard, et al., Quantization and training of neural networks for efficient integer-arithmetic-only inference, in: Proceedings of the IEEE Conference on Computer Vision and Pattern Recognition, 2018, pp. 2704−2713.

[30] R. Chandra, A. Jain, D. Singh Chauhan, Deep learning via LSTM models for COVID-19 infection forecasting in India, PLoS One 17 (1) (2022) e0262708.

[31] M.S. Pandianchery, V. Sowmya, E.A. Gopalakrishnan, K.P. Soman, Long short-term memory-based recurrent neural network model for COVID-19 prediction in different states of India, in: Emerging Technologies for Combatting Pandemics, Auerbach Publications, 2022, pp. 245−270.

[32] A.I. Middya, S. Roy, Spatio-temporal variation of Covid-19 health outcomes in India using deep learning based models, Technological Forecasting and Social Change 183 (2022) 121911.

[33] B. Tang, N.L. Bragazzi, Q. Li, S. Tang, Y. Xiao, J. Wu, An updated estimation of the risk of transmission of the novel coronavirus (2019-nCov), Infectious Disease Modelling 5 (2020) 248−255.

[34] S. Zhao, Q. Lin, J. Ran, S.S. Musa, G. Yang, W. Wang, et al., Preliminary estimation of the basic reproduction number of novel coronavirus (2019-nCoV) in China, from 2019 to 2020: a data driven analysis in the early phase of the outbreak, International Journal of Infectious Diseases 92 (2020) 214–217.

[35] D. Fanelli, F. Piazza, Analysis and forecast of COVID-19 spreading in China, Italy and France, Chaos, Solitons & Fractals 134 (2020) 109761.

[36] Z. Ibrahim, P. Tulay, J. Abdullahi, Multi-region machine learning-based novel ensemble approaches for predicting COVID-19 pandemic in Africa, Environmental Science and Pollution Research (2022) 1–23.

[37] M. Chetoui, M.A. Akhloufi, Explainable vision transformers and radiomics for COVID-19 detection in chest X-rays, Journal of Clinical Medicine 11 (11) (2022) 3013.

[38] N. Sharma, S. Yadav, M. Mangla, A. Mohanty, S. Satpathy, S.N. Mohanty, T. Choudhury, Geospatial multivariate analysis of COVID-19: a global perspective, GeoJournal (2021) 1–15.

[39] K. Shankar, S.N. Mohanty, K. Yadav, T. Gopalakrishnan, A.M. Elmisery, Automated COVID-19 diagnosis and classification using convolutional neural network with fusion based feature extraction model, Cognitive Neurodynamics (2021) 1–14.

[40] S. Satpathy, M. Mangla, N. Sharma, H. Deshmukh, S. Mohanty, Predicting mortality rate and associated risks in COVID-19 patients, Spatial Information Research 29 (4) (2021) 455–464.

[41] A. Khadidos, A.O. Khadidos, S. Kannan, Y. Natarajan, S.N. Mohanty, G. Tsaramirsis, Analysis of covid-19 infections on a ct image using deepsense model, Frontiers in Public Health 8 (2020) 599550.

[42] H.A. Hassan, Coronavirus COVID-19: current situation in Nigeria, Journal of Ongoing Chemical Research 5 (1) (2020) 32–34.

[43] S. Rauch, E. Jasny, K.E. Schmidt, B. Petsch, New vaccine technologies to combat outbreak situations, Frontiers in Immunology 19 (9) (September 2018) 1963.

[44] World Health Organization (WHO), World Health Organization DRAFT Landscape of COVID-19 Candidate Vaccines, July 14, 2020.

[45] J.S. Tregoning, E.S. Brown, H.M. Cheeseman, K.E. Flight, S.L. Higham, N.M. Lemm, et al., Vaccines for COVID-19, Clinical and Experimental Immunology 202 (2020) 162–192.

[46] R. Bhatia, P. Abraham, The enigmatic COVID-19 pandemic, Indian Journal of Medical Research 152 (2020) 1–5.

[47] Pfizer BioNTech, Pfizer and BioNTech Celebrate Historic First Authorization in the US of Vaccine to Prevent COVID-19. Available from: https://www.pfizer.com/news/press-release/press-release-detail/pfizer-and-biontech-celebrate-historicfirst-authorization. (Accessed 10 December 2020).

[48] F. Liu, Respiratory Sound Analysis Using Deep Learning and Machine Learning for Detection of COVID-19 Symptoms (Doctoral Dissertation), Faculty of the Graduate School of the University at Buffalo, State University of New York, 2022.

[49] E. Seydoux, L.J. Homad, A.J. MacCamy, K.R. Parks, N.K. Hurlburt, M.F. Jennewein, et al., Analysis of a SARS-CoV-2-infected individual reveals development of potent neutralizing antibodies with limited somatic mutation, Immunity 53 (1) (2020) 98–105.

[50] L.B. Schonberger, D.J. Bregman, J.Z. Sullivan-Bolyai, R.A. Keenlyside, D.W. Ziegler, H.F. Retailliau, et al., Guillain-Barre syndrome following vaccination in the national influenza immunization program, United States, 1976–1977, American Journal of Epidemiology 110 (1979) 105–123.

[51] A. Koirala, Y.J. Joo, A. Khatami, C. Chiu, P.N. Britton, Vaccines for COVID-19: the current state of play, Paediatric Respiratory Reviews 35 (2020) 43–49.

[52] B. Sahu, S. Mohanty, S. Rout, A hybrid approach for breast cancer classification and diagnosis, EAI Endorsed Transactions on Scalable Information Systems 6 (20) (2019).

[53] S.S. Haykin, Neural Networks and Learning Machines/Simon Haykin, 2009.

[54] F. Kratzert, D. Klotz, C. Brenner, K. Schulz, M. Herrnegger, Rainfall–runoff modelling using long short-term memory (LSTM) networks, Hydrology and Earth System Sciences 22 (11) (2018) 6005–6022.

[55] X.H. Le, H.V. Ho, G. Lee, S. Jung, Application of long short-term memory (LSTM) neural network for flood forecasting, Water 11 (7) (2019) 1387.

[56] A. Akandeh, F.M. Salem, Simplified Long Short-Term Memory Recurrent Neural Networks: Part II, 2017 arXiv preprint arXiv:1707.04623.

[57] T. Chai, R.R. Draxler, Root mean square error (RMSE) or mean absolute error (MAE)?—arguments against avoiding RMSE in the literature, Geoscientific Model Development 7 (2014) 1247–1250.

[58] J. Benesty, J. Chen, Y. Huang, I. Cohen, Pearson Correlation Coefficient, Noise Reduction in Speech Processing, Springer, Berlin, Heidelberg, 2009, pp. 1–4.

[59] A.B. Nassif, I. Shahin, I. Attili, M. Azzeh, K. Shaalan, Speech recognition using deep neural networks: a systematic review, IEEE Access 7 (2019) 19143–19165.

[60] C. Olah, Understanding LSTM Networks, 2015. http://colah.github.io/posts/2015-08-Understanding-LSTMs/.

Chapter 12

Post-COVID-19 Indian healthcare system: Challenges and solutions

Sanjeev Kumar Mathur[1], Akash Saxena[2], Ali Wagdy Mohamed[3,4], Karam M. Sallam[5] and Shivani Mathur[6]

[1]*Faculty of Management & Commerce, Poornima University, Jaipur, Rajasthan, India;* [2]*School of Engineering and Technology, Central University of Haryana, Mahendergarh, India;* [3]*Operations Research Department, Faculty of Graduate Studies for Statistical Research, Cairo University, Giza, Egypt;* [4]*Department of Mathematics and Actuarial Science, School of Sciences Engineering, The American University, Cairo, Egypt;* [5]*School of IT and Systems, University of Canberra, Canberra, ACT, Australia;* [6]*Human Development and Family Studies, University of Rajasthan, Jaipur, Rajasthan, India*

1. Creation of robust healthcare system—A nationwide priority and its emergence

India has been a conservative nation as far as expenditure on the healthcare of its people is concerned. The percentage contribution of country's gross domestic product (GDP) on healthcare was always inconsequential. Hence, India was not able to grapple with COVID-19 pandemic due to its unpreparedness when it entered the country.

The nation's economy and healthcare system got a severe jolt due to pandemic and its consequences were immense and got etched forever on the heart, mind, and soul of the countrymen. Prior to pandemic, no concrete measures were taken up. Somehow the words of the policymakers were not complimenting the actions and only concerns were raised without any action for a smooth and prolific healthcare system to cater to the needs of more than one billion population. Hence, historically country's healthcare system was weak, and no earlier efforts were made by the previous governments to think of creation of a system that was self-reliant and self-sufficient to address to the healthcare needs of the people living in the country.

2. Pandemonium scenes

The intensity with which India got struck with COVID-19 pandemic was profound enough to spread a pall of gloom in all ranks. The nation's census at a regular interval of 10 years was consistently cautioning the policy makers about the rapid increase in the population and the need to build solid health-related infrastructure at a fast pace to meet the future challenges. A point to ponder over is that India had less than 1% expenditure on healthcare system till 2019 when the average global healthcare expenditure was 6%. In other words, no match with the average global standards. The healthcare sector became a priority sector overnight after the pandemic, and the government announced an increase in the year 2021 in the healthcare expenditure to the tune of 137% over preceding year in a jiffy which was still 1.8% of its gross domestic product (GDP) [1]. The callousness and nonchalance can be understood from the fact that as per a Human Development Report−2020, the country was pathetically placed at 155th rank in the world as far as the availability of beds in the hospitals were concerned. Here, we are not qualified to comment on the availability of Intensive Care Unit (ICU) beds, ventilators, oxygen, essential medicines, well-equipped surgical theaters etc. in the hospitals. The reality was that India was offering a meager 5 beds on a population of 10,000. The strength of healthcare system of any nation across the world is being evaluated based on number of beds per 10,000. The policies were existing on papers, and they were never thought of, for any implementation. The first lockdown announced by the central government pan India later got four back-to-back extensions to buy some time to revamp the entire healthcare system. At that point of time, a lot of emphasis was given to social distancing of six feet and the use of N 95 masks. It was perhaps the best possible mitigation strategy to slow down the pace of the transmission rate of pandemic. The diligent maintenance of social distancing was a herculean task in a nation like India that was densely populated. The loss of jobs, economic downturn, increased mortality rate due to

Deep Learning in Personalized Healthcare and Decision Support. https://doi.org/10.1016/B978-0-443-19413-9.00025-4

pandemic, paucity of hospitals for indoor admissions, etc. were the immediate impact. There was a spurt in the demand of medical oxygen in the first wave to the level of 2800 metric tons per day—an incremental rise of 2100 metric tons per day in the prepandemic times. It was somehow managed staggeringly. This demand for medical oxygen went up to mammoth levels of 5000 metric tons per day during the second wave. The fact is that India was producing 7000 metric tons per day (MTPD) at that point of time. The major issue was not the nonavailability of medical oxygen, but the transportation blues and its storage. This mismanagement along with shortages of Remdesivir proved to be the nemesis of the country. The second wave of COVID 19 pandemic left 80% COVID patients gasping for oxygen and its uneven distribution and logistics issues created artificial shortages which resulted in the colossal loss of human life in the country. The nation witnessed long queues of people at the cremation grounds pan India for performing the last rites of their loved ones who died of COVID-19. On May 7, 2021, India reported highest 0.414 million cases and the number of active cases shot up to 21.5 million. The months of April, May, and June 2020 will go down in the annals of Indian history as the darkest phase of the country. The states were divided into red, orange, and green zones indicating the relative severity of caseloads and a yardstick to open up economy and provide means of livelihood to millions of people who were having difficulty in arranging the two square meals a day for their households. All these miseries the nation was subjected to.

The abrupt declaration of lockdown broke the backbone of Indian economy due to a halt in the economic activities of the nation. This accentuated mass movement of hapless labor class from urban to rural areas which augmented the number of COVID-19 cases in the densely populated 739 districts of the country. The socio-economic multilayers were a major hindrance to provide efficient and effective medical treatment to large population in India. There were millions of Indians who lived below the poverty line, and they did not have the financial strength to pick up the massive cost of COVID-19 treatment. The private healthcare players who account for treating 60% patients in India were clearly instructed by the government to give admissions to COVID-19 patients and treat them without fail. The outpatient door (OPD) facilities were ceased. The sources of revenue dried up for private hospitals except for COVID-19 patients' treatment. The government hospitals were also converted into 100% COVID-dedicated hospitals. The poor masses were given free-of-cost treatment in the COVID-dedicated government run hospitals. Still, the government hospitals were running short of adequate medical infrastructure in the form of beds, ICUs, and essential medicines to treat the ever-increasing numbers of patients in the pandemic.

3. Efforts for healthcare system development

Although the Indian healthcare system had undergone a sea change in the 21st century, the deficient and ineffective healthcare system of the last century had shown signs of improvement through private hospitals' phenomenal growth to provide healthcare solutions to millions of people as the central and state governments felt the pressure on their existing multilayered healthcare system comprising public healthcare centers (PHCs) equipped with one physician to provide initial medical treatment in the OPDs and performing minor surgeries, district-level hospitals, and government hospitals associated with medical colleges in the cities. The government healthcare facilities were not adequate to bear the patient load due to scarcity of doctors earlier. It was at that time the state and central governments encouraged the participation of private hospitals in the 8-metropolitan tier-1 cities, 104 cities under tier-2, and remaining 3-tier cities to work in tandem with the public healthcare system and reduced increasing pressure of quality treatment to the patients through PHCs and district-level hospitals. It was a well thought out sequential growth plan of the healthcare system to provide healthcare solutions to the people of the country. All possible efforts have been made in the last few decades to create adequate infrastructure in the country to extend basic healthcare services in the remote areas too.

Undoubtedly, India made tremendous progress to deliver a robust healthcare system to the country since 2020, and as on date, this is one of the most promising sectors of Indian economy offering myriad job opportunities and substantial revenues. Both the public and private healthcare providers have worked together to create an efficient and effective healthcare delivery system. The major responsibility of state governments has been to build up the public healthcare system so that primary healthcare gets delivered through beneficial community coverage health programs. The objective of such programs is to mitigate both the mortality and morbidity among the people arising out of multifarious communicable and noncommunicable diseases. The infrastructure creation has been intended to cure "not so serious diseases" with the help of sub-centers and primary health centers and avoid the influx of patients from rural to urban areas. The presence of specialists in the district hospitals and medical colleges has been strengthened to take care of referral cases of primary health centers.

The WHO has classified nations into four categories based on their per capita gross national income (GNI) which is the benchmark of a nation's development. These categories are low-income, low-middle income, upper-middle income, and high-income nations. India has been a nation with high dispersion rate in the income of its people. The chasm between the rich and poor has been widening over a period.

4. Providing treatment to all amidst difficulties

The financial problems of a vast population in the pandemic were a major concern. This is where the concept of crowdfunding came into being. Not so popular concept for Indian community till then! Soon, this crowdfunding involving millions of people contributing, be it small or large money donations to various NGOs became a movement to extend selfless financial help to the riffraff of the society to get their treatment in private healthcare hospitals. In India, those who are a part of sports and entertainment industries are highly revered and paid handsomely. They along with prominent business conglomerates came forward to contribute handsomely to the NGOs to combat the pandemic. The PM CARES fund was constituted on March 28, 2020, to accept generous contributions to up and about healthcare system and straighten up the long-standing flaws. According to an estimate by IndiaSpend, the donations for PM CARES fund were amounted to nearly Rs. 9678 crore (1.27 billion US dollars) between the period March 28 and May 13, 2020 [2]. Out of this collected amount, Rs. 3100 crore (32%) were channelized for COVID-19 works. The deep wounds of second wave made the central government release funds of Rs. 23,000 crore to negotiate the third wave and make the states and hospitals self-reliant for the production of medical oxygen gas. The dearth of hospital beds for the admission of patients found a unique and innovate response in the creation of 100/200/500/1000 or more bedded COVID-19 temporary air-conditioned domes erection through volunteers where nongovernmental organizations (NGOs) and state governments became partners to provide COVID treatment along with essentials such as medical oxygen, medicines, injections, food, etc. A positive inference in this narrative is that Indian citizens rose to the occasion in support of their fellow countrymen to protect their lives and attempted to create better medical infrastructure in the battle against COVID-19. The age-old Indian ethos triumphed! Mankind again surfaced and flourished. Indian ethics tested and emboldened. Kindness rekindled. Camaraderie spread. This way the shortcomings of decades old poor healthcare system were taken upfront.

It was a horrendous phase in the 21st century with no one ever had previous experience to handle this ticklish situation.

Implementing lessons from the seven major preaching of Guru Gobind Singh for a meaningful life, the Sikh community of India opened their shrines (gurudwaras) to the doctors' boarding and lodging so that they could take shelter in their shrines close to the hospital they were associated with and attend the pandemic emergencies. Suddenly, India had become a land of generosity.

5. Corona warriors and their woes

The Indian healthcare workers were overworked. They were handling cases of COVID-19 24*7. Their physical and mental health also took a heavy toll. The mental stress levels went up, and they were also to be treated through counseling sessions to get over the human miseries they had been to witness and reduce their stress and anxiety levels. There has been a long list of doctors pan India who also contracted COVID-19 and succumbed to the virus. There have been some unfortunate sporadic incidences where the excessive stress resulted in doctors ending their lives. All this put together was a colossal loss of precious jewels of the society who gallantly fought against the pandemic and finally lost the battle of life. Still, the nation owes to the doctors and nursing staff for the invaluable contribution they gave to the society during the corona pandemic in 2020 and 2021. It is not that the nation has underlined their importance post-COVID-19 pandemic. The respect for doctors in the country and world has always been high. When the world's advanced medical systems known for their quintessential medical infrastructure were collapsing, it was the sheer grit and determination insurmountable of health workers that gave the country semblance to stood tall in the mid of century's worst human misery.

6. Transformation of Indian healthcare sector post COVID-19

Indian healthcare industry is made up of the following constituents given below with their areas of contribution.

Hospitals	Comprises government (public health centers, district hospitals), and private players (nursing homes, mid- and top-tier private hospitals—corporate, local investors')
Pharmaceutical	Comprises manufacturing drugs as per British Pharmacopeia/Indian Pharmacopeia standards for use as medications for human beings or animals
Medical equipment suppliers	Comprises manufacturing and supply of medical equipment such as orthopedics, surgery, dental, cardiac, etc.
Diagnostics	Includes laboratories using reagents, machines, etc. to provide diagnostic services to assist doctors to clinically diagnose disease
Telemedicine	Promising Internet of Things (IOT)-based technological intervention to meet up challenges of effective treatment solutions to the rural masses
Medical insurance	Comprises health insurance wherein an individual gets hospitalized for treatment and all medical bills are reimbursed
Medical tourism	Includes foreign patients getting world class treatment at affordable rates

The above-mentioned constituents in India were very poorly handled prior to the COVID-19 pandemic except for pharmaceutical industry growth. However, the overall situation of Indian healthcare industry is a quite different as on date. The stakeholders are fully aware of their roles and responsibilities for the establishment of an efficient and effective healthcare system.

It will be apt to comment that COVID-19 caught the world's finest and advanced healthcare systems by surprise, exposing chinks in the armour to the hilt. The COVID-19 pandemic was a reality check for world healthcare systems for the readiness to handle any health-related crisis on this planet. Indeed, it was an unprecedented crisis the world over, and India's healthcare system was not an exception to that. The COVID-19 pandemic exposed the shortfalls and pitfalls of the Indian healthcare system which was at the crossroads when the pandemic became rampant and audacious [3]. The socio-economic, political, and healthcare fabric of the nation experienced tumultuous changes. The loss of human lives was huge, and the flourishing gross domestic product of the country since 2004 got savagely damaged. The Indian healthcare system was not robust in the pre-COVID-19 times [4]. The onslaught of COVID-19 completely shook the multilayered healthcare system of the country.

In 2020, India had 1.2 million doctors to provide healthcare solutions to 1.38 billion population [5]. The World Health Organization (WHO) expects 44 doctors per 10,000 population (4.4 per 1000) and just double the nursing staff, that is, 88 on 10,000 population. This ratio (0.87 per 1000) is too less than expected norms of WHO, and India is today an awakened nation that has been up from the deep slumber to make every possible bit toward a self-reliant and self-sufficient (Atmanirbhar) India that can deal effectively all the healthcare related issues.

7. Healthcare component 1—Hospitals

The hospital industry has a sizable contribution of 80% to the overall healthcare market of India (Fig. 12.1).

Taking lessons from the pandemic, the resolute government has decided to create more world-class healthcare facilities in India. The flagship brand of autonomous government public medical universities under the aegis of the Ministry of Medical and Family Welfare is all set to add five more institutes to its existing base of 19 institutes by 2025 and a further expansion of six more later.

Speaking with figures, India boasts of currently having a massive health-related infrastructure with more than 23,000 primary health centers and approximately 0.15 million subcenters to cover nearly three fourth population (72.2% precisely) of India which resides in rural areas to offer health services. For every four PHCs, there is one referral center which offers consultation of specialists and obstetrics and gynecology care. Hence, an overall 5335 community healthcare centers are operational. There were 542 medical colleges in India and 64 autonomous postgraduate institutions as on May 21, 2020 [7].

The Center has given green signal to establish 157 medical colleges in the backward districts in a bid to strengthen the healthcare infrastructure and be future ready to face any natural calamity. The private sector has exponentially grown in the major cities of the country, and its services to the citizens may be taken into consideration from the referral services point of view, as they hire the services of specialists and superspecialists who are extremely competent. The growth of private medical colleges has also given an opportunity to eminent specialist doctors who retired from government medical colleges to provide their services to the private medical colleges. The management of latter wholeheartedly welcomed them post retirement. However, this public—private partnership which was meant to develop a holistic and impregnable healthcare

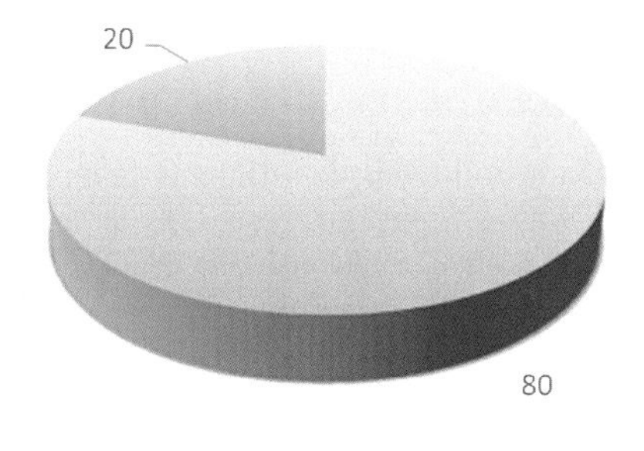

FIGURE 12.1 Ratio of % contribution of hospital industry to healthcare market in India [6].

system to address societal health care issues causes discriminations and disparities on account of different objectives through their services. The central and state governments have been introducing new healthcare schemes to ensure treatment to all including state sponsored medical policies to treat patients in both the private and public hospitals for all kinds of surgeries and indoor admissions.

8. Healthcare component 2—Pharmaceutical industry

The Indian pharmaceutical industry is the pride of the nation, and it has become the epicenter of the world supply of generic medicines to the global markets. A decade ago, the Indian pharmaceutical market was pegged at US Dollars (USD) 73 billion, and it is likely to touch USD 372 billion in 2022 (Fig. 12.2).

The likely win that appears over COVID-19 has been possible due to Serum Institute of India's (SII) massive production of Covishield vaccine. The press release of SII dated 26th November stated the company surpassed 1.25 billion doses mark. The severity of infection has mitigated among the COVID-contracted patients now. The way this world has got respite from the untimely deaths arising out of pandemic due to COVID vaccine jabs may be attributed to the Indian pharmaceutical company, the Serum Institute of India which not only owned up the responsibility to manufacture mass scale production of Covishield but managing the stringent criteria of cold chain maintenance successfully, which was a major challenge for the success of this vaccine through efficacy preservation.

9. Healthcare component 3—Medical devices and equipment

The medical devices are a promising sector of healthcare system in India. It has relatively lesser barriers. This sunshine market is like to grow fivefold from the current USD 11 billion to USD 50 billion by the end of year 2025 [6]. In terms of competition with other Asian markets, it is ranked fourth after Japan, China and South Korea. Indian market has nearly 6000 medical equipment devices, of which 86% devices are imported to meet out the domestic demand. There are problems in the manufacturing of high price medical equipment, especially those required for the diagnosis of cancer, magnetic resonance imaging (MRI) machines, ultrasound machines, Computed Tomography (CT) scan machines, etc. India reported demand of medical devices to the tune of USD 1.77 billion as compared to USD 0.99 billion exports, The Government of India has been generous to the medical devises segment and have permitted 100% Foreign Direct Investment (FDI) (Table 12.1).

Indian dependency is very high for high-end medical devices on the supply of multinational organizations. The Prime Minister's call to promote local manufacturers for a self-reliant India under "Vocal for Local" mantra is likely to give a boost to this industry of healthcare system in the future.

The domestic manufacturers are strong enough to produce Class A and B medical equipment where technology is not in an advanced stage. However, same is not true for high-cost medical devices.

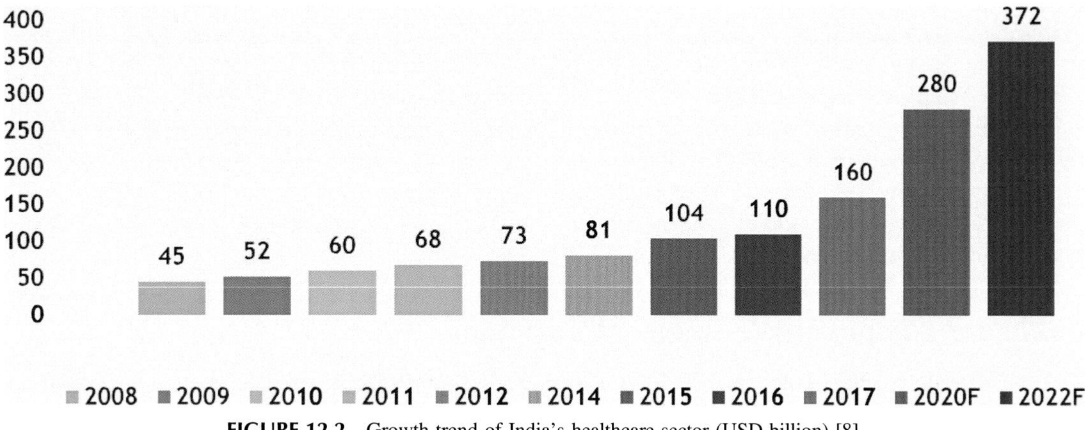

FIGURE 12.2 Growth trend of India's healthcare sector (USD billion) [8].

TABLE 12.1 Imports and exports in India's medical devices sector [9].

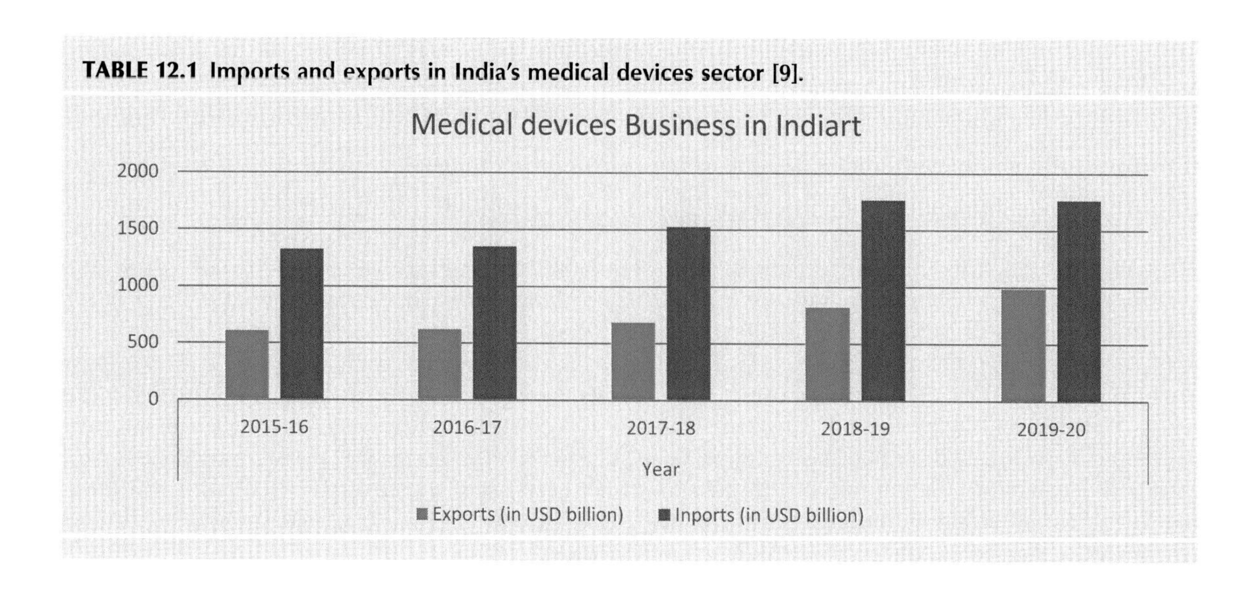

10. Healthcare component 4—Diagnostics

India' diagnostic market value is USD 4 billion, of which 15% goes to laboratories and 10% is into radiology. This market is likely to experience exponential rise to the extent of USD 32 billion with a compound annual growth rate (CAGR) pegged at 20.4% [6].

There were issues of testing kits for novel coronavirus in 2020 and later in 2021. The diagnostic labs were not equipped adequately to perform routine tests for assessing the lungs damage due to the virus. In due course of time, things were resolved and free testing facility by state and central government hospitals were provided to the vast population of the country. The diagnostic labs are getting digitalized, and there has been a deluge of fully automated new diagnostic centers establishment across the country now.

11. Healthcare component 5—Telemedicine

On a positive note, the COVID-19 pandemic opened new vistas for the digital intervention into the system and brought focus on the importance and revamping of healthcare solutions to be future ready for unseen endemic and pandemic conditions. The technological push became inevitable and a boon [10].

The policy makers have made teleconsultation legal, and they have started bridging the gap between specialist doctors and rural population health-related issues. The future of India lies in the phenomenal growth of telemedicine. This will reduce the pressure on the urban healthcare system. From the comforts of home, anyone from rural or urban area may take specialist advice and start medication.

The telemedicine market worth till the year 2019 was USD 830 billion, and it is estimated to touch USD 5.5 billion by 2025 growing at a cumulative average growth rate of 20.4% during the period 2020–25 [11].

The Indian telemedicine market for the period 2010–19, and it estimates up to 2025 are given below [12] (Fig. 12.3):

In early 2020, Arogya Setu mobile app proved to be a panacea in India for assisting in the syndromic mapping, contact tracing, and self-assessment [13]. The web-based technology was effectively used to complement the response management, ensuring smooth supply of essential items in containment zones, tele-consultations with patients, bed management, and real-time monitoring and review by the authorities [10].

12. Healthcare component 6—medical insurance

The contribution of health insurance stands at 20% to the business of nonlife insurance. This is the second largest portfolio. The India Brand Equity Foundation has revealed that the gross direct premium income from health insurance sector has risen by 17.16% year-on-yearly basis to touch the value of INR 516.37 billion (approx. USD 6.87 Billion) in the financial year 2020. This is attributable to the overall rise in the awareness level among urban class and growing lifestyle-related

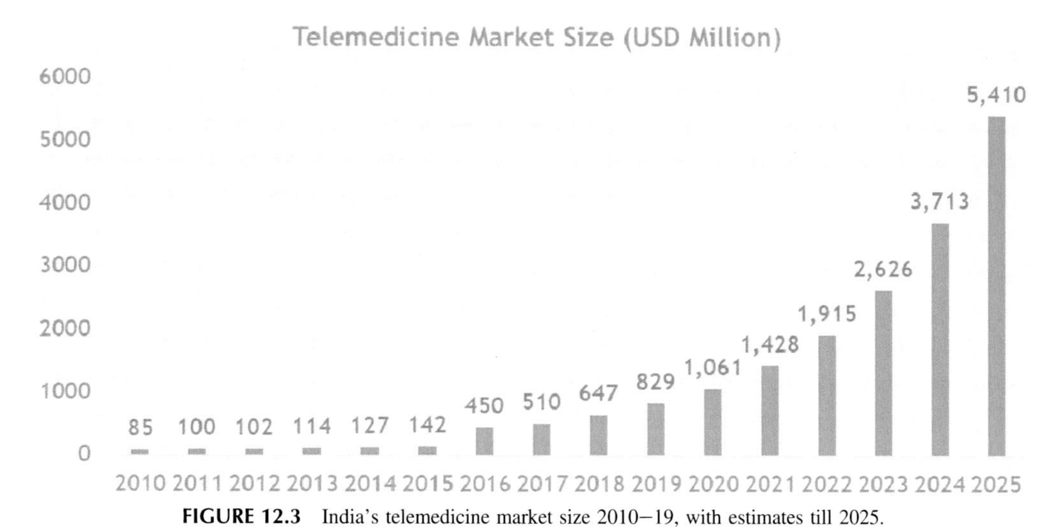

FIGURE 12.3 India's telemedicine market size 2010–19, with estimates till 2025.

health issues. The determination of market size of the health insurance sector was based on number of lives covered and the price per life. The private health insurance sector has not yet been able to penetrate the market in India, and it is still less than 10% [6,14].

Thus, this sector has started strengthening and emerging as a key constituent of the healthcare market of India. This has drawn attention of millions of Indians who have now started the importance of buying an insurance policy to meet out the hospital expenditures. Earlier an average Indian had no qualms to put aside the importance of getting all the family members insured medically. Now, India comprehends how significant it is to first insure the entire family against a particular sum of money. The flourishing medical insurance sector is a testimony in the last 2 years. It will continue to grow by leaps and bounds in near future. It was a deep in slumber sector, now fully awakened.

India's health insurance ecosystem has multifarious stakeholders such as insurance companies, beneficiaries, pharmaceutical companies, druggists, diagnostic centers, hospitals providing services, Government funded insurance schemes, third-party intermediaries, etc.

13. Healthcare component 7—Medical tourism

Medical tourism is again a silver lining to the Indian healthcare industry with its worth of 9 billion till 2020 and an estimated value of USD 13 billion in 2022 [6,14]. In the Asia–Pacific region, it is 3rd ranked and growing at a faster pace.

The reasons for its growth are apparent. The medical tourism sector in India offers low-cost treatment to the offshore patients. India has proved to the world that there are frugal ways by which treatment can be done of low-income groups. The world has taken a cognizance of it. There are habitants of African nations and middle-east countries who have acknowledges since long the medical expertise and low cost of treatment. The opening up of economy will be a gateway to many nations where cost of medical treatment is exorbitantly high.

14. No need for complacency

The Global Health Security (GHS) Index (2021) which evaluates the readiness of 195 nations across the world in its study pronounced the unpreparedness of nations for future endemics and pandemics onslaughts that might be more lethal on humanity than COVID-19 pandemic. The GHS index has put India at rank number 66 with a score of 42.8 in 2021, which is a further dip by 0.8 points in comparison to 2019 scores. The assessment of Global Health Security Index is based on six-point criteria, namely, prevention, detection and reporting, rapid response, healthcare system, and compliance with international norms and risk environment. The GHS study acknowledged India's giant strides into detection and reporting factors, but no headway was observed for the prevention practices, improved healthcare system and rapid response procedures. The study reported newly acquired capacity building but deemed to be temporary [15].

The way the private Indian healthcare players have risen to the call of the nation to extend all possible support to the central and state governments is praiseworthy. It was a major challenge for the nation to provide adequate quantities of N-95 masks, surgical masks, medicines, sanitizers, disinfectant sprays, etc. besides impeccable COVID-19 testing kits. The nation responded with sincerity and commitment. The isolation beds for treatment, ventilators, ICUs, and medical staff were in great demand in both the government and private hospitals and together they were able to address these teething issues with a sense of responsibility [16]. The nation saluted the corona warriors from time to time expressing its gratitude to those healthcare workers who put their lives in danger to protect the lives of millions of countrymen.

India's private healthcare sector took up the challenge to treat and cure COVID-19 patients, and nearly 60% patients were admitted to the private hospitals during the first wave of COVID-19 pandemic. The culture of encouraging establishment of private medical colleges, corporate hospitals, and private entrepreneurs' hospitals came to the rescue of the nation. The government funding to create more beds in the hospitals at tehsildar level and district level not only ensured timely treatment and indoor hospitalization but also proved to be a game changer to reduce the mortality rate. At tier-1 and tier-2 levels, the private hospitals made huge investments to prevent further damage and ran mammoth vaccination program with Serum Institute of India Limited and Bharat Biotech working day in and day out to ensure uninterrupted supply of Covishield and Covaxin, respectively, to the Indian population on a war footing note. The large investments at all levels were there to handle and prevent the infection, infrastructure creation for quarantining and isolating patients, providing treatment, besides offering pathological lab facilities at government prescribed rates to check the vital parameters, and supplying simple yet important equipment like oximeters, BP instruments, glucometers, etc. [17].

The second wave of COVID-19 pandemic came like a Tsunami and proved to be devastating to the humongous population of the country causing widespread COVID-19-related deaths in million in the months of April and May 2021. The country's healthcare system collapsed for a while and witnessed pandemonium scenes all over. The game changer during the first COVID-19 wave, Remdesivir, was also in acute shortage. Everything happened suddenly. The major hospitals of the country were having acute shortages of oxygen, the demand for ICU beds and general beds with oxygen raised alarmingly, and the country's healthcare systems became helpless.

India learned its lessons the hard way quickly during the second wave of COVID-19 pandemic. Once the second wave got over, the country's healthcare system got refurbished to be future ready to tackle probable third wave which India witnessed in the beginning of 2022. That was the time by which the government established pressure swing absorption (PSA) plants in each district hospital to be self-sufficient to produce oxygen to meet future requirements. The far-flung areas were on priority for oxygen production [18].

The COVID-19 management was emboldened with the establishment of COVID Drugs Management Cell under the aegis of the Department of Pharmaceuticals, Government of India, to ensure smooth flow of drugs needed for the treatment of COVID-19 even in distant places.

The State Health Index of NITI Aayog has recently pronounced that only five major and two minor states of the country have been able to improve their overall healthcare index from Base Year of 2018–91 to the Reference Year 2019–20. The need of the hour for the Indian healthcare industry is to do preparations on a war-footing note to tackle future challenges to the system with aplomb and complete resourcefulness.

One more thing that emerged prominently was the paucity of trained nursing staff to use ventilators in the COVID-19-dedicated hospitals and do effective and skillful utilization of oxygen to prevent any wastage. Keeping this aspect on mind, the Indian healthcare system is getting ready to work on knowledge enhancement and skill acquisition of paramedical staff in every district hospital. The endeavor is to train and create at least one such "oxygen manager" in every hospital to do proper management of "scarce resources."

The exceptional pandemic crisis and the absence of standard treatment protocol compelled the government to announce nationwide lockdown which has impacted the profitability of operations of private hospitals in 2020–21. According to a report published in Business Standard (October 20, 2020), it is the estimation of Crisil that the profits of private hospitals are likely to take a plunge to the extent of 40% [19]. The reasons are obvious. The government converted most of the private hospitals into COVID1-19 centers, and the general OPDs were closed for a long duration which seemed to have adversely impacted the revenue collection and thus, the profits of private sector hospitals.

It is also very true that the treatment of COVID-19 patients also made a substantial contribution to the revenues of the private hospitals.

15. Strength weakness threat opportunity (SWOT) analysis of Indian healthcare industry in the post-COVID-19 era

There were initial hiccups in the Indian healthcare system to combat pandemic in 2020, and after a clueless start to take COVID-19 upfront, the Indian healthcare industry has come of age. The system is on the right track and getting ahead smoothly.

Some positive things that have happened in Indian healthcare system in the pandemic era are:

- ☐ The state governments are scouting lands to build satellite hospitals in all directions of the districts with adequate bed facilities and proper facilities for the treatment.
- ☐ There has been an announcement in many states to open up medical colleges in every district which will provide strength to the healthcare system.
- ☐ The existing bed capacity in each hospital is getting expanded. The ventilators are getting produced and supplied to the hospitals. The staff is getting trained to operate ventilators.
- ☐ Every major hospital is becoming self-reliant in the production of oxygen to meet its demand without looking at other sources.
- ☐ India has emerged as the world capital of vaccine production.
- ☐ The Indian pharmaceutical industry is future ready to step up production of emergency medicines as and when required. They are on high alert.

Also, India has emerged as a major force to be heard at all platforms. Recently, India has raised its voice at the global COVID-19 summit to reform the World Health Organization's approval processes for vaccines and therapeutic treatment. The benefit would be in the form of creating a resilient global health security architecture. The building of a resilient supply chain is a must and equitable access to vaccines and medicines is the need of the hour.

Having talked of so much happening at the government level, let us do the SWOT analysis of the private healthcare system of India, without which, it would be an uphill task to accomplish the healthcare system—related objectives. The SWOT analysis of private healthcare system is given below:

Strengths

- ☐ Emergence of corporate hospitals such as Fortis, Apollo, Narayan, Manipal Hospital etc. in tier-1 and tier-2 cities
- ☐ Better facilities and expansion due to increased investments in the private hospitals
- ☐ Recruitment of high caliber specialty doctors by HR teams
- ☐ Regular training and knowledge updation of doctors and supporting staff
- ☐ Excellent patient care through wide array of services
- ☐ Increased NABH accreditations—a hallmark of quality services among private hospitals promoting healthy competition
- ☐ Use of web based applications for consultation is bridging the gap between rural and urban areas with a possibility of patients getting indoor treatment in private hospitals
- ☐ Collaborations with foreign hospitals to provide world class opinion to the patients
- ☐ Stiff competition among the private hospitals has been raising the quality of services

Weakness

- ☐ Domestic expansion for growth of private healthcare system. Not going global is a lacuna.
- ☐ The cost of medical services is getting beyond the reach of a common man in India. The GDP per capita of India stands at USD 2000, which makes majority of Indian population inaccessible to these hospitals.

Opportunities

- ☐ Growing health concerns and the use of ICT enablers especially after COVID-19 pandemic has made a common man aware of the importance of health in life. Thus, a huge potential exists in the market to be tapped.
- ☐ The phenomenal rise in the medical insurance in India has made the treatment of a common man possible in the private hospitals.
- ☐ Medical tourism boom is a great opportunity. The reasons are high caliber professionals, world-class infrastructure, and economical cost of treatment/complicated surgeries. India is a destination for dental tourism since the start of 21st century. It is now expanding into other areas of medical treatments.

Threats

- ☐ The private service providers are being governed by the Department of Pharmaceuticals under the Ministry of Chemicals and Fertilizers. Their policies are people centric. In difficult times, they can exercise complete control over the operations of private healthcare service providers with restriction over cost of treatment as has been observed during COVID-19. This can impact profits of private hospitals.
- ☐ The ever increasing healthcare cost is a matter of concern for private players. If the private service providers pass on raised cost of treatment to patients every time, then very soon private hospitals will be inaccessible due to raised costs.

16. Future of Indian healthcare system

Going by the old saying, "once bitten, twice shy"; the government is not in a mood to undermine the importance of establishment of a robust healthcare system. The lessons are painful and profound. It has become a priority sector, and it is not only the government-created healthcare facilities that are getting a facelift, but the private sector healthcare system is also making huge investments in healthcare infrastructure.

In order to fulfill motto of "health for all," the central and state governments have already floated Ayushman Bharat, Chiranjeevi Yojana, etc. schemes to provide medical treatment to one and all. The state-owned government hospitals are providing free of cost treatment to the visiting patients. The legalization of telemedicine practice by the government is a welcome move. There has been a shift in the approach of government to promote telemedicine practice in India.

The "future of healthcare report" of Healthcare Information and Management System Society suggests an incremental rise of 80% in the digital healthcare tools in the forthcoming 5 years from now.

India's healthcare industry turnover is expected to be pegged at USD 372 billion in the year 2022 as per a report published by Invest India. The contribution of hospital industry would be USD 132.84 billion this year, which was struggling in the year 2017 at USD 61.79 billion. In other words, the hospital industry growth is likely to be at 16%−17% compound annual growth rate (CAGR) [20].

The healthcare system of India would continue to be under enormous pressure due to the country's vast population in the forthcoming years. There is every likelihood that the concept of homecare would gain a steep rise in the coming years as COVID-19 has not gone as yet and the probability is that it may continue to bother the mankind in the next 5−6 years.

An effective means to tackle COVID-19 pandemic and work toward its eradication from this planet is to promote vaccination at the global level. As regards to Indian population, 80% population has had two jabs and a whopping 97% at least one jab. It would not be erroneous to say that vast population coverage in the country for vaccination has allowed the country to thwart the perils of hospitalization and mortality to a great extent in the third wave when Omicron, a new variant, was reported in the country.

Is this fourth wave of COVID-19 in India in May 2022? Even, if the numbers are on the rise, the hospitalization and mortality rate, two major concerns, are still not discernible.

The bitter COVID-19 experience will continue to give sleepless nights to the policy makers of the country to take the leverage of technology-based numerous applications at all levels to set free this country and the world from this dreadful pandemic.

One good news is that the Indian government has taken a considerate decision to raise the level of expenditure to 2.5% of the GDP in the next 5 years [21].

In the last, we are in the era of digital and technological innovations, which are all set to provide healthcare solutions to the countrymen at a faster pace.

17. Conclusion

There is a big lesson for India in the aftermath of pandemic crisis. The life of every citizen is invaluable, and the progress of this nation would largely depend on the creation of robust healthcare system. There is no scope to undermine its importance or feel any complacency. The healthcare progression should be made an ongoing process. The technological push into healthcare system through Internet of Things (IOTs) must be intensified to facilitate medical treatment. The Computers and information technology (ICT) enablers would reduce pressure on both the private and government hospitals in the urban areas and facilitate the process of indoor treatments and surgeries in the referral hospitals at the district level.

References

[1] B. Jain, et al., Health budget in light of pandemic: health reforms from mirage to reality, Journal of Family Medicine and Primary Care 11 (1) (January 2022) 1−4.
[2] An Article in IndiaSpend Dated, May 20, 2020.
[3] D. Roy, et al., Study of knowledge, attitude, anxiety & perceived mental healthcare need in Indian population during COVID-19 pandemic, Asian Journal of Psychiatry 51 (June 2020) 102083.
[4] Joint Statement by ILO, FAO, IFAD and WHO, Impact of COVID-19 on People's Livelihoods, Their Health and Our Food Systems, October 2020.
[5] Healthcare, Invest India. https://www.investindia.gov.in/sector/healthcare.
[6] Note on Health and Pharmaceutical Sector, Invest India.
[7] List of Medical Colleges in India—Wikipedia.

[8] Healthcare: The Neglected GDP Driver. Need for a paradigm shift. KPMG and FICCI. Retrieved January 10, 2021, from http://ficci.in/study-page.asp?spid=20634§orid=18.
[9] Note on Home Healthcare. Nidhi Saxena. Founder and CEO, Zoctr.com.
[10] R.P. Singh, et al., Internet of Things (IoT) Applications to Fight against COVID-19 Pandemic.
[11] Recommendations to Reinvigorate the Private Health Sector in India, NATHEALTH.
[12] Telemedicine Market Size in India 2010–2025, Statista Research Department. Retrieved January 7, 20201 from https://www.statista.com/statistics/605179/india-telemedicine-market/.
[13] S. Basu, Effective contact tracing for COVID-19 using mobile phones: an ethical analysis of the mandatory use of the Aarogya Setu application in India, Cambridge Quarterly of Healthcare Ethics (September 30, 2020) 1–10.
[14] National Health Policy, Ministry of health and family welfare, Government of India (2017). Retrieved February 23, 2021 from, https://www.nhp.gov.in/nhpfiles/national_health_policy_2017.pdf.
[15] World Unprepared for Future Pandemics: Global Health Security Index 2021, December 9, 2021.
[16] J.-L. Vincent, et al., Ethical aspects of the COVID-19 crisis: how to deal with an overwhelming shortage of acute beds, European Heart Journal: Acute Cardiovascular Care 9 (3) (April 1, 2020) 248–252.
[17] Key Policy Responses from OECD; The Territorial Impact of COVID-19: Managing the Crisis and Recovery Across Levels of Government, May 10, 2021.
[18] B. Srinivas, et al., Medical oxygen supply during COVID-19: a study with specific reference to State of Andhra Pradesh, India, Materials Today Proceedings (January 26, 2021).
[19] Pandemic to Shave Off 40% Operating Profit of Private Hospitals, Business Standard, October 20, 2020.
[20] Indian Healthcare Market to Hit $372 Billion by 2022, The Economic Times, December 3, 2017.
[21] Committed to Raise Health Expenditure to 2.5% of GDP: Harsh Vardhan, Business Standard, September 28, 2020.

Chapter 13

SWOT Perspective of the Internet of Healthcare Things

M.S. Sadiq[1], I.P. Singh[2] and M.M. Ahmad[3]
[1]Department of Agricultural Economics and Extension, FUD, Dutse, Jigawa, Nigeria; [2]Department of Agricultural Economics, SKRAU, Bikaner, Rajasthan, India; [3]Department of Agricultural Economics and Extension, BUK, Kano, Nigeria

1. Introduction

A sophisticated network of intelligent gadgets that frequently exchange data via the Internet is known as the Internet of Things (IoT) [1]. It has cleverly transformed real-world objects into virtual counterparts. Everything in our world should be brought together under a mutually beneficial arrangement as part of the Internet of Things (IoT), which will assist users in both controlling the objects around them and keeping them informed about their current conditions [2]. Without requiring human-to-human or human-to-machine interaction, IoT devices sense the surroundings and transfer the collected data to the Internet cloud. In the current modern era of communication, when tens of millions of devices are connected via the IoT and the number is increasing quickly [3], the IoT has become an essential component. The IoT has the potential to be fundamental in a number of areas of life, including health systems, autonomous vehicles, home and industrial automation, intelligent transportation, smart grids, etc. [2]. In order to send information to the appropriate body or organization, sensors collect data and related information from the environment and do so via the Internet cloud [4]. The implementation of seamless communication between numerous devices is the basic idea behind the IoT. Better resource utilization, cost savings, and a reduction in manual contact are all expected benefits of this system.

It has been inevitable to talk about and explain the IoT potential during pandemics as the 2019 coronavirus disease (COVID-19) spreads across the globe. The number of confirmed COVID-19 cases had surpassed 8.5 million as of June 21, 2020, with a 3.7% fatality rate [5]. Diverse solutions are being developed by researchers in various domains that may aid in the fight against COVID-19 [6]. The elements required to assist countries in reducing the impact of COVID-19 are developed through the IoT. There are numerous IoT applications that can be used to ensure that all safety and precautionary measures advised by health professionals are followed [7]. The IoT provides a scalable network that has the capacity to handle massive amounts of data gathered from sensors utilized by numerous applications to combat COVID-19. Additionally, dependable IoT networks speed up the distribution of vital information, which can aid in delivering prompt assistance during the global COVID-19 epidemic [8]. Because of the coronavirus outbreak, the IoT has never been as important as it is today.

The IoT has recently established itself as a compelling study issue across a wide range of academic and industry fields, particularly in health care. Modern healthcare systems are being transformed by the IoT revolution by embracing technological, economic, and social prospects. Healthcare systems are changing from traditional to more individualized ones so that patients can be examined, cured, and monitored more simply. The unique acute respiratory syndrome coronavirus 2, which is causing the current epidemic, is the biggest threat to global public health since the influenza pandemic of 1918. IoT technology represents one of the trailblazers in this field. Since the epidemic began, there has been a vigorous push in various research communities to use a wide array of technologies to tackle this global menace. In the current pandemic crisis, all countries are battling COVID-19 and are still searching for a workable and affordable solution to deal with the issues that are emerging in various ways. Physical scientists and engineers are working to meet these challenges, develop new theories, characterize new research problems, produce user-centered explanations, and enlighten ourselves and the general public. By utilizing early diagnosis, patient monitoring, and the deployment of predetermined protocols following

Deep Learning in Personalized Healthcare and Decision Support. https://doi.org/10.1016/B978-0-443-19413-9.00006-0

patient recovery, IoT-enabled/linked devices/applications are used in the context of COVID-19 to reduce the possibility of COVID-19 spreading to others. This succinct study is intended to raise awareness of this ground-breaking technology and its important uses in the healthcare industry, particularly in light of the COVID-19 epidemic.

At the time of writing, the number of sick patients was rising daily in the pandemic, and there is a huge need to use the adequate and well-organized facilities provided by the IoT. The Internet of Healthcare Things (IoHT) and the Internet of Medical Things (IoMT) are related to the current challenges, and the IoT has already been used to serve the sought-after aims in several areas. The number of settled cases can be increased and improved by adhering to the recommendations and resources of the IoHT/IoMT. We can expect to witness an increase in the productivity of medical staff and a decrease in their workload with the successful adoption of this technology. With less costs and errors, it also can be used for the COVID-19 pandemic.

2. The IoT in healthcare systems

Because it directly affects people's social welfare and quality of life, the healthcare sector is a crucial concern for both emerging and established nations. Since it can aid in reducing numerous health problems and diseases, research and development in the healthcare industry should be a continuous activity. The healthcare industry can readily improve thanks to recent and advanced technological breakthroughs. The implementation of cutting-edge computer technology in the healthcare sector can further enhance the already-existing capabilities of the healthcare and medical sectors. These cutting-edge computer technologies can help physicians and other healthcare professionals identify numerous ailments at an early stage. They can also significantly increase the accuracy of early disease detection.

Different cutting-edge and ground-breaking computer technologies are already having amazing effects in other industries. IoT, blockchain, machine learning, data mining, natural language processing (NLP), image processing, cloud computing, and many other technologies are among these. The IoT refers to the Internet of Everything. Here, everything is referred to as being embedded with electronics, software, sensors, actuators, connections, and other components that allow it to connect, gather data, and exchange information. This includes household appliances, cars, and other goods. The IoT, which encompasses Internet connectivity beyond standard devices like desktops, laptops, smartphones, and tablets to any variety of often simple or non-Internet-enabled physical items and daily objects, is credited to Kevin Ashton. In the Internet of Things, sensors, cloud computing, wireless technology, and security are the most important technologies.

The fundamental IoT life cycle has four stages: (1) data collection through devices using sensors; (2) data storage in the cloud for analysis; (3) data analysis and subsequent data transmission to the device; and (4) device action. Our lives are made more comfortable by IoT's applicability across numerous industries. Smart homes, smart cities, agriculture, smart retail, driverless cars, and health care are the primary IoT applications. Security remains a key component of all technologies and is essential to the efficient operation of IoT networks. Methods for ensuring data confidentiality and authentication, access control inside the IoT network, privacy, and trust among users and things, and the enforcement of security and privacy regulations are some active projects for improving IoT security. Careless program design creates vulnerabilities, which are a major cause of network security problems and the security issue within IoT.

So that any unauthorized receiver cannot access the system, correct initialization of the IoT is performed at the physical level in IoT architecture. The perception layer, network layer, middleware layer, application layer, and business layer are the five levels that make up the IoT architecture [3]. Each layer has a goal and problems. Confidentiality, integrity, and availability are the primary security objectives that are essential in IoT. IoT attacks fall into four types based on vulnerabilities: "physical attack," "software attack," "network attack," and "encryption attack."

1. Physical attack
  - **i.** ***Node tempering***: By modifying the compromised node, the attacker is able to access the encryption key.
  - **ii.** ***Physical damage***: When an attacker physically affects an IoT system component, it results in a denial of service (DOS) attack.
  - **iii.** ***Injecting malicious code***: Using this approach, the attacker can take complete control of the Internet of Things system.
  - **iv.** ***RF interference on RFID***: The attacker delivers noise signals across radio frequency signals that are used for RFID communication.
  - **v.** ***Social engineering***: The attacker uses IoT system users to gather sensitive information in order to further his objectives.
  - **vi.** ***Sleep deprivation attack***: The attacker's primary goal is to shut down nodes.
  - **vii.** ***Node jamming in WSNs***: This attack uses a jammer to disrupt wireless communication.

2. Software attack
   i. ***Phishing attacks are a common type of attack***: To obtain the user's confidential information, the attacker makes use of bogus websites.
   ii. ***Viruses, worms, Trojan horses, spyware, and adware***: The appearance of these entities can harm the system by dispersing harmful code through e-mail attachments and from the Internet. The worm is capable of self-replication without the assistance of people.
   iii. ***Malicious scripts***: Malicious scripts are used in this attack to gain access to the system.
   iv. ***DOS***: The adversary's primary objective is to stop the users.
3. Network attack
   i. ***Traffic analysis attacks***: The attacker intercepts and examines messages to get network information.
   ii. ***RFID spoofing***: An attacker manipulates RFID signals to modify the message and provide the system with false information. The attacker modifies the information, which the system accepts.
   iii. ***Sinkhole attacks are a particularly frequent kind of attack***: This attack's main goal is to trick nearby nodes into believing they are on a particular route.
   iv. ***Sybil attack***: The attacker introduces a malicious node into the network, and that node assumes the identities of numerous other nodes.
4. Encryption attack

   The primary goal of this attack is to gain the private key, which is required for communication between two devices.

   i. ***Side-channel***: In this attack, the attacker discloses some additional information when the message is sent from the user to the server or vice versa.
   ii. ***Cryptanalysis attack***: In these attacks, the attacker converts the message from an unintelligible format to an understandable one without having access to the key.
   iii. ***Man in the middle attack***: To steal sensitive information, the attacker continuously monitors the communication between the nodes. Various security solutions have been made in the literature.

Due to current issues such as centralization, single points of failure, etc., security remains a worry in IoT networks. Therefore, to increase the security of IoT, a new and growing technology called blockchain can be deployed. By addressing the difficulties and problems of centralization in the current security procedures and introducing the idea of decentralization using the blockchain, the strength of blockchain technology can be implemented in the IoT to strengthen its security and make it a more secure network. No third party is needed for transactions or communication in a point-to-point distributed network like blockchain [4]. Each transaction is separate from the others and operates independently. Blockchain is the technology that underpins the well-known and ground-breaking idea of cryptocurrencies. The security and hack-proof qualities of cryptocurrencies are well regarded. Other networks can improve security by utilizing the exact same blockchain idea.

A public distributed ledger system is available to everyone in blockchain. Blockchain is a database of records that keeps data in chronological order and is accessible to the public. Details about transactions are contained in a block. In addition to data, each block also carries the hash of the block before it and the block in question. Header and transaction information are its two components. Information about the block is contained in the header. The time the block was produced is recorded by "timestamp." The "difficulty level" determines how challenging mining a block will be. The fingerprints of all the transactions contained in a block are represented by "Merkle Root," while the answer to the proof-of-work algorithm's mathematical conundrum is represented by "NONCE."

3. Key merits of the IoT during the COVID-19 pandemic

The IoT is a cutting-edge technology that makes sure that all those who have contracted this infection are quarantined. A good monitoring system is helpful when under quarantine. Through the use of an Internet-based network, all high-risk patients may be easily tracked. Biometric parameters like blood pressure, heart rate, and glucose level are taken using this technique. The major benefits of the IoT for the COVID-19 pandemic are shown in Fig. 13.1. We can expect to witness an increase in the productivity of medical staff and a decrease in their workload with the successful adoption of this technology. With reduced costs and errors, the same can be applied to the COVID-19 pandemic.

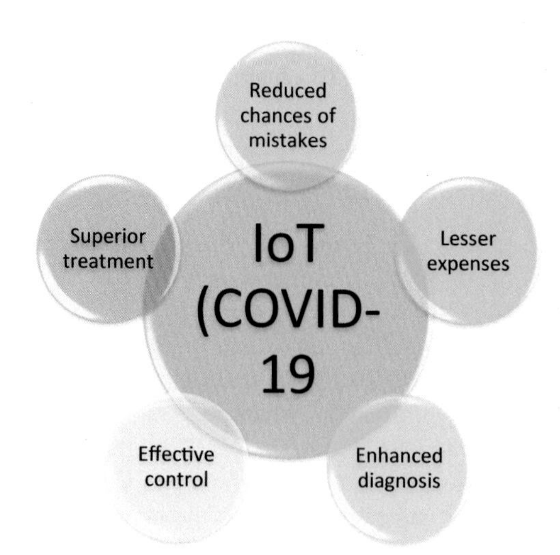

FIGURE 13.1 Key merits of using the IoT for curbing the spread of COVID-19.

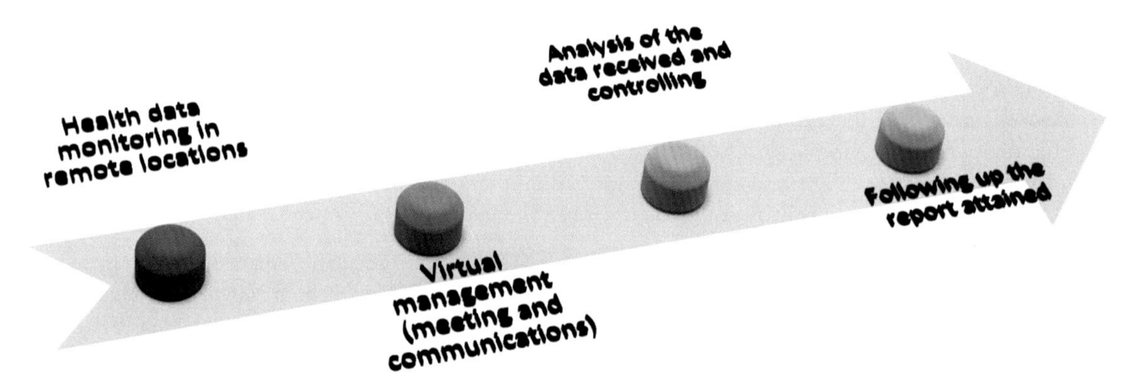

FIGURE 13.2 Step-up flow for using the IoT to curb COVID-19.

4. IoT processes for combatting COVID-19

The IoT is a cutting-edge technological platform that can meet considerable problems in a lockdown situation. The real-time data and other necessary information about the infected patient can be collected using this method. Fig. 13.2 displays the key IoT procedures performed for COVID-19. The IoT is employed in the first stage to collect health data from the infected patient's numerous locations and manage all of the data using a virtual management system. This technology aids in data control and report completion monitoring.

5. The IoT's overall impacts in relation to COVID-19 concerns

The concept of the IoT makes use of the connected network for efficient data flow and exchange, as was noted previously. Additionally, it enables anyone to connect with the service beneficiaries to cooperate and address any issues, including social workers, patients, and other members of the public. Therefore, the effective tracing of patients as well as suspicious cases may be entirely assured by using the proposed IoT technique in the COVID-19 pandemic. Most of the general public are now aware of the coronavirus symptoms.

The identification of a cluster can be considerably improved by creating an informed group within a connected network. It is also possible to create a specific smartphone application for the benefit of those in need. In order to opt out of the impressive move and shorten the overall quarantine period, the symptoms and recovery must be properly reported to the controller, i.e., doctors, physicians, caretakers, etc.

6. Global technology developments to quickly treat COVID-19 cases

Therefore, in an effort to combat the COVID-19 pandemic and raise public awareness of it, the Indian government has introduced e ArogyaSetu, a smartphone application that aims to connect Indians with the most vital potential healthcare services. The mobile application e Close Contact, which translates as "close contact," was also released for civilian use in China. This app notifies the user of how close they are to a corona-positive individual in order to allow for greater caution to be used when moving outside. At the end of April 2020, the US government introduced a comparable smartphone application for its citizens. Taiwan was the country where COVID-19 cases were most likely to increase after China. To protect the public's health, Taiwan swiftly militarized the country and put in place particular procedures for coronavirus case detection, eradication, and resource distribution. In order to start the creation of big data for analytics, Taiwan provided and integrated its national health insurance database with its immigration department and cataloged the details; it generated real-time warnings during a clinical visit based on travel time and medical symptoms to help in case identification. In order to possibly identify those who are infected, they have also utilized the most recent technology, which includes QR code scanning, connected reporting of travel history, etc.

7. IoT applications to fight COVID-19

The IoT is one of the promising technologies that will revolutionize our lives because of its seamless connections and impressive integration with other technologies [9]. In order to effectively tackle this worldwide pandemic, IoT solutions can be applied to a number of industries, which can significantly lower the danger of coronavirus recurrence [10].

1. Digital telehealth and the Internet of Health Things
   IoT's extension, the Internet of Healthcare Things (IoHT), uses the communication infrastructure to link people to healthcare facilities so that they may monitor and regulate the vital indicators of the human body [11]. In rural locations where finding a qualified doctor is difficult for a variety of reasons, telemedicine is becoming increasingly popular [12]. Examples of important body signs that can be remotely monitored without the patients' physical presence include electrocardiography, diabetes, and heart rate. Using a local gateway, the sensors and actuators communicate the data they collect from the patient to the cloud. Utilizing any mobile or desktop tool made available to them, the physician reviews the data and informs the patient or the medical team caring for the patient about the findings.
   A critical role is played by digital telehealth during the COVID-19 outbreak. Patients can communicate with clinicians through a portal, and treatment is delivered remotely. Employing a secure IoHT system in COVID-19 has the advantage of preventing the spread of viruses because doctors do not have direct contact with patients [13]. In these circumstances, digital telehealth has been working in several nations. Patients who use Health Arc's [14] IoT-based medical devices have their data continuously checked by medical personnel. Patients receive recommendations and prescriptions on their smartphones or tablets when the data are evaluated. Additionally, it helps to reduce the frequency of hospitalizations, readmissions, and patient density in hospitals, all of which improve patient care and quality of life for COVID-19 patients.
   a. IoT-enabled ambulances
      The medical personnel who work in ambulances frequently face extremely stressful and error-prone circumstances. Situations for medical workers caring for COVID-19 patients have been even more stressful and difficult during the present COVID-19 pandemic. An efficient answer is provided by IoT-enabled ambulances, which can advise appropriate steps to the medical personnel caring for the patient inside. This results in prompt action and efficient patient management. Emergency cars are provided by WAS vehicles [15], which use smart solutions. The equipment that uses radio-frequency identification (RFID) is linked to a wireless local area network (WLAN). The relevant medical staff has remote access to the patient's information.
   b. Smart devices for tracking one's health
      IoHT-enabled devices are divided into two groups: personal and clinical. IoHT devices for personal use are used to track health. The most widely utilized technologies include the Fitbit and Apple Watch. Using these devices, the user keeps tabs on their weight, exercise, and heart rate. These are helpful in the fight against COVID-19 as well because, for those with this illness, sleep and rest are crucial components. Getting enough sleep boosts the body's ability to fight off viruses. The patient can access the portals supplied by these device manufacturers to view their results and, if necessary, send information to the relevant doctors. If specific algorithms are added to the current

devices, wearable technology based on the IoT can aid in limiting the spread of the coronavirus. This wearable technology alerts in real time if any of the following occur:

➢ The social distancing protocol is broken;
➢ A COVID-19 patient is present; and,
➢ The region has been deemed a danger zone by the government due to an epidemic of the coronavirus.

c. Social distance and AI-based forecasting

One of the most promising technologies, artificial intelligence (AI), is bringing about a revolution in a variety of industries. The IoT's adoption of AI algorithms has created new opportunities in this field. AI gives users the chance to learn from their data and discover useful patterns. It is simple to use a database of IoT device data to predict the spread of coronavirus, its consequences, and ways to counteract it. The information from COVID-19 patients is used to forecast the behavior of the virus in the future and to compare its impacts by geographic area. Additionally, it facilitates a quick and efficient match of COVID-19 symptoms. Quick recuperation and patient monitoring are two benefits of AI-based therapy. The patient's medical history and the results produced can be used to forecast better treatment options using AI and machine learning (ML) algorithms. One of the earliest AI-powered companies, BlueDot, predicted the spread of the coronavirus and recognized the global threat. They offered data on the virus's movement and the propensity for an epidemic. In addition to Deargen, Insilico Medicine, SRI Biosciences, Iktos, Benevolent AI, DeepMind, Nanox, Baidu, Alibaba, and EndoAngel Medical Technology Company are other AI-powered businesses that teamed together and cooperated to combat COVID-19.

They did this by detecting individuals within 2 m or 6 feet of one another, computer vision, and deep learning, which enabled real-time social distance assessment. This is done using a pretrained model that runs in real time on a GPU. When a breach of the social distance protocol is discovered, the information is continuously sent via the Internet, where the approving authority acts immediately. This is one of the most effective methods for reducing coronavirus spread and for battling the ongoing COVID-19 pandemic. Ambulances and other emergency service providers can enter with ease thanks to AI-based emergency traffic control. Life First Emergency Traffic Control (LiFE), an algorithm created by Red Ninja, a Liverpool-based business, enables paramedics to use real-time data about congestion to regulate traffic [17].

2. Internet of Things for Industry

Machine-to-machine (M2M) communication is covered by the Industrial Internet of Things (IIoT), a subset of the Internet of Things that provides automation to industrial communication technologies [18]. It has a wide range of applications and significantly aids in a nation's economic development. The IoT can assist in preserving the economy during the present COVID-19 outbreak. Additionally, it helps industries run their operations remotely.

a. Smart infrastructure

In many countries, cameras that are integrated with sensors in autonomous human body temperature-detecting devices, sending real-time data to the server. The system also makes use of AI to identify faces and compare them to the central database. The use of these tools facilitates the tracking of COVID-19 patients. Enabling smart infrastructure that senses the environment and generates reports for law enforcement authorities in real time enforces social separation. For continuous monitoring, the sensor data are continuously recorded on an online database. The Environmental Protection Agency is informed of the presence of harmful gases or carbon content, and the server is updated so that it may be accessed online. Researchers are attempting to detect the coronavirus, which can be used similarly. Some of the major firms that offer smart infrastructure solutions include IBM, Microsoft, Huawie, and Cisco.

b. Automation in industry

Designing safety systems for businesses, infrastructure, and transportation systems has been significantly influenced by the emergence of autonomous systems and their rapid evolution [19]. By giving early warnings and implementing the appropriate precautions, these safety systems have greatly reduced the occurrence of abnormalities caused by human error or environmental causes. The advantages and beneficial impacts of integrating IoT with the current autonomous systems can be seen to a significant extent. Due to the coronavirus pandemic, industries have currently halted operations in numerous nations. The use of industrial automation makes remote operation possible. Additionally, all functions and risks are monitored by the industries, and utilizing smartphones, operations are carried out with a single tap.

c. Telemart

According to WHO, the recommended social distance is 1 meter, or three feet. It can be challenging to maintain this distance in supermarkets and shopping centers. As a result, the IoT is used to introduce the idea of telemart. A popular illustration is Amazon Go, where customers scan a QR code at the entrance to the store to begin buying. The user finishes their purchasing, packs their belongings, and departs. The sum is deducted from their Amazon account. The IoT-based shopping system communicates with the sensors on the shopping cart and maintains a track of all the things that are selected from the store. The best way to maintain social distance is to avoid standing in line to make a purchase.

d. Modern grid

The significance of a constant power supply has grown during these times of global crises. There should be no power interruptions for any hospitals, emergency service providers, or domestic customers. Transducers and sensors are employed in smart grids to monitor and manage the demand for electricity [20]. The advanced metering infrastructure (AMI) is a crucial component of the smart grid. The key functions of the modernized traditional power grids are communication and information systems. In order to analyze, monitor, and control system devices, these functions are crucial. Energy conservation, cost cutting, and improved dependability and transparency are the objectives of these updated networks. AMI, also known as non-technical losses or theft, is installed by electricity service providers (ESPs). AMI is crucial for the identification, localization, and prevention of any malicious activity in the network. Smart meters (SMs) are part of AMI in the smart grid. All of the data from these SMs are transmitted via the Internet to the data concentrator unit (DCU), which then feeds it to the Meter Data Management System (MDMS). The IoT is used to remotely calculate things like energy use, bill creation, and theft detection. The system requires this application to reduce interactions between ESPs and customers in order to decrease the spread of COVID-19. AMI assists in remotely accessing the meter/consumer data and refreshing the system if any error develops; also keeping in mind the safety of numerous workers who manually correct any faults that arise in the metering infrastructure. Additionally, it aids in keeping track of any local power outages so that any necessary action can be taken as soon as possible. The quality of service (QoS) provided by AMI is prioritized. Information is transferred to the MDMS via DCU in the event of a network outage [21]. The government-designated danger zone during the coronavirus outbreak gets priority over all other places. The fastest possible fault recovery and continuous power supply are offered.

e. IoT for online learning and conferencing

All educational institutions ceased operations to slow the spread of the coronavirus. It is suggested that the professors run lectures remotely. The majority of corporations have also encouraged their employees to work remotely. Virtual conferencing and online learning are more necessary than ever. Digital tools, a fast Internet connection, and uninterrupted connectivity are requirements for e-learning and e-conferencing. Google Meet, Zoom, Microsoft Teams, and Skype are the platforms that are most frequently used by staff members to communicate with one another while working from home and by students to take online lectures. End-to-end communication is being made possible in large part via IoT.

8. IoT challenges in the aftermath of COVID-19

IoT implementation is never a simple process. There are numerous obstacles to overcome when deploying IoT for COVID-19, some of which are listed below.

1. **Scalability**: The number of IoT devices is increasing rapidly as a result of the development of digital technologies [22]. Due to the fact that there are currently numerous IoT applications in use, they are not restricted to just one. A recent survey found that the adoption of home automation appliances will significantly expand between 2018 and 2022. The implementation of IoT to combat the global COVID-19 pandemic faces significant scaling challenges. In the IoHT alone, a sizable number of sensors are needed to precisely sense the patients' vital signs and transmit those to the Internet cloud. There are currently over 3.7 million active cases worldwide. The number of sensors in an IoT device can be many. It is difficult to implement IoT in this massively scaled environment. Large amounts of data will circulate around these small IoT nodes and a lot of devices are needed. In addition, scalability causes an increase in energy needs.
2. **Spectrum and bandwidth constraints:** More bandwidth is needed to transport all the data from sensors to the cloud as the number of IoT devices grows [23]. The majority of IoT devices already in use do so using licensed spectra provided by mobile operators. The need for bandwidth has grown along with the use of these gadgets. Data transfers can occasionally go wrong because of data latency. Operators use WiFi for fixed IoT, but as the number of IoT devices grows inside its coverage area, WiFi becomes unreliable. Many IoT gadgets currently use 4G/LTE networks to carry out their functions. It will not be long before the number of IoT devices outnumbers the 3G, LTE, and 4G spectra's capacity. Timely data flow from IoT devices to the concerned body is crucial during

the COVID-19 pandemic. Data delays or errors could cost human lives. The issues of latency and slow data rates can be solved if the bandwidth is large.

3. **Privacy and security concerns:** IoT security cannot be implemented using typical cryptographic algorithms because of the scalability and energy requirements of IoT devices. In order to provide complete data protection, user privacy, and safe authentication, security systems must be energy-efficient and their established algorithms should be less computationally complex. To implement security in the IoT, lightweight security algorithms must be created. The coronavirus outbreak has raised the bar for network security for IoT-enabled systems. The following security issues with IoT implementation in relation to COVID-19 apply:
   - ➢ The data supplied by the sensors attached to the COVID-19 patent's body should be accurate;
   - ➢ arrive at its intended location without error;
   - ➢ not be falsified;
   - ➢ shouldn't be intercepted during communication; and,
   - ➢ no-one should have access to the data kept in the IoT device's memory.

IoT devices should be considered while using security primitives because they have limited processing power. The necessary security algorithms must be accurate and able to maintain user trust in addition to being lightweight.

4. **Big data hubs**: A vast amount of data is kept at data centers as each IoT device delivers data to the cloud via a predefined application program interface (API). Implementing the IoT to tackle COVID-19 presents one of its toughest challenges since it necessitates large storage facilities where all the necessary data may be stored without overloading.

9. Strengths, weaknesses, opportunities, and threats (SWOT) analysis

Table 13.1 displays the results of a SWOT analysis of the IoT. The internal elements, which are confined to enterprises or researchers who seek to deploy the IoT in any sector, are made up of strengths and weaknesses. Internal factors are subject to alteration over time. Opportunities and threats are viewed as external forces that are dependent on the market and immutable [2].

1. Strengths
   The accuracy of data in the IoT is one of its benefits when implemented, using COVID-19 as a test case. The sensors collect environmental data in real time and upload them to the cloud. As a result, patients are assisted in receiving

TABLE 13.1 SWOT analysis of the IoT from the perspective of a global pandemic.

Internal elements	
Strengths	**Weaknesses**
Information about safety precautions	IoT device scalability
IoT-based systems are in high demand	Maintenance of privacy and security
Dependable forecasting	Insufficient spectral resources
Prompt diagnosis	It is necessary to have high-processing server/fusion centers
Data accuracy	Hefty bandwidth demands
Prompt therapy	Data aggregation and massive data centers
External elements	
Opportunities	**Threats**
The raising of awareness of IoT needs	Using unpermitted bands
In the direction of mmWave communication for greater bandwidths	Device compatibility
Cooperative dialogue	
Employment creation	
Programmable radios	

timely care, which has the potential to save many lives. If someone is experiencing COVID-19 symptoms and needs to see a doctor, the IoT can help by offering a telemedicine platform that allows them to do so without physically going to a hospital or clinic. This is in reference to the early detection of COVID-19. The IoT can aid in raising knowledge of the facts and precautions that should be taken to prevent the spread of coronavirus. There is a great demand for IoT-based systems because of how crucial it is to battle the current global coronavirus crisis. Incorporating AI and IoT can improve forecasting of future requirements to combat COVID-19.

2. Weaknesses
When considering the use of the IoT to tackle this infection, it is impossible to disregard the drawbacks and vulnerabilities. The data-processing units should have high processing power because scalability and a large number of IoT devices are necessary. To maintain tracking of patients and related information, data centers need be expanded. The security algorithms should be created in a way that keeps complexity to a minimum while still providing high levels of security for the entire IoT network. The need for high bandwidth cannot be disregarded because numerous devices will be frequently transferring data to the cloud. The method should be created such that the scarce spectrum is utilized effectively. Frequency planning and reuse mechanisms can be used to accomplish this.
3. Opportunities
The potential benefits of using the IoT to address the COVID-19 global situation are enormous. The awareness of employing IoT apps specifically for containing the coronavirus outbreak is not a difficult task to carry out due to the paradigm change toward the use of digital technologies and smartphones. Additionally, the IoT sector can contribute significantly to the growth of a country's economy by creating jobs in local marketplaces. IoT networks, which offer massive bandwidth and high data rates, do not yet make use of millimeter wave (mmWave) or 5G technology. The adoption of the IoT may persuade IT titans to employ these high-bandwidth mmWave technologies, which run between 3 and 300 GHz. In several sectors of wireless communication networks, this will create new opportunities. Software-defined radios, cognitive radio networks, and cooperative communication are currently applicable in existing IoT networks to optimally use the spectrum by sensing the open spaces in licensed bands and using them for their operations.
4. Threats
Threats from outside sources are rather limited. IoT devices are currently interoperable with the same vendor's manufacturer. In order to foster vendor rivalry, compatibility is absolutely necessary. As a result, IoT activities will be of higher quality, and the apps will develop over time. Furthermore, there are not many unlicensed bands. The Industrial, Scientific, and Medical (ISM) (2.4 GHz) frequency band, which is used for the majority of IoT communication, might introduce interference if sufficient planning is not done.

10. An answer to the problems of the COVID-19 battle

Unquestionably, creating scalable IoT networks is difficult, but there are answers to these problems in the literature that can aid in effectively installing IoT networks. The subsections that follow list some of the well-known solutions.

1. Simple security techniques
The majority of IoT devices are compact and simple to use because of their scalability. It is necessary to take precautions to guarantee the safety of the data and their timely delivery to the intended location. Since IoT nodes are typically not physically secure, data security and provenance form the foundation of IoT network implementation. If the appropriate security primitives are not employed, data can be easily fabricated. Specific attack detection, channel state masking, intrusion detection, geolocation, and data provenance are a few examples of security primitives. Discovering the data's source is provenance. Major issues can arise from a single data modification. During the global COVID-19 pandemic, for instance, IoT-generated medical health reports of COVID-19 patients sent to doctors and smart grid power outages can both be very problematic. The energy constraints of IoT devices prevent the use of conventional encryption methods as a feasible alternative. The basic building blocks for end-to-end content protection, user authentication, and customer confidentiality in the IoT era are energy-efficient security primitives that occupy less memory and have lower computational complexities. Simple encryption methods are the foundation of most lightweight security algorithms. IoT security algorithms can be created using a variety of metrics, including angle of arrival, time of arrival, phasor information, and received signal strength indicators (RSSIs). IoT devices that communicate produce link fingerprints. The symmetric key used to encrypt these link fingerprints is then used to send the decoded data to the server, which uses the link fingerprints of linked IoT devices to compute the Pearson correlation coefficient. To identify any adversary in the IoT network, a fairly straightforward technique called Pearson correlation coefficient computation is used [9].

a. Blockchain for privacy protection and connected healthcare units

The virtual currency known as bitcoin helped make blockchain, a rapidly developing technology, prominent. Numerous fields are now utilizing blockchain technology. Sharing data is made secure and private thanks to blockchain. The private key is kept on the IoT device in a blockchain-based IoT system, while the public key is kept on the Ethereum. Healthcare facilities that are all interconnected can use blockchain technology. By establishing a blockchain-based IoT network, each healthcare unit functions as a block and correct data transfer is made feasible. Data may be fully secured with blockchain technology. When IoT and blockchain are combined, security and anonymity are guaranteed. Any official will verify crucial information from medical records as well as the list of all accessible medical supplies and other resources by determining whether or not the document is authentic.

2. IoT with cognitive radio

Cognitive radio IoT (CRIoT) is the result of the fusion of cognitive radio with the IoT. The traditional method of allocating spectrum involves obtaining licenses. It has been noted that the majority of the licensed spectrum is not being used entirely. The frequency reuse issue may be successfully solved by cognitive radios. IoT devices are capable of sensing the surroundings and automatically adjusting the configuration parameters while employing cognitive radio parameters. IoT devices detect free spectrum availability, also known as holes in the spectrum, and communicate in the detected holes without interfering with the licensed user known as the primary user (PU). This promotes continuous data transmission and effective use of the permitted spectrum.

a. Spectrum sensing

Spectrum sensing is the method used to find an unutilized spectrum. The fundamental component of a cognitive radio network is spectrum sensing. Free holes in the PU's spectrum must be found in order for IoT devices to communicate using a licensed band. In order to identify the available holes, the IoT device scans the spectrum of PUs that are active. The following stage for an IoT device is to choose the most suitable spectrum in accordance with their QoS requirements after the spectrum has been sensed. The radio environment and the statistical behavior of the PUs are taken into consideration while choosing the spectrum.

b. Spectrum sharing

Spectrum sharing is the process of allocating radio waves to PU and IoT devices. When licensed PUs and unlicensed IoT devices share a spectrum, it is essential for reliable communication to occur. Spectrum sharing is essential in order to share the spectrum by reducing collisions in overlapping portions of the spectrum because there may be more IoT devices attempting to access it. Spectrum sharing offers the ability to flexibly share the spectrum resources with various IoT devices, including resource allocation to prevent interference to the PU.

c. Spectrum movement

Spectrum mobility is the process of keeping in touch while the spectrum is changing. When a licensed user is discovered, the IoT device closes the channel. Another part of the spectrum that is open is where the IoT device will continue to transmit. With minimal quality loss, the signal is transferred to a new path or band. To detect if the PU is present or not in the frequency spectrum, there are numerous techniques available. These techniques include feature detection, energy detection, matching filter (MF) detection, and cyclostationary detection. The energy detection method is more energy efficient since it does not require the system to have any prior knowledge about the principal signal, unlike the other three methods.

3. Toward greater bandwidth with millimeter wave

The emergence of the IoT has increased the need for bandwidth. The lack of available bandwidth has encouraged companies developing IoT devices to investigate the underutilized millimeter wave (mmWave) frequency spectrum for IoT networks in the future. The frequency range of mmWave is 3–300 GHz. For next-generation cellular networks, the spectrum at 28 GHz, 38 GHz, and 70–80 GHz appears to be very promising. Large bandwidths enable the attainment of many gigabits per second. Other application scenarios, such as wearable networks, automotive communications, autonomous robotics, etc., show promise for mmWave communication.

4. IoT networks powered by artificial intelligence

It is undeniable that the Internet-connected communication gadgets that link to one another provide a lot of relevant data. The communication device's decisions can be influenced by these data. AI adds context to these data in order to make sense of them, so supplying more information to support the decision made by a communication endpoint. Pattern recognition is made easier by AI, which also gives terminals the ability to learn from previous patterns. This may occur in one of two ways:

i. ***Predictive analysis***: The information is employed to foretell potential decision outcomes.

ii. ***Adaptive analysis***: In order to maximize the decision-making process, what decisions may be made based on prior experiences?

The ability to automatically recognize patterns and spot anomalies in the data produced by smart sensors and devices is provided by machine learning, an AI technology [24]. Operational predictions are up to 20 times faster and more accurate when made using machine learning techniques. Insights from data that needed human validation can be extracted with the use of other AI technologies, such as voice recognition and computer vision [25]. In order to combat DOS attacks and replay attacks on IoT networks, machine learning offers solutions. When very small IoT devices with less RAM are installed, it also offers options for resource-efficient IoT networks [26,27].

11. Conclusion

The use of technology like IoT, AI, blockchain, big data analytics, and cloud computing has expanded as a result of the global COVID-19 pandemic outbreak. By providing platforms that aid in adhering to the guidelines set forth by the WHO, the IoT significantly lowers the risks of coronavirus propagation. Medical staff can respond quickly to COVID-19 patients thanks to IoT-based healthcare units. IoT and AI integration improve future situational predictions. Professionals in business and academics can do their work remotely thanks to the IoT. IoT networks built on blockchain technology make it easier to oversee the supply chain and spot data fraud. While it will take some time to fully recognize and quantify the full health, social, and economic effects of this pandemic and its limitations, there are numerous ongoing efforts in the research and industrial communities to use various technologies to detect, treat, and trace the virus to reduce its effects. Although early identification, quarantine, and recovery from COVID-19 have all benefited from IoT technology, as we learn more about the virus and its behavior, we need to modify and improve our strategies at various stages. The IoT appears to be a great tool for screening sick patients. With real-time information, this technology helps the healthcare industry maintain quality supervision. The IoT can be useful for forecasting an approaching scenario of this disease by employing a statistically based method. Researchers, medical professionals, governments, and academics can improve the environment in which this disease is fought by properly implementing this technology.

References

[1] T.R. Gadekallu, Q.V. Pham, D.C. Nguyen, P.K.R. Maddikunta, N. Deepa, B. Prabadevi, W.J. Hwang, Blockchain for edge of things: applications, opportunities, and challenges, IEEE Internet of Things Journal 9 (2) (2021) 964–988.

[2] M. Al-Emran, S.I. Malik, M.N. Al-Kabi, A survey of internet of things (IoT) in education: opportunities and challenges, Toward Social Internet of Things (Siot): Enabling Technologies, Architectures and Applications (2020) 197–209.

[3] M.J. Baucas, P. Spachos, S. Gregori, Internet-of-Things devices and assistive technologies for health care: applications, challenges, and opportunities, IEEE Signal Processing Magazine 38 (4) (2021) 65–77.

[4] S. Anmulwar, A.K. Gupta, M. Derawi, Challenges of IoT in healthcare, in: IoT and ICT for Healthcare Applications, Springer, Cham, 2020, pp. 11–20.

[5] M. Nasajpour, S. Pouriyeh, R.M. Parizi, M. Dorodchi, M. Valero, H.R. Arabnia, Internet of Things for current COVID-19 and future pandemics: an exploratory study, Journal of Healthcare Informatics Research 4 (4) (2020) 325–364.

[6] M. Kamal, A. Aljohani, E. Alanazi, IoT Meets COVID-19: Status, Challenges, and Opportunities, 2020 *arXiv preprint arXiv:2007.12268*.

[7] M.A. Khan, Challenges facing the application of IoT in medicine and healthcare, International Journal of Computer Integrated Manufacturing 1 (1) (2021).

[8] S. Badotra, D. Nagpal, S.N. Panda, S. Tanwar, S. Bajaj, IoT-enabled healthcare network with SDN, in: 2020 8th International Conference on Reliability, Infocom Technologies and Optimization (Trends and Future Directions)(ICRITO), IEEE, June 2020, pp. 38–42.

[9] A. Abugabah, N. Nizamuddin, A. Abuqabbeh, A review of challenges and barriers implementing RFID technology in the Healthcare sector, Procedia Computer Science 170 (2020) 1003–1010.

[10] P. Ratta, A. Kaur, S. Sharma, M. Shabaz, G. Dhiman, Application of blockchain and internet of things in healthcare and medical sector: applications, challenges, and future perspectives, Journal of Food Quality (2021) 1–20.

[11] A.K. Mourya, B. Alankar, H. Kaur, Blockchain technology and its implementation challenges with IoT for healthcare industries, in: Advances in Intelligent Computing and Communication, Springer, Singapore, 2021, pp. 221–229.

[12] V.S. Naresh, S.S. Pericherla, P.S.R. Murty, R. Sivaranjani, Internet of things in healthcare: architecture, applications, challenges, and solutions, Computer Systems Science and Engineering 35 (6) (2020) 411–421.

[13] S. Goyal, N. Sharma, B. Bhushan, A. Shankar, M. Sagayam, Iot enabled technology in secured healthcare: applications, challenges and future directions, in: Cognitive Internet of Medical Things for Smart Healthcare, Springer, Cham, 2021, pp. 25–48.

[14] A. Hassan, D. Prasad, M. Khurana, U.K. Lilhore, S. Simaiya, Integration of internet of things (IoT) in health care industry: an overview of benefits, challenges, and applications, Data Science and Innovations for Intelligent Systems (2021) 165–180.

[15] M. Usak, M. Kubiatko, M.S. Shabbir, O. Viktorovna Dudnik, K. Jermsittiparsert, L. Rajabion, Health care service delivery based on the Internet of things: a systematic and comprehensive study, International Journal of Communication Systems 33 (2) (2020) e4179.

[16] F.J. Dian, R. Vahidnia, A. Rahmati, Wearables and the internet of things (IoT), applications, opportunities, and challenges: a survey, IEEE Access 8 (2020) 69200–69211.

[17] B. Lin, S. Wu, COVID-19 (coronavirus disease 2019): opportunities and challenges for digital health and the internet of medical things in China, OMICS: A Journal of Integrative Biology 24 (5) (2020) 231–232.

[18] T. Car, L.P. Stifanich, M. Šimunić, Internet of things (iot) in tourism and hospitality: opportunities and challenges, Tourism in South East Europe 5 (2019) 163–175.

[19] M.R. Naqvi, M. Aslam, M.W. Iqbal, S.K. Shahzad, M. Malik, M.U. Tahir, Study of block chain and its impact on internet of health things (IoHT): challenges and opportunities, in: 2020 International Congress on Human-Computer Interaction, Optimization and Robotic Applications (HORA), IEEE, 2020, pp. 1–6.

[20] S.P. Amaraweera, M.N. Halgamuge, Internet of things in the healthcare sector: overview of security and privacy issues, Security, Privacy and Trust in the IoT Environment (2019) 153–179.

[21] N.N. Thilakarathne, M.K. Kagita, T.R. Gadekallu, The role of the internet of things in health care: a systematic and comprehensive study, SSRN (2020), https://doi.org/10.2139/ssrn.3690815.

[22] R. De Michele, M. Furini, Iot healthcare: benefits, issues and challenges, in: Proceedings of the 5th EAI International Conference on Smart Objects and Technologies for Social Good, September 2019, pp. 160–164.

[23] B. Al-Shargabi, S. Abuarqoub, IoT-enabled healthcare: benefits, issues and challenges, in: The 4th International Conference on Future Networks and Distributed Systems (ICFNDS), 2020, pp. 1–5.

[24] A. Rghioui, A. Oumnad, Challenges and opportunities of internet of things in healthcare, International Journal of Electrical and Computer Engineering 8 (5) (2018). ISSN: 2088-8708.

[25] A.M. Longva, M. Haddara, How can IoT improve the life-quality of diabetes patients?, in: MATEC Web of Conferences 292 EDP Sciences, 2019, p. 03016.

[26] P. Singh, Internet of things based health monitoring system: opportunities and challenges, International Journal of Advanced Research in Computer Science 9 (1) (2018) 224–228.

[27] A. Zubiaga, R. Procter, C. Maple, A longitudinal analysis of the public perception of the opportunities and challenges of the Internet of Things, PLoS One 13 (12) (2018) e0209472.

Further reading

[1] E. Chukwu, L. Garg, R. Zahra, Internet of Health Things: Opportunities and Challenges, Artificial Intelligence and the Fourth Industrial Revolution, 2022, pp. 105–131.

[2] A. Darwish, A.E. Hassanien, M. Elhoseny, A.K. Sangaiah, K. Muhammad, The impact of the hybrid platform of internet of things and cloud computing on healthcare systems: opportunities, challenges, and open problems, Journal of Ambient Intelligence and Humanized Computing 10 (10) (2019) 4151–4166.

Chapter 14

Deep learning for clinical decision-making and improved healthcare outcome

Russell Kabir[1], Haniya Zehra Syed[1], Divya Vinnakota[2], Madhini Sivasubramanian[2], Geeta Hitch[2], Sharon Akinyi Okello[1], Sharon-Shivuli-Isigi[1], Amal Thomas Pulikkottil[1], Ilias Mahmud[3], Leila Dehghani[4] and Ali Davod Parsa[1]

[1]*School of Allied Health, Faculty of Health, Education, Medicine and Social Care, Anglia Ruskin University, Essex, United Kingdom;* [2]*Faculty of Health Sciences and Wellbeing, University of Sunderland, London, United Kingdom;* [3]*Department of Public Health, College of Public Health and Health Informatics, Qassim University, Al Bukairiyah, Saudi Arabia;* [4]*CRN Eastern (NIHR), Cambridge University Hospitals NHS Foundation Trust, Cambridge, United Kingdom*

1. Introduction

Technologies based on machine learning (ML) are being developed more and more for use in the medical field. Concurrent advances in clinical genetics, bioinformatics, artificial intelligence (AI), and statistics have made it possible to create tools that help clinical decision-making. Applications of ML have been successful in a number of medical fields, including disease prediction [1] using various data modalities, including speech signals and medical imaging [2,3], as well as clinical outcome prediction to detect deterioration, such as cardiac arrest, mortality, or admission to an intensive care unit (ICU) [4,5].

Future cardiovascular problems can be predicted. For example, using the left atrium's enlargement (LAE) that can help doctors predict and avert unfavorable outcomes in many clinical situations. According to a study [6], deep learning method (DLM) was applied to an ECG to detect on going changes such as LAE, making it the use of ECG as a reliable screening tool for diagnosing LAE and predicting its severity. The results of this investigation proved that a DLM-ECG could also offer additional predictive data on newly developed hypertension (HTN), stroke, mitral regurgitation (MR), and atrial fibrillation (AF).

An important condition known as left ventricular dysfunction (LVD) causes significant mortality [7] and considerable medical expenses [8]. Around 3%–6% of the general population has asymptomatic LVD [9].

Chen et al. [10] developed a DLM algorithm to precisely identify LVD and predict ejection fraction (EF) changes using a large number of ECGs and echocardiographic data. DLM could be used in a variety of applications, such as wearable technology and remote health care systems, to identify asymptomatic LVD patients and enable doctors start the management for high-risk patients. Additionally, DLM-based ECG-EF analysis improves outcomes prediction for cardiovascular (CV) diseases.

For example, Yao et al. [11] have created a screening system that can identify people who have influenza based on three clinical criteria (heart rate, respiration rate, and facial temperature). Their technology is particularly interesting since it employs contactless technologies that make it particularly suitable for clinical use with the infectious patients [11]. The same three clinical parameters were used to create a screening technique more recently [12]. Their method, which can be represented as a flowchart, predicts the patient's infection condition using a random tree algorithm [12].

A smartphone app that offers doctors access to clinical guidelines is another interesting ML-based decision support solution. In order to forecast the mortality risk of Ebola patients [13,14], the ML-based models have been created. Their models were later put into a smartphone app to assist in supporting a physician judgment in the isolated clinical care

Deep Learning in Personalized Healthcare and Decision Support. https://doi.org/10.1016/B978-0-443-19413-9.00004-7

situations [13,14]. As new biomedical information becomes available, these systems will be frequently updated to make them particularly appealing during an epidemic.

Deep learning technology employs the information contained in Electronic Medical Record (EMR) software to respond to healthcare-related questions, such as lowering the rate of diagnosis and outcome prediction errors. By improving the diagnosis and raising patients outcome standards, deep learning (DL) in healthcare is having a significant impact on the healthcare system. DL aids clinicians in data analysis and multiple conditions identification including: identifying cardiac problems, image analysis is used to find cancers, identifying malignant cells in the body and making a cancer diagnosis, monitoring glucose levels in people with diabetes, examining blood samples, and cancer detection in blood samples.

Deep learning techniques have not, however, been thoroughly examined for a wide range of medical issues that can benefit from their capabilities. DL has numerous features that could be used in the healthcare industry, including its superior performance, end-to-end learning model with integrated feature learning, ability to handle complicated and multimodality data, and more. The DL research community as a whole needs to address a number of issues related to the characteristics of healthcare data (i.e., sparse, noisy, heterogeneous, and time-dependent), as well as the need for improved techniques and tools that allow DL to interface with clinical decision support workflows.

In this book chapter, we will discuss the application of DL in healthcare, highlighting the critical elements that will have a big impact on health care. We do not want to give a thorough foundation on the technical aspects or widespread applications of DL. Instead, we focus on DLM for clinical decision-making and improved healthcare outcome in diagnosis of different diseases, patient care plan in chronic disease management and rehabilitation, healthcare data analytics and modeling, mental health support and personalized care, research and development in healthcare, diagnosis and treatment in healthcare, population level future disease prediction, and screening and referral of clinical cases in remote and poor access areas. According to authors best knowledge, there has been no extensive research on DLM in healthcare on the above-mentioned aspects. This is a comprehensive piece of work on DLM application in healthcare for clinical decision-making and improved healthcare outcome. The key findings of DLM use in healthcare are highlighted along with advantages and its limitations. The aim of this chapter is to provide the reader with a good understanding of applications of DLM in healthcare and how to improve healthcare outcome. The epidemic, an increase in diseases linked to modern lifestyles, and an expanding global population are all putting enormous strain on our modern healthcare system. The good news is that healthcare might become more affordable, efficient, individualized, and equitable by leveraging DLM to develop intelligent systems and workflows.

2. Application of deep learning machine (DLM) in clinical decision-making in the diagnosis of different diseases

Both communicable and noncommunicable diseases are a huge threat to public health. Therefore, it is vital that early detection techniques and treatments are put in place in order to prevent outbreaks. More recently, over 6.4 million deaths have been recorded due to COVID-19 globally [15]. The global burden of disease indeed places chronic diseases at the top of the list (Fig. 14.1).

Due to the huge explosion of the "genomics" technologies, there has been a rapid development in the development of biomarkers at the DNA level to enable detection and diagnosis of many complex diseases. A single nucleotide polymorphism (SNP) is a DNA sequence variation that occurs when a single nucleotide (adenine, thymine, cytosine, or guanine) in the genome sequence is altered, and the particular alteration is present in at least 1% of the population [16]. Epistasis is the interaction between different SNPs that influences a phenotype by suppressing the effect of nonalleles genes by another. However, underlying biological mechanisms are quite complex and this is where DLM has been used since the 1990s and more recently, in the diagnosis of diseases in oncology, rheumatology, cardiology, craniostenosis syndrome liver pathology, thyroid diseases, dermatoglyptic diagnosis, neuropsychology, gynecology, and perinatology [17]. For many diseases, biobanks of data are created initially whereby MRI or CT scans are used to produce 3D impressions of diseased affected areas for diagnosis [18]. Then, various machines and DLM are applied with exacted algorithms based upon these impressions of specific disease states enabling diagnosis of the disease in question. Examples are techniques such as Boltzmann machine, K-nearest neighbor (kNN), support vector machine (SVM), decision tree, logistic regression, fuzzy logic, and artificial neural network, which have been applied in deep and ML to diagnose numerous diseases include diseases of the skin. Wang et al. [19], Rodrigues et al. [20], liver [21,22], urology [23,24], Alzheimer's disease· [25–27] arrhythmia [28], kidney [29], gastrointestinal [30], cardiovascular [31–33], tuberculosis [34], retinopathy [35], cancers [23,24,36], and COVID-19 disease [37,38]. Recent studies have shown the use of video-based convolutional neural network model with spatiotemporal convolutions to predict the cause of left ventricular hypertrophy (LVH) aiding in the

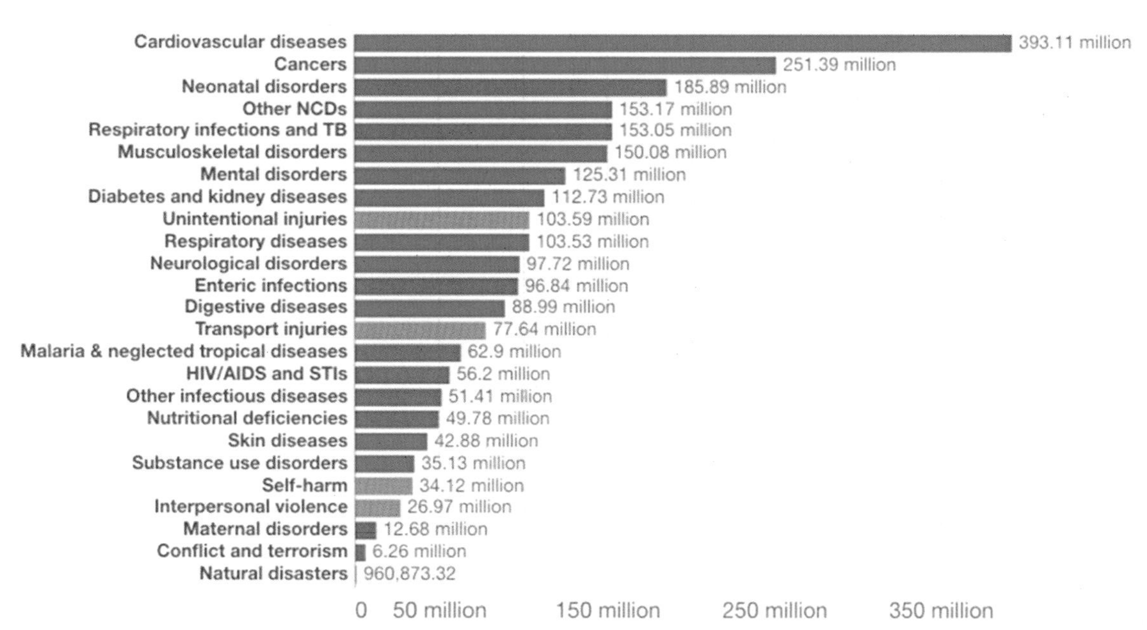

FIGURE 14.1 Burden of disease by cause, World, 2019. *Source: M. Roser, H. Ritchie, Burden of Disease, Published online at OurWorldInData.org. 2021, https://ourworldindata.org/burden-of-disease. (Accessed August 9, 2022). NB: Communicable, maternal, neonatal, and nutritional diseases are shown in red; other noncommunicable diseases including injuries are shown in blue and gray, respectively.*

diagnosis of hypertrophic cardiomyopathy and cardiac amyloidosis [39]. This deep ML technology has enabled rapid diagnosis of tumors without waiting for biopsies to be extracted and sent to labs for analysis thereby increasing survival chances for patients in the event of cancer related diseases in a study in China [23,24]. Another example of the use of DML technology is in the diagnosis of liver viral hepatitis, which has been diagnosed with an accuracy of over 97% by application of RNN [40].

Deep learning algorithms have also been used in other recent studies to identify several retinal diseases. The retinal artificial intelligence diagnosis system (RAIDS) has been used by researchers in China to identify 10 retinal diseases based on over 120,000 ocular fundus images [41].

When making clinical decisions in the field of orthodontics, the doctor mostly uses diagnostic tools. AI-based models, including convolutional neural networks (CNNs), artificial neural networks (ANNs), and DL applications, have shown to be effective and beneficial for the dentist to be more exact in diagnostic and clinical decision-making [42]. These systems can speed up the process and deliver results, saving the dentist time and allowing them to carry out the duties more effectively.

Rapid diagnosis of disease by application of DL and ML has enabled physicians to make rapid life-saving decisions and spend precious time with their patients, as well as reducing hospital stays and spending on healthcare. DL and ML have the capability of dealing with "big complex data" and have gained enormous traction over the past decade and are finding their niche within the healthcare sector.

3. Application of DLM in patient care plan in chronic disease management and rehabilitation

In recent years, advances in computer hardware/software and algorithms have accelerated DL use in e-commerce, finance, and voice and image recognition [43]. The application of DL in medical diagnosis, lesion detection and segmentation, image registration, and disease classification has made a great progress [44].

Deep learning has proven to be a viable solution for both reducing the rising cost of healthcare and improving the patient–doctor relationships in the management of chronic conditions such as diabetes, cardiac problems, pulmonary diseases, and cancer. The phrase chronic disease management refers to an approach to illness management that involves screenings, check-ups, monitoring, coordinating, and educating the patients [45].

4. Application in cardiovascular medicine

- **Ischemic and structural heart disease imaging**
  The use of DL to classify images has been extended to cardiac imaging using CT and MRI in a variety of medical specialties [43]. It is predicted that ML models will improve the diagnostic capacity of echocardiography and the predominant cardiovascular imaging modality which are heavily dependent on human expertise.
  Deep learning for outcomes prediction in heart failure.
  In addition to improving cardiovascular risk prediction, this technique will identify the patients who could benefit from preventive treatment, as well as reducing unnecessary treatment for patients [45]
- **Deep learning for arrhythmia detection and phenotyping**
  To predict cardiac arrhythmia based on ECG, DNN architectures include autoencoders, DBNs, CNNs, and RNNs [46].
  Application in pulmonary disease

5. Application in Pulmonary Disease

- **Asthma**
  Asthma management system (AMS) experts have developed the Bluetooth-enabled inhalers, which collect time stamps for every time they are used. Taking advantage of the patient monitoring capability of the AMS, the authors developed a data analytics framework for detecting an abnormal inhaler use that is out of the patient's normal usage pattern [45].
- **COPD**
  With an application on a smartphone, wearable devices and air quality sensors combined with supervised prediction algorithms will lead to an excellent prediction accuracy of acute exacerbations of chronic obstructive pulmonary disease (AECOPD) within the upcoming 7 days. AECOPD predictions were based on lifestyle and environmental data collected by the prediction system and were reliable [47]
- **Application in hepatic and renal diseases**
  Deep learning models have been used extensively to facilitate the diagnosis of multiple types of liver disease, most of which are life threatening. Interest has been primarily focused on the automated detection of nonalcoholic fatty liver disease (NAFLD), as most patients remain asymptomatic until the development of liver cirrhosis. NAFLD can be diagnosed with 97.2% accuracy using a recently developed AI neural network [48]. Additionally, the same model can distinguish between patients with NAFLD and those with nonalcoholic steato-hepatitis (NASH). In patients with chronic kidney damage, ML is able to accurately predict how long their kidneys will function adequately. Specifically, DL-based CNN was trained on renal biopsies by Boston University researchers. As compared to traditional pathologist-estimated scoring systems, CNN algorithms were more precise and accurate [49]. A clinical decision support system based on AI could significantly enhance nephrologists' clinical decision-making
- **Application in cancer prognosis and survival**
  Clinical oncology relies heavily on prognosis prediction as it can guide treatment choices by predicting the course of disease and the likelihood of survival [50]. DL can predict prognosis and patient's survival when used with genomic, transcriptomic, and other types of data [51]. The Cox proportional hazard regression model is the approach used to predict survival. In addition to transcriptome prognosis predictions, CNN models trained on histopathology pictures have been used to predict survival in a variety of malignancies, including brain, colorectal, renal cell, liver, and mesothelioma tumors [52].
- **Application in ophthalmology**
  Vision care is already being revolutionized by ML and DL in many ophthalmology practices. An AI-based device has already been FDA approved to detect diabetic retinopathy [53]. Using DL, this system automatically detects and quantifies intraretinal cystoid fluid (IRC) and subretinal fluid (SRF) [50].
- **Application of deep learning models in health system decision-making** via **healthcare data analytics and modeling**
  When ML is applied to clinical decision-making, it implies that the system will perceive a specific individual by gathering and analyzing data relevant to that individual's health, and it will then use the data to explain about the best method that should be taken to maintain or improve the individual's health [54]. DL enables neural network-based computational models to learn descriptions of data at various levels of abstraction [55]. By suitably valuing connections between network nodes, DL systems use training data to create neural networks that can detect even weak correlations in data. Through training, hidden "features" in the data are identified, enabling a neural network to recognize similar features in the future. In other words, it enables such networks to identify and weigh such correlations appropriately in

additional cases that are not included in the initial training data that it is provided with [56]. Use of secondary electronic health records facilitates clinical predictive modeling. Deep models make it possible to find high-level characteristics, which enhance performance compared to typical models, increases interpretability, and offers more insight into the structure of biological data. By optimizing a local unsupervised criterion, a DL system creates a representation of the observed patterns depending on the data it gets as inputs from the layer below [57]. These models can perform differential diagnosis with similar accuracy to physicians. CNNs, RNNs, restricted Boltzmann machines (RBMs), and autoencoders (AEs) make up the majority of the deep architectures used in the healthcare industry. The application of DL technique with these models are listed below [55].

	Applications	Model
Clinical imaging	Diabetic retinopathy detection in retinal fundus images [58].	Convolutional neural networks
	Using brain MRIs to detect Alzheimer disease early [59].	Autoencoders
	Multiple brain MRIs to identify Alzheimer disease characteristics [60].	Restricted Boltzmann machines
	Automatic knee cartilage segmentation to identify osteoarthritis risk [61].	Convolutional neural networks
	Using deep learning to determine survival from colorectal cancer histology slides [30].	Convolutional neural networks
	Multiple sclerosis lesions in multichannel 3D MRIs are segmented [62]	
	Prediction of the prognosis of human breast cancer using multidimensional data integration [63].	Recurrent neural network
	Classification of skin cancer at dermatologist level [64].	Autoencoders
Electronic health records	Congestive heart failure and chronic obstructive pulmonary disease prediction using longitudinal electronic health records [59].	Convolutional neural networks
	Prediction of future diseases from the patient clinical status[55].	Autoencoders
	Predict suicide risk of mental health patients by low-dimensional representations of the medical concepts embedded in the HER [65].	Restricted Boltzmann machines
	Discovering and detection of characteristic patterns of physiology in clinical time series [66].	Autoencoders
	Use the history of patients to predict diagnoses and medications for a subsequent visit [67].	Recurrent neural network
	Prediction of disease onsets from longitudinal lab tests [68].	Recurrent neural network
	A dynamic memory model for predictive medicine based on patient history [69].	Recurrent neural network
	Diagnosis classification based on clinical parameters of infants receiving care in intensive care units [70].	Recurrent neural network
Mobile data	Health monitoring using photoplethysmography signal identification [71].	Restricted Boltzmann machines
	Analysis of electroencephalogram and local field potentials signals [72].	Convolutional neural networks
	Analyze wearable activity data from physical activity to predict the quality of sleep at night [73]	Convolutional neural networks
	HAR to detect freezing of gait in PD patients [74].	Convolutional neural networks
Genomics	Predict chromatin marks from DNA sequences [75]	Convolutional neural networks
	Predict methylation states in single-cell bisulfite sequencing studies [76].	Convolutional neural networks
	Determine the specifics of proteins that bind to DNA and RNA [77].	Convolutional neural networks
	Cancer classification using gene expression patterns [78].	Autoencoders

- **The DLM role in mental health support and personalized care**
  Being healthy is not just being free of a disease state physically but also being mental well health wise as defined by World Health Organisation, "*Health is a state of complete physical, mental and social well-being and not merely the absence of disease or infirmity*" [79: 77].
  This includes **emotional and psychological well-being** since this affects how we think, feel, and act. It also helps determine how we handle stress, relate to others, and make healthy choices. Mental health is important at every stage of life, from childhood and adolescence through adulthood. An A to Z guide on mental health (MH) definition ranges from conditions such as anger to anxiety, depression, schizophrenia, psychosis, self-esteem, paranoia, and stress to

suicidal feelings including many more serious problems. The WHO report on mental health and COVID-19: early evidence of the pandemic's impact (2022) showed a "significant increase in mental health problems in the general population in the first year of the pandemic" [15: 3]. Barriers that stand in the way of patients seeking help include fears of stigmatization to lack of access and affordability for treatment. Additionally, it is often quite complex to diagnose MH conditions as many depression disorders present with similar symptoms and research has shown schizophrenia to comprise of biologically diverse subgroups [80].

The many psychological depression elements, such as negative and ambiguity, affect how emotions are extracted. The most difficult duty for any psychiatrist treating their patients is figuring out these factors. To capture the text sequence of data, this also contains DL models using global vector representations (GloVe) embeddings. Authors suggested a model called MHA-BCNN, which is a preeminent mechanism that outperforms previous research works for capturing the negative text-based emotions [81]. It combines multihead attention with bidirectional long short-term memory and convolutional neural network. The numerous mental health disorders like addiction, anxiety, sadness, insomnia, stress, and obsessive compulsive disorder were retrieved using DL. By utilizing GloVe embeddings, the ambiguity elements, such as numerous emotion words in a specific order, were managed [81].

Current treatment and management of MH conditions include use of evidence-based clinical guidelines such as National Institute for Health and Care Excellence (NICE) in the UK of which there are many ranging from "Mental health and Wellbeing at Work" to "Antenatal and Postnatal Mental Health Clinical Management and Service Guidance" [82]. Standardized questionnaires and/or patient interviews are used to diagnose as an assessment tool in the clinical setting. Research has shown that ML has the potential to aid and improve diagnosis and patient outcomes suffering from many depressive illnesses [83]. A recent scoping review of ML and its application in MH identified four domains of application of MH. These include (1) detection and diagnosis (2) prognosis, treatment and support (3) public health applications, and (4) and research and clinical administration. For example, one study used random forest models of ML to predict substance abuse in patients from data derived in the United States and psychotic mental health disorders in patients from Egypt. The researchers used two models of random forest techniques to increase the accuracy in these groups of patients who presented with diverse risks for substance abuse and psychotic disorders. Use of RF techniques improved the diagnosis of these conditions [84].

Different types of ML techniques have been applied to improve outcomes for MH patients These include random forest; SVM; Naive Bayes; neural networks; latent Dirichlet allocation; k-nearest neighbors (kNN); Hidden Markov Model; Bayesian network; and association rule mining and principal component analysis to improve outcomes for MH patients [83]. For example, improved treatment outcomes have resulted due to the use of supervised ML has been used to trial responses to a drug for Parkinson disease [85]. ML has been used to predict prognosis of patient outcome in conditions such as schizophrenia [86], Alzheimer's disease [87,88], posttraumatic stress disorder [89], depression [90–93], and psychosis [94–96].

A systematic review [97] found that random forest model has shown to present with a lower error rate of around 73% to predict onset of childhood onset schizophrenia. In other studies where random forest model and multinomial naive Bayes presented with an accuracy of 68.6% and 66.9%, respectively, in predicting schizophrenia. When other models such as the XG Boost accuracy score and the SVM were used, these presented with an accuracy of 66.3% and 58.2%, respectively [98].

Hence, the use of ML still has some way to go in the field of mental health. In patients with SNP-induced schizophrenia, the SVM algorithm showed an accuracy of 74% on classification of this disorder [99].

Just-in-time adaptive interventions (JITAI) which allows for "on demand" personalized support has been delivered by mobile applications (mHealth) for addictive behaviors [100]. JITAIs intervention is directed at coping strategies, emotional support, and feedback just in time of need [100]. Psychotherapies comprise behavioral activation and dialectical behavioral therapy which are considered to be gold standard treatment for the support and treatment of depression disorders and are face-to-face interventions which are complex and layered and personalized approaches for supporting MH patients. ML algorithms need to be adapted to include such numerous complex aspects of the treatment such as the intricacy of the problem, timing, and treatment dose, and frequency and length of treatment, and this lends itself to lack of "personalization" and the gold standard approach. Indeed studies have shown patient engagement with mobile applications to be less than desirable [101–103]. Hence, researchers face several challenges when developing digitalized therapies. ML has a niche role to play according to Lewis's team [104] who propose using the Collaborative filtering (CF) algorithms (matrix factorization and k-nearest neighbor) and the recommender systems (RS) to improve the user experience of digital mental health apps. These ML algorithm models have shown their potential use personalizing support for patients by generating recommendations derived algorithmically from a group of patients with similar preferences. In this way, for future work, more scenarios can be added to the algorithms to improve accuracy for supporting patients digitally [104]. ML has a fair way to develop and is already rapidly developing but the human personal element still lacks as this review has shown in terms of personally supporting patients.

- **DLM facilitated research and development in healthcare:**
The application of AI into healthcare research has gathered immense attention and is the most rapidly growing idea of the 21st century. AI-related research and publications have increased significantly over the past decade, uncovering the vast potential of AI in healthcare [105]. It is also suggested that in the area of healthcare research methodology, use of AI is inevitable and will soon be ubiquitous [106]. Standard research methodology such as case-control, cohort, and randomized control trails are the foundation of evidence-based medicine. However, these methods are also laborious, expensive, and prone to biases. ML, which falls under the umbrella of AI, can be used to tackle this problem. ML provides the ability to manage the large and multidimensional data that are specific to healthcare research [107]. Health data accumulated through electronic health records (EHRs), gene sequencing, and digital health wearable devices have resulted in a robust and ever-growing biomedical information, commonly referred to as the "big data." ML, more specifically, DL, techniques have exhibited the ability to transform this "big data" into actionable knowledge [108]. Developments in ML and AI, in healthcare research, aims to translate patient data from multiple sources in a more effective and efficient way, which will pave way for the use of these data in a more interpretable, ethical, and transparent manner. Furthermore, it will lead to improved patient safety, better quality of care, and decreased healthcare costs [107,108].
Identifying risk factors, etiology, effective treatment options, and variants of diseases comes under the horizon of epidemiology. Methods such as randomized control trails, cohort, and case-control studies are used for the same. However, these methods are difficult to use on large patient populations due to operational challenges and patient follow-up. ML using data collected from EHRs to address these epidemiological issues and to increase precision in healthcare delivery are now widely used [109]. EHRs grant access to a large quantity and quality of variables, enabling high-end classification and prediction. Application of ML to EHR data analysis is the front runner of modern clinical informatics [110].
An important step in systematic reviews in healthcare research is the recognition of relevant randomized controlled trials (RCTs) from enormous research databases. A standard practice to recognize these RCTs is using a database filter, which automatically removes a number of irrelevant documents, while the rest need to be screened manually [111]. According to a study conducted by Marshall et al. [112], RCT identification using state-of-the-art ML approaches outperform standard database filters. ML is shown to have substantially high specificity compared to Cochrane Highly Sensitive Search Strategy, for systematic reviews. Moreover, ML has shown greater sensitivity than PubMed for rapid reviews and high precision searches [112].
- **DLM facilitated diagnosis and treatment in healthcare:**
Other examples of AI algorithm application in healthcare include screening and triage, diagnosis, prognostication, decision support, and treatment recommendation [105]. DL is being used extensively for speech recognition in the form of Natural Language Processing (NLP). In healthcare, NLPs are dominantly used to create, interpret, and classify clinical documentation and published research. These NLP systems can further translate unorganized clinical notes on patients, formulate reports, transcribe patient communications, and run conversational AI [113]. DL algorithms have exhibited vast potential in automated analysis of medical data across various medical disciplines [114]. For example, DL algorithms have been used in ophthalmology to detect ocular diseases such as glaucoma and retinopathy [115]. ML algorithms have been known to correctly diagnose osteoarthritis on basic radiographs, making these algorithms useful in both clinical and research settings, for confirmation of osteoarthritis [107]. An AI algorithm, known as faster R−CNN, uses DL and advanced computer vision to accurately detect pressure ulcers. It is designed to augment clinical practice, reduce diagnostic errors, and standardize analysis and reporting of pressure ulcers [116]. Auto Prognosis, is a recent ML program, which provides clinicians with association rules which links patients to predicted risk groups. This program has exhibited higher accuracy in predicting cardiovascular disease risk, as compared to traditional scoring systems based on standard risk factors [108]. ML techniques have the potential to improve patient risk score systems, predict disease onset, and streamline hospital operations [109].
- **Clinical trial management:**
ML, when applied to clinical trials, has the potential to enhance the generalizability, patient-centeredness, accuracy of the process and hence the trail success. ML works across the spectrum of clinical trials in healthcare, from preclinical drug discovery to pretrial planning including study implementation to data management and analysis. Robust preclinical investigation and planning is essential for a successful clinical trial. ML can help researchers to identify relevant previous and ongoing research to reduce inefficiencies of the preclinical process [117]. Recruitment of subjects, who would prove most beneficial for the trail, is one of the principal barriers that affect the success of a clinical trial.

Enrollment of sufficient well-matched research participants poses a big challenge, with wrong/delay in selection process leading to extension of deadlines, delayed submission of trial protocols for approval, with subsequent delay in product/research presentation, and increased costs of the trial [118,119]. ML and advanced data analytics can be utilized to understand the biology and clinical status of subjects in the trial, which in turn will facilitate participant matching for the trials [120,121]. AI-driven clinical trials allow the researcher to identify and randomize participants using AI-assisted randomization techniques and can also be used to reduce biases and challenges when prescreening populations for participation in clinical trials [122].

Big-data clinical trials (BCT) will enhance and match RCTs by making robust data available for researchers. Amalgamating genomic and clinical data, and analyzing it, brings us closer to the goal of personalized medicine [120]. In a study conducted by Vazquez et al., supervised ML classifiers and DL models were applied to data from an online registry which had people showing a prior interest to participate in clinical trials. The classifiers in the study were shown to perform relatively well, with the DL model performing better than traditional approaches, in determining interested participants for the clinical trial [123].

RCTs have an associated complex issue. The statistical analysis of RCT data focuses mainly on the relationship between baseline factors and an intended clinical outcome, without considering the multiple confounding variables that affect individuals from the point of disease progression to death. These confounding interventions need to be considered in the final analysis. To overcome this drawback, a more empirical analysis of RCT is required, which will depend mainly on the analysis of big data using ML, commonly referred to as "Big Data Clinical Trials" (BCTs) [121].

Moreover, ML plays an essential role in preclinical drug discovery and development research. ML can simplify the process of drug target identification and candidate molecule generation through production of large amounts of existing research, interpretation of drug mechanisms, and predictive modeling of protein structures and future drug−target interactions [124]. This type of targeted approach helps increase the possibility that the drugs are tested in populations who are most likely to benefit from the drug, for example, in case of a drug evaluated by Madhukar et al., a more detailed understanding of its mechanism paved way for new clinical trials in pheochromocytoma, which was more likely to benefit from that specific drug, than other cancer types [117]. Newer drugs developed using ML have approached the phase of human testing, for example, an obsessive-compulsive personality disorder (OCPD) drug, enabled by AI, was scheduled to begin phase 1 trial in 2021. The drug was selected from among only 250 candidates and was developed within a year. This is in contrast to typical drug development with requires 2000+ candidates and nearly 5 years to fully develop [117].

There has been a surge in the big data approaches over the years, which is believed to have a positive impact on the efficacy of clinical trials. Standardization of data and nomenclature will improve exchange of data and trial design. It will also allow multistudy analysis and provide benefit to data curation for RCTs. The ML-based approach is expected to elevate the level for evidence threshold required from the RCTs and inspire more validation studies [118].

6. Role of DLM in population level future disease prediction

Four models; random forest, SVM, ANN, and logistic regression data mining technique with 10-fold cross validation have been experimented to predict fatty liver disease. The performance was evaluated based on accuracy, sensitivity, specificity, negative predictive value, and positive predictive value. According to Islam et al. [125] logistic regression provides higher prediction of early fatty liver disease based on medical data from electronic medical records providing 76.3% accuracy, 64.90% specificity, and 74.10% sensitivity the view has been supported by Ref. [126] that logistic regression has an accuracy of 77.40% in lung cancer prediction. Consideration of different easy-to-use and interpret models is important for clinical decision-making; as supported by Zhu et al. [127,128], the model predicted fatty liver disease without abdominal ultrasonography providing fast, easy, low cost, and noninvasive procedures for better diagnosis and treatment of fatty liver disease [125].

Immobile stroke patients suffer infections of up to 30%, UTI 2%−27%, and pneumonia 7%−23%. Predictive comprehensive models have been tested to study different variables in identifying UTI risk. A study conducted by Zhu et al. [127,128] adopted different models including; SMOTE Tomek, ensemble learning pipeline that blended results from (random forest, gradient boosting, regularized logistic regression, and multilayer perception algorithms) and neural network intelligence. The ensemble learning model had highest performance with a sensitivity of 80.9% and specificity of 81.1% in internal and external validation, respectively, comparable to existing predictive models for populations with increased UTI prevalence. Combination of complex models in predicting poststroke UTI complications is important in immobile patients for improved clinical decision-making [127,128].

Deep learning has been adopted in brain research and neurodegenerative diseases such as Alzheimer's disease [129]. CNN or recurrent neural network that use neuro-imaging data have an accuracy of up to 96.0% for Alzheimer's disease and 84.2% for MCI conversion prediction with best classification performance achieved when fluid biomark and multimodal neuroimaging data are combined. Multiple models have been used for early detection and progression of Alzheimer's disease such as linear discrimination analysis, linear program boosting, logistic regression, and SVM [129,130]. DL methods such as DNN, RBM, DBM, DBN, sparse AE, and stacked AE have been used to classify Alzheimer's disease patients from cognitively normal controls or mild cognitive impairments each predicting conversion of MCI to Alzheimer's disease using multi modal neuroimaging data that identify structural and molecular function biomarkers for Alzheimer's disease. The volume of cortical thickness in preselected Alzheimer's disease specific regions.

Hippocampus and entorhinal cortex could be feature enhancers of classification accuracy in ML [130].

Anil Kumar et al. [131] found that support vector machine classifies patients based on their symptoms and enables real-time treatment, cost-effectiveness with least amount of effort, and latency from any location at any time. SVM analyses data and recognizes patterns to detect and calculate size, shape, and location. ANNs analyses breast cancer data from microarrays and UCI ML detects emergence of lung cancer using multilay feed forward neural networks similar to artificial neural networks. Support vector machine and SMOTE used together increases lung cancer detection accuracy to 98.8% [131]. Another study by Tirzīte et al. [132] demonstrated the effectiveness of SVM in discriminating lung cancer patients from healthy subjects using exhale breath analysis by electronic nose resulting to detection of 87.3% of cases. Evaluation of SVM has produced positive results for predicting development of lung cancer encouraging oncologists to use in identification of lung cancer [131].

7. Effectiveness of DLM in screening and referral of clinical cases in remote and poor access areas

The expanding accuracy and booming of deep ML models in working out numerous problems of human have made ventures to establish DL designs in the health care institutions [133]. DL in the recent years has grown as top technique for imaging and computer vision tasks [134]. Basically, ML is divided into three basic categories namely: unsupervised learning, supervised learning, and reinforcement learning [135]. Although healthcare disparities exist in the rural areas, ML supports in identification, diagnosis and management of the diseases, clinical trial research, radiotherapy, personalized medicine, behavioral issues, radiology, and outbreak predictions [135]. According to Ref. [134], ML in medicine field especially cardiology has been involved in various research studies. Therefore, ML in rural areas has the ability power to predict and prevent healthcare discrepancy at high speed and decreased cost [135]. With all these technologies, ML has the capability of generating predictive results, which are more accurate and can facilitate intelligent ways which are human centered [133]. The population in the rural areas has increased number of benefits through use of ML [135]. Using ML algorithms, physicians provide quick and inexpensive services without involving costly specialists who outlay long hours to offer their services to the patients [135]. Similarly, supervised learning algorithms are used in identification of problems such as prediction, diagnosis, image analysis, and heart diseases treatment [134]. AI has stretched out its extent to enhance smart healthcare strategies by use of DL concepts [136]. Moreover, DL methods serve in the analysis and computation of the generated data from image segmentation of MRI [136]. In rural areas of China, DL machine has been used in diagnosing ophthalmologic diseases whereby selection of diseases is done manually and detected through ML [127]. Though DL machine has not been discovered in many rural areas in the world, Thailand has the widest nationwide screening for diabetic retinopathy program in both rural and urban areas with DL machine having an increased efficiency in screening and accuracy [137]. Furthermore, DL is beneficial for the identification of any medical issues [138]. On the other hand, AI offers high satisfaction, accuracy, and efficacy based on the progress of medical practices [139]. In a matter of fact, DL machine will not replace consultants but rather improve the ability to diagnose abnormal lesions in an easy and understandable context in remote setups [139].

8. Conclusion

ML-based medical theologies are becoming a widely used aid in clinical practice. Clinical data and parameters have been used to create a predictive model for both diagnosis and prognosis of disease. ML has opened a new horizon to the medical science and clinical practice to maximize benefits from data and patients records and improve the accuracy of clinical decision-making and ultimately enhance the patient and population health outcomes.

Sustainable chronic disease management has been a challenge for every health service provider because of longevity of services required and increasing need for a personalized care plan. DLM has offered its potential to easily design and respond to the individual care while put the least pressure of the resources (staff and costs) and hence making health service delivery for existing and emerging health needs more viable.

DLM could be a new hope and affordable alternative for the health services providers in low income countries with a diverse, low-density demographic distribution, to benefit from technological advances in developed nations and get relatively comparable standard care within affordable range of costs while investing on expansion of their care provision and overcoming on healthcare professional staff shortage for their universal health coverage (UHC) target.

Combination of DLM and E-health (mHealth) would make community service users more involved in managing their own personalized long-term care plan and potentially more satisfaction and higher health improvement rate at potentially lower costs. Finally, there is an optimistic view now that DLM would be capable to support global health strategic planners to meaningfully model and predict the emerging global population level health risks and identify an accurate timely action.

The advantages of using DLM is, it can learn better representations of unstructured (or) unlabeled data with several levels of abstraction, to solve the data handling problems. To perform the exact procedure of ML, DLM makes use of hierarchical level artificial neural networks (designed like a human brain). One of the main benefits of DLM is that it processes data nonlinearly, as opposed to other technologies that process data linearly. Due to its ability to analyze data more rapidly and precisely, the system is able to quickly adapt to the healthcare industry. It also has the added advantage of enabling decision-making with a far lower involvement of human trainers. Compared to ML, DL requires less pre-processing of the data. Filtering and normalization activities, which are carried out by human programmers in other ML techniques, can be completed by the DLM network itself.

However, there are a few limitations of DLM that include privacy, interpretability of DL models, and developing efficient models to handle diverse and constantly changing healthcare data. Additionally, understanding diseases and their variations is a lot more difficult than other tasks, like speech or image recognition. As a result, from the standpoint of big data, the quantity of medical data required to train an efficient and reliable DL model would be substantially higher when compared to other media. It is difficult to train a robust DL model with such large and diverse datasets and must take into account a number of factors, including data sparsity, redundancy, and missing values. All of these challenges present a number of chances and potential areas for future study that could advance the field.

References

[1] M. Chen, Y. Hao, K. Hwang, L. Wang, L. Wang, Disease prediction by machine learning over big data from healthcare communities, IEEE Access 5 (2017) 8869–8879.

[2] H. Nishi, N. Oishi, A. Ishii, I. Ono, T. Ogura, T. Sunohara, H. Chihara, R. Fukumitsu, M. Okawa, N. Yamana, H. Imamura, Deep learning–derived high-level neuroimaging features predict clinical outcomes for large vessel occlusion, Stroke 51 (5) (2020) 1484–1492.

[3] A. Tsanas, M.A. Little, P.E. McSharry, J. Spielman, L.O. Ramig, Novel speech signal processing algorithms for high-accuracy classification of Parkinson's disease, IEEE Transactions on Biomedical Engineering 59 (5) (2012) 1264–1271.

[4] H. Lee, S.Y. Shin, M. Seo, G.B. Nam, S. Joo, Prediction of ventricular tachycardia one hour before occurrence using artificial neural networks, Scientific Reports 6 (1) (2016) 1–7.

[5] F. Shamout, T. Zhu, D.A. Clifton, Machine learning for clinical outcome prediction, IEEE Reviews in Biomedical Engineering 14 (2020) 116–126.

[6] Y.S. Lou, C.S. Lin, W.H. Fang, C.C. Lee, C.L. Ho, C.H. Wang, C. Lin, Artificial intelligence-enabled electrocardiogram estimates left atrium enlargement as a predictor of future cardiovascular disease, Journal of Personalized Medicine 12 (2) (2022) 315.

[7] P. Ponikowski, S.D. Anker, K.F. AlHabib, M.R. Cowie, T.L. Force, S. Hu, T. Jaarsma, H. Krum, V. Rastogi, L.E. Rohde, U.C. Samal, Heart failure: preventing disease and death worldwide, ESC Heart Failure 1 (1) (2014) 4–25.

[8] A.P. Ambrosy, G.C. Fonarow, J. Butler, O. Chioncel, S.J. Greene, M. Vaduganathan, S. Nodari, C.S. Lam, N. Sato, A.N. Shah, M. Gheorghiade, The global health and economic burden of hospitalizations for heart failure: lessons learned from hospitalized heart failure registries, Journal of the American College of Cardiology 63 (12) (2014) 1123–1133.

[9] Z.I. Attia, S. Kapa, F. Lopez-Jimenez, P.M. McKie, D.J. Ladewig, G. Satam, P.A. Pellikka, M. Enriquez-Sarano, P.A. Noseworthy, T.M. Munger, S.J. Asirvatham, Screening for cardiac contractile dysfunction using an artificial intelligence–enabled electrocardiogram, Nature Medicine 25 (1) (2019) 70–74.

[10] H.Y. Chen, C.S. Lin, W.H. Fang, Y.S. Lou, C.C. Cheng, C.C. Lee, C. Lin, Artificial intelligence-enabled electrocardiography predicts left ventricular dysfunction and future cardiovascular outcomes: a retrospective analysis, Journal of Personalized Medicine 12 (3) (2022) 455.

[11] Y. Yao, G. Sun, T. Matsui, Y. Hakozaki, S. van Waasen, M. Schiek, Multiple vital-sign-based infection screening outperforms thermography independent of the classification algorithm, IEEE Transactions on Biomedical Engineering 63 (5) (2015) 1025–1033.

[12] S. Dagdanpurev, S. Abe, G. Sun, H. Nishimura, L. Choimaa, Y. Hakozaki, T. Matsui, A novel machine-learning-based infection screening system via 2013–2017 seasonal influenza patients' vital signs as training datasets, Journal of Infection 78 (5) (2019) 409–421.

[13] A. Colubri, M.A. Hartley, M. Siakor, V. Wolfman, A. Felix, T. Sesay, J.G. Shaffer, R.F. Garry, D.S. Grant, A.C. Levine, P.C. Sabeti, Machine-learning prognostic models from the 2014–16 Ebola outbreak: data-harmonization challenges, validation strategies, and mHealth applications, EClinicalMedicine 11 (2019) 54–64.

[14] A. Colubri, T. Silver, T. Fradet, K. Retzepi, B. Fry, P. Sabeti, Transforming clinical data into actionable prognosis models: machine-learning framework and field-deployable app to predict outcome of Ebola patients, PLoS Neglected Tropical Diseases 10 (3) (2016) e0004549.

[15] World Health Organization, Coronavirus (COVID-19) dashboard, Covid19.who.int (2022).

[16] National Cancer Institute. https://www.cancer.gov/publications/dictionaries/genetics-dictionary/def/snp. (Accessed August 9, 2022).

[17] I. Kononenko, Machine learning for medical diagnosis: history, state of the art and perspective, Artificial Intelligence in Medicine 23 (1) (2001) 89–109.

[18] Biobanking.com (2021). Data from 10 largest biobanks in the world .[Online] Available from: https://www.biobanking.com/10-largest-biobanks-in-the-world/ [Accessed 9thAugust 2022].

[19] W.C. Wang, L.B. Chen, W.J. Chang, Development and experimental evaluation of machine-learning techniques for an intelligent hairy scalp detection system, Applied Sciences 8 (6) (2018) 853.

[20] D.D.A. Rodrigues, R.F. Ivo, S.C. Satapathy, S. Wang, J. Hemanth, P.P. Reboucas Filho, A new approach for classification skin lesion based on transfer learning, deep learning, and IoT system, Pattern Recognition Letters 136 (2020) 8–15.

[21] C.L. Chuang, Case-based reasoning support for liver disease diagnosis, Artificial Intelligence in Medicine 53 (1) (2011) 15–23.

[22] M.M. Musleh, E. Alajrami, A.J. Khalil, B.S. Abu-Nasser, A.M. Barhoom, S.A. Naser, Predicting liver patients using artificial neural network, International Journal of Academic Information Systems Research (IJAISR) 3 (10) (2019).

[23] J. Chen, D. Remulla, J.H. Nguyen, Y. Liu, P. Dasgupta, A.J. Hung, Current status of artificial intelligence applications in urology and their potential to influence clinical practice, BJU International 124 (4) (2019) 567–577.

[24] P.H.C. Chen, K. Gadepalli, R. MacDonald, Y. Liu, S. Kadowaki, K. Nagpal, T. Kohlberger, J. Dean, G.S. Corrado, J.D. Hipp, C.H. Mermel, An augmented reality microscope with real-time artificial intelligence integration for cancer diagnosis, Nature Medicine 25 (9) (2019) 1453–1457.

[25] F.C. Morabito, M. Campolo, C. Ieracitano, J.M. Ebadi, L. Bonanno, A. Bramanti, S. Desalvo, N. Mammone, P. Bramanti, Deep convolutional neural networks for classification of mild cognitive impaired and Alzheimer's disease patients from scalp EEG recordings, in: 2016 IEEE 2nd International Forum on Research and Technologies for Society and Industry Leveraging a Better Tomorrow (RTSI), IEEE, September 2016, pp. 1–6.

[26] S. Sathitratanacheewin, P. Sunanta, K. Pongpirul, Deep learning for automated classification of tuberculosis-related chest X-Ray: dataset distribution shift limits diagnostic performance generalizability, Heliyon 6 (8) (2020) e04614.

[27] A.M. Shabut, M.H. Tania, K.T. Lwin, B.A. Evans, N.A. Yusof, K.J. Abu-Hassan, M.A. Hossain, An intelligent mobile-enabled expert system for tuberculosis disease diagnosis in real time, Expert Systems with Applications 114 (2018) 65–77.

[28] Ö. Yõldõrõm, P. Pławiak, R.S. Tan, U.R. Acharya, Arrhythmia detection using deep convolutional neural network with long duration ECG signals, Computers in Biology and Medicine 102 (2018) 411–420.

[29] A. Nithya, A. Appathurai, N. Venkatadri, D.R. Ramji, C.A. Palagan, Kidney disease detection and segmentation using artificial neural network and multi-kernel k-means clustering for ultrasound images, Measurement 149 (2020) 106952.

[30] J.N. Kather, A.T. Pearson, N. Halama, D. Jäger, J. Krause, S.H. Loosen, A. Marx, P. Boor, F. Tacke, U.P. Neumann, H.I. Grabsch, Deep learning can predict microsatellite instability directly from histology in gastrointestinal cancer, Nature Medicine 25 (7) (2019) 1054–1056.

[31] H. Kanegae, K. Suzuki, K. Fukatani, T. Ito, N. Harada, K. Kario, Highly precise risk prediction model for new-onset hypertension using artificial intelligence techniques, Journal of Clinical Hypertension 22 (3) (2020) 445–450.

[32] A.S. Kasasbeh, S. Christensen, M.W. Parsons, B. Campbell, G.W. Albers, M.G. Lansberg, Artificial neural network computer tomography perfusion prediction of ischemic core, Stroke 50 (6) (2019) 1578–1581.

[33] E.K. Oikonomou, M.C. Williams, C.P. Kotanidis, M.Y. Desai, M. Marwan, A.S. Antonopoulos, K.E. Thomas, S. Thomas, I. Akoumianakis, L.M. Fan, S. Kesavan, A novel machine learning-derived radiotranscriptomic signature of perivascular fat improves cardiac risk prediction using coronary CT angiography, European Heart Journal 40 (43) (2019) 3529–3543.

[34] R.O. Panicker, K.S. Kalmady, J. Rajan, M.K. Sabu, Automatic detection of tuberculosis bacilli from microscopic sputum smear images using deep learning methods, Biocybernetics and Biomedical Engineering 38 (3) (2018) 691–699.

[35] V. Sarao, D. Veritti, P. Lanzetta, Automated diabetic retinopathy detection with two different retinal imaging devices using artificial intelligence: a comparison study, Graefe's Archive for Clinical and Experimental Ophthalmology 258 (12) (2020) 2647–2654.

[36] E. Shkolyar, X. Jia, T.C. Chang, D. Trivedi, K.E. Mach, M.Q.H. Meng, L. Xing, J.C. Liao, Augmented bladder tumor detection using deep learning, European Urology 76 (6) (2019) 714–718.

[37] W. Gouda, R. Yasin, COVID-19 disease: CT pneumonia analysis prototype by using artificial intelligence, predicting the disease severity, Egyptian Journal of Radiology and Nuclear Medicine 51 (1) (2020) 1–11.

[38] S. Vasal, S. Jain, A. Verma, COVID-AI: an artificial intelligence system to diagnose COVID 19 disease, Journal of Engineering Research & Technology 9 (2020) 1–6.

[39] G. Duffy, P.P. Cheng, N. Yuan, B. He, A.C. Kwan, M.J. Shun-Shin, K.M. Alexander, J. Ebinger, M.P. Lungren, F. Rader, D.H. Liang, High-throughput precision phenotyping of left ventricular hypertrophy with cardiovascular deep learning, JAMA Cardiology 7 (4) (2022) 386–395.

[40] S. Ansari, I. Shafi, A. Ansari, J. Ahmad, S.I. Shah, Diagnosis of liver disease induced by hepatitis virus using artificial neural networks, in: 2011 IEEE 14th International Multitopic Conference, IEEE, December 2011, pp. 8–12.

[41] L. Dong, W. He, R. Zhang, Z. Ge, Y.X. Wang, J. Zhou, J. Xu, L. Shao, Q. Wang, Y. Yan, Y. Xie, Artificial intelligence for screening of multiple retinal and optic nerve diseases, JAMA Network Open 5 (5) (2022) e229960-e229960.
[42] S.B. Khanagar, A. Al-Ehaideb, S. Vishwanathaiah, P.C. Maganur, S. Patil, S. Naik, H.A. Baeshen, S.S. Sarode, Scope and performance of artificial intelligence technology in orthodontic diagnosis, treatment planning, and clinical decision-making-a systematic review, Journal of Dental Sciences 16 (1) (2021) 482–492.
[43] C. Krittanawong, K.W. Johnson, R.S. Rosenson, Z. Wang, M. Aydar, U. Baber, J.K. Min, W.W. Tang, J.L. Halperin, S.M. Narayan, Deep learning for cardiovascular medicine: a practical primer, European Heart Journal 40 (25) (2019) 2058–2073.
[44] Y. Cao, Z. Liu, P. Zhang, Y. Zheng, Y. Song, L. Cui, Deep learning methods for cardiovascular image, Journal of Artificial Intelligence and Systems 1 (1) (2019) 96–109.
[45] I. Bardhan, H. Chen, E. Karahanna, Connecting systems, data, and people: a multidisciplinary research roadmap for chronic disease management, MIS Quarterly 44 (1) (2020) 185–200.
[46] R. Amin, M.A. Al Ghamdi, S.H. Almotiri, M. Alruily, Healthcare techniques through deep learning: issues, challenges and opportunities, IEEE Access 9 (2021) 98523–98541.
[47] C.T. Wu, G.H. Li, C.T. Huang, Y.C. Cheng, C.H. Chen, J.Y. Chien, P.H. Kuo, L.C. Kuo, F. Lai, Acute exacerbation of a chronic obstructive pulmonary disease prediction system using wearable device data, machine learning, and deep learning: development and cohort study, JMIR mHealth and uHealth 9 (5) (2021) e22591.
[48] T. Okanoue, T. Shima, Y. Mitsumoto, A. Umemura, K. Yamaguchi, Y. Itoh, M. Yoneda, A. Nakajima, E. Mizukoshi, S. Kaneko, K. Harada, Artificial intelligence/neural network system for the screening of nonalcoholic fatty liver disease and nonalcoholic steatohepatitis, Hepatology Research 51 (5) (2021) 554–569.
[49] A.S. Ahuja, The impact of artificial intelligence in medicine on the future role of the physician, PeerJ 7 (2019) e7702.
[50] M. Nair, S. Singh Sandhu, A. K Sharma, Prognostic and predictive biomarkers in cancer, Current Cancer Drug Targets 14 (5) (2014) 477–504.
[51] Y.H. Lai, W.N. Chen, T.C. Hsu, C. Lin, Y. Tsao, S. Wu, Overall survival prediction of non-small cell lung cancer by integrating microarray and clinical data with deep learning, Scientific Reports 10 (1) (2020) 1–11.
[52] K.A. Tran, O. Kondrashova, A. Bradley, E.D. Williams, J.V. Pearson, N. Waddell, Deep learning in cancer diagnosis, prognosis and treatment selection, Genome Medicine 13 (1) (2021) 1–17.
[53] FDA News Release (2018). FDA permits marketing of artificial intelligence-based device to detect certain diabetes-related eye problems. [Online] Available from: https://www.fda.gov/news-events/press-announcements/fda-permits-marketing-artificial-intelligence-based-device-detect-certain-diabetes-related-eye [9 August 2022]
[54] S.M.D.A.C. Jayatilake, G.U. Ganegoda, Involvement of machine learning tools in healthcare decision making, Journal of Healthcare Engineering (2021), 2021.
[55] R. Miotto, F. Wang, S. Wang, X. Jiang, J.T. Dudley, Deep learning for healthcare: review, opportunities and challenges, Briefings in Bioinformatics 19 (6) (2018) 1236–1246.
[56] K. Begley, C. Begley, V. Smith, Shared decision-making and maternity care in the deep learning age: acknowledging and overcoming inherited defeaters, Journal of Evaluation in Clinical Practice 27 (3) (2021) 497–503.
[57] Y. Bengio, Learning deep architectures for AI, Foundations and Trends® in Machine Learning 2 (1) (2009) 1–127.
[58] V. Gulshan, L. Peng, M. Coram, M.C. Stumpe, D. Wu, A. Narayanaswamy, S. Venugopalan, K. Widner, T. Madams, J. Cuadros, R. Kim, Development and validation of a deep learning algorithm for detection of diabetic retinopathy in retinal fundus photographs, JAMA 316 (22) (2016) 2402–2410.
[59] S. Liu, S. Liu, W. Cai, S. Pujol, R. Kikinis, D. Feng, Early diagnosis of Alzheimer's disease with deep learning, in: 2014 IEEE 11th International Symposium on Biomedical Imaging (ISBI), April 2014, pp. 1015–1018. IEEE.
[60] J. Brosch, M. Farlow, Early-onset dementia in adults, Uptodate 1 (1) (2018) 1–4.
[61] A. Prasoon, K. Petersen, C. Igel, F. Lauze, E. Dam, M. Nielsen, Deep feature learning for knee cartilage segmentation using a triplanar convolutional neural network, in: International Conference on Medical Image Computing and Computer-Assisted Intervention, Springer, *Berlin,* Heidelberg, September 2013, pp. 246–253.
[62] Y. Yoo, T. Brosch, A. Traboulsee, D.K. Li, R. Tam, Deep learning of image features from unlabeled data for multiple sclerosis lesion segmentation. In: Machine Learning in Medical Imaging: 5th International Workshop, MLMI 2014, Held in Conjunction with MICCAI 2014, Boston, MA, USA, September 14, 2014. Proceedings 5 2014 (pp. 117-124). Springer International Publishing
[63] D. Sun, M. Wang, A. Li, A multimodal deep neural network for human breast cancer prognosis prediction by integrating multi-dimensional data, IEEE/ACM Trans Comput Biol Bioinform 16 (3) (2018) 841–850.
[64] A. Esteva, B. Kuprel, R.A. Novoa, J. Ko, S.M. Swetter, H.M. Blau, S. Thrun, Dermatologist-level classification of skin cancer with deep neural networks, Nature 542 (7639) (2017) 115–118.
[65] T. Tran, T.D. Nguyen, D. Phung, S. Venkatesh, Learning vector representation of medical objects via EMR-driven nonnegative restricted Boltzmann machines (eNRBM), Journal of biomedical informatics 54 (2015) 96–105.
[66] Z. Che, D. Kale, W. Li, M.T. Bahadori, Y. Liu, Deep computational phenotyping, in: Proceedings of the 21th ACM SIGKDD International Conference on Knowledge Discovery and Data Mining, August 2015, pp. 507–516.
[67] E. Choi, M.T. Bahadori, A. Schuetz, W.F. Stewart, J. Sun, Doctor ai: predicting clinical events via recurrent neural networks, in: Machine Learning for Healthcare Conference, PMLR, December 2016, pp. 301–318.

[68] N. Razavian, J. Marcus, D. Sontag, Multi-task prediction of disease onsets from longitudinal laboratory tests, in: Machine Learning for Healthcare Conference, PMLR, December 2016, pp. 73–100.

[69] T. Pham, T. Tran, D. Phung, S. Venkatesh. Deepcare, A deep dynamic memory model for predictive medicine, in: Advances in Knowledge Discovery and Data Mining, 20th Pacific-Asia Conference, PAKDD 2016, Springer International Publishing, 2016, pp. 30–41.

[70] Z.C. Lipton, D.C. Kale, C. Elkan, R. Wetzel, Learning to Diagnose with LSTM Recurrent Neural Networks, 2015 *arXiv preprint arXiv:1511.03677.*

[71] V. Jindal, J. Birjandtalab, M.B. Pouyan, M. Nourani, An adaptive deep learning approach for PPG-based identification, in: 2016 38th Annual International Conference of the IEEE Engineering in Medicine and Biology Society (EMBC), IEEE, August 2016, pp. 6401–6404.

[72] E. Nurse, B.S. Mashford, A.J. Yepes, I. Kiral-Kornek, S. Harrer, D.R. Freestone, Decoding EEG and LFP signals using deep learning: heading TrueNorth, in: Proceedings of the ACM International Conference on Computing Frontiers, May 2016, pp. 259–266.

[73] A. Sathyanarayana, S. Joty, L. Fernandez-Luque, F. Ofli, J. Srivastava, A. Elmagarmid, T. Arora, S. Taheri. Sleep quality prediction from wearable data using deep learning. JMIR mHealth and uHealth. 4 (4): (2016) 6562.

[74] N.Y. Hammerla, S. Halloran, T. Plötz, Deep, Convolutional, and Recurrent Models for Human Activity Recognition Using Wearables, 2016 *arXiv preprint arXiv:1604.08880.*

[75] J. Zhou, OG. Troyanskaya, Predicting effects of noncoding variants with deep learning–based sequence model, Nature methods 12 (10) (2015) 931–934.

[76] C. Angermueller, H.J. Lee, W. Reik, O. Stegle, DeepCpG: accurate prediction of single-cell DNA methylation states using deep learning, Genome Biology 18 (1) (2017) 1–13, https://doi.org/10.1101/052118.

[77] B. Alipanahi, A. Delong, M.T. Weirauch, B.J. Frey, Predicting the sequence specificities of DNA-and RNA-binding proteins by deep learning, Nature Biotechnology 33 (8) (2015) 831–838.

[78] R. Fakoor, F. Ladhak, A. Nazi, M. Huber, Using deep learning to enhance cancer diagnosis and classification, in: Proceedings of the International Conference on Machine Learning vol 28, ACM, New York, USA, June 2013, pp. 3937–3949.

[79] D. Callahan, The WHO Definition of 'health', Hastings Center Studies, 1973, pp. 77–87.

[80] O.D. Howes, S. Kapur, A neurobiological hypothesis for the classification of schizophrenia: type A (hyperdopaminergic) and type B (normodopaminergic), The British Journal of Psychiatry 205 (1) (2014) 1–3.

[81] K. Dheeraj, T. Ramakrishnudu, Negative emotions detection on online mental-health related patients texts using the deep learning with MHA-BCNN model, Expert Systems with Applications 182 (2021) 115265.

[82] NICE(2022). Mental Health and Well-being. 2022. https://www.nice.org.uk/guidance/lifestyle-and-wellbeing/mental-health-and-wellbeing. [Accessed 14th August 2022]

[83] A.B. Shatte, D.M. Hutchinson, S.J. Teague, Machine learning in mental health: a scoping review of methods and applications, Psychological Medicine 49 (9) (2019) 1426–1448.

[84] H. Abou-Warda, N.A. Belal, Y. El-Sonbaty, S. Darwish, A random forest model for mental disorders diagnostic systems, in: InInternational Conference on Advanced Intelligent Systems and Informatics, Springer, Cham, October 2016, pp. 670–680.

[85] Z. Ye, C.L. Rae, C. Nombela, T. Ham, T. Rittman, P.S. Jones, P.V. Rodríguez, I. Coyle-Gilchrist, R. Regenthal, E. Altena, C.R. Housden, Predicting beneficial effects of atomoxetine and citalopram on response inhibition in Parkinson's disease with clinical and neuroimaging measures, Human Brain Mapping 37 (3) (2016) 1026–1037.

[86] N. Bak, B.H. Ebdrup, B. Oranje, B. Fagerlund, M.H. Jensen, S.W. Düring, M.Ø. Nielsen, B.Y. Glenthøj, L.K. Hansen, Two subgroups of antipsychotic-naive, first-episode schizophrenia patients identified with a Gaussian mixture model on cognition and electrophysiology, Translational Psychiatry 7 (4) (2017) e1087-e1087.

[87] T. Chen, D. Zeng, Y. Wang, Multiple kernel learning with random effects for predicting longitudinal outcomes and data integration, Biometrics 71 (4) (2015) 918–928.

[88] F. Zhu, B. Panwar, H.H. Dodge, H. Li, B.M. Hampstead, R.L. Albin, H.L. Paulson, Y. Guan, COMPASS: a computational model to predict changes in MMSE scores 24-months after initial assessment of Alzheimer's disease, Scientific Reports 6 (1) (2016) 1–12.

[89] G.N. Saxe, S. Ma, J. Ren, C. Aliferis, Machine learning methods to predict child posttraumatic stress: a proof of concept study, BMC Psychiatry 17 (1) (2017) 1–13.

[90] T.T. Erguzel, N. Tarhan, Machine learning approaches to predict repetitive transcranial magnetic stimulation treatment response in major depressive disorder, in: Proceedings of SAI Intelligent Systems Conference, Springer, Cham, September 2016, pp. 391–401.

[91] J.P. Guilloux, S. Bassi, Y. Ding, C. Walsh, G. Turecki, G. Tseng, J.M. Cyranowski, E. Sibille, Testing the predictive value of peripheral gene expression for nonremission following citalopram treatment for major depression, Neuropsychopharmacology 40 (3) (2015) 701–710.

[92] R. Iniesta, K. Malki, W. Maier, M. Rietschel, O. Mors, J. Hauser, N. Henigsberg, M.Z. Dernovsek, D. Souery, D. Stahl, R. Dobson, Combining clinical variables to optimize prediction of antidepressant treatment outcomes, Journal of Psychiatric Research 78 (2016) 94–102.

[93] R.C. Kessler, H.M. van Loo, K.J. Wardenaar, R.M. Bossarte, L.A. Brenner, T. Cai, D.D. Ebert, I. Hwang, J. Li, P. de Jonge, A.A. Nierenberg, Testing a machine-learning algorithm to predict the persistence and severity of major depressive disorder from baseline self-reports, Molecular Psychiatry 21 (10) (2016) 1366–1371.

[94] G.P. Amminger, A. Mechelli, S. Rice, S.W. Kim, C.M. Klier, R.K. McNamara, M. Berk, P.D. McGorry, M.R. Schäfer, Predictors of treatment response in young people at ultra-high risk for psychosis who received long-chain omega-3 fatty acids, Translational Psychiatry 5 (1) (2015) e495-e495.

[95] N. Koutsouleris, R.S. Kahn, A.M. Chekroud, S. Leucht, P. Falkai, T. Wobrock, E.M. Derks, W.W. Fleischhacker, A. Hasan, Multisite prediction of 4-week and 52-week treatment outcomes in patients with first-episode psychosis: a machine learning approach, The Lancet Psychiatry 3 (10) (2016) 935–946.
[96] A. Mechelli, A. Lin, S. Wood, P. McGorry, P. Amminger, S. Tognin, P. McGuire, J. Young, B. Nelson, A. Yung, Using clinical information to make individualized prognostic predictions in people at ultra high risk for psychosis, Schizophrenia Research 184 (2017) 32–38.
[97] J. Chung, J. Teo, Mental Health Prediction Using Machine Learning: Taxonomy, Applications, and Challenges, Applied Computational Intelligence and Soft Computing, 2022, p. 2022.
[98] Y.T. Jo, S.W. Joo, S.H. Shon, H. Kim, Y. Kim, J. Lee, Diagnosing schizophrenia with network analysis and a machine learning method, International Journal of Methods in Psychiatric Research 29 (1) (2020) e1818.
[99] H. Yang, J. Liu, J. Sui, G. Pearlson, V.D. Calhoun, A hybrid machine learning method for fusing fMRI and genetic data: combining both improves classification of schizophrenia, Frontiers in Human Neuroscience 4 (2010) 192.
[100] I. Nahum-Shani, S.N. Smith, B.J. Spring, L.M. Collins, K. Witkiewitz, A. Tewari, S.A. Murphy, Just-in-time adaptive interventions (JITAIs) in mobile health: key components and design principles for ongoing health behavior support, Annals of Behavioral Medicine 52 (6) (2018) 446–462.
[101] J. Lipschitz, C.J. Miller, T.P. Hogan, K.E. Burdick, R. Lippin-Foster, S.R. Simon, J. Burgess, Adoption of mobile apps for depression and anxiety: cross-sectional survey study on patient interest and barriers to engagement, JMIR Mental Health 6 (1) (2019) e11334.
[102] O. Perski, A. Blandford, R. West, S. Michie, Conceptualising engagement with digital behaviour change interventions: a systematic review using principles from critical interpretive synthesis, Translational Behavioral Medicine 7 (2) (2017) 254–267.
[103] C. Qu, C. Sas, C.D. Roquet, G. Doherty, Functionality of top-rated mobile apps for depression: systematic search and evaluation, JMIR Mental Health 7 (1) (2020) e15321.
[104] R. Lewis, C. Ferguson, C. Wilks, N. Jones, R.W. Picard, Can a recommender system support treatment personalisation in digital mental health therapy? A quantitative feasibility assessment using data from a behavioural activation therapy app, in: CHI Conference on Human Factors in Computing Systems Extended Abstracts, April 2022, pp. 1–8.
[105] H. Ibrahim, X. Liu, A.K. Denniston, Reporting guidelines for artificial intelligence in healthcare research, Clinical and Experimental Opthamology 49 (2021) 470–476.
[106] T.B. Murdoch, A.S. Detsky, The inevitable application of big data to healthcare, JAMA 309 (13) (2013) 1351–1352.
[107] L. Rubinger, A. Gazendam, S. Ekhtiari, M. Bhandari, Machine learning and artificial intelligence in research and healthcare, Injury 3 (37) (2022).
[108] J. Waring, C. Lindvall, R. Umeton, Automated machine learning: review of the state-of-the-art and opportunities in healthcare, Artificial Intelligence in Medicine Vol 104 (2020).
[109] A. Callahan, N.H. Shah, Machine learning in healthcare, in: A. Sheikh, D. Bates, A. Wright, K. Cresswell (Eds.), Key Advances in Clinical Informatics: Transforming Health Care through Health Information Technology, Elsevier Scinece & Technology, Stanford, CA, 2017, pp. 279–291.
[110] B. Goldstein, A. Navar, M. Pencina, J. Ioannidis, Opportunities and challenges in developing risk prediction models with electronic health records data: a systemic review, Journal of the American Medical Informatics Association 27 (1) (2016) 198–208.
[111] C. Lefebvre, E. Manheimer, J. Glanville, Searching for studies, in: J. Higgins, S. Green (Eds.), Cochrane Handbook for Systematic Reviews of Interventions. 5.1.0, The Cochrane Collaboration, Chichester, 2011.
[112] I.J. Marshall, et al., Machine learning for identifying randomized controlled trials: an evaluation and practitioner's guide, Research Synthesis Methods Volume 9 (2018) 602–614.
[113] T. Davenport, R. Kalakota, The potential for artificial intelligence in healthcare, Future Healthcare Journal 6 (2019) 94–98.
[114] P. Lakhani, B. Sundaram, Deep Learning at chest radiography: automated classification of pulmonary tuberculosis by using convolutional neural networks, Radiology 284 (2017) 574–582.
[115] T. Tan, et al., Retinal photograph-based deep learning algorithms for myopia and a blockchain platform to facilitate artificial intelligence and a blockchain platform to facilitate artificial intelligence medical research: a retrospective multicohort study, Lancet Digital Health 3 (2021) e317–e329.
[116] P. Fergus, et al., Pressure Ulcer Categorisation Using Deep Learning: A Clinical Trial to Evaluate Model Performance, 2022 *(arXiv)*.
[117] E.H. Weissler, et al., The role of machine learning in clinical research: transforming the future of evidence generation, Trials 22 (537) (2021).
[118] C. Mayo, et al., Big data in designing clinical trials: opportunities and challenges, Frontiers in Oncology 7 (187) (2017).
[119] A. Sen, et al., Correlating eligibility criteria generalizability and adverse events using Big Data for patients and clinical trials, Annals of the New York Academy of Sciences 1387 (2016) 34–43.
[120] N. Mehta, A. Pandit, S. Shukla, Transforming healthcare with big data analytics and artificial intelligence: a systemic mapping study, Journal of Biomedical Informatics 100 (2019).
[121] G. Taglang, D.B. Jackson, Use of "big data" in drug discovery and clinical trials, Gynecologic Oncology 14 (2016) 17–23.
[122] E.N. Ngayua, J. He, K. Ageyei-Boahene, Applying advanced technologies to improve clinical trials: a systematic mapping study, Scientometrics 126 (2021) 1217–1238.
[123] J. Vezquez, et al., Using supervised machine learning classifiers to estimate likelihood of participating in clinical trials of a de-identified version of Research Match, Journal of Clinical and Translational Science 5 (42) (2020) 1–7.
[124] A. Senior, et al., Improved protien structure prediction using potentials from deep learning, Nature 577 (7792) (2020) 706–710.
[125] M. Islam, C. Wu, T.N. Poly, H. Yang, Y.J. Li, Applications of machine learning in fatty live disease prediction, in: Building Continents of Knowledge in Oceans of Data: The Future of Co-created eHealth, IOS Press, 2018, pp. 166–170, 2018.

[126] A. Hazra, N. Bera, A. Mandal, Predicting lung cancer survivability using SVM and logistic regression algorithms, International Journal of Computer Applications 174 (2) (2017) 19–24.
[127] C. Zhu, Z. Xu, Y. Gu, S. Zheng, X. Sun, J. Cao, B. Song, J. Jin, Y. Liu, X. Wen, Prediction of post-stroke urinary tract infection risk in immobile patients using machine learning: an observational cohort study, Journal of Hospital Infection 122 (2022) 96–107.
[128] S. Zhu, B. Lu, C. Wang, M. Wu, B. Zheng, Q. Jiang, R. Wei, Q. Cao, W. Yang, Screening of common retinal diseases using six-category models based on EfficientNet, Frontiers of Medicine (2022) 130.
[129] S. Tabarestani, M. Eslami, M. Cabrerizo, R.E. Curiel, A. Barreto, N. Rishe, D. Vaillancourt, S.T. DeKosky, D.A. Loewenstein, R. Duara, A tensorized multitask deep learning network for progression prediction of Alzheimer's disease, Frontiers in Aging Neuroscience 14 (2022).
[130] T. Jo, K. Nho, A.J. Saykin, Deep learning in Alzheimer's disease: diagnostic classification and prognostic prediction using neuroimaging data, Frontiers in Aging Neuroscience 11 (2019) 220.
[131] C. Anil Kumar, S. Harish, P. Ravi, M. Svn, B.P. Kumar, V. Mohanavel, N.M. Alyami, S.S. Priya, A.K. Asfaw, Lung cancer prediction from text datasets using machine learning, BioMed Research International 2022 (2022) 10, 6254177.
[132] M. Tirzīte, M. Bukovskis, G. Strazda, N. Jurka, I. Taivans, Detection of lung cancer in exhaled breath with an electronic nose using support vector machine analysis, Journal of Breath Research 11 (3) (2017) 036009.
[133] P.U. Eze, C.O. Asogwa, Deep machine learning model trade-offs for malaria elimination in resource-constrained locations, Bioengineering 8 (11) (2021) 150.
[134] B.C. Loh, P.H. Then, Deep learning for cardiac computer-aided diagnosis: benefits, issues & solutions, mHealth 3 (2017).
[135] A.A. Cecchetti, Why introduce machine learning to rural health care? Marshall Journal of Medicine 4 (2) (2018) 3.
[136] S. Bhattacharya, S.R.K. Somayaji, T.R. Gadekallu, M. Alazab, P.K.R. Maddikunta, A review on deep learning for future smart cities, Internet Technology Letters 5 (1) (2022) e187.
[137] P. Ruamviboonsuk, R. Tiwari, R. Sayres, V. Nganthavee, K. Hemarat, A. Kongprayoon, R. Raman, B. Levinstein, Y. Liu, M. Schaekermann, Real-time diabetic retinopathy screening by deep learning in a multisite national screening programme: a prospective interventional cohort study, The Lancet Digital Health 4 (4) (2022) e235–e244.
[138] D. Bordoloi, V. Singh, S. Sanober, S.M. Buhari, J.A. Ujjan, R. Boddu, Deep learning in healthcare system for quality of service, Journal of Healthcare Engineering (2022) 2022.
[139] I. Wijesinghe, C. Gamage, I. Perera, C. Chitraranjan (Eds.), 2019 Moratuwa Engineering Research Conference (MERCon), IEEE, 2019.

Further reading

[1] M. Alsharqi, W.J. Woodward, J.A. Mumith, D.C. Markham, R. Upton, P. Leeson, Artificial intelligence and echocardiography, Echo Research and Practice 5 (4) (2018) R115–R125.
[2] T.O. Ayodele, Types of machine learning algorithms, New Advances in Machine Learning 3 (2010) 19–48.
[3] Y. Cheng, F. Wang, P. Zhang, J. Hu, Risk prediction with electronic health records: a deep learning approach, in: Proceedings of the 2016 SIAM International Conference on Data Mining, *Society for* Industrial and Applied Mathematics, June 2016, pp. 432–440.
[4] Data from 10 Largest Biobanks in the World https://www.biobanking.com/10-largest-biobanks-in-the-world/(Accessed August 9, 2022).
[5] M. Fatima, M. Pasha, Survey of machine learning algorithms for disease diagnostic, Journal of Intelligent Learning Systems and Applications 9 (01) (2017) 1.
[6] Mental Health and COVID-19: Early Evidence of the Pandemic's Impact: Scientific Brief, March 2, 2022. WHO Report 2022, https://apps.who.int/iris/handle/10665/352189?search-result=true&query=WHO+and+mental+health+due+to+covid-19&scope=&rpp=10&sort_by=score&order=desc. (Accessed 15 August 2022).
[8] P.A. Pattanaik, M. Mittal, M.Z. Khan, Unsupervised deep learning cad scheme for the detection of malaria in blood smear microscopic images, IEEE Access 8 (2020) 94936–94946.
[9] R. Punithavathi, M. Sharmila, T. Avudaiappan, I. Raj, S. Kanchana, S.A. Mamo, Empirical investigation for predicting depression from different machine learning based voice recognition techniques, Evidence-based Complementary and Alternative Medicine (2022).
[10] M. Roser, H. Ritchie, Burden of Disease, 2021. Published online at OurWorldInData.org, https://ourworldindata.org/burden-of-disease. (Accessed 9 August 2022).
[11] T. Schlegl, S.M. Waldstein, H. Bogunovic, F. Endstraßer, A. Sadeghipour, A.M. Philip, D. Podkowinski, B.S. Gerendas, G. Langs, U. Schmidt-Erfurth, Fully automated detection and quantification of macular fluid in OCT using deep learning, Ophthalmology 125 (4) (2018) 549–558.
[12] Types of Mental Health Problems, 2022. https://www.mind.org.uk/information-support/types-of-mental-health-problems/. (Accessed 14 August 2022).
[13] Coronavirus disease (COVID-19).https://covid19.who.int/(Accessed August 9, 2022).

Chapter 15

Development of a no-regret deep learning framework for efficient clinical decision-making

Yamuna Mundru[1], Manas Kumar Yogi[1], Jyotir Moy Chatterjee[2], Madhur Meduri[3] and Ketha Dhana Veera Chaitanya[3]

[1]CSE Department, Pragati Engineering College, Surampalem, Andhra Pradesh, India; [2]Department of Information Technology, Lord Buddha Education Foundation (Asia Pacific University), Kathmandu, Nepal; [3]Pragati Engineering College, Surampalem, Andhra Pradesh, India

1. Introduction

Decision-making is performed by analyzing the solution to a problem connected to a large amount of data. We already know while taking making decisions we need to check a variety of gathered data [1].

From different relevant information sources from various areas support the decision-making process. Data-driven systems (decision systems based on databases, e.g., external and internal databases) are modeled on the following factors:

- Choice at the time of testing
- Choice of treatment
- Prognosis
- Diagnosis

With the help of decision support we can provide a solution with different options for a patient.

For example, for eye sight we can replace spectacles with contact lenses which are easily useable. However, we need to know more about the advantages and disadvantages of contact lenses. We can also get different options for lenses. With the help of artificial intelligence we have upgraded technology to choose the option with the help of multineural architecture, feed forward neural network, which has a high degree of interpredictability [2]. They store information on the entire network and have the ability to work with incomplete knowledge, working with incomplete work fault-tolerance and distributed memory. It also uses CNN to provide high accuracy in image recognition problems and is capable of automatically detecting important features without any human supervision when problems occur. Decision trees are also quite useful (Fig. 15.1).

While making decisions it always performs a sum of product is done by using disjunctive normal form (DNF) [3]. In order to increase the degree of homogeneity,the decision tree splits the decision node using more than pne algorithm. As shown in Fig. 15.2, a branch of machine learning algorithm has been developed called a reinforcement algorithm.

A calculation in clinical terms ought to be utilized basically as a doctor collaborator to reduce the burden of doctors, and not to replace them. The advancement of calculation medication in medical care presents an amazing and promising future.

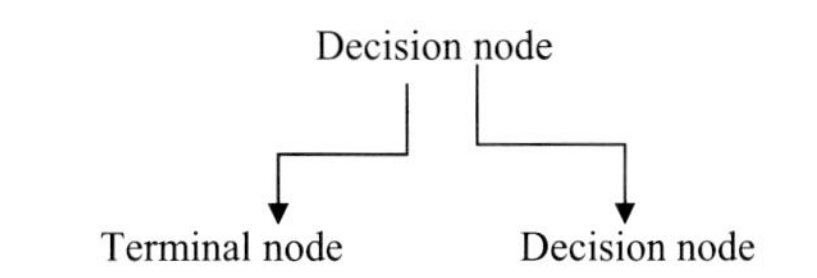

FIGURE 15.1 Operational logic of a decision tree.

Deep Learning in Personalized Healthcare and Decision Support. https://doi.org/10.1016/B978-0-443-19413-9.00007-2

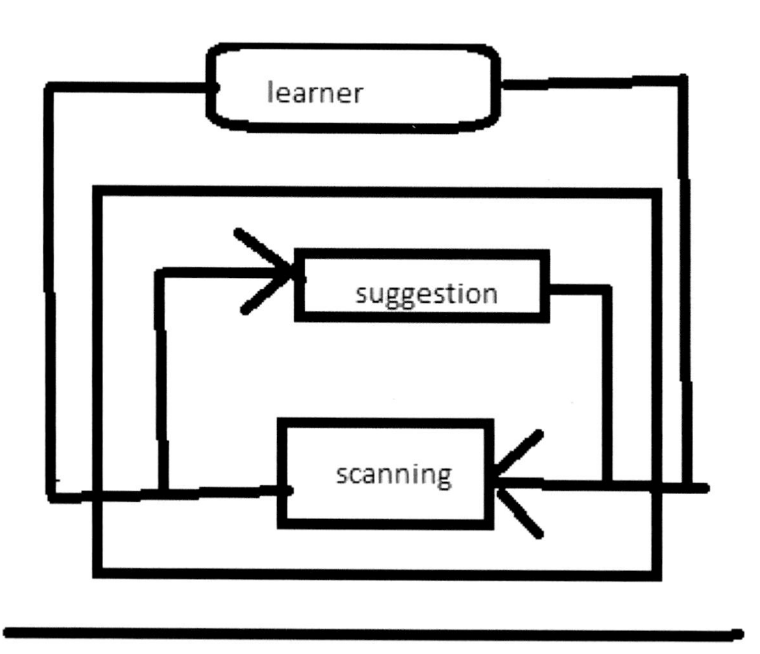

FIGURE 15.2 Basic mechanism of reinforcement learning.

A time wherein significant clinical advances can be made open, and more doctors and patients can profit from it [4]. Programming works where you want it, at the place of care. The proposed operational mechanism equipped with machine learning properties tries to use the patient data stored in the most meaningful manner.

Some of the disadvantages to the use of clinical calculations include a shortage of information, unknown capacities, trouble in understanding the outcomes communicated in the calculation, and absence of access at the point of care. Computerization of a therapy calculation can serve to both offer clinical data, as well as help in the correct utilization of those data with less errors, saving staff long periods of managing complaints and administrative bottlenecks [5]. The algorithms are an essential format for sharing medically pertinent information, and the sharing of that information can be all the more comprehensively utilized in the event that they are effectively accessible in a provision for using calculations assessing clinical parameters concerned with clinicians, teachers, and specialists (Fig. 15.3) [6]. Providing result by observing test results and the previous history for a diagnostic approach process can be high-pressured because the situation can be serious. Dealing with uncertainty is difficult. Things are typically not straightforward when patients' disease conditions are constantly progressing and evolving over time. Some techniques used to solve these problems are described below.

The naive Bayes algorithm uses Bayes theorem to arrive at a conclusion which is beneficial to the user. It uses Bayes theorem to predict the occurrence of an event using the given factors. It calculates the probability of an event and gives out a decision based on the obtained probability [7]. The principle of conditional probability is used, where a variable X is taken for an outcome of whether the event will happen or not and we take the variable Y for the actual condition itself. Then, we calculate the probabilities of the individual events and also the probability of the condition over the event and multiply it by the probability of the event taking place and divide the obtained value with the probability of the occurrence of the condition to get the final value which will tell us the probability of the event happening under the given condition. We can also consider the probability of the event not happening under the given condition and then compare the probabilities of the event happening or not happening under the given conditions to obtain a conclusion. If the probability of the event happening over the given condition is more than the probability of the event not happening under the given condition, we can conclude that the event is likely to occur. If the probability of the event not happening over the given condition is more than the probability of the event happening over the given condition, we can conclude that the event is unlikely to occur. Finally, if the probability of the event happening over the given condition and the probability of the event not happening under the given condition are equal, we can conclude that the event has a 50% chance of either happening or not. Based on the likelihood of the problem, we can make a decision regarding our problem. If the event is likely, we can

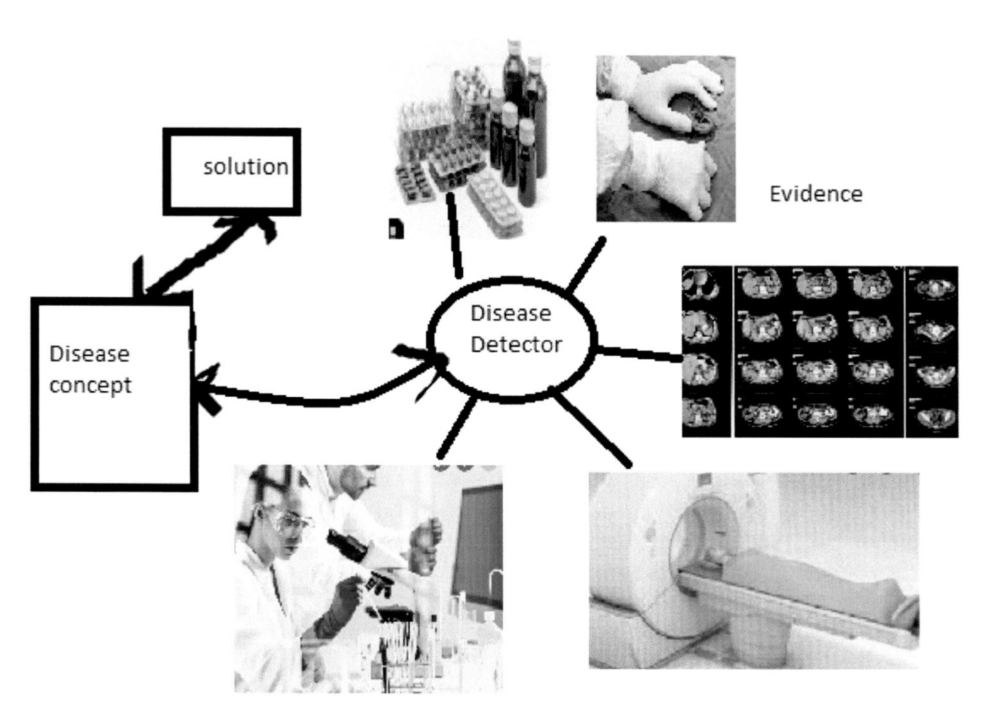

FIGURE 15.3 Aspects of a disease detector for clinical decision-making.

continue with the current decision. If the event is unlikely, we can think about the conditions and make some changes to the initial condition and run the algorithm again to get the probability with the new conditions and, once we are satisfied with the outcome, we can make a decision. If the probability of the event occurring is 50%, we can either continue to make the decision or we can upgrade the condition so that we get an optimal outcome.

The naive Bayes algorithm is an easy-to-use and fast algorithm since it is relatively fast as it uses the previously observed probabilities or takes the probabilities directly from the user [8]. The speed of the algorithm depends on the size of the sample data given to the algorithm as it has to consider the individual probabilities of each type of element. The naive Bayes algorithm is useful in many scenarios and is flexible, which makes it a very useful decision-making algorithm in almost all situations. The naive Bayes algorithm also has its own disadvantages. It only takes in the probabilities of events, so it does not consider all the factors that are or may be affecting the outcome, therefore we won't have the possibilities of the tasks which need additional consideration. The algorithm also does not know the relations between the conditions and the event occurrences, and so it becomes somewhat unreliable when we have a complex problem. The naive Bayes algorithm also struggles whenever there is no record of the previous cases and it cannot understand the problem, further resulting in an error and we are unable to come to a conclusion. Even if there is a single instance or a couple of instances, the naive Bayes algorithm will not perform at optimal efficiency as some problems need multiple instances in order to include all the possible outcomes.

Decision trees are used in machine learning algorithms for decision-making. In decision trees, we select the specific conditions that are required for the problem we are trying to solve [9]. In decision trees, we usually prepare multiple conditions, which lead to further execution of the algorithm. We take an initial condition and pass the problem through the initial condition node which is called the decision node. The result of the decision node will be in a yes/no format. We then have left and right subtrees from the decision node. We call the nodes below a decision node the leaf nodes of that decision node. The decision tree will be traversed depending on the output of the decision node, we can select yes for the left subtree and no for the right subtree, or vice versa, depending on the problem or the understanding of the user. When the condition is passed through the decision node, it will enter either the left or right subtree, depending on the obtained result, and then the condition will be processed again in the right or left subtree, which will be the current decision node. The process of the condition passing through a decision node and then moving onto one of the subtrees depends on the output of that specific decision node and continues till there are no further nodes in the decision tree. At that point, we will have arrived at a final conclusion at the end of the tree, where we can conclude the result of the decision tree. The decision tree is a supervised algorithm. It is relatively efficient at many problems since it considers different solutions for the given problem statement and changes the path of considering the solution accordingly and can include possibilities that may rise in the future based on future conditions and can divert a problem toward the optimal solution [2]. The decision trees are understandable by the users as the users provide the required conditions for a specific problem statement. They are simple for execution, the system just checks whether the requirements for an output are met and sends the problem statement to the next node. This can save a lot of system resources since it rules out unnecessary conditions and proceeds in a path that is optimal for the problem statement. For a decision tree of "N" levels, we can get the conclusion with a worst case scenario at "N" executions. For example, if we have a decision tree with four levels, with a worst case scenario (where the whole tree will be traversed), we will only have four executions as the problem will be checked once with the condition at every level. We also have a couple of disadvantages with the decision tree algorithm. With the decision tree algorithm, the user has to include every condition that they want to check with the problem statement. As such, the decision tree will be more complex to create than to execute. It still is highly efficient whenever we have to apply the decision tree on a large database. The decision trees are useful on large-scale industrial use cases, where the same conditions might be applied to a vast number of entities. The decision tree algorithm will become a bottleneck for some complex problems where we do not just have two outcomes at a certain decision node since most of the decision trees that are used in real life are binary and we may have some difficulty if we have to go back and forth for a special problem statement to check the conditions on other binary trees [10]. Therefore, decision trees are complex at the user's end and are not very suitable for complex problems and need thorough verification of the placement of the decision nodes since placing a node or two at the wrong place might cause the entire decision tree to malfunction and the decision made by the decision tree will be rendered useless. In rare cases, we can end up with faulty decisions that might lead to fatal losses which are clearly not beneficial to the user.

1.1 Research gaps

Some researchers have examined the doctor's grasp of basic concepts and decision-making abilities in issues related to coronary heart disease and hypercholesterolemia [11]. The investigation was implemented in two stages:

(1) An interview in which two clinical issues were discussed;
(2) A session in which patients answered to a sequence of queries.

The queries pertained to the discussion on a theme related to factors which may give rise to risk, diagnostic criteria for finding a high level of lipid, and continuous and accurate diagnoses for lipid disorders. The outcome showed that many doctors had gaps in their conceptual understanding. Specifically, most of the physicians illustrated a lower level of knowledge on the primary genetic ailments that lead to coronary heart problems, as well as deficiencies in their understanding of the secondary causes of hypercholesterolemia [12]. Many of the doctors overestimated the lipid value intervals for determining patients at high risk. Physicians had no difficulty in diagnosing the first patient problem of familial hypercholesterolemia, but failed to identify the problem of elevated lipids secondary to hypothyroidism. This implies that either empirical knowledge of the cooccurrence of hypercholesterolemia and thyroid issue or a mechanistic understanding of the way in which the two factors are connected with each other.

1.2 Major findings

1. The process of healthcare decision-making is affected by human biases and heuristics chosen.
2. The traditional mechanisms and research propelled by heuristics and biases do not adequately support the decision-making process [13].
3. Clinical decision-making and problem-solving research deploy various types of methodological and theoretical strategies ranging from multiple historical conventions to diagnose the same issue and resulting in hugely different conclusions.
4. Clinical decision-making in a "real-world" context imposes unique demands (e.g., stress and time pressure) on the decision process and these features are not completely captured in most laboratory decision investigations [14].
5. Clinical decision-making in realistic settings is mostly attributed by the sequence of examinations of a single choice rather than the determination of a fixed set of alternative options [15].
6. Clinical decision-making by a team is due to emergent characteristics that cannot be made by individual decision makers.
7. Technologies facilitate the decision-making approach in different and often counterintuitive aspects that can deliver unintended outcomes.
8. Clinical decision-making technology does not just mediate or augment the decision-making mechanism rather it reorganizes the practice of clinical decision-making.

1.3 Motivation

The main motivation for employing ML algorithms in clinical decision-making is to rule out limitations in terms of efficiency or any type of bias occurring during clinical decision-making [16]. Irrespective of the traditional framework with robust underlying principles to make clinical decisions, the cost and effort related to these are considerable, including technologically. Popular e-healthcare systems aim to lower the cost of treatment for a patient and also infer efficient healthcare decisions. As more and more medical institutions are employing data scientists to leverage the usage of technology while making clinical decisions, we are motivated to exploit the aspects of deep learning-based approaches so that the amount of data does not affect our operating principles [17]. The no-regret deep learning mechanism enforces an approach where huge improvements in decision-making policies are possible.

2. Related works

Nowadays, most medical-related works are carried out with the help of computer technology, and in the future there is the opportunity to operate on patients with the help of robots through the rapid improvements in technology. For this approach we need to be able to solve problems with strong and efficient techniques with the help of diagnosis, medical test selection, therapy, and prognosis, to name a few.

Why are algorithms important in health care?

There are many algorithms used in health care, but most practitioners use only a small amount of medicine algorithms routinely (Fig. 15.4). A medical treatment algorithm can assist in standardizing the selection of patient care plans and answering a number of queries.

During monitoring,decisions are taken if changes occur.

The logical parameters concerned with the patient are verified by the data aggregators and results are compared with respect to the patient history. If any other information is needed, it is also retrieved to start the diagnosis. It provides heuristic models as an outcome.

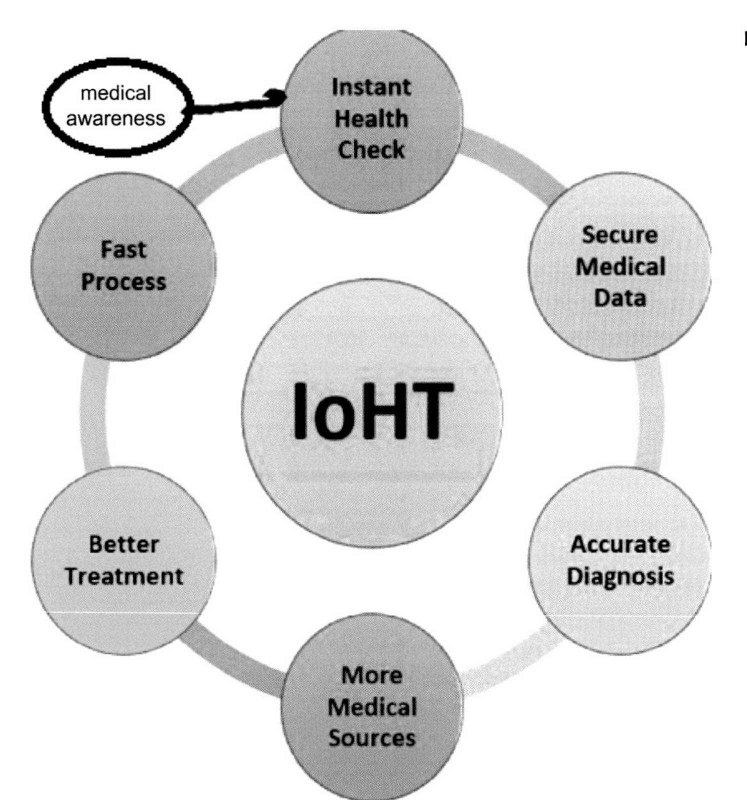

FIGURE 15.4 Ecosystem of the Internet of Healthcare Things.

The efficiency of the proposed mechanism is based on generation of probability models and it is obtained from below steps [18]:

(a) Trigger medication orders
(b) Input current medications
(c) The model performs forecasting based on the changes observed in the patient health record.

The role of the Internet of Things (IoT) for decision systems for healthcare treatment is discussed next.

The tremendous growth of the IoT in health systems has been enabled through IoT advancements with newly emerging technologies such as artificial intelligence, machine learning, big data, etc.

The IoT will not only help to cure patients, but also help workers in the healthcare system [19]. This type of revolution in health care enhances various areas as it reduces the costs and prevents medical errors, saving patients lives in critical situations. For example, if we take any IoT device as an example, they helps to monitor the patient's health condition constantly, and automatically update the health reports to medical staff. In the case of an emergency the medication can be immediately given by the IoT device itself.

The main purpose of the IoT in health systems is to diagnose health issues earlier, as many people die because they do not receive the proper treatment at the right time. As a current example, we can take COVID-19, where many people lost their lives due to the panic situations and without proper guidance. Therefore, in order to deal with such circumstances in future the IoT can help a great deal [20].

Telecare can be used as the primary form of IoT in a healthcare example. With this technological advancement, patients can be monitored and, according to the situation, treatment can be provided remotely through cameras, sensors, and with such electronic actuators (for example, endoscopy, MRI).

Examples of IoT applications include:

- Using sensors to upload the patient health status to a nearby hospital from the home or an ambulance;
- IoT wearables can be introduced for use at home by patients and elderly persons to communicate with the healthcare communities;
- Temperature sensors can be used to detect temperature levels;
- Pressure sensors are used to sense the human body pressure and are mainly based on piezo resistive technology, and other sensors include heartbeat sensors;
- IoT devices that are tagged with sensors are used for tracking monitoring equipment such as oxygen pumps, wheelchairs, defibrillators, nebulizers, etc.;
- Robotic surgeries can be performed using the IoT.

With a view to deploying machine learning clinical decision support systems (ML CDSS), there are multiple crucial viewpoints to analyze [21]. Primarily, if the application of an ML CDSS results in any risks to the diagnostic or therapeutic procedure, these risks have to be made transparent and the patient should know about the risks beforehand. Therefore, the risks may result in a high rate of false-positive or false-negative outcomes, or problematic secondary findings. In the case of adverse situations arising due to application of ML CDSS, an alternative procedure should be easily accessible to the patient. For example, due to the secondary findings, provisions must be available to use other techniques, such as medical genetics and genetic counseling, where a matured professionalized practice where the patient's "right not to know" has been intensively discussed and thoroughly implemented [22]. Also, in case of a privacy risk to the patient being elevated due to usage of ML CDSS, the patient's stress should also be resolved in a professional manner. In addition, this patient tension due to privacy leakage depends on the trust in the used framework on ML and AI in general, i.e., techniques of trustworthy or explainable AI. The physician should also verify and validate (and reaffirm at regular intervals) that the ML CDSS represents (or outperforms) her or his own medico-scientific knowledge to a fair degree and that its use improves the outcome of the medical practice. If this is not confirmed, then it is not suitable to use ML CDSS. Yet another threat to the validity of ML CDSS includes areas such as the patient's life situation, daily routine, cultural background, and daily healthcare practices, which must also be taken into consideration [23]. These factors are collectively described as patient autonomy.

3. Proposed mechanism

For a decision-making system, we intend to use interactive no-regret learning. We model the decision-making problem in a scenario where there are two entities. The learner who wants to initialize a secure policy so as to minimize their security loss and an adversary who generates an accurate result with a better option [24]. This function value indicates how the best decision system is chosen. Naturally, the objective is to select the best decision system with minimum error value. In the no-regret learning approach, the learner updates the decision system as and when they observe a minimum error function value.

We formulated the decision policy as shown below.

Table 15.1 shows the effect of the single value attribute in a patient health record on the performance of the algorithm using such dataset. The benefit increases for multivalued attributes of a patient data. Such dataset helps in training the model used in the algorithm to a greater extent.

We can observe that in none of the rounds do we obtain an optimal result. Therefore, we proceed to formulate the factor of regret as given below.

$$\text{Regret} = \sum_{t=1}^{T}(\text{Lt}(\text{D}\phi\text{t})) - \min\sum_{t=1}^{T}((\text{Lt}(\text{D}\phi*)) \tag{15.1}$$

In other words, the regret for the learner is the difference between the sum of all security loss function values of all the experts and the minimum loss function chosen from all the rows with hindsight.

This minimum value of the security loss becomes the best expert in the game. For our example, we can see that expert 3 is the best expert with hindsight or for the chosen decision.

We are now in a position to calculate the average regret as:

Average Regret = (1/T) * Regret

As, T->∞, Average Regret->0 (No Regret)

To select the best expert in hindsight, The Follow the Leader (FTL) technique is used to select the best expert (decision policy) at every round of the game. So that,

$$\text{D}\phi\text{t} = \arg\left(\min\sum_{i=1}^{t-1}((\text{Li})(\text{D}\phi))\right) \tag{15.2}$$

In the above equation, $\text{D}\phi$ represents the lowest total decision value.

An illustration of the FTL mechanism is provided next.

In Table 15.2, we keep track of the cumulative loss of each expert in all the rounds of the game, and we also keep track of the average regret.

After round 1, we choose the expert with minimum options, i.e., expert 2 with a result of 0.2 units. Now, in round 2, we get the below function (Table 15.3).

We stay with expert 2 for round 3 as the leader. Proceeding to round 3, we get Table 15.4.

We now change the leader; expert 3 becomes the new leader. Subsequently in rounds 4 and 5 we get Table 15.5.

We can observe that, more or less, the average regret values do not change after rounds 3, 4, and 5; it is still 0.53. Hence, we can infer that expert 3 has the best decision policy with no regret. Now we try to generalize the decision policy in the FTL algorithm after all rounds of the game as:

TABLE 15.1 Initial decision values for each expert.

		Decision values at each round			
No. of experts	**Decision policies**	**L1**	**L2**	**L3**	…
E1	Dϕ1	0.1	1.0	1.0	
E2	Dϕ2	1.0	0.1	0.1	
E3	Dϕ3	0.1	0.1	0.5	
…	…	…	…	…	
E4	DϕN	1.0	1.0	1.0	

In round 1, we choose expert 1 (E1) and we get a lower error of 1.0. In round 2, if we choose expert 1, we get the result = 1.0 and in round 3, we choose expert 3 and obtain a result = 0.5.

TABLE 15.2 Regret values after the FTL mechanism is applied.

		Decision system matched at each round			
No. of experts	**Decision policies**	**L1**	**L2**	**L3**	**Total regret**
E1	Dϕ1	1.0			1.0
E2	Dϕ2	0.2			0.2
E3	Dϕ3	0.5			0.5

Average regret = 0.80.

TABLE 15.3 Regret values after round 2.

		Decision system matched at each round			
No of experts	**Decision system**	**L1**	**L2**	**L3**	**Total regret**
E1	Dϕ1	1.0	0.5		1.5
E2	Dϕ2	0.2	0.5		0.7
E3	Dϕ3	0.5	0.2		0.7

Average regret = 0.40.

TABLE 15.4 Regret values after round 3.

		Decision system matched at each round			
No of experts	**Decision system**	**L1**	**L2**	**L3**	**Total regret**
E1	Dϕ1	1.0	0.5	0.5	2.0
E2	Dϕ2	0.2	0.5	1.0	1.7
E3	Dϕ3	0.5	0.2	0.2	0.9

TABLE 15.5 Regret values after rounds 4 and 5.

		Decision system matched at each round					
No of experts	**Decision system**	**L1**	**L2**	**L3**	**L4**	**L5**	**Total regret**
E1	Dϕ1	1.0	0.5	0.5	1.0	0.5	3.5
E2	Dϕ2	0.2	0.5	1.0	0.2	1.0	2.9
E3	Dϕ3	0.5	0.2	0.2	0.5	0.2	1.6

Average regret = 0.53.

$$\sum_{t=1}^{T}(\mathrm{IT}(\mathrm{D}\phi t+1)) \leq \sum_{t=1}^{T}(\mathrm{IT}(\mathrm{D}\phi *)) \tag{15.3}$$

In the above equation, the left-hand-side term denotes the decision system result of the leader after all rounds of the game, and the right-hand-side term denotes the result of the best expert.

4. Experimental results

As observed from Fig. 15.5, the proposed mechanism performs decently when the dataset size increases. It can be observed that for nearly 800 records, the NRDL (No Regret Deep Learning) approach achieves a magnificent learning rate of 0.8, which is noteworthy and such types of support in terms of decision-making provide a huge amount of trust and confidence among patients.

From Fig. 15.6, we can infer that among the current popular methods used in ML for making clinical decisions, the proposed approach outperforms the other approaches. The logistic regression method is least popular as is evident from the graph, and the other two techniques, naïve Bayes and support vector machine-based technique, perform well when trained

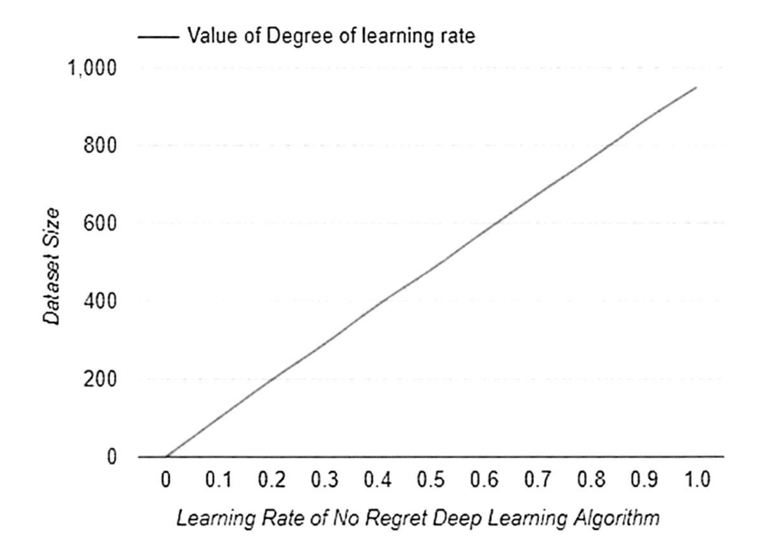

FIGURE 15.5 Plot of learning rate versus size of dataset.

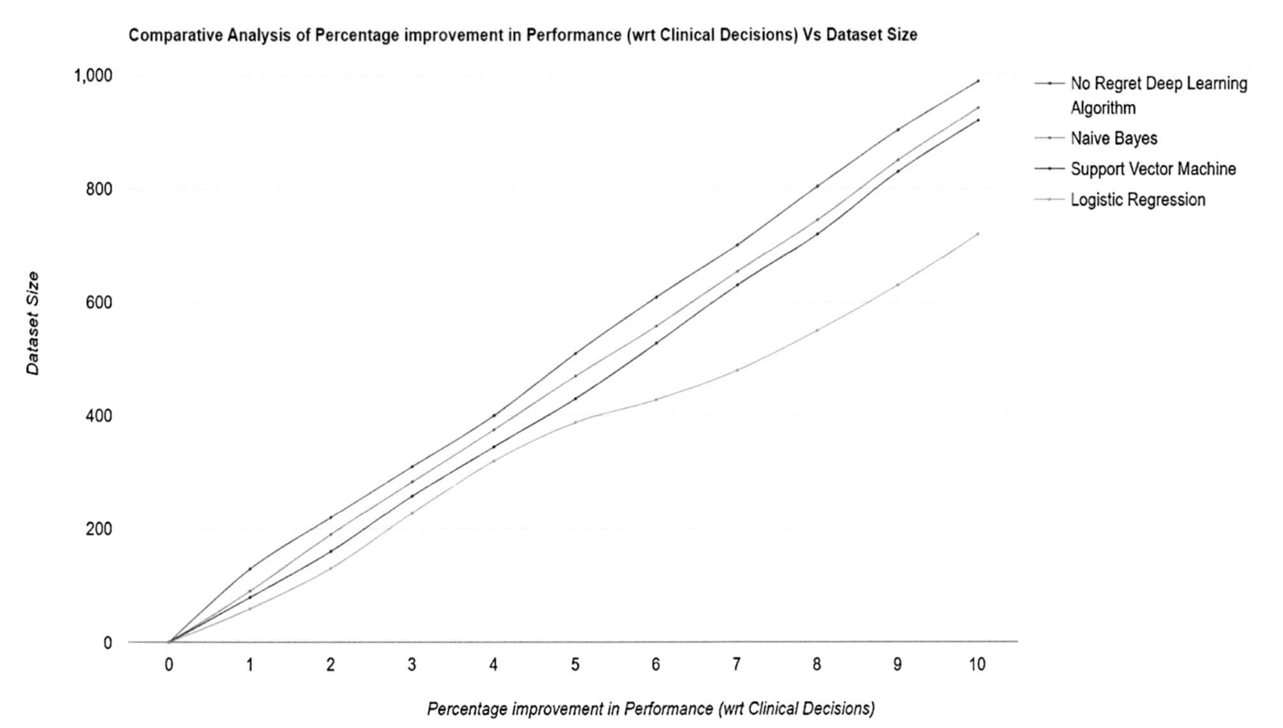

FIGURE 15.6 Comparative analysis of popular algorithms used for clinical decision-making.

with a dataset of over 700 records. We can observe that our proposed mechanism is almost 2% better than its nearest competitor mechanism.

The merits and features of the proposed approach include:

1. The proposed mechanism has the ability to solve a complex task with less prior knowledge due to its derivative ability to learn from various low-level abstractions of data.
2. The cost of feature engineering across the massive amount of data is reduced considerably.
3. Imbalanced clinical tasks for making decisions are handled efficiently by the NRDL algorithm due to the iterative nature of the mechanism.
4. Our proposed mechanism advocates the formation of clinical groups to decrease the cardinality of both diagnoses and medications during the decision-making process.

5. Future directions

Due to the assumptions we have considered while employing the FTL mechanism there may be an issue with the stability of FTL predictions. Suppose that the adversary is powerful enough to propel the regret of any deterministic algorithm to O(T) by assigning the maximum accuracy results, then we must face this challenge. We propose to work with a hedging technique to address this. In this process, instead of following the leader among the experts, we use the following steps.

Step 1: At the first round, $t = 1$, set weight of an expert E_i as $W1_i = 1$
Step 2: At round t, we choose expert E_i with probability $W_{ti}/\sum W_{ti}$
Step 3: Update weight for expert I (if decision result is more, decrease the number of test results)

$$W_{ti} + 1 = W_{ti}\exp(-\eta Li)(D\phi i)$$

Stability is a key aspect. If the decision system of all the expert policies oscillates then we cannot obtain an average regret value of zero. Therefore, simply selecting the best hindsight result may be unsuitable. In future, we plan to work on the area where optimal hedging is possible against an adversary who is shifting the decision values. Also, we need to formalize approaches where the tradeoff between exploration and exploitation is balanced. Because ,in real-life scenarios, the policies are finite, we may or may not ensure an average regret value of zero. In future, this aspect should be worked on.

6. Conclusion

In our chapter, we have introduced the novel mechanism of no-regret learning in deep architecture to minimize errors while making clinical decisions. The no-regret algorithm provides sublinear growth with every input sequence in the deep architecture, thereby increasing the overall prediction rate of the learning model. In this chapter, we have modeled the decision system to solve problems in healthcare systems as an interactive no-regret learning approach using a game setup. In a decision system, a change to the states which force the user to input some sensitive information in turn changes the results with different options. Due to this uncertainty in data distribution, expected regret values may not be minimal. Therefore, the no-regret learning approach is suitable and the experimental results have shown it performs better than the popular methods such as clinical decision systems using machine learning techniques, etc. Our chapter provides a pathway to researchers working in the area of game theory and designing decision systems with highly accurate decision results using no-regret learning. This system also provides awareness of the different option we have available. This area is nascent and the future potential is immense due to the highly adaptable nature of the algorithm we have proposed.

The major key findings of our proposed approach are as follows:

1. A deep learning no-regret learning framework will help in improving the effectiveness of clinical decisions to a significant degree.
2. Improving the healthcare decisions through value-added decision support with minimum cost and low training data for the learning agent is possible.
3. The performance of deep learning with a no-regret learning approach is better than its peer algorithms by almost 2% −3%.

The advantages of the proposed approach are manifold as stated below.

1. Auditing the training-learning procedure using deep learning is highly transparent.
2. The efficiency of decision-making is established by applying the no-regret learning algorithm to suit the dynamic context of the problem statement.
3. The proposed technique can be integrated with e-health analytical systems for rendering better healthcare services.

Limitations:

1. The empirical frequency of the decision-making agent affects the amount of regret, which is not included as a learning parameter in our proposed algorithm.
2. The approachability of the algorithm is constrained due to the FTL (Follow the leader) principle. Due to this, the dynamic decision-making ability may be affected.
3. The training data available for making clinical decisions are not purely longitudinal data and are curated to a reach a possible fair degree of accuracy for efficient learning purposes, so it is biased in a certain way.

References

[1] A.M. Antoniadi, Y. Du, Y. Guendouz, L. Wei, C. Mazo, B.A. Becker, C. Mooney, Current challenges and future opportunities for XAI in machine learning-based clinical decision support systems: a systematic review, Applied Sciences 11 (11) (January 2021) 5088.

[2] H. Haick, N. Tang, Artificial intelligence in medical sensors for clinical decisions, ACS Nano 15 (3) (February 23, 2021) 3557–3567.

[3] N. Panigrahi, I. Ayus, O.P. Jena, An expert system-based clinical decision support system for Hepatitis-B prediction & diagnosis, Machine Learning for Healthcare Applications (April 12, 2021) 57–75.

[4] A. Karthikeyan, A. Garg, P.K. Vinod, U.D. Priyakumar, Machine learning based clinical decision support system for early COVID-19 mortality prediction, Frontiers in Public Health 9 (May 12, 2021) 626697.

[5] T. Chen, C. Shang, P. Su, E. Keravnou-Papailiou, Y. Zhao, G. Antoniou, Q. Shen, A decision tree-initialised neuro-fuzzy approach for clinical decision support, Artificial Intelligence in Medicine 111 (January 1, 2021) 101986.

[6] F. Shaikh, J. Dehmeshki, S. Bisdas, D. Roettger-Dupont, O. Kubassova, M. Aziz, O. Awan, Artificial intelligence-based clinical decision support systems using advanced medical imaging and radiomics, Current Problems in Diagnostic Radiology 50 (2) (March 1, 2021) 262–267.

[7] C. Panigutti, A. Perotti, A. Panisson, P. Bajardi, D. Pedreschi, FairLens: auditing black-box clinical decision support systems, Information Processing & Management 58 (5) (September 1, 2021) 102657.

[8] J.M. Schwartz, A.J. Moy, S.C. Rossetti, N. Elhadad, K.D. Cato, Clinician involvement in research on machine learning–based predictive clinical decision support for the hospital setting: a scoping review, Journal of the American Medical Informatics Association 28 (3) (March 1, 2021) 653–663.

[9] D. Wang, L. Wang, Z. Zhang, D. Wang, H. Zhu, Y. Gao, et al., "Brilliant AI doctor" in rural clinics: challenges in AI-powered clinical decision support system deployment, in: Proceedings of the 2021 CHI Conference on Human Factors in Computing Systems, May 6, 2021, pp. 1–18.

[10] S.M. Shortreed, E. Laber, D.J. Lizotte, T.S. Stroup, J. Pineau, S.A. Murphy, Informing sequential clinical decision-making through reinforcement learning: an empirical study, Machine Learning 84 (1) (July 2011) 109–136.

[11] S. Liu, K.C. See, K.Y. Ngiam, L.A. Celi, X. Sun, M. Feng, Reinforcement learning for clinical decision support in critical care: comprehensive review, Journal of Medical Internet Research 22 (7) (July 20, 2020) e18477.

[12] J. Futoma, M.C. Hughes, F. Doshi-Velez, Popcorn: partially observed prediction constrained reinforcement learning, arXiv preprint arXiv:2001.04032 (January 13, 2020).

[13] C. Yu, J. Liu, S. Nemati, G. Yin, Reinforcement learning in healthcare: a survey, ACM Computing Surveys 55 (1) (November 23, 2021) 1–36.

[14] J.A. Li, D. Dong, Z. Wei, Y. Liu, Y. Pan, F. Nori, X. Zhang, Quantum reinforcement learning during human decision-making, Nature Human Behaviour 4 (3) (March 2020) 294–307.

[15] O. Gottesman, F. Johansson, J. Meier, J. Dent, D. Lee, S. Srinivasan, et al., Evaluating reinforcement learning algorithms in observational health settings, arXiv preprint arXiv:1805.12298 (May 31, 2018).

[16] W.Y. Ahn, N. Haines, L. Zhang, Revealing neurocomputational mechanisms of reinforcement learning and decision-making with the hBayesDM package, Computational Psychiatry 1 (October 2017) 24.

[17] M.L. Pedersen, M.J. Frank, G. Biele, The drift diffusion model as the choice rule in reinforcement learning, Psychonomic Bulletin & Review 24 (4) (August 2017) 1234–1251.

[18] S. Nemati, M.M. Ghassemi, G.D. Clifford, Optimal medication dosing from suboptimal clinical examples: a deep reinforcement learning approach, in: 2016 38th Annual International Conference of the IEEE Engineering in Medicine and Biology Society (EMBC), IEEE, August 16, 2016, pp. 2978–2981.

[19] M.L. Littman, Reinforcement learning improves behaviour from evaluative feedback, Nature 521 (7553) (May 2015) 445–451.

[20] S. Levine, A. Kumar, G. Tucker, J. Fu, Offline reinforcement learning: tutorial, review, and perspectives on open problems, arXiv preprint arXiv:2005.01643 (May 4, 2020).

[21] A. Raghu, M. Komorowski, L.A. Celi, P. Szolovits, M. Ghassemi, Continuous state-space models for optimal sepsis treatment: a deep reinforcement learning approach, in: Machine Learning for Healthcare Conference 6, PMLR, November 2017, pp. 147–163.

[22] Y. Liu, B. Logan, N. Liu, Z. Xu, J. Tang, Y. Wang, Deep reinforcement learning for dynamic treatment regimes on medical registry data, in: 2017 IEEE International Conference on Healthcare Informatics (ICHI), IEEE, August 23, 2017, pp. 380–385.

[23] Z. Zhu, K. Lin, J. Zhou, Transfer learning in deep reinforcement learning: a survey, arXiv preprint arXiv:2009.07888 (September 16, 2020).

[24] V. François-Lavet, P. Henderson, R. Islam, M.G. Bellemare, J. Pineau, An introduction to deep reinforcement learning, Foundations and Trends in Machine Learning 11 (3–4) (December 19, 2018) 219–354.

Chapter 16

Symptom-based diagnosis of diseases for primary health check-ups using biomedical text mining

B. Rakesh[1] and Sukanta Nayak[1,2]

[1]*School of Advanced Sciences, VIT-AP University, Amaravati, Andhra Pradesh, India;* [2]*Center of Excellence, AI and Robotics, VIT-AP University, Amaravati, Andhra Pradesh, India*

1. Introduction

Recent development in technology provides a sustainable solution to many complicated social problems. Decades back, due to the lack of advancement in information technology, the problems are supplied with small data. As such, many techniques [1–5] were adopted to handle the same. But, over the past two decades, huge data are collected and kept in the warehouse as information, knowledge, and values. So, there is a need for different tools to handle big data. Hence, the current developments provide various tools with the existing ones which can ease our task. In this context, one can address important social issues, that is, population. There are plenty of social problems that arise with the rise in population. One of those is poor management of health check-ups in health centers.

A current report [6] says India possesses second place in the world based on a population with more than 1.38 billion people. Overpopulation creates both challenges and opportunities. In the healthcare system, the challenges are more than opportunities. The proper availability and management of primary health care are one of them. Due to the huge population, the number of patients increases and makes an infinite queuing system. To take an appointment with the doctors, patients wait a long in queue at the health center which may hamper the conditions of the patients. As such, there is an essence of automation method which can solve the problem to reduce the waiting time with precise appointment order. Moreover, the costly health care facility and the high price of doctors' consulting fees impact revenue generation. The increased cost discourages many patients to acquire routine health follow-ups. Therefore, sometimes it creates dangerous consequences. The main victims are poor people who cannot afford the same. It is observed that the ignorance of early body check-ups based on symptoms is the cause of more than a quarter percent of the population leads death. Besides, the unavailability of hospitals and doctors in rural areas is one of the other concerns. From various sources, it is predicted that a shortage of approximately 100,000 doctors by the year 2030. The same may be alternatively managed with the help of technology. Telehealth is one of the technologies. Besides, live streaming, store-and-forward imaging, and biomedical text mining techniques can be used to analyze the data of patients as well as remote location patient diagnostics approach shall be helpful to reach the healthcare in remote places.

In a broader sense, a disease is defined as a type of disposition regarded as adversely affecting a person [7,8]. An organism is said to be diseased if it shows a sign of abnormal state. To know the status of the same, primary diagnoses of health play a key role [9,10]. This is a condition made after study that mainly responsible for the admission of the patient to the hospital for care. Usually, healthcare systems use this information to deduce the complication occur in the treatment of a patient. The information contains symptoms reported by the patients, the type of medications, treatment procedures, and resources available at hospital. This information helps to provide a precise Mediclaim. A symptom is an indication of a disease that appears to the patient. Whereas, a sign is a display of a disease that the physician perceives. If severe symptoms are noted, then the patient is referred for doctor's consultancy.

Deep Learning in Personalized Healthcare and Decision Support. https://doi.org/10.1016/B978-0-443-19413-9.00023-0

In light of this, the biomedical text mining technique is one of the best analyzing tools to recognize hidden relationships and drift in the data. Biomedical text mining includes the study and methodology with the application of text mining in the field of molecular biology and biomedical. Further, research point of view, it hybridizes the concepts of natural language processing (NLP) with bioinformatics and medical informatics. Various strategies of the same are frequently used in biomedical. Based on the theme and contents of the topics, biomedical documentation are either classified or clustered. A manual procedure is followed for document classification [11], whereas algorithm-dependent and unique groups are followed in clustering [12]. Supervised and unsupervised procedures are adopted to perform these two tasks. The main objective of these tasks is the formation of documentation subsets referring to the distinct features. As such, the biomedical clustering of documents mainly depends on the k-means clustering algorithm [12]. In this context, many types of research are reported by the researchers, and few works are reported in Refs. [13,14]. Isolating texts, phrases, and sentences depending on the core arguments are the basis of their investigation. Due to the importance and utility of biomedical text mining, it got popular and is needed by practitioners for clinical use [15]. In Ref. [16], a concise study of the development and application of NLP is mentioned with free-text clinical notes related to chronic diseases.

The above literature review reveals that an automation process has an edge over the traditional fixing of the long queuing system. As such, NLP plays a vital role to identify the diseases from the symptoms that ease the referral processes to the particular doctor. Hence, in this chapter, Section 2 provides a concise knowledge of NLP with the procedure presented in flow charts. Then, Section 3 includes the architecture of the primary diagnosis of diseases based on symptoms. A sample case study is discussed in Section 4 to show the effectiveness of machine learning approach to diagnosis of the disease. Finally, the results and discussions are reported in Section 5.

2. Natural language processing

This section includes NLP as a tool to handle linguistic information in the use of the primary health care system to manage the infinite queuing system of patients. NLP is one of the core areas of artificial intelligence (AI) that makes the human language capable of being understood by machines. As such, the task of an interpreter becomes easier. NLP possesses the potential of computer science skills and linguistics to study the structure and rules of different languages and create intelligent systems to apprehend, analyze, and reveal the meaning of a paragraph and/or a phrase. In this regard, text classification plays a major role in NLP as a foundation in various complex tasks. Text classification classifies a full text using predefined labels [17]. The broad applications of text classification can be found in the biomedical field such as automatic diagnosis [18,19], biomedical indexing [20,21], tweets classification [22–24], safety reports classification [25], etc. During the course of applications, text classification can be named multinomial or multiclass text classification and multilabel text classification. Multinomial or multiclass text classification is associated with only one of the labels. In other words, in this process, labels are mutually exclusive. Whereas, in multilabel text classification, one or more labels can be used to assign each text. In Medical Subject Headings (MeSH) indexing, generally, new publications are provided with a good number of relevant MeSH terms [26]. Usually, multilabel text classification is more difficult than multinomial classification due to the assignment of an indeterminate number of labels for each textual document [27]. Binary relevance is one of the traditional techniques to solve the multilabel text classification problems. It divides the problem into a number of independent binary tasks (one for each label). However, the independence of each label is assumed in this approach [26,28,29]. Label power set, which creates binary classifiers for each label combination, is able to model potential correlations between labels [30]. However, both two approaches could have low throughout when the number of different labels becomes extremely large. There are also some other algorithms for multilabel text classification, including learning to rank [26], classifier chains [31], etc. A review of multilabel learning algorithms can be found in Ref. [32].

A schematic diagram of NLP is shown in Fig. 16.1. The process involves texts and paragraphs as input which are undergone through a feature extractor and different features are extracted. Then, the same is sent to the machine learning algorithm execution. Based on the systems, labels are presumed. Using the training dataset with labels and machine learning algorithm a system is built which can be useful to predict the new data and its class.

As such, Fig. 16.2 describes the prediction process for NLP. For new data, the texts and paragraphs are collected, and then features are extracted through extractors. The same are classified through classifiers to give labels. Altogether, the procedures given in Figs. 16.1 and 16.2 compete with the process of prediction based on the model.

Common types of NLP tasks include syntactic and semantic analysis. These analyses break down the human languages into machine-understandable chunks. Syntactic analysis is also known as syntax analysis. It helps to find the syntactic structure of a text and/or paragraphs. Further, it identifies relationships between words and their dependency. Generally, the syntactic analysis is represented on a parse tree diagram. Whereas, the semantic analysis identifies the meaning of a

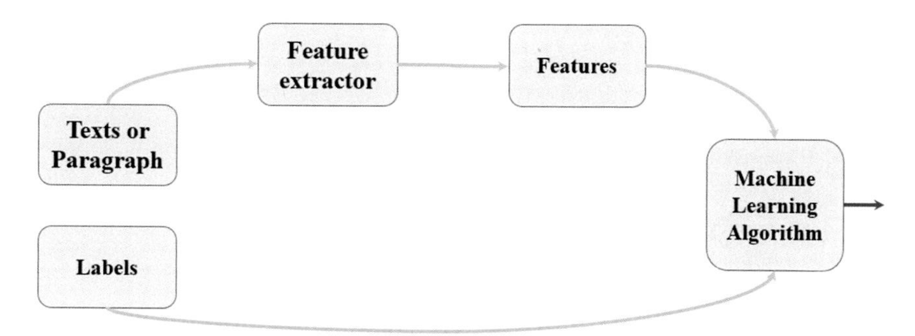

FIGURE 16.1 Schematic diagram of natural language processing (NLP).

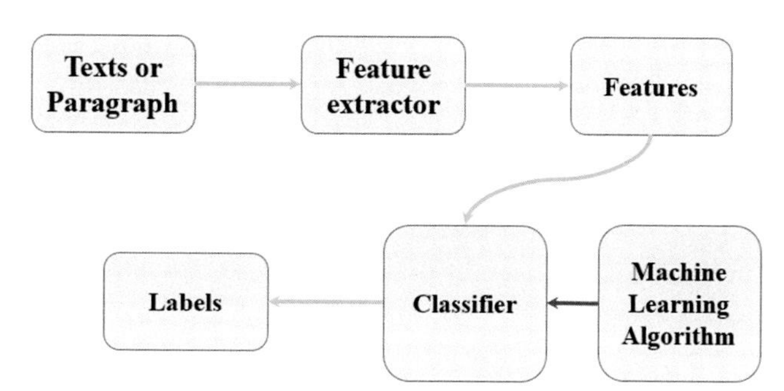

FIGURE 16.2 Prediction process in natural language processing (NLP).

language. It analyses the structure of a phrase, sentences, words, and their interactions to discover the meaning of the theme. In this context, one can consider the task tokenization that possesses both semantic and syntactic analysis.

The above concepts are useful for easy demonstration of NLP through many application problems. Hence, primary diagnosis of diseases based on the symptoms may be achieved. The next section presents the use of NLP in the primary diagnosis of diseases and its system architecture.

3. System architecture of primary diagnosis of diseases

The system architecture of primary diagnosis of diseases based on symptoms is provided in Fig. 16.3. The system architecture involves the following steps.

1. Based on the patient's report, the data are collected.
2. The collected data are processed and the data frame is created.
3. Features are extracted using feature extractor viz. count vectorization.
4. Then the same is executed through a machine learning algorithm viz. random forest.
5. New sets of data are collected for the prediction that undergoes different stages viz. feature vector, predictive model, and labels.
6. The predicted labels are verified by the doctor and incorporated into the library for further use.

The detailed implementation is described below.

3.1 Data collection

For the training set, data are collected based on the basic symptoms of the diseases from various data repositories and created a library. The main focus is to collect a huge data set for better prediction. Accordingly, the library is prepared as well as the new verified data are incorporated. It is noticed that few of the data are sentences and/or long phrases. Hence, data cleaning is implemented to get a structured data frame. In this process, high-frequency words that have no semantic values are filtered. A sample data subset is presented in Table 16.1. Different symptoms of patients with the type of diseases as labels are reported here. In this case, only two labels viz. fever and skin infection are considered. But depending on the user, a greater number of labels can be used for the investigation.

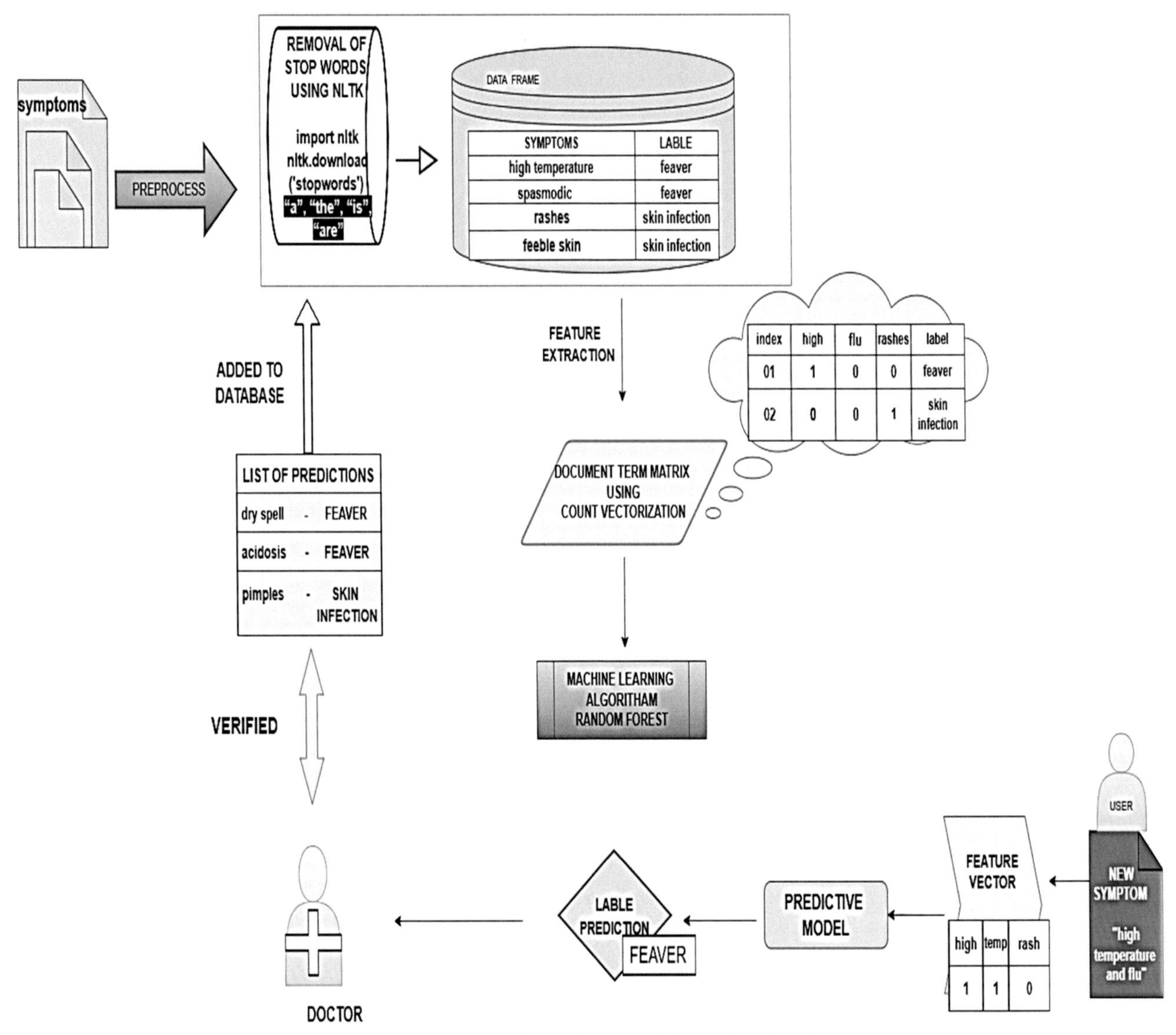

FIGURE 16.3 Flow chart of primary diagnosis of diseases based on the symptoms.

3.2 Lemmatization and stemming

The lemmatization procedure uses morphological analysis and vocabulary of words. It focuses to delete inflectional endings and giving the dictionary form of a word, which is called a lemma. Trivially, lemmatization is performed with a simple dictionary lookup. Lemmatization works fine for straightforward inflected forms. Otherwise, a rule-based system is needed. For example, a language containing long compound words needs a rule-based system. Those rules can be either manually created or learned automatically from an annotated corpus. Whereas, stemming is a method that withdraws the base form of words by deleting affixes. The words which are send out after or before the processing of natural language data from the stop lists are called stop words. There is no unique list of stop words by all NLP tools, rather, it is user-defined. Hence, depending on the purpose, a group of words is chosen as the stop words.

3.3 Feature extraction

Feature extraction is a tool to extract the features from the texts. Here, the features are extracted with the help of NLP. The count vectorization technique is adopted for the same. With the help of count vectorization, numerical feature vectors are obtained from the texts. Bunch of words takes a document from a corpus and converts it into a numeric vector. Count

TABLE 16.1 Sample data subset of the library.

Index	Symptoms	Diseases
01	Bathed	Fever
02	Drenched	Fever
03	Dripping	Fever
04	Glowing	Fever
05	Pulverizable	Skin infection
06	Rotted	Skin infection
07	Rotten	Skin infection
08	Rusted	Skin infection
09	Shivery	Skin infection
10	Short	Skin infection
11	Worn	Skin infection

vectorization involves counting the number of occurrences each word appears in a document. For execution purposes, Python's Sci-kit learn library is used to get the numeric vectors. Count vectorization uses the concept of frequency to convert a given text to a vector. The importance of it lies wherever a multiple given text occurs. In this regard, the count vectorizer forms a matrix that includes unique words. These unique words are represented by a column and each text sample from the document is a row. Each cell gives a count of the word in that particular text sample. Inside the count vectorizer, these words are given a particular index value. In the present study, "at" is indexed 0, "each" is indexed 1, "four" is indexed 2, and so on. A sample representation is shown in Table 16.2.

Use of optimization algorithm with count vectorization can make the process faster in the sense of computation than the nonvectorized implementation. The main focus is to get unique features out of the text to train with the numerical vectors. In addition, word embeddings can be used to find the word predictions, semantics and/or word similarities (Table 16.3).

Some of the popular methods to accomplish text vectorization are listed below.

1. Word2Vec
2. Binary Term Frequency
3. Normalized Term Frequency
4. Bag of Words (BoW) Term Frequency
5. Normalized TF-IDF

3.4 Algorithms

The next big task is the implementation of algorithms. Here, a decision tree classifier is used for the given data set, mainly on three things such as the predictability of each attribute, predictability of the whole dataset, and information gain. The root node is selected based on the attribute having lower predictability (or higher information gain). Then the dataset is split into subsets using attributes for which the predictability is minimum. The process goes on till substantial information is retrieved from the dataset.

TABLE 16.2 Representation of index values.

0	1	2	3	4	5	6	7	8	9	10	11
0	0	0	1	1	0	0	1	0	0	0	1
0	0	1	0	2	0	1	0	0	0	0	1
1	1	0	1	1	1	0	1	1	0	1	0

TABLE 16.3 Representation of index values with labels.

Index	High	Flu	Rash	Label
01	1	0	0	Fever
02	0	0	1	Skin infection

3.4.1 Random forest

Random forest is one of the supervised machine learning algorithms that is constituted of individual decision trees. This algorithm is popularly used in classification problems. Due to the higher accuracy across cross-validation, random forest is a well-known tool to handle classification problems. This classifier manages the missing values and maintains a good accuracy of a huge proportion of data. As the random forest is simple and easy to use, it is the most used machine learning algorithm. The step-by-step procedure is mentioned below.

Step 1: First, n number of random records are considered from the dataset. The data set possesses k number of records.
Step 2: For each sample particular decision trees are created.
Step 3: Each decision tree produces an output.
Step 4: Final output is decided on the basis of majority voting. Sometimes, averaging is also considered for classification.

3.4.2 Naive Bayes

In this approach, frequency plays an important role. Here, the term frequency (TF) is the total counts of a token that appears in a document. For practical purposes, usually, it is normalized. The mathematical representation of the same is

$$\text{Normalized term frequency} = \frac{TF(t,d)}{n_d} \tag{16.1}$$

where t is the count of the tokens, d is the document, and n_d is the total number of terms in document d.

In the multinominal model, the TFs are used to compute the maximum likelihood to estimate conditional probabilities for different classes. The class-conditional probabilities are represented in Eq. (16.2).

$$P(x_i|\omega_j) = \frac{\sum tf(x_i, d \in \omega_j) + \alpha}{\sum N_{d\in\omega_j} + \alpha V} \tag{16.2}$$

where x_i is a word in the feature vector x of an individual sample; $\sum tf(x_i, d \in \omega_j)$ is defined as the sum of term frequencies of x_i from all documents in the training sample that belong to class ω_j; $\sum N_{d\in\omega_j}$ is defined as the sum of all term frequencies in the training dataset for class ω_j; α is an additive smoothing parameter; V is the size of the vocabulary (number of different words in the training set).

The product of likelihoods of particular words is known as the class-conditional probability of text x. The class-conditional probability for the same can be calculated as follows. These computations are performed under the naive assumption.

$$P(x|\omega_j) = \prod_{i=1}^{m} P(x_i|\omega_j) \tag{16.3}$$

Besides, another approach is term frequency-inverse document frequency that is denoted as $Tf - idf$. It is usually preferred for ranking the documents based on its relevance with different text mining tasks. For example, page ranking based on hits by search engines. Other important application includes text classification using naive Bayes. Mathematically, $Tf - idf$ can be defined as below.

$$Tf - idf = tfn(t,d) \cdot idf(t) \tag{16.4}$$

where idf is the inverse document frequency, and $tfn(d,f)$ is denoted as the normalized term. The idf can be evaluated as follows.

$$idf(t) = \log\frac{n_d}{n_d(t)} \tag{16.5}$$

where $n_d(t)$ is defined as the number of documents contain the term t.

In this context, one of the important tasks that is needed in decision-making is the text classification. It is a classic case of categorical data. As such, naive Bayes may be used for continuous data. For example, the Iris flower dataset is a supervised classification task possessing continuous features. For the same, dataset contains the data as length of the petals and width of the petals in centimeters. Naive Bayes classification can be used to partition the different features and creates unique categories to obtain the class-conditional probabilities. With the reference of Gaussian naive Bayes model, the probability distributions can be represented as follows

$$P(x_{ik}|\omega) = \frac{1}{\sqrt{2\pi\sigma_\omega^2}}e^{\frac{-(x_{ik}-\mu_\omega)^2}{2\sigma_\omega^2}} \tag{16.6}$$

where μ is the sample mean and σ is the standard deviation of the training data.

Within this frame of reference, the class-conditional probability may be defined as the product of the particular probabilities in Eq. (16.7).

$$P(x_i|\omega) = \prod_{k=1}^{d} P(x_{ik}|\omega) \tag{16.7}$$

The next section explains the demonstration of above-mentioned procedure through a simple example.

4. Case study

A model is developed to predict the diseases from the symptoms reported by the patients. For easy understanding two labeled model is developed using the concepts of NLP. First the symptoms were collected and stored, then the same is processed. For example, a patient reported that "he is suffering from high temperature with headache and flu." Then, the texts are stored after removing the stop words. Therefore, the texts are changed to "suffering high temperature headache flu." The through tokenization, tokens viz. "suffering," "high," "temperature," "headache," and "flu" are created (Fig. 16.4).

As such, the library is created. For this study prior to start the implementation, a database is prepared with the symptoms and corresponding levels. Then, the new data is accessed for prediction. After each prediction, the data is verified by the doctor and the same is included in the library with proper allocation. This helps the system more effective to predict the disease. In this course of time frequency plays a major role for identification. Each inclusion of verified data increases the surety through frequency change.

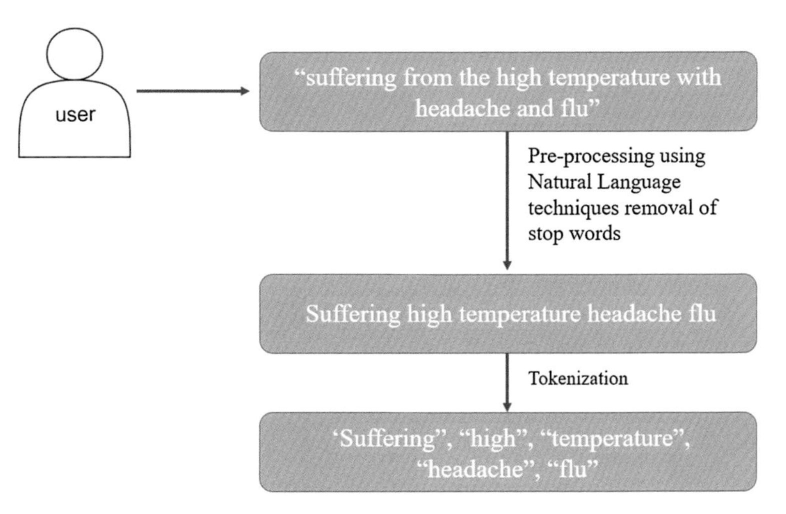

FIGURE 16.4 Schematic diagram of the model used in case study.

5. Results and discussion

It is observed that the model had implemented with an accuracy of 78.5%, on the given symptoms given by the various patients. The present system covers the all symptoms related to fever and the skin infection. The same model can be used for more number of labeled disease with a huge symptom data base. The early prediction of disease can be done using the model, so that the patient can go for the proper treatment. This helps in the welfare of the patient and their family.

References

[1] S. Nayak, Fundamentals of Optimization Techniques with Algorithms, Academic Press, San Diego, California, USA, 2020.
[2] S. Nayak, Uncertain quantification of field variables involved in transient convection diffusion problems for imprecisely defined parameters, International Communications in Heat and Mass Transfer 119 (2020) 104894.
[3] S. Nayak, S. Chakraverty, Interval Finite Element Method with MATLAB, Academic Press, San Diego, California, USA, 2018.
[4] S. Nayak, T.D. Rao, S. Chakraverty, Nonprobabilistic analysis of thermal and chemical diffusion problems with uncertain bounded parameters, Mathematical Methods in Interdisciplinary Sciences (2020) 99–113.
[5] A.K. Sahoo, A.K. Gupta, New analytical exact solutions of time fractional (2+1)-dimensional Calogero–Bogoyavlenskii–Schiff (CBS) equations, in: S.S. Ray, H. Jafari, T.R. Sekhar, S. Kayal (Eds.), Applied Analysis, Computation and Mathematical Modelling in Engineering, Lecture Notes in Electrical Engineering, vol 897, 2022.
[6] India has an estimated population of 1.38 billion inhabitants - Prensa Latina, www.plenglish.com, December 31, 2021. Retrieved August 16, 2022.
[7] Disease, Dorland's Medical Dictionary, www.dorlandsonline.com.
[8] T. White, What is the difference between an 'injury' and 'disease' for commonwealth injury claims? Tindall Gask Bentley (2014). Archived from the original on October 27, 2017. Retrieved November 6, 2017.
[9] Beyond Intuition: Quantifying and Understanding the Signs and Symptoms of Fever, clinicaltrials.gov, October 5, 2017. Retrieved January 9, 2021.
[10] Symptoms and Self-Help Guides by Body Part | NHS Inform, www.nhsinform.scot. Retrieved January 9, 2021.
[11] A.M. Cohen, An effective general-purpose approach for automated biomedical document classification, AMIA Annual Symposium Proceedings (2006) 161–165.
[12] R. Xu, D.C. Wunsch, Clustering algorithms in biomedical research: a review, IEEE Reviews in Biomedical Engineering 3 (2010) 120–154.
[13] A. Alamri, M. Stevensony, Automatic identification of potentially contradictory claims to support systematic reviews, IEEE International Conference on Bioinformatics and Biomedicine (BIBM) (2015).
[14] C. Blake, Beyond genes, proteins, and abstracts: identifying scientific claims from full-text biomedical articles, Journal of Biomedical Informatics 43 (2) (2010) 173–189.
[15] L. Ohno-Machado, P. Nadkarni, K. Johnson, Natural language processing: algorithms and tools to extract computable information from EHRs and from the biomedical literature, Journal of the American Medical Informatics Association 20 (5) (2013) 805.
[16] S. Sheikhalishahi, R. Miotto, J.T. Dudley, A. Lavelli, F. Rinaldi, V. Osmani, Natural language processing of clinical notes on chronic diseases: systematic review, JMIR Medical Informatics 7 (2) (2019) e12239.
[17] D. Jurafsky, J.H. Martin, Speech and Language Processing, Pearson, London, 2014.
[18] T. Baumel, J. Nassour-Kassis, M. Elhadad, N. Elhadad, Multi-label classification of patient notes a case study on ICD code assignment, 2017 arXiv Prepr. arXiv1709.09587.
[19] A. Perotte, R. Pivovarov, K. Natarajan, N. Weiskopf, F. Wood, N. Elhadad, Diagnosis code assignment: models and evaluation metrics, Journal of the American Medical Informatics Association 21 (2) (2013) 231–237.
[20] M. Huang, A. Névéol, Z. Lu, Recommending MeSH terms for annotating biomedical articles, Journal of the American Medical Informatics Association: JAMIA 18 (5) (2011) 660–667.
[21] S. Peng, R. You, H. Wang, C. Zhai, H. Mamitsuka, S. Zhu, Bioinformatics 32 (12) (2016) i70–i79.
[22] B. Jiang, Y. Zhao, R.G. Salloum, Y. Guo, M. Wang, M. Prosperi, H. Zhang, X. Du, L.J. Ramirez-Diaz, Z. He, Y. Sun, Using social media data to understand the impact of promotional information on Laypeople's discussions: a case study of Lynch syndrome, Journal of Medical Internet Research 19 (12) (2017) e414.
[23] J. Du, L. Tang, Y. Xiang, D. Zhi, J. Xu, H.-Y. Song, T. Cui, Public perception analysis of tweets during the 2015 measles outbreak: comparative study using convolutional neural network models, Journal of Medical Internet Research 20 (7) (2018) e236.
[24] J. Du, Y. Zhang, J. Luo, Y. Jia, Q. Wei, C. Tao, M.Q. Wang, Extracting psychiatric stressors for suicide from social media using deep learning, BMC Medical Informatics and Decision Making 18 (S2) (2018).
[25] C. Liang, Y. Gong, Automated classification of multi-labeled patient safety reports: a shift from quantity to quality measure, Studies in Health Technology and Informatics 245 (2017) 1070–1074.
[26] Y. Mao, Z. Lu, MeSH now: automatic MeSH indexing at PubMed scale via learning to rank, Journal of Biomedical Semantics 8 (2017) 1–9.
[27] F. Gargiulo, S. Silvestri, M. Ciampi, Deep convolution neural network for extreme multi-label text classification, International Workshop on Artificial Intelligence for HealthAt: Funchal, Madeira, Portugal 5 (2018).
[28] Y. Li, Y. Song, J. Luo, Improving pairwise ranking for multi-label image classification, Proceedings of IEEE Conference on Computer Vision and Pattern Recognition (2017) 1837–1845.

[29] J. Nam, J. Kim, E.L. Mencía, I. Gurevych, J. Fürnkranz, Large-scale multi-label text classification - revisiting neural networks, Lecture Notes in Computer Science (2014) 437−452.

[30] M.R. Boutell, J. Luo, X. Shen, C.M. Brown, Learning multi-label scene classification, Pattern Recognition 37 (2004) 1757−1771.

[31] J. Read, B. Pfahringer, G. Holmes, E. Frank, Classifier chains for multi-label classification, Machine Learning 85 (2011) 333.

[32] Z. Min-Ling, Z. Zhi-Hua, A review on multi-label learning algorithms, IEEE Transactions on Knowledge and Data Engineering 26 (2014) 1819−1837.

Chapter 17

"Deep learning" for healthcare: Opportunities, threats, and challenges

Russell Kabir[1], Madhini Sivasubramanian[2], Geeta Hitch[2], Saira Hakkim[2], John Kainesie[2], Divya Vinnakota[2], Ilias Mahmud[3], Ehsanul Hoque Apu[4,5], Haniya Zehra Syed[1] and Ali Davod Parsa[1]

[1]School of Allied Health, Faculty of Health, Education, Medicine and Social Care, Anglia Ruskin University, Essex, United Kingdom; [2]Faculty of Health Sciences and Wellbeing, University of Sunderland, London, United Kingdom; [3]College of Public Health and Health Informatics, Qassim University, Buraydah, Saudi Arabia; [4]Department of Biomedical Engineering, Institute of Quantitative Health Science and Engineering, Michigan State University, East Lansing, MI, United States; [5]Division of Hematology and Oncology, Department of Internal Medicine, Michigan Medicine, University of Michigan, Ann Arbor, MI, United States

1. Introduction

Machine learning (ML) is being used in many entities already and proven to be helpful. In many industries it is a great way to increase the productivity. However "deep learning" (DL) has additional advantages in healthcare by mimetic process and absorbs information. Artificial intelligence (AI) offers the chance to optimize routes for diagnosis and prognosis as well as to generate individualized treatment plans. For instance, studies that include potential risk factors, such as underlying genetics and particular surroundings, may help with the creation of preventative measures and more precise diagnosis using massive datasets which is used in ML. Additionally, the use of structural and functional imaging tools can help healthcare professionals better understand present condition and plan and execute their treatment pathways. Biomedical and healthcare sectors are increasingly applying "big data" for medical data analysis, which, in turn, is critical for early disease diagnosis, treatment, and community healthcare. However, when the quality of the medical data is lacking, the analysis's accuracy suffers. For example, distinct regional diseases in different places have their own features, which could make it harder to forecast when a disease would spread. In this chapter, we streamline ML techniques and how effective is the DL techniques are and necessary to make accurate prediction in healthcare practices [1]. The neural network–driven DL mimics the human brain. It employs a multilayered neural network that generates results without the need for preparing the input data. The algorithm receives the raw data from data scientists, evaluates it based on what it already knows and what it can deduct from the new data, and then produces a decision. Contemporary issues need contemporary solutions. There is a global shortage for healthcare professionals and ML was helpful with limitations. DL could help to face some challenges. This chapter examines the modernization of healthcare industry through DL. It provides an insight into what DL is and how it has revolutionized healthcare by proving beneficial in nearly every sector of this mega-industry. From digitalized patient data to remote healthcare, DL has proven to be significant and advantageous to patients and healthcare providers. Moreover, the chapter highlights the various opportunities, challenges, and threats, which accompany integration of DL and "big data" into healthcare.

2. Machine learning and deep learning

2.1 Machine learning

Computers that can be taught to assist humans with little to no ongoing human work are referred to as machine-learning system [2]. As a logical progression from conventional statistical methods, ML has been increasingly used in medical decision-making [3–5]. Machine-learning techniques range from wholly automated algorithms that process huge data without human participation to completely machine-driven analysis and machine-learning algorithms [6]. ML depends on

Deep Learning in Personalized Healthcare and Decision Support. https://doi.org/10.1016/B978-0-443-19413-9.00017-5

"big data" availability and processes. The more meaningful data influences the efficiency and accuracy of the ML. Accuracy is increased by preprocessing data using exploratory data analysis and feature engineering. Since it is more crucial to precisely identify the percentage of sick people in the healthcare business than it is to identify the percentage of healthy people accurately, research should be done to make algorithms more sensitive (specificity). Healthcare is considered one of ML's most challenging applications. ML could never replace the human touch in the delivery of compassionate healthcare. Still, it could speed up the process of investigation, diagnosis, and treatment decision-making of healthcare delivery in all levels [7].

Application of ML surely enhances healthcare delivery. However, the ML focuses on biomedical applications only, and DL minimizes and almost eliminates the underlying undetectable various as it creates reliable outcomes from multiple data sources [8]. For example, healthcare professionals diagnose and plan a treatment that is effective with obtained information from ML that must be combined and properly interpreted. To achieve this, ML algorithms need to be learned and used well. Supervised learning (SL) and unsupervised learning (UL) play a huge role. Through UL, there are a number of hidden data that can be looked at and learned from for important cues which are otherwise detectable.

2.2 Deep learning

The essentials of DL are the same as ML, and it is widely used safely from medical imaging, medical records (electronic medical records [EMRs]), genomics medicine, and molecular design. DL best fits with healthcare as this has UL and creating hierarchical models through neural networks. DL shares the same methodological algorithms as ML (Holzinger, 2017). DL needs less human interaction in fields such as medical image analysis [9]. Fetal developmental analysis has been made lot accurate by DL [10]. For diagnosis of chronic respiratory diseases such as chronic obstructive pulmonary disease (COPD), specific biomarkers created using DL are used for early diagnosis and treatment in cases which may otherwise be asymptomatic [11]. DL capabilities allow analyzing complex datasets such as images, comparison between the biomarkers, and many more variables otherwise undetectable. DL is inevitable wherever healthcare needs exceeding the capacity of the specialists and resources [12]. Using DL in healthcare is not without its challenges. This chapter looks into the various opportunities, challenges, and threats caused by using DL.

3. Deep learning and its opportunities in healthcare

A new phase in healthcare is rapidly approaching, one in which the abundance of biological data will play an increasingly significant role [13]. The health sector could benefit from the various DL features, including its superior execution, end-to-end learning structure amalgamated with feature learning, the capability to process complex and high volumes of data, and more. The DL framework is a live model which gets updated in real time to reflect any changes, for example, when implemented into the EHR system of a hospital, the framework would indicate any changes in patient data. Multilayer neural networks when used for medical data enhanced the prognostic capacity for multiple clinical applications [14].

High rates of low-income and middle-income countries' (LMICs') under-5 mortality are frequently a syndromic sign of a weakening healthcare system [15]. DL deployment in the healthcare system is becoming more and more popular in high-income countries due to its advantages such as computer vision, automatic speech recognition, natural language processing, audio recognition, and bioinformatics [16,17]. There is a need for capacity building across the board and cooperative use of technological resources between LMICs in order to achieve equity in the application of this technology [18]. Different DL opportunities are showin in Fig. 17.1.

3.1 "Big data" management

The term "big data" in the context of healthcare refers to a variety of data sources, including genomics-driven experiments, smart web data from the Internet of Things (IoT), and payer-provider data from the healthcare industry such as EMRs, prescription and insurance records, and pharmacy records [19]. The abundance of biomedical data presents both enormous potential and difficulties for healthcare research [20].

The knowledge of epidemiology, diagnosis, treatment, etc., of numerous diseases available today has been made possible by the collective data contribution of healthcare organizations and biomedical researchers. This information has paved way for a more efficient and healthier model toward the development of "individualized healthcare" [21].Various medical ontologies used to generalize the data, such as the International Classification of Disease-ninth version, Systematized Nomenclature of Medicine-Clinical Terms (SNOMED-CT), Unified Medical Language System (UMLS), and others, frequently contain inconsistencies and conflicts, making full use of the biomedical data difficult due to their high-dimensionality heterogeneity, temporal dependency, sparsity, and irregularity [22,23].

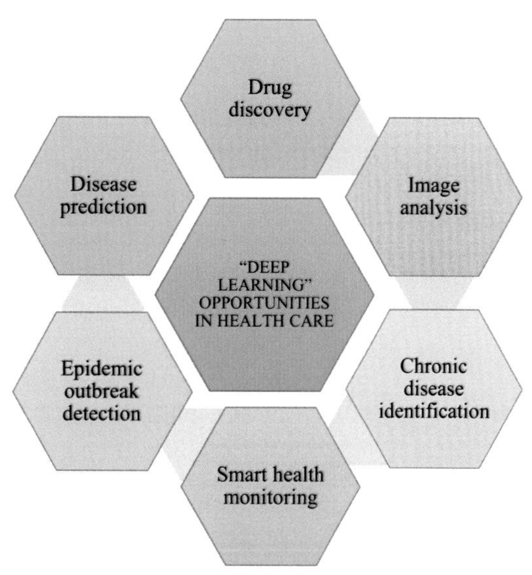

FIGURE 17.1 Deep learning opportunities in healthcare system.

A fresh and original approach to the analysis of healthcare "big data" includes clinical trials, the combined analysis of pharmacy and insurance claims, and the development of biomarkers. The discrepancy between structured and unstructured data sources is exploited by "big data" analytics. Medical practice management software (MPM), personal health records (PHR), and electronic health records (EHRs), among other healthcare data components show the capacity to reduce healthcare costs while being more effective and providing superior services [19].

4. Data analysis and statistical modules

"Big data" from the healthcare industry holds promise for enhancing health outcomes and reducing expenses. The data must be kept in a file format that is simple to obtain and understandable for an effective. Translational bioinformatics is the application of bioinformatics methods to turn biomedical and genetic data into predictive and preventive health. It pioneers the use of data in healthcare. It is possible to combine and use many quantitative data sources from the healthcare industry, such as genomic profiles, drug data, and laboratory measures, to find new meta-data that can support precision therapies [24]. "Big Data Research and Development Initiative" aims to improve the quality of data tools and techniques used, which, in turn, will provide enhanced organization, better access and skillful analysis of "big data" [25].

Such a solution is cloud computing, which offers dependable services and virtualized storage technologies. It provides all-encompassing access, dynamic resource discovery, high dependability, scalability, and autonomy, as well as composability. These platforms can operate as a computer to analyze and interpret the data, a receiver of data from ubiquitous sensors, and a source of user-friendly web-based visualization. Hadoop and Apache Spark are two of the more popular technologies for using "big data" [26].

Convolutional neural networks, recurrent neural networks, deep belief networks, auto-encoders, and deep Boltzmann machines are a few of the variations of DL models. Several researchers have recently shown that these DL models work well when used for machine health monitoring. DL uses numerous layers of information processing modules stacked in hierarchical architectures to attempt to model the hierarchical representations underlying the data and identify (predict) patterns [27].

4.1 "Deep learning" and advancement

Deep learning algorithms have proven to be more sophisticated in multiple fields, including object detection, speech recognition, visual recognition, drug discovery, genomics, proteomics, and drug discovery [18]. Then, by leveraging clinician actions in various domains and applications, including as illness risk prediction, individualized prescriptions, treatment recommendations, clinical trial recruitment, as well as research and data analysis, deep representations can be leveraged [27]. Deep representations can then be used to support therapeutic activities across a range of domains and applications, including illness risk assessment, individualized medication recommendations, therapy suggestions, clinical trial enrollment, and data analysis [28].

There are initiatives in progress which aim to transfer patient records from pre-EHR notes to a standardized digital format which can be recognized by the machines. Optical character recognition (OCR) software is one such method that promotes digitization and can recognize both handwriting and computer fonts. Identifying important grammatical patterns in free text, assisting with speech recognition, and extracting the meaning behind a narrative are all capabilities of the emerging field of ML known as natural language processing (NLP). NLP can be beneficial for recognizing unstructured text which include patient records from nurses, reception reports, clinical text, and discharge records. Radiology and MRI reports can also be converted by NLP to predict cancer progression and classify various pathologies [29].

In clinical areas, DL can act as a guiding principle for organizing both hypothesis-driven research and exploratory investigation (e.g., clustering, visualization of patient cohorts, stratification of disease populations). All levels of statistical and medical duties, such as research design, experiment planning, model development and refinement, and data interpretation, must be integrated [14].

It is imperative that published papers be evaluated and their quality determined [30]. Various techniques have been implemented by the researchers to overcome the issue of text plagiarism, which depends on linguistic analysis detecting similarity using semantic factors. . To build the DL models for detecting text plagiarism, a new text similarity feature (TSF) database was created [29,31,32]. The number of citations is a significant indicator because it is frequently used to gauge a paper's impact and because it has been the foundation for other systems of measurement, including the h-index, impact factor, and i-10 index, which provide as a reference for reviewing academic writings, discussions, and research institutions [33,34]. The proposed neural network is trained to anticipate future paper citation counts using the citation patterns of numerous existing scientific papers. Deep neural networks have been successfully used in recent years to solve a variety of issues, including voice recognition, object recognition, image processing, and text processing [30,35–37].

5. Deep learning and immunization management

The mechanism of SARS-CoV-2 transmission was unpredictable and unknown during the first wave of impact, which made it difficult to develop and implement an immediate and effective public health response [38]. Digitalization of information is being heavily used to help the global public health response to "COVID-19" [39]. A global real-time surveillance study emphasizes the significance of DL-based social media monitoring as a quick and efficient method for spotting emerging trends of "COVID-19" vaccination intention and confidence to guide timely interventions, especially in environments with limited resources and pressing timelines [40].

When there is no obvious cure for a global pandemic, reinforcement learning is very helpful. Countries without recorded cases can benefit from other nations' public initiatives, including their effectiveness, timing, and intensity. Reinforcement learning identified the best strategies to decrease the "COVID-19" load based on the temporal and population features particular to each nation and region [41]. In case of "COVID-19" vaccination, text messages, email-based communication, frequent webinars, the distribution of digital newsletters and toolkits, smartphone apps, and other digital interventions are being utilized more and more to encourage the uptake of vaccinations across all age groups, including in low- and middle-income countries [42].

6. Role in "COVID-19" vaccination

The use of a live model to improve vaccine rollout has been of significant importance during "COVID-19" vaccination campaign. The digital technology used has made the data analysis and exchange more efficient and accessible, which in turn has improved the campaign quality. Digital interventions were used to manage the lists and reminders for immunizations. Population monitoring, case identification, contact tracing, and evaluation of interventions based on mobility data and public communication all depend on instant phone messaging services including social media and other connected computer devices, and advancements in ML. In order to find out what issues are being addressed, what information is being given, and whether it is correct, health organizations use social listening to find and comprehend posts concerning vaccines and vaccination on social media [39,43].

Hesitancy and attitude of mass population: Immunization reluctance is not a fresh phenomenon. Since the first human immunization campaigns, people have had a negative attitude about vaccinations. The WHO views vaccine hesitancy as one of the greatest challenges to global health, which is defined as people's hesitation to receive vaccinations for a variety of reasons. It has a negative effect on efforts to maintain a decent level of vaccination coverage as well as vaccine demand. Vaccine reluctance varies across time, geography, and vaccine types and is complex and context specific. It frequently takes the stage in discussions or when only a partial picture of a condition is provided. People may exhibit reluctance to receive vaccinations for a variety of reasons, such as lack of confidence, complacency, a lack of understanding of the benefits of the vaccine, and difficulty accessing the vaccine [43,44].

The UK's vaccine reluctance is more likely due to mistrust of the nation's healthcare system [45]. The reluctance toward vaccination can also be attributed to factors such as general fear and disgust of medical equipment, media and misinformation, personal choice and unpredictability of viral mutations [46]. In developing nations, the influence of customs and beliefs, alternative health beliefs, and ignorance might be added to the list [43].

Attempts to persuade people to take vaccines or alter their attitudes toward them frequently meet with resistance. The tendency for health messaging to be succinct and to ignore the majority of potential issues is one reason why people may be so hesitant to change their beliefs. Application of DL, giving participants access to a chatbot that responds to the most frequent queries about the "COVID-19" vaccines, have been implemented to resolve the issue [47].

7. Role in future disease prediction

Due to its effectiveness across various domains and the speed at which methodological advancements are being made, DL paradigms present fascinating new potential for biomedical informatics. Enlitic is utilizing DL intelligence to recognize health concerns on radiographic scans such as X-rays and CT scans [14,48].

Healthcare companies are using mobile health and wellness services more and more to create cutting-edge and creative approaches to caregiving and wellness coordination. Mobile platforms can advance healthcare by facilitating faster interactive communication between patients and healthcare professionals. Apple and Google have created specific platforms for developing research applications for fitness and health statistics, such as Apple's Research Kit and Google Fit [49]. These available applications use customer devices and inbuilt sensors which offer easy data integration. The apps provide doctors simple and up to date access to health data and health status of an individual [39,43].

Siah et al. [50], conducted a study to detect air leakage from respirators worn during Fit checks, using AI tools such as DL algorithms and thermal infrared imaging. This study proved the potential of DL to accurately detect the air leakage location and its application in manufacturing safe respirators [50].

The analysis of brain magnetic resonance imaging (MRI) scans to forecast Alzheimer disease and its variations was one of the earliest clinical data uses of DL in image processing [51,52]. In order to automatically partition cartilage and predict the likelihood of osteoarthritis in low-field knee MRI data, CNNs were applied. High-throughput biology uses DL to identify the internal organization of increasingly vast and high-dimensional data sets (e.g., DNA sequencing, RNA measurements). Deep models make it possible to find high-level characteristics, which enhances performance compared to typical models, increases interpretability, and offers more insight into the structure of biological data [14].

To perform medical image analysis and uncover hidden information, a number of software solutions have been created based on functions as generic, registration, segmentation, visualization, reconstruction, simulation, and diffusion. For instance, SPM can process and analyze five various types of brain pictures, including MRI, fMRI, PET, CT-scan, and EEG, whereas Visualization Toolkit is a free program that enables powerful processing and analysis of 3D images from medical tests. MITK, Elastix, and GIMIAS are a few other programs that support various picture kind [53].

Hospitals, providers, and researchers can share and analyze health data using IBM's Watson Health AI platform. Flatiron Health offers technology-driven healthcare analytics services that are particularly geared on cancer research. Other large corporations, such Oracle Corporation and Google Inc., are concentrating on creating platforms for distributed computing and cloud-based storage [19].

The most significant factors affecting U5MR according to random forest are duration of breastfeeding, family wealth index, and level of mother education. Additionally, for the prediction of U5MR, DL algorithms are more sensitive and specific. The most significant implication for the future is that DL algorithms can be used in production to identify and flag children who are most at risk and not likely to survive until the age of five, allowing for the targeting of necessary interventions to the communities where those children live [18,54].

8. Public health misinformation

Internet users are becoming increasingly dependent on social media, which has created an ideal environment for spreading fake news, misleading information, fake reviews, rumors, etc [55,56]. The term misinformation refers to information that is either factually inaccurate or lacking evidence [57]. Health-related misinformation on social media is a growing and urgent problem, particularly in the field of health [58].

It has been reported that patients' health outcomes can be influenced by health information obtained from social media, including online health communities [59]. As health information is sought on social media platforms, concerns about health misinformation have increased. Online communities often lack gatekeepers that prevent the spread of health misinformation [60].

An online health misinformation model integrates central-level (including topic features) and peripheral-level (linguistic, sentiment, and user behavior) features into a single model. Online health misinformation is classified into two levels: central-level and peripheral-level [61]. An ML-based model combining linguistic, topic, sentiment, and behavioral features to detect health misinformation appearing in online health communities is essential due to the exacerbation of health misinformation spread on social media. In turn, this will help curb the spread of health misinformation through peer-to-peer communication and the Elaboration Likelihood Model (ELM), hence, facilitates the detection of health misinformation in online community settings and provides solutions on how to detect it accurately and automatically [61,62].

9. Contact tracing and pandemic management

Management of pandemic crises brought on by a virus like "COVID-19," which has nearly brought the world to a standstill and claimed millions of lives, has never before been so critical [63]. The "COVID-19" virus mutation and new variants made it impossible for the immunization campaign to keep up with the development. It led to the implementation of contact tracing software by many nations to lessen "COVID-19" virus exposure [64–67].

Deep learning neural networks were used to screen, track, and anticipate future events during the "COVID-19" pandemic because they had a higher computational power [68,69]. The "COVID-19" pandemic may be mitigated and reduced by using ML methods, such as the DL neural network or random forest classifier. The prediction of perceived usability on contract tracing applications has a high accuracy of 97.32%, thanks to the use of DL neural networks. Because it selects the best tree from among the several decision trees created, the random forest classifier can be a potent predictive ML algorithm tool for human behavior with improved accuracy [68,69]. ML and artificial intelligence have played an integral part in the smartphone-based contract tracing approaches to control the "COVID-19" pandemic [70].

10. Remote healthcare

E-healthcare, often known as remote healthcare, includes remote health monitoring. It is a method for remote, automated health monitoring. This has been made possible by improvements in sensors, WSN, and other more recent technologies like WBAN and IoT. The body's various internal and external health sensors collect various physiological data, including body temperature, heart rate, blood pressure, blood sugar, brainwaves, blood oxygen saturation, and more. The relevant medical specialists get these data and interpret them to determine the patient's condition, diagnosis, and treatment [71,72]. Due to the rapid advancement of wireless communication technologies and computer processing capacity, mobile healthcare may now deliver services that are quick, affordable, pleasant, and hassle-free [71].

Estimates regarding ML algorithms and the "Internet of Things" powered by deep neural networks were made in analyses on digital epidemiological surveillance, intelligent telemedicine diagnosis systems, and ML-based real-time data sensing and processing in "COVID-19" remote patient monitoring. DL algorithms and intelligent networked medical devices can be used to improve the diagnostic' precision rate. Wearable technology and the Internet of Things powered by deep neural networks are essential to patient-focused real-time medical analytics and smart healthcare. Real-time data sensing and processing based on ML and artificial intelligence have been integrated into massive healthcare data analytics [73].

11. Telemedicine

Telemedicine is the idea of employing technology to provide clinical services and medical care remotely. Through the use of a telecommunications link, doctors can provide remote patient care through telemedicine. A significant component of remote monitoring is introduced as telemedicine, which enables a variety of chronic patients to receive straightforward therapy at home. The reach of telemedicine has increased as a result of the Internet's extensive use and omnipresence [72,74].

The concept of telemedicine is implemented in the following three ways.

- **i.** Store and forward: The patient's health information and medical records are provided to the doctors for review. The physicians review the reports at their convenience and provide comments or directions to the local medical personnel.
- **ii.** Remote monitoring: A doctor keeps track on a patient's vital signs.
- **iii.** Real-time communication: A doctor and patient communicate live from a distance [72].

To provide effective and quick healthcare services, telecommunications techniques and algorithms must be used, potentially saving the patient's life. The triaging and prioritization procedures grow more difficult as the number of patients rises. This is seen as a serious issue that might even endanger human life. To this purpose, the difficulties associated with triaging and prioritizing processes could be resolved by utilizing the valuable instrument of ML. Applying medical standards to the patient's vital signs determines the patient's triage level. The patients who are given priority receive the appropriate care. This strategy makes it simple to monitor and manage patient conditions both indoors and outdoors, including those of nursing home residents [75,76].

12. Virtual healthcare platforms

The telecommunication link used in various forms. Technologies, such as mobile communication, video conferencing, fax, scanners, etc., are the primary facilitators of the virtual healthcare platform, to communicate and exchange medical records (such as X-ray and sonography images, photographs of infections, previous prescriptions, and pathological reports, ECG analysis, etc.). Doctors evaluate patients' conditions and offer recommendations based on these records [72].

Patients can obtain supportive care through telemedicine without physically visiting a hospital, for instance, by conversing with a conversational assistant powered by artificial intelligence that offers treatment recommendations [77]. Chatbots can assist in telehealth delivery by giving chronic patients free primary healthcare education, information, and guidance as well as facilitating virtual connection while socially isolating them [78,79]. Similar to this, a mobile clinical decision support system can offer a thorough framework for identifying and tracking diseases. In telemedicine, ML approaches were used to enhance "COVID-19" patient identification and tracking for usage by health systems in real-time [79,80]. Patients can receive care in their homes thanks to community paramedicine or mobile integrated healthcare programs, with higher-level medical support being delivered virtually [81].

13. Online healthcare education: opportunities and challenges

The need to create alternate learning methods to give healthcare students enough patient exposure has arisen as a result of rising student numbers and declining clinical teaching opportunities [82]. Internationally, there have been several e-learning efforts in medical education that have been tried out with varying degrees of success. Those that have fallen short frequently did so due to deficient funding or low quality, which resulted in a lack of acceptance among students and staff. In medical schools abroad, similar e-learning efforts that incorporate online lectures, notes, structured learning modules, and case reports have been successful [83,84].

There is no much proof that computer-assisted instruction (CAI) is more effective than other conventional methods, but it works best in a blended learning environment where it can go before traditional didactic teaching and complement clinical experiences, laying the groundwork for other types of instruction [85,86].

E-learning has its limitations, though. In contrast to traditional education, e-learning frequently denies pupils the chance to raise questions in person. Additionally, it depends on students having reliable access to computers and the internet. Using both traditional clinical attachments and online learning would help overcome each teaching method's inherent challenges [87].

Importantly, e-learning permits evaluation linked to a defined curriculum and uniformity in the subject covered. Additionally, an online learning tool can adjust to changes more quickly than traditional textbooks can in the fast-evolving field of medical knowledge [85]. Especially, during the "COVID-19" pandemic period, online-based clinical learning was the only mode of education which was optimized with ML, DL, high-speed internet, and "big data" management systems [88].

14. Hospital, clinical research, and healthcare facilities

Many healthcare organizations employ ML and have a huge vested interest in how ML can provide a safe, cost-effective, and efficacious patient care with improving efficiency. Artificial intelligence (AI) or DL or ML has erupted in the recent years leaving healthcare stakeholders and users with an improbable array and task in the choice of technologies. But what exactly is artificial intelligence (AI) or DL or ML?

The definition of AI has changed over the course of time and there are many versions [89,90].

Artificial intelligence can be loosely described as "any human-like behavior displayed by a machine or a system." For example, *"computers can be programmed to* "mimic" *human behaviour using extensive data from past examples of similar behavior such as recognizing differences between a human and an animal, to performing complex robotic activities in industry*" [91].

AI is powerful in generating real-time analytics from huge volumes of data for the purposes of automating and predicting to enable the completion of tasks free from human error and with precision and reliability [91]. Both DL and ML are branches of AI, with DL being a branch of ML. ML has advanced in leaps and bounds since Alan Turning created his "Turning test" to check a machine's intelligence in 1950 [92]. The approach by which how computers are used to simulate human behavior is called ML [93,94].

A definition of ML by Ref. [94]; stated of ML: *"A computer program is said to learn from experience E with respect to some task T and some performance measure P, if its performance on T, as measured by P, improves with experience E."* most certainly ML has found its niche in many areas within the healthcare industry.

Deep learning is also a type of ML based upon the body's nervous system where the body's neurons connect and function with automated precision across layers of nodes which when stimulated by electrical signals from other neurons in the network, become activated to ultimately process information in the brain. Likewise, in DL, similar to the nervous system, there are layers of artificial neural networks (ANN) through which data are filtered from each successive structured layer to inform the output for the next layer. Through this structured learning, DL models perform increasingly accurately using algorithms for data analysis. Due to ability of data to traverse across layers multiple times, there is efficiency as the raw data becomes more refined and meaningful to the user to find meaning in complex patterns in large data sets.

Hence, deep neural networks (DNNs) are artificial neural networks (ANN) with multiple layers between the input and output layers (Fig. 17.2). Within ML there are artificial neural networks (ANN), convolution neural networks (CNN) and recurrent neural networks (RNN), all capable of identifying highly complex patterns within huge datasets [95–97]. Each type of NN works in a different way.

Today ML has many applications, which include solving complex data using Bayesian Methods for example, from auto-correction to predicting mortality in injured patients in the medical field [92]. The application of ML in healthcare, which has gained momentum over the last decade, has made a noteworthy impact on healthcare. Machines are enabled with the ability to detect pattern recognition, that is, an algorithm called the training set or training model, of which exist three main groups which include (1) supervised learning, (2) unsupervised learning, and (3) reinforcement learning [98].

Supervised learning is when humans label input and out data to train algorithms; unsupervised learning is when there is no labeled data to provide the algorithm with an underlying structure from within its input data. Reinforcement learning (RL) does not require labeled input and out data and, therefore, no supervision. Instead, in RL, the focus is on learning by trial and error, whereby outcomes are both random and under the control of a decision-maker to maximize a reward in a dynamic situation [99]. There are several architectural models of DL (Fig. 17.3).

With these features of ML, there are numerous application for ML in Healthcare as shown in Fig. 17.4.

15. Electronic health records (EHRs)

A patient's journey to the hospital starts with taking a patient medication and drug history. Healthcare professionals (HCPs) need to take accurate information from the patients to provide quality healthcare services and treat complex

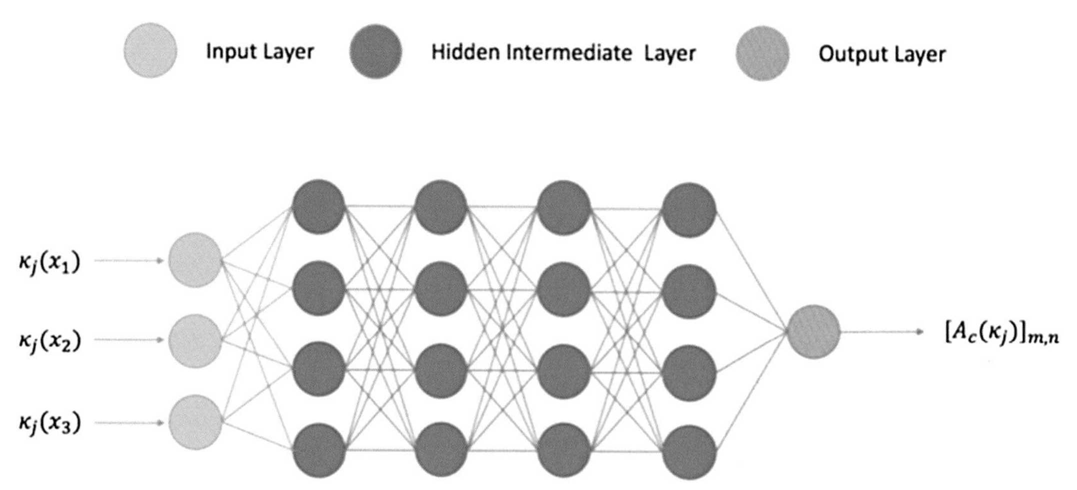

FIGURE 17.2 An illustration of deep neural network. *Courtesy of Wang, M., Cheung, S.W., Chung, E.T., Efendiev, Y., Leung, W.T. and Wang, Y., 2019. Prediction of discretization of gmsfem using "Deep Learning".* Mathematics, *7(5), p.412.*

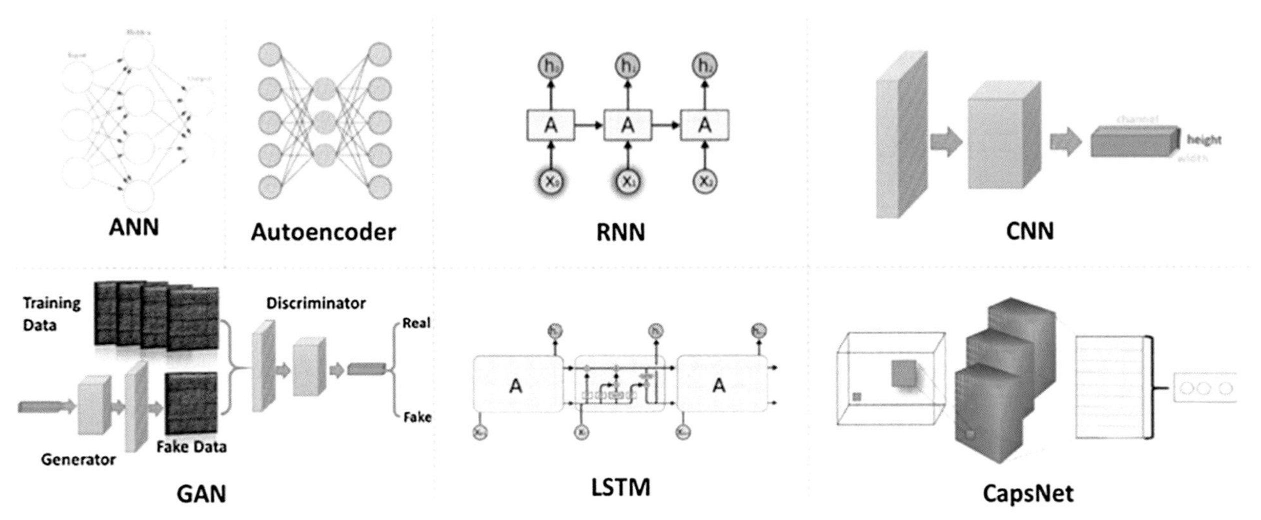

FIGURE 17.3 Architecture of main "deep learning" models. *Taken from Koumakis, L., 2020. "Deep Learning" models in genomics; are we there yet?.* Computational and Structural Biotechnology Journal, *18, pp.1466–1473. (This is an open access article under the CC BY-NC-ND license (http://creativecommons.org/licenses/by-nc-nd/4.0/).*

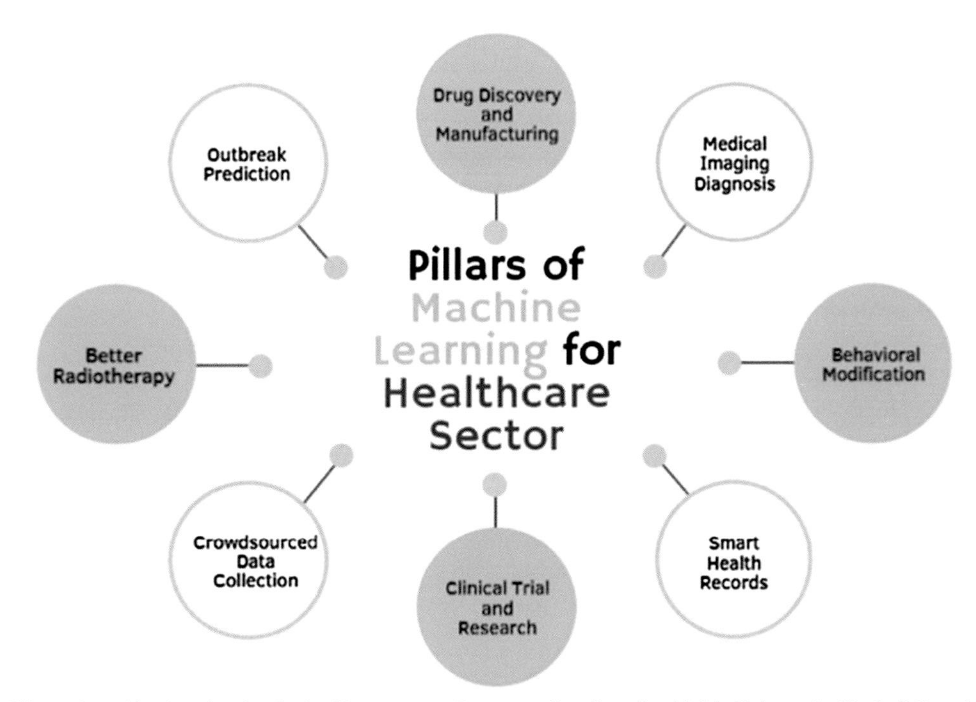

FIGURE 17.4 Pillars of machine learning for the healthcare sector. *Diagram taken from Javaid, M., Haleem, A., Singh, R.P. and Suman, R., 2022. Artificial intelligence applications for industry 4.0: a literature-based study.* Journal of Industrial Integration and Management, *7(01), pp.83–111.*

diseases and conditions. ML technologies provide automation and rapid turnaround in the provision of services for HCPs. Recent publications have shown that the use of ML modeling analysis to estimate patient exposure to antimicrobials and the number of days of therapy from numerous EHRs using random algorithm forests showed a robust ability to identify antibiotic exposures. Although more development is needed with more sophisticated algorithms, ML offers more meaningful comparisons via evaluation of high-volume encounter-level data to *"develop an antimicrobial stewardship program strategy and improve the efficiency of antimicrobial stewardship practice assessments"* as concluded by the authors [100].

Other areas where EHRs have been employed include determining mortality after myocardial infarction, which has shown that ML models for HF mortality where race was used as a covariate were "superior" compared with traditional and rederived logistic regressions models [101].

Among many ML applications in healthcare, the organization of patients' EHRs is automated with ease and accuracy, not to mention speed. It removes the risk for the more invasive processes of human interventions that may, for example, require diagnosis of a medical condition via a biopsy by the use of more streamlined noninvasive algorithms to detect the presence of a condition by analysis of "big data" in real-time from thousands of patients. Indeed, this has been recently applied to diagnosing skin diseases, (Liu et al., 2020), and Kawasaki disease [102] and in the characterisation of immune-related adverse events in cancers where immune-checkpoint inhibitors (ICIs) are used to treat certain cancers and cause organ-specific inflammatory toxicities [103].

16. Clinical imaging

Deep learning techniques are employed in medical imaging where diagnostic accuracy that informs clinical decision is of utmost importance in healthcare in order to improve the health outcomes of patients. During a period spanning nearly 11 years, over 1.5 million computed tomography (CT) and magnetic resonance imaging (MRI) scans were evaluated at one institution alone, concluding that one image is interpreted every 3–4 s in an 8-h workday by a radiologist to meet workload demands, leading to an increase in potential errors due to fatigue and overworking [104]. Clinical data are essential for informing a treatment decision together with accurate medical imaging. Recently, medical imaging has taken the form of a "fusion-paradigm" whereby it is used in conjunction with EHRs to leverage multiple modalities to diagnose and inform treatment decisions in the medical field. This is called multimodal fusion, of which there are three main types, early, joint, and late fusion [105].

In clinical imaging, multimodal fusion has been employed to detect "COVID-19" in medical images [106,107], whereby several different platforms are used. For example, the NiftyNet platform is the CNNs platform for medical image analysis and image-guided therapy [108]. This is used for CT segmentation of abdominal organs; regression of brain images for prediction of CT maps; and the use of ultrasound anatomical images [35,109]; applied deep CNN to classify the 1.2 million high-resolution images into the 1000 different classes [35] as shown in Fig. 17.5.

Supervised learning: Datasets labeled or annotated are fed into the computer to use these labels to generate new correct labels from unseen datasets. Malignant skin lesions have been classified using labeled human image annotation data [111]. Although the accuracy is still low, CNN-based classification has been applied to identify bacterial and viral pneumonia in chest X-ray datasets [112,113]. Many different modalities are used to identify and treat diseases. Radiologists use MRI, CT, ultrasound, and X-rays to interpret diseases in various parts of the human anatomy. Some DL networks include Alexnet, VGG16, GoogleNet, and ResNet deep network [114].

The GoogleNet Inception v3 DL model has been used to classify skin lesions from skin conditions such as keratinocyte carcinomas, seborrheic keratoses, malignant melanomas, and benign nevi [111].

The ResNet-152 DL model has been used for the classification of basal cell carcinoma, squamous cell carcinoma, intraepithelial carcinoma, actinic keratosis, seborrheic keratosis, malignant melanoma, melanocytic nevus, lentigo, pyogenic granuloma, hemangioma, dermatofibroma, and warts [115] and for melanoma detection in seborrheic keratoses and benign nevi skin lesions [116]. Since Esteva's [111], use of the DL model for the classification of skin conditions, DL models have advanced rapidly to the use of transfer learning and pretrained VGG19, Inception V3, ResNet50, and SqueezeNet models by Ref. [117] for benign and malignant skin lesion classification.

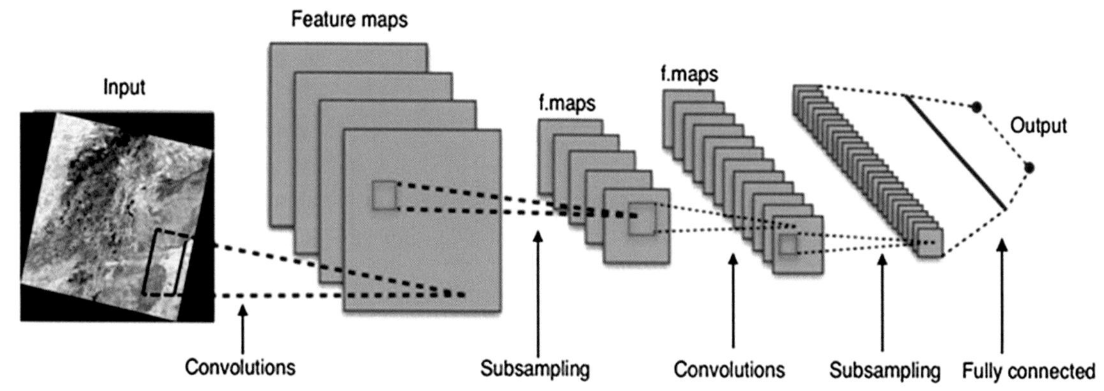

FIGURE 17.5 General CNN architecture. *Taken from Traore, B.B., Kamsu-Foguem, B. and Tangara, F., 2018. Deep convolution neural network for image recognition. Ecological Informatics, 48, pp. 257–268* [110].

Deep learning models are also employed in chest X-ray recognition for the classification of thorax diseases such as Cardiomegaly, pneumothorax, pneumonia, and pleuro-thickening, as was shown by Wang et al., [118] who used the ChestNet model to achieve this classification along with 14 thorax diseases in total [119].

More recently, with the advent of the "COVID-19" pandemic, researchers have used the multilayer perceptron algorithm (MLP) to detect, diagnose and treat this disease. It is difficult to differentiate normal pneumonia from "COVID-19" pneumonia since both present similar X-ray patterns on the lungs. The inception C-net (IC-Net) has been used to detect and differentiate COVID-19 from other diseases [120].

Researchers such as [121] have used the DeTrac Deep CNNs to further enable differentiation between the healthy and COVID-19 causes of pneumonia. **De**compose, **tra**nsfer, and **c**ompose lend to the name "DeTrac," whereby the CNNs are used to extract the deeper information from the images (Fig. 17.6).

Schematic depicting how a convolutional neural network trained on the ImageNet dataset of 1000 categories can be adapted to significantly increase the accuracy and shorten the training duration of a network trained on a novel dataset of

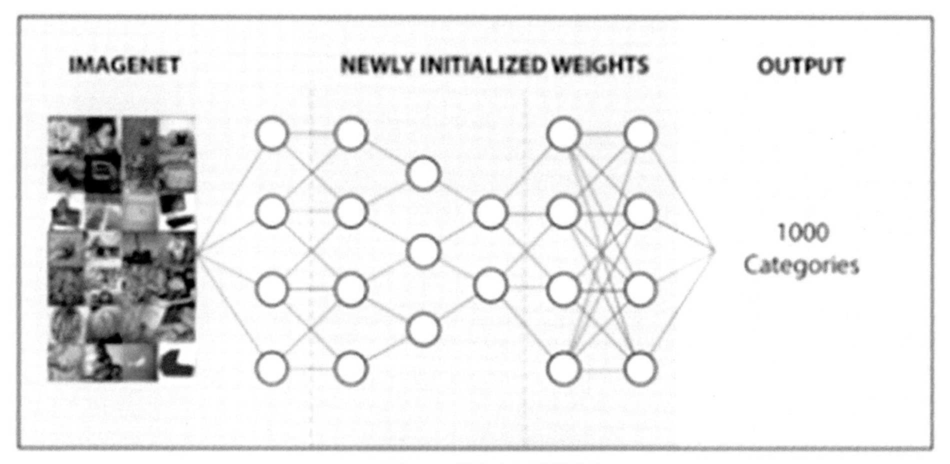

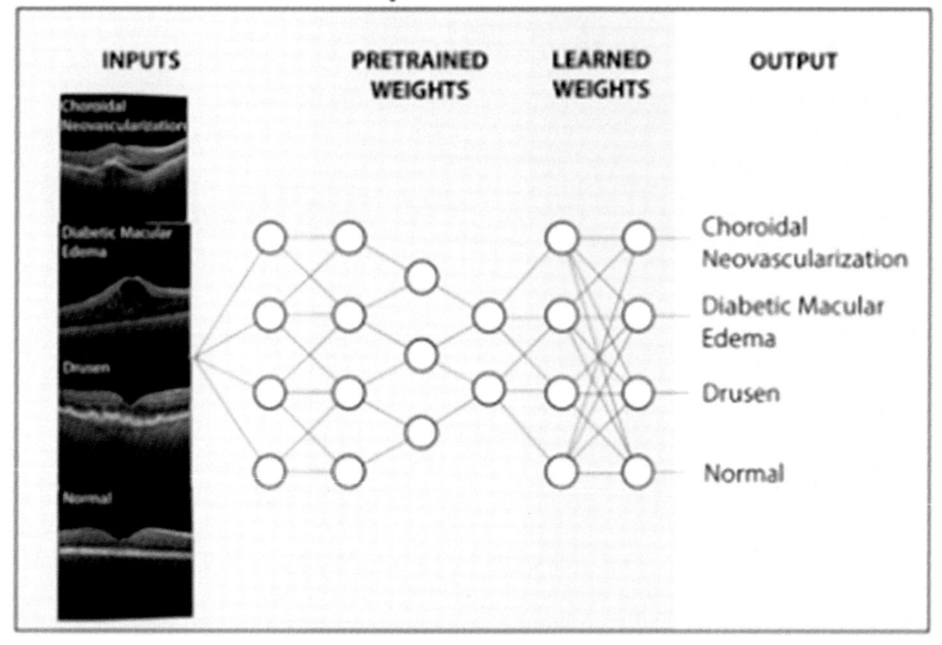

FIGURE 17.6 Schematic of a convolutional neural network. *Source: Taken from Kermany, D., Zhang, K. and Goldbaum, M., 2018a. Labeled optical coherence tomography (oct) and chest x-ray images for classification.* Mendeley Data, *2(2); Kermany, D.S., Goldbaum, M., Cai, W., Valentim, C.C., Liang, H., Baxter, S.L., McKeown, A., Yang, G., Wu, X., Yan, F. and Dong, J., 2018b. Identifying medical diagnoses and treatable diseases by image-based "Deep Learning".* Cell, *172(5), pp.1122–1131.*

OCT images. The locally connected (convolutional) layers are frozen and transferred into a new network. In contrast, the final, fully connected layers are recreated and retrained from random initialization on top of the transferred layers.

Applications of CNNs in medical imaging include diagnosing severe vision loss associated with diabetic macular edema and choroidal neovascularization.

17. Genomics

In genomics, research is focused on various areas, such as high throughput sequencing techniques which generate "big data." Therefore, it is no surprise that DL has found its way into the research of genomics, and the era of "-omics" brings with it its own set of challenges in DL. With DL, the data are transformed through as many as 200 layers making this truly DL. High-throughput platforms in the "omics" areas produce a high volume of data, which gives us a better understanding of the diseases by which the data are generated. The Cancer Genome Atlas (TCGA) is a repository of cancer samples from over 11,000 patients collected over 12 years. Sequencing and molecular characterization have been used to characterize these samples and are accessible by the research community [122]. Data from cancer tumor samples include clinical, single nucleotide polymorphism (SNP) SNP microarray data, Low-Pass DNA sequencing, whole exome, whole genome (proteomic data and epigenetic data (i.e., methylation and other chromosomal modifications), expression data (i.e., mRNA and miRNA data) such as reverse-Phase Protein Array and total RNA Sequencing among some of the data types collected [123]. This data, together with DL model application for cancer prognosis prediction, for example, for low-grade glioma and glioblastoma where H&E images, genomics data, and clinical data have been used alongside VGG19 as a base model and cox regression used as output for prognosis and thus survival from this cancer type [124–126].

There is a lack of biomarkers for the diagnosis and prognosis of soft tissue sarcoma. Using the molecular profile and morphology applying a random forest analysis, [127]; identified subtype-specific genes that could be used as diagnostic markers for soft tissue sarcoma [127,128]. The US National Institutes of Health (NIH) National Center for Biotechnology Information (NCBI) in Bethesda, Maryland, has developed the database ClinVar, which stores data on disease-causing variants. Using this database and the application of DL approaches, it is possible to predict mutations that affect pre-mRNA splicing. RNA splicing is a regulated process that removes introns (a segment of a DNA or RNA molecule which does not code for proteins and interrupts the sequence of genes), from pre-mRNA transcripts to generate mature mRNA; if there is a mutation in this process, this then leads to human monogenic diseases which include spinal muscular atrophy, neurofibromatosis type 1, cystic fibrosis, familial dysautonomia, Duchenne muscular dystrophy, and myotonic dystrophy (NIH, 2022) [126,129–133].

18. Drug discovery and precision medicine

It takes between 10 and 12 years to develop a new drug from bench to bedside with an average cost of some $314 million to $2.6 billion [134,135]. Around 90% of the drugs in phase 1 trials fail or get approval from health regulatory bodies due to lack of efficacy or toxicity induced by the novel compounds.

> *"Pharmacogenomics is a branch of genetics that is concerned with the way in which an individual genetic attributes affect the likely response to therapeutic drugs.*
>
> Ref. [136]

Pharmacogenomics screen enable associations to be found via high-throughput screens (HTS) in *in vitro* and in vivo disease models [137,138].

Drug discovery and development is now informed through computer-based approaches. AI has found its niche regarding efficacy and safety in preclinical studies in disease models.

Precision medicine has become more promising since the completion of the human genome project in 2003 [139]. Precision medicine tailor's treatments based upon the patient's disease specific biomarkers. Personalized medicine is based upon the patient's genotype, phenotype, lifestyle, and the environmental factors in which they live. With such precision, the success of clinical trials can also be increased. Indeed, personalized cancer vaccines have been developed with the use of ML to identify neoepitopes. In fact an AI-based vaccine discovery framework has been proposed by Refs. [106,107]. These researchers have proposed a combination of the in silico immuno-informatics and deep neural network strategies, into the DeepVacPred, which is a novel AI-based in silico multiepitope vaccine design computational framework, enabling 26 potential vaccine subunits from the SARS-CoV-2 spike protein sequence for a **worldwide human population coverage.** They also hope to construct a multiepitope vaccine for SARS-CoV-2 virus in the future. DeepVacPred, is a

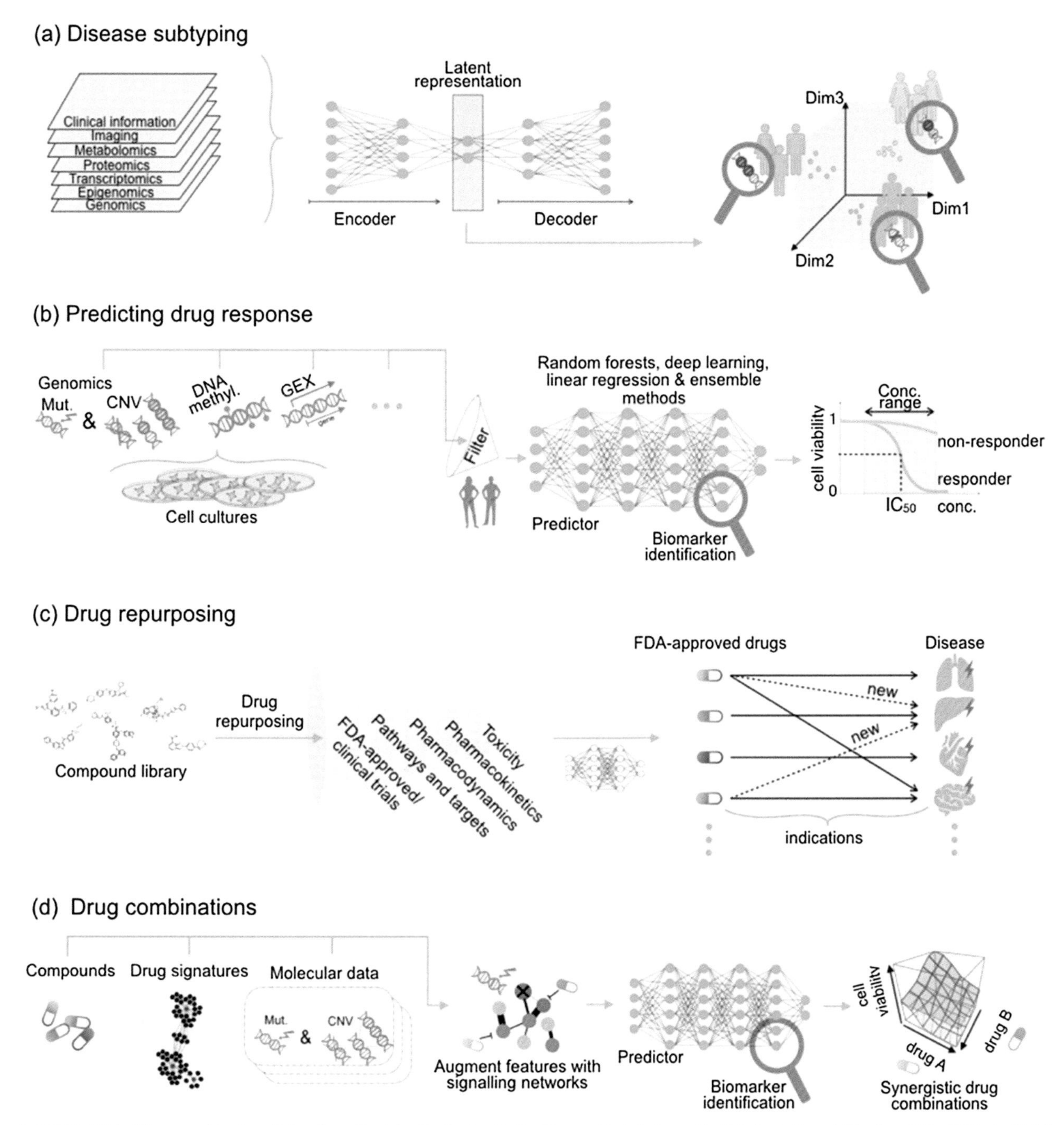

FIGURE 17.7 Disease subtyping. *Taken from Boniolo, F., Dorigatti, E., Ohnmacht, A.J., Saur, D., Schubert, B. and Menden, M.P., 2021. Artificial intelligence in early drug discovery enabling precision medicine. Expert Opinion on Drug Discovery, 16(9), pp.991–1007.*

novel AI-based in silico multiepitope vaccine design framework. The use and application of neural networks have been applied in many areas of drug discovery [140–144] (Fig. 17.7).

19. Biobank

There are several Biobanks globally, which have been initiated to capture and store is large-scale biomedical/clinical/imaging/genomic data (www.biobanking.com 2022). These provide a rich research resource for all medical practitioners and computer scientists to enable that new scientific discoveries such as new drugs and treatment targets in order to

TABLE 17.1 compiled from data on Data from 10 largest biobanks in the world https://www.biobanking.com/10-largest-biobanks-in-the-world/(www.biobanking.com, 2022) [146].

Biobank (established year)	Region	Size	Types of samples	Website
Biobank Graz [147],	Austria, Europe	20 million samples from over 30 years	Fresh frozen tissues plus fluid samples	https://biobank.medunigraz.at
Shanghai Zhangjiang Biobank [148],	Shanghai, China	10 million human derived samples	Human tissue, cells, blood, and intestinal microflora, tumor tissues	http://www.shbiobank.com
All of Us" Biobank (2018)	USA	1 million target samples	Health surveys and electronic health records. Blood and urine samples and other physical measurements.	https://www.joinallofus.org/
IARC [149],	France, Europe	5.1 million biological samples from 562,000 individuals.	Body fluids, including plasma, serum, urine and extracted DNA samples.	https://ibb.iarc.fr/
China Kadoorie Biobank [150],	China	510,000 participants	Global resource for the investigation of lifestyle, environmental, blood biochemical and genetic factors as determinants of common diseases in the Chinese population	https://www.ckbiobank.org/
UK Biobank [151]	UK (Europe)	500,000 people aged between 40 and 69 years	Genetics, lifestyle, imaging, and health records	https://www.ukbiobank.ac.uk
FINNGEN biobanks (2017)	Finland, Europe	500,000 participants	Blood samples—genome information with digital healthcare data from national registries	https://www.finngen.fi/en
Canadian Partnership for Tomorrow Project (CPTP) (2008)	Canada	Over 300,000 participants aged 30–74 years	Physical measures, mental health measures, magnetic resonance imaging (MRI) data and biological samples such as blood and urine.	https://partnershipfortomorrow.ca/
Estonian Biobank [152],	Estonia, Eastern Europe	200,000 participants	Descriptions of the state of health and lifestyle (phenotype data) and genetic data which include biological samples such as DNA	https://genomics.ut.ee/en/access-biobank
EuroBioBank [153],	Europe	150,000 biological samples	Human DNA, cell and tissue samples	http://www.eurobiobank.org/

improve the health of the population. Biobanks have to be officially registered in accordance with the International Organization for Standardization ISO 20,387 to ensure all biobanks conform to their standards and are competent and in alignment with their regulations. Table 17.1 shows a variety of Biobanks established globally (www.biobanking.com 2022). Recent publication from the UK Biobank used genome-wide association studies (GWASs) with VEGAS2 DL model and computational prediction algorithms, PolyPhen-2, CADD, DANN, PROVEAN, REVEL, VEST3, and Eigen, that consider genetic, evolutionary, structural, and biochemical information to infer variant pathogenicity and deleteriousness to identify 48 loci of which 18 harbor genes associated with hearing loss [145].

20. Conclusion

Alongside AI, both ML and DL techniques have revolutionized the healthcare system. Unlike the previous pandemics or healthcare disasters, ML and DL provided the tools to combat "COVID-19" besides the healthcare professionals. They supported sharing real-time viral sequencing, updates on contract tracing management, predicting the patient flow, and controlling the "COVID-19" vaccination efforts. The densely populated low-middle-income and low-income countries are widely using technology platforms for monitoring and managing healthcare facilities. From drug development to hands-on clinical education system, both ML and DL have proved their efficiency against deadly diseases. Their potential applications will enhance the possibilities of predicting possible disease outbreaks, tracking the spread, warning potentially vulnerable populations, and providing real-time information exchange among healthcare professionals, researchers, policymakers, media, local authorities, and government agencies. Further studies should be conducted to reduce the risk of data breaches and increase online security for future "big data"-based healthcare management systems.

References

[1] M. Chen, Y. Hao, K. Hwang, L. Wang, L. Wang, Disease prediction by machine learning over "big data" from healthcare communities, IEEE Access 5 (2017) 8869–8879.

[2] E. Topol, High-performance medicine: the convergence of human and artificial intelligence, Nature Medicine 25 (1) (2019) 44–56.

[3] A. Beam, I. Kohane, "big data" and machine learning in health care, JAMA 319 (13) (2018) 1317.

[4] D. Char, N. Shah, D. Magnus, Implementing machine learning in health care — addressing ethical challenges, New England Journal of Medicine 378 (11) (2018) 981–983.

[5] R. Delahanty, D. Kaufman, S. Jones, Development and evaluation of an automated machine learning algorithm for in-hospital mortality risk adjustment among critical care patients, Critical Care Medicine 46 (6) (2018) e481–e488.

[6] Z. Obermeyer, E. Emanuel, Predicting the future—"big data", machine learning, and clinical medicine, New England Journal of Medicine 375 (13) (2016) 1216–1219.

[7] F. Khan, Educating and engaging junior doctors, Future Healthcare Journal 6 (Suppl. 1) (2019), 189-189.

[8] P. Baldi, Deep learning" in biomedical data science, Annual Review of Biomedical Data Science 1 (1) (2018) 181–205.

[9] J. Wang, H. Zhu, S.H. Wang, Y.D. Zhang, A review of deep learning on medical image analysis, Mobile Networks and Applications 26 (2021), 351–80.

[10] R. Qu, G. Xu, C. Ding, W. Jia, M. Sun, Standard plane identification in fetal brain ultrasound scans using a differential convolutional neural network, IEEE Access 8 (2020) 83821–83830.

[11] J. Nam, H. Kang, S. Lee, H. Kim, C. Rhee, J. Goo, Y. Oh, C. Lee, C. Park, "Deep learning" prediction of survival in patients with chronic obstructive pulmonary disease using chest radiographs, Radiology (2022).

[12] L. Zhang, H. Wang, Q. Li, M. Zhao, Q. Zhan, "big data" and medical research in China, BMJ (2018) j5910.

[13] F.S. Collins, H. Varmus, A new initiative on precision medicine, New England Journal of Medicine 372 (9) (2015) 793–795.

[14] R. Miotto, F. Wang, S. Wang, X. Jiang, J.T. Dudley, "Deep Learning" for healthcare: review, opportunities and challenges, Briefings in Bioinformatics 19 (6) (2018) 1236–1246.

[15] M. Maruthappu, K.Y.B. Ng, C. Williams, R. Atun, T. Zeltner, Government health care spending and child mortality, Pediatrics 135 (4) (2015) e887–e894.

[16] Y. LeCun, Y. Bengio, G. Hinton, Deep learning, Nature 521 (7553) (2015) 436–444.

[17] M.K. Leung, H.Y. Xiong, L.J. Lee, B.J. Frey, "Deep Learning" of the tissue-regulated splicing code, Bioinformatics 30 (12) (2014) i121–i129.

[18] A.E. Adegbosin, B. Stantic, J. Sun, Efficacy of "Deep Learning" methods for predicting under-five mortality in 34 low-income and middle-income countries, BMJ Open 10 (8) (2020) e034524.

[19] S. Dash, S.K. Shakyawar, M. Sharma, S. Kaushik, "big data" in healthcare: management, analysis and future prospects, Journal of Big Data 6 (1) (2019) 1–25.

[20] B. Wang, A.M. Mezlini, F. Demir, M. Fiume, Z. Tu, M. Brudno, B. Haibe-Kains, A. Goldenberg, Similarity network fusion for aggregating data types on a genomic scale, Nature Methods 11 (3) (2014) 333–337.

[21] K. Shameer, M.A. Badgeley, R. Miotto, B.S. Glicksberg, J.W. Morgan, J.T. Dudley, Translational bioinformatics in the era of real-time biomedical, health care and wellness data streams, Briefings in Bioinformatics 18 (1) (2017) 105–124.

[22] R. Cornet, C.G. Chute, Health concept and knowledge management: twenty-five years of evolution, Yearbook of medical informatics 25 (S 01) (2016) S32–S41.

[23] A. Mohan, D.M. Blough, T. Kurc, A. Post, J. Saltz, Detection of conflicts and inconsistencies in taxonomy-based authorization policies, in: 2011 IEEE International Conference on Bioinformatics and Biomedicine, IEEE, November 2011, pp. 590–594.

[24] L. Li, W.Y. Cheng, B.S. Glicksberg, O. Gottesman, R. Tamler, R. Chen, E.P. Bottinger, J.T. Dudley, Identification of type 2 diabetes subgroups through topological analysis of patient similarity, Science Translational Medicine 7 (311) (2015), 311ra174-311ra174.

[25] A. Belle, R. Thiagarajan, S.M. Soroushmehr, F. Navidi, D.A. Beard, K. Najarian, "big Data" Analytics in Healthcare, BioMed research international, 2015.

[26] S. Gopalani, R. Arora, Comparing apache spark and map reduce with performance analysis using k-means, International Journal of Computer Applications 113 (1) (2015).

[27] R. Zhao, R. Yan, Z. Chen, K. Mao, P. Wang, R.X. Gao, "Deep Learning" and its applications to machine health monitoring, Mechanical Systems and Signal Processing 115 (2019) 213–237.

[28] Y. Bengio, A. Courville, P. Vincent, Representation learning: a review and new perspectives, IEEE Transactions on Pattern Analysis and Machine Intelligence 35 (8) (2013) 1798–1828.

[29] M.A. El-Rashidy, R.G. Mohamed, N.A. El-Fishawy, M.A. Shouman, Reliable plagiarism detection system based on "Deep Learning" approaches, Neural Computing & Applications (2022) 1–22.
[30] A. Abrishami, S. Aliakbary, Predicting citation counts based on deep neural network learning techniques, Journal of Informetrics 13 (2) (2019) 485–499.
[31] F. Sánchez-Vega, E. Villatoro-Tello, M. Montes-y-Gómez, P. Rosso, E. Stamatatos, L. Villasenor-Pineda, Paraphrase plagiarism identification with character-level features, Pattern Analysis & Applications 22 (2) (2019) 669–681.
[32] E. Stamatatos, Plagiarism detection using stopword n-grams, Journal of the American Society for Information Science and Technology 62 (12) (2011) 2512–2527.
[33] G. Eugene, The Agony and the Ecstasy—The History and Meaning of the Journal Impact Factor, 2005, p. 2006. *Retrieved December, 28.*
[34] H.F. Moed, L. Colledge, J. Reedijk, F. Moya-Anegon, V. Guerrero-Bote, A. Plume, M. Amin, Citation-based metrics are appropriate tools in journal assessment provided that they are accurate and used in an informed way, Scientometrics 92 (2) (2012) 367–376.
[35] A. Krizhevsky, I. Sutskever, G.E. Hinton, Imagenet classification with deep convolutional neural networks, Communications of the ACM 60 (6) (2017) 84–90.
[36] G.E. Dahl, D. Yu, L. Deng, A. Acero, Context-dependent pre-trained deep neural networks for large-vocabulary speech recognition, IEEE Transactions on Audio Speech and Language Processing 20 (1) (2011) 30–42.
[37] G. Hinton, L. Deng, D. Yu, G.E. Dahl, A.R. Mohamed, N. Jaitly, A. Senior, V. Vanhoucke, P. Nguyen, T.N. Sainath, B. Kingsbury, Deep neural networks for acoustic modeling in speech recognition: the shared views of four research groups, IEEE Signal Processing Magazine 29 (6) (2012) 82–97.
[38] Y. Xiao, M.E. Torok, Taking the right measures to control "COVID-19", The Lancet Infectious Diseases 20 (5) (2020) 523–524.
[39] J. Budd, B.S. Miller, E.M. Manning, V. Lampos, M. Zhuang, M. Edelstein, G. Rees, V.C. Emery, M.M. Stevens, N. Keegan, M.J. Short, Digital technologies in the public-health response to "COVID-19", Nature Medicine 26 (8) (2020) 1183–1192.
[40] X. Zhou, A. de Figueiredo, Q. Xu, L. Lin, P.E. Kummervold, H. Larson, M. Jit, Z. Hou, Monitoring Global Trends in "COVID-19" Vaccination Intention and Confidence: A Social Media-Based "Deep Learning" Study, medRxiv, 2021.
[41] G.H. Kwak, L. Ling, P. Hui, Deep reinforcement learning approaches for global public health strategies for "COVID-19" pandemic, PLoS One 16 (5) (2021) e0251550.
[42] A. Odone, V. Gianfredi, S. Sorbello, M. Capraro, B. Frascella, G.P. Vigezzi, C. Signorelli, The use of digital technologies to support vaccination programmes in Europe: state of the art and best practices from experts' interviews, Vaccines 9 (10) (2021) 1126.
[43] S. Nyawa, D. Tchuente, S. Fosso-Wamba, "COVID-19" vaccine hesitancy: a social media analysis using "Deep Learning", Annals of Operations Research (2022) 1–39.
[44] E. Dubé, M. Vivion, N.E. MacDonald, Vaccine hesitancy, vaccine refusal and the anti-vaccine movement: influence, impact and implications, Expert Review of Vaccines 14 (1) (2015) 99–117.
[45] D. Freeman, F. Waite, L. Rosebrock, A. Petit, C. Causier, A. East, L. Jenner, A.L. Teale, L. Carr, S. Mulhall, E. Bold, Coronavirus conspiracy beliefs, mistrust, and compliance with government guidelines in England, Psychological Medicine 52 (2) (2022) 251–263.
[46] L.J. Kay, The Enterprise, 252, RED, 2020, pp. 265–8117.
[47] S. Altay, A.S. Hacquin, C. Chevallier, H. Mercier, Information delivered by a chatbot has a positive impact on "COVID-19" vaccines attitudes and intentions, Journal of Experimental Psychology: Applied (2021).
[48] L. Deng, X. Li, Machine learning paradigms for speech recognition: an overview, IEEE Transactions on Audio Speech and Language Processing 21 (5) (2013) 1060–1089.
[49] R. Sayeed, D. Gottlieb, K.D. Mandl, SMART Markers: collecting patient-generated health data as a standardized property of health information technology, NPJ Digital Medicine 3 (1) (2020) 1–8.
[50] C.J. Siah, S.T. Lau, S.S. Tng, C.H. Chua, Using infrared imaging and deep learning in fit-checking of respiratory protective devices among healthcare professionals, Journal of Nursing Scholarship 54 (3) (2022) 345–354.
[51] T. Brosch, R. Tam, Alzheimer's Disease Neuroimaging Initiative, Manifold learning of brain MRIs by "deep learning", in: International Conference on Medical Image Computing And Computer-Assisted Intervention, Springer, Berlin, Heidelberg, September 2013, pp. 633–640.
[52] S. Liu, S. Liu, W. Cai, S. Pujol, R. Kikinis, D. Feng, Early diagnosis of alzheimer's disease with "deep learning", in: 2014 IEEE 11th International Symposium on Biomedical Imaging (ISBI), IEEE, April 2014, pp. 1015–1018.
[53] W. Schroeder, K.M. Martin, W.E. Lorensen, The Visualization Toolkit an Object-Oriented Approach to 3D Graphics, Prentice-Hall, Inc, 1998.
[54] K.J. Friston, Statistical parametric mapping, in: Neuroscience Databases, Springer, Boston, MA, 2003, pp. 237–250.
[55] A.B. Shams, E. Hoque Apu, A. Rahman, M.M. Sarker Raihan, N. Siddika, R.B. Preo, M.R. Hussein, S. Mostari, R. Kabir, Web search engine misinformation notifier extension (SEMiNExt): a machine learning based approach during "COVID-19" Pandemic, Healthcare 9 (2) (February 2021) 156 (MDPI).
[56] X. Zhang, A.A. Ghorbani, An overview of online fake news: characterization, detection, and discussion, Information Processing and Management 57 (2) (2020) 102025.
[57] L. Bode, E.K. Vraga, In related news, that was wrong: the correction of misinformation through related stories functionality in social media, Journal of Communication 65 (4) (2015) 619–638.
[58] A. Ghenai, Y. Mejova, Fake cures: user-centric modeling of health misinformation in social media, Proceedings of the ACM on Human-Computer Interaction 2 (2018) 1–20. CSCW.

[59] Y. Zhao, J. Zhang, Consumer health information seeking in social media: a literature review, Health Information and Libraries Journal 34 (4) (2017) 268–283.

[60] L. Bode, E.K. Vraga, See something, say something: correction of global health misinformation on social media, Health Communication 33 (9) (2018) 1131–1140.

[61] Y. Zhao, J. Da, J. Yan, Detecting health misinformation in online health communities: incorporating behavioral features into machine learning based approaches, Information Processing and Management 58 (1) (2021) 102390.

[62] R.E. Petty, J.T. Cacioppo, The elaboration likelihood model of persuasion, in: Communication and Persuasion, Springer, New York, NY, 1986, pp. 1–24.

[63] R.A. Raghavendra, D. Samanta, A real-time approach with "deep learning" for pandemic management, in: Healthcare Informatics For Fighting "COVID-19" and Future Epidemics, Springer, Cham, 2022, pp. 113–139.

[64] Y.T. Prasetyo, S.A.R. Tumanan, L.A.F. Yarte, M.C.C. Ogoy, A.K.S. Ong, Blackboard E-learning system acceptance and satisfaction among Filipino high school students: an extended technology acceptance model (TAM) approach, in: 2020 IEEE International Conference on Industrial Engineering and Engineering Management (IEEM), IEEE, December 2020, pp. 1271–1275.

[65] E.A. Rashed, A. Hirata, Infectivity upsurge by "COVID-19" viral variants in Japan: evidence from "deep learning" modeling, International Journal of Environmental Research and Public Health 18 (15) (2021) 7799.

[66] I. Rodríguez-Rodríguez, J.V. Rodríguez, N. Shirvanizadeh, A. Ortiz, D.J. Pardo-Quiles, Applications of artificial intelligence, machine learning, "big data" and the internet of things to the "COVID-19" pandemic: a scientometric review using text mining, International Journal of Environmental Research and Public Health 18 (16) (2021) 8578.

[67] N. Yuduang, A.K.S. Ong, Y.T. Prasetyo, T. Chuenyindee, P. Kusonwattana, W. Limpasart, T. Sittiwatethanasiri, M.J.J. Gumasing, J.D. German, R. Nadlifatin, Factors influencing the perceived effectiveness of "COVID-19" risk assessment mobile application "MorChana" in Thailand: UTAUT2 approach, International Journal of Environmental Research and Public Health 19 (9) (2022) 5643.

[68] Y.T. Prasetyo, R.S. Dewi, N.M. Balatbat, M.L.B. Antonio, T. Chuenyindee, A.A.N. Perwira Redi, M.N. Young, J.F.T. Diaz, Y.B. Kurata, The evaluation of preference and perceived quality of health communication icons associated with "COVID-19" prevention measures, Healthcare 9 (9) (2021a) 1115 (MDPI), August.

[69] Y.T. Prasetyo, T. Maulanti, S.F. Persada, A.A.N. Perwira Redi, M.N. Young, J.F.T. Diaz, Factors influencing job satisfaction among dentists during the new normal of the "COVID-19" pandemic in Indonesia: a structural equation modeling approach, Work (2021b) 1–12.

[70] R. Jalabneh, H.Z. Syed, S. Pillai, E.H. Apu, M.R. Hussein, R. Kabir, S.M. Arafat, M. Majumder, A. Azim, S.K. Saxena, Use of mobile phone apps for contact tracing to control the "COVID-19" pandemic: a literature review, Applications of Artificial Intelligence in "COVID-19" (2021) 389–404.

[71] M. Alhussein, G. Muhammad, Voice pathology detection using "Deep Learning" on mobile healthcare framework, IEEE Access 6 (2018) 41034–41041.

[72] D.J. Hemanth, V.E. Balas (Eds.), Telemedicine Technologies: "Big Data", "Deep Learning", Robotics, Mobile and Remote Applications for Global Healthcare, 2019 (Academic Press).

[73] M. Woods, R. Miklencicova, Digital epidemiological surveillance, smart telemedicine diagnosis systems, and machine learning-based real-time data sensing and processing in "COVID-19" remote patient monitoring, American Journal of Medical Research 8 (2) (2021) 65–78.

[74] O.H. Salman, Z. Taha, M.Q. Alsabah, Y.S. Hussein, A.S. Mohammed, M. Aal-Nouman, A review on utilizing machine learning technology in the fields of electronic emergency triage and patient priority systems in telemedicine: coherent taxonomy, motivations, open research challenges and recommendations for intelligent future work, Computer Methods and Programs in Biomedicine 209 (2021) 106357.

[75] N. Kalid, A.A. Zaidan, B.B. Zaidan, O.H. Salman, M. Hashim, H.J.J.O.M.S. Muzammil, Based real time remote health monitoring systems: a review on patients prioritization and related" "big data"" using body sensors information and communication technology, Journal of Medical Systems 42 (2) (2018) 1–30.

[76] O.H. Salman, M.I. Aal-Nouman, Z.K. Taha, Reducing waiting time for remote patients in telemedicine with considering treated patients in emergency department based on body sensors technologies and hybrid computational algorithms: toward scalable and efficient real time healthcare monitoring system, Journal of Biomedical Informatics 112 (2020) 103592.

[77] I.R. Mendo, G. Marques, I. de la Torre Díez, M. López-Coronado, F. Martín-Rodríguez, Machine learning in medical emergencies: a systematic review and analysis, Journal of Medical Systems 45 (10) (2021) 1–16.

[78] U. Bharti, D. Bajaj, H. Batra, S. Lalit, S. Lalit, A. Gangwani, Medbot: conversational artificial intelligence powered chatbot for delivering telehealth after "COVID-19", in: 2020 5th International Conference on Communication and Electronics Systems (ICCES), IEEE, June 2020, pp. 870–875.

[79] E. Meinert, M. Milne-Ives, S. Surodina, C. Lam, Agile requirements engineering and software planning for a digital health platform to engage the effects of isolation caused by social distancing: case study, JMIR Public Health and Surveillance 6 (2) (2020) e19297.

[80] N. El-Rashidy, S. El-Sappagh, S.M. Islam, H.M. El-Bakry, S. Abdelrazek, End-to-end "Deep Learning" framework for coronavirus ("COVID-19") detection and monitoring, Electronics 9 (9) (2020) 1439.

[81] J.E. Hollander, B.G. Carr, Virtually perfect? Telemedicine for "COVID-19", New England Journal of Medicine 382 (18) (2020) 1679–1681.

[82] L.G. Olson, S.R. Hill, D.A. Newby, Barriers to student access to patients in a group of teaching hospitals, Medical Journal of Australia 183 (9) (2005) 461–463.

[83] S. Jenkins, R. Goel, D.S. Morrell, Computer-assisted instruction versus traditional lecture for medical student teaching of dermatology morphology: a randomized control trial, Journal of the American Academy of Dermatology 59 (2) (2008) 255–259.

[84] T. Lüdert, A. Nast, H. Zielke, W. Sterry, B. Rzany, E-learning in the dermatological education at the Charité: evaluation of the last three years, JDDG: Journal der Deutschen Dermatologischen Gesellschaft 6 (6) (2008) 467–472.
[85] N.B. Berman, L.H. Fall, C.G. Maloney, D.A. Levine, Computer-assisted instruction in clinical education: a roadmap to increasing CAI implementation, Advances in Health Sciences Education 13 (3) (2008) 373–383.
[86] J.G. Ruiz, M.J. Mintzer, R.M. Leipzig, The impact of e-learning in medical education, Academic Medicine 81 (3) (2006) 207–212.
[87] D.G. Singh, N. Boudville, R. Corderoy, S. Ralston, C.P. Tait, Impact on the dermatology educational experience of medical students with the introduction of online teaching support modules to help address the reduction in clinical teaching, Australasian Journal of Dermatology 52 (4) (2011) 264–269.
[88] M.I. Islam, S.S. Jahan, M.T.H. Chowdhury, S.N. Isha, A.K. Saha, S.K. Nath, M.S. Jahan, M.H. Kabir, E. Hoque Apu, R. Kabir, N. Siddika, Experience of Bangladeshi dental students towards online learning during the "COVID-19" pandemic: a web-based cross-sectional study, International Journal of Environmental Research and Public Health 19 (13) (2022) 7786.
[89] J.N. Kok, E.J. Boers, W.A. Kosters, P. Van der Putten, M. Poel, Artificial intelligence: definition, trends, techniques, and cases, Artificial Intelligence 1 (2009) 270–299.
[90] A.B. Simmons, S.G. Chappell, Artificial intelligence-definition and practice, IEEE Journal of Oceanic Engineering 13 (2) (1988) 14–42.
[91] Hewlett Packard Enterprise Development LP, What Is Artificial Intelligence?, 2022. Available at: https://www.hpe.com/uk/en/what-is/artificial-intelligence.html?jumpid=ps_3hmavknz8b_aid-520061736&ef_id. (Accessed 14 July 2022).
[92] J. Alzubi, A. Nayyar, A. Kumar, Machine learning from theory to algorithms: an overview, Journal of Physics: Conference Series 1142 (1) (November 2018) 012012 (IOP Publishing).
[93] H. Lv, H. Tang, Machine learning methods and their application research, in: 2011 2nd International Symposium on Intelligence Information Processing and Trusted Computing, IEEE, October 2011, pp. 108–110.
[94] T.M. Mitchell, T.M. Mitchell, Machine Learning, No. 9 vol 1, McGraw-Hill, New York, 1997.
[95] B. Mahesh, Machine learning algorithms-a review, International Journal of Science and Research 9 (2020) 381–386.
[96] Royal Society, Machine Learning: The Power and Promise of Computers that Learn by Example, Royal Society, London, UK, 2017. Available at: https://royalsociety.org/~/media/policy/projects/machine-learning/publications/machine-learning-report.pdf. (Accessed 14 July 2022).
[97] N. Sandhya, K.R. Charanjeet, A review on machine learning techniques, International Journal on Recent and Innovation Trends in Computing and Communication 4 (3) (2016) 451–458.
[98] R. Kashyap, Machine learning for internet of things, in: Research Anthology on Artificial Intelligence Applications in Security, IGI Global, 2021, pp. 976–1002.
[99] M. Naeem, S.T.H. Rizvi, A. Coronato, A gentle introduction to reinforcement learning and its application in different fields, IEEE Access 8 (2020) 209320–209344.
[100] R.W. Moehring, M. Phelan, E. Lofgren, A. Nelson, E.D. Ashley, D.J. Anderson, B.A. Goldstein, Development of a machine learning model using electronic health record data to identify antibiotic use among hospitalized patients, JAMA Network Open 4 (3) (2021) e213460-e213460.
[101] M.W. Segar, J.L. Hall, P.S. Jhund, T.M. Powell-Wiley, A.A. Morris, D. Kao, G.C. Fonarow, R. Hernandez, N.E. Ibrahim, C. Rutan, A.M. Navar, Machine learning–based models incorporating social determinants of health vs traditional models for predicting in-hospital mortality in patients with heart failure, JAMA Cardiology (2022).
[102] E. Xu, S. Nemati, A.H. Tremoulet, A deep convolutional neural network for Kawasaki disease diagnosis, Scientific Reports 12 (1) (2022) 1–6.
[103] Y. Jing, J. Yang, D.B. Johnson, J.J. Moslehi, L. Han, Harnessing "big data" to characterize immune-related adverse events, Nature Reviews Clinical Oncology 19 (4) (2022) 269–280.
[104] R.J. McDonald, K.M. Schwartz, L.J. Eckel, F.E. Diehn, C.H. Hunt, B.J. Bartholmai, B.J. Erickson, D.F. Kallmes, The effects of changes in utilization and technological advancements of cross-sectional imaging on radiologist workload, Academic Radiology 22 (9) (2015) 1191–1198.
[105] S.C. Huang, A. Pareek, S. Seyyedi, I. Banerjee, M.P. Lungren, Fusion of medical imaging and electronic health records using "deep learning": a systematic review and implementation guidelines, NPJ digital medicine 3 (1) (2020) 1–9.
[106] D. Yang, C. Martinez, L. Visuña, H. Khandhar, C. Bhatt, J. Carretero, Detection and analysis of "COVID-19" in medical images using "Deep Learning" techniques, Scientific Reports 11 (1) (2021a) 1–13.
[107] Z. Yang, P. Bogdan, S. Nazarian, An in silico "Deep Learning" approach to multi-epitope vaccine design: a SARS-CoV-2 case study, Scientific Reports 11 (1) (2021b) 1–21.
[108] E. Gibson, W. Li, C. Sudre, L. Fidon, D.I. Shakir, G. Wang, Z. Eaton-Rosen, R. Gray, T. Doel, Y. Hu, T. Whyntie, NiftyNet: a deep-learning platform for medical imaging, Computer Methods and Programs in Biomedicine 158 (2018) 113–122.
[109] J. Muschelli, Recommendations for processing head CT data, Frontiers in Neuroinformatics 13 (2019) 61.
[110] B.B. Traore, B. Kamsu-Foguem, F. Tangara, Deep convolution neural network for image recognition, Ecological Informatics 48 (2018) 257–268.
[111] A. Esteva, B. Kuprel, R.A. Novoa, J. Ko, S.M. Swetter, H.M. Blau, S. Thrun, Dermatologist-level classification of skin cancer with deep neural networks, Nature 542 (7639) (2017) 115–118.
[112] D. Kermany, K. Zhang, M. Goldbaum, Labeled optical coherence tomography (oct) and chest x-ray images for classification, Mendeley Data 2 (2) (2018a).
[113] D.S. Kermany, M. Goldbaum, W. Cai, C.C. Valentim, H. Liang, S.L. Baxter, A. McKeown, G. Yang, X. Wu, F. Yan, J. Dong, Identifying medical diagnoses and treatable diseases by image-based "Deep Learning", Cell 172 (5) (2018b) 1122–1131.
[114] S. Serte, A. Serener, F. Al-Turjman, "Deep Learning" in medical imaging: a brief review, Transactions on Emerging Telecommunications Technologies (2020) e4080.

[115] S.S. Han, M.S. Kim, W. Lim, G.H. Park, I. Park, S.E. Chang, Classification of the clinical images for benign and malignant cutaneous tumors using a "Deep Learning" algorithm, Journal of Investigative Dermatology 138 (7) (2018) 1529–1538.
[116] A. Menegola, J. Tavares, M. Fornaciali, L.T. Li, S. Avila, E. Valle, RECOD Titans at ISIC Challenge 2017, 2017 *arXiv preprint arXiv:1703.04819*.
[117] A. Khamparia, P.K. Singh, P. Rani, D. Samanta, A. Khanna, B. Bhushan, An internet of health things-driven "Deep Learning" framework for detection and classification of skin cancer using transfer learning, Transactions on Emerging Telecommunications Technologies 32 (7) (2021) e3963.
[118] M. Wang, S.W. Cheung, E.T. Chung, Y. Efendiev, W.T. Leung, Y. Wang, Prediction of discretization of gmsfem using "Deep Learning", Mathematics 7 (5) (2019) 412.
[119] H. Wang, Y. Xia, Chestnet: A Deep Neural Network for Classification of Thoracic Diseases on Chest Radiography, 2018 arXiv preprint arXiv:1807.03058.
[120] N. Darapaneni, T. Gupta, A.R. Paduri, A. Banerji, S. Sharma, D. Sharma, N. Gupta, Inception C-net (IC-net): altered inception module for detection of "COVID-19" and pneumonia using chest X-rays, in: 2020 IEEE 15th International Conference on Industrial and Information Systems (ICIIS), IEEE, November 2020, pp. 393–398.
[121] A. Abbas, M.M. Abdelsamea, M.M. Gaber, Classification of "COVID-19" in chest X-ray images using DeTraC deep convolutional neural network, Applied Intelligence 51 (2) (2021) 854–864.
[122] C. Hutter, J.C. Zenklusen, The cancer genome atlas: creating lasting value beyond its data, Cell 173 (2) (2018) 283–285.
[123] National Cancer Institute. NCI's Genome Characterization Pipeline. Available at: https://www.cancer.gov/about-nci/organization/ccg/research/genomic-pipeline#genome-characterization [Accessed 17th July 2022].
[124] K. Kourou, T.P. Exarchos, K.P. Exarchos, M.V. Karamouzis, D.I. Fotiadis, Machine learning applications in cancer prognosis and prediction, Computational and Structural Biotechnology Journal 13 (2015) 8–17.
[125] P. Mobadersany, S. Yousefi, M. Amgad, D.A. Gutman, J.S. Barnholtz-Sloan, J.E. Velázquez Vega, D.J. Brat, L.A. Cooper, Predicting cancer outcomes from histology and genomics using convolutional networks, Proceedings of the National Academy of Sciences 115 (13) (2018) E2970–E2979.
[126] W. Zhu, L. Xie, J. Han, X. Guo, The application of "Deep Learning" in cancer prognosis prediction, Cancers 12 (3) (2020) 603, https://doi.org/10.3390/cancers12030603.
[127] D.G. van IJzendoorn, K. Szuhai, I.H. Briaire-de Bruijn, M. Kostine, M.L. Kuijjer, J.V. Bovée, Machine learning analysis of gene expression data reveals novel diagnostic and prognostic biomarkers and identifies therapeutic targets for soft tissue sarcomas, PLoS Computational Biology 15 (2) (2019) e1006826.
[128] M. Baker, One-stop shop for disease genes, Nature 491 (7423) (2012) 171.
[129] M. Civelek, A.J. Lusis, Systems genetics approaches to understand complex traits, Nature Reviews Genetics 15 (1) (2014) 34–48.
[130] D. Gao, E. Morini, M. Salani, A.J. Krauson, A. Chekuri, N. Sharma, A. Ragavendran, S. Erdin, E.M. Logan, W. Li, A. Dakka, A "deep learning" approach to identify gene targets of a therapeutic for human splicing disorders, Nature Communications 12 (1) (2021) 1–15.
[131] G. Graham, N. Csicsery, E. Stasiowski, G. Thouvenin, W.H. Mather, M. Ferry, S. Cookson, J. Hasty, Genome-scale transcriptional dynamics and environmental biosensing, Proceedings of the National Academy of Sciences 117 (6) (2020) 3301–3306.
[132] V. Prasad, T. Fojo, M. Brada, Precision oncology: origins, optimism, and potential, The Lancet Oncology 17 (2) (2016) e81–e86.
[133] E. Trivizakis, G.Z. Papadakis, I. Souglakos, N. Papanikolaou, L. Koumakis, D.A. Spandidos, A. Tsatsakis, A.H. Karantanas, K. Marias, Artificial intelligence radiogenomics for advancing precision and effectiveness in oncologic care, International Journal of Oncology 57 (1) (2020) 43–53.
[134] M.J. Waring, J. Arrowsmith, A.R. Leach, P.D. Leeson, S. Mandrell, R.M. Owen, G. Pairaudeau, W.D. Pennie, S.D. Pickett, J. Wang, O. Wallace, An analysis of the attrition of drug candidates from four major pharmaceutical companies, Nature Reviews Drug Discovery 14 (7) (2015) 475–486.
[135] O.J. Wouters, M. McKee, J. Luyten, Estimated research and development investment needed to bring a new medicine to market, 2009–2018, JAMA 323 (9) (2020) 844–853.
[136] Oxford English Dictionary. Available at: https://www.lexico.com/definition/pharmacogenomics [Accessed 17 July 2022].
[137] H. Gao, J.M. Korn, S. Ferretti, J.E. Monahan, Y. Wang, M. Singh, C. Zhang, C. Schnell, G. Yang, Y. Zhang, O.A. Balbin, High-throughput screening using patient-derived tumor xenografts to predict clinical trial drug response, Nature Medicine 21 (11) (2015) 1318–1325.
[138] F. Iorio, T.A. Knijnenburg, D.J. Vis, G.R. Bignell, M.P. Menden, M. Schubert, N. Aben, E. Gonçalves, S. Barthorpe, H. Lightfoot, T. Cokelaer, A landscape of pharmacogenomic interactions in cancer, Cell 166 (3) (2016) 740–754.
[139] J.C. Denny, S.L. Van Driest, W.Q. Wei, D.M. Roden, The influence of big (clinical) data and genomics on precision medicine and drug development, Clinical Pharmacology and Therapeutics (St. Louis) 103 (3) (2018) 409–418.
[140] N. Hilf, S. Kuttruff-Coqui, K. Frenzel, V. Bukur, S. Stevanović, C. Gouttefangeas, M. Platten, G. Tabatabai, V. Dutoit, S.H. van der Burg, F. Martínez-Ricarte, Actively personalized vaccination trial for newly diagnosed glioblastoma, Nature 565 (7738) (2019) 240–245.
[141] D.B. Keskin, A.J. Anandappa, J. Sun, I. Tirosh, N.D. Mathewson, S. Li, G. Oliveira, A. Giobbie-Hurder, K. Felt, E. Gjini, S.A. Shukla, Neoantigen vaccine generates intratumoral T cell responses in phase Ib glioblastoma trial, Nature 565 (7738) (2019) 234–239.
[142] P.A. Ott, S. Hu-Lieskovan, B. Chmielowski, R. Govindan, A. Naing, N. Bhardwaj, K. Margolin, M.M. Awad, M.D. Hellmann, J.J. Lin, T. Friedlander, A phase Ib trial of personalized neoantigen therapy plus anti-PD-1 in patients with advanced melanoma, non-small cell lung cancer, or bladder cancer, Cell 183 (2) (2020) 347–362.
[143] P.A. Ott, Z. Hu, D.B. Keskin, S.A. Shukla, J. Sun, D.J. Bozym, W. Zhang, A. Luoma, A. Giobbie-Hurder, L. Peter, C. Chen, An immunogenic personal neoantigen vaccine for patients with melanoma, Nature 547 (7662) (2017) 217–221.

[144] U. Sahin, E. Derhovanessian, M. Miller, B.P. Kloke, P. Simon, M. Löwer, V. Bukur, A.D. Tadmor, U. Luxemburger, B. Schrörs, T. Omokoko, Personalized RNA mutanome vaccines mobilize poly-specific therapeutic immunity against cancer, Nature 547 (7662) (2017) 222–226.
[145] N. Trpchevska, M.B. Freidin, L. Broer, B.C. Oosterloo, S. Yao, Y. Zhou, B. Vona, C. Bishop, A. Bizaki-Vallaskangas, B. Canlon, F. Castellana, Genome-wide association meta-analysis identifies 48 risk variants and highlights the role of the stria vascularis in hearing loss, The American Journal of Human Genetics 109 (6) (2022) 1077–1091.
[146] Data from 10 largest biobanks in the world. Available at: https://www.biobanking.com/10-largest-biobanks-in-the-world/[Accessed 14 July 2022].
[147] Biobank Graz, Medical University of Graz, 2022. Available at: https://biobank.medunigraz.at. (Accessed 14 July 2022).
[148] Shanghai Zhangjiang Biobank, Available at: https://www.shbiobank.com/, 2022. (Accessed 14 July 2022).
[149] The International Agency for Research on Cancer (IARC) 2022. Available at: https://ibb.iarc.fr/[Accessed 14 July].
[150] China Kadoorie Biobank 2022. Available at: https://www.ckbiobank.org//. [Accessed 14 July 2022].
[151] UK Biobank. 2022. Available at: https://www.ukbiobank.ac.uk [Accessed 14 July 2022].
[152] Estonian Biobank, Institute of Genomics, 2022. Available at: https://genomics.ut.ee/en/access-biobank. (Accessed 14 July 2022).
[153] EuroBioBank 2022. Available at: http://www.eurobiobank.org/[Accessed 14 July 2022].
[154] L. Koumakis, "Deep Learning" models in genomics; are we there yet? Computational and Structural Biotechnology Journal 18 (2020) 1466–1473.
[155] M. Javaid, A. Haleem, R.P. Singh, R. Suman, Artificial intelligence applications for industry 4.0: a literature-based study, Journal of Industrial Integration and Management 7 (01) (2022) 83–111.
[156] F. Boniolo, E. Dorigatti, A.J. Ohnmacht, D. Saur, B. Schubert, M.P. Menden, Artificial intelligence in early drug discovery enabling precision medicine, Expert Opinion on Drug Discovery 16 (9) (2021) 991–1007.

Further reading

[1] K. Allel, F. Salustri, H. Haghparast-Bidgoli, A. Kiadaliri, The contributions of public health policies and healthcare quality to gender gap and country differences in life expectancy in the UK, Population Health Metrics 19 (1) (2021) 1–11.
[2] A. Bender, I. Cortés-Ciriano, Artificial intelligence in drug discovery: what is realistic, what are illusions? Part 1: ways to make an impact, and why we are not there yet, Drug Discovery Today 26 (2) (2021) 511–524.
[3] CanPath—The Canadian Partnership for Tomorrow's Health, 2022. Available at: https://partnershipfortomorrow.ca/. (Accessed 14 July 2022).
[4] J. Chen, Q. Li, H. Wang, M. Deng, A machine learning ensemble approach based on random forest and radial basis function neural network for risk evaluation of regional flood disaster: a case study of the Yangtze River Delta, China, International Journal of Environmental Research and Public Health 17 (1) (2020) 49.
[5] Institute for Molecular Medicine Finland (FIMM) at the University of Helsinki. 2022. Available at: https://www.finngen.fi/en. [Accessed 14 July 2022].
[6] Journal of Xidian University, Medical image analysis using machine learning and "deep learning", The Review 14 (5) (2020).
[7] L.E. Juarez-Orozco, O. Martinez-Manzanera, S.V. Nesterov, S. Kajander, J. Knuuti, The machine learning horizon in cardiac hybrid imaging, European Journal of Hybrid Imaging 2 (1) (2018) 1–15.
[8] G.F. Murphy, A.M. Hanken, K.A. Waters (Eds.), Electronic Health Records: Changing the Vision, WB Saunders Company, 1999.
[9] National Center for Biotechnology Information (NCBI), National Library of Medicine (US), National Center for Biotechnology Information, Bethesda (MD), 2022. Available at: https://www.ncbi.nlm.nih.gov/clinvar/intro/. (Accessed 17 July 2022).
[10] S.S. Yadav, S.M. Jadhav, Deep convolutional neural network based medical image classification for disease diagnosis, Journal of Big Data 6 (1) (2019) 1–18.
[11] C.A. Zaouiat, A. Latif, Internet of things and machine learning convergence: the e-healthcare revolution, in: Proceedings of the 2nd International Conference on Computing and Wireless Communication Systems, November 2017, pp. 1–5.

Chapter 18

Deep learning IoT in medical and healthcare

Ashwani Sharma[1], Anjali Sharma[1], Reshu Virmani[1], Girish Kumar[1], Tarun Virmani[1] and Nitin Chitranshi[2]

[1]*School of Pharmaceutical Sciences, MVN University, Palwal, Haryana, India;* [2]*Faculty of Medicines, Health and Human Sciences, Macquarie University, Sydney, NSW, Australia*

1. Introduction

Health is a condition of overall physical, mental, and social well-being, not only the absence of disease and disability. Health improves one's quality of life by utilizing smart healthcare applications such as disease diagnosis and treatment, smart medicines, remote patient monitoring, biosensors, emergency healthcare, robots, and healthcare facilities available 24 h a day, 7 days a week. The abundance of biomedical data is playing an increasingly crucial role in healthcare. Precision medicine, for example, aims to ensure that the appropriate therapy is administered to the right patient at the right time by considering numerous components of a patient's data, such as molecular trait variability, genomic variation, electronic health records (EHRs), ongoing drugs, and lifestyle. The vast amount of biomedical data available presents both constraints and benefits for healthcare research. Exploring the relationships between all of the many bits of information in large datasets is a critical difficulty for developing trustworthy medical solutions based on data-driven techniques and machine learning (ML) [1,2]. The vast amount of data in the healthcare industry has been effectively analyzed over the past 10 years using a variety of artificial intelligence (AI) and ML approaches. For instance, a prediction model based on logistic regression was used to automatically diagnose heart disease. Medical imaging has also used ML to automatically identify object attributes. In the current era of big data, scientific and technological research produces huge data volumes. The necessity to switch from conventional methodology to more sophisticated big data analysis methodologies arises from the particular difficulties that come with studying and understanding massive data. Deep learning (DL), a unique approach to AI, is one of these sophisticated techniques [3]. DL is a branch of ML conceptually. A DL model, in contrast to conventional methods, is made up of numerous deep networking layers and is capable of learning from enormous amounts of data. Convolutional neural networks (CNN) [4], deep belief networks (DBN) [5], autoencoders (AE) [6], long short-term memory networks (LSTM) [7], deep feedforward neural networks, and physiological signal analysis are some examples of deep learning models that are frequently used. It has also been used to manage or assist in the diagnosis of deadly infectious diseases such as COVID-19. AI-based models, including both traditional ML and sophisticated deep models [8,9]. Deep neural network (DNN)-based techniques are drawing a lot of attention among the many ML models, especially when it comes to the analysis of large datasets. Data is filtered through a cascade of layers using DL techniques, which involve numerous phases in the learning process. DL is regarded as a novel field in ML and, as a result, is a crucial component of AI. DNN models surpass many traditional ML models because they grow increasingly precise while processing vast amounts of data. DNN-based methods have also shown excellent results in the processing of images and natural language processing (NLP) [10—12]. These models are emerging as cutting-edge and exciting new tools to analyze healthcare data in light of the success of DL (also sometimes referred to as DNN) techniques in several fields and their quick ongoing methodological improvement. Using DL models on biological and healthcare data, a variety of activities have been carried out [13]. For instance, a computer-based support system for analyzing healthcare data has been developed by IBM Watson [14] and Google DeepMind [15,16]. The left ventricle's (LV) segmentation using short-axis cardiac magnetic resonance imaging (MRI) has been effectively accomplished by encoding parameters of a deformable model using DL [17]. The identification of biomarkers from MRI scans was accomplished using a different DL model based on Restricted Boltzmann

Deep Learning in Personalized Healthcare and Decision Support. https://doi.org/10.1016/B978-0-443-19413-9.00027-8

Machine (RBM) [18]. Currently, a significant portion of data is produced by the Internet of Things (IoT) devices, used as big data, and fed into DL algorithms as input to provide useful information. DL has a wide range of applications and is practically present in every part of our lives. The DL has significantly influenced how images are categorized in the healthcare industry, and it will soon serve as an effective tool for doctors to evaluate and assess the issue. Solutions based on the IoT and ML are effective because of improvements in processing, sensing, spectrum usage, and machine intelligence. These solutions are made possible by the development of microelectronics, which has produced small, affordable medical sensing devices and transformed medical services. In light of this, healthcare organizations categorize these remedies as preventive treatment and symptomatic treatment. People today place a high priority on disease prevention, early disease identification, and the most effective treatment for numerous chronic illnesses [19]. As a result, it has become clear that national healthcare monitoring systems are a necessary trend today. In recent times, telemedicine has paid a lot of attention to ML algorithms and devices to remotely monitor people for accurate diagnosis. Healthcare systems with IoT capabilities frequently use DL techniques and ML nowadays [20]. IoT refers to physical objects that are linked to the Internet to facilitate simple data sharing and collecting needed for sound decision-making, without the need for a great deal of human involvement. An IoT system is made up of five layers, including web-based smart devices with embedded systems like communication, sensors, and processor tools that help with data collecting and exchange. These layers consist of the network, business, application, middleware, and perception levels. Physical objects like sensors and barcodes make up the lowest perception layer, which collects data for transmission to the network layer [21,22]. The network layer allows data to be transmitted over a wired or wireless connection, such as Wi-Fi, Bluetooth, or mobile communication protocols (4G, 5G). The next middleware layer analyses the information that has been gathered and determines the best course of action based on the output from all available computing devices. The application layer makes use of the information that has been processed to offer users application-specific services. The application layer's data and statistics are used in the top-level business layer to plan out future objectives and strategies [23,24]. The rapid growth of medical data and the enormous number of medical devices have increased the risk to the current hospital information systems [25] due to the rapid improvement of medical knowledge. The healthcare industry produces a diverse range of data, including records of patients, treatments, prescription drugs, diagnosis, and more. The underlying issue is that, as a result of the restricted data processing, the organization of healthcare is impacted by the veracity of some reports [26]. Due to such a problem, the data is meaningless without an accurate result that has been generated through the processing of massive amounts of data. Utilizing systems that gather hundreds of actual medical records and have them reviewed by professionals and AI tools can enhance the outcome [27]. Some hospitals have already begun utilizing big data in the healthcare sector, particularly in radiology employing cutting-edge algorithms like DL [28]. DL techniques are representation-learning algorithms with multiple representational levels that are created by composing straightforward but nonlinear modules that each transform the representation at one level (beginning with the raw input) into a representation at a higher, marginally more abstract level [29]. These models excelled at computer vision, speech recognition, and NLP tasks, showing significant promise. DL paradigms offer fascinating new potential for biomedical informatics because of their shown efficacy in numerous fields and the quick advancement of methodological developments. DL initiatives are already in the works or planned for the healthcare industry. For instance, Google DeepMind has stated that it will use its knowledge in the field of healthcare, and Enlitic is utilizing DL intelligence to identify health issues on Computed Tomography (CT) scans and X-rays [10,11,30]. X-rays, clinical blood test results, and computed tomography (CT) are all utilized in the screening of patients by ML and AI. Radiology Images, such as those from CT scans and X-rays, can be used by medical professionals to extend the scope of current screening and diagnosing methods [8]. The rapid advancement of wearable technology, the IoT, and cloud computing has impacted every part of people's lives and created substantial development prospects. The implementation of IoT in the healthcare industry could lead to continuous health monitoring [31]. This chapter details the various algorithms of Deep learning and applications of IoT used in the healthcare system and medical devices.

2. Review of literature

Mohamed et al. [32] concluded that the analysis of patient behavior is one area where smart healthcare has been found to be very important. For improved health self-management, patients can gain from smart healthcare. Compared to traditional healthcare, smart healthcare systems can reduce staff workload, reduce costs, and improve patient care. In order to ascertain the patient's precise health status, the deep belief neural network analyses the patient's specifics from health data. The system that was constructed demonstrated an average error rate of 0.04% and 99% accuracy in assessing patient behavior.

Mohana et al. [33] showed that In order to identify the occurrence of heart failure and arrhythmia from the image collection, CNN analyses the input signal from the IoT devices, where ECG data are categorized. Testing packages and faster training are made possible by the classification of the dataset using two distinct heartbeat signals. A simulation is run to evaluate the effectiveness of the CNN against ECG picture datasets. The use of cloud resources to perform the CNN classification demonstrates the fastest classification procedure without lags. The simulation results demonstrate an enhanced accuracy rate of 98% over image ECG classification algorithms, demonstrating that the suggested CNN identifies the instances better than other current approaches.

Sharma et al. [34] concluded that by analyzing the data at hand and assisting in the creation of effective treatment plans as well as a novel vaccine candidate, AI techniques like ML and DL have proven to be helpful in the research of the virus. Due to the vast amount of readily available data, AI algorithms have become more crucial for advanced analysis and turning fundamental discoveries into novel vaccination candidates. By analyzing the data at hand and assisting in the creation of effective treatment plans as well as novel vaccine candidates, AI techniques like ML and DL have proven to be helpful in the research of the virus. Due to the vast amount of readily available data, AI algorithms have become more crucial for advanced analysis and turning fundamental discoveries into novel vaccination candidates. Although AI methods cannot completely replace time-consuming processes like laboratory research and clinical trials, they can help with trial preparation, monitoring, and risk factor forecasts.

Guo et al. [35] demonstrated that the CNN model is tested using the test data set, and the prediction accuracy is over 77.6% after training. The CNN model does good work in recognizing the health state. By enhancing the diagnosis of health status in clinical practice, the Smart Healthcare System is anticipated to help physicians.

Hussain et al. [36] used the linear binary pattern histogram (LBPH) face recognition method for classification purposes, in conjunction with pretrained CNN models, namely VGG-16 and ResNet-50 with an SVM classifier. They analyzed the outcomes of these and discovered that ResNet-50 performed more successfully, providing a recognition accuracy of 99.56% as opposed to the other two algorithms. In contrast to VGG-16 and LBPH algorithms, ResNet-50 demonstrated superior performance while requiring less computing time.

Nweke et al. [37] suggested a hybrid strategy that can enhance feature learning: a discriminative and generative model. The benefits of using mobile sensor data to generate efficient vectors include faster computation and accurate recognition performance.

Jiao et al. [38] concluded that for classifying breast masses, a parasitic metric learning method was contrasted with a CNN method. In terms of accuracy, CNN's method was compared with CNN's-SVM and the CNN-based Stochastic Gradient Descent (SGD) techniques created by other researchers.

Zhao et al. [39] concluded the systems that use deep learning-based machines to monitor human health (MHMS) are designed to extract hierarchical indicators from input data by building deep neural networks with various nonlinear change layers.

Pasluosta et al. [40] researched the idea of the Internet of Health Things (IoHT). By gathering and combining information on critically important signals in hospitals, switching objects help control patients' physical states. IoHT consists of four stages: gathering, storing, processing, and presenting.

Islam et al. [41] proposed various medical network platforms and architectures that make the IoT foundation accessible and facilitate the transfer and collection of data in medicine. Reduced costs and improved life quality are benefits of IoT-based medical services.

3. Deep learning: Algorithms and its application

AI is currently the field of study most commonly pursued by researchers. DL and traditional ML are the two subfields of AI as shown in Fig. 18.1. The technique that has most recently generated widespread interest is DL, which is currently the fastest-growing area of ML [29,42]. DL has been used in a variety of applications, including analysis of images, drug discovery [43], object detection [44], speech recognition [42], health monitoring [45,46], text mining, computer vision, and many more [47].

The most common DL models that have been utilized to evaluate and analyze clinical data used in medical and healthcare systems have been discussed below.

1. **Convolutional Neural Network (CNN):** It is an architecture for supervised DL. Applications for image analysis are its primary use cases. Other domains where CNN is commonly used are NLP, voice recognition, computer vision, and text detection and recognition. Convolutional, pooling, and fully linked layers are the three types of layers used in CNN. The input image is processed by kernels or filters in the convolutional layer to produce various feature maps. To keep the number of weights low in the pooling layer, the size of each feature map is decreased. This method is often referred to as subsampling or downsampling. Following these layers, the fully linked layer is utilized to convert two-dimensional feature maps into a one-dimensional vector for the purpose of the final classification. The basic

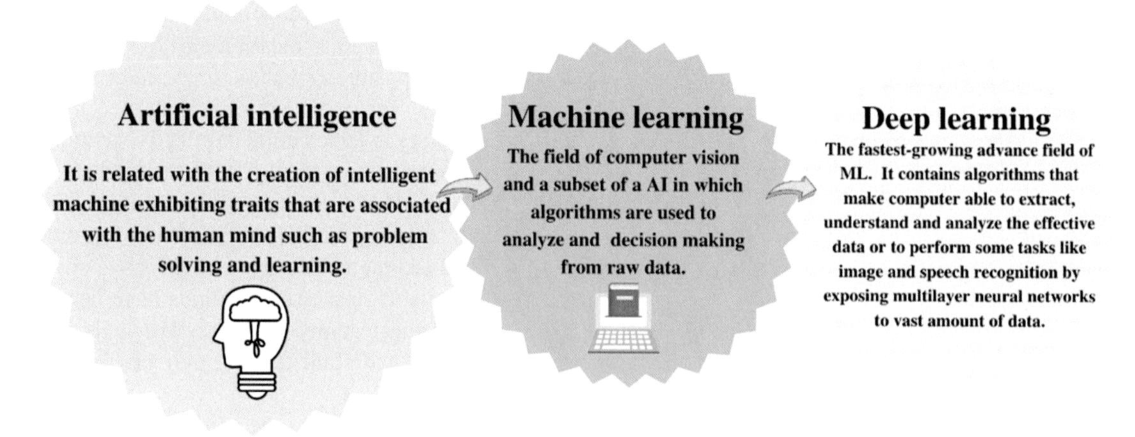

FIGURE 18.1 Subfields of artificial intelligence.

architecture of CNN is shown in Fig. 18.2A and 1D representation of CNN architecture is shown in Fig. 18.2B. The most prevalent CNN architectures include ResNet [48], GoogleNet [49], AlexNet, ZFNet [50], and VGGNet [51]. A DL model named Deepr was suggested to extract significant information from medical records and predict the anomaly [52]. The investigation of several NLP tasks on biological texts, including named entity recognition [53] and mesh indexing [54,55], has been done using CNN. A DL approach was used by Cheng et al. [56] to investigate the temporal aspects of patient EHR [56].

2. **Recurrent Neural Network (RNN):** Another representative form of deep learning architecture is the RNN which is used to recognize patterns in streams of input, including handwriting, speech, and text. Extraction of features from clinical texts is one of the key uses of RNN. The architecture of RNN was shown in Fig. 18.3. RNN has also been used to recognize chemical name entities [57,58] and is used to predict diagnosis codes based on medical literature. Having to keep the old data around for a very long time is a problem with RNN long-term dependencies. So, in order to solve this issue, a different RNN variation called long short-term memory is used (LSTM). Using memory cells and gate

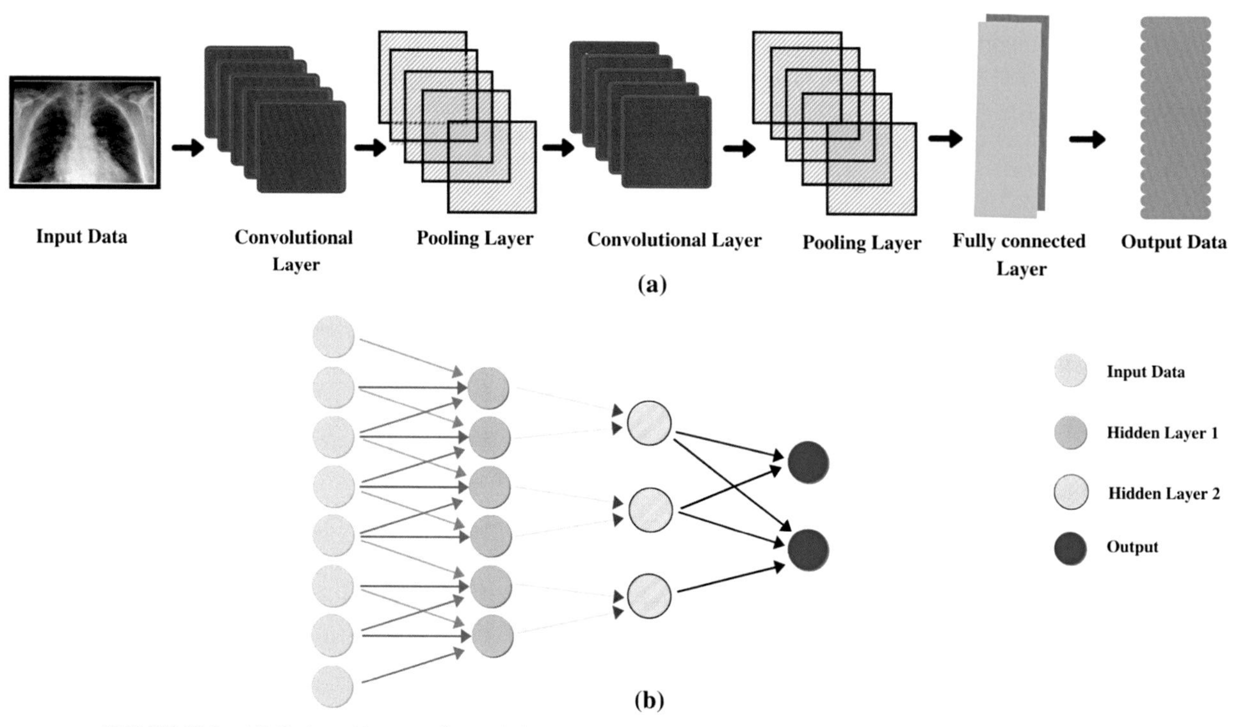

FIGURE 18.2 (A) Basic architecture of convolutional neural network (CNN). (B) 1D representation of CNN architecture.

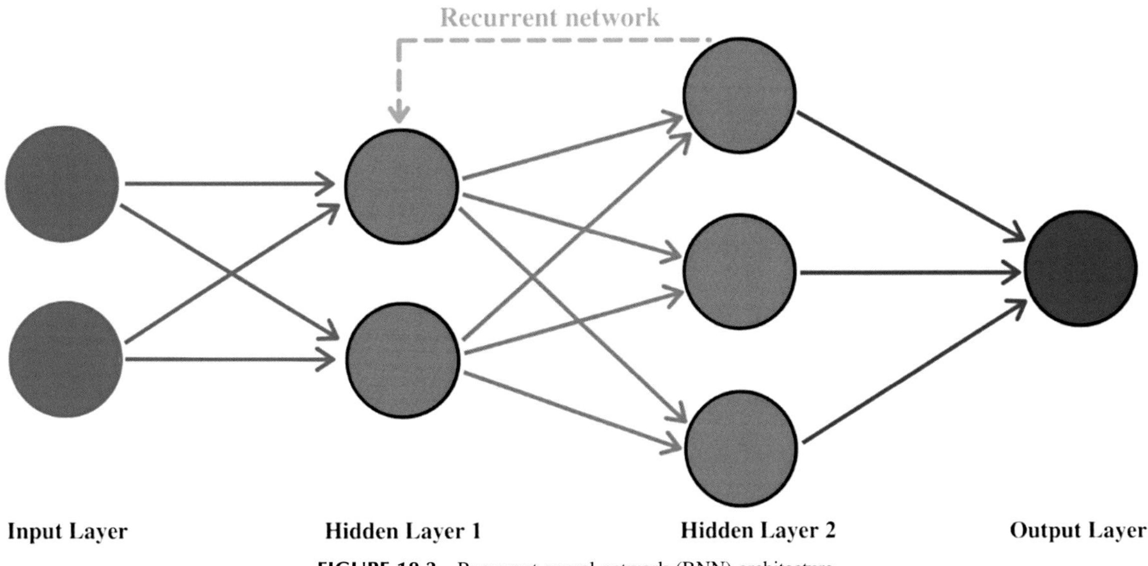

FIGURE 18.3 Recurrent neural network (RNN) architecture.

mechanisms, it has been able to find solutions to the issues. The long-term memory of the parameters in an input form is achieved by LSTM using hidden layers or nodes [7]. The gated recurrent unit (GRU) of LSTM is a further simplified version. Clinical decision-making issues were reportedly addressed by the LSTM model [59]. When analyzing extended sequences, LSTM and GRU both outperform alternative plain models. Additionally, LSTM performed better in speech recognition systems. Additionally, in the field of image retrieval, LSTM is just as successful and well-liked as CNN, and it is used with CNN in AI to automatically generate image descriptions.

3. **Restricted Boltzmann Machine (RBM):** RBM is a term used to describe a generative design that depicts a probability distribution [60]. There are input and hidden layers in RBM (also called visible layers). The RBM architecture is depicted in Fig. 18.4. In comparison to before, RBMs are currently used to solve more intriguing issues thanks to the development of novel learning algorithms and advances in computer technology. It has been established that there exist X-ray image categorization models based on RBM. DBN, which is a multi-layer learning task, has been proposed as using RBM as its building blocks [60,61], and this has lately attracted attention. It is a hierarchical stack of RBM that addresses nodule classification for supervised learning applications such as computed tomography (CT) imaging.
4. **Deep Neural Network (DNN):** A DNN is an artificial neural network (ANN) with multiple hidden layers between the input and output layers [62]. It is possible to think of each hidden layer as a feature extraction technique. Higher layers may mix functions from lower layers, enabling more complicated data processing with fewer units than a shallow network with a comparable level of performance. As a result, DNN can model complex non-linear connections using training data. To address various problems, various extensions and versions have been created [57]. Fig. 18.5A depicts the ANN architecture and Fig. 18.5B depicts the DNN architecture.
5. **Auto encoder (AE):** An ANN called the Auto-Encoder AE seeks to efficiently code the data. As a result, it can be used for feature reduction or network initialization. Fig. 18.6 shows Deep autoencoder architecture. To achieve this, it translates the input through an interconnected neural network to itself. Unsupervised learning is categorized as AE, which encompasses Variational Autoencoder (VAE), Denoising Autoencoder (DAE), and, Stacked Autoencoder (SAE) [63].

The various DL algorithms that are being used in healthcare systems to distinguish between healthy and ill people are discussed below with the models specific to disease models.

1. **Parkinson's disease**: In order to better understand Parkinson's disease (PD), a DL-based FPCIT SPECT interpretation system [59] was developed by Choi et al. [59]. In this work, it was demonstrated that the problem of interobserver variability can be solved using this strategy. DL was employed to classify PD more recently [64] in 2019 and they applied 9-layered CNN on the set of vocal (speech) features to achieve this. They provided evidence of the usefulness of the classifier they used to address this issue.
2. **Diagnosis of Epilepsy**: Acharya et al. [65] presented a novel computer-aided diagnosis system for epilepsy diagnosis using CNN analysis of the encephalogram data. The specificity, accuracy, and sensitivity of this CNN method which

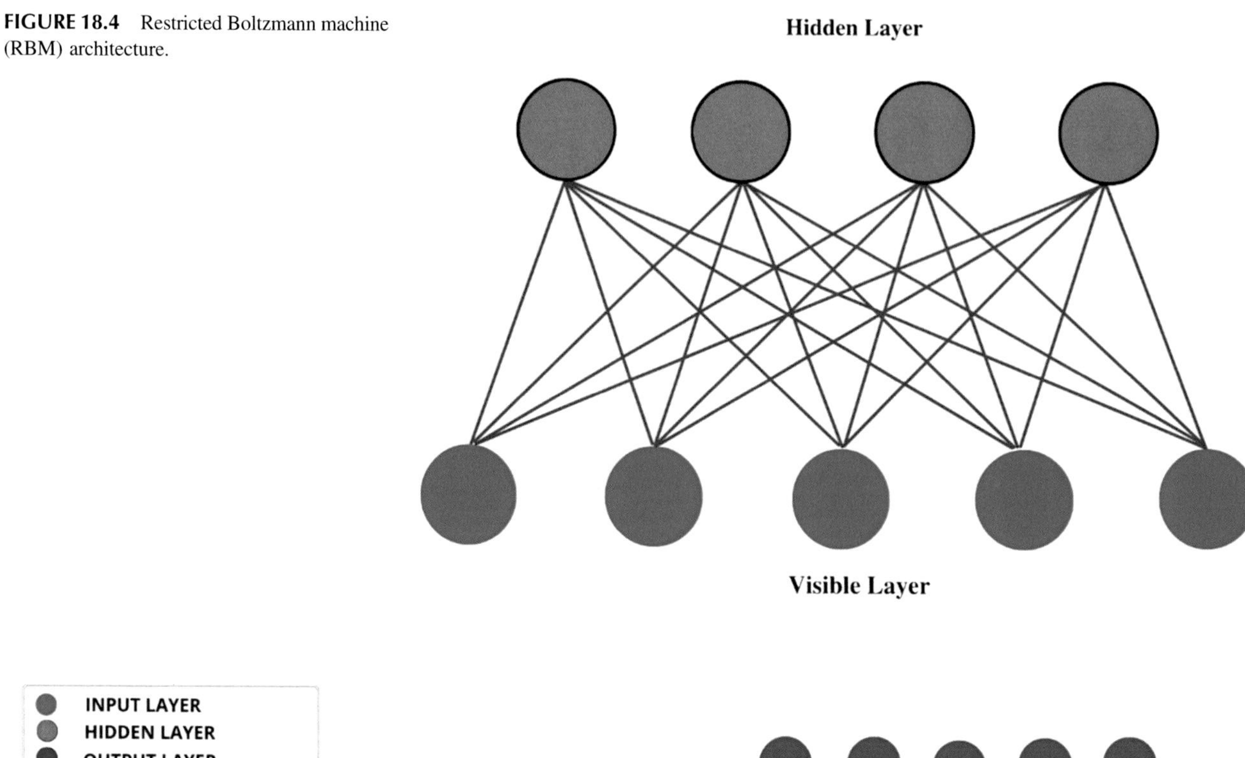

FIGURE 18.4 Restricted Boltzmann machine (RBM) architecture.

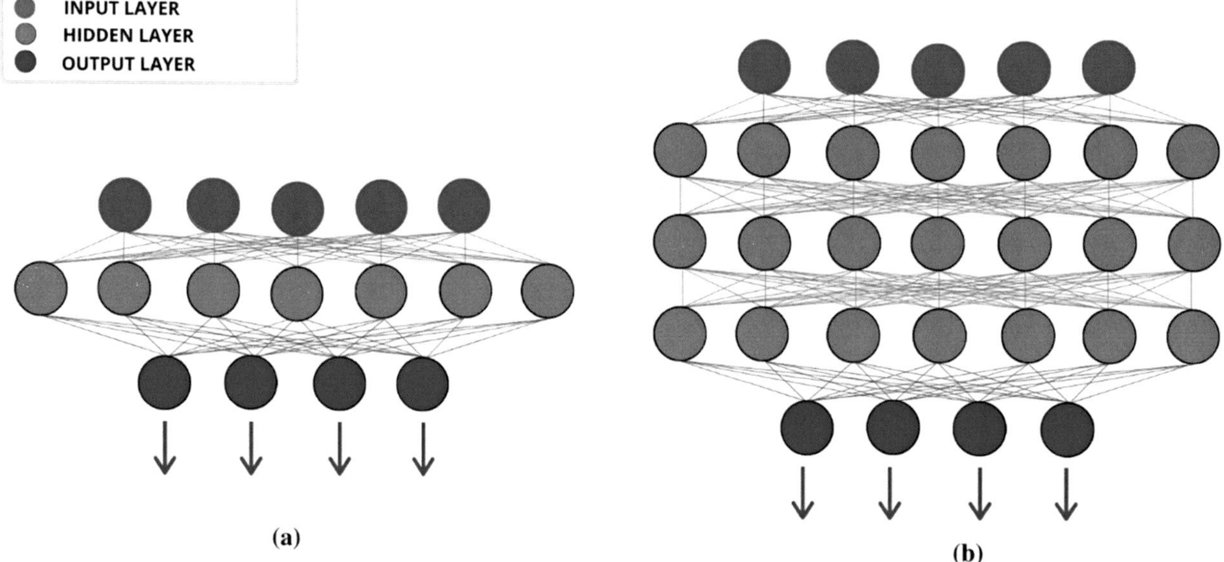

FIGURE 18.5 (A) depicts artificial neural network (ANN) architecture and (B) depicts deep neural network (DNN) architecture.

consists of 13 layers are compared to those of other ML techniques for detecting seizures (both normal and preictal classes). This is the only substantial application of DL for the diagnosis of epilepsy [65].

3. **Multiple sclerosis:** Latent multiple sclerosis (MS) lesion patterns were extracted from baseline pictures using a CNN technique [66]. According to Yoo et al. [67], it was suggested to use a DBN and random forest (RF) combined with myelin and T1W images to detect MS pathology in brain tissue that appeared normal on MRI. A latent feature representation for the 3D image patches of NAWM and NAGM was learned using a four-layer DBN in this model [67].
4. **Brain cancer**: Moeskops et al. [68] suggested a multiscale CNN-based tissue segmentation technique for brain MRI. Various levels of abnormalities were used to test the approach. According to the findings, it can precisely segment brain tissues [68]. A new D model for brain tumor segmentation was proposed by Zhao et al. [69] by combining conditional random fields (CRF) and fully convolutional neural networks (FCNN). For classifying brain images, DL approaches (such as ResNets and CNNs) are gaining ground over ML approaches [69]. In order to achieve more precise

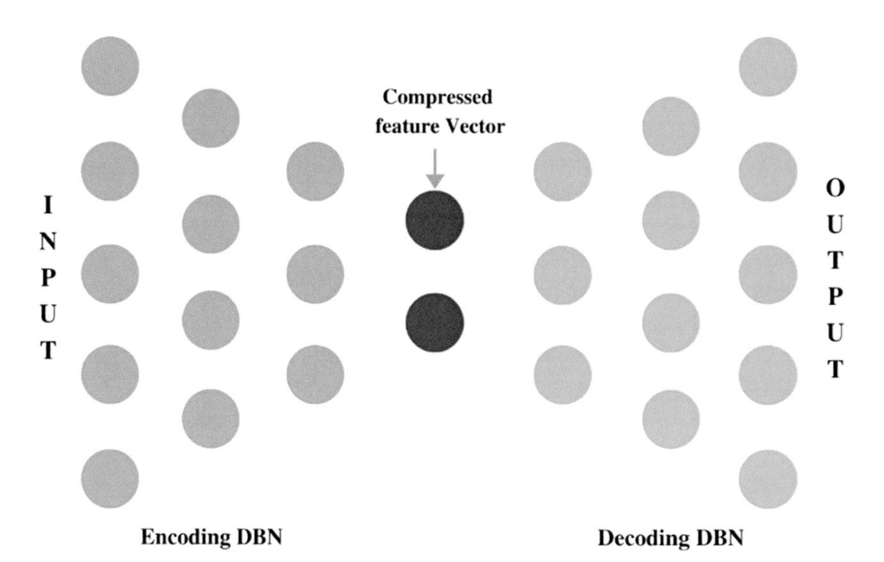

FIGURE 18.6 Deep autoencoder architecture.

characterization and segmentation, DL algorithms like CNN extract features from images that are translation invariant and stable to deformations. It is advised to use DL models for grading brain tumors in addition to characterizing and segmenting the brain [70].

5. **Eye disease:** Using a cutting-edge AI framework, DeepMind and the Moorfields Eye Hospital have integrated neural network power for segmentation and classification tasks. The initial step in this method is to employ a segmentation network to separate a variety of 15 different retinal morphological features and optical coherence tomography (OCT) acquisition artifacts [71].
6. **Heart disease**: In order to detect heart failures, 70 proposed an RNN model with gated recurrent units that examined the relationships between time-stamped events using an observation window of cases and controls.
7. **Breast cancer:** Dhungel et al. [72] described a computer-aided diagnosis system that uses DL algorithms to identify, segment, and classify masses in mammograms [72]. They thought about three steps: detection, segmentation, and classification to do. For mass detection, a series of DL methods including CNN, DBN, and CRF were proposed. The highest accuracy of 88.2% was reported by the multilayer perceptron (MLP) model. Support vector machine (SVM), RF, and decision tree (DT) performed well in breast cancer survival prediction in previous studies [73−75].
8. **Alzheimer's disease (AD)**: A CNN-based approach for diagnosing AD using structural MRI [76] was put forth by Yue et al. [76]. In order to diagnose AD in its early stages using structural MRI, Gunawardena et al. [77] developed a procedure based on CNN [77]. The CNN model outperformed the SVM when the performance of the proposed method was compared to that of the SVM in the study. For anticipating the transformation of Mild cognitive impairment (MCI) participants into AD patients, Lee et al. [78] suggested an RNN-based model that retrieved temporal variables from multimodal data [78].

Early disease diagnosis or identification is crucial to saving lives. Early notice or discovery of a disease can help one avoid it from developing into something more serious or worse. A DE network that was built on resting-state functional magnetic resonance imaging (RfMRI) data and some pertinent text information was utilized to detect AD in its early stages. The overall prediction accuracy achieved was 86.47%. Another study used a multiscale and multimodal DNN and demonstrated success in differentiating between participants with different AD trajectories [79]. The ability to measure the structural and functional alterations connected to autism spectrum disorder (ASD) is a result of advancements in neuroimaging technologies over the past 10 years [80]. In the field of ASD biomarker identification and classification, fMRI, a method for detecting biomarker patterns in the brain, has attracted a lot of interest [81,82]. Additionally, it has been reported that ML algorithms have been successful in detecting biomarkers using fMRI datasets for the identification and categorization of biomarkers for a variety of brain disorders [83,84]. The detection and classification of ASD biomarkers have drawn a lot of interest in DL techniques like CNN, DNN, and AE. In order to classify the images from the clinical exam, various deep learning models have been applied. DBNs have reportedly been used to categorize people with nervous system problems [85]. MCI and AD were classified using SAEs. CNN and its derivatives, though, are the models that are most effective at identifying diseases from medical images. Additionally, CNN was used to identify diabetic retinopathy and diabetic macular edema in retinal fundus images, and the model was pretrained on the ImageNet dataset [86]. According to other research, it is possible to classify skin lesions by utilizing a single CNN and a pretrained instance of Google's Inception v3 architecture on medical data [87]. DNNs can also be used to identify enlarged lymph nodes, colonic polyps, and cerebral microhemorrhage in MRI and CT images [88]. Numerous research studies concentrating on estimating the patient's risk of mortality have been reported. For instance in the case of COVID-19, to identify certain patients by mortality risk and predict mortality 20 days in advance, a COVID-19 mortality risk prediction model (MRPMC) based on clinical data has been proposed [89]. They developed an ensemble framework using a neural network (NN), an SVM, a gradient-boosted decision tree, and logistic regression. They promptly and efficiently categorized the mortality risk of COVID-19 patients using their method [90]. A potent statistical ML model called Deep-Neo-V was developed for clinical use in patients admitted with COVID-19 infection that was confirmed by RT-PCR in order to forecast mortality risk [91]. Both LSTM and Bi-LSTM models were created using clinical EHR data from a large population to identify the best hypertension drug regimens to achieve the best blood pressure (BP) outcomes. Based on the prediction outcomes, LSTM models can determine a patient's best treatment options, assisting doctors in better clinical decision-making, lowering the risk of serious adverse events, enhancing early prevention and detection of hypertension-related risks, raising patient awareness, and promoting dietary and lifestyle changes [92]. In all, the different DL algorithms have been suggested for the analysis of radiological images (X-rays, MRI, CT), and are helpful in disease predictions, identifying the risk factors, early detection and diagnosis of disease, and predicting mortality risk.

4. IoT and its applications

DL and IoT are rapidly expanding technologies that have important applications in the medical and healthcare industries [85]. IoT is the most popular, advanced, and up-to-date technology that can automate intelligent administration, remote control, data monitoring, and management over a real-time network. Highly creative linked health technology is driven by the positioning of the healthcare sector, which includes IoT services, IoT apps, and IoT solutions. The main objective of digital health is to enhance healthcare quality while reducing expenses. It has been reported that smart wearables can provide real-time health information in everyday conditions [93]. The IoT revolution is reshaping contemporary healthcare systems, with ramifications for the economy, society, and technology. The traditional healthcare system is being transformed into one that is more customized, making it simpler to diagnose, treat, and monitor patients. IoT is gradually becoming a crucial technology in the healthcare industry, where it is used to enhance service quality, potentially cut costs, and offer cutting-edge user experiences [41]. With the most recent developments in the fields of AI and big data, there is a significant need for improvement in the field of computer vision for face recognition systems, particularly in medical and healthcare environments and locations to prevent the spread of infectious diseases such as COVID-19, a recent outbreak. IoT can potentially be used in epidemic circumstances to monitor infected patients. This technology is made to be able to be activated in the event of a pandemic and offers strategies for dealing with pandemic breakouts in addition to the capacity to supply computerized data. An extensive program for early detection of any sickness, whether it is incurable or not, is essential for early treatment to save more lives. The spread of pandemic diseases like COVID-19 can be stopped by quick screening and detection techniques. For early treatment to save more lives, it is extremely important for early discovery of any sickness, whether it is incurable or not. Effective screening and detection techniques can stop the spread of pandemic diseases like COVID-19 [94]. The condition of the patient is evaluated utilizing devices that have an IoT-based sensor integrated into them. The patient's smartphone is used by wearable devices to transmit the patient's blood glucose, oxygen saturation, and pulse rate to the caregiver. Instead of physically visiting a clinic or hospital, physicians can consult with individuals online through telehealth. By changing bad behavior and embracing healthier lifestyles, behavior change is a technique that may aid people in enhancing their health [32,95]. Wearable technology based on IoT enables users to monitor and record their own physiological data, such as respiration rate, body temperature, heart rate, and other vital signs. Even when they are in an isolated state, the person is able to recognize irregularities in their vital signs right away [94,96]. IoT-based approaches have inspired a growing interest among developers of healthcare applications with the advent of communication and information technology. Without a doubt, IoT applications present a very practical solution for COVID-19 pandemic management. IoT technology has been used to detect COVID-19 patients, diagnose COVID-19 infections, deliver telemedicine services, and integrate these applications with wearables. Wearable sensors, medical equipment, and portable devices have all improved in price and usability as a result of the IoT's quick development. These technologies can be used to gather information about patients, identify diseases, keep track of patients' health, and send out notifications in the event of a medical emergency. Patients in medical clinics are extremely concerned about disease transmission. Devices for IoT-enabled cleanliness monitoring assist in keeping patients safe. IoT tools also support resource managers in fields like nature observation, such as analyzing colder temperatures and temperature and moisture control and pharmacy store stock control [97,98]. Mohammed et al. [32] proposed the use of a smart helmet with a thermal imaging system to automatically identify coronavirus using the thermal imaging technique, reducing human-to-human interaction. To monitor smart houses and heartbeats, Desai et al. [99] created a wireless sensor network (WSN) [99]. Smart helmets have a positioning system. The system will react right away if it notices temperatures that are higher than usual. Doctors can keep a close check on a patient's condition, thanks to wearable technology that tracks their whereabouts using Global Positioning System (GPS) data. IoT-enabled wearable systems were also used to assess biosignals such as an electrocardiogram (ECG) and electromyography (EMG) signals to obtain crucial patient data [100]. Patients who utilize wearable technology use them to monitor their health at all times, provide remote medical services, and get medical treatment via virtual assistants; doctors employ a variety of sophisticated clinical judgments to support systems to improve diagnosis. Table 18.1 illustrates the IoT-linked devices used in COVID-19 and in the healthcare system [101]. The electronic medical record (EMR) and communication systems, as well as photo archiving, laboratory information management systems, and other tools, are handled by doctor's forms. The usage of surgical robots that utilize mixed reality technologies is beneficial. The hospital's supply chain and personnel are both tracked via radio frequency identification (RFID). To collect data and facilitate decision-making, RFID is utilized in conjunction with many management systems [32]. RFID makes it possible for medical professionals to easily find and monitor medical devices. The key benefit of RFID is that it does not require an external power supply. It is a rather insecure protocol, though, and connecting with a smartphone may cause compatibility concerns. In order to track patients' health records and other physiological parameters, mobile computers, sensors, communication technologies, and cloud computing are combined. This is known as mobile IoT or m-IoT. Because of the widespread usage of mobile devices, healthcare professionals now have easier access to healthcare IoT services, allowing them to quickly diagnose and treat patients by quickly accessing their data. An IoT-based real-time monitoring system has been described in previous studies [102] that detects an irregularity in heart activity and notifies the patient when the heart rate exceeds 60−100 beats per minute. A major concern in an m-IoT system is the privacy and security of the user and user data. The pathological condition of the patient is determined by cognitive computing in Amin et al. [103], which is an EEG-based smart healthcare monitoring system [103]. The state of a patient was evaluated using EEG data as well as other sensor data, including voice, gesture, bodily movement, and facial expressions. Furthermore, it makes immediate assistance possible in cases of pathogenic situations [104]

A brand-new IoT-based architecture for cancer treatment was put forward [105], and it included chemotherapy and radiotherapy as well as different stages of cancer treatment. For online doctor consultations, a smartphone app was utilized. Healthcare professionals have access to patient lab test results that were saved on a cloud server to make decisions about when and how much medication to provide. In the past several years, a lot of IoT-based solutions for monitoring asthma have been suggested [106]. A smart sensor was utilized to monitor the respiratory rate as part of a smart healthcare IoT solution that was presented for asthma sufferers [36,107]. Fig. 18.7 depicts the classification of deep learning and IoT in healthcare and also the percentage of various deep learning applications for IoT in healthcare. Healthcare providers may now assess a variety of health metrics, monitor and diagnose a number of health conditions, and provide remote diagnostic services because of IoT technology.

5. Conclusion

This chapter concludes the applications of DL and IoT in the medical and healthcare system. Techniques based on DL have been demonstrated to be effective in dealing with disease detection in preprocessing, feature selection and extraction, classification, and clustering processes. DL is advancing swiftly and is commonly used in voice and image processing. This chapter also entails about the various DL techniques in the management of various diseases. The existing healthcare services where IoT-based solutions have been investigated are also included in this study. A DBN network examines the patient's specifics from health data in order to discover the patient's precise health state. Patient data are transmitted by the Internet, where they are seen and evaluated using ML methods. The healthcare system is more adaptable as a result of the IoT. Incorporating the IoT is essential to providing patients with an accurate diagnosis. By utilizing these ideas, IoT technology has assisted healthcare practitioners in monitoring and diagnosing a variety of health concerns, measuring a variety of health factors, and offering diagnostic services in remote areas. The healthcare sector has changed from being primarily focused on hospitals to becoming more patient-centric. It has been concluded that different DL algorithms and IoT have enormous potential for solving complex issues in the healthcare sector.

Future Scope: Data security is the issue that has embroiled the healthcare sector. Healthcare data are extremely sensitive, highly individualized, and required to be dealt with the greatest care and secrecy. While the data have significant value in terms of numerous paradigms, this is the most significant aspect of it all. Along with conducting research on deep learning and its integration with IoT devices, substantial work needs to be done on developing safe hardware, securing the

TABLE 18.1 Illustrates the IoT-linked devices used in COVID-19 and in the healthcare system [101].

Technology	Description	Types	Uses	Examples (products)	Trial phase	Advantages	Disadvantages
Wearables	A device with an app that is attached to the body or worn on the person to receive and process information	Smart glasses	- Measuring and recording temperature - Less human contact	Rokid in China, Vuzix and Onsight	I	- Continuous observation - Enhancing the standard of patient care under medicare - Hospitals that are more effective and safe - Reducing hospitalizations	- Data privacy and security - A poor battery life
		Smart helmet	- Temperature observation - Location and face picture capture	KC N901 in China	I		
		Smart thermometers	- Increasing the percentage of diagnoses - Temperature observation	Ran's Night, Temp-drop, iFever, Kinsa, iSense	I		
		EasyBand	- Tracking social distancing of individuals - LED warning of approaching peril	Pact wristband	III		
		IoT-Q-Band	- Keeping track of confined patients in case they escape	USA electronic ankle bracelet and Hong Kong electronic wristband	II		
		Proximity trace	- Observing employees for social distance - Tracking down compromised workers' contacts	Instant Trace and Hardhat TraceTag	III		
Robots	A programmed device capable of performing complicated activities like a living being	Autonomous robots	- Identifying symptoms - Managing social isolation - Preventing infection among medical personnel - Hospitals cleaning and sanitising polluted areas - Administering patient care - Examining the patients' respiratory symptoms - Gathering swab test data	Intelligent care robot and spot robot	I, II, III	- Reducing interactions through online diagnosis and care - Lowering the problems with mental health - Upkeep like cleaning and sanitizing	- Injustice and privacy issues
		Telerobots	- Reducing the risk of infection for medical staff	DaVinci surgical robots	II		
		Collaborative robots	- Reducing tiredness in healthcare workers - Disinfecting difficult-to-reach areas	Asimov Robotics and eXtremeDisinfection Robot	II		
		Social robot	- Lowering mental fatigue	Paro	II		

Smartphone applications	A software program created for mobile devices that can only do specific functions	DetectaChem	- Doing low-cost COVID-19 tests with a kit linked to a smartphone app	DetectaChem, USA	I	- Tracking and monitoring - Cost effective	- Security and privacy of collected data
		Stop Corona	- Obtaining daily health reports with names, addresses, and places - Creating a map with high-risk areas	Stop Corona, Croatia	I		
		Social monitoring	- Monitor individuals with COVID-19 diagnoses - The government's access to the user's information (privacy concern)	Social monitoring, Russia	II		
		Selfie app	- Asking patients for selfies at random to monitor them	Selfie app, Poland	II		
		Civitas	- Determining the ideal departure time for suspects to go get necessities	Civitas, Canada	II		
		StayHomeSafe	- Using a bracelet and a smartphone app, the airport can be watched for arrivals.	StayHomeSafe, Hong Kong	II		
		Aarogya Setu	- Better connecting individuals with health services	Aarogya Setu, India	III		
		TraceTogether	- Using encrypted IDs to capture the user's friends and family - Notifying those who have direct touch with the sick person if they are affected	TraceTogether, Singapore	III		

Continued

TABLE 18.1 Illustrates the IoT-linked devices used in COVID-19 and in the healthcare system [101].—cont'd

Technology	Description	Types	Uses	Examples (products)	Trial phase	Advantages	Disadvantages
Drones	An aircraft with sensors, cameras, GPS, and communication systems that requires little to no human input while flying	Thermal imaging drone	- Detecting the temperature in a crowd	Pandemic Drone	I	- Perform a range of duties, such as delivery, monitoring, and searching. - Reach difficult-to-reach areas - Reduce employee contacts like maintaining	- Security concern (large unstructured data) - The level of service - Slender connections
		Disinfectant drone	- Sterlizing up polluted places - Preventing infection among medical personnel	DJI	II		
		Medical/delivery drone	- Lowering hospital visits - Increasing treatment availability	Delivery Drone Canada	II, III		
		Multipurpose drone	- Disinfecting areas - Temperature capturing - Broadcasting information - Crowd monitoring	Corona Combat	I, II, III		

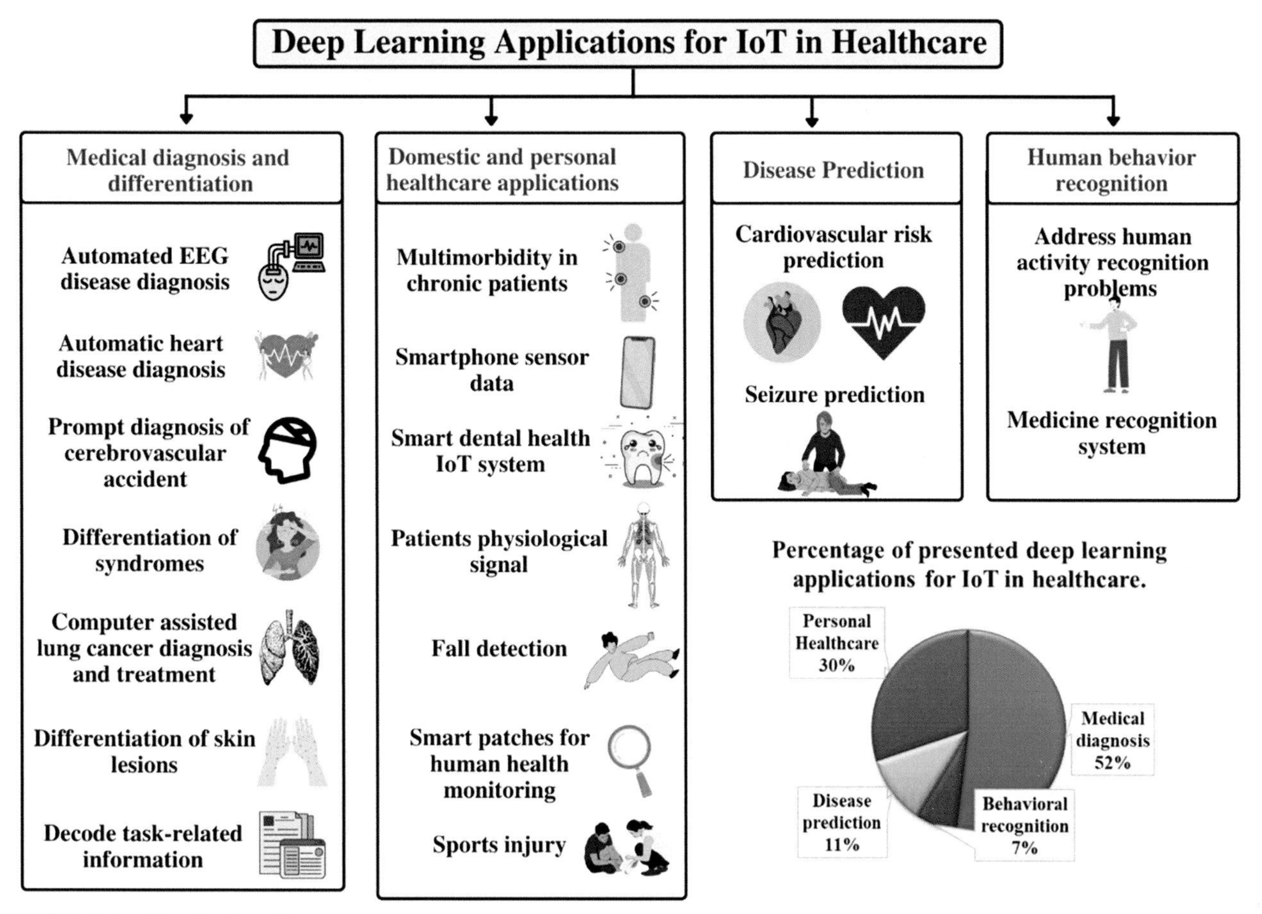

FIGURE 18.7 Classification of deep learning applications for Internet of things (IoT) in the healthcare with percentage of various deep learning applications for IoT in healthcare.

network over which the data will be incorporated, and using the right data encryption to store and process this sensitive data. The availability of data for research purposes is one of the major difficulties that will face us in the future. Data collection is significantly more difficult for healthcare because the data are more private and individualized. The data's interpretability is one of the subproblems with deep learning architecture. While the model's output is clear, there are still many unanswered questions about the internal layers that led to the final outcome.

References

[1] F.S. Collins, H. Varmus, A new initiative on precision medicine, New England Journal of Medicine 372 (9) (2015) 793–795, https://doi.org/10.1056/NEJMp1500523.

[2] G.H. Lyman, H.L. Moses, Biomarker tests for molecularly targeted therapies—the key to unlocking precision medicine, New England Journal of Medicine 375 (1) (2016) 4–6, https://doi.org/10.1056/NEJMp1604033.

[3] F. Emmert-Streib, Z. Yang, H. Feng, S. Tripathi, M. Dehmer, An introductory review of deep learning for prediction models with big data, Frontiers in Artificial Intelligence 3 (2020) 4.

[4] S. Indolia, A.K. Goswami, S.P. Mishra, P. Asopa, Conceptual understanding of convolutional neural network—a deep learning approach, Procedia Computer Science 132 (2018) 679–688, https://doi.org/10.1016/j.procs.2018.05.069.

[5] Y. Hua, J. Guo, H. Zhao, Deep belief networks and deep learning, in: Proceedings of 2015 International Conference on Intelligent Computing and Internet of Things, 2015, pp. 1–4, https://doi.org/10.1109/ICAIOT.2015.7111524.

[6] X. Lu, Y. Tsao, S. Matsuda, C. Hori, Speech Enhancement Based on Deep Denoising Autoencoder, Vol 2013, Interspeech, August 2013, pp. 436–440.

[7] S. Hochreiter, J. Schmidhuber, Long short-term memory, Neural Computation 9 (8) (1997) 1735–1780, https://doi.org/10.1162/neco.1997.9.8.1735.

[8] A.A. Ardakani, A.R. Kanafi, U.R. Acharya, N. Khadem, A. Mohammadi, Application of deep learning technique to manage COVID-19 in routine clinical practice using CT images: results of 10 convolutional neural networks, Computers in Biology and Medicine 121 (2020) 103795, https://doi.org/10.1016/j.compbiomed.2020.103795.
[9] T. Ozturk, M. Talo, E.A. Yildirim, U.B. Baloglu, O. Yildirim, U. Rajendra Acharya, Automated detection of COVID-19 cases using deep neural networks with X-ray images, Computers in Biology and Medicine 121 (2020) 103792, https://doi.org/10.1016/j.compbiomed.2020.103792.
[10] K. Cho, A. Courville, Y. Bengio, Describing multimedia content using attention-based encoder—decoder networks, IEEE Transactions on Multimedia 17 (11) (2015) 1875—1886, https://doi.org/10.1109/TMM.2015.2477044.
[11] O. Abdel-Hamid, A. Mohamed, H. Jiang, L. Deng, G. Penn, D. Yu, Convolutional neural networks for speech recognition, IEEE/ACM Transactions on Audio, Speech, and Language Processing 22 (10) (2014) 1533—1545, https://doi.org/10.1109/TASLP.2014.2339736.
[12] A. Hannun, C. Case, J. Casper, B. Catanzaro, G. Diamos, E. Elsen, et al., Deep Speech: Scaling Up End-to-End Speech Recognition, arXiv:1412.5567; Version 2, 2014, https://doi.org/10.48550/arXiv.1412.5567.
[13] S. Shamshirband, M. Fathi, A. Dehzangi, A.T. Chronopoulos, H. Alinejad-Rokny, A review on deep learning approaches in healthcare systems: taxonomies, challenges, and open issues, Journal of Biomedical Informatics 113 (2021) 103627, https://doi.org/10.1016/j.jbi.2020.103627.
[14] IBM Watson Health | AI Healthcare Solutions, IBM Watson Health, July 26, 2021. https://www.ibm.com/in-en/watson-health.
[15] DeepMind. (n.d.). Retrieved August 31, 2022, from https://www.deepmind.com/.
[16] M. Cabrita, H. op den Akker, M. Tabak, H.J. Hermens, M.M.R. Vollenbroek-Hutten, Persuasive technology to support active and healthy ageing: an exploration of past, present, and future, Journal of Biomedical Informatics 84 (2018) 17—30, https://doi.org/10.1016/j.jbi.2018.06.010.
[17] M.R. Avendi, A. Kheradvar, H. Jafarkhani, A combined deep-learning and deformable-model approach to fully automatic segmentation of the left ventricle in cardiac MRI, Medical Image Analysis 30 (2016) 108—119, https://doi.org/10.1016/j.media.2016.01.005.
[18] F. Li, L. Tran, K.-H. Thung, S. Ji, D. Shen, J. Li, A robust deep model for improved classification of AD/MCI patients, IEEE Journal of Biomedical and Health Informatics 19 (5) (2015) 1610—1616, https://doi.org/10.1109/JBHI.2015.2429556.
[19] J. Sun, F. Khan, J. Li, M.D. Alshehri, R. Alturki, M. Wedyan, Mutual authentication scheme for the device-to-server communication in the internet of medical things, IEEE Internet of Things Journal (2021), https://doi.org/10.1109/JIOT.2021.3078702.
[20] M.A. Jan, F. Khan, R. Khan, S. Mastorakis, V.G. Menon, M. Alazab, P. Watters, Lightweight mutual authentication and privacy-preservation scheme for intelligent wearable devices in industrial-CPS, IEEE Transactions on Industrial Informatics 17 (8) (2020) 5829—5839.
[21] L. Zheng, O. Wang, S. Hao, C. Ye, M. Liu, M. Xia, et al., Development of an early-warning system for high-risk patients for suicide attempt using deep learning and electronic health records, Translational Psychiatry 10 (1) (2020) 1—10, https://doi.org/10.1038/s41398-020-0684-2.
[22] A.R. Sfar, Z. Chtourou, Y. Challal, A systemic and cognitive vision for IoT security: a case study of military live simulation and security challenges, in: 2017 International Conference on Smart, Monitored and Controlled Cities (SM2C), 2017, pp. 101—105, https://doi.org/10.1109/SM2C.2017.8071828.
[23] S. Kumar, P. Tiwari, M. Zymbler, Internet of things is a revolutionary approach for future technology enhancement: a review, Journal of Big Data 6 (1) (2019) 111, https://doi.org/10.1186/s40537-019-0268-2.
[24] V. Jahmunah, V.K. Sudarshan, S.L. Oh, R. Gururajan, R. Gururajan, X. Zhou, et al., Future IoT tools for COVID-19 contact tracing and prediction: a review of the state-of-the-science, International Journal of Imaging Systems and Technology 31 (2) (2021) 455—471, https://doi.org/10.1002/ima.22552.
[25] J. Latif, C. Xiao, A. Imran, S. Tu, Medical Imaging Using Machine Learning and Deep Learning Algorithms: A Review, 2019, https://doi.org/10.1109/ICOMET.2019.8673502.
[26] Y. Yang, T. Chen, Analysis and visualization implementation of medical big data resource sharing mechanism based on deep learning, IEEE Access 7 (2019) 156077—156088, https://doi.org/10.1109/ACCESS.2019.2949879.
[27] H. Zhao, G. Li, W. Feng, Research on application of artificial intelligence in medical education, in: 2018 International Conference on Engineering Simulation and Intelligent Control (ESAIC), 2018, pp. 340—342, https://doi.org/10.1109/ESAIC.2018.00085.
[28] F. Al-Turjman, M.H. Nawaz, U.D. Ulusar, Intelligence in the internet of medical things era: a systematic review of current and future trends, Computer Communications 150 (2020) 644—660.
[29] Y. LeCun, Y. Bengio, G. Hinton, Deep learning, Nature 521 (7553) (2015) 436—444.
[30] R. Miotto, F. Wang, S. Wang, X. Jiang, J.T. Dudley, Deep learning for healthcare: review, opportunities and challenges, Briefings in Bioinformatics 19 (6) (2018) 1236—1246, https://doi.org/10.1093/bib/bbx044.
[31] W. Shi, J. Cao, Q. Zhang, Y. Li, L. Xu, Edge computing: vision and challenges, IEEE Internet of Things Journal 3 (5) (2016) 637—646, https://doi.org/10.1109/JIOT.2016.2579198.
[32] R.M. Mohamed, O.R. Shahin, N.O. Hamed, H.Y. Zahran, M.H. Abdellattif, Analyzing the patient behavior for improving the medical treatment using smart healthcare and IoT-based deep belief network, Journal of Healthcare Engineering 2022 (2022) e6389069, https://doi.org/10.1155/2022/6389069.
[33] J. Mohana, B. Yakkala, S. Vimalnath, P.M. Benson Mansingh, N. Yuvaraj, K. Srihari, et al., Application of internet of things on the healthcare field using convolutional neural network processing, Journal of Healthcare Engineering 2022 (2022) e1892123, https://doi.org/10.1155/2022/1892123.
[34] A. Sharma, T. Virmani, V. Pathak, A. Sharma, K. Pathak, G. Kumar, D. Pathak, Artificial intelligence-based data-driven strategy to accelerate research, development, and clinical trials of COVID vaccine, BioMed Research International 2022 (2022) 7205241, https://doi.org/10.1155/2022/7205241.
[35] B. Guo, Y. Ma, J. Yang, Z. Wang, Smart healthcare system based on cloud-internet of things and deep learning, Journal of Healthcare Engineering 2021 (2021) e4109102, https://doi.org/10.1155/2021/4109102.

[36] T. Hussain, D. Hussain, I. Hussain, H. AlSalman, S. Hussain, S.S. Ullah, S. Al-Hadhrami, Internet of things with deep learning-based face recognition approach for authentication in control medical systems, Computational and Mathematical Methods in Medicine 2022 (2022) e5137513, https://doi.org/10.1155/2022/5137513.

[37] H.F. Nweke, Y.W. Teh, M.A. Al-garadi, U.R. Alo, Deep learning algorithms for human activity recognition using mobile and wearable sensor networks: state of the art and research challenges, Expert Systems with Applications 105 (2018) 233–261, https://doi.org/10.1016/j.eswa.2018.03.056.

[38] Z. Jiao, X. Gao, Y. Wang, J. Li, A parasitic metric learning net for breast mass classification based on mammography, Pattern Recognition 75 (C) (2018) 292–301, https://doi.org/10.1016/j.patcog.2017.07.008.

[39] R. Zhao, R. Yan, Z. Chen, K. Mao, P. Wang, R.X. Gao, Deep Learning and its Applications to Machine Health Monitoring: A Survey, 2016, https://doi.org/10.48550/arXiv.1612.07640 arXiv:1612.07640.

[40] C.A. da Costa, C.F. Pasluosta, B. Eskofier, D.B. da Silva, R. da Rosa Righi, Internet of health things: toward intelligent vital signs monitoring in hospital wards, Artificial Intelligence in Medicine 89 (2018) 61–69, https://doi.org/10.1016/j.artmed.2018.05.005.

[41] S.M.R. Islam, D. Kwak, M.D.H. Kabir, M. Hossain, K.-S. Kwak, The internet of things for health care: a comprehensive survey, IEEE Access 3 (2015) 678–708, https://doi.org/10.1109/ACCESS.2015.2437951.

[42] D. Yu, L. Deng, Deep learning and its applications to signal and information processing [exploratory DSP], Signal Processing Magazine, IEEE 28 (2011) 145–154, https://doi.org/10.1109/MSP.2010.939038.

[43] E. Gawehn, J.A. Hiss, G. Schneider, Deep learning in drug discovery, Molecular Informatics 35 (1) (2016) 3–14, https://doi.org/10.1002/minf.201501008.

[44] X. Chen, S. Xiang, C.-L. Liu, C.-H. Pan, Vehicle detection in satellite images by hybrid deep convolutional neural networks, IEEE Geoscience and Remote Sensing Letters 11 (10) (2014) 1797–1801, https://doi.org/10.1109/LGRS.2014.2309695.

[45] D. Ravì, C. Wong, F. Deligianni, M. Berthelot, J. Andreu-Perez, B. Lo, G.-Z. Yang, Deep learning for health informatics, IEEE Journal of Biomedical and Health Informatics 21 (1) (2017) 4–21, https://doi.org/10.1109/JBHI.2016.2636665.

[46] S. Min, B. Lee, S. Yoon, Deep learning in bioinformatics, Briefings in Bioinformatics 18 (5) (2017) 851–869, https://doi.org/10.1093/bib/bbw068.

[47] I. Lenz, H. Lee, A. Saxena, Deep learning for detecting robotic grasps, The International Journal of Robotics Research 34 (4–5) (2015) 705–724, https://doi.org/10.1177/0278364914549607.

[48] K. He, X. Zhang, S. Ren, J. Sun, Deep residual learning for image recognition, in: 2016 IEEE Conference on Computer Vision and Pattern Recognition (CVPR), 2016, pp. 770–778, https://doi.org/10.1109/CVPR.2016.90.

[49] C. Szegedy, W. Liu, Y. Jia, P. Sermanet, S. Reed, D. Anguelov, et al., Going deeper with convolutions, in: 2015 IEEE Conference on Computer Vision and Pattern Recognition (CVPR), 2015, pp. 1–9, https://doi.org/10.1109/CVPR.2015.7298594.

[50] M.D. Zeiler, R. Fergus, Visualizing and understanding convolutional networks, in: D. Fleet, T. Pajdla, B. Schiele, T. Tuytelaars (Eds.), Computer Vision—ECCV 2014, Springer International Publishing, 2014, pp. 818–833, https://doi.org/10.1007/978-3-319-10590-1_53.

[51] K. Simonyan, A. Zisserman, Very Deep Convolutional Networks for Large-Scale Image Recognition, 2015, https://doi.org/10.48550/arXiv.1409.1556 arXiv:1409.1556.

[52] A. Khatami, A. Khosravi, T. Nguyen, C.P. Lim, S. Nahavandi, Medical image analysis using wavelet transform and deep belief networks, Expert Systems with Applications 86 (2017) 190–198, https://doi.org/10.1016/j.eswa.2017.05.073.

[53] Q. Zhu, X. Li, A. Conesa, C. Pereira, GRAM-CNN: a deep learning approach with local context for named entity recognition in biomedical text, Bioinformatics (Oxford, England) 34 (9) (2018) 1547–1554, https://doi.org/10.1093/bioinformatics/btx815.

[54] K. Liu, S. Peng, J. Wu, C. Zhai, H. Mamitsuka, S. Zhu, MeSHLabeler: improving the accuracy of large-scale MeSH indexing by integrating diverse evidence, Bioinformatics (Oxford, England) 31 (12) (2015) i339–i347, https://doi.org/10.1093/bioinformatics/btv237.

[55] S. Peng, R. You, H. Wang, C. Zhai, H. Mamitsuka, S. Zhu, DeepMeSH: deep semantic representation for improving large-scale MeSH indexing, Bioinformatics (Oxford, England) 32 (12) (2016) i70–i79, https://doi.org/10.1093/bioinformatics/btw294.

[56] Y. Cheng, F. Wang, P. Zhang, J. Hu, Risk prediction with electronic health records: a deep learning approach, in: Proceedings of the 2016 SIAM International Conference on Data Mining, Society for Industrial and Applied Mathematics, June 2016, pp. 432–440.

[57] Y. Yu, M. Li, L. Liu, Y. Li, J. Wang, Clinical big data and deep learning: applications, challenges, and future outlooks, Big Data Mining and Analytics 2 (4) (2019) 288–305, https://doi.org/10.26599/BDMA.2019.9020007.

[58] Y. Wu, M. Jiang, J. Lei, H. Xu, Named entity recognition in Chinese clinical text using deep neural network, Studies in Health Technology and Informatics 216 (2015) 624–628.

[59] H. Choi, S. Ha, H.J. Im, S.H. Paek, D.S. Lee, Refining diagnosis of Parkinson's disease with deep learning-based interpretation of dopamine transporter imaging, NeuroImage: Clinical 16 (2017) 586–594, https://doi.org/10.1016/j.nicl.2017.09.010.

[60] G.E. Hinton, Learning multiple layers of representation, Trends in Cognitive Sciences 11 (10) (2007) 428–434, https://doi.org/10.1016/j.tics.2007.09.004.

[61] G.E. Hinton, R.R. Salakhutdinov, Reducing the dimensionality of data with neural networks, Science (New York, N.Y.) 313 (5786) (2006) 504–507, https://doi.org/10.1126/science.1127647.

[62] J. Schmidhuber, Deep learning in neural networks: an overview, Neural Networks 61 (2015) 85–117, https://doi.org/10.1016/j.neunet.2014.09.003.

[63] V. Gulshan, L. Peng, M. Coram, M.C. Stumpe, D. Wu, A. Narayanaswamy, et al., Development and validation of a deep learning algorithm for detection of diabetic retinopathy in retinal fundus photographs, JAMA 316 (22) (2016) 2402–2410, https://doi.org/10.1001/jama.2016.17216.

[64] H. Gunduz, Deep learning-based Parkinson's disease classification using vocal feature sets, IEEE Access 7 (2019) 115540–115551, https://doi.org/10.1109/ACCESS.2019.2936564.

[65] U.R. Acharya, S.L. Oh, Y. Hagiwara, J.H. Tan, H. Adeli, Deep convolutional neural network for the automated detection and diagnosis of seizure using EEG signals, Computers in Biology and Medicine 100 (2018) 270–278, https://doi.org/10.1016/j.compbiomed.2017.09.017.

[66] Y. Yoo, L.Y.W. Tang, D.K.B. Li, L. Metz, S. Kolind, A.L. Traboulsee, R.C. Tam, Deep learning of brain lesion patterns and user-defined clinical and MRI features for predicting conversion to multiple sclerosis from clinically isolated syndrome, Computer Methods in Biomechanics and Biomedical Engineering: Imaging & Visualization 7 (3) (2019) 250–259, https://doi.org/10.1080/21681163.2017.1356750.

[67] Y. Yoo, L.Y.W. Tang, T. Brosch, D.K.B. Li, S. Kolind, I. Vavasour, et al., Deep learning of joint myelin and T1w MRI features in normal-appearing brain tissue to distinguish between multiple sclerosis patients and healthy controls, NeuroImage: Clinical 17 (2018) 169–178, https://doi.org/10.1016/j.nicl.2017.10.015.

[68] P. Moeskops, J. de Bresser, H.J. Kuijf, A.M. Mendrik, G.J. Biessels, J.P.W. Pluim, I. Išgum, Evaluation of a deep learning approach for the segmentation of brain tissues and white matter hyperintensities of presumed vascular origin in MRI, NeuroImage: Clinical 17 (2017) 251–262, https://doi.org/10.1016/j.nicl.2017.10.007.

[69] X. Zhao, Y. Wu, G. Song, Z. Li, Y. Zhang, Y. Fan, A deep learning model integrating FCNNs and CRFs for brain tumor segmentation, Medical Image Analysis 43 (2018) 98–111, https://doi.org/10.1016/j.media.2017.10.002.

[70] G.S. Tandel, M. Biswas, O.G. Kakde, A. Tiwari, H.S. Suri, M. Turk, et al., A review on a deep learning perspective in brain cancer classification, Cancers 11 (1) (2019) 111, https://doi.org/10.3390/cancers11010111.

[71] D.S.W. Ting, L.R. Pasquale, L. Peng, J.P. Campbell, A.Y. Lee, R. Raman, et al., Artificial intelligence and deep learning in ophthalmology, British Journal of Ophthalmology 103 (2) (2019) 167–175, https://doi.org/10.1136/bjophthalmol-2018-313173.

[72] N. Dhungel, G. Carneiro, A.P. Bradley, A deep learning approach for the analysis of masses in mammograms with minimal user intervention, Medical Image Analysis 37 (2017) 114–128, https://doi.org/10.1016/j.media.2017.01.009.

[73] E.Y. Kalafi, N.A.M. Nor, N.A. Taib, M.D. Ganggayah, C. Town, S.K. Dhillon, Machine learning and deep learning approaches in breast cancer survival prediction using clinical data, Folia Biologica 65 (5–6) (2019) 212–220.

[74] M.D. Ganggayah, N.A. Taib, Y.C. Har, P. Lio, S.K. Dhillon, Predicting factors for survival of breast cancer patients using machine learning techniques, BMC Medical Informatics and Decision Making 19 (1) (2019) 48, https://doi.org/10.1186/s12911-019-0801-4.

[75] S.M.H. Hosseini, S.R. Kesler, Multivariate pattern analysis of FMRI in breast cancer survivors and healthy women, Journal of the International Neuropsychological Society: JINS 20 (4) (2014) 391–401, https://doi.org/10.1017/S1355617713001173.

[76] L. Yue, X. Gong, K. Chen, M. Mao, J. Li, A.K. Nandi, M. Li, Auto-detection of alzheimer's disease using deep convolutional neural networks, in: 2018 14th International Conference on Natural Computation, Fuzzy Systems and Knowledge Discovery (ICNC-FSKD), 2018, pp. 228–234, https://doi.org/10.1109/FSKD.2018.8687207.

[77] K.A.N.N.P. Gunawardena, R.N. Rajapakse, N.D. Kodikara, Applying convolutional neural networks for pre-detection of alzheimer's disease from structural MRI data, in: 2017 24th International Conference on Mechatronics and Machine Vision in Practice (M2VIP), 2017, pp. 1–7, https://doi.org/10.1109/M2VIP.2017.8211486.

[78] G. Lee, K. Nho, B. Kang, K.-A. Sohn, D. Kim, Predicting Alzheimer's disease progression using multi-modal deep learning approach, Scientific Reports 9 (2019) 1952, https://doi.org/10.1038/s41598-018-37769-z.

[79] R. Ju, C. Hu, Q. Li, Early diagnosis of Alzheimer's disease based on resting-state brain networks and deep learning, IEEE/ACM Transactions on Computational Biology and Bioinformatics 16 (1) (2017) 244–257.

[80] M.A. Just, V.L. Cherkassky, T.A. Keller, R.K. Kana, N.J. Minshew, Functional and anatomical cortical underconnectivity in autism: evidence from an FMRI study of an executive function task and corpus callosum morphometry, Cerebral Cortex 17 (4) (2007) 951–961.

[81] G.S. Dichter, Functional magnetic resonance imaging of autism spectrum disorders, Dialogues in Clinical Neuroscience 14 (3) (2012) 319–351.

[82] C. Wang, Z. Xiao, B. Wang, J. Wu, Identification of autism based on SVM-RFE and stacked sparse auto-encoder, IEEE Access 7 (2019) 118030–118036, https://doi.org/10.1109/ACCESS.2019.2936639.

[83] T. Eslami, F. Saeed, Similarity based classification of ADHD using singular value decomposition, in: Proceedings of the 15th ACM International Conference on Computing Frontiers, May 2018, pp. 19–25.

[84] Q. Yao, H. Lu, Brain functional connectivity augmentation method for mental disease classification with generative adversarial network, in: Chinese Conference on Pattern Recognition and Computer Vision (PRCV), Springer, Cham, November 2019, pp. 444–455.

[85] S.M. Plis, D.R. Hjelm, R. Salakhutdinov, E.A. Allen, H.J. Bockholt, J.D. Long, et al., Deep learning for neuroimaging: a validation study, Frontiers in Neuroscience 8 (2014) 229.

[86] R. Wason, D. Goyal, V. Jain, S. Balamurugan, A. Baliyan, (1 C.E.). Applications of Deep Learning and Big IoT on Personalized Healthcare Services. https://services.igi-global.com/resolvedoi/resolve.aspx?doi=10.4018/978-1-7998-2101-4. IGI Global. https://www.igi-global.com/book/applications-deep-learning-big-iot/www.igi-global.com/book/applications-deep-learning-big-iot/235494.

[87] V. Jain, J.M. Chatterjee, Machine Learning with Health Care Perspective, Springer, Cham, 2020, pp. 1–415.

[88] Q. Dou, H. Chen, L. Yu, L. Zhao, J. Qin, D. Wang, et al., Automatic detection of cerebral microbleeds from MR images via 3D convolutional neural networks, IEEE Transactions on Medical Imaging 35 (5) (2016) 1182–1195, https://doi.org/10.1109/TMI.2016.2528129.

[89] Y. Gao, G.-Y. Cai, W. Fang, H.-Y. Li, S.-Y. Wang, L. Chen, et al., Machine learning based early warning system enables accurate mortality risk prediction for COVID-19, Nature Communications 11 (1) (2020) 5033, https://doi.org/10.1038/s41467-020-18684-2.

[90] F. Khozeimeh, D. Sharifrazi, N.H. Izadi, J.H. Joloudari, A. Shoeibi, R. Alizadehsani, et al., Combining a convolutional neural network with autoencoders to predict the survival chance of COVID-19 patients, Scientific Reports 11 (1) (2021) 1–18.

[91] M. Naseem, H. Arshad, S.A. Hashmi, F. Irfan, F.S. Ahmed, Predicting mortality in SARS-COV-2 (COVID-19) positive patients in the inpatient setting using a novel deep neural network, International Journal of Medical Informatics 154 (2021) 104556, https://doi.org/10.1016/j.ijmedinf.2021.104556.

[92] X. Ye, Q.T. Zeng, J.C. Facelli, D.I. Brixner, M. Conway, B.E. Bray, Predicting optimal hypertension treatment pathways using recurrent neural networks, International Journal of Medical Informatics 139 (2020) 104122, https://doi.org/10.1016/j.ijmedinf.2020.104122.

[93] A. Lanata, G. Valenza, M. Nardelli, C. Gentili, E.P. Scilingo, Complexity index from a personalized wearable monitoring system for assessing remission in mental health, IEEE Journal of Biomedical and Health Informatics 19 (1) (2015) 132−139, https://doi.org/10.1109/JBHI.2014.2360711.

[94] T. Ai, Z. Yang, H. Hou, C. Zhan, C. Chen, W. Lv, Q. Tao, Z. Sun, L. Xia, Correlation of chest CT and RT-PCR testing for coronavirus disease 2019 (COVID-19) in China: a report of 1014 cases, Radiology 296 (2) (2020) E32−E40, https://doi.org/10.1148/radiol.2020200642.

[95] B. Al-Shargabi, S. Abuarqoub, IoT-Enabled Healthcare: Benefits, Issues and Challenges, 2020, https://doi.org/10.1145/3440749.3442596.

[96] S.W. Chen, X.W. Gu, J.J. Wang, H.S. Zhu, AIoT used for COVID-19 pandemic prevention and control, Contrast Media & Molecular Imaging 2021 (2021).

[97] R. Sitharthan, S. Krishnamoorthy, P. Sanjeevikumar, J.B. Holm-Nielsen, R.R. Singh, M. Rajesh, Torque ripple minimization of PMSM using an adaptive Elman neural network-controlled feedback linearization-based direct torque control strategy, International Transactions on Electrical Energy Systems 31 (1) (2021) e12685.

[98] M. Ramamurthy, I. Krishnamurthi, S. Vimal, Y.H. Robinson, Deep learning based genome analysis and NGS-RNA LL identification with a novel hybrid model, Biosystems 197 (2020) 104211, https://doi.org/10.1016/j.biosystems.2020.104211.

[99] M.R. Desai, S. Toravi, A smart sensor interface for smart homes and heart beat monitoring using WSN in IoT environment, in: 2017 International Conference on Current Trends in Computer, Electrical, Electronics and Communication (CTCEEC), IEEE, September 2017, pp. 74−77.

[100] A. Kelati, I.B. Dhaou, H. Tenhunen, Biosignal Monitoring Platform Using Wearable IoT 6, 2018.

[101] M. Nasajpour, S. Pouriyeh, R.M. Parizi, M. Dorodchi, M. Valero, H.R. Arabnia, Internet of Things for current COVID-19 and future pandemics: an exploratory study, Journal of Healthcare Informatics Research 4 (4) (2020) 325−364.

[102] L. Chuquimarca, D. Roca, W. Torres, L. Amaya, J. Orozco, D. Sánchez, Mobile IoT device for BPM monitoring people with heart problems, in: 2020 International Conference on Electrical, Communication, and Computer Engineering (ICECCE), 2020, pp. 1−5, https://doi.org/10.1109/ICECCE49384.2020.9179293.

[103] S.U. Amin, M.S. Hossain, G. Muhammad, M. Alhussein, M.A. Rahman, Cognitive smart healthcare for pathology detection and monitoring, IEEE Access 7 (2019) 10745−10753, https://doi.org/10.1109/ACCESS.2019.2891390.

[104] B. Pradhan, S. Bhattacharyya, K. Pal, IoT-based applications in healthcare devices, Journal of Healthcare Engineering 2021 (2021) e6632599, https://doi.org/10.1155/2021/6632599.

[105] M. Heshmat, A.-R.S. Shehata, A Framework about Using Internet of Things for Smart Cancer Treatment Process 6, 2018.

[106] B. Li, Q. Dong, R.S. Downen, N. Tran, J.H. Jackson, D. Pillai, et al., A wearable IoT aldehyde sensor for pediatric asthma research and management, Sensors and Actuators B: Chemical 287 (2019) 584−594.

[107] S.T.U. Shah, F. Badshah, F. Dad, N. Amin, M.A. Jan, Cloud-assisted IoT-based smart respiratory monitoring system for asthma patients, in: F. Khan, M.A. Jan, M. Alam (Eds.), Applications of Intelligent Technologies in Healthcare, Springer International Publishing, 2019, pp. 77−86.

Chapter 19

Deep learning in drug discovery

Meenu Bhati[1], Tarun Virmani[1], Girish Kumar[1], Ashwani Sharma[1] and Nitin Chitranshi[2]
[1]*School of Pharmaceutical Sciences, MVN University, Palwal, Haryana, India;* [2]*Faculty of Medicines, Health and Human Sciences, Macquarie University, Sydney, NSW, Australia*

1. Introduction

The development of a new drug can take up to 12 years, and it is estimated that its average cost until it reaches the market is approximately one billion euros. The time and costs involved are largely associated with the large number of molecules that fail at one or more stages of their development, as it is estimated that only 1 in 5000 drugs finally reach the market [1]. Drug discovery is an expensive process that takes on average more than a decade from discovery to approval. However, current advanced technologies not only hastened the discovery process but also work to enable the development of personalized therapeutics. Consequently, drug discovery has witnessed rapid changes that over the last few decades produced significant advances in technological innovation and scientific studies that enabled the generation of novel drug candidates and rapid translation to useable entities. The area of AI/ML has witnessed many changes recently, particularly concerning deep learning (DL) methods [2]. DL has made significant progress in many artificial intelligence research fields over the last decade. In recent years, there has been the emergence of the first wave of DL applications in pharmaceutical research. These applications go earlier projections of bioactivity and show promise for solving a variety of issues in drug discovery [3]. Artificial intelligence is the tool used for the discovery of drugs. AI consists of the different tools, techniques, and algorithms such as symbolic AI, neural networks, DL, and genetic algorithms [4]. The ability of a machine to mimic intelligent human behavior is known as AI. Smart machines (Siri, Alexa, Google Assistant, etc.), automated public transportation, airplanes, and electronic games are just a few examples of how AI is incorporated into our daily lives. AI has more recently started to be used in medicine to enhance patient care by accelerating procedures and obtaining higher accuracy, paving the way for improved healthcare overall. Machine learning (ML) is evaluating radiological pictures, pathology slides, and patient electronic medical records (EMR), assisting in the process of patient diagnosis and treatment, and enhancing doctors' abilities. Herein, we describe the current status of AI-based DL in medicine, the way it is used in the different disciplines, and future trends [5]. Since the 1960s, medicinal chemistry has used AI in a variety of ways to create molecules, with variable degrees of success. Supervised learning, where labeled training datasets are used to train models, is extensively applied. An example is the QSAR approach, which is widely used to predict properties, such as logP, solubility, and bioactivity, for given chemical structures [6]. More than 1060 molecules constitute the vast chemical space, which promotes the development of many various pharmacological compounds. However, the drug development process is constrained by a lack of cutting-edge technologies; consequently, it is time-consuming and costly endeavor that can be rectified by applying AI. AI can identify hit and lead compounds, as well as accelerate therapeutic target validation and structural design optimization [7]. The exhibited information by computers is referred to as machine intelligence, which is typically referred to as artificial intelligence. In the evolution of rational drug development history, a variety of machine intelligence techniques have been used to direct costly and time-consuming traditional studies. QSAR modeling is one of the machine-learning technologies that have been created over the past few decades that may efficiently and cheaply discover possible biologically active molecules from millions of candidate compounds [8]. After the advent of ML, AI was the most studied concept in September 2015. Some refer to ML as the main application of AI, while others refer to it as a subclass of AI [9]. ML approaches and recent developments in DL provide many opportunities to increase efficiency across the drug discovery and development pipeline. As such, we expect to see increasing numbers of applications for well-defined problems across the industry in the coming years. With available data becoming "bigger," at least in the sense of more thoroughly covering the relevant variability of the whole data space, and as computers become increasingly more

Deep Learning in Personalized Healthcare and Decision Support. https://doi.org/10.1016/B978-0-443-19413-9.00013-8

powerful, ML algorithms are going to systematically generate improved outputs, and new, interesting applications are expected to follow. The previous sections, where we discussed some ML applications for target discovery and validation, drug design, and development, serve as the main example of this, biomarker identification, and pathology for disease diagnosis and therapy [10]. The use of a DL technique has increased recently which draws on years of research into artificial neural networks [11], which has demonstrated significant benefits in learning from images and languages [12]. This may indicate the beginning of the next phase of cheminformatics and pharmaceutical research, which will be centered on the analysis of heterogeneous big data, which is accumulating using more sophisticated algorithms such as DL [13]. Thus, the last 2 decades developed many techniques and tools for computational drug discovery and implemented in favor of drug synthesis and drug discovery as depicted in Table 19.1.

TABLE 19.1 List of a toolset and methods based on machine learning with their descriptions in favor of drug synthesis and drug discovery.

S. No.	Toolbox and techniques	Characteristics	References
1.	OCED	Prediction of toxicity	Yang et al. [14]
2.	PaDEL-descriptor	Determine the pharmacodynamics, pharmacokinetic properties, and toxic effects of a substance	He et al. [15]
3.	OpenBabel	Prediction of molecule 3D geometry	Yoshikawa et al. [16]
4.	Google's DeepMind	Predict the 3D structure of proteins from their amino acid sequences	Powles and Hodson [17]
5.	Cloud 3D-QSAR	Development of QSAR models in drug discovery	Wang et al. [18]
6.	ChemSAR	Generating molecular SAR modeling	Dong et al. [19]
7.	HNet-DNN	Prediction of the new drug–disease associations	Liu et al. [20]
8.	Gaussian interaction profile kernel and autoencoder (GIPAE)	Prediction of the drug-disease associations	Jian et al. [21]
9.	DeepConv-DTI	Prediction of drug-target interaction	Lee et al. [22]
10.	METADOCK	Molecular docking methodology	Imbernon et al. [23]
11.	Chembench	Used for curation, visualization, analysis, and modeling of chemogenomics data	Capuzzi et al. [24]
12.	TargetNet	Prediction of the potential drug target interaction profiling	Yao et al. [25]
13.	Vienna LiverTox P	Pharmacokinetic property identification	Montanari et al. [26]
14.	dendPoint	A web resource for exploring the pharmacokinetics of dendrimers and prediction.	Kaminskas et al. [27]
15.	Lazar	A modular predictive toxicology framework	Maunz et al. [28]
16.	DeepPurpose	Library for drug-target interaction prediction.	Huang et al. [29]
17.	IDDkin	Prediction of kinase inhibitors	Shen et al. [30]
18.	RosENet	Prediction of the absolute binding affinity of protein ligand complexes	Hussein Hassan-Harrirou et al. [31]
19.	MDeePred	Target protein interaction prediction system	Rifaioglu et al. [32]
20.	ProTox-II	Webserver for the prediction of toxicity of chemicals	Banerjee et al. [33]
21.	PSBP-SVM (polystyrene binding peptides)	Prediction of polystyrene binding peptide	Meng et al. [34]
22.	CSM-lig	A web server for assessing and comparing protein small molecule affinities.	Pires and Ascher [35]
23.	ChemGrapher	Chemical compound recognition using optical graphs	Oldenhof et al. [36]
24.	HeteroDualNet	Prediction of drug-disease association prediction	Jian et al. [37]

2. Literature review

Fu et al. discussed the implementation and design of the probabilistic and dynamic neural inference (PRIME) method to replace conventional point estimation with such a high-dimensional distribution to reflect drug and disease. Such probabilistic modeling can choose certain confident predictions for task repurposing and enhance the performance. Uncertainty estimation allows thorough exploration of the chemical space for de-novo design. Models are usually based on either drug—target or drug—disease interaction for drug repurposing [38].

Jimenez Luna et al. focused on quantitative structure—activity/property relationship (QSAR/QSPR) approaches have transitioned from the use of simpler models, such as linear regression and k-nearest neighbors, toward more universally applicable ML techniques, such as support vector machines (SVM) and gradient boosting methods (GBM), aiming to address more complex and potentially nonlinear relationships between the chemical structure and its physicochemical/biological properties, often at the expenses of interpretability [39].

Jamshidi et al. reported that to overcome the challenges faced by conventional approaches, AI and its emerging branches, such as DL and ML, can be used to design drugs. Additionally, using smart methods will give researchers the chance to apply some simple and effective services to solve such issues. In this regard, a formal DL framework has been investigated to show the efficacy and utility of these techniques [40].

Arpaci et al. found that most of the research that were undertaken tended to use straightforward classification algorithms, making few attempts to utilize more advanced methods. These outcomes open the door for future research and encourage scholars to work on more advanced techniques, such as developing new algorithms or improving the existing ones. In addition, the limited access to patient's data hinders the use of ML in an appropriate manner. Hence, the availability of patients' clinical features is another challenge to the application of ML algorithms [41].

Muniz Castro et al. stated that AI is a technology that could revolutionize pharmaceutical 3-dimensional printed (3DP) through evaluating huge data. Here, key elements of the 3DP formulation pipeline and in vitro dissolving parameters were projected using literature-mined data for creating AI-ML models. From 114 papers, a total of 968 formulas were evaluated. The ML algorithms were able to learn and offer accuracies as high as 93% for values in the filament hot melt extrusion process. Additionally, ML systems were able to forecast the release of 3DP medications using information from the formulations' composition and extra input variables [42].

Boso et al. stated that artificial neural networks are a mathematical-computational model that is inspired by the structure and functional aspects of biological neural networks. They are composed of a collection of nodes also called neurons gathered into layers. The nodes in the first layer contain the input data of the problems; the nodes in the last layer contain the desired output. The number of nodes in those two layers depends upon the physics of the problem at hand. Between the input and output layers, nodes are organized in the so-called "hidden layers": the number of hidden layers and the number of nodes per each hidden layer is decided by the user [43].

Fields et al. reported that ML algorithms, including NNs-based techniques and SVM models, were used to discover novel antimicrobial peptides, also known as bacteriocins, from bacteria could ultimately be used as compelling antibiotic candidates. These input bacteriocins were represented as complicated vector sums. The machine-learning algorithm then took the inputs and generated new vector structure outputs that preserved the original inputs key features. 676 bacteriocins that were not similar to the input bacteriocins were synthesized from these outputs. A sliding window method was used to create 28,895 peptides from the output bacteriocins; these peptides spanned 20-mers and were placed through biophysical parameters [44].

Urban et al. recently introduced a CNN to automatically classify images of vasculature networks formed in our managed print service (MPS) into no-hit, soft-hit, and hard-hit categories. The accuracy of our best model is significantly better than our minimally trained human raters and requires no human intervention to operate. This model is a first step toward automation of data analysis for high-throughput drug screening [45].

Cai et al. on the attempt to forecast the site of metabolism for UDP-glucuronosyl transferase (UGT) catalyzed events using tree form ML techniques. Membrane permeation of drug molecules can also be considered and tackled by traditional ML methods [46].

Dey et al., a neural fingerprint technique in a concurrent DL framework for prediction of drug-related adverse events, so that the label information can be used in the feature generation stage of ML process and evaluate the characteristics to determine which subsets within the bioactive compounds are specifically associated with a given ADR [47].

Pang et al. focused on Naive Bayes (NB) models and additional techniques as classifiers for active and inactive compounds, with possible activity as antagonists for estrogen receptors in breast cancer. The researchers utilized the ability of NB algorithms to process vast quantities of information while having a unique tolerance to random noise. The technique, in combination with other tools such as extended-connectivity fingerprint-6, is able to collect excellent outputs [48].

Popova et al. proposed a method called Reinforcement Learning for Structural Evolution (ReLeaSE) for creating chemical compounds and focused chemical libraries with desired physical, chemical, and/or bioactivity properties that are based on deep reinforcement learning (RL). Based on deep and reinforcement learning (RL) approaches, ReLeaSE integrates two deep neural networks, namely, generative and predictive that are trained separately but are used jointly to generate novel targeted chemical libraries. ReLeaSE uses the simplified molecular depiction by their simplified molecular-input line-entry system (SMILES) strings only [49].

Mamoshina et al. compared the several supervised ML methods and revealed SVMs with linear kernel and deep feature selection to be best suited to the identification of aging biomarkers. In each of these examples, ML generated a set of predictions of targets that have properties that suggest they are likely to bind drugs or be involved in disease [50].

Lo et al. developed the shape align program that combines 2D and 3D metrics based on the open babel path-based fingerprint (PF2) fingerprint, shapes and pharmacophoric points for unsupervised 3D chemical similarity clustering [51].

Rahman et al. utilized multivariate radio frequency (RF) by including information relating to genomic sequencing, which helped sustain error and achieve drug responses based on genomic characterizations. Data incorporating genetic and epigenetic characterization combinations are fed into the computational framework, enabling the framework to forecast the mean and confidence interval of the medication reactions., an important quality essential for analyzing any drug to be processed in clinical trials [52].

Gomez et al. introduced an algorithm named atomic convolutional neural network (ACNN) for predicting the energy gap between a bounded drug–target complex and an unbounded state using spatial convolutional neural network (CNN) directly from atomic coordinates. To process the input, the authors employed radial pooling filters with learnable mean and variance [53].

Olivecrona et al. found that the pretrained RNNs for creating molecules with specified user defined features; a policy-based reinforcement learning approach has been presented. In one test case for adjusting the model to produce compounds predicted to be active against the dopamine receptor type 2, the model produced structures of which $>95\%$ were predicted to be active, including experimentally confirmed actives not included in either the generative model or the activity prediction model [3].

Glaab et al. examined most current breakthroughs in virtual screening (VS) methods based on ligands and structures. The four primary phases that built up the overall process were collection of data, preprocessing, screening, selectivity, and ADMETox (i.e., absorption, distribution, metabolism, excretion, and toxicity). With a focus on relevant open-access tools and datasets, each stage was described in detail. Additionally, the author integrated open-access screening tools into executable cross-platform software using the virtual machine platform [54].

Qiu et al. introduced advanced proteochemometric techniques and mentioned its advantages by referring to studies in which in deep tensor imaging (DTI) modeling, PCM models perform better than traditional QSAR models. The researcher concentrated on the most latest developments in PCM modeling with regard to target descriptors, crossterm descriptors, and PCM's application breadth, which included interactions between proteins and small and large molecules. It was also stated that it may be possible to develop PCM models for more complex systems, like ligand catalyst–target reactions, with further developments in representations of molecules, ML techniques, and the available pharmacodynamic data. This could help to more accurately identify biochemical reactions [55].

Wan et al. proposed a deep neural network (DNN) for DTI prediction. Additionally, by locating low-dimensional representations of the original input features, their method incorporated unsupervised representation learning for feature development. Natural language processing (NLP) methods were used to embed the initial input features, which consisted of Morgan fingerprints for chemicals and protein sequences for targets, in a predetermined low-dimensional space (i.e., 200 dimensions for compounds and 100 for proteins) (i.e., latent semantic analysis and Word2vec) [56].

Aliper et al. demonstrated the ability of DL algorithms to classify different drugs into therapeutic categories merely based on their transcriptional profiles in diverse cell lines combined with pathway information [57].

Lusci et al. reported a method that employed a variant of RNN, called undirected Graph (UGRNN), which, in order to generate models, first converts molecular structures into vectors of the same length as the molecular representation. Bit values in the vectors are learned from the dataset. It was demonstrated that models created using the UGRNN approach could predict solubility with an accuracy that was on par with models created using molecular descriptors [58].

Wang et al. showed that the oral bioavailability and blood–brain barrier (BBB) models can be improved by including biological descriptors of membrane transportations. The differences in transporter interaction activities are able to differentiate the two structurally similar compounds in chemical space but with different bio-activities, thus correct the prediction for the query compound. Therefore, including meaningful biological descriptors (e.g., transporter descriptors in this study) can improve the resulting models [59].

3. Approaches of machine and deep learning for drug target

Cheminformatics and bioinformatics are two emerging computer-assisted computational tools that have already been applied to the discovery of new drugs [60]

I. **Cheminformatics:** In 2015, the term "epi-informatics" was introduced and conceptualized to summarize advances in epigenetic drug and chemical probe discovery driven by computational methods [61]. In addition, due to increased emphasis on "Big Data," ML and artificial intelligence, not only in the society in general, but also in drug discovery, it is expected that the cheminformatics field will be even more important in the future. Besides traditional cheminformatics tasks as outlined above, the use of ML in cheminformatics has become a very hot topic and will be extensively discussed in the article [62]. Compounds and targets must be digitally recorded as quantitative matrices (i.e., representations and descriptors) in accordance with their molecular characteristics in order to do this. These features of vectors are used as input by VS techniques to model the interactions between substances and their target molecules. VS methods can be divided into various groups based on the employed input features which are as:

- **Structure-based virtual screening (SBVS)** employs 3D structure of targets and compounds to model the interactions. These methods have augmented the drug discovery process since the inception of in-silico molecular modeling. The SBVS aims to prioritize small molecules potentially targeting a protein target of interest, subsequently leading to the reduction of experimental costs and increase of successful outcomes. The typical approach includes the prediction of the binding affinity (or related affinity scores) between small molecules and the target of interest. This binding affinity/score is then used to rank-order compounds for prioritization, and the general rule is that the greater the predicted affinity the greater the likelihood of a true small molecule-target interaction.
- **Ligand-based virtual screening** uses the molecular properties of compounds (mostly nonstructural) to model the interactions with targets. These methods do not require target protein-derived information.
- **Molecular docking using ML**, the adoption and use of DL methods for molecular docking has been extremely successful mainly due to the accessibility to DL computational frameworks such as Tensorflow. Typically, these methods are applied to postdocking drug–target complexes in order to more accurately predict binding poses and drug activity as depicted in Fig. 19.1. Such techniques are essential for drug repurposing because they enable fast virtual screening of existing medications against a large number of novel therapeutically significant targets.
- **EHR-based ML:** In addition to the utilization of cheminformatic, phenotypic, and transcriptomic datasets, electronic health records (EHRs) also contain a wealth of information that can be mined for drug discovery, repurposing and adverse drug event (ADE) predictions. Recent advances in NLP have enabled investigators to take the largely unstructured data sources (i.e., clinic notes, web forum posts) and translate them into a machine-interpretable format, thus enabling the application of machine and DL analytics. For example, in an effort to improve upon the underreporting of ADEs through current means such as the FDA Adverse Events Reporting System.
- **Proteochemometric modeling (PCM) approach models** the interactions at the input level by combining nonstructural features of both compounds and targets. The combination of information collected from the drug and the target protein has proved very helpful for creating prediction models of drug activity as a logical extension of the QSAR methodology. Proteochemometrics are these combinatorial techniques that can reveal target characteristics relevant to binding, such as crucial amino acids. Numerous new inhibitors, specifically for adenosine receptors, have been discovered using PCM [39,40].

For example: One of the main examples of VS is molecular docking. It is routinely used to process virtual libraries containing millions of molecular structures against a variety of drug targets with known three-dimensional structures [65] as shown in Fig. 19.1. Some of the additional parameters are discussed as.

a. **Evolutionary clustering:** Evolutionary computation (EC) is a sub-set of AI that contains a family of nature-inspired algorithms. These are population-based algorithms, which maintain a population of possible solutions (individuals) and evolve toward good/optimal solutions. By evolving multiple solutions simultaneously, EC techniques are well known for their good global search ability. Herein, researchers present a novel evolutionary chemical binding similarity

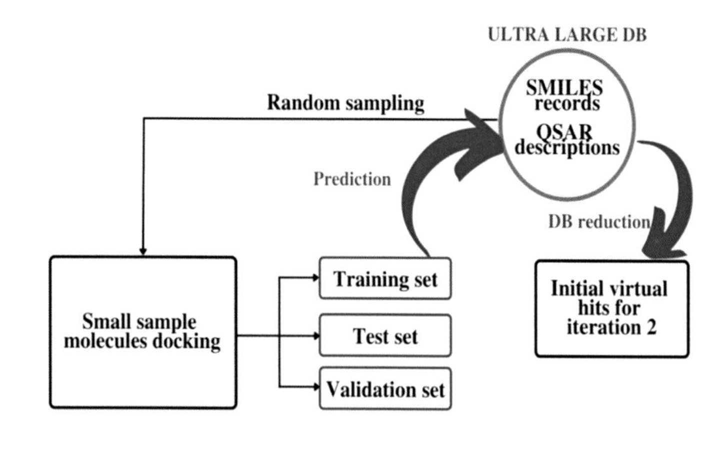

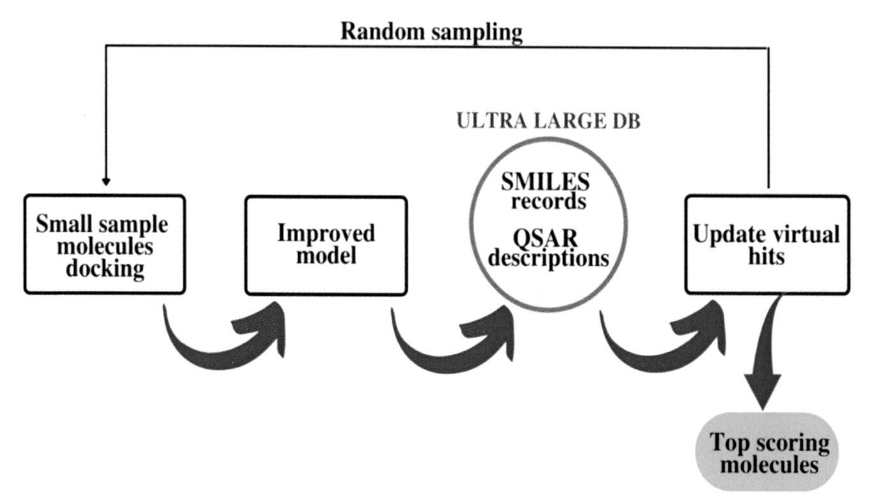

FIGURE 19.1 Schematic view of the DD pipeline. (Top) DD initialization: Small samples of molecules that have been docked to a target under consideration and were randomly chosen from an enormous docking database are used to train a QSAR deep model. (Bottom) DD screening: The QSAR-predicted virtual hits from the previous DD iteration are randomly chosen and added to the training set, gradually improving the deep model.

(ECBS) method using a classification similarity learning framework defined with paired chemical data and target's evolutionary relationship depicted as Fig. 19.2.

The ECBS method is designed to encode molecular features enriched in evolutionarily conserved chemical−target binding relationships and formulated by the likelihood of chemical compounds binding to identical targets [66].

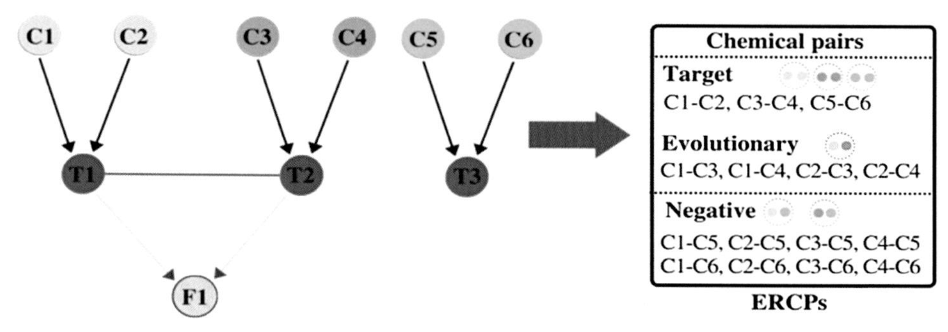

FIGURE 19.2 The simplified chemical targeted evolutionary relationship for the description of evolutionary chemical binding similarity (ECBS) method.

The field of evolutionary clustering algorithms can be split into two categories: fixed algorithms that require K that is known and automatic algorithms, which discover K themselves. Fixed clustering algorithms are prevalent historically, whereas most recent work tackles the more difficult automatic clustering problem. The third category of algorithms which has emerged recently uses feature reduction to improve clustering performance [67].

b. De-novo drug designing: De-novo drug design is an approach that develops new chemical compounds solely from knowledge of a biological target (receptor) or its recognized active binders (ligands found to possess good binding or inhibitory activity against the receptor). The synthesis of the molecules (sampling), evaluation of the created compounds, and description of the receptor active site or ligand pharmacophore modeling are the main elements of de-novo drug design. There are two main de-novo drug design methodologies: structure-based and ligand-based design. X-ray crystallography, NMR, or electron microscopy is frequently used to determine the three-dimensional (3D) structure of receptors. When the structure of the receptor is unclear, homology modeling can be utilized to determine a suitable structure for de-novo drug development. However, the quality of a homology model depends on the template structure and sequence similarity. The ligand-based method is typically used when the biological target lacks structural information but has one or more known active binders [68]. The ML -based de-novo drug design, put out as a solution to the issue of the limited variety of therapeutic candidate molecules offered by the conventional prediction models, is one computational drug discovery topic that is quickly gaining interest. The goal of de-novo drug design is finding innovative therapeutic drug that are structurally considerably different from those currently on the market or in the development stage. Classical de-novo drug design methods follow a rather manual procedure, where the researcher carries out a series of intensive computational processes such as docking and molecular dynamics simulations [63]. Overall, docking has proved to be a very effective drug repurposing tool for many diseases such as HIV, Dengue virus, and multiple cancer types [64] and the creation of new chemical structures using neural networks (NNs) is an intriguing use of DL in cheminformatics. Goomez-Bombarelli et al. introduced a novel method using variational autoencoder (VAE) to generate chemical structures. Medicinal chemists are exploring to what extent an RNN with long short-term memory (LSTM) cells can figure out sensible chemical rules and generate synthetically feasible molecules after being trained on existing compounds encoded as SMILES [69]. The initial phase is to map chemical structures using unsupervised learning using VAE as SMILES strings in the ZINC database into latent space. Once the training of the VAE is complete, the latent vector in the latent space represents molecular structure continuously and may be reversely translated into an SMILES string using the trained VAE. By using any optimization technique, such as Bayesian optimization, to find the best latent solutions in the continuous latent space, and then decoding those latent solutions into SMILES, it is possible to create new structures with desirable attributes. Following on from Gomez-Bombarelli's work, Kadurin et al. [70] used VAE as a molecular descriptor generator coupled with a generative adversarial network (GAN) a special NN architecture, to generate new structures that were claimed to have promising specific anticancer properties. Blaschke et al. [71] utilized VAE to generate novel structures with predicted activity against dopamine receptor type II. De-novo drug design (DNDD) refers to the design of novel chemical entities that fit a set of constraints using computational growth algorithms.

The machine-learning approaches such as deep reinforcement learning and its application in the development of novel de novo drug design methods works together to make the process easy and efficient which is summarized in Fig. 19.3. Future directions for this important field, including integration with toxicogenomics and opportunities in vaccine development, are presented as the frontiers for machine-learning-enabled de-novo drug design [72].

c. Quantitative structure−activity relationships (QSAR): In the past, different QSAR models for VS have been produced using machine-learning techniques, which are one of the most crucial elements of artificial intelligence. Some of the QSAR strategies used in drug discovery can be divided into linear and nonlinear methods. For instance, Belhumeur created linear discriminant analysis (LDA) in 1996 for pattern identification and artificial intelligence. LDA is a supervised machine-learning technique that works well with small datasets as depicted in Fig. 19.4.

LDA is a classifier that considers a linear equation to maximize the between-class distance and minimize the within-class distance. LDA has been used to predict drug−drug interactions, identify new compounds [73], and detect adverse drug events, among others. LDA is a basic methodology, but when combined with innovative descriptors, it is still regarded as an effective modeling approach as depicted in Fig. 19.4 [74].

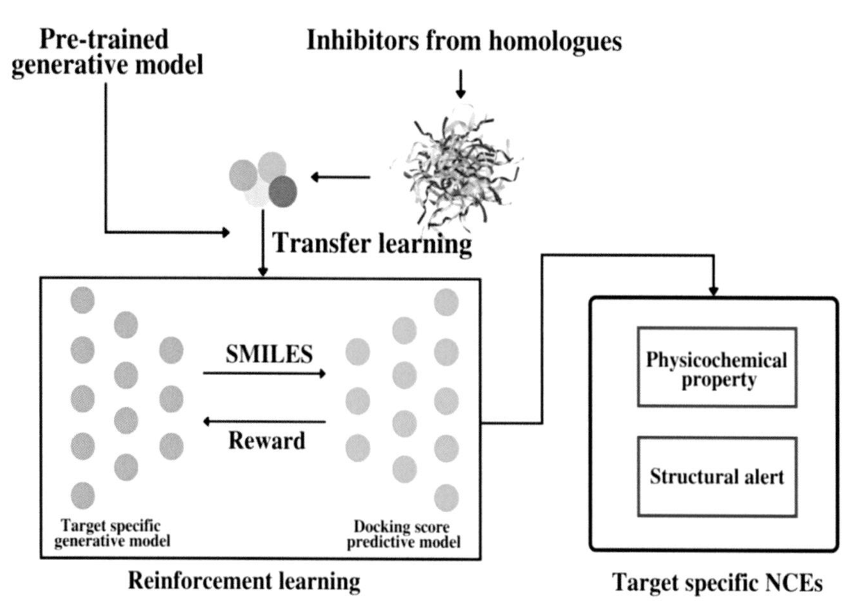

FIGURE 19.3 Accelerating de-novo drug design against novel proteins using deep learning.

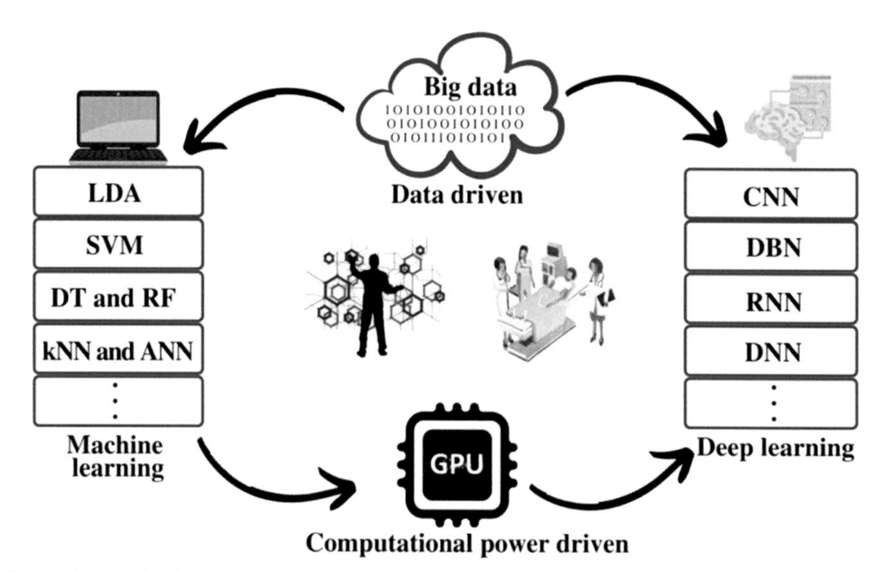

FIGURE 19.4 The combination of linear discriminant analysis (LDA) and novel descriptors for automatic change detection.

4. Practical applications of QSAR-based virtual screening

QSAR is still seen as a complementary analysis to studies of synthesis and biological evaluation, often introduced in the study without any justification or additional perspective. Despite the small number of virtual screening (VS) applications available in the literature, most of them led to the discovery of promising hits and lead candidates several effective uses of QSAR-based VS for hit-to-lead optimization and finding new hits discussed below [75]

a. **Malaria:** Due to the ongoing evolution of parasite resistance to available antimalarial medications, malaria is a contagious disease, making the search for novel treatment candidates a top global health priority. Researchers designed binary and continuous QSAR models implementing DL for predicting antiplasmodial activity and cytotoxicity of untested compounds [76].
b. **Schistosomiasis:** Schistosomiasis is a disease caused by flatworms from the schistosoma genus that affects 206 million of people worldwide [77]. QSAR models for *Schistosoma mansoni* thioredoxin glutathione reductase (SmTGR), a

validated target for schistosomiasis to find new structurally dissimilar compounds with antischistosomal activity [78]. For the achievement of the goal, it was designed in a study with the following steps:

(i) Curation of the largest possible data set of SmTGR inhibitors,
(ii) Development of rigorously validated and mechanistically interpretable models,
(iii) Application of generated models for VS of ChemBridge library [75].

c. Chagas disease: Chagas disease (CD), also known as american trypanosomiasis, is a multisystemic disorder [79]. It was also reported herein the identification of novel potent and selective hits against *T. cruzi* intracellular stage. Validated binary QSAR models are developed for prediction of mitigation of trypanosomiasis and cytotoxicity against mammalian cells using the ideal procedures for QSAR modeling. These models were then used for virtual screening of a commercial database, leading to the identification of 39 virtual hits [80].

II. Bioinformatics: Here are also some problems in the bioinformatics field as follows which need to be tackled. First, the interpretability of model is essential to biologists to understand how model helps solve the biological problem, fr example, predicting DNA−protein binding [81]. In bioinformatics, ML has been a popular and effective way for obtaining knowledge from huge data. Furthermore, the bioinformatics domain classified it (i.e., omics, biomedical imaging, biomedical signal processing) and DL architecture (i.e., deep neural networks, convolutional neural networks, recurrent neural networks, emergent architectures) and present brief descriptions of each study [82].

a. Omics: Owing to the fast growth of bioinformatics knowledge and large-scale "omics" data, drug repositioning has decreased significantly the time cost for the drug development process [83]. With improvements in omics technology, it is now possible to systematically assess each of these areas. Technologies such as RNA sequencing (RNA-Seq), chromatin immunoprecipitation (ChIP-Seq), and mass spectrometry provide extensive measurements of gene expression, chromatin accessibility, metabolite expression, protein expression, and posttranslational modifications.

The integration of these omics data can provide a more comprehensive view of the compounds and allow for discoveries that could be overlooked in the analysis of any individual dataset. Biological samples are analyzed using multiomics methodologies like genomes, transcriptomics, proteomics, and metabolomics; each type of analysis has the potential to provide and add significant information [84]. There are several distinct types of methodology with omics data that would be useful for translational research as depicted in Fig. 19.5.

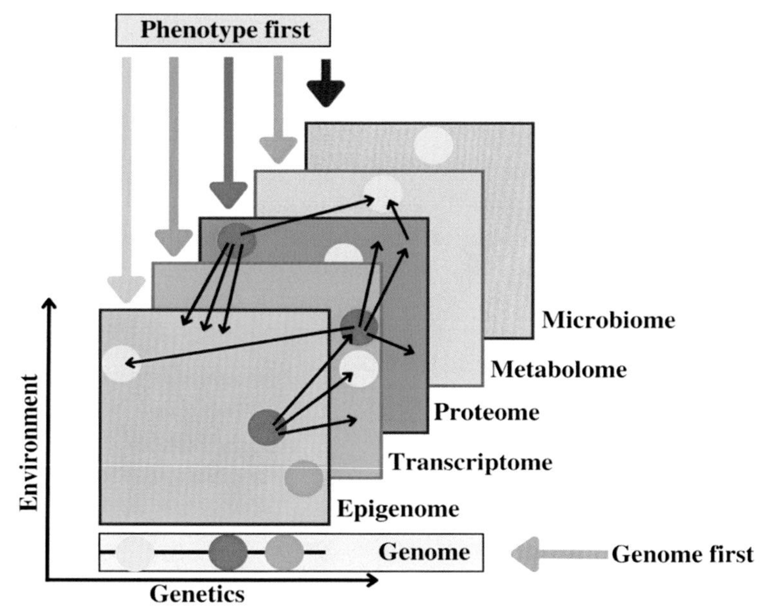

FIGURE 19.5 Various omics data types and disease research methodologies.

Omics information is gathered for the complete collection of molecules, which are shown as circles. With the exception of the genome, all data layers reflect both genetic regulation and environment, each of which may have a different impact on a particular molecule. Potential molecular interactions or correlations are represented by the thin red arrows in distinct layers—the red transcript, for instance, can be linked to numerous proteins. Although common, interactions within layers are not shown. Several alternative beginning points or conceptual frameworks for combining various omics data to explain disease are indicated by thicker arrows. The genome first strategy suggests starting from a linked locus, whereas the phenotype first approach suggests starting from any other layer.

b. **Deep learning architecture:** Deep architecture is necessary to process, analyze, and inference such a large volume of data, while the CNN and RNN for extracting the features from the traffic flow information [85]. As a result, one may encounter difficulties in choosing optimal architectures, interpreting their working mechanisms, and initializing the parameters and deep architecture allow a larger feature learning capacity compared to conventional sparse coding [86].

➢ **Convolutional neural networks (CNNs):** CNN is one of the most popular and used of DL networks [87]. CNNs are composed of connected neural nodes with learnable parameters. We applied a CNN for image analysis with a large number of layers to establish a hierarchical representation of magnetic resonance images. The CNN that we developed for the purpose of this study consists of an input layer, convolutional layer, fully connected layer, and output layer or loss layer with input dimension of $150 \times 3 \times 3$ and two output nodes. This network architecture was adapted from architecture available in the public domain [88] CNN is primarily designed to extract 2D spatial features from still image, and videos are naturally viewed as 3D spatiotemporal signals, the core issue of extending the CNN from image to video is temporal information exploitation. The methodologies and CNN architectures extended from image to video have undergone significant advancements. Therefore, CNN-based action recognition has become an active research field, with numerous CNN-based approaches, which have also achieved great success, recently emerging [89]. And it also extracts the latent association information between drug and target [90].

➢ **Recurrent neural network (RNN):** The term "RNN" refers to the generic class of neural networks that originates and includes the LSTM network as a specific case. RNN is frequently mentioned without context and the concept of "unrolling" is introduced without explanation. Moreover, the training equations are often omitted altogether, leaving the reader puzzled and searching for more resources, while having to reconcile disparate notation used therein [91].

➢ **Deep belief network (DBN):** The popular DL architecture, that is, DBN, can learn the hidden structure of the patterns through a layered structure, which also be able to extract the significant features from the data [92]. DBN algorithm has high learning and generalization ability for data and has achieved a lot of achievements in the field of image processing and network applications. For example, Dahl et al., improved context-dependent pretrained deep neural networks for large vocabulary speech recognition [93].

5. Conclusion and future scope

Machine intelligence has been applied in the drug discovery field for decades. Traditional ML modeling has evolved into a variety of new methods, such as combi-QSAR and hybrid QSAR, and remains a popular approach to study various drug-related topics. There are various tools and techniques designed with help of ML or computational methods in favor of discovery and synthesis of drug.

Despite the advantages and popularity of using ML approaches (e.g., QSAR) in modeling studies, machine intelligence has, in some instances, been replaced by DL in recent years. The development of DL methods is driven by the accumulation of massive amounts of biomedical data and the powerful parallel computing capacity of graphics processing units (GPUs). Importantly, DL methods can deal with complex tasks based on large, heterogeneous, and high-dimensional datasets without the need for human input. These methods have been shown to be useful in many practical and commercial applications, including drug discovery studies.

Although they have been shown to have high prediction accuracies, DL models still perform as “black boxes” that are difficult to use to reveal the biological mechanisms integrated in the data used for modeling. Overall, DL has shown promise for usage in the new big data era of drug development as a newly developed machine intelligence technique. With more data becoming available and new approaches being developed, DL methods will become a major computer-aided drug design (CADD) approach in the near future.

References

[1] P. Carracedo-Reboredo, J. Liñares-Blanco, N. Rodríguez-Fernández, F. Cedrón, F.J. Novoa, A. Carballal, V. Maojo, A. Pazos, C. Fernandez-Lozano, A review on machine learning approaches and trends in drug discovery, Computational and Structural Biotechnology Journal 19 (2021) 4538–4558, https://doi.org/10.1016/j.csbj.2021.08.011.

[2] A. Zhavoronkov, Q. Vanhaelen, T.I. Oprea, Will artificial intelligence for drug discovery impact clinical pharmacology? Clinical Pharmacology & Therapeutics 107 (4) (2020) 780–785, https://doi.org/10.1002/cpt.1795.

[3] H. Chen, O. Engkvist, Y. Wang, M. Olivecrona, T. Blaschke, The rise of deep learning in drug discovery, Drug Discovery Today 23 (6) (2018) 1241–1250, https://doi.org/10.1016/j.drudis.2018.01.039.

[4] A. Sharma, T. Virmani, V. Pathak, A. Sharma, K. Pathak, G. Kumar, D. Pathak, Artificial intelligence-based data-driven strategy to accelerate research, development, and clinical trials of COVID vaccine, BioMed Research International 2022 (2022) 1–16, https://doi.org/10.1155/2022/7205241.

[5] Y. Mintz, R. Brodie, Introduction to artificial intelligence in medicine, Minimally Invasive Therapy and Allied Technologies 28 (2) (2019) 73–81, https://doi.org/10.1080/13645706.2019.1575882.

[6] M.A. Sellwood, M. Ahmed, M.H. Segler, N. Brown, Artificial intelligence in drug discovery, Future Medicinal Chemistry 10 (17) (2018) 2025–2028, https://doi.org/10.4155/fmc-2018-0212.

[7] D. Paul, G. Sanap, S. Shenoy, D. Kalyane, K. Kalia, R.K. Tekade, Artificial intelligence in drug discovery and development, Drug Discovery Today 26 (1) (2021) 80–93, https://doi.org/10.1016/j.drudis.2020.10.010.

[8] L. Zhang, J. Tan, D. Han, H. Zhu, From machine learning to deep learning: progress in machine intelligence for rational drug discovery, Drug Discovery Today 22 (11) (2017) 1680–1685, https://doi.org/10.1016/j.drudis.2017.08.010.

[9] R. Gupta, D. Srivastava, M. Sahu, S. Tiwari, R.K. Ambasta, P. Kumar, Artificial intelligence to deep learning: machine intelligence approach for drug discovery, Molecular Diversity 25 (3) (2021) 1315–1360, https://doi.org/10.1007/s11030-021-10217-3.

[10] J. Vamathevan, D. Clark, P. Czodrowski, I. Dunham, E. Ferran, G. Lee, B. Li, A. Madabhushi, P. Shah, M. Spitzer, S. Zhao, Applications of machine learning in drug discovery and development, Nature Reviews Drug Discovery 18 (6) (2019) 463–477, https://doi.org/10.1038/s41573-019-0024-5.

[11] I.I. Baskin, D. Winkler, I.V. Tetko, A renaissance of neural networks in drug discovery, Expert Opinion on Drug Discovery 11 (8) (2016) 785–795, https://doi.org/10.1080/17460441.2016.1201262.

[12] Y. LeCun, Y. Bengio, G. Hinton, Deep learning, Nature 521 (7553) (2015) 436–444, https://doi.org/10.1038/nature14539.

[13] S. Ekins, The next era: deep learning in pharmaceutical research, Pharmaceutical Research 33 (11) (2016) 2594–2603, https://doi.org/10.1007/s11095-016-2029-7.

[14] H. Yang, L. Sun, W. Li, G. Liu, Y. Tang, Silico prediction of chemical toxicity for drug design using machine learning methods and structural alerts, Frontiers of Chemistry 6 (2018) 30, https://doi.org/10.3389/fchem.2018.00030.

[15] Y. He, C.Y. Liew, N. Sharma, S.K. Woo, Y.T. Chau, C.W. Yap, PaDEL-DDPredictor: open-source software for PD-PK-T prediction, Journal of Computational Chemistry 34 (7) (2013) 604–610, https://doi.org/10.1002/jcc.23173.

[16] N. Yoshikawa, G.R. Hutchison, Fast, efficient fragment-based coordinate generation for Open Babel, Journal of Cheminformatics 11 (1) (2019) 49, https://doi.org/10.1186/s13321-019-0372-5.

[17] J. Powles, H. Hodson, Google DeepMind and healthcare in an age of algorithms, Health Technology 7 (4) (2017) 351–367, https://doi.org/10.1007/s12553-017-0179-1.

[18] Y.-L. Wang, F. Wang, X.-X. Shi, C.-Y. Jia, F.-X. Wu, G.-F. Hao, G.-F. Yang, Cloud 3D-QSAR: a web tool for the development of quantitative structure–activity relationship models in drug discovery, Briefings in Bioinformatics 22 (4) (2021) bbaa276, https://doi.org/10.1093/bib/bbaa276.

[19] J. Dong, Z.-J. Yao, M.-F. Zhu, N.-N. Wang, B. Lu, A.F. Chen, A.-P. Lu, H. Miao, W.-B. Zeng, D.-S. Cao, ChemSAR: an online pipelining platform for molecular SAR modeling, Journal of Cheminformatics 9 (1) (2017) 27, https://doi.org/10.1186/s13321-017-0215-1.

[20] H. Liu, W. Zhang, Y. Song, L. Deng, S. Zhou, HNet-DNN: inferring new drug–disease associations with deep neural network based on heterogeneous network features, Journal of Chemical Information and Modeling 60 (4) (2020) 2367–2376, https://doi.org/10.1021/acs.jcim.9b01008.

[21] H.-J. Jiang, Y.-A. Huang, Z.-H. You, Predicting drug-disease associations via using Gaussian interaction profile and kernel-based autoencoder, BioMed Research International 2019 (2019) 1–11, https://doi.org/10.1155/2019/2426958.

[22] I. Lee, J. Keum, H. Nam, DeepConv-DTI: prediction of drug-target interactions via deep learning with convolution on protein sequences, PLoS Computational Biology 15 (6) (2019) e1007129, https://doi.org/10.1371/journal.pcbi.1007129.

[23] B. Imbernón, J.M. Cecilia, H. Pérez-Sánchez, D. Giménez, METADOCK: a parallel metaheuristic schema for virtual screening methods, International Journal of High Performance Computing Applications 32 (6) (2018) 789–803, https://doi.org/10.1177/1094342017697471.

[24] S.J. Capuzzi, I.S.-J. Kim, W.I. Lam, T.E. Thornton, E.N. Muratov, D. Pozefsky, A. Tropsha, Chembench: a publicly accessible, integrated cheminformatics portal, Journal of Chemical Information and Modeling 57 (2) (2017) 105–108, https://doi.org/10.1021/acs.jcim.6b00462.

[25] Z.-J. Yao, J. Dong, Y.-J. Che, M.-F. Zhu, M. Wen, N.-N. Wang, S. Wang, A.-P. Lu, D.-S. Cao, TargetNet: a web service for predicting potential drug–target interaction profiling via multi-target SAR models, Journal of Computer-Aided Molecular Design 30 (5) (2016) 413–424, https://doi.org/10.1007/s10822-016-9915-2.

[26] F. Montanari, B. Knasmüller, S. Kohlbacher, C. Hillisch, C. Baierová, M. Grandits, G.F. Ecker, Vienna LiverTox workspace—a set of machine learning models for prediction of interactions profiles of small molecules with transporters relevant for regulatory agencies, Frontiers of Chemistry 7 (2020) 899, https://doi.org/10.3389/fchem.2019.00899.

[27] L.M. Kaminskas, D.E.V. Pires, D.B. Ascher, dendPoint: a web resource for dendrimer pharmacokinetics investigation and prediction, Scientific Reports 9 (1) (2019) 15465, https://doi.org/10.1038/s41598-019-51789-3.

[28] A. Maunz, M. Gütlein, M. Rautenberg, D. Vorgrimmler, D. Gebele, C. Helma, lazar: a modular predictive toxicology framework, Frontiers in Pharmacology 4 (2013), https://doi.org/10.3389/fphar.2013.00038.

[29] K. Huang, T. Fu, L.M. Glass, M. Zitnik, C. Xiao, J. Sun, DeepPurpose: a deep learning library for drug—target interaction prediction, Bioinformatics 36 (22—23) (2021) 5545—5547, https://doi.org/10.1093/bioinformatics/btaa1005.

[30] C. Shen, J. Luo, W. Ouyang, P. Ding, X. Chen, IDDkin: network-based influence deep diffusion model for enhancing prediction of kinase inhibitors, Bioinformatics 36 (22—23) (2021) 5481—5491, https://doi.org/10.1093/bioinformatics/btaa1058.

[31] H. Hassan-Harrirou, C. Zhang, T. Lemmin, RosENet: improving binding affinity prediction by leveraging molecular mechanics energies with an ensemble of 3D convolutional neural networks, Journal of Chemical Information and Modeling 60 (6) (2020) 2791—2802, https://doi.org/10.1021/acs.jcim.0c00075.

[32] A.S. Rifaioglu, R. Cetin Atalay, D. Cansen Kahraman, T. Doğan, M. Martin, V. Atalay, MDeePred: novel multi-channel protein featurization for deep learning-based binding affinity prediction in drug discovery, Bioinformatics 37 (5) (2021) 693—704, https://doi.org/10.1093/bioinformatics/btaa858.

[33] P. Banerjee, A.O. Eckert, A.K. Schrey, R. Preissner, ProTox-II: a webserver for the prediction of toxicity of chemicals, Nucleic Acids Research 46 (W1) (2018) W257—W263, https://doi.org/10.1093/nar/gky318.

[34] C. Meng, Y. Hu, Y. Zhang, F. Guo, PSBP-SVM: a machine learning-based computational identifier for predicting polystyrene binding peptides, Frontiers in Bioengineering and Biotechnology 8 (2020) 245, https://doi.org/10.3389/fbioe.2020.00245.

[35] D.E.V. Pires, D.B. Ascher, CSM-lig: a web server for assessing and comparing protein—small molecule affinities, Nucleic Acids Research 44 (W1) (2016) W557—W561, https://doi.org/10.1093/nar/gkw390.

[36] M. Oldenhof, A. Arany, Y. Moreau, J. Simm, ChemGrapher: optical graph recognition of chemical compounds by deep learning, Journal of Chemical Information and Modeling 60 (10) (2020) 4506—4517, https://doi.org/10.1021/acs.jcim.0c00459.

[37] H.-J. Jiang, Y.-A. Huang, Z.-H. You, SAEROF: an ensemble approach for large-scale drug-disease association prediction by incorporating rotation forest and sparse autoencoder deep neural network, Scientific Reports 10 (1) (2020) 4972, https://doi.org/10.1038/s41598-020-61616-9.

[38] T. Fu, C. Xiao, C. Qian, L.M. Glass, J. Sun, Probabilistic and dynamic molecule-disease interaction modeling for drug discovery, in: Proceedings of the 27th ACM SIGKDD Conference on Knowledge Discovery and Data Mining, 2021, pp. 404—414, https://doi.org/10.1145/3447548.3467286.

[39] J. Jiménez-Luna, F. Grisoni, N. Weskamp, G. Schneider, Artificial intelligence in drug discovery: recent advances and future perspectives, Expert Opinion on Drug Discovery 16 (9) (2021) 949—959, https://doi.org/10.1080/17460441.2021.1909567.

[40] M.B. Jamshidi, J. Talla, A. Lalbakhsh, M.S. Sharifi-Atashgah, A. Sabet, Z. Peroutka, A conceptual deep learning framework for COVID-19 drug discovery, in: 2021 IEEE 12th Annual Ubiquitous Computing, Electronics and Mobile Communication Conference (UEMCON), 2021, pp. 00030—00034, https://doi.org/10.1109/UEMCON53757.2021.9666715.

[41] I. Arpaci, M. Al-Emran, M. Al-Sharafi, G. Marques (Eds.), Emerging Technologies during the Era of COVID-19 Pandemic, vol 348, Springer International Publishing, 2021, https://doi.org/10.1007/978-3-030-67716-9.

[42] B. Muñiz Castro, M. Elbadawi, J.J. Ong, T. Pollard, Z. Song, S. Gaisford, G. Pérez, A.W. Basit, P. Cabalar, A. Goyanes, Machine learning predicts 3D printing performance of over 900 drug delivery systems, Journal of Controlled Release 337 (2021) 530—545, https://doi.org/10.1016/j.jconrel.2021.07.046.

[43] D.P. Boso, D. Di Mascolo, R. Santagiuliana, P. Decuzzi, B.A. Schrefler, Drug delivery: experiments, mathematical modelling and machine learning, Computers in Biology and Medicine 123 (2020) 103820, https://doi.org/10.1016/j.compbiomed.2020.103820.

[44] F.R. Fields, S.D. Freed, K.E. Carothers, M.N. Hamid, D.E. Hammers, J.N. Ross, V.R. Kalwajtys, A.J. Gonzalez, A.D. Hildreth, I. Friedberg, S.W. Lee, Novel antimicrobial peptide discovery using machine learning and biophysical selection of minimal bacteriocin domains, Drug Development Research 81 (1) (2020) 43—51, https://doi.org/10.1002/ddr.21601.

[45] G. Urban, K. Bache, D.T.T. Phan, A. Sobrino, A.K. Shmakov, S.J. Hachey, C.C.W. Hughes, P. Baldi, Deep learning for drug discovery and cancer research: automated analysis of vascularization images, IEEE/ACM Transactions on Computational Biology and Bioinformatics 16 (3) (2019) 1029—1035, https://doi.org/10.1109/TCBB.2018.2841396.

[46] Y. Cai, H. Yang, W. Li, G. Liu, P.W. Lee, Y. Tang, Computational prediction of site of metabolism for UGT-catalyzed reactions, Journal of Chemical Information and Modeling 59 (3) (2019) 1085—1095, https://doi.org/10.1021/acs.jcim.8b00851.

[47] S. Dey, H. Luo, A. Fokoue, J. Hu, P. Zhang, Predicting adverse drug reactions through interpretable deep learning framework, BMC Bioinformatics 19 (S21) (2018) 476, https://doi.org/10.1186/s12859-018-2544-0.

[48] X. Pang, W. Fu, J. Wang, D. Kang, L. Xu, Y. Zhao, A.-L. Liu, G.-H. Du, Identification of estrogen receptor α antagonists from natural products via *in vitro* and *in silico* approaches, Oxidative Medicine and Cellular Longevity 2018 (2018) 1—11, https://doi.org/10.1155/2018/6040149.

[49] M. Popova, O. Isayev, A. Tropsha, Deep reinforcement learning for de novo drug design, Science Advances 4 (7) (2018) eaap7885, https://doi.org/10.1126/sciadv.aap7885.

[50] P. Mamoshina, M. Volosnikova, I.V. Ozerov, E. Putin, E. Skibina, F. Cortese, A. Zhavoronkov, Machine learning on human muscle transcriptomic data for biomarker discovery and tissue-specific drug target identification, Frontiers in Genetics 9 (2018) 242, https://doi.org/10.3389/fgene.2018.00242.

[51] Y.-C. Lo, S. Senese, B. France, A.A. Gholkar, R. Damoiseaux, J.Z. Torres, Computational cell cycle profiling of cancer cells for prioritizing FDA-approved drugs with repurposing potential, Scientific Reports 7 (1) (2017) 11261, https://doi.org/10.1038/s41598-017-11508-2.
[52] R. Rahman, J. Otridge, R. Pal, IntegratedMRF: random forest-based framework for integrating prediction from different data types, Bioinformatics 33 (9) (2017) 1407−1410, https://doi.org/10.1093/bioinformatics/btw765.
[53] J. Gomes, B. Ramsundar, E.N. Feinberg, V.S. Pande, Atomic Convolutional Networks for Predicting Protein-Ligand Binding Affinity, 2017 (arXiv:1703.10603). arXiv, http://arxiv.org/abs/1703.10603.
[54] M. Olivecrona, T. Blaschke, O. Engkvist, H. Chen, Molecular de-novo design through deep reinforcement learning, Journal of Cheminformatics 9 (1) (2017) 48, https://doi.org/10.1186/s13321-017-0235-x.
[55] T. Qiu, J. Qiu, J. Feng, D. Wu, Y. Yang, K. Tang, Z. Cao, R. Zhu, The recent progress in proteochemometric modelling: focusing on target descriptors, cross-term descriptors and application scope, Briefings in Bioinformatics 18 (1) (2017) 125−136, https://doi.org/10.1093/bib/bbw004.
[56] F. Wan, J. Zeng, Deep learning with feature embedding for compound-protein interaction prediction, BioRxiv 7 (2016) 086033, https://doi.org/10.1101/086033.
[57] A. Aliper, S. Plis, A. Artemov, A. Ulloa, P. Mamoshina, A. Zhavoronkov, Deep learning applications for predicting pharmacological properties of drugs and drug repurposing using transcriptomic data, Molecular Pharmaceutics 13 (7) (2016) 2524−2530, https://doi.org/10.1021/acs.molpharmaceut.6b00248.
[58] A. Lusci, G. Pollastri, P. Baldi, Deep architectures and deep learning in chemoinformatics: the prediction of aqueous solubility for drug-like molecules, Journal of Chemical Information and Modeling 53 (7) (2013) 1563−1575, https://doi.org/10.1021/ci400187y.
[59] W. Wang, M.T. Kim, A. Sedykh, H. Zhu, Developing enhanced blood−brain barrier permeability models: integrating external bio-assay data in QSAR modeling, Pharmaceutical Research 32 (9) (2015) 3055−3065, https://doi.org/10.1007/s11095-015-1687-1.
[60] L. Patel, T. Shukla, X. Huang, D.W. Ussery, S. Wang, Machine learning methods in drug discovery, Molecules 25 (22) (2020) 5277, https://doi.org/10.3390/molecules25225277.
[61] Z. Sessions, N. Sánchez-Cruz, F.D. Prieto-Martínez, V.M. Alves, H.P. Santos, E. Muratov, A. Tropsha, J.L. Medina-Franco, Recent progress on cheminformatics approaches to epigenetic drug discovery, Drug Discovery Today 25 (12) (2020) 2268−2276, https://doi.org/10.1016/j.drudis.2020.09.021.
[62] H. Chen, T. Kogej, O. Engkvist, Cheminformatics in drug discovery, an industrial perspective, Molecular Informatics 37 (9−10) (2018) 1800041, https://doi.org/10.1002/minf.201800041.
[63] A.S. Rifaioglu, H. Atas, M.J. Martin, R. Cetin-Atalay, V. Atalay, T. Doğan, Recent applications of deep learning and machine intelligence on in silico drug discovery: methods, tools and databases, Briefings in Bioinformatics 20 (5) (2019) 1878−1912, https://doi.org/10.1093/bib/bby061.
[64] N.T. Issa, V. Stathias, S. Schürer, S. Dakshanamurthy, Machine and deep learning approaches for cancer drug repurposing, Seminars in Cancer Biology 68 (2021) 132−142, https://doi.org/10.1016/j.semcancer.2019.12.011.
[65] F. Gentile, V. Agrawal, M. Hsing, A.-T. Ton, F. Ban, U. Norinder, M.E. Gleave, A. Cherkasov, Deep docking: a deep learning platform for augmentation of structure based drug discovery, ACS Central Science 6 (6) (2020) 939−949, https://doi.org/10.1021/acscentsci.0c00229.
[66] K. Park, Y.-J. Ko, P. Durai, C.-H. Pan, Machine learning-based chemical binding similarity using evolutionary relationships of target genes, Nucleic Acids Research 47 (20) (2019), https://doi.org/10.1093/nar/gkz743 e128−e128.
[67] H. Al-Sahaf, Y. Bi, Q. Chen, A. Lensen, Y. Mei, Y. Sun, B. Tran, B. Xue, M. Zhang, A survey on evolutionary machine learning, Journal of the Royal Society of New Zealand 49 (2) (2019) 205−228, https://doi.org/10.1080/03036758.2019.1609052.
[68] M. Batool, B. Ahmad, S. Choi, A structure-based drug discovery paradigm, International Journal of Molecular Sciences 20 (11) (2019) 2783, https://doi.org/10.3390/ijms20112783.
[69] D. Dana, S. Gadhiya, L. St Surin, D. Li, F. Naaz, Q. Ali, L. Paka, M. Yamin, M. Narayan, I. Goldberg, P. Narayan, Deep learning in drug discovery and medicine; scratching the surface, Molecules 23 (9) (2018) 2384, https://doi.org/10.3390/molecules23092384.
[70] A. Kadurin, S. Nikolenko, K. Khrabrov, A. Aliper, A. Zhavoronkov, druGAN: an advanced generative adversarial autoencoder model for de Novo generation of new molecules with desired molecular properties in silico, Molecular Pharmaceutics 14 (9) (2017) 3098−3104, https://doi.org/10.1021/acs.molpharmaceut.7b00346.
[71] X. Liu, T. Blaschke, B. Thomas, S. De Geest, S. Jiang, Y. Gao, X. Li, E. Buono, S. Buchanan, Z. Zhang, S. Huan, Usability of a medication event reminder monitor system (MERM) by providers and patients to improve adherence in the management of tuberculosis, International Journal of Environmental Research and Public Health 14 (10) (2017) 1115, https://doi.org/10.3390/ijerph14101115.
[72] V.D. Mouchlis, A. Afantitis, A. Serra, M. Fratello, A.G. Papadiamantis, V. Aidinis, I. Lynch, D. Greco, G. Melagraki, Advances in de novo drug design: from conventional to machine learning methods, International Journal of Molecular Sciences 22 (4) (2021) 1676, https://doi.org/10.3390/ijms22041676.
[73] R. Medina Marrero, Y. Marrero-Ponce, S.J. Barigye, Y. Echeverría Díaz, R. Acevedo-Barrios, G.M. Casañola-Martín, M. García Bernal, F. Torrens, F. Pérez-Giménez, QuBiLs-MAS method in early drug discovery and rational drug identification of antifungal agents, SAR and QSAR in Environmental Research 26 (11) (2015) 943−958, https://doi.org/10.1080/1062936X.2015.1104517.
[74] S. Vilar, N.P. Tatonetti, G. Hripcsak, 3D pharmacophoric similarity improves multi adverse drug event identification in pharmacovigilance, Scientific Reports 5 (1) (2015) 8809, https://doi.org/10.1038/srep08809.
[75] B.J. Neves, R.C. Braga, C.C. Melo-Filho, J.T. Moreira-Filho, E.N. Muratov, C.H. Andrade, QSAR-based virtual screening: advances and applications in drug discovery, Frontiers in Pharmacology 9 (2018) 1275, https://doi.org/10.3389/fphar.2018.01275.

[76] B.J. Neves, R.C. Braga, V.M. Alves, M.N.N. Lima, G.C. Cassiano, E.N. Muratov, F.T.M. Costa, C.H. Andrade, Deep learning-driven research for drug discovery: tackling malaria, PLoS Computational Biology 16 (2) (2020) e1007025, https://doi.org/10.1371/journal.pcbi.1007025.

[77] K. Yang, H. Mehlhorn (Eds.), Sino-African Cooperation for Schistosomiasis Control in Zanzibar: A Blueprint for Combating Other Parasitic Diseases, vol 15, Springer International Publishing, 2021, https://doi.org/10.1007/978-3-030-72165-7.

[78] B.J. Neves, R.F. Dantas, M.R. Senger, C.C. Melo-Filho, W.C.G. Valente, A.C.M. de Almeida, J.M. Rezende-Neto, E.F.C. Lima, R. Paveley, N. Furnham, E. Muratov, L. Kamentsky, A.E. Carpenter, R.C. Braga, F.P. Silva-Junior, C.H. Andrade, Discovery of new anti-schistosomal hits by integration of QSAR-based virtual screening and high content screening, Journal of Medicinal Chemistry 59 (15) (2016) 7075−7088, https://doi.org/10.1021/acs.jmedchem.5b02038.

[79] L.E. Echeverria, C.A. Morillo, American trypanosomiasis (Chagas disease), Infectious Disease Clinics of North America 33 (1) (2019) 119−134, https://doi.org/10.1016/j.idc.2018.10.015.

[80] C.C. Melo-Filho, R.C. Braga, E.N. Muratov, C.H. Franco, C.B. Moraes, L.H. Freitas-Junior, C.H. Andrade, Discovery of new potent hits against intracellular Trypanosoma cruzi by QSAR-based virtual screening, European Journal of Medicinal Chemistry 163 (2019) 649−659, https://doi.org/10.1016/j.ejmech.2018.11.062.

[81] H. Li, S. Tian, Y. Li, Q. Fang, R. Tan, Y. Pan, C. Huang, Y. Xu, X. Gao, Modern deep learning in bioinformatics, Journal of Molecular Cell Biology 12 (11) (2021) 823−827, https://doi.org/10.1093/jmcb/mjaa030.

[82] S. Min, B. Lee, S. Yoon, Deep learning in bioinformatics, Briefings in Bioinformatics (2016) bbw068, https://doi.org/10.1093/bib/bbw068.

[83] M. Koromina, M.-T. Pandi, G.P. Patrinos, Rethinking drug repositioning and development with artificial intelligence, machine learning, and omics, OMICS: A Journal of Integrative Biology 23 (11) (2019) 539−548, https://doi.org/10.1089/omi.2019.0151.

[84] A. Kedaigle, E. Fraenkel, Turning omics data into therapeutic insights, Current Opinion in Pharmacology 42 (2018) 95−101, https://doi.org/10.1016/j.coph.2018.08.006.

[85] B. Vijayalakshmi, K. Ramar, N. Jhanjhi, S. Verma, M. Kaliappan, K. Vijayalakshmi, S. Vimal, Kavita, U. Ghosh, An attention-based deep learning model for traffic flow prediction using spatiotemporal features towards sustainable smart city, International Journal of Communication Systems 34 (3) (2021), https://doi.org/10.1002/dac.4609.

[86] Z. Wang, S. Chang, J. Zhou, M. Wang, T.S. Huang, Learning a task-specific deep architecture for clustering, in: Proceedings of the 2016 SIAM International Conference on Data Mining, 2016, pp. 369−377, https://doi.org/10.1137/1.9781611974348.42.

[87] A. Dhillon, G.K. Verma, Convolutional neural network: a review of models, methodologies and applications to object detection, Progress in Artificial Intelligence 9 (2) (2020) 85−112, https://doi.org/10.1007/s13748-019-00203-0.

[88] S.J. Esses, X. Lu, T. Zhao, K. Shanbhogue, B. Dane, M. Bruno, H. Chandarana, Automated image quality evaluation of T_2-weighted liver MRI utilizing deep learning architecture: automated image quality evaluation, Journal of Magnetic Resonance Imaging 47 (3) (2018) 723−728, https://doi.org/10.1002/jmri.25779.

[89] G. Yao, T. Lei, J. Zhong, A review of convolutional-neural-network-based action recognition, Pattern Recognition Letters 118 (2019) 14−22, https://doi.org/10.1016/j.patrec.2018.05.018.

[90] K. Shao, Z. Zhang, S. He, X. Bo, DTIGCCN: prediction of drug-target interactions based on GCN and CNN, in: 2020 IEEE 32nd International Conference on Tools with Artificial Intelligence (ICTAI), 2020, pp. 337−342, https://doi.org/10.1109/ICTAI50040.2020.00060.

[91] A. Sherstinsky, Fundamentals of recurrent neural network (RNN) and long short-term memory (LSTM) network, Physica D: Nonlinear Phenomena 404 (2020) 132306, https://doi.org/10.1016/j.physd.2019.132306.

[92] M.M. Hassan, M.G.R. Alam, M.Z. Uddin, S. Huda, A. Almogren, G. Fortino, Human emotion recognition using deep belief network architecture, Information Fusion 51 (2019) 10−18, https://doi.org/10.1016/j.inffus.2018.10.009.

[93] C. Ying, Z. Huang, C. Ying, Accelerating the image processing by the optimization strategy for deep learning algorithm DBN, EURASIP Journal on Wireless Communications and Networking 2018 (1) (2018) 232, https://doi.org/10.1186/s13638-018-1255-6.

Chapter 20

Avant-garde techniques in machine for detecting financial fraud in healthcare

S. Geetha[1], G. Soniya Priyatharsini[1], N. Ethiraj[2] and G. Victo Sudha George[1]
[1]*Department of Computer Science and Engineering, Dr.M.G.R Educational and Research Institute of Technology, Madhuravoyal, Chennai, Tamil Nadu, India;* [2]*Department of Mechanical Engineering, Dr.M.G.R Educational and Research Institute of Technology, Madhuravoyal, Chennai, Tamil Nadu, India*

1. Introduction

Machine learning (ML) is everywhere nowadays, from gaming technology to all the working environments where the computer is related to in our daily life. So, it is also equally important to secure those particular data which we are handling; this is because we have the possibility of losing the data and the hackers can hack our personal data. With the quick blooming progress in the industry of science and technologies, ML also accompanies the advanced progressing opportunities. All the technologies which ever accompany the ML techniques comprises multidisciplinary knowledge theoretically. This includes the complexity of an algorithm and statistics. These functional attributes of ML will reinforce additionally [1]. To attain the enhancement in the ML process, it is necessary to do an appropriate analysis of ML algorithms. This will furnish the guidance to the successive processing in ML. This process leads to the productive expansion in an industry when the appropriateness of the ML algorithms is enhanced.

These ML algorithms are excessively useful in healthcare industries. The healthcare industry is a broad area where millions of customers consume. This industry never stops contributing to economic growth since it always cares for millions of people. ML is the processing area in computing mechanisms. Even though it is based on computer systems, human intelligence is required to do some specific tasks. Those are decision-making, detecting any objects, and solving any problems which are complex and even more. The incorporation of technology development in the healthcare industry area is efficacious for medical professionals [2].It is used to keep the data organized and to enhance the business-oriented services. The impact of the ML in healthcare is enormous.

Broadly they are useful the following areas.

1. Data management in the hospitals
   Maintaining the patient record and hospital data is one of the most important tasks in healthcare management. By improving the technology support, this can be attained easily.
2. Diagnosing the disease
   To identify a person whether he is ill is easy. But to pinpoint what the disease is tough work. By using the various ML algorithms, it is easy to compare with the previous symptoms and is able to do this task easily.
3. Early stage identification
   Nowadays, there are many wearables that give warnings regarding health issues. Example is some person's BP is getting high, this wearable which was designed with ML knowledge indicates them.
4. Virtual assistance
   A 24/7 nursing assistance can be available for the patients virtually. This can be done only with the help of the ML algorithms
5. Decision-making
   Surgical robots which are very useful in minimizing the errors during critical operations are used to make decisions in the crucial situations.

Deep Learning in Personalized Healthcare and Decision Support. https://doi.org/10.1016/B978-0-443-19413-9.00020-5

6. Personalized treatment
   The user can get the personalized treatment with the help of the predictive analysis of the ML. These treatments are done with the help of the user's record. So this will give the personalized treatment.
7. Flaws in prescriptions
   Once a patient is prescribed with the prescription, ML intelligence can detect whether the prescription is correct or not. This can be done easily by examining the health record of the patient.
8. Payment in healthcare industries
   Due to the numerous patients being taken care of by the healthcare industry, the records to be maintained is huge. The abundance of money is rolling over in the field of medical science. With the help of ML intelligence, the money can be used securely.

Thus, in an enormous number of ways, the ML techniques help the healthcare industries [3]. Even though there are multiple ways to save a user's time and money through ML techniques, there are also hackers and scammers there to steal the user's credentials for their own well-being. This usage of the credit cards is increased for the people to buy online and pay their bills with ease methods. But nowadays, these methods are found by the scammers, and they capitalize on other people's credit card details. As mentioned above, the credit card credentials are used for the payment of the medical bill and also for the insurance claim. Sometimes fabricated reports are submitted for the insurance with this credit card data. E-commerce websites and the banking system are aware of these types of fraud and they increase the security. E-commerce websites and Banking systems were making efforts to avoid such fraudulent transactions. So that they avoid the fraud transactions. Deep learning and ML methods are used to filter the fraudsters whenever the transaction is happening. By doing this, the fraudulent uses can be identified easily. ML, deep learning, and artificial intelligence are the hot topics IN these past decades. When a service is satisfied, users are ready to start using that particular service so the companies are ready to spend on those services, and the efforts for that particular service can be high. Once the services get improved automatically, the client numbers get increased. Without hard coding, the computer can perform tasks. In this type, ML is the fusion of statistical modeling and computer algorithms. The repository of the knowledge stored is used to predict and with the help of the prediction's actions will be performed. ML comprises the deep learning techniques, and it also contains the artificial neural networks [4,5].There are also various methods such as auto encoders, restricted Boltzmann machine, convolutional neural networks, recurrent neural network, and deep belief network. Thus, it is represented with the total dataset. With these datasets, the uniqueness in the relationship can be captured only by the properly trained neural networks.

In the financial domain, fraud is happening in various fields. Credit card fraud is growing like anything. It is a serious issue while considering all the other frauds in the domain of finance. Nowadays, people prefer to buy things online from vegetables to stock exchange markets. They use credit cards for the payment. This technique is captured by the business persons, and they encourage the consumers to use the credit card consistently. This results in the dark web people finding the credentials of the customers and starting stealing the cash flow. There are a huge number of fraudsters out there. There are two types of credit cards available in the market. One is virtual and the other one is physical. Virtual cards can be easily stolen compared to the Physical credit card. Fraudsters here hide their original identity and location. When the user is saving the credit card details in the system or any other places, this fraudster can easily get those details and hack the particular place. With this particular credential and the card details, the fraudster can steal the amount very easily. There are various methods for this fraudster to steal the data. The second card type is the physical card type. When the customer keeps his/her credit card carelessly, there is a chance of missing the card. Or purposely the fraudster can follow some person and he can steal the card [6].When the credit card details are entered by the user and when the credentials are shared with the unsecured medium, then the details can be leaked. Thus, using the credit card, the fraudster can steal the data. This is how a particular person's details are leaked, and his data are transferred to the network. So, to avoid these types of unsecured transactions, a secure detection mechanism is needed. This can reduce the fraud rate and eliminate the fraudsters entering the network. Most monotonous or endless task in online progressing is the fraud detection and the security for the same. It is an expensive process too.

Even though technology develops, there are still people over there who are not aware of the online fraudsters. When a person through an unknown call is asking for credentials, they are ready to provide. Such practices should be put to an end. All around the world is turning to the digital format and still some of these illiterates do not know how to handle the credit card details. Banks are conveying proper message transformations to their customers regarding these fraudster activities. Still such information is not taken seriously by some percentage of the people [7]. In this digitalized world, humans are expecting the daily routine in an easier way. That is everything should be done within the fingertip. Still the fraud attacks are happening because of some of the illiterates, and they are the victims too. From the past four or 5 years, everyone started using online systems habitually. Every individual person's personal information is scattered across the globe in various forms. Whether they paid through online, they are buying through online or selling through online. That information is in digital languages. So, if anyone is good in those languages, they can hack the details very easily. Thus, it is a threat with the various attacks. There are various types of attacks which are menace to the digital world [8]. To avoid these types of severe attacks, it is a prerequisite to develop more secured algorithms for developing the secured digital world. It is also important to avoid phishing to the dark web world.

In recent times, artificial intelligence, ML, and deep learning are in the limelight for the digital world. Hence, they can provide various technologies for the productivity and safer environment with the enhanced technical settings. This helps the customers and the service providers to work on the developed and highly technical background.

Various applications are provided by artificial intelligence. One such type is ML. It is learning of various means. There are several types of learning. They are listed as semi-supervised learning, unsupervised learning and unsupervised learning. In semisupervised learning, partial data are assigned as labels, whereas other partial data are not at all assigned with anyone. This leads to some types of complex computations. In unsupervised learning models, the real-time applications are used mostly. In supervised learning cases, the labels are assigned along with all the data. It is very easy to predict the details in detail. In ML, there are some of the commonly used algorithms for the betterment of any technology [9].They are linear regression, decision tree classifier, Naive Bayesian classifier, logistic regression, etc.

In this work, it is concentrated on credit card fraud during the health care payment process. The datasets which are considered from are identified from the publicly available domain. With the help of the datasets, the credit card details are extracted, and then it is preprocessed. Thus, the algorithm processes the datasets, and it confirms whether the card holder is genuine or fraud. This is tested with nearly 400 datasets. The results are discussed in the discussion part.

2. Survey of frequently used algorithms for machine learning

2.1 Random forest algorithm

To control unreasonable data, the random tree forest algorithm is very much useful. This can be used for data processing. It is used in the data calculation process and can be used in the extended processing of any data. This results in the split results of the data which give high accuracy in data analysis and can be effectively calculated [10]. There are also some complex issues here, that is, when the multiple set is getting calculated, the classification trees are also created. This can be done by the unified algorithm. Thus, the regression is getting processed. Let as assume a set which is independent, and it contains the following values (i.e.,) i = {1,2,3 … n}. The set of random forests contains the total forest set. This gives the values of R= (r1, r2, r3, … rn). Here r = 1,2,3 … n. Here, the random distribution is in state, and every set will remain in an independent state.

2.2 Decision tree algorithm

This algorithm pertains to the content of classic algorithms. There is a working principle of this algorithm which states that the data information is processed first. The transformation starts from the root which is having the collection instance. It reaches the position in which the completion of the nodes takes place. This leads to the practical example of division of scientific datasets. Here, the branches are split on the basis of the decision number algorithm. This happens to facilitate the data analysis of the information. After that, the data content will be trimmed to improve the integrity. Here, the algorithm acts as the top-down algorithm. This is from the point of calculation method. The optimal attributes are analyzed, and the contents are expanded on the basis of the node. These expanded nodes can be of two or more with respect to the content [11].Thus, the split can be provided the data information with the comprehensive analysis. The number of samples can be increased to get branching like the tree method. These samples can be analyzed. The sample number of statistics can be analyzed with the increase in the number. The decision tree can be named with the large amount of the information with the data. There are some limits in the branching method with the upper format. This happens while splitting in the upper limit. For example, if the limit is set to the value of 10, it stops the continuation of splitting the value. It started to use cropping technology to trim the content. Thus, the larger tree value can be refined. This will happen to improve the analysis result of the data.

2.3 Artificial neural network algorithm

Artificial neural network algorithm handles the processing of the human information transmission. It classifies one neuron from various data. With the help of the network, it connects the neurons and accomplishes different memory activities.

Moreover, this algorithm is based on the process of data analyzing, the data are unfolded. Every data have a degree of high authenticity. Thus, the data can be completed with the process of output. For example, if a person is running or moving. In the artificial neural network (ANN) algorithm, a variety of application characteristics are presented. It corresponds to the process of analyzing. It completes according to the needs. Recently, the ANN comprises a multilayer forward network. This included the calculation of the data, self-organized neural networks, etc, [12]. The coefficient should be weighed with respect to that of the threshold output. When the output is calculated with the sum, it exceeds the value. It is used to improve some of the orderliness of the process of numerical analysis.

2.4 SVM algorithm

This algorithm is commonly used in the ML process. In some of the specified application processes, it is commonly identified as the application process. This algorithm mainly depends on the vector method. The data analysis task is finished using the vector machine method. The processing of the data in the data information is supported by this algorithm. As the data are sent, it gets optimized. For the purpose of improvising the technique format, the scientific data being acquired are processed with the real data. As a result, this method aids in obtaining samples from various sets together with the analysis process, improving the data analysis of the end results [13].This details how the scientific data were actually analyzed. For the purpose of determining the boundary value, samples are gathered here. The Hd examples are provided in the processing of the data information. Here, the data are handled initially with complete use of the SVM techniques. In order to fully disperse, the SVM technology processed the revised data. To estimate the greatest distance with the complete aircraft, the Hd is then calculated for the plane. In order to accurately determine the value of the plane, the Hd is studied after the vector's content has been determined for the entire plane. While extracting plane value, this lowers the vector values.

3. Background study

This describes the actual analysis of the scientific data. Here, samples are gathered in order to determine the boundary value. The processing of the data information includes the HD samples. The first handling of the data in this case makes full use of SVM algorithms. The amended data was processed by the SVM technology in order to fully disperse [14]. The Hd for the aircraft is then computed in order to determine its maximum range. After determining the content of the vector for the whole plane, the Hd is studied to obtain the value of the plane. This decreases the vector values while extracting the plane value [15]. There are numerous methods for detecting fraud that are now in use, including artificial immune system, logistic regression, decision tree classifier, genetic algorithm, Bayesian belief network, neural network, support vector machine, and others [16]. Each of these strategies is still the subject of ongoing research. A paradigm for ensemble learning based on training set partitioning and clustering is provided. The integrity of the sample features has been guaranteed [17]. The dataset is split into a training and testing set, and the training set is further segmented into samples from the majority and minority groups. They balanced the dataset using the random sampling strategy by including an equal number of samples from the minority and majority classes in the training set. Now that each base classifier has been trained with a balanced training set, each estimator casts a vote to determine the ensemble model that will be used, according to the majority voting rule [18]. The idea of detecting credit card fraud by combining ML models is presented. Data preprocessing is done on the credit card dataset by deleting duplicate data, filling in blank areas, and other data cleaning procedures. K-Fold to divide the dataset into a training and testing dataset, cross-validation is done. The creation of models for logistic regression, decision tree classifier, and SVM is combined with the computation of metrics like accuracy, precision, etc., to arrive at a comparison [19]. The use of ML to identify financial fraud in mobile payment systems is suggested [20]. Using filter-based approaches, feature selection is performed on the preprocessed data.

4. Materials and methods

4.1 Working model

From the public websites like Kaggle, a real-world credit card dataset is gathered. The collection includes cardholder transactions. A total of 500 transactions have been made, of which 92 are fraudulent. There are 12 features total, including time and amount that are not transformed and 28 numerical input variables that are derived. The final component of the class accepts the class variable and sends the datasets. It responds to fraud as 1 and nonfraud as 0. Imbalanced credit card dataset, Fig. 20.1 explains the working model of the proposed work. Since there are more occurrences of the majority class (nonfraud) than the minority class that the dataset is severely unbalanced (fraud). As a result, the dataset must be balanced.

FIGURE 20.1 Working model of the proposed work.

TABLE 20.1 Total number of times supported for the given inputs.

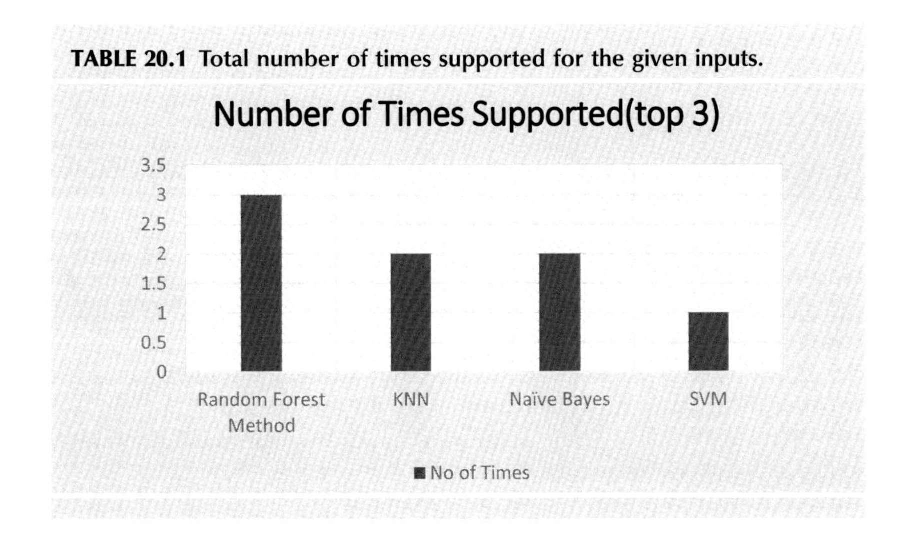

4.2 Data preprocessing

The approach of extra trees classifier (ET) is used to preprocess the throughput of the preprocessing. This step is relevant with each of the features. This determines the extractions of the data from the datasets. This classifier is highly unpredictable. For each, feature index is computed for the feature. There are 30 features totally and the top 12 features are chosen for the extraction. Descending order is followed for the classification. Table 20.1 shows the comparative analysis. Compared to the other algorithms the proposed algorithms are given as good as the output. Here, the top three outputs are listed with the times.

4.3 Partitioning of datasets

Here the cross validation (CV) K fold method is used. In this process, for the training process and the testing methods, the data sets are divided. The total number of datasets given with the parameters are divided by the specified parameter(k). Fig. 20.2 explains in detail. Using the random validation, the dataset is shuffled randomly. Next, it is separated into groups(k). This is done by the cross validation. The distinct group for which the mean squared error is computed. This is carried out by the calculation of the lowest mean squared error.

4.4 Sampling of the collected data

The unbalanced dataset is used here. There is one easiest method for dealing the unbalanced and balanced dataset. This is done for the sampling process. In this work, the under sampling technique of random sampling is employed. The process of choosing a particular class from the given instances is referred to as the random under sampling. It is used for the classes of minorities. The generated void is named as random.

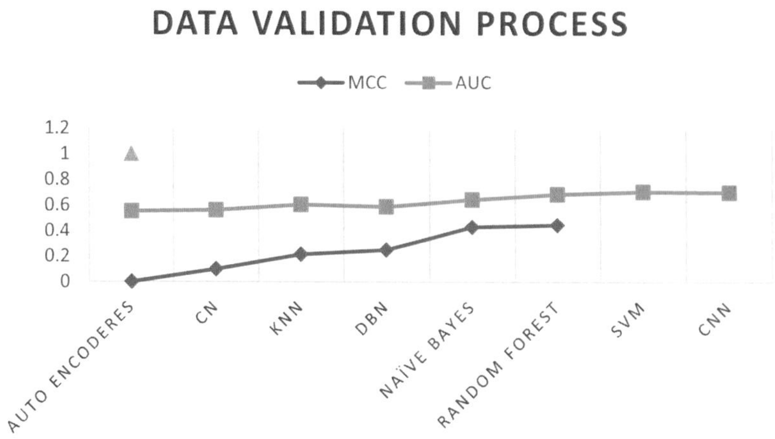

FIGURE 20.2 Process of the data validation.

4.5 Standard machine learning models

To begin with, the dataset is modeled using four common ML models: Naive Bayesian classifier, logistic regression, decision tree classifier, and random forest classifier.

(a) Random forest classifier: With the exception of the fact that only a random subset of features is available, random tree functions as a decision tree operator. An ensemble of random trees is produced by RF. The number of trees is user-configurable. The resulting model uses voting on each tree that has been generated to determine the categorization result.
(b) Naive Bayes classifier: Naive Bayes classifies data using the Bayes theorem. It is believed that some characteristics are not related to others. For classification, only a minimal training dataset is required.
(c) Decision tree classifier: A group of nodes called a decision tree is used to make decisions about features. For a feature, each node stands in for a split rule. Until the stopping requirement is attained, new characteristics are established. The class that a certain feature belongs to is specified by the leaf nodes.
(d) Logistic regression: Data having nominal and numerical properties can be handled using LR. On the basis of one or more predictor features, it calculates the likelihood.

From the datasets collected, the preprocessing techniques are carried out with the format of above explained methods. This can be applied for every dataset to check whether the data are consuming more cost or not. This process is explained in the following figure (Fig. 20.3).

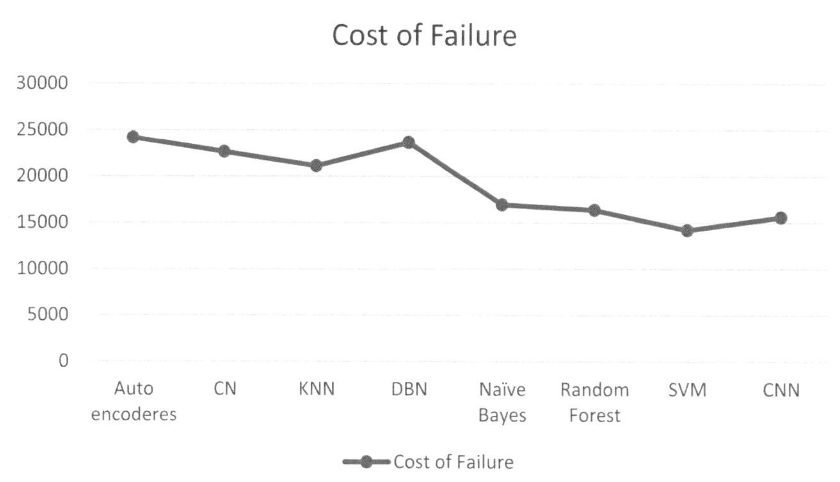

FIGURE 20.3 Cost of failure with the given methods.

5. Conclusion

Various methods, such as manual inspection and consumer end authentication, have been used in fraud detection studies during the past 20 years. Recent advancements in better computation power and lower computing costs have allowed deep learning models to be used in a variety of applications. This paper includes empirical research comparing different ML and deep learning models on various data sets in order to detect fraudulent transactions. The main objective of this effort is to determine which approaches are most suited for various dataset types. Given that many firms are investing in new technologies these days to grow their operations, this research may help practitioners and businesses better understand how various approaches work on various sorts of datasets. According to this study, SVMs are the best methods for identifying fraud in bigger datasets. They can be used with CNNs to deliver a more trustworthy performance. For smaller datasets, SVM, random forest, and KNN ensemble algorithms can provide good improvements. Convolutional neural networks (CNNs) consistently outperform DBN, autoencoders, RBM, and other deep learning methods. The limitation of this study is that it only deals with fraud detection in the setting of supervised learning. Even while supervised learning methods like CNN, KNN, and random forest have a good performance and look appealing, they are useless in dynamic environments. Fraudulent behaviors often change over time and are difficult to spot. Fresh data sets would need to be gathered, and ML algorithms would need to be retrained. Autoencoders provide a good solution in that circumstance because they are only trained on routine traffic. Fraudulent transactions are recognized as being out of the ordinary. Although initially quite expensive, autoencoder training can be helpful for labeling data. Once enough data have been labeled, it may be used to either build new supervised models or retrain ones that already exist.

6. Future work

It might be a good idea to look at fraud detection using neural networks in more detail. The main barrier to developing successful NN models is the lack of adequate training data. We are continually looking into the availability of new datasets and larger fraud transaction sizes to be able to train larger models. Additionally, we are looking into other deep learning network models. Generative adversarial networks (GANs) have been used for anomaly detection to uncover cyberattacks, especially in the field of cybersecurity. Some studies have used GANs to simultaneously train the generator and discriminator using data from healthy eyes in order to discover anomalies in human eyes. This could also include the usage of fraud detection. Consequently, there will be a lot of unstudied areas in the future connected to the creation of dynamic, hybrid, and adaptable fraud detection systems.

References

[1] K. Li, S. Jiang, Machine learning and cultural production reform based on the perspective of the development of AI technology, Journal of Xiangtan University—Philosophy and Social Sciences 44 (01) (2020) 74–79.

[2] H. Wang, P. Zhu, X. Zou, S. Qin, An ensemble learning framework for credit card fraud detection based on training set partitioning and clustering, in: IEEE Conference on Smart World, Science Indexed, December 2018.

[3] C. Bycroft, C. Freeman, D. Petkova, G. Band, L.T. Elliott, K. Sharp, The UK biobank resource with deep phenotyping and genomic data, Nature 562 (7726) (2018) 203–209.

[4] S. Geetha, S. Ramamoorthy, N. Kanya, G.S. Anandha Mala, Extraction of sequence of actions from unstructured requirements specification document, Journal of Computational and Theoretical Nanoscience 16 (2019) 2108–2112.

[5] T. Sweers, Autoencoding Credit Card Fraud". Bachelor Thesis, Radboud University, June 2018, pp. 214–220.

[6] C.K. Loo, K. Randhawa, C.P. Lim, M.S. Nandi, Credit Card Fraud Detection Using AdaBoost and Majority Voting", vol. 6, IEEE Access, 2018, pp. 14277–14284.

[7] Y. Liu, A. Pumsirirat, Credit card fraud detection using deep learning based on auto-encoder and restricted Boltzmann machine, International Journal of Advanced Computer Science and Applications 9 (1–2) (2018) 18–25.

[8] Z. Chen, C.K. Yeo, B.S. Lee, C.T. Lau, Autoencoder based network anomaly detection, Wireless Telecommunications Symposium (2018) 1–5.

[9] B. Manderick, Credit card fraud detection using Bayesian and neural networks, International Journal of Neural Networks 4 (2002) 140–144.

[10] A. Chouiekha, E.L. Hassane, I.E.L. Haj, Conv nets for fraud detection analysis, Procedia Computer Science 127 (2018) 133–138.

[11] W. Wang, Fraud detection using deep learning, Restricted Boltzmann Machine. GitHub 4 (August 2017) 36–40.

[12] E.W.T. Ngai, Y. Hu, Y.H. Wong, Y. Chen, X. Sun, The application of data mining techniques in financial fraud detection: a classification framework and an academic review of literature, Decision Support Systems 50 (2011) 559–569.

[13] N. Kanya, P.S. Rani, S. Geetha, M. Rajkumar, G. Sandhiya, An efficient damage relief system based on image processing and deep learning techniques, Revista Geintec-Gestao Inovacao e Tecnologias 11 (2) (2021) 2124–2131.

[14] Albashrawi, Detecting financial fraud using data mining techniques: a decade review from 2004 to 2015, Journal of Data Science 14 (2016) 553–570.

[15] M. Jans, jan martijn van der werf, N. Lybaert, K. Vanhoof, A business process mining application for internal transaction fraud mitigation, Expert Systems with Applications 38 (2011) 13351–13359.

[16] D. Ashokkumar Patel, A data mining with hybrid approach based transaction risk score generation model (TRSGM) for fraud detection of online financial transaction, International Journal of Computer Application 16 (2012) 18–25.

[17] G.S. Priyatharsini, N. Malarvizhi, Efficient utilization of energy consumption in cloud environment, International Journal of Engineering and Technology 7(1):- (2018) 189–193, https://doi.org/10.14419/ijet.v7i1.7.10649.

[18] Yusufsahin, S. Bulkan, E. Duman, A cost-sensitive decision tree approach for fraud detection, expert systems with applications, International journal of Engineering and applications 40 (2013) 5916–5923.

[19] S.M. Rubio, S.I. Perez, Solving the false positives problem in fraud prediction automated data science at an industrial scale, International Journals of Applied Engineering and Sciences (2017).

[20] W. James Max Kanter, K. Veeramachaneni, Solving the false positives problem in fraud prediction using automated feature engineering, machine learning and knowledge discovery in databases, International Journal of Applied Engineering and Sciences 4 (2018) 372–388.

Chapter 21

Predicting mental health using social media: A roadmap for future development

Ramin Safa[1], S.A. Edalatpanah[2] and Ali Sorourkhah[3]
[1]*Department of Computer Engineering, Ayandegan Institute of Higher Education, Tonekabon, Mazandaran, Iran;* [2]*Department of Applied Mathematics, Ayandegan Institute of Higher Education, Tonekabon, Mazandaran, Iran;* [3]*Department of Management, Ayandegan Institute of Higher Education, Tonekabon, Mazandaran, Iran*

1. Introduction

The World Health Organization (WHO) explains mental disorders as various issues with a broad range of symptoms. A general characteristic of these disorders is the presence of abnormal behavior, thoughts, and emotions. The following disorders are a few examples: Depression, anxiety, bipolar, eating, post-traumatic stress, schizophrenia, problems associated with drug abuse, and intellectual disabilities. Approximately, 970 million people worldwide suffer from mental disorders, according to the Mental Disorders Fact Sheet[1] published by the WHO in June 2022. Since the beginning of the COVID-19 pandemic in 2020, many people suffering from anxiety and depression have increased significantly. Approximately 83 million European Economic Area (EEA) citizens (between 18 and 65 years) have experienced one or more mental disorders over the past year [1]. In addition, without appropriate treatment, someone can experience psychotic episodes, disability, self-harm attitudes, or commit suicide. Symptoms of these mental disturbances must therefore be recognized at early stages to avoid each unwanted consequence.

As reported by the Lancet Commission on global mental health and sustainable development in 2018, mental health disorders contribute to an increasing number of diseases worldwide. While social services often fail to meet the same standards as physical health services [2]. Depressive disorders were ranked the most prevalent mental health problem worldwide in the 2013 GBD[2] study, continued by bipolar, schizophrenia, and anxiety disorder. According to the GBD 2019[3] reports, 1.2% (more than 815,000 cases) of the United Kingdom's population suffer from bipolar disorder [3].

Social networks offer unique opportunities to study interpersonal relationships and the social context of modern societies, particularly among the under-25s, their primary consumers [4]. As, among youth, suicide is a primer ground of death, this examination is statistically crucial [5], even though suicide rates generally increase with age. In addition, the number of suicide attempts among young people is also exceptionally high. In adolescence, extensive brain changes will shape cognitive development forced by social media [6]. It is also possible to reduce intervention costs using social workers in mental health services [7].

Social media platforms have become increasingly widespread among people nowadays to express their feelings and moods. As a result of this phenomenon, researchers and healthcare professionals can classify linguistic indicators related to mental illnesses like schizophrenia, depression, and suicide. By examining an individual's language use, we can gain valuable insight into their mental state, personality, and social and emotional conditions [8]. Studies have shown that using language attributes to describe one's mental state, nature, and even personal values can be a strong indicator of their current

1. https://who.int/news-room/fact-sheets/detail/mental-disorders.
2. Global Burden of Disease.
3. https://healthdata.org/gbd/2019.

Deep Learning in Personalized Healthcare and Decision Support. https://doi.org/10.1016/B978-0-443-19413-9.00014-X

mental state, character, and values. The interaction of language and clinical disorders has been investigated in several studies. According to conclusions, "speech content can provide a unique window into thoughts" [1], making it possible to directly diagnose mental disorders, for instance, resulting from addiction. As people increasingly turn to platforms such as Twitter, Facebook, and Reddit to express their opinions, feelings, and moods, tools may be developed to diagnose various mental health issues [9,10]. The language and emotions used in social media posts can illuminate feelings such as worthlessness, guilt, and helplessness. Thus, the symptoms of psychological disorders can be characterized this way. Coppersmith et al. [11] argued that these kinds of concerns are often disclosed on social media for a variety of reasons, including pursuing or providing encouragement, changing society's stigma or taboo against mental illness, or explaining certain behaviors [1].

Researchers have been studying the possibility of detecting mental state alterations through language for several years. As a result of social data, health specialists can identify people and communities at risk. For example, experts can identify individuals who need immediate care through a large-scale monitoring program. As the subject has received considerable attention lately, it has become necessary to organize and summarize conventional approaches and the latest trends in a scoping review. This chapter outlines a roadmap and current advances for evaluating mental states and identifying disorders based on user digital footprints on social platforms. The assessment approaches and feature extraction process were also categorized. Thus, young researchers can understand the latest trends and developments in the area.

Below is a summary of the remainder of the chapter. Following the literature on mental disorders, social platforms, and social data analysis, Section 2 discusses these concepts. This chapter demonstrates how social data, information retrieval, NLP,[4] and machine learning techniques are used to build mental health assessment tools and develop DSSs[5] in psychiatry. Section 3 will present various data collection methods and evaluations in the field of research. Feature engineering and preprocessing big social data are discussed in Section 4. Prediction methods and evaluation methods are also discussed in Sections 5 and 6. In Section 7, there are concluding remarks including challenges and possible directions for future research. Finally, Section 8 summarizes the findings and presents future directions.

2. Mental disorders and big social data

Based on WHO, mental health involves understanding yourself, coping with everyday stress, working productively, and positively contributing to society. According to estimates, the global economy loses trillions of dollars to mental disorders [12]. Some of the most known disorders are summarized in Table 21.1. There is, however, a significant decline in treatment and quality of care for those with mental illnesses because of resource shortages [13–15]. It is also declared that many countries are suffering from a lack of psychiatrists [16]. Furthermore, mental health professionals have insufficient tools and methods for decision-making on care-related issues, such as accurate diagnoses [17].

Recent pandemics have also worsened the global mental health crisis [14,18]. Therefore, governments, policymakers, and mental health professionals (like psychiatrists and counselors) require innovative tools to help them in diagnostic decisions made with greater efficiency and accuracy [19,20].

Even with various diagnostic guidelines and tools available, diagnosing patients accurately and efficiently remains challenging. As a result, recent literature suggests that more research should be done on how cutting-edge technologies (such as data science) can be used to create clinical DSSs assisting psychologists and directing health informatics developers [21]. According to Fig. 21.1, social media-based e-mental health research has the following conceptual framework.

This chapter precisely responds to social media-based e-mental health assessment and addresses diagnosing mental disorder challenges concerning the guidelines and tools used. We aim to show a road map to build a social media-based e-mental health assessment tool and present the requirements to develop an artificial intelligence–based DSS.

The development of data science has been driven by increasing data volumes, variety, analytics, visualization, and computational power. Using information retrieval and machine learning techniques, researchers are extracting information from massive amounts of data and using them to expound and construct classification models in numerous fields. Researchers have detected meaningful patterns in healthcare datasets using data science methods. Section 3 discusses how the data of social media mining are applied to detect mental disorder symptoms.

Social media is social by nature. As a result, raw data may readily reveal and quantify social interaction patterns, which are crucial to mental state assessments. Because social media is (semi) anonymous and open, people are more likely to socialize and share their information [22]. Sharing content concerning your everyday life and announcing prominent life

4. Natural Language Processing.
5. Decision Support System.

TABLE 21.1 Common mental disorders definition.

Mental disorder	Definition
Depression	There is a difference between depression and mood swings or short-lived emotional reactions to daily experiments; a mental state causing painful symptoms adversely disrupts normal activities (e.g., sleeping). For at least 2 weeks, the person experiences depressive moods (sad, irritable, empty) or a lack of interest in activities for most of the day or the week.
Anxiety	Several behavioral disturbances are associated with anxiety disorders, including excessive fear and worry. Severe symptoms cause significant impairment in functioning cause considerable distress. Anxiety disorders come in many forms, such as social anxiety, generalized anxiety, panic, etc.
Bipolar disorder	An alternating pattern of depression and manic symptoms is associated with bipolar disorder. An individual experiencing a depressive episode may feel sad, irritable, empty, or lose interest in daily activities. Emotions of euphoria or irritability, excessive energy, and increased talkativeness can all be signs of manic depression. Increased self-esteem, decreased sleep need, disorientation, and reckless behavior may also be signs of manic depression.
Post-traumatic stress disorder (PTSD)	In PTSD, persistent mental and emotional stress can occur after an injury or severe psychological shock, characterized by sleep disturbances, constant vivid memories, and dulled response to others and the outside world. People who re-experience symptoms may have difficulties with their everyday routines and experience significant impairment in their performance.
Seasonal affective disorder (SAD)	In most cases, SAD occurs in the fall/winter and enters remission in the spring/summer; although in some cases, it may happen in the summer and remit in the autumn and winter. A majority of the cause-and-effect mechanisms of SAD have not yet been discovered. However, several hypotheses have been posed regarding the disease, and they promise to deliver new information to scientists.
Schizophrenia	A schizophrenic disorder is characterized by episodes of psychosis that occur continuously or recur continuously. Disorganized thinking, hallucinations, and delusions are some of the significant symptoms. Other symptoms are apathy, social withdrawal, and a decreased expression of emotions. Most symptoms develop gradually, begin during young adulthood, and do not resolve in most cases.
Eating disorders	In eating disorders, there is a persistent disturbance in eating behavior, along with distressing thoughts and feelings. These conditions negatively impact physical, psychological, and social functioning. Disorders like anxiety and obsessive-compulsive disorders often coexist with eating disorders.

landmarks on social media is expected. The way the subject expresses herself can provide a considerable amount of information about their mental state and emotional conditions, as introduced in the introduction. Thus, a user's social data content can convey feelings such as helplessness, worthlessness, or guilt due to the language employed and the emotions conveyed. Different mental disorders can thus be identified and characterized in this way. Data collected from social platforms occurs nonreactive and can complement conventional data in a valuable way.

With billions of users, Facebook is the leading social media network, followed by Twitter and Instagram. Many features are available on Facebook, including creating a profile, uploading files, sending messages, and being in contact. Users can post information about themself, whether about their occupation, religion, political views, or favorite movies and musicians. With Twitter, users can tweet, short stream up to 280 characters, and follow other individuals' updates. Young adults between 18 and 29 years tend to be more attracted to it [23]. Based on Statista,[6] in 2019, Twitter had 290 million active users (monthly), and the population is anticipated to be over 340 million by 2024. Users can upload photos and videos on Instagram while exposing their current location. Reddit is also another social network that can be analyzed for a variety of reasons. In recent years, these platforms have become increasingly popular for expressing opinions, interacting with others, and sharing feelings. Hootsuite's well-established social media management platform claims that social media

6. https://statista.com/statistics/303681/twitter-users-worldwide/.

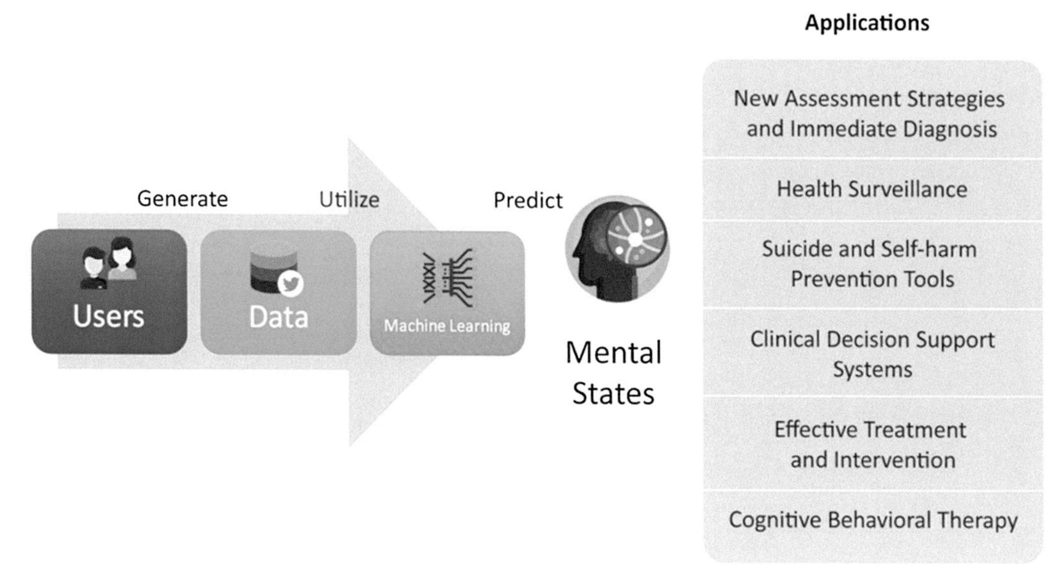

FIGURE 21.1 Research on e-mental health through social media: a conceptual framework.

users have recently grown by more than 200 million. Globally, social media usage has increased by more than 5%, reaching 59% of the population. In addition, unique users have risen by some 520 million, representing an annual growth rate of over 13%. Fig. 21.2 illustrates the growing trend of social media users in the last 11 years.

It results in social big data, which contains worthwhile information about people's behavior, moods, and interests [24–26], which covers a considerable number of fields containing machine learning, data and graph mining, statistics, information retrieval, linguistics, NLP, and text mining. In addition to health and e-commerce, their applications can be extended to many other areas [27,28].

In a study of a large set of tweets, self-reported signs were found to be the most trustworthy signal for predicting syndrome outbreaks [29]. Similarly, studies have found that social networks can detect trends in disease outbreaks, such as flu spread investigations [30] or depression [23]. Using these data, researchers can deeply understand the users' behavior.

Monitoring population and mental health are increasingly being conducted through social media [31]. Online screening tools, like medical DSSs, are practical, and may serve as more common assessment strategies in the future [32]; likewise,

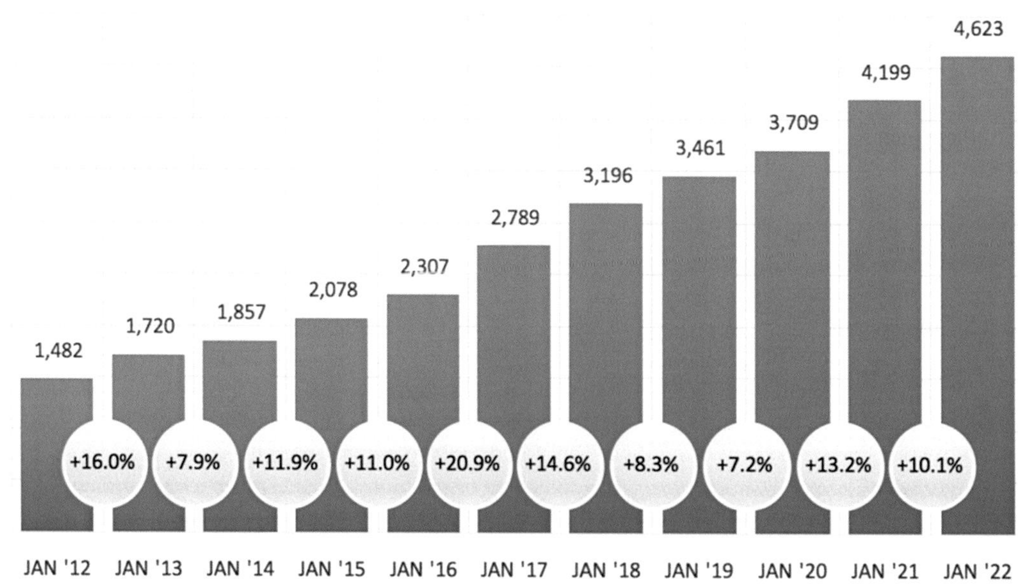

FIGURE 21.2 Users of social media over time (in millions).

health surveillance tools detect mental disorders signs, despite inefficient traditional approaches relying heavily on interviews and surveys. Distinguishing recognized signals from user-generated social media content might shape new mental disorders screening forms. A potential advantage of automated social data analysis is the ability to detect early warning signs. According to previous studies, machine learning may identify depression early using language patterns as indicators [26,33–35].

3. Assessment strategies

By using an automated process, it would be possible to identify the signs of an individual's disorder in their behavior and target them for a more thorough assessment. There has been an increase in the number of researchers investigating mental health within the context of social media in recent years, examining the association between social media use and behavioral patterns and disorders such as anxiety, depression, suicidality, and stress. This section discusses methods to predict mental disorders and five commonly used approaches. A discussion of the differences is then provided, along with directions for future research.

To collect social data with associated information about people's mental health, several approaches have been studied. Fig. 21.3 shows how data are gathered from sources by either recruiting participants to take a survey and/or sharing their social network profile information. Users' tweets can be searched for specific keywords to detect (and collect all tweets from) those who have shared diagnoses, user language on tweets, or mental illness–related forums mentioning mental disease keywords can be collected. Using public data can be more cost-effective and faster than administering surveys if a much larger sample is collected. However, clinical information and survey-based assessments generally have a higher degree of validity [9]. A more detailed examination of these methods will follow in the next sections.

3.1 Questionnaires and surveys

The psychological characteristics of users are identified through questionnaires and surveys, which are highly used in the social and behavioral sciences [36]. Using psychometric self-report surveys to assess mental illness is highly reliable and valid. The use of self-report surveys in psychology and epidemiology is second only to clinical interviews [37]. CES-D,[7]

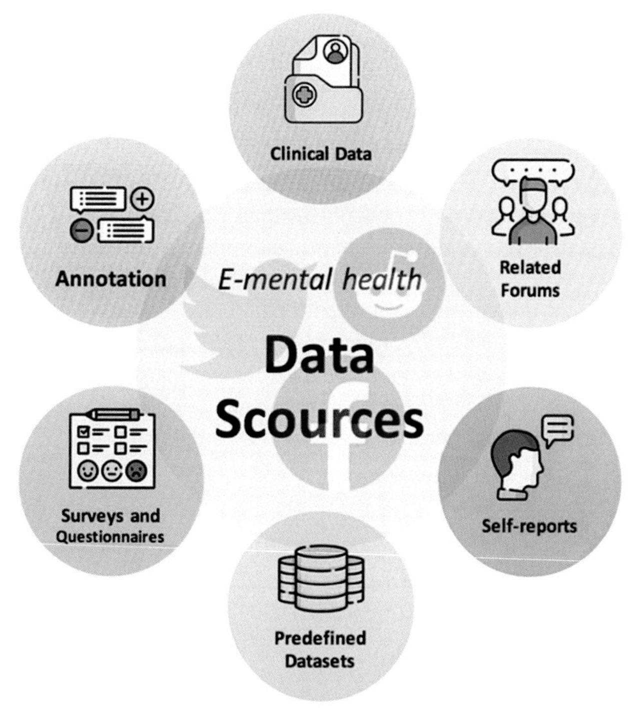

FIGURE 21.3 The data source for e-mental health research.

7. Center for Epidemiologic Studies Depression Scale.

PHQ-9,[8] and BDI[9] are the most common feedback form to assess participants' depression levels. Among the relevant instruments for detecting suicidal ideation, and measuring well-being, are the SPS[10] and SWLS[11] [38]. Participants could participate in the research by answering questionnaires and following data collection guidelines by visiting data donation websites or crowdsourcing platforms, like OurDataHelps[12] or MTurk[13] [33,39,40].

Predictive models in this emerging literature can be most accurately built using survey responses. As a result of the high costs associated with this method, more publicly accessible assessment criteria are being used, such as those outlined in the following sections. Combining social data with surveys is additionally not without challenges in practice. Social media identities can be asked for in surveys or questionnaires as part of participants' contact information. Whether or not participants volunteer this information is up to them [36].

3.2 Self-declared mental health status

Public data are used in a number of studies. Social media APIs[14] are used to collect related posts using keywords/phrases or regular expressions. Because there were no standards for data collection, there was a need for a custom data capture mechanism for certain data sources. As an example, custom tools or web apps developed to connect to the Facebook APIs were used to gather datasets for Facebook-based experiments [41,42]. Likewise, several studies have explored mental disorder cues using Twitter APIs [38,43,44]. There have been similar approaches taken for Reddit [45], Sina Weibo [46], and Instagram APIs [47].

The leading queries for post-retrieval are "suicide," "self-harm," "kill yourself," "want to die," and "I was diagnosed with [disorder]" [48–50]. An example of publicly available data is self-declared mental illness diagnoses on Twitter. Self-reported diagnosis is based on regular expressions of this type. API extraction leads to a set of tweets that must be assessed before analysis. Note that posts negate suicide ideation, discuss the suicide of others, or report news of suicide removed from the collection. In the same way, posts without hypothetical statements, negations, or quotations are selected as positive samples in the case of self-report diagnosis.

Twitter users' self-declared diagnoses were used to predict whether they had PTSD or depression to facilitate collaboration between computer scientists and clinical psychologists in the 2015 CLPsych[15] workshop [51]. During the study, participating teams constructed language topic models [52], identified words most related to disorder status, considered character sequences as features, and built relative counts of N-grams present in all disorders statuses using a rule-based approach [53]. Prediction performance was highest for the latter. Researchers have been interested in this method, which has recently been used to develop automatic tools for detecting depression [38].

3.3 Forum membership

Another source of available mental health information is online forums and discussion websites. Generally, they enable people to discuss stigmatized mental health issues in an open space, receive and provide emotional support, and ask for advice. This approach commonly uses Reddit and Facebook for data collection.

De Choudhury et al. [54] looked at posts of Reddit users (via subreddits) who addressed mental health issues and then spoke about suicidal ideation in the future. It was discovered that these shifts were characterized by poor linguistic coordination, reduced social engagement, heightened self-attentional focus, and a sense of hopelessness, impulsiveness, anxiety, and loneliness in shared content. A study by Tadesse et al. [55] looked for ways to detect depression on Reddit social media and solutions for detecting depression through effective performance increases. Language usage and depression were found to be closely related using NLP and text classification techniques. In summary, depressing accounts were more likely to contain language predictors of depression, with an emphasis on the present and the future, referring to feelings of sadness, anger, anxiety, or suicidal thoughts.

8. Patient Health Questionnaire.
9. Beck Depression Inventory.
10. Suicide Probability Scale.
11. Satisfaction with Life Scale.
12. https://ourdatahelps.org.
13. https://mturk.com.
14. Application Programming Interface.
15. https://clpsych.org.

3.4 Posts annotation

Manually reviewing and annotating posts that contain mental health keywords is the next source of publicly available data. The language of social media posts can predict annotations. Social media posts are coded concerning pre-established categories by annotations [56]. Annotation studies on depression typically look for posts where users discuss their own experiences with depression [57]. In addition to guidelines on recognizing depression symptoms, annotators are provided with a reduced set of symptoms, including disturbed sleep, depressed mood, and fatigue, as described in clinical assessment manuals such as the DSM-5[16] [58]. Furthermore, annotations have been applied to distinguish between stigmatizing and insulting mentions of mental illness from expressing, sharing, or supporting helpful information with those with mental disorders. The annotations of posts are generally used as a supplementary method for revealing life conditions accompanying mental illness (such as education, employment, housing, or weather problems) not apprehended by conventional depression diagnostic indicators [59].

An innovative dataset for CAMS[17] has been developed by Garg et al. [60]; using two separate datasets: crawling and annotating 3155 Reddit posts and re-annotating 1896 instances from the available SDCNL dataset[18] [61] for interpretable causal analysis, the authors present a causal analysis annotation schema. Their experimental results showed that a classic logistic regression model performed better than a CNN-LSTM[19] model on the CAMS dataset.

3.5 Other methods

To provide access to individuals with mental illnesses, computer scientists should collaborate with physicians and psychologists. Predictive models could be built based on real patients' data. It is possible to create very reliable documents based on clinical reports along with social data. Social network platforms can be analyzed using user-generated content and behavior to discover meaningful patterns by having the mental state of individuals and the appropriate permissions.

Another option is to use predefined datasets, in which data are collected by other researchers and pass the initial evaluation phase to identify class labels. The myPersonality[20] project, CLPsych [49], and eRisk[21] workshops [62] provide users' social data and psychometric test scores for academic purposes and are three well-known datasets in the field in question. AutoDep[22] is another new dataset that is provided automatically and used to examine more features in the social data produced by Twitter users with symptoms of depression [38].

4. Social data configuration

Building predictive models using extracted data is the process of automating the analysis of social networks. The social data collected on social media networks come from users' online activity. The content can include text, images, videos, context information (such as location tags), user biographical information, connections, and interests (Fig. 21.4).

Numerous research applies textual content and linguistic patterns to recognize the most significant consequence of mental disorder prediction. Posting frequencies, hashtags, and times of posts are among users' common features.

LIWC[23] [63] analysis obtained some of these results. In related studies, the LIWC is a notable text analysis application for finding linguistic patterns [34,50,64,65]. Various psychologically meaningful categories are covered by dictionaries that psychologists manually construct. In addition to extracting positive or negative sentiments and personal pronouns from the textual content, it can extract potential signs from the text. As popular sentiment analysis tools, OpinionFinder [66] and SentiStrength [67] were frequently used in selected studies to quantify textual sentiment [68,69]. In many attempts to disclose latent topics from user posts, topic modeling techniques such as LDA[24] [70] have also been used as part of the content analysis [71,72].

N-gram language models are also tools used to determine the probability of the occurrence of certain characters and word sequences. Traditional word-based approaches are not ideal in social media texts because of shortening and spelling

16. Diagnostic and Statistical Manual of Mental Disorders.
17. Causal Analysis of Mental Health Issues in Social Media Posts.
18. https://ayaanzhaque.github.io/SDCNL.
19. Convolutional Long Short-Term Memory Neural Networks.
20. https://sites.google.com/michalkosinski.com/mypersonality.
21. https://erisk.irlab.org.
22. https://github.com/rsafa/autodep.
23. Linguistic Inquiry and Word Count.
24. Latent Dirichlet Allocation.

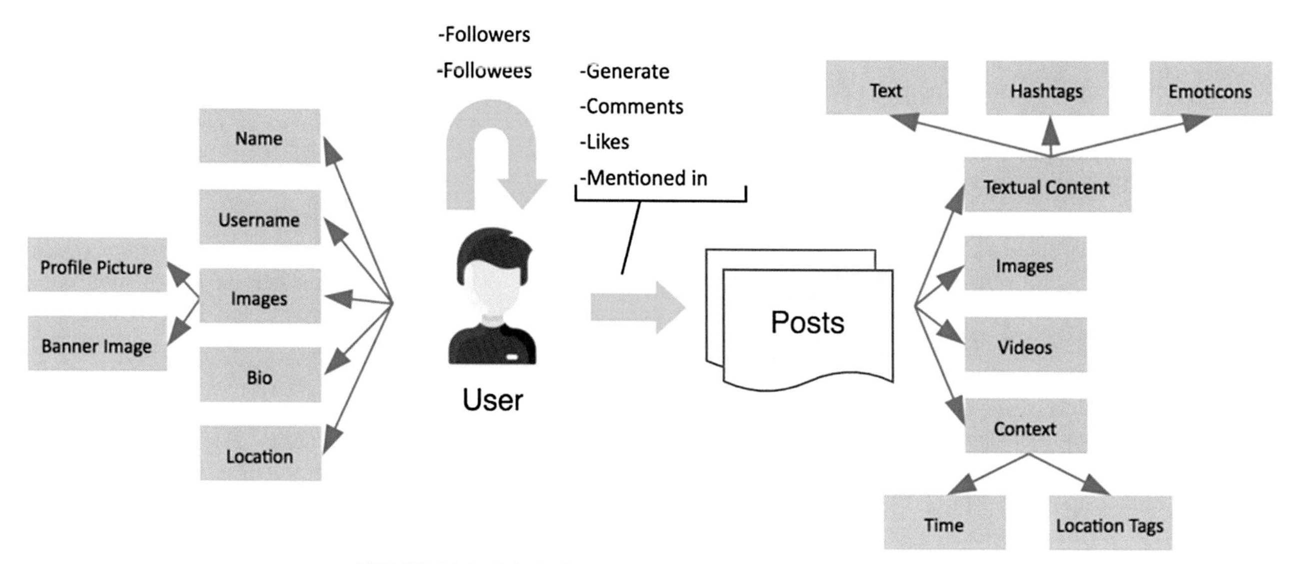

FIGURE 21.4 Principal components and relationships in social data.

errors [73]. Hence, to extract features, we can employ character/word n-grams and use the tf-idf technique [38]. After the features have been extracted, they are preserved as independent variables in a model, such as linear regression [37] or SVM[25] [74], to signify the dependent variable (mental disorder). A machine learning technique is utilized for training a predictive model using training data and evaluated by the test data to avoid overfitting (cross-validation). As one of the different metrics, the prediction performances, which we will discuss in the evaluation section, are presented.

An example of Twitter data analysis using surveys involves collecting information on social network accounts while administering the survey to a well-defined study population. It is possible to sample social data systematically using Twitter account identifiers (User_IDs). According to Fig. 21.4, we can access various observational data types using the Twitter API. Each class must be queried separately. The most recent tweets of a user are available at the time of querying (limited to 3200 recent tweets). Typical data collection can conduct at regular intervals to resume longitudinal observation. To reach the target dataset, several standard steps must be done with the users' data. The first step was to clean and preprocess the data to ensure they were suitable for the analysis algorithms. Feature engineering selects a subset of features in the next step. In order to fit the model, the desired characteristics and information extracted from the questionnaires are analyzed. Following the testing phase, an evaluation is conducted based on the test set [38]. In the following, we will take a quick look at these steps.

4.1 Preprocessing

It is typical for the corpus of data to be preprocessed to remove unsuitable samples and to clean and prepare the data for analysis. Participants' questionnaires and pieces of information, usually extracted from studies, can contain unworkable and vague details to enhance prediction and classification accuracy. The dataset excludes individuals whose profiles lack sufficient information. A number of people with low activity levels were removed since they published fewer than a defined number of posts. The questionnaires excluded individuals whose completion time was abnormally short or long [75]. Those whose responses to two different questionnaires were not correlated were also excluded. Each post was checked for written language during the data cleaning [38]. As a result, the available tools were suitable for analyzing the posts.

In the preprocessing step, stop-words are usually removed, sentences are segmented, retweets, duplicates, URLs, special characters, mentioned usernames, and hashtags are handled, and lowercasing is done [64,76,77]. Additionally, emojis were converted to ASCII to ensure that machines could read the data. For ethical reasons, any potentially identifying usernames were also anonymized. The user may upload images and videos in a different format on some social networks, so it is essential to consider format conversion for visual content.

25. Support Vector Machines.

4.2 Feature engineering

Social network users may show signs of mental health problems based on a number of different features that can be extracted from their social networking profiles. Several studies have examined textual content to determine which factors are associated with mental health. Alternative research techniques, such as image analysis and social interaction analysis, have been used in other research projects.

Sentiment analysis is a popular tool in NLP and text mining, which classifies a given text's polarity into positive, negative, and neutral categories to understand the emotional expression [78]. Several studies used LIWC [63] to draw out mental disorders' signals from textual data (e.g., the occurrence of using pronouns "I" or "me" as determinants). OpinionFinder [66] was used by Bollen et al. [68], and SentiStrength [67] was used by Kang et al. [25] to carry out sentiment analysis. The literature likewise used VADER to exploit the benefits of rule-based modeling and lexicon-based characteristics for social media messages [38]. Moreover, topic modeling was employed in many studies [40,79] to extract topics from social content.

It is common for social media posts to contain a variety of emoticons. Consequently, some studies [25] looked into their use's meaning and mood states. Users of social network platforms can also post visual content in addition to text messages; researchers have examined these data for mental disorder cues in some studies [38].

However, most of these studies rely heavily on textual content and only a few image-processing techniques [80,81]. Kang et al. [25] used color composition and SIFT descriptors to discover emotional meaning from Twitter images. Reece et al. [82] predict depression in Instagram users based on the image's hue, saturation, and brightness. Sharath et al. [83] have demonstrated that utilizing VGG-Net [84] image classifier together with image features like facial, aesthetics, color, and content are used to predict depression. More recently, Safa et al. [38] used user profile photos and header image content to discover depression's latent patterns. They used Imagga[26] tagging API to represent image content and generate a Bag-of-Visual-Words (BoVW) as a part of the analysis.

Users on social network platforms interact and form relationships with each other millions of times each day. A graph structure containing information about friendships, relationships, and interactions was analyzed to determine how mental disorders can be detected (e.g., assortative mixing patterns and interactions among depressed users) [43]. It should also be noted that incorporating hashtags and context information (e.g., post time) can improve prediction accuracy, which demonstrated a difference in a study on the biggest social media platforms in China (Sina Weibo) [85]).

Following preprocessing, the next step is feature selection, where the key features associated with the research domain are prepared for classification. In feature selection, relevant subsets of features are selected so that they are able to predict mental disorder symptoms or accurately mark participants while bypassing overfitting. Analyzing statistics aims to identify parameters that differentiate nondisordered users from disordered users. Previously, Pearson correlation coefficients and Spearman rank correlation coefficients were applied [38,42], as well as Mann–Whitney U tests [47,86]. PCA[27] [87], gain ratio [88], forward greedy stepwise [89], relief technique [88], and convolutional neural network with cross-auto encoder technique [90] are also used to reduce the dimensionality of features. Finally, the model will be constructed using the training dataset, and the performance will be analogized and assessed. Consequently, the model can predict new user states (unseen data) with gained accuracy. Fig. 21.5A and B depict the overall survey and self-report-based analysis frameworks.

5. Prediction algorithms

Machine learning algorithms learn patterns from data by using selected features as training sets. Five broad categories of machine learning techniques are presented here: supervised, semi-supervised, unsupervised, reinforcement, and deep learning.

5.1 Supervised learning

Supervised learning involves presenting algorithms with examples for training. Instances consist of inputs and outputs. Inputs are often represented as numerical arrays that describe examples numerically. The feature vector is usually referred to as this array. A classification problem results in an output called a class, a value we want to predict, such as a mental state. By training, the model can predict incognito output values from unexplored, unseen input values [91]. The model can

26. https://docs.imagga.com.
27. Principal component analysis.

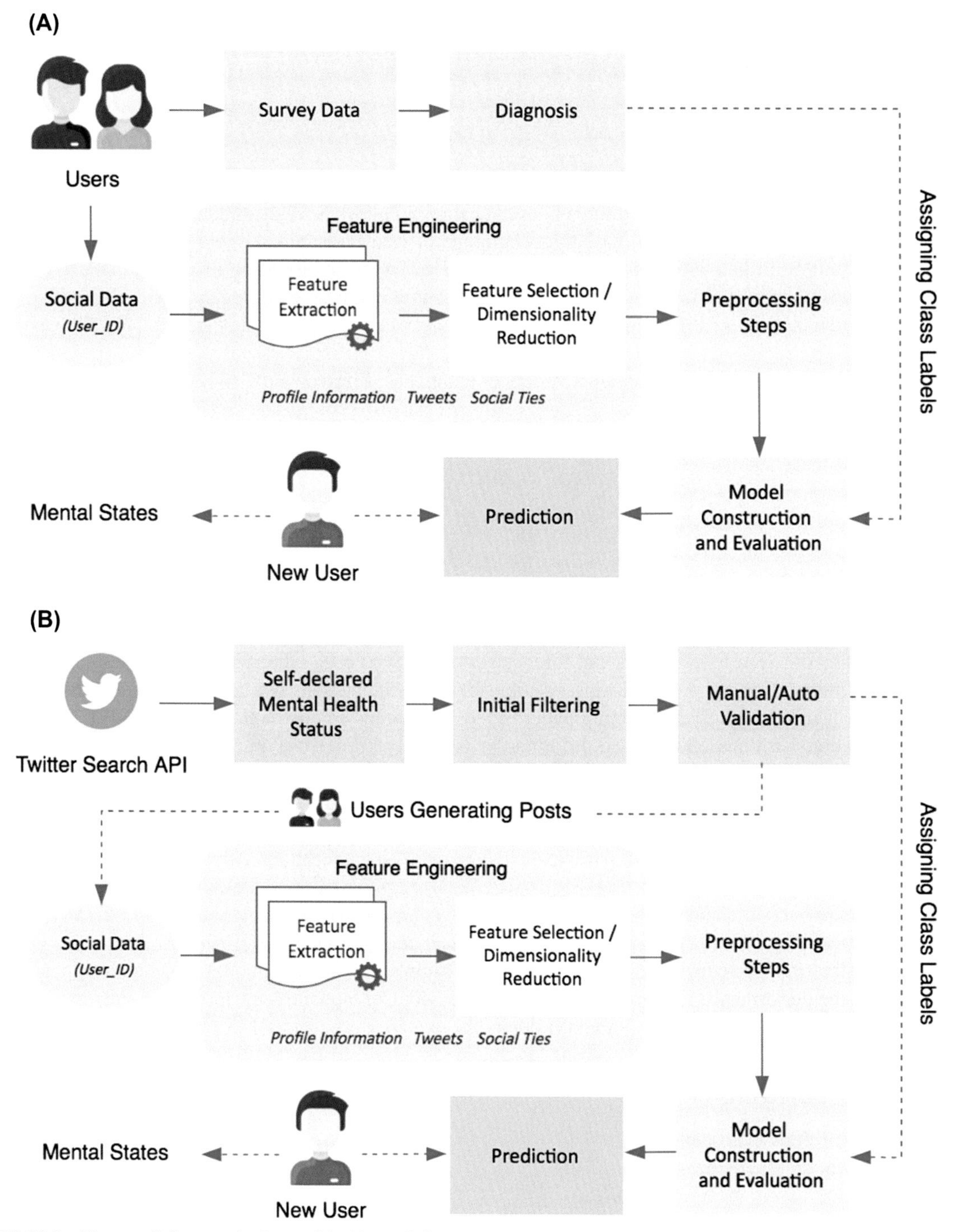

FIGURE 21.5 The overall framework of mental health prediction via social data. (A) Survey-based analysis. (B) Self-report-based analysis.

be tested for performance when the instances' output values are hidden. After that, the predictions can be compared to the ground truth, which is the actual output.

Classification refers to categorical output predictions; Regression refers to quantitative output predictions. A model can be trained using a set of signals from user behavior on social media with labels corresponding to mental states. Due to the categorical output (e.g., depressed and not depressed), a classifier can be applied. Therefore, most machine learning models used to detect mental states are classifiers. Allowing the categorical output variable to take a given desirable mental state as a value makes it possible to detect the presence/absence of a particular mental state. It is possible to determine the final category from probabilities rather than the final class from some classifiers. Cognitive state detection has been achieved through supervised learning classifiers such as decision tree [92,93], random forest [38,55,64], AdaBoost [55], support vector machines (SVM) together with other kernels like linear [38,55,94], and radial basis function (RBF) [25,43,87,92,95–97], Naïve Bayes [43,92,93], different types of regression [48,89,98–100], artificial neural networks [38,101], etc. In Fig. 21.6, a part of the survey conducted by Wongkoblap et al. [102] is visualized as a diagram to understand better how supervised learning methods classify users according to mental disorders are distributed in the literature.

5.2 Unsupervised learning

Unlike supervised learning, it is necessary to find the classes that can naturally arise from a similarity in input data instead of knowing the output variable at training time. Identifying groups of users with similar characteristics, for example, unsupervised learning algorithms used for clustering are common methods for identifying groups or hierarchies within data. The unsupervised learning process can be considered a preprocessing step before the supervised method is applied. A number of standard clustering algorithms have been used for stress evaluation tasks [103,104], including k-medoids clustering, k-means clustering, hierarchical clustering, fuzzy c-means clustering, and density-based spatial clustering of applications with noise (DBSCAN) [91].

Finding a numerical textual content representation is crucial since clustering methods rely on numerical data as input. In more precise terms, the posts would have to be mapped into a latent space with an inherent structure based on contextual similarity. According to this continuous space, similar tweets have similar number vectors, and different tweets are distant. Embedding models can help solve this problem. Representing words as numerical vectors to capture the semantics can be equated to representing text as numerical vectors with meaning [105]. Using word embeddings, which are real-valued vectors, similar words are represented similarly. Embeddings of words have become an important trend in NLP research since the development of Word2Vec [106]. A single sentence or document embedding can then be obtained by further processing the words of a sentence or document.

5.3 Semi-supervised learning

Semi-supervised learning occurs when a large number of training samples are available, but the output labels are only known for a small percentage of samples. Models are trained by applying labeled and unlabeled instances in semi-

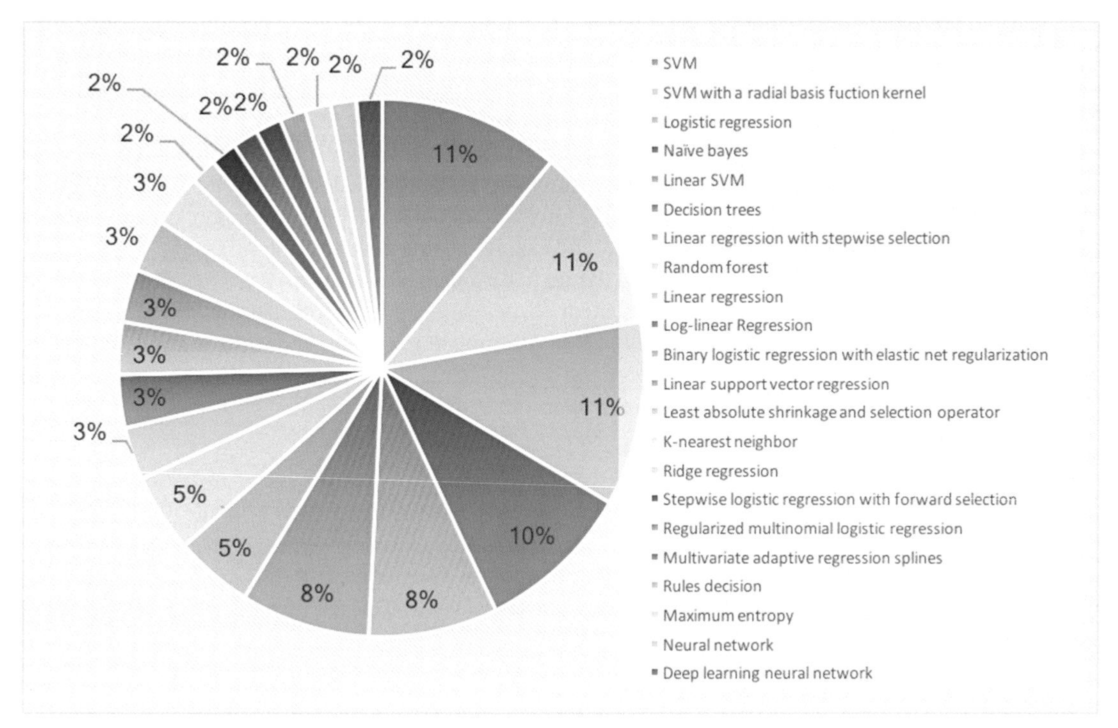

FIGURE 21.6 Distribution of supervised learning techniques in the selected articles based on Wongkoblap [102] research.

supervised algorithms [107]. Since ground truth classes are difficult to tag, semi-supervised learning helps detect mental states. For example, questionnaires are usually used to tag daily mood states; nevertheless, it is not uncommon for participants to forget to answer them, resulting in multifold days going untagged. Identifying the current state of a bipolar patient requires an expert's evaluation, usually clinical, to classify the data. This type of issue likely limits data tags.

According to Tariq et al. [108], data extracted from Reddit can be used to classify those who suffer from chronic mental illnesses. The proposed method uses semi-supervised learning (co-training) by incorporating widely used machine learning techniques. According to the experimental results, co-training-based classification appears more effective than typical approaches.

5.4 Reinforcement learning

Reinforcement learning occurs when agents learn through trial-and-error interactions with their environment. Depending on the current state, an agent will decide what action to take in order to maximize its rewards. A reinforcement learning algorithm explores a wide range of activities and chooses one that maximizes reward instead of being presented with inputs and targets [91]. Agents are rewarded or punished when they perform the right or wrong actions. Over time, the system tries to figure out what actions result in the gilt-edged rewards. Its advantage is that no human expertise is required to understand the problem domain with reinforcement learning.

According to Gui et al. [109], depression can be detected using only contextual information by analyzing what users post. From users' recorded posts, they suggested a reinforcement learning-based approach for selecting indicator posts automatically. Compared to feature-based and neural network-based methods, the proposed method outperforms them.

5.5 Deep learning

Research on e-mental health has been boosted recently by deep learning, a rapidly expanding field of machine learning. In contrast to statistical methods, deep learning uses neural networks, usually with many hidden layers, to learn various abstraction levels. The values of these layers are not included in the input data, making them referred to as hidden layers. It is, therefore, necessary for the network to determine which concepts help explain observed relationships. The MLP[28] architecture is used in some studies [110]. In contrast, more complex network architectures are used in others [111], including LSTM [112], GRU[29] [113], and CNN [114]. Various RNNs,[30] including LSTM and GRU, have been presented over the past few years, with the main difference being how the input maps to the output. They hold significant promise for detecting mental health conditions in clinical notes and social data because they demonstrate state-of-the-art implementation in plenty of applications, including NLP. It has been investigated whether deep learning could detect depression [115,116] or recognize suicide-related psychiatric stressors [76]. A study by Gkotsis et al. [117] analyzed posts on Reddit and developed classifiers for identifying and categorizing mental illness posts. In the dataset, they automatically recognized mental illness-related posts using neural networks and deep learning.

In RNN, input elements are prioritized using an attention mechanism, providing some interpretability of results [118]. An attention mechanism was included in a hierarchical RNN architecture introduced to predict the classes of posts, such as anxiety, depression, etc. [119]. With the attention mechanism, the authors observed that the model effectively predicts risk text and extracts elements critical for decision-making. Users' mental health status can be inferred from the properties of the shared images, interactions, and relationships within the textual content. Researchers have increasingly endeavored to compound these two information types with textual content to develop predictive models in recent years. Using psychological and art theories, researchers analyzed text, image, and comment representations at low and middle levels [120]. Using CNN's hybrid model, they integrated post content and social interactions [121]. Based on the results, it was found that stressed people tended to have less connected friends than those who were not.

5.5.1 Transfer learning

Transfer learning refers to the process of using a model developed for a task to develop a model for another one. As an initial point for NLP and computer vision, using pretrained algorithms is a popular approach in deep learning. As a result of the extended computing and time resources needed to build neural network models and the significant improvements they provide on related topics, this approach is prevalent in deep learning. Due to the lack of millions of labeled data points to

28. Multilayer Perceptron.
29. Gated Recurrent Unit.
30. Recurrent neural networks.

train such complex models, data science is an advantageous field. Basically, by combining knowledge from similar or other areas, transfer learning allows us to learn novel things in a diverse domain without or with limited labeled sets [122]. It is possible, for instance, to identify cats from images in the same way that dogs are recognized from images. A typical example is a system that understands how to differentiate whether an image contains signs of a particular disease based on the categories [123]. Besides overcoming the problem of insufficient labeled data, this method can also be used in conjunction with other methods. In particular, it seems to work quite well regarding image-related use cases. The results are not very good if the domains are too different. Moreover, most research on this topic has been limited to images and videos. It is important to note that transfer learning has the potential to address the challenge of reducing the amount of labeled data required by e-mental health applications.

6. Evaluation

Analyzing the generalizability of a trained model refers to its estimated performance on unseen data, which is the purpose of model evaluation. The train and test sets are two subs of a dataset that can be applied to estimate a model's generalization ability. Holdout validation uses a training set for model training, while a testing set is used to assess its performance. A random assignment is often made to a training or testing subset. Machine learning models often perform superbly if evaluated with the same data they were trained with, but they may not generalize well when evaluated with new data. As a result, holdout validation allows for better estimations of generalization since the trained model doesn't include information about sample sizes from the testing set.

Parameter tuning is required for some models; Continuously tuning such parameters based on the testing set performance leads to the model overfitting risk. This can be avoided by dividing the dataset into training, validation, and testing. Hence, models are built using training data, and parameters are tuned with validation data. Generalization performance is then assessed based on the testing set. Depending on the application, the three sets are typically split between 60% (train set), 20% (validation set), and 20% (test set) [124]. When there is a large amount of data, holdout validation makes sense. Cross-validation with k folds is preferred when data is limited. A random division of the data into k equal-sized subsets is used in this method; After that, k iterations are performed. One subset is used to test the model, while the remaining is used to train it. All iterations are averaged to determine the performance. The estimate's variance can be reduced as k increases, but the computational load increases as well. Typically, k is considered 10 in most cases. It is also known as LOOCV,[31] where the number of samples equals k.

Another commonly used method in social media monitoring of mental health is to apply the training set, all other users' data, and the data from a new user as a testing set to assess the performance of a model. Users do not need to train data for this user-independent or general model [125]. A classification model's reliability is assessed after prediction by an evaluation mechanism. The problem's type (regression, classification, etc.) determines the metrics to use. In the literature, the most applied metrics and visualization tools helping to examine the proposed models' performance are classification accuracy (proportion of correctly classified instances), precision (the positive predictive value), recall (the true positive rate or sensitivity), F1-score (the harmonic mean of the precision and recall), and ROC[32] curve [38]. According to Table 21.2 (the confusion matrix), these metrics are as follows.

$$\text{Accuracy} = \frac{\text{TP} + \text{TN}}{\text{TP} + \text{TN} + \text{FP} + \text{FN}} \tag{21.1}$$

$$\text{Precision} = \frac{\text{TP}}{\text{TP} + \text{FP}} \tag{21.2}$$

$$\text{Recall} = \text{TPR} = \frac{\text{TP}}{\text{TP} + \text{FN}} \tag{21.3}$$

$$\text{F1} = 2 \times \frac{\text{Precision} \times \text{Recall}}{\text{Precision} + \text{Recall}} \tag{21.4}$$

ROC is a probability curve and AUC[33] measures separability. It determines the degree to which the model will be able to distinguish between different classes. Increasing the AUC increases the model's ability to predict 0 classes as 0 and 1

31. Leave-one-out cross-validation.
32. Receiver Operating Characteristics.
33. Area Under the ROC Curve.

TABLE 21.2 Confusion matrix.

	Relevant	Non-relevant
Retrieved	True positive (TP)	False positive (FP)
Not retrieved	False negative (FN)	True negative (TN)

classes as 1. Analogously, users with the disorder and control group with a higher AUC can be distinguished more easily. ROC curves are plotted with TPR[34] versus FPR,[35] which are computed as follows. Specificity refers to the true negative rate, i.e., how many negatives are categorized precisely.

$$\text{Specificity} = \frac{\text{TN}}{\text{TN} + \text{FP}} \tag{21.5}$$

$$\text{FPR} = 1 - \text{Specificity} = \frac{\text{FP}}{\text{TN} + \text{FP}} \tag{21.6}$$

In evaluating performance, it is essential to avoid using only one metric. When several different metrics are used simultaneously, a classifier or prediction approach's real performance and robustness can be better assessed [125]. As a result, reporting all metrics is good practice when classes are imbalanced, as precision may be deceptive. A sample imbalance occurs when only a tiny percentage of samples belong to a particular category.

Typical performance metrics for regression problems include mean square, root means square, and mean absolute error, as well as correlation coefficients [126]. Mental state detection often produces ordinal output classes. In other words, it has a natural order; It is possible to model depression levels with an ordinal variable that takes the values "high," "medium," and "low." In regular performance metrics, errors are treated equally: Confusion between low and medium has equivalent error weight as confusion between low and high; obviously, the latter mistake should be penalized intensively. Metrics like linear correlation, mean squared or absolute errors, and accuracy within *n* can be used to measure ordinal variables.

7. Status, challenges, and future direction

In mental health diagnostics and classifications, mental health professionals use DSM[36] and the ICD[37] to classify mental disorders based on their type, intensity, and duration [127]. Although these tools are frequently used, they have certain limitations. Many diseases are diagnosed using the same criteria; it is, therefore, impossible to distinguish between diagnostic groups. Further, these tools do not take into account additional factors like demographics and biochemistry, patient interview records, mental disorder background in the family, and medication response [20]. In spite of the fact that screening surveys that have been validated and are reliable are closest to clinical practice, they are time-consuming to administer and rely on the self-selection of crowd workers, introducing biases in the sampling [128]. The assessments of this approach also suffer from significant temporal gaps, as identifying risk factors for mental illness often requires immediate action, limiting the development of effective interventions [129]. It is also possible that mentally ill people are less likely to cooperate with researchers; Social media analysis can detect cases that would otherwise go undiagnosed. In contrast, public data approaches (i.e., selecting users based on self-report analysis, or specific forum membership) have a larger sample size and bring additional information. As of yet, no studies have focused explicitly on effectively recognizing someone oblivious to their mental health condition, so the chances of capturing users unaware of their diagnosis are low [9].

For social media-based screening to be effective and distinguish between mental illnesses, gold-standard structured clinical interviews, and further screening methods must be combined with social media data collection in ecologically valid samples [11]. It remains challenging to gather, process, fusion, and analyze big social data from unstructured, semi-structured sources to exploit valuable knowledge. Due to this issue, several challenges and problems have emerged in

34. True Positive Rate.
35. False Positive Rate.
36. The Diagnostic and Statistical Manual of Mental Disorders.
37. International Classification of Diseases.

the social big data domain, including knowledge representation, data processing, data analysis, data management, and data visualization [27,28].

However, most former works investigated textual/visual features from social media platforms without considering users' social networks. Mental health problems are highly associated with friends' circles, suggesting that social network analysis may be an effective method for studying the prevalence of such disorders [130]. It remains challenging, however, to model text information and network structure comprehensively. A convolutional graph network has been developed to address networked data mining in this context.

In practice, the final prediction model is often derived by combining different types of algorithms. Before building supervised learning models, for insistence, unsupervised learning models are often applied as a preprocessing phase [124]. It is also important to differentiate between user-dependent and user-independent (known as general) training schemes. Training the former is founded on the data from the particular user under consideration. All other users' data are used to train the latter, except the target user (who is using the system). User-dependent models are advantageous since they capture the behavior of each individual user and produce better results, but they require a substantial amount of training time. Users who are 'atypical' might not benefit from user-independent models since they need no data from the target user.

Although social media analyses are applied to detect online users with mental disorders, the importance of translating this innovation into practical application and providing users with real-time assistance, for instance, cannot be overstated [131]. In addition, the lack of large training data is one of the main obstacles to the widespread application of deep learning approaches in e-mental health. It is worth noting that the training data used for these deep architectures is extensive and hand-labeled.

There are some ethical concerns associated with social-media-based mental illness assessment. Privacy has been a concern for a long time. People with mental illness may be disadvantaged by these policies when their employers or insurers use them. A data protection and ownership framework are essential for ensuring that mental illnesses are not stigmatized or discriminated against [132]. Most people are unaware that their digital traces contain mental-health-related information. Transparency is essential regarding who generates which health indicators and why. Mandatory reporting guidelines need to be clarified from a mental health perspective. Because of misclassifications, derived mental health indicators are difficult to integrate into systems of care responsibly [9].

8. Conclusion

Social networks are becoming increasingly popular as a platform for sharing opinions, feelings, and thoughts. Recently, social networks have been exploited as a public health tool with increasing interest. This chapter discussed the implications related to big social data and explained how these data could be used to analyze and predict mental disorders. We reviewed the recent approaches addressing mental state evaluation and disorders diagnosis using users' digital footprints. The studies, as mentioned earlier, were organized concerning the assessment strategy, data configuration (including data preprocessing and the feature engineering phase), and prediction algorithms. Furthermore, we presented a series that examines the language and behavior of people enduring mental disorders (like PTSD or depression) and discussed diverse aspects associated with developing experimental frameworks. Thus, this chapter's primary contributions are comprehensively analyzing mental state assessment methods on social data, structurally categorizing them by considering their design principles and lessons learned during their development, and discussing challenges and achievable directions for future studies.

References

[1] E.A. Ríssola, D.E. Losada, F. Crestani, A survey of computational methods for online mental state assessment on social media, ACM Transactions on Computing for Healthcare 2 (2) (2021) 1−31.

[2] V. Patel, S. Saxena, C. Lund, G. Thornicroft, F. Baingana, P. Bolton, et al., The Lancet Commission on global mental health and sustainable development, The Lancet 392 (10157) (2018) 1553−1598.

[3] D. Harvey, F. Lobban, P. Rayson, A. Warner, S. Jones, Natural language processing methods and bipolar disorder: scoping review, JMIR Mental Health 9 (4) (2022) e35928.

[4] F.S. Bersani, B. Barchielli, S. Ferracuti, A. Panno, G.A. Carbone, C. Massullo, et al., The association of problematic use of social media and online videogames with aggression is mediated by insomnia severity: a cross-sectional study in a sample of 18-to 24-year-old individuals, Aggressive Behavior 48 (3) (2022) 348−355.

[5] W.H. Organization, Suicide Worldwide in 2019: Global Health Estimates, 2021.

[6] E.A. Crone, E.A. Konijn, Media use and brain development during adolescence, Nature Communications 9 (1) (2018) 1−10.

[7] S. Rice, J. Robinson, S. Bendall, S. Hetrick, G. Cox, E. Bailey, et al., Online and social media suicide prevention interventions for young people: a focus on implementation and moderation, Journal of the Canadian Academy of Child and Adolescent Psychiatry 25 (2) (2016) 80.

[8] A.-S. Uban, B. Chulvi, P. Rosso, An emotion and cognitive based analysis of mental health disorders from social media data, Future Generation Computer Systems 124 (2021) 480–494.

[9] S.C. Guntuku, D.B. Yaden, M.L. Kern, L.H. Ungar, J.C. Eichstaedt, Detecting depression and mental illness on social media: an integrative review, Current Opinion in Behavioral Sciences 18 (2017) 43–49.

[10] R. Thorstad, P. Wolff, Predicting future mental illness from social media: a big-data approach, Behavior Research Methods 51 (4) (2019) 1586–1600.

[11] G. Coppersmith, M. Dredze, C. Harman, K. Hollingshead, From ADHD to SAD: analyzing the language of mental health on Twitter through self-reported diagnoses, in: Proceedings of the 2nd Workshop on Computational Linguistics and Clinical Psychology: From Linguistic Signal to Clinical Reality, 2015.

[12] H. Whiteford, A. Ferrari, L. Degenhardt, Global burden of disease studies: implications for mental and substance use disorders, Health Affairs 35 (6) (2016) 1114–1120.

[13] S. Docrat, D. Besada, S. Cleary, E. Daviaud, C. Lund, Mental health system costs, resources and constraints in South Africa: a national survey, Health Policy and Planning 34 (9) (2019) 706–719.

[14] P. Lee, A. Abernethy, D. Shaywitz, A.V. Gundlapalli, J. Weinstein, P.M. Doraiswamy, et al., Digital health COVID-19 impact assessment: lessons learned and compelling needs, in: NAM Perspectives, 2022, 2022.

[15] I. Petersen, A. Bhana, L.R. Fairall, O. Selohilwe, T. Kathree, E.C. Baron, et al., Evaluation of a collaborative care model for integrated primary care of common mental disorders comorbid with chronic conditions in South Africa, BMC Psychiatry 19 (1) (2019) 1–11.

[16] F. Hanna, C. Barbui, T. Dua, A. Lora, M. van Regteren Altena, S. Saxena, Global mental health: how are we doing? World Psychiatry 17 (3) (2018) 367.

[17] A.M. Kilbourne, K. Beck, B. Spaeth-Rublee, P. Ramanuj, R.W. O'Brien, N. Tomoyasu, H.A. Pincus, Measuring and improving the quality of mental health care: a global perspective, World Psychiatry 17 (1) (2018) 30–38.

[18] M. Johnson, A. Albizri, A. Harfouche, S. Tutun, Digital Transformation to Mitigate Emergency Situations: Increasing Opioid Overdose Survival Rates through Explainable Artificial Intelligence, Industrial Management & Data Systems, 2021.

[19] A. Thieme, D. Belgrave, G. Doherty, Machine learning in mental health: a systematic review of the HCI literature to support the development of effective and implementable ML systems, ACM Transactions on Computer-Human Interaction 27 (5) (2020) 1–53.

[20] S. Tutun, M.E. Johnson, A. Ahmed, A. Albizri, S. Irgil, I. Yesilkaya, et al., An AI-based decision support system for predicting mental health disorders, Information Systems Frontiers (2022) 1–16.

[21] L. Balcombe, D. De Leo, Digital mental health challenges and the horizon ahead for solutions, JMIR Mental Health 8 (3) (2021) e26811.

[22] M. De Choudhury, S. Counts, E. Horvitz, Major life changes and behavioral markers in social media: case of childbirth, in: Proceedings of the 2013 Conference on Computer Supported Cooperative Work, 2013.

[23] J. Lopez-Castroman, B. Moulahi, J. Azé, S. Bringay, J. Deninotti, S. Guillaume, E. Baca-Garcia, Mining social networks to improve suicide prevention: a scoping review, Journal of Neuroscience Research 98 (4) (2020) 616–625.

[24] S. Javadi, R. Safa, M. Azizi, S.A. Mirroshandel, A recommendation system for finding experts in online scientific communities, Journal of AI and Data Mining 8 (4) (2020) 573–584.

[25] K. Kang, C. Yoon, E.Y. Kim, Identifying depressive users in Twitter using multimodal analysis, in: 2016 International Conference on Big Data and Smart Computing (BigComp), 2016.

[26] R. Martínez-Castaño, J.C. Pichel, D.E. Losada, A big data platform for real time analysis of signs of depression in social media, International Journal of Environmental Research and Public Health 17 (13) (2020) 4752.

[27] G. Bello-Orgaz, J.J. Jung, D. Camacho, Social big data: recent achievements and new challenges, Information Fusion 28 (2016) 45–59.

[28] A. Kumar, S.R. Sangwan, A. Nayyar, Multimedia social big data: mining, in: Multimedia Big Data Computing for IoT Applications, Springer, 2020, pp. 289–321.

[29] M. Krieck, J. Dreesman, L. Otrusina, K. Denecke, A new age of public health: identifying disease outbreaks by analyzing tweets, in: Proceedings of Health Web-Science Workshop, ACM Web Science Conference, 2011.

[30] J. Sooknanan, N. Mays, Harnessing social media in the modelling of pandemics—challenges and opportunities, Bulletin of Mathematical Biology 83 (5) (2021) 1–11.

[31] M. Conway, D. O'Connor, Social media, big data, and mental health: current advances and ethical implications, Current Opinion in Psychology 9 (2016) 77–82.

[32] D.D. Ebert, M. Harrer, J. Apolinário-Hagen, H. Baumeister, Digital interventions for mental disorders: key features, efficacy, and potential for artificial intelligence applications, in: Frontiers in Psychiatry, Springer, 2019, pp. 583–627.

[33] G. Coppersmith, R. Leary, P. Crutchley, A. Fine, Natural language processing of social media as screening for suicide risk, Biomedical Informatics Insights 10 (2018), 1178222618792860.

[34] K. Loveys, P. Crutchley, E. Wyatt, G. Coppersmith, Small but mighty: affective micropatterns for quantifying mental health from social media language, in: Proceedings of the Fourth Workshop on Computational Linguistics and Clinical Psychology—From Linguistic Signal to Clinical Reality, 2017.

[35] F.M. Plaza-del-Arco, M.T. Martín-Valdivia, L.A. Ureña-López, R. Mitkov, Improved emotion recognition in Spanish social media through incorporation of lexical knowledge, Future Generation Computer Systems 110 (2020) 1000–1008.

[36] E.S. Spiro, Research opportunities at the intersection of social media and survey data, Current Opinion in Psychology 9 (2016) 67–71.
[37] J. Neter, M.H. Kutner, C.J. Nachtsheim, W. Wasserman, Applied Linear Statistical Models, 1996.
[38] R. Safa, P. Bayat, L. Moghtader, Automatic detection of depression symptoms in twitter using multimodal analysis, The Journal of Supercomputing 78 (4) (2022) 4709–4744.
[39] S.R. Braithwaite, C. Giraud-Carrier, J. West, M.D. Barnes, C.L. Hanson, Validating machine learning algorithms for Twitter data against established measures of suicidality, JMIR Mental Health 3 (2) (2016) e21.
[40] H.A. Schwartz, M. Sap, M.L. Kern, J.C. Eichstaedt, A. Kapelner, M. Agrawal, et al., Predicting individual well-being through the language of social media, in: Biocomputing 2016: Proceedings of the Pacific Symposium, 2016.
[41] M. De Choudhury, S. Counts, E.J. Horvitz, A. Hoff, Characterizing and predicting postpartum depression from shared Facebook data, in: Proceedings of the 17th ACM Conference on Computer Supported Cooperative Work & Social Computing, 2014.
[42] S. Park, I. Kim, S.W. Lee, J. Yoo, B. Jeong, M. Cha, Manifestation of depression and loneliness on social networks: a case study of young adults on Facebook, in: Proceedings of the 18th ACM Conference on Computer Supported Cooperative Work & Social Computing, 2015.
[43] T. Wang, M. Brede, A. Ianni, E. Mentzakis, Detecting and characterizing eating-disorder communities on social media, in: Proceedings of the Tenth ACM International Conference on Web Search and Data Mining, 2017.
[44] S.R. Braithwaite, C. Giraud-Carrier, J. West, M.D. Barnes, C.L. Hanson, Validating machine learning algorithms for Twitter data against established measures of suicidality, JMIR Mental Health 3 (2) (2016) e4822.
[45] N. Boettcher, Studies of depression and anxiety using Reddit as a data source: scoping review, JMIR Mental Health 8 (11) (2021) e29487.
[46] M. Lv, A. Li, T. Liu, T. Zhu, Creating a Chinese suicide dictionary for identifying suicide risk on social media, PeerJ 3 (2015) e1455.
[47] B. Ferwerda, M. Tkalcic, You are what you post: what the content of Instagram pictures tells about users' personality, in: The 23rd International on Intelligent User Interfaces, March 7–11, Tokyo, Japan, 2018.
[48] G. Coppersmith, M. Dredze, C. Harman, Quantifying mental health signals in Twitter, in: Proceedings of the Workshop on Computational Linguistics and Clinical Psychology: From Linguistic Signal to Clinical Reality, 2014.
[49] D.E. Losada, F. Crestani, A test collection for research on depression and language use, in: International Conference of the Cross-Language Evaluation Forum for European Languages, 2016.
[50] E.A. Ríssola, M. Aliannejadi, F. Crestani, Beyond modelling: understanding mental disorders in online social media, in: European Conference on Information Retrieval, 2020.
[51] G. Coppersmith, M. Dredze, C. Harman, K. Hollingshead, M. Mitchell, CLPsych 2015 shared task: depression and PTSD on Twitter, in: Proceedings of the 2nd Workshop on Computational Linguistics and Clinical Psychology: From Linguistic Signal to Clinical Reality, 2015.
[52] P. Resnik, W. Armstrong, L. Claudino, T. Nguyen, V.-A. Nguyen, J. Boyd-Graber, Beyond LDA: exploring supervised topic modeling for depression-related language in Twitter, in: Proceedings of the 2nd Workshop on Computational Linguistics and Clinical Psychology: From Linguistic Signal to Clinical Reality, 2015.
[53] T. Pedersen, Screening Twitter users for depression and PTSD with lexical decision lists, in: Proceedings of the 2nd Workshop on Computational Linguistics and Clinical Psychology: From Linguistic Signal to Clinical Reality, 2015.
[54] M. De Choudhury, E. Kiciman, M. Dredze, G. Coppersmith, M. Kumar, Discovering shifts to suicidal ideation from mental health content in social media, in: Proceedings of the 2016 CHI Conference on Human Factors in Computing Systems, 2016.
[55] M.M. Tadesse, H. Lin, B. Xu, L. Yang, Detection of depression-related posts in Reddit social media forum, IEEE Access 7 (2019) 44883–44893.
[56] M.L. Kern, G. Park, J.C. Eichstaedt, H.A. Schwartz, M. Sap, L.K. Smith, L.H. Ungar, Gaining insights from social media language: methodologies and challenges, Psychological Methods 21 (4) (2016) 507.
[57] P.A. Cavazos-Rehg, M.J. Krauss, S. Sowles, S. Connolly, C. Rosas, M. Bharadwaj, L.J. Bierut, A content analysis of depression-related tweets, Computers in Human Behavior 54 (2016) 351–357.
[58] Association, A. P., Diagnostic and Statistical Manual of Mental Disorders (DSM-5®), American Psychiatric Pub, 2013.
[59] D.L. Mowery, C. Bryan, M. Conway, Towards developing an annotation scheme for depressive disorder symptoms: a preliminary study using twitter data, in: Proceedings of the 2nd Workshop on Computational Linguistics and Clinical Psychology: From Linguistic Signal to Clinical Reality, 2015.
[60] M. Garg, C. Saxena, V. Krishnan, R. Joshi, S. Saha, V. Mago, B.J. Dorr, CAMS: an annotated corpus for causal analysis of mental health issues in social media posts, arXiv preprint arXiv:2207.04674 (2022).
[61] A. Haque, V. Reddi, T. Giallanza, Deep learning for suicide and depression identification with unsupervised label correction, in: International Conference on Artificial Neural Networks, 2021.
[62] D.E. Losada, F. Crestani, J. Parapar, eRisk 2020: self-harm and depression challenges, in: European Conference on Information Retrieval, 2020.
[63] J.W. Pennebaker, R.L. Boyd, K. Jordan, K. Blackburn, The Development and Psychometric Properties of LIWC2015, 2015.
[64] X. Chen, M. Sykora, T. Jackson, S. Elayan, F. Munir, Tweeting Your Mental Health: An Exploration of Different Classifiers and Features with Emotional Signals in Identifying Mental Health Conditions, 2018.
[65] X. Chen, M.D. Sykora, T.W. Jackson, S. Elayan, What about mood swings: identifying depression on twitter with temporal measures of emotions, Companion Proceedings of the the Web Conference 2018 (2018).
[66] T. Wilson, P. Hoffmann, S. Somasundaran, J. Kessler, J. Wiebe, Y. Choi, et al., OpinionFinder: a system for subjectivity analysis, in: Proceedings of HLT/EMNLP 2005 Interactive Demonstrations, 2005.
[67] M. Thelwall, K. Buckley, G. Paltoglou, D. Cai, A. Kappas, Sentiment strength detection in short informal text, Journal of the American Society for Information Science and Technology 61 (12) (2010) 2544–2558.

[68] J. Bollen, B. Gonçalves, G. Ruan, H. Mao, Happiness is assortative in online social networks, Artificial Life 17 (3) (2011) 237–251.
[69] A.O. Durahim, M. Coşkun, # iamhappybecause: Gross National Happiness through Twitter analysis and big data, Technological Forecasting and Social Change 99 (2015) 92–105.
[70] D.M. Blei, A.Y. Ng, M.I. Jordan, Latent dirichlet allocation, Journal of Machine Learning Research 3 (2003) 993–1022.
[71] S. Ji, C.P. Yu, S.-f. Fung, S. Pan, G. Long, Supervised learning for suicidal ideation detection in online user content, Complexity 2018 (2018).
[72] M.J. Paul, M. Dredze, You are what you tweet: analyzing twitter for public health, in: Fifth International AAAI Conference on Weblogs and Social Media, 2011.
[73] C.E.H. Chua, V.C. Storey, X. Li, M. Kaul, Developing insights from social media using semantic lexical chains to mine short text structures, Decision Support Systems 127 (2019) 113142.
[74] C. Cortes, V. Vapnik, Support-vector networks, Machine Learning 20 (3) (1995) 273–297.
[75] A. Wongkoblap, Multiple Instance Learning for Detecting Depression Markers in Social Media Content, 2020.
[76] J. Du, Y. Zhang, J. Luo, Y. Jia, Q. Wei, C. Tao, H. Xu, Extracting psychiatric stressors for suicide from social media using deep learning, BMC Medical Informatics and Decision Making 18 (2) (2018) 43.
[77] L. Ma, Z. Wang, Y. Zhang, Extracting depression symptoms from social networks and web blogs via text mining, in: International Symposium on Bioinformatics Research and Applications, 2017.
[78] S. Kiritchenko, X. Zhu, S.M. Mohammad, Sentiment analysis of short informal texts, Journal of Artificial Intelligence Research 50 (2014) 723–762.
[79] C. Margus, N. Brown, A.J. Hertelendy, M.R. Safferman, A. Hart, G.R. Ciottone, Emergency physician Twitter use in the COVID-19 pandemic as a potential predictor of impending surge: retrospective observational study, Journal of Medical Internet Research 23 (7) (2021) e28615.
[80] C.Y. Chiu, H.Y. Lane, J.L. Koh, A.L. Chen, Multimodal depression detection on instagram considering time interval of posts, Journal of Intelligent Information Systems (2020) 1–23.
[81] A. Kumar, G. Garg, Sentiment analysis of multimodal twitter data, Multimedia Tools and Applications 78 (17) (2019) 24103–24119.
[82] A.G. Reece, C.M. Danforth, Instagram photos reveal predictive markers of depression, EPJ Data Science 6 (1) (2017) 1–12.
[83] S.C. Guntuku, D. Preotiuc-Pietro, J.C. Eichstaedt, L.H. Ungar, What twitter profile and posted images reveal about depression and anxiety, in: Proceedings of the International AAAI Conference on Web and Social Media, 2019.
[84] K. Simonyan, A. Zisserman, Very deep convolutional networks for large-scale image recognition, in: International Conference on Learning Representations, 2015.
[85] K. Mao, J. Niu, H. Chen, L. Wang, M. Atiquzzaman, Mining of marital distress from microblogging social networks: a case study on Sina Weibo, Future Generation Computer Systems 86 (2018) 1481–1490.
[86] S. Park, S.W. Lee, J. Kwak, M. Cha, B. Jeong, Activities on Facebook reveal the depressive state of users, Journal of Medical Internet Research 15 (10) (2013) e2718.
[87] M. De Choudhury, M. Gamon, S. Counts, E. Horvitz, Predicting depression via social media, Icwsm 13 (2013) 1–10.
[88] V.M. Prieto, S. Matos, M. Alvarez, F. Cacheda, J.L. Oliveira, Twitter: a good place to detect health conditions, PLoS One 9 (1) (2014) e86191.
[89] Q. Hu, A. Li, F. Heng, J. Li, T. Zhu, Predicting depression of social media user on different observation windows, in: 2015 IEEE/WIC/ACM International Conference on Web Intelligence and Intelligent Agent Technology (WI-IAT), 2015.
[90] H. Lin, J. Jia, Q. Guo, Y. Xue, J. Huang, L. Cai, L. Feng, Psychological stress detection from cross-media microblog data using deep sparse neural network, in: 2014 IEEE International Conference on Multimedia and Expo (ICME), 2014.
[91] G. Bonaccorso, Machine Learning Algorithms, Packt Publishing Ltd, 2017.
[92] P. Burnap, W. Colombo, J. Scourfield, Machine classification and analysis of suicide-related communication on twitter, in: Proceedings of the 26th ACM Conference on Hypertext & Social Media, 2015.
[93] X. Huang, L. Zhang, D. Chiu, T. Liu, X. Li, T. Zhu, Detecting suicidal ideation in Chinese microblogs with psychological lexicons, in: 2014 IEEE 11th International Conference on Ubiquitous Intelligence and Computing and 2014 IEEE 11th International Conference on Autonomic and Trusted Computing and 2014 IEEE 14th International Conference on Scalable Computing and Communications and Its Associated Workshops, 2014.
[94] M.R. Islam, M.A. Kabir, A. Ahmed, A.R.M. Kamal, H. Wang, A. Ulhaq, Depression detection from social network data using machine learning techniques, Health Information Science and Systems 6 (1) (2018) 1–12.
[95] M. De Choudhury, S. Counts, E. Horvitz, Social media as a measurement tool of depression in populations, in: Proceedings of the 5th Annual ACM Web Science Conference, 2013.
[96] D. Preoţiuc-Pietro, M. Sap, H.A. Schwartz, L. Ungar, Mental illness detection at the World Well-Being Project for the CLPsych 2015 shared task, in: Proceedings of the 2nd Workshop on Computational Linguistics and Clinical Psychology: From Linguistic Signal to Clinical Reality, 2015.
[97] S. Tsugawa, Y. Kikuchi, F. Kishino, K. Nakajima, Y. Itoh, H. Ohsaki, Recognizing depression from twitter activity, in: Proceedings of the 33rd Annual ACM Conference on Human Factors in Computing Systems, 2015.
[98] G. Coppersmith, K. Ngo, R. Leary, A. Wood, Exploratory analysis of social media prior to a suicide attempt, in: Proceedings of the Third Workshop on Computational Linguistics and Clinical Psychology, 2016.
[99] T. Nguyen, D. Phung, B. Dao, S. Venkatesh, M. Berk, Affective and content analysis of online depression communities, IEEE Transactions on Affective Computing 5 (3) (2014) 217–226.
[100] Z. Yin, L.M. Sulieman, B.A. Malin, A systematic literature review of machine learning in online personal health data, Journal of the American Medical Informatics Association 26 (6) (2019) 561–576.

[101] J. Kim, J. Lee, E. Park, J. Han, A deep learning model for detecting mental illness from user content on social media, Scientific Reports 10 (1) (2020) 1–6.
[102] A. Wongkoblap, M.A. Vadillo, V. Curcin, Researching mental health disorders in the era of social media: systematic review, Journal of Medical Internet Research 19 (6) (2017) e7215.
[103] E. Garcia-Ceja, V. Osmani, O. Mayora, Automatic stress detection in working environments from smartphones' accelerometer data: a first step, IEEE Journal of Biomedical and Health Informatics 20 (4) (2015) 1053–1060.
[104] Q. Xu, T.L. Nwe, C. Guan, Cluster-based analysis for personalized stress evaluation using physiological signals, IEEE Journal of Biomedical and Health Informatics 19 (1) (2014) 275–281.
[105] M. Bayer, M.-A. Kaufhold, C. Reuter, Information Overload in Crisis Management: Bilingual Evaluation of Embedding Models for Clustering Social Media Posts in Emergencies, ECIS, 2021.
[106] T. Mikolov, K. Chen, G. Corrado, J. Dean, Efficient estimation of word representations in vector space, arXiv preprint arXiv:1301.3781 (2013).
[107] X. Zhu, A.B. Goldberg, Introduction to semi-supervised learning, Synthesis lectures on artificial intelligence and machine learning 3 (1) (2009) 1–130.
[108] S. Tariq, N. Akhtar, H. Afzal, S. Khalid, M.R. Mufti, S. Hussain, et al., A novel co-training-based approach for the classification of mental illnesses using social media posts, IEEE Access 7 (2019) 166165–166172.
[109] T. Gui, Q. Zhang, L. Zhu, X. Zhou, M. Peng, X. Huang, Depression Detection on Social Media with Reinforcement Learning, China National Conference on Chinese Computational Linguistics, 2019.
[110] D. Maupomé, M.-J. Meurs, Using topic extraction on social media content for the early detection of depression, CLEF (Working Notes) (2018) 2125.
[111] K. Cho, B. Van Merriënboer, C. Gulcehre, D. Bahdanau, F. Bougares, H. Schwenk, Y. Bengio, Learning phrase representations using RNN encoder-decoder for statistical machine translation, arXiv preprint arXiv:1406.1078 (2014).
[112] S. Paul, S.K. Jandhyala, T. Basu, Early detection of signs of anorexia and depression over social media using effective machine learning frameworks, CLEF (Working Notes) (2018).
[113] F. Sadeque, D. Xu, S. Bethard, UArizona at the CLEF eRisk 2017 pilot task: linear and recurrent models for early depression detection, CEUR Workshop Proceedings (2017).
[114] M. Trotzek, S. Koitka, C.M. Friedrich, Word embeddings and linguistic metadata at the CLEF 2018 tasks for early detection of depression and anorexia, CLEF (Working Notes) (2018).
[115] A.H. Orabi, P. Buddhitha, M.H. Orabi, D. Inkpen, Deep learning for depression detection of twitter users, in: Proceedings of the Fifth Workshop on Computational Linguistics and Clinical Psychology: From Keyboard to Clinic, 2018.
[116] Y. Wang, Z. Wang, C. Li, Y. Zhang, H. Wang, A multitask deep learning approach for user depression detection on Sina Weibo, arXiv preprint arXiv:2008.11708 (2020).
[117] G. Gkotsis, A. Oellrich, S. Velupillai, M. Liakata, T.J. Hubbard, R.J. Dobson, R. Dutta, Characterisation of mental health conditions in social media using Informed Deep Learning, Scientific Reports 7 (1) (2017) 1–11.
[118] C. Su, Z. Xu, J. Pathak, F. Wang, Deep learning in mental health outcome research: a scoping review, Translational Psychiatry 10 (1) (2020) 1–26.
[119] J. Ive, G. Gkotsis, R. Dutta, R. Stewart, S. Velupillai, Hierarchical neural model with attention mechanisms for the classification of social media text related to mental health, in: Proceedings of the Fifth Workshop on Computational Linguistics and Clinical Psychology: From Keyboard to Clinic, 2018.
[120] H. Lin, J. Jia, Q. Guo, Y. Xue, Q. Li, J. Huang, et al., User-level psychological stress detection from social media using deep neural network, in: Proceedings of the 22nd ACM International Conference on Multimedia, 2014.
[121] H. Lin, J. Jia, J. Qiu, Y. Zhang, G. Shen, L. Xie, et al., Detecting stress based on social interactions in social networks, IEEE Transactions on Knowledge and Data Engineering 29 (9) (2017) 1820–1833.
[122] S.J. Pan, Q. Yang, A survey on transfer learning, IEEE Transactions on Knowledge and Data Engineering 22 (10) (2009) 1345–1359.
[123] K. Pogorelov, M. Riegler, S.L. Eskeland, T. de Lange, D. Johansen, C. Griwodz, et al., Efficient disease detection in gastrointestinal videos–global features versus neural networks, Multimedia Tools and Applications 76 (21) (2017) 22493–22525.
[124] E. Garcia-Ceja, M. Riegler, T. Nordgreen, P. Jakobsen, K.J. Oedegaard, J. Tørresen, Mental health monitoring with multimodal sensing and machine learning: a survey, Pervasive and Mobile Computing 51 (2018) 1–26.
[125] E. Garcia-Ceja, R. Brena, Building personalized activity recognition models with scarce labeled data based on class similarities, in: International Conference on Ubiquitous Computing and Ambient Intelligence, 2015.
[126] A. Botchkarev, Performance metrics (error measures) in machine learning regression, forecasting and prognostics: properties and typology, arXiv preprint arXiv:1809.03006 (2018).
[127] Statistics, N. C. f. H., Administration, U. S. H. C. F., The International Classification of Diseases, 9th Revision, Clinical Modification: Diseases: Alphabetic Index vol. 2, US Department of Health and Human Services, Public Health Service, Health …, 1980.
[128] K.A. Arditte, D. Çek, A.M. Shaw, K.R. Timpano, The importance of assessing clinical phenomena in Mechanical Turk research, Psychological Assessment 28 (6) (2016) 684.
[129] M. De Choudhury, Role of social media in tackling challenges in mental health, in: Proceedings of the 2nd International Workshop on Socially-Aware Multimedia, 2013.
[130] J.N. Rosenquist, J.H. Fowler, N.A. Christakis, Social network determinants of depression, Molecular Psychiatry 16 (3) (2011) 273–281.
[131] S.M. Rice, J. Goodall, S.E. Hetrick, A.G. Parker, T. Gilbertson, G.P. Amminger, et al., Online and social networking interventions for the treatment of depression in young people: a systematic review, Journal of Medical Internet Research 16 (9) (2014) e3304.
[132] S.J. Luna Ansari, Q. Chen, E. Cambria, P.M. Pekka, Ensemble hybrid learning methods for automated depression detection, IEEE Transactions on Computational Social Systems (2022).

Chapter 22

Applied picture fuzzy sets with its picture fuzzy database for identification of patients in a hospital

Van Hai Pham[1], Quoc Hung Nguyen[2], Kim Phung Thai[2] and Le Phuc Thinh Tran[2]
[1]*Hanoi University of Science and Technology, Hanoi, Viet Nam;* [2]*University of Economics, Ho Chi Minh City (UEH), Ho Chi Minh City, Viet Nam*

1. Introduction

Fuzzy set theory (FST) was developed by Zadeh in 1965 [1]. It plays an important role in decision-making in an uncertain environment. FST methods have been made and applied in many circumstances related to real-world problems. For instance, intuitionistic fuzzy set (IFS), introduced by Atanassov in 1986 [2], was explained as the generalization of a fuzzy set. On the one hand, a fuzzy set shows the rate of membership of a set in an element; however, an intuitionistic fuzzy set shows both a membership degree and a nonmembership degree of an element in a specific set. In 2013, Cuong et al. [3] introduced the picture fuzzy set, containing three levels membership for each element of a particular set: positive, negative, and neutral. Picture fuzzy set (PFS) has been used widely in various applications such as: weather forecasting and health prediction. Furthermore, PFS can be presented for the quantification of uncertain information while identifying objectives with multiple features in an uncertain environment.

Tanaka et al. [4] investigated a picture fuzzy database, and represented members combined directly to a single relationship. Picture fuzzy database is the generalization of a classically relational database since it allows uncertain information to be demonstrated and controlled. In addition, a relational database presents a table, with field (columns), and record (rows). Field (columns) is an attribute of entity, while record (rows) displays the information of each entity. Roy et al. [5] introduced a new concept called intuitionistic fuzzy database, where the relationship between values of any attribute in the relational database have an intuitionistic fuzzy relationship. Chunxin et al. [6] proposed some new operations and basic properties of picture fuzzy relations that have been intensively studied, and a new inclusion relation (called type-2 inclusion relation) of picture fuzzy relations. Dutta et al. [7] investigated a picture fuzzy set which is applied to an application domain by dealing with uncertainty which is a direct extension of an intuitionistic fuzzy set. Furthermore, PFS can be presented under uncertainty in such responses as: yes, abstain, and no. The basics of fuzzy set, relational database, and fuzzy database can be represented as a picture fuzzy database that inherits their features as it is an extension of these concepts. To identify noises of human including hair coverage, hair texture, face shape, and skin color in various data types, the PFS approach is used combined with a criminal database for human identification.

In related works [8—10], applied PFSs were used for group decision-making in the evaluation of learners' behaviors and tracking people using historical learning resources [8]. PFS is also applied to decision-making for quantifying sensibilities together with human decisions in an uncertain environment [9,10]. To identify humans in an uncertain environment such as forests, rule area, and fire disasters, researchers have applied rule-based methods for decision supports [11]. Recently, a fuzzy knowledge graph using fuzzy rules in pairs has been used to aid decision-making in human identification in healthcare domains [12]. Lan et al. investigated a fuzzy knowledge graph to ensure the right decision-making in a human profile in the healthcare domain [13]. In large digital society data sets, studies [14] using knowledge graph and graph deep learning [15] for improvement of recognition of user behavior in the digital society has been attempted. The forensic science approach has been applied to find a person in the American criminal justice system [16]. An investigation used fuzzy-rough uncertainty measure [17] and picture fuzzy sets with unsupervised training [18] to

Deep Learning in Personalized Healthcare and Decision Support. https://doi.org/10.1016/B978-0-443-19413-9.00011-4

quantify human features in order to find the matched person with COVID-19. Other studies used type-2 fuzzy temporal integrated with convolutional model for improvement of human identification in an uncertain environment [19]. Some studies [20,21] used spherical fuzzy sets, a context matching algorithm, and decision-making [22,23] to quantify an uncertain environment to make the right decisions. Identifying human records in a framework of big data [24] and big data in health care 4.0 [25] are also significant in current research. All of these studies have considered partially solutions for human identification such as lack of information, insufficient feacture, and sensing data.

This chapter presents picture fuzzy sets based on human features, combined with a database of personal attributes. First, the picture fuzzy set is an extension of the fuzzy and intuitionistic fuzzy set combined with a picture fuzzy database aiming to improve human recognition in an uncertain environment. Second, the picture fuzzy database stores all the human features while developing an application. The final stage is implemented as an application to test human recognition together with their identification. The technical contributions in this study include uncertain information of people for the identification of patients in a case study in a hospital.

2. Research background

2.1 Terms and definitions

Let U be a nonempty set which is called the universe set. Then $P(U)$ represents a class of all subsets of U and $F(U)$ represents a class of all fuzzy subsets of U, respectively.

Definition 22.1. A picture fuzzy set A on a universe set U is defined as follows.

$$A = \{(x, \mu_A(x), \eta_A(x), \gamma_A(x) | x \varepsilon U\} \tag{22.1}$$

where:

- $\mu_A(x)\varepsilon[0, 1]$: a positive membership degree of x in A
- $\eta_A(x)\varepsilon[0, 1]$: a neutral membership degree of x in A
- $\gamma_A(x)\varepsilon[0, 1]$: a negative membership degree of x in A

And $\mu_A(x)$, $\eta_A(x)$, and $\gamma_A(x)$ should be satisfied with the condition following.

$$\mu_A(x) + \eta_A(x) + \gamma_A(x) \le 1, \forall x \in X \tag{22.2}$$

The member of all picture fuzzy sets in U is denoted as PFS(U). The complement of a picture fuzzy set is presented as follows:

$$A = \{(x, \gamma_A(x), \eta_A(x), \mu_A(x)) | x \in U\} \tag{22.3}$$

Normally, a picture fuzzy set represents three fuzzy sets: $\mu_A : U \rightarrow [0, 1]$, $\eta_A : U \rightarrow [0, 1]$, and $\gamma_A : U \rightarrow [0, 1]$ that are associated as $(\mu_A, \eta_A, \gamma_A)$. $A = \{(x, \mu_A(x), \gamma_A(x))\}$ is an intuitionistic fuzzy set which can be defined by a picture fuzzy set as $A = \{(x, \gamma_A(x), 0, \mu_A(x)) | x \varepsilon U\}$.

Operator **PFS(U)** was introduced: $\forall A, B \in PFS(U)$, we then have:

- $A \subseteq B$ if $\mu_A(x) \le \mu_B(x), \eta_A(x) \le \eta_B(x)$ and $\gamma_A(x) \ge \gamma_B(x) \forall x \in U$
- $A = B$ if $A \subseteq B$ and $B \subseteq A$
- $A \cup B = \{(x, \max(\mu_A(x), \mu_B(x)), \min(\eta_A(x), \eta_B(x), \min(\gamma_A(x), \gamma_B(x)))) | x \varepsilon U\}$
- $A \cap B = \{(x, \min(\mu_A(x), \mu_B(x)), \min(\eta_A(x), \eta_B(x), \max(\gamma_A(x), \gamma_B(x)))) | x \varepsilon U\}$

Note that some special picture fuzzy sets are as follows: a constant picture fuzzy set is a picture fuzzy set in which $\left(\widehat{\alpha, \beta, \theta}\right) = \{(x, \alpha, \beta, \theta) | x \varepsilon U\}$; a picture fuzzy universe set is defined by $U = 1_U = \left(\widehat{1, 0, 0}\right) = \{(x, 1, 0, 0) | x \varepsilon U\}$, and the picture fuzzy empty set is $\phi = 0_U = \left(\widehat{0, 1, 0}\right) = \{(x, 0, 1, 0) | x \varepsilon U\}$.

For any $x \in U$, picture fuzzy sets 1_x and $1_{U-\{x\}}$, respectively, are defined by:

$$\mu_{1_x}(y) = \begin{cases} 1, \text{if } y = x \\ 0, \text{if } y \ne x \end{cases}$$

$$\gamma_{1_x}(y) = \begin{cases} 0, \text{if } y = x \\ 1, \text{if } y \neq x \end{cases}$$

$$\eta_{1_x}(y) = \begin{cases} 0, \text{if } y = x \\ 0, \text{if } y \neq x \end{cases}$$

$$\mu_{1_{U-\{x\}}}(y) = \begin{cases} 0, \text{if } y = x \\ 1, \text{if } y \neq x \end{cases}$$

$$\gamma_{1_{U-\{x\}}}(y) = \begin{cases} 1, \text{if } y = x \\ 0, \text{if } y \neq x \end{cases}$$

$$\eta_{1_{U-\{x\}}}(y) = \begin{cases} 0, \text{if } y = x \\ 0, \text{if } y \neq x \end{cases}$$

Definition 22.2. Let **U** be an infinite nonempty universe set. A relation of the picture fuzzy set (PFS) from **U** to **V** represents a PFS of **U × V**, denoted by $R(U \rightarrow V)$, which is a function given by:

$$R = \{((x, y), \mu_R(x, y), \eta_R(x, y), \gamma_R(x, y)) | (x, y) \in U \times V\} \tag{22.4}$$

where μ_R, η_R, γ_R are relations in the range [0,1] from $U \times V$ to that of:

$$\mu_R(x, y) + \eta_R(x, y) + \gamma_R(x, y) \leq 1 \forall (x, y) \in U \times V \tag{22.5}$$

While $U \equiv V$, we have $R(U \rightarrow V)$, called the picture fuzzy relation.

Definition 22.3. We have $P(U \rightarrow V)$ and $Q(U \rightarrow V)$. The max−min picture fuzzy relation **P** with **Q** picture fuzzy relation can be presented by $\boldsymbol{P} \circ \boldsymbol{Q}$, called the picture fuzzy relation on **U × W**, expressed for all by $(x, z) \in U \times W$:

$$\mu_{P \circ Q}(x, z) = \max_{y \in V}\{\min\{\mu_P(x, y), \mu_Q(y, z)\}\} \tag{22.6}$$

$$\eta_{P \circ Q}(x, z) = \min_{y \in V}\{\min\{\eta_P(x, y), \eta_Q(y, z)\}\} \tag{22.7}$$

$$\gamma_{P \circ Q}(x, z) = \min_{y \in V}\{\max\{\gamma_P(x, y), \gamma_Q(y, z)\}\} \tag{22.8}$$

Definition 22.4. A picture fuzzy relation **R** on **U** is:

- if $\forall x \varepsilon U, \mu_R(x, x) = 1$ called reflexive.
- if $\forall x, y \, \varepsilon \, U, \mu_R(x, y) = \mu_R(y, x), \eta_R(x, y) = \eta_R(y, x), \gamma_R(x, y) = \gamma_R(y, x)$ called symmetric.
- if $R^2 \subset R$, where $R^2 = R \circ R$ called transitive.
- If R is reflexive and symmetric it is called a picture tolerance.
- If R is reflexive and transitive it is called a picture preorder.
- If R is reflexive, symmetric, and transitive, it is called as picture similarity (picture fuzzy equivalence).

2.2 Picture fuzzy database for applications

A criminal patient database called picture fuzzy database is represented by tables, where each entity is a table, containing multiple rows (records, rows, or tuples) and columns (columns, fields, or domains), where columns correspond to attributes of entity, while rows display detailed information about each individual on that entity.

The proposed model is applied to use a picture fuzzy database which represents relation R which is a subset expressed by:

$$2^{D_1} \times 2^{D_2} \times \ldots \times 2^{D_m} \tag{22.9}$$

Thus, $R \subset 2^{D_1} \times 2^{D_2} \times \ldots \times 2^{D_m}$ presents a picture fuzzy database relation. A picture fuzzy tuple is an element of R.
A picture fuzzy tuple $t_i = (d_{i1}, d_{i2}, \ldots, d_{im})$, where $d_{ij} \in D_j$.
Let $\theta = (a_1, a_2, \ldots, a_m)$, where $a_i \in d_{ij}$ for each domain D_j.
For each domain D_j, if R_j is a picture fuzzy database, tolerance in relationships is defined as:

- $\mu_{R_j} : D_j \times D_j \rightarrow [0, 1]$ for positive membership degree
- $\eta_{R_j} : D_j \times D_j \rightarrow [0, 1]$ for neutral membership
- $\gamma_{R_j} : D_j \times D_j \rightarrow [0, 1]$ for negative membership degree.

Procedure for picture fuzzy tolerance relation R
Input: + Let vector U represent (μ,η,ν)
+ Let vector V represent (μ,η,ν)
Output: A matrix in a fuzzy relation
```
Begin:
k = 0, l = 0
Initialize matrix R
For (u ∈ U)
For (v ∈ V)
If (k==l) R[k][l] = (1,0,0);
Else R[k][l] = Smax(n(u),v);
   l++;
  end for
k++;
end for
return R;
End.
```
Procedure for inference matrix relation
Input: + relation R
+ a vector U′ represents (μ,η,ν) of universe X in linguistic term.
Output: a vector V′ containing (μ,η,ν) of each element
```
Begin
Let c = No. column in R
Let r = No. row in R
  For i = 0 to c do
    For j = 0 to r do
  V′ [j] = U′∘R = ∨_i (U′ [i]∧R[i][j]))
    End for
End for
Return V′
End.
```
Satisfy with: $\mu_{R_j}(x, y) + \eta_{R_j}(x, y) + \gamma_{R_j}(x, y) \leq 1, (x, y) \in D_j \times D_j$

3. The proposed model

3.1 Process steps in a summary of the proposed model

The process steps in a summary of the proposed model are as follows:

- **Step 1.** Create a table with all needed information, including attributes and records.
- **Step 2.** For each attribute, create a picture fuzzy tolerance relation whose table has the first row and column being the values in its domain, as described in Section 2.2.
- **Step 3.** From the tables in step 2, calculate the clustering on each attribute (suppose that each attribute is a universe set).
- **Step 4.** Give the final result based on step 3.

3.2 The proposed model and its application

In this section, we have applied the theory in a specific real-world problem which is investigating patient behavior. As clinic's claim to a patient, who has a strange behavior in a hospital. While carrying out an investigation, an inspector tries to recognize a patient who has different "hair color," "hair style," and "body," as shown in Table 22.1. It considers the picture fuzzy tolerance relation R1 on the attribute domain "hair color" in Table 22.2, and similarly R2 on domain "hair style" in Table 22.3 and R3 on domain "body" in Table 22.4.

TABLE 22.1 Patient information (brief patient data).

Name	Hair style	Hair color	Body
Hung	Stc.	Full small (FS)	Large
Ban	Short	Rec.	Very small (VS)
Chien	Straight (Str.)	Full small (FS)	Small (S)
Dita	Curly	Beauty	Medium (M)
Elsa	Short	Beauty	Medium (M)
Minh	Stc.	Full big (FB)	Very large (VL)
Quynh	Straight (Str.)	Full small (FS)	Small (S)
Hong	Curly	Rec.	Medium (M)

TABLE 22.2 The weights of picture fuzzy tolerance relation R_1 "hair color."

R_1	FB	Rec.	Beauty	FS
FB	(1,0,0)	(0.4,0.1,0.4)	(0,0,1)	(0.8,0.1,0.1)
FS	(0.8,0.1,0.1)	(0.5,0.1,0.4)	(0,0.1,0.9)	(1,0,0)
Rec.	(0.4,0.1,0.4)	(1,0,0)	(0.4,0.1,0.4)	(0.5,0.1,0.4)
Beauty	(0,0,1)	(0.4,0.1,0.4)	(1,0,0)	(0,0.1,0.9)

TABLE 22.3 The weights of picture fuzzy tolerance relation R_2 "hair style."

R_2	Short	Curly	Stc.	Str.
Str.	(0.1,0.1,0.7)	(0.1,0,0.7)	(0.6,0.1,0.3)	(1,0,0)
Stc.	(0.3,0.1,0.4)	(0.5,0.1,0.2)	(1,0,0)	(0.6,0.1,0.3)
Short	(1,0,0)	(0.4,0.1,0.4)	(0.5,0.1,0.4)	(0.1,0.1,0.7)
Curly	(0.4,0.1,0.4)	(1,0,0)	(0.5,0.1,0.2)	(0.1,0,0.7)

TABLE 22.4 The weights of picture fuzzy tolerance relation R_3 "body."

R_3	VL	VS	M	L	S
VL	(1,0,0)	(0,0,1)	(0.4,0.1,0.4)	(0.7,0.1,0.2)	(0.3,0.1,0.6)
L	(0.7,0.1,0.2)	(0,1,0.9)	(0.5,0.1,0.4)	(1,0,0)	(0.4,0,0.5)
A	(0.5,0.1,0.4)	(0.3,0.1,0.6)	(1,0,0)	(0.5,0.1,0.4)	(0.5,0.1,0.3)
S	(0.3,0.1,0.6)	(0.7,0.1,0.2)	(0.5,0.1,0.3)	(0.4,0,0.5)	(1,0,0)
VS	(0,0,1)	(1,0,0)	(0.3,0.1,0.6)	(0,1,0.9)	(0.7,0.1,0.2)

- **Step 1: Create an information table about a particular problem, consisting of attributes and records.**
  As described in Table 22.1, this is depicted for all suspected patients.
- **Step 2: Create a picture fuzzy set in the relation which is stated to the table in its domain.**
  Apply the knowledge of a picture fuzzy set in relation to the attribute domain "hair color" in Table 22.2, "hair style" in Table 22.3, and "body" in Table 22.4.
- **Step 3: Calculate the clustering on all attributes.**
  From the tables in step 2, calculate the clustering on all attributes (suppose that each attribute is a universe set). From Tables 22.1–22.4, we can identify the criminal. Since there is a tolerance, it is also reflexive and symmetric (definition 4), we have just considered the upper half of the diagonal from top to bottom, and left to right in Tables 22.2–22.4, respectively.
  First, it considers Table 22.2: this is a picture fuzzy in relation R_1 on the attribute domain "hair color."
  - For Table 22.2, we use definition 5, 6, 7. For $\alpha = 0.8$. We have $\gamma_{R_1}(x,y) \le 1 - \alpha = 1 - 0.8 = 0.2$. Therefore, we will have:

    $$\{(FB, FB), (FS, FS), (Rec., Rec.), (Bald, Bald), (FB, FS)\}$$

    From there we can take the partition $R_1 = \{FB,\ FS,\ Rec.,\ Beauty\}$ as:

    $$\{\{FB, FS\}, \{Rec.\}, \{Bald\}\}$$

  - Do the same thing with Table 22.3 as Table 22.2 and we have $\alpha = 0.8$. So we have $\gamma_{R_1}(x,y) \le 1 - \alpha = 1 - 0.8 = 0.2$. Therefore, we will have these pairs:

    $$\{(Str., Str.), (Stc., Stc.,), (Wavy, Wavy), (Curly, Curly), (Stc., Curly)\}$$

    From there we can take the partition R_2 as:

    $$\{\{Str.\}, \{Stc., Curly\}, \{Wavy\}\}$$

  - Similarly with Table 22.4 for $\alpha = 0.7$ we have the partition R_3:

    $$\{\{VL, L\}, \{A, S, VS\}\}$$

- **Step 4: Give the final result based on step 3.**
  From the results in all tables and people's descriptions, the patient has: *more or less full big hair color, more or less curly hair style, moderately large body.*
  We can choose attributes based on the results mentioned above.
  - For Table 22.2: $\{FB, FS\}$
  - For Table 22.3: $\{Stc., Curly\}$
  - For Table 22.4: $\{VL, L\}$

From the above result, combining with Table 22.1, we can find two of the most suspected patients among them all (Table 22.5).

In summary, it is much easier for the police's investigation because they only need to question two people instead of all of those present.

4. Experimental results

4.1 Case study of a criminal investigation problem

The input data contain suspected criminal information and attributes: name, body, hair coverage, hair texture, face shape, and skin color in thousands of records. To find a criminal in a database, we can consider the list of all suspected people. Let

TABLE 22.5 Relation "find the right patient".

Name	Hair color	Body	Hair color	Hair style
{Hung, Minh}ti	{Full Big, Full Small}	{Large, Very Large}	{Full Big, Full Small}	{Curly, Stc.}

a witness select all features of the suspect that they think are similar for the evaluation of each attribute for each feature. Then press the search button, the program runs and shows a list of suspects in the Suspect box on the screen, as shown in Fig. 22.1.

As shown in Fig. 22.1, all of those on the list of suspected patients in the hospital have unusual behaviors. Based on the features of human descriptions, the proposed system automatically recognizes the correct patients in the list. Hence, it is easy for the decision maker to select the correct patient from the list. In addition, the decision maker can freely select the other attributes of patients with priority in the list.

4.2 Evaluation

To evaluate our proposed system and establish a performance study, we tested simulations of the proposed model performance with respect to expert responses in criminal persons in a case study of a hospital for the list, as shown in Fig. 22.2.

We selected nine persons for testing of the proposed model. In the evaluation, the feedbacks of experts were calculated from the correct responses. Fig. 22.2 shows the correct answers of the proposed model as compared to the experts' responses.

In the evaluation, the proposed model has initially been implemented only partially to evaluate the identifying persons with respect to the correct answers. In the simulation results, the proposed model has figured out in large data sets in data-driven in decision making while tracking patient behaviors in real-time of the hospital domain. This supports the hospital in finding quickly the correct person who matches the description in the warning. Furthermore, it is also applied to find the HIS (Health Information System) in a hospital in order to manage patients by tracking them at any time and any location. It is indicated that the proposed system is able to recognize patients/persons correctly in an uncertain environment for the hospital domain.

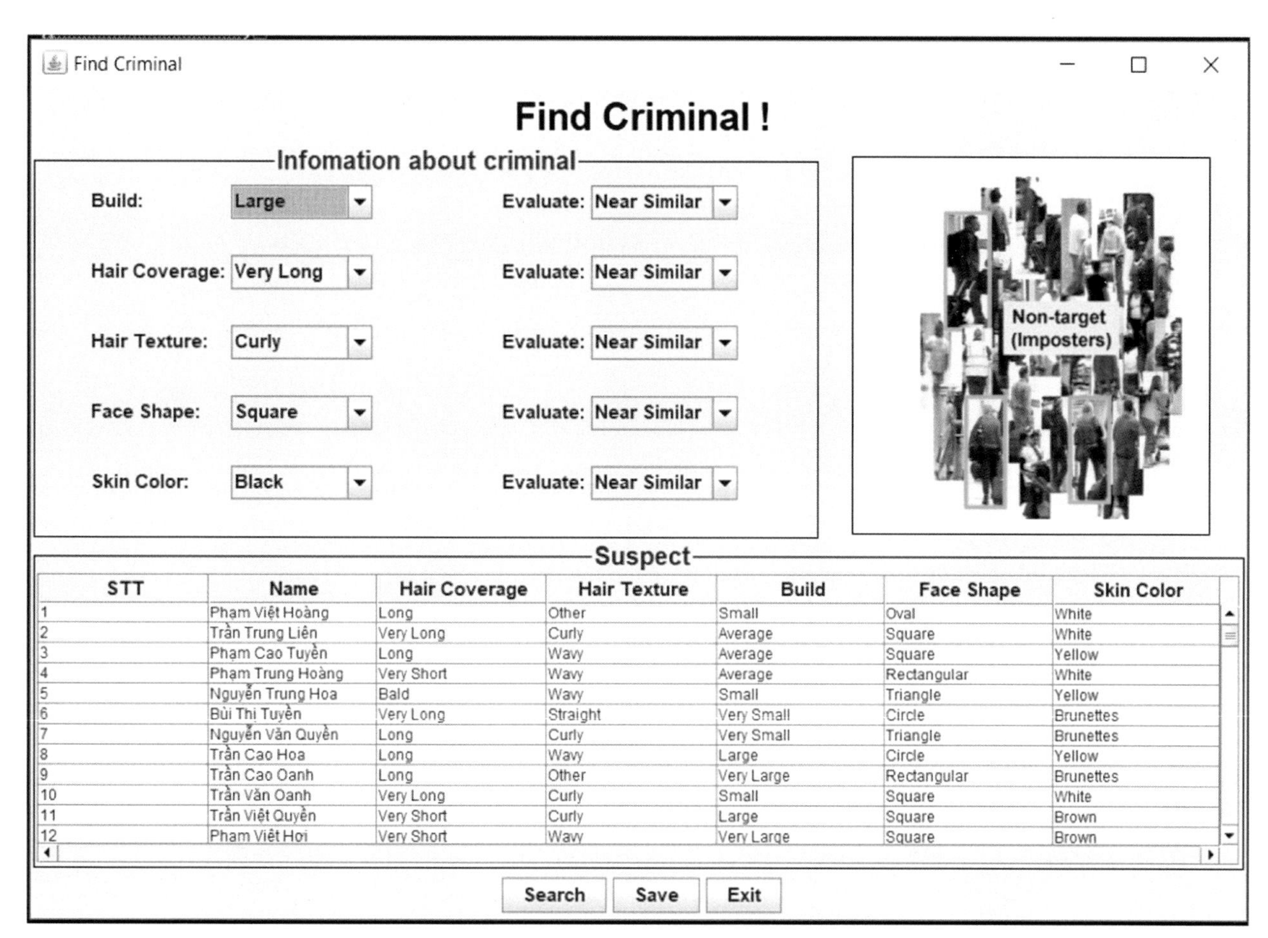

STT	Name	Hair Coverage	Hair Texture	Build	Face Shape	Skin Color
1	Phạm Việt Hoàng	Long	Other	Small	Oval	White
2	Trần Trung Liên	Very Long	Curly	Average	Square	White
3	Phạm Cao Tuyền	Long	Wavy	Average	Square	Yellow
4	Phạm Trung Hoàng	Very Short	Wavy	Average	Rectangular	White
5	Nguyễn Trung Hoa	Bald	Wavy	Small	Triangle	Yellow
6	Bùi Thị Tuyền	Very Long	Straight	Very Small	Circle	Brunettes
7	Nguyễn Văn Quyền	Long	Curly	Very Small	Triangle	Brunettes
8	Trần Cao Hoa	Long	Wavy	Large	Circle	Yellow
9	Trần Cao Oanh	Long	Other	Very Large	Rectangular	Brunettes
10	Trần Văn Oanh	Very Long	Curly	Small	Square	White
11	Trần Việt Quyền	Very Short	Curly	Large	Square	Brown
12	Phạm Việt Hơi	Very Short	Wavy	Very Large	Square	Brown

FIGURE 22.1 The screen of criminal persons as provided by the proposed system.

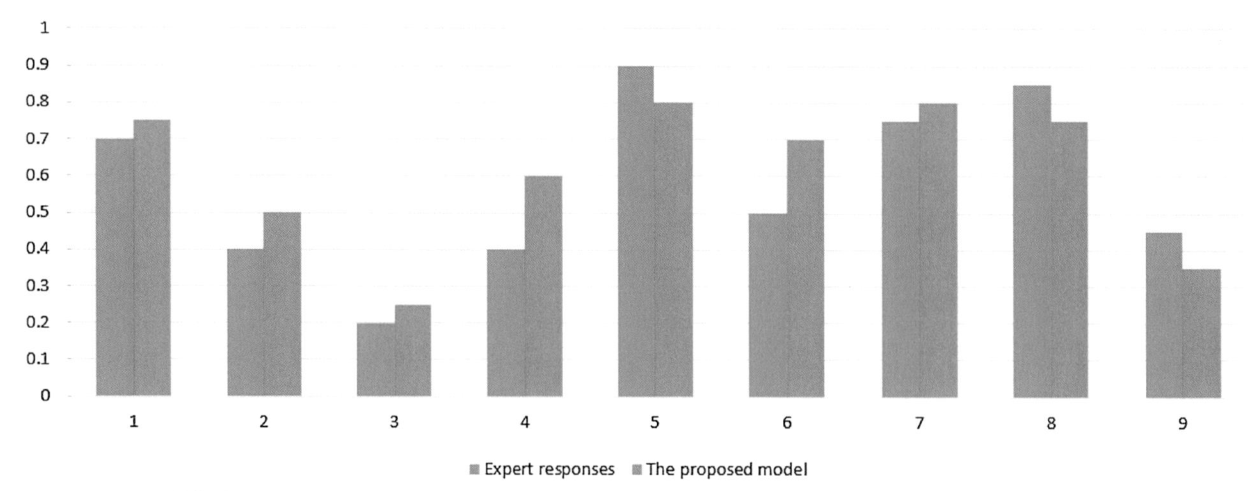

FIGURE 22.2 The screening for identifying criminal persons by expert responses and by the proposed model.

5. Conclusions

This chapter uses the picture fuzzy set as an extension of the fuzzy and intuitionistic fuzzy set for solving real-world problems in identified people in an uncertain environment. Experimental results show that the picture fuzzy set is an extension of the fuzzy and intuitionistic fuzzy set combined with the criminal database for reasoning to improve human recognition in an uncertain environment. In addition, the picture fuzzy sets are based on features of criminal information and attributes such as name, body, hair coverage, hair texture, face shape, and skin color of the patient, and building applications in the hospital have been investigated in real-world applications.

To improve the quality of the automated responses obtained in the proposed model and in its effectiveness and efficiency, further investigations into the model will be developed with other functions implemented while updating it with new knowledge automated systems in future works.

Acknowledgments

This work was supported by the University of Economics Ho Chi Minh City (UEH), Ho Chi Minh City, Vietnam.

References

[1] L.A. Zadeh, Electrical engineering at the crossroads, IEEE Transactions on Education 8 (2) (1965) 30–33.
[2] A.I. Atanassov, Sugar beet (beta vulgaris L.), in: Y.P.S. Bajaj (Ed.), Crops I, Springer, Berlin, Heidelberg, 1986, pp. 462–470.
[3] B.C. Cuong, V. Kreinovich, Picture fuzzy sets - a new concept for computational intelligence problems, in: 2013 Third World Congress on Information and Communication Technologies (WICT 2013), 2013, pp. 1–6.
[4] K. Kobayashi, T. Sirao, H. Tanaka, On the growing up problem for semilinear heat equations, Journal of the Mathematical Society of Japan 29 (3) (1977) 407–424.
[5] K.T. Atanassov, New topological operator over intuitionistic fuzzy sets, Journal of Computational and Cognitive Engineering 1 (3) (2022) 94–102.
[6] C. Bo and X. Zhang, "New operations of picture fuzzy relations and fuzzy comprehensive evaluation," Symmetry, vol. 9, no. 11. doi: 10.3390/sym9110268
[7] P. Dutta, S. Ganju, Some aspects of picture fuzzy set, Transactions of A. Razmadze Mathematical Institute 172 (2) (2018) 164–175.
[8] H. Van Pham, N.D. Khoa, T.T.H. Bui, N.T.H. Giang, P. Moore, Applied picture fuzzy sets for group decision-support in the evaluation of pedagogic systems, International Journal of Mathematical, Engineering Management Sciences 7 (2) (2022) 243.
[9] P. Moore, H.V. Pham, On context and the open world assumption, in: 2015 IEEE 29th International Conference on Advanced Information Networking and Applications Workshops, 2015, pp. 387–392.
[10] L. Kim, H. Pham, An integrated picture fuzzy set with TOPSIS-AHP approach to group decision-making in policymaking under uncertainty, International Journal of Mathematical, Engineering and Management Sciences 6 (2021) 1578–1593.
[11] H. Van Pham, Q.H. Nguyen, Intelligent IoT monitoring system using rule-based for decision supports in Fired forest images, in: Industrial Networks and Intelligent Systems, Springer International Publishing, Cham, 2021, pp. 367–378.
[12] C.K. Long, P. Van Hai, T.M. Tuan, L.T.H. Lan, P.M. Chuan, L.H. Son, A novel fuzzy knowledge graph pairs approach in decision making, Multimedia Tools and Applications 81 (18) (2022) 26505–26534.

[13] L. Luong Thi Hong, et al., A new complex fuzzy inference system with fuzzy knowledge graph and extensions in decision making, IEEE Access 8 (2020) 164899–164921.

[14] H. Van Pham, D.N. Tien, Hybrid louvain-clustering model using knowledge graph for improvement of clustering user's behavior on social networks, in: Intelligent Systems and Networks, Springer Singapore, Singapore, 2021, pp. 126–133.

[15] X.T. Dinh, H. Van Pham, Social network analysis based on combining probabilistic models with graph deep learning, in: Communication and Intelligent Systems, Springer Singapore, Singapore, 2021, pp. 975–986.

[16] H. Swofford, C. Champod, Probabilistic reporting and algorithms in forensic science: Stakeholder perspectives within the American criminal justice system, Forensic Science International: Synergy 4 (2022) 100220.

[17] G. Nápoles, L. Koutsoviti Koumeri, A fuzzy-rough uncertainty measure to discover bias encoded explicitly or implicitly in features of structured pattern classification datasets, Pattern Recognition Letters 154 (2022).

[18] H.V. Pham, Q.H. Nguyen, The clustering approach using SOM and picture fuzzy sets for tracking influenced COVID-19 persons, in: Artificial Intelligence in Data and Big Data Processing, Springer International Publishing, Cham, 2022, pp. 531–541.

[19] W. Ding, M. Abdel-Basset, H. Hawash, N. Moustafa, Interval type-2 fuzzy temporal convolutional autoencoder for gait-based human identification and authentication, Information Sciences 597 (2022) 144–165.

[20] V.H. Pham, Q.H. Nguyen, V.P. Truong, L.P.T. Tran, The proposed context matching algorithm and its application for user preferences of tourism in COVID-19 pandemic, in: International Conference on Innovative Computing and Communications, Springer Nature Singapore, Singapore, 2023, pp. 285–293.

[21] M. Olgun, E. Türkarslan, M. Ünver, J. Ye, A cosine similarity measure based on the choquet integral for intuitionistic fuzzy sets and its applications to pattern recognition, Informatica (2021) 1–16.

[22] H. Garg, R. Krishankumar, K.S. Ravichandran, Decision framework with integrated methods for group decision-making under probabilistic hesitant fuzzy context and unknown weights, Expert Systems with Applications 200 (2022) 117082.

[23] C. Jana, H. Garg, M. Pal, Multi-attribute decision making for power Dombi operators under Pythagorean fuzzy information with MABAC method, Journal of Ambient Intelligence and Humanized Computing (2022) 1–18.

[24] C.K. Long, R. Agrawal, H.Q. Trung, H.V. Pham, A big data framework for E-Government in Industry 4.0, Journal Open Computer Science 11 (1) (2021) 461–479.

[25] M. Karatas, L. Eriskin, M. Deveci, D. Pamucar, H. Garg, Big data for healthcare industry 4.0: Applications, challenges and future perspectives, Expert Systems with Applications: International Journal 200 (2022) 15.

Chapter 23

A deep learning framework for surgery action detection

Prabu Selvam and Joseph Abraham Sundar K
SASTRA Deemed University, Thanjavur, India

1. Introduction

Minimally Invasive Surgery (MIS) is a very sensitive medical procedure. Degree of coordination between the surgeons, expertise and attentiveness are key factors for the success of MIS procedures. A study from Johns Hopkins University stated that 15% of deaths in the USA are due to medical faults. According to a survey by Lancet Commission, more than 4.2 million people die every year within a month after surgery. Since there is no precise way to quantify and anticipate the risk factor linked to surgeon behavior, it is crucial to track the sequence of activities carried out by surgeons throughout an operation in real-time to prevent any unfortunate event. Artificial intelligence (AI) is frequently hired in applications where human error needs to be reduced. In order to reduce the risks of human mistakes in diagnostic imaging and electrodiagnosis, various applications are developed have been in the past few years based on AI. Exploring the use of AI in several other areas, including patient monitoring, clinical decision support, healthcare delivery, interventions, and administration, is becoming increasingly important.

Action detection (AD) and action recognition (AR) are well-established fields in computer vision [1]. It combines two main tasks: locating the action and recognizing an action. AR describes what kind of action is performed. AD describes the action's location, typically on the image plane; it can be represented in the form of a rectangular bounding box containing the action instance. Both AD and AR provide complete information on what is happening and where. Surgical activity detection (SAD) is a critical topic in the medical field [2]. Autonomous surgeries, navigation during surgery, and automated real-time feedback are open research problems. AI technology became an essential driving force for SAD, which can identify the surgical actions of surgeons and the types of surgical tools used in surgery. The identification of surgical activities serves as a reference for several automatic intraoperative or postoperative operations, including surgical procedure optimization, assessment of the surgical proficiency of the surgeons, and creation of surgical reports. These techniques can significantly reduce surgeons' burden and enhance patients' safety.

2. Literature review

Surgeon action detection/recognition from images or videos is less explored. Nevertheless, some existing works deserve to be mentioned. Petlenkov et al. [2] computed the surgeon's hand movement during the endoscopic surgery using the Kohonen map. The primary advantage of the Kohonen map-based segmentation method is that it easily adapts the dynamic properties of various motions and allows the detection algorithm to track motions. It also helps to reduce the time required to perform etalon segmentation for model adaptation and tuning. Voros et al. [3] presented a tool-tissue interaction spotting approach from live surgical data, notably in the scenario of robotic-assisted laparoscopy adopting the daVinciTM system along with visual and kinematic information. Kocev et al. [4] used a Microsoft Kinect camera to record the interactions between the surgeon and the patient, and it also helped to track the surface changes over time. The Kinect camera and projector are calibrated to detect finger movement automatically. The interaction of the surgeon with the projected virtual information is determined via a point cloud surface model. Projector-Kinect calibration data and the distance between the surgeon's fingertips and the interaction zone are used to recognize the interaction. Multitouch gestures are used for interaction.

Deep Learning in Personalized Healthcare and Decision Support. https://doi.org/10.1016/B978-0-443-19413-9.00008-4

Van Amsterdam et al. [5] proposed a surgical gesture recognition approach based on a weakly supervised algorithm to overcome state-of-the-art methods' limitations, such as being time-consuming and error-prone. The authors use Gaussian Mixture Models (GMM) to detect and recognize surgical actions in medical images. Meanwhile, this approach is not trained with original surgical images; trained was performed with simple, homogeneous artificial surfaces, and it can recognize only one action per frame. It cannot recognize multiple actions in a medical image is the main drawback of this approach. Azari et al. [6] designed a surgical maneuvers prediction system using the surgeon's hand motion. This system uses three popular machine learning algorithms, including hidden Markov model (HMM), decision trees, and random forest, to predict the surgeon's action every 2 s in a video frame. Random forest aided by HMM achieves better performance than stand-alone algorithms. However, the surgeon restricted this technique to direct hand-to-hand motions and did not allow for teleoperations.

Li et al. [7] proposed an early hand action recognition model using a subaction-based approach from a video sequence. This model is divided into two stages: quasi-start point identification and early action recognition. In the first stage, a cascade histogram feature is generated by computing the shortest distance between each subaction model's start portion and the test video present within the sliding window. Then, the histogram feature within the sliding window at each position is fed into a support vector machine (SVM) classifier to identify probable start locations of subactions. Candidates are screened to determine quasi-start points. In the second stage, the likelihood that the test video corresponds to that subaction is calculated for each time instance based on the dynamic matching output between the subaction model and the test video sequence. If that likelihood is higher than the cutoff, the subaction is identified instantly. Early recognition of actions is achieved by looking at the outcomes of subactions. The main drawback of this model is that all the experiments are performed only on artificial data.

Ameur et al. [8] employed the leap motion controller and different classification methods to recognize 11 hand gestures in medical images. This algorithm utilizes statistical characteristics from finger and hand data that were standardized and fed into various classifiers, including the SVM, AdaBoost, linear discriminant analysis (LDA), and multi-layer perceptron (MLP). In an operating room, Sa-nguannarm et al. [9] demonstrated a method for manipulating medical imaging with just one hand. The proposed method comprises five important phases: data acquisition, color coding and hand visualization, open and close hand categorization, identification of hand gestures, and interpretation of commands [10]. This method works based on 10 commands: six commands correspond to the program screen and four to button clicks. This approach is limited by sensor distance and commands to decrease the risk of contamination during surgical procedures.

To improve the robustness of gesture detection for contactless interfaces, Cho et al. [11] combined leap motion with a customized automatic classifier. Training gestures for each user meant that this program was developed and tested individually. In order to predict and train five different types of gestures, including hover, grasp, click, one peak and two peaks, the authors employed 30 different features that were chosen and given as input to a multiclass SVM and Naive Bayes classifiers. Forestier et al. [12] analyze the performance of surgical action detection based on discriminative interpretable patterns. In this method, continuous kinematic data are decomposed into a series of overlapping gestures represented by strings and relative geometric statistics are combined to discover the discriminative gestures. Khalid et al. [13] trained various deep learning (DL) models to identify surgeon actions such as needle passing, suturing, and knot tying. These DL models are trained and predicted at different activity levels, including simple, intermediate, and complex.

Huang [14] developed a transformer-based surgical action prediction and recognition system. This system can identify 20 different surgeon actions in a continuous video. This self-attention-based model follows three relationships: relationships between video frames and actions, relations among surgical actions, and relations among endoscopic images. This system is not effective in real-time surgical action detection and recognition. Zhang et al. [15] combine multiple convolutional neural networks (CNNs) with a transformer network to identify the surgical workflow.

Similarly, Nwoye et al. [16] utilize a transformer network for surgical interaction detection. Xu et al. [17] introduced a system to improve the efficiency of surgical actions and instrument detection. The authors developed a new feature pyramid network (FPN) called multiscale fusion feature pyramid network (MSF-FPN) to combine high-level and low-level semantic data. In the bottom layer, the feature maps collect the semantic data via the pyramid network and diverse the feature data via cross-transmission in the intermediate layer. Finally, the output layer generates a robust semantic feature map. The performance of MSF-FPN is evaluated using Endoscopic Surgeon Action Detection (ESAD) dataset.

Li et al. [18] developed a surgical interaction recognition network (SIRNet) to predict surgical interaction triplets. A multi-head cross-attention (MHCA) mechanism in SIRNet learns the associations between triplet and endoscopic images and understands the relationships among surgical interaction triplets before training starts. The bipartite matching loss in SIRNet is used to create formal learning and prediction for each component of the surgical interaction triplet. This loss takes permutation and a combination of tools, verbs, and targets into consideration to avoid false predictions. Additionally, a weight attention module is developed to determine the significance of each anticipated surgical interaction triplet and each triplet element for detecting legitimate surgical interaction triplets. To obtain discriminative characteristics from the kinematic data and improve recognition scores, Van Amsterdam et al. [19] designed a multi-task recurrent neural network (MTRNN) for the detection of surgical gestures and to estimate a unique formulation of surgical task progress. Huynhnguyen et al. [20] designed a two-stage surgical gesture segmentation and recognition approach. The first stage discovers the transition gestures. The second stage classifies the frames into the appropriate class. Finally, both stages are combined and trained with additional samples to improve recognition accuracy.

3. Proposed approach

Fig. 23.1 shows the overall pipeline of the proposed framework. It has been divided into three-step processes [21]: in the first step, the most popular object detection algorithms YOLOv5 and EfficientDet are combined to identify the surgeon's action in the OR. In the second step, a deep learning model, EfficientNet, is employed to remove false detection. Finally, the surgeon's action detection results are based on three deep learners' decisions.

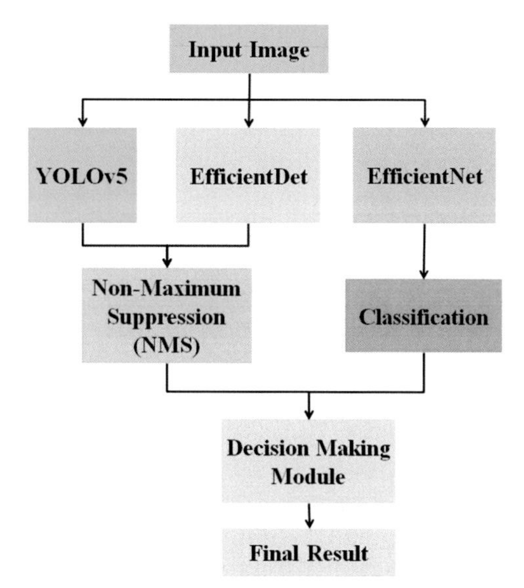

FIGURE 23.1 Proposed framework for surgical action detection.

4. YOLOv5

YOLOv5 is a modern, real-time object detector built on YOLOv1-YOLOv4. It has achieved top results on benchmark datasets, including Microsoft COCO and Pascal VOC. Fig. 23.2 shows the architecture of the YOLOv5 algorithm. The authors find three aspects for choosing the YOLOv5 model as one of the learners. First, YOLOv5 uses CSPDarknet as its backbone. CSPDarknet was constructed initially by integrating cross stage partial network (CSPNet) with Darknet. CSPNet is designed to resolve the problem of repeating gradients in network backbones. It incorporates the gradient changes into the feature maps that reduce the number of hyperparameters required for the model. Speed and accuracy of detection are crucial in surgical action detection. Hence, CSPNet is forced to make attention to inference accuracy, speed, and model size.

The YOLOv5 algorithm employed path aggregation network (PANet) as its neck block to increase information flow between the top and bottom layers [22]. It introduced a novel FPN topology with an improved bottom-up path to maximize the propagation of low-level features. Adaptive feature pooling in the FPN allows direct propagation of important information from one feature level to the subnetwork and vice-versa. PANet can increase object detection accuracy by improving precise localization signals in lower layers. In order to handle small, medium, and large objects, the YOLO layer, the head of the YOLOv5 algorithm, generates feature maps in three distinct scales (20×20, 40×40, and 80×80) in the third step. The multiscale detection mechanisms can track each action of the surgeon. It produces three vectors such as class name, confident score, and bounding box as its output.

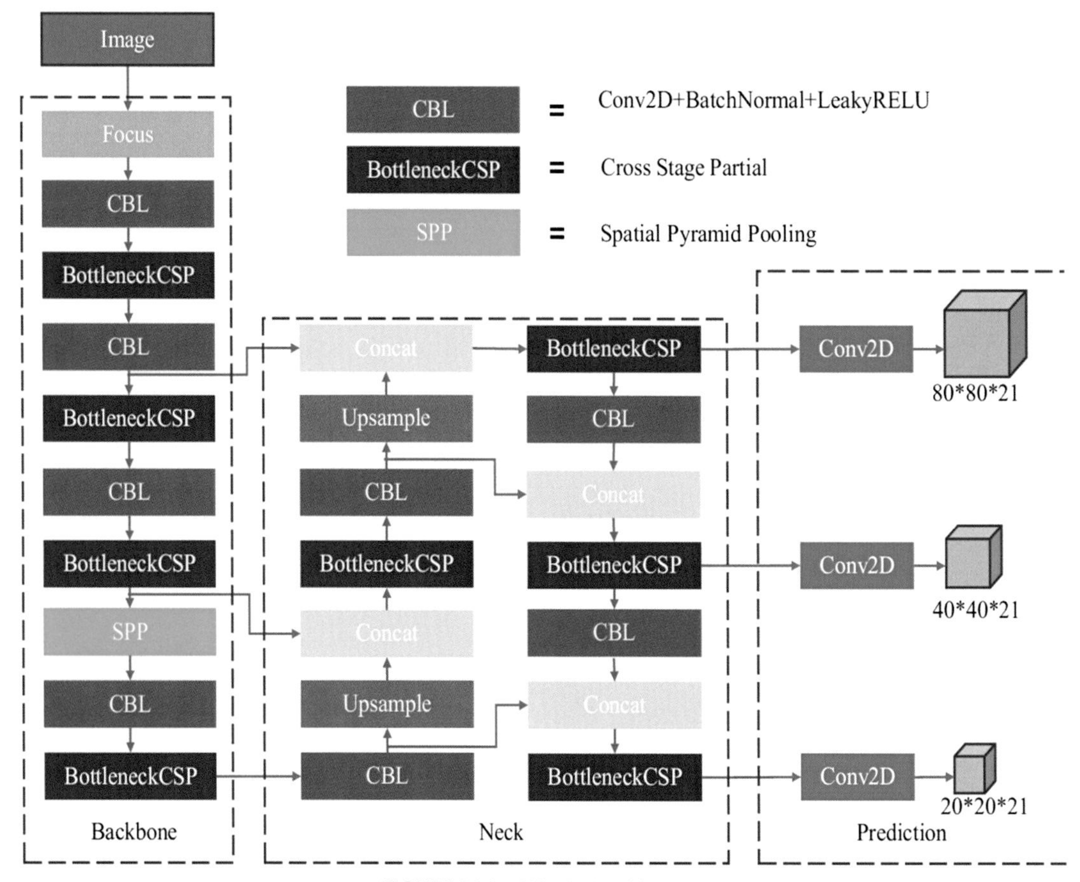

FIGURE 23.2 YOLOv5 architecture.

5. EfficientDet

Google has created a new class of object detectors called EfficientDet, and it reliably outperforms most object detection algorithms effectively under various resource limitations [23]. EfficientDet, which is often used in real-world applications, has similar achievements to YOLOv5 in terms of impressive performances in Microsoft COCO and Pascal VOC datasets.

Fig. 23.3 depicts the architecture of EfficientDet. There are three motivations for choosing EfficientDet as a succeeding network. First, it uses the cutting-edge network, EfficientNet as its backbone, supporting the model with the necessary

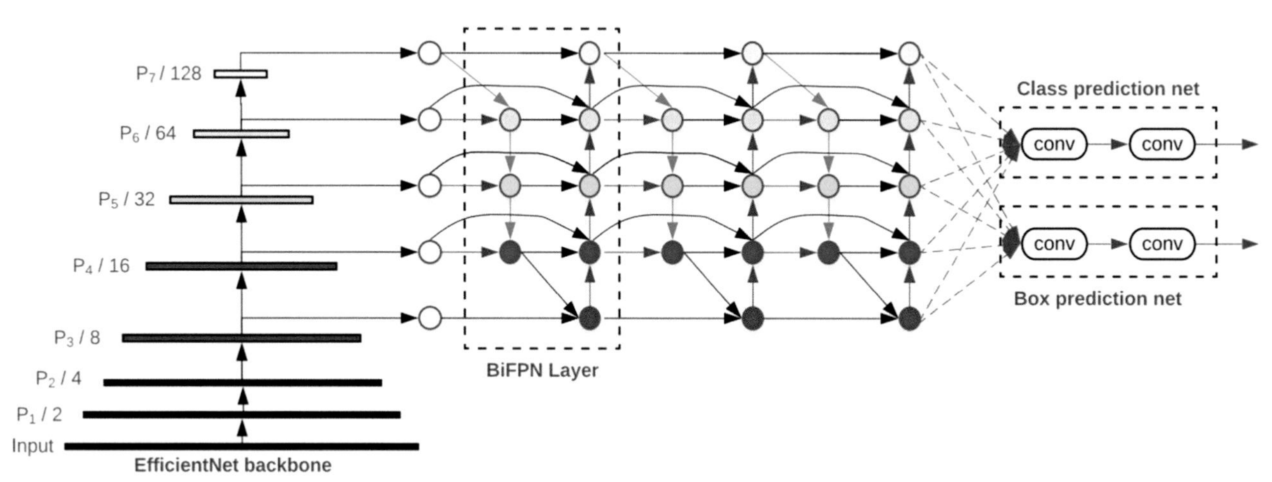

FIGURE 23.3 EfficientDet architecture.

capacity to understand the complicated features of various surgery actions. Second, to provide quick and simple multiscale feature fusion, EfficientDet also employed an upgraded PANet called the bi-directional feature pyramid network (Bi-FPN) as its neck. Bi-FPN repeatedly employs top-down and bottom-up multiscale feature fusion and includes learnable weights, enabling the network to learn the significance of various inputs. Bi-FPN performs efficiently with fewer hyperparameters and floating-point operations per second than PANet. Distinct feature fusion strategies provide various semantic information that helps to produce different detection outputs. Thirdly, like EfficientNet, EfficientDet also incorporates a composite scaling mechanism that simultaneously scales the breadth, depth, and resolution of all feature maps, backbone network and prediction network, ensuring the highest accuracy and efficiency obtained with the available computational resources. Accuracy will steadily increase as more resources become available. YOLOv5 has a different backbone, neck, and head. It cannot learn the same knowledge as the second learner (EfficientDet).

6. EfficientNet

Google has proposed a brand-new efficient network called EfficientNet. It used a cutting-edge model sizing technique called the compound scaling approach to balance image quality, network depth, and breadth for higher accuracy with available computational resources [24]. In the ImageNet image classification challenge, EfficientNet achieves the Top-1 accuracy by outperforming other popular networks, including ResNet, DenseNet, and ResNeXt.

Fig. 23.4 depicts the EfficientNet network architecture. The main reason for choosing EfficientNet as a third learner is because it delivers a higher accuracy-to-efficiency trade-off. It has the greatest impact on the proposed methodology. Decision-making ability has a direct impact on the outcome. Meanwhile, EfficientNet must be extremely effective; otherwise, the speed of the entire 'model will slow down.

7. Proposed ensemble network

Recognizing different surgeon actions such as needle passing, suturing, and knot tying is very important in a real-world surgery action detection task. These actions in the medical images are dominated by the surroundings, such as shape, color, and texture add additional complexity for the individual networks to capture robust features. After conducting multiple experiments, the authors find that YOLOv5 is better at learning complex surgeon actions, but it sometimes misses actions happening far away. Even though the EfficientDet is not sensitive to complex actions, it can perform complementary detection. Therefore, the authors integrated two efficient learners (YOLOv5 and EfficientDet) to perform surgeon action detection that improves recognition accuracy. Some limitations to the object detector's ability add more complexity. Generally, object detection algorithms ignore background information and only learns the action movement zone [25]. The object detector can mistakenly identify medical instruments in OR as surgeon action. Hence, a competent leader like EfficientNet is integrated into the framework to analyze the entire image.

Three learners are combined to jointly make decisions to overcome the abovementioned problems and ensure that the model is resilient to many circumstances. The two individual networks, YOLOv5 and EfficientDet, are combined and serve as an object detector to find the location information of different surgeon actions in the medical image by creating candidate boxes. After that, redundant boxes are removed using the nonmaximum suppression (NMS) technique, which keeps the highest confidence boxes. EfficientNet, the third learner, serves as a binary classifier and is in charge of learning the entire image to detect different surgeon actions. The classification and detection results are given as input to a decision-making module (DMM). DMM will retain the image if there is any surgeon action on it. Otherwise, it will discard. In addition, all three learners are designed to work independently, and the entire network is implemented in numerous stages. Each learner has a distinct role in this process; integrating multiple learners does not affect the model's overall performance.

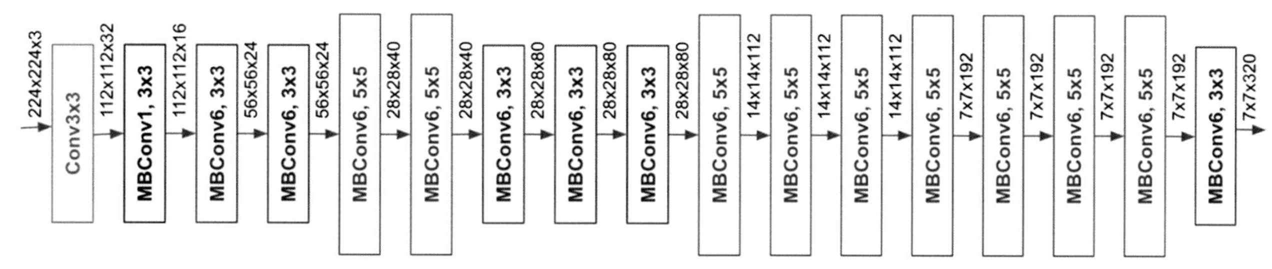

FIGURE 23.4 EfficientNet architecture.

8. Experiment

8.1 Dataset

The ESAD dataset [26] was released in the SARAS-ESAD challenge as part of the medical imaging and deep learning (MIDL) international conference held in 2020. ESAD is the first benchmark dataset created to test and evaluate approaches for recognizing surgeon actions from endoscopic images and recordings. Fig. 23.5 illustrates the sample images of the ESAD dataset.

The ESAD dataset was developed with the help of skilled surgeons and medical experts. It includes four complete radical prostatectomy surgery videos. Each video is about 4 h long, annotated with 46,325 action occurrences in the form of bounding boxes with the corresponding action labels. It includes 21 action classes unique to radical prostatectomy (see Table 23.1). The SARAS-ESAD dataset is available at https://saras-esad.grand-challenge.org/.

8.2 Training environment

The proposed ensemble model is developed using Python programming. All the experiments were conducted in a DELL Precision Tower P7810 workstation with a 64 GB NVIDIA graphics card. The proposed model is trained with the SARAS-ESAD benchmark dataset. The dataset has been divided into three sets: training set, validation set, and test set. The learning rate and training batch size are set to 0.01 and eight for all three learners. The YOLOv5 and EfficientNet use stochastic gradient descent (SGD) optimizer, and EfficientDet uses an Adam optimizer. The YOLOv5 and EfficientNet are trained with 200 epochs with a batch size of 8. Similarly, EfficientDet is trained with 200 epochs with a batch size of 16. The detailed training strategy of the proposed framework is presented in Table 23.2.

9. Result and discussion

9.1 Ablation study

The performance of the proposed framework was evaluated using the ESAD dataset. To analyze the performance of the proposed framework, the authors conducted an ablation study (see Table 23.3) before comparing the framework with existing methods (see Table 23.4). In the first experiment, the authors utilized the YOLOv5 object detection algorithm to perform this action detection task. Similarly, in the second and third experiments, the authors employed EfficientDet and EfficientNet. The individual learners (YOLOv5, EfficientDet, and EfficientNet) perform better than most state-of-the-art algorithms by achieving a recognition accuracy of 82.4%, 80.7%, and 79.8%, respectively. Compared with single learners, two learners (YOLOv5 + EfficientDet) produced a satisfactory F1-Score of 88.2% and Recognition accuracy of 86.9%.

The authors integrated all three learners into the final experiment and created a framework. However, that framework achieves superior performance than individual learners and two learners. The three learners achieve an F1-Score of 94.1% and a recognition accuracy of 92.6%. To conclude, with the assistance of the final network, EfficientNet, and false-positive values were reduced below 0.2%, and it eliminated incorrect detection results accurately. The model's (3 learners) absolute latency is 54.3 ms which demonstrates the model's adaptability for real-time detection tasks and the great trade-off between detection performance and efficiency.

9.2 Comparison with state-of-the-art approaches

In this section, the authors compared the performance of the proposed framework with the current state-of-the-art approaches (see Table 23.4). Li et al. [7] used a threshold-based algorithm for hand gesture recognition to achieve an accuracy of 56.4%, which is the least percentage in Table 23.4. It proves that threshold-based approaches are not effective for this problem. Forestier et al. [12] extract discriminative interpretable patterns from medical images, but extracting patterns from real-time videos is a challenging problem. Unfortunately, Forestier et al. [12] failed to extract discriminative patterns from the complex ESAD dataset, achieving an F1-score of 54.8% and recognition accuracy of 50.0%, which is the second least percentage in Table 23.4. The methods of Azari et al. [6] and Van Amsterdam et al. [5] are trained with artificial image samples; when these methods are utilized to experiment on benchmarks, they fail to produce satisfactory results.

On the ESAD dataset, Azari et al. [6] and Van Amsterdam et al. [5] achieved an F1-score of 64.7% and 68.3% and recognition accuracy of 62.8% and 66.9%, respectively. Methods of Ameur et al. [8], Sa-nguannarm et al. [9] and Khalid et al. [13] employed machine algorithms like SVM, LDA, and Adaboost to accomplish surgeon hand action recognition

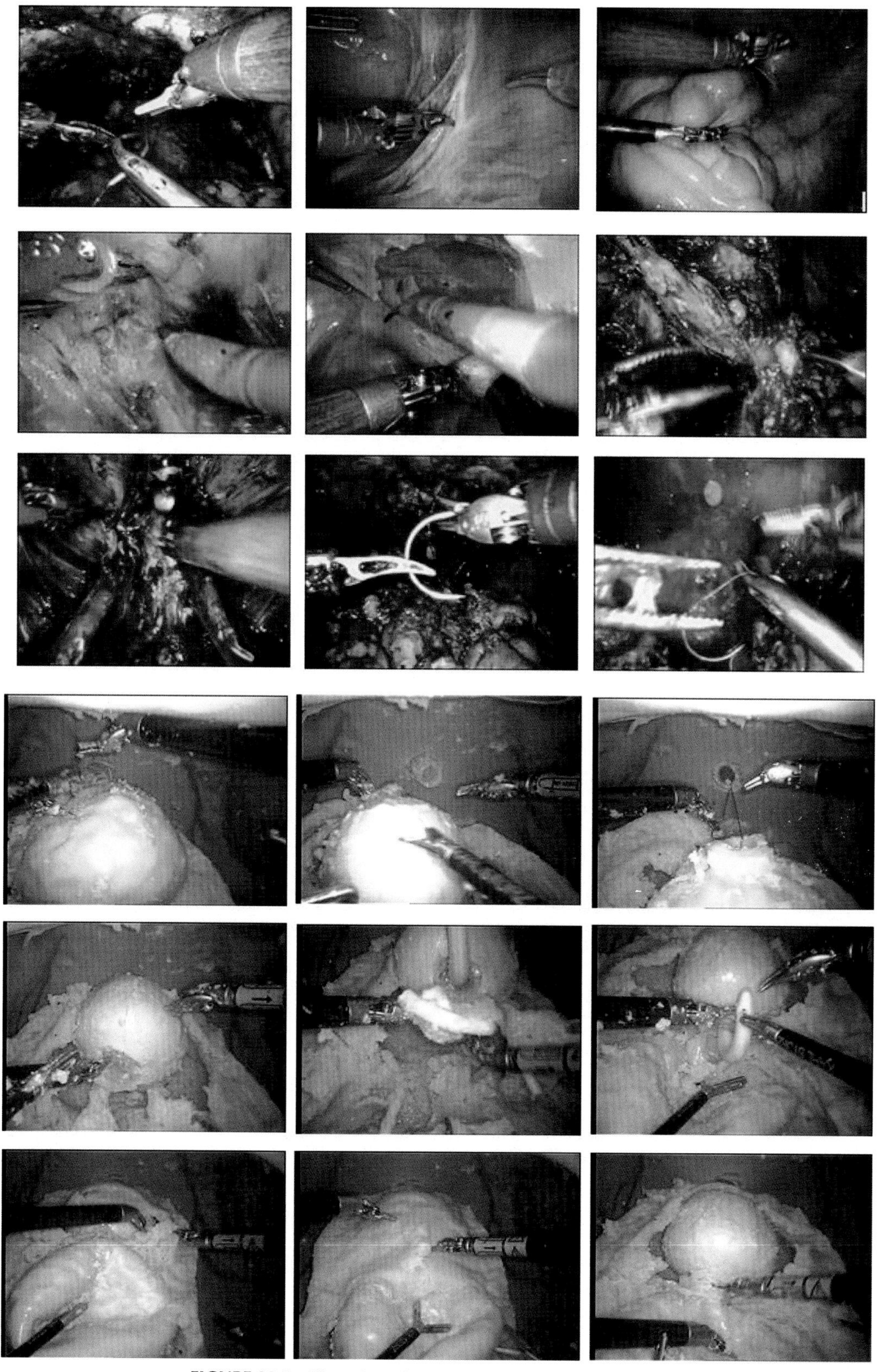

FIGURE 23.5 Illustration of sample SARAS-ESAD dataset images.

TABLE 23.1 List of actions for the SARAS-ESAD dataset.

Sucking blood	Clipping vas deferens	Clipping bladder neck
Clipping seminal vesicle	Cutting prostate	Cutting mesocolon
Bladder anastomosis	Cutting tissue	Pulling bladder neck
Bagging prostate	Pulling prostate	Cutting vas deferens
Cutting thread	Sucking smoke	Bladder neck dissection
Clipping tissue	Urethra dissection	Pulling seminal vesicle
Cutting seminal vesicle	Pulling tissue	Pulling vas deferens

TABLE 23.2 Detailed training strategy.

Model	Train: Val: Test	Learning rate	Optimizer	Epoch	Batch size
YOLOv5	80:10:10	10^{-2}	SGD	200	8
EfficientDet	80:10:10	10^{-2}	Adam	200	16
EfficientNet	80:10:10	10^{-2}	SGD	200	8

TABLE 23.3 Performance comparison of single learners with multiple learners.

Methods	Precision (%)	Recall (%)	F1-score (%)	Accuracy (%)	Latency (ms)
YOLOv5	84.2	87.6	85.9	82.4	24.0
EfficientDet	81.3	83.6	82.4	80.7	66.8
EfficientNet	82.7	80.9	81.8	79.8	88.4
Proposed (YOLOv5 + EfficientDet)	88.9	87.5	88.2	86.9	59.5
Proposed (YOLOv5 + EfficientDet + EfficientNet)	**93.6**	**94.7**	**94.1**	**92.6**	**54.3**

TABLE 23.4 Comparison of the proposed framework with existing methods.

Methods	Precision (%)	Recall (%)	F1-score (%)	Accuracy (%)
Li et al. [7]	60.0	53.8	56.7	56.4
Forestier et al. [12]	57.1	52.6	54.8	50.0
Cho et al. [11]	64.2	58.6	61.3	62.1
Azari et al. [6]	65.4	64.0	64.7	62.8
Van Amsterdam et al. [5]	67.8	68.9	68.3	66.9
Ameur et al. [8]	71.0	68.3	69.6	67.5
Sa-nguannarm et al. [9]	68.7	61.3	64.8	65.5
Khalid et al. [13]	73.6	69.9	71.7	70.0
Nwoye et al. [16]	75.3	70.4	72.8	71.4
3Huynhnguyen et al. [20]	76.5	75.7	76.1	73.8
Zhang et al. [15]	81.5	83.6	82.5	80.3
Xu et al. [17]	80.7	84.6	82.6	83.7
Huang [14]	82.3	81.1	81.7	82.2
Li et al. [18]	85.2	84.0	84.6	83.6
Van Amsterdam et al. [19]	87.3	83.4	85.3	86.5
Ours (3 learners)	**93.6**	**94.7**	**94.1**	**92.6**

tasks and achieved an accuracy of 67.5%, 65.5%, and 70.0%. Transformers were initially designed for natural language processing (NLP). Nwoye et al. [16] utilized them to perform the action detection task achieving an F1-score of 72.8% and recognition accuracy of 71.4%. The ensemble network outperforms Nwoye et al. [16] by a margin of around +20%. The proposed framework surpassed current state-of-the-art algorithms Zhang et al. [15], Xu et al. [17] and Huang [14] by achieving an accuracy of 92.6% v.s. 80.3%, 83.7%, and 82.2%. Fig. 23.6 illustrates the sample results obtained by the proposed framework.

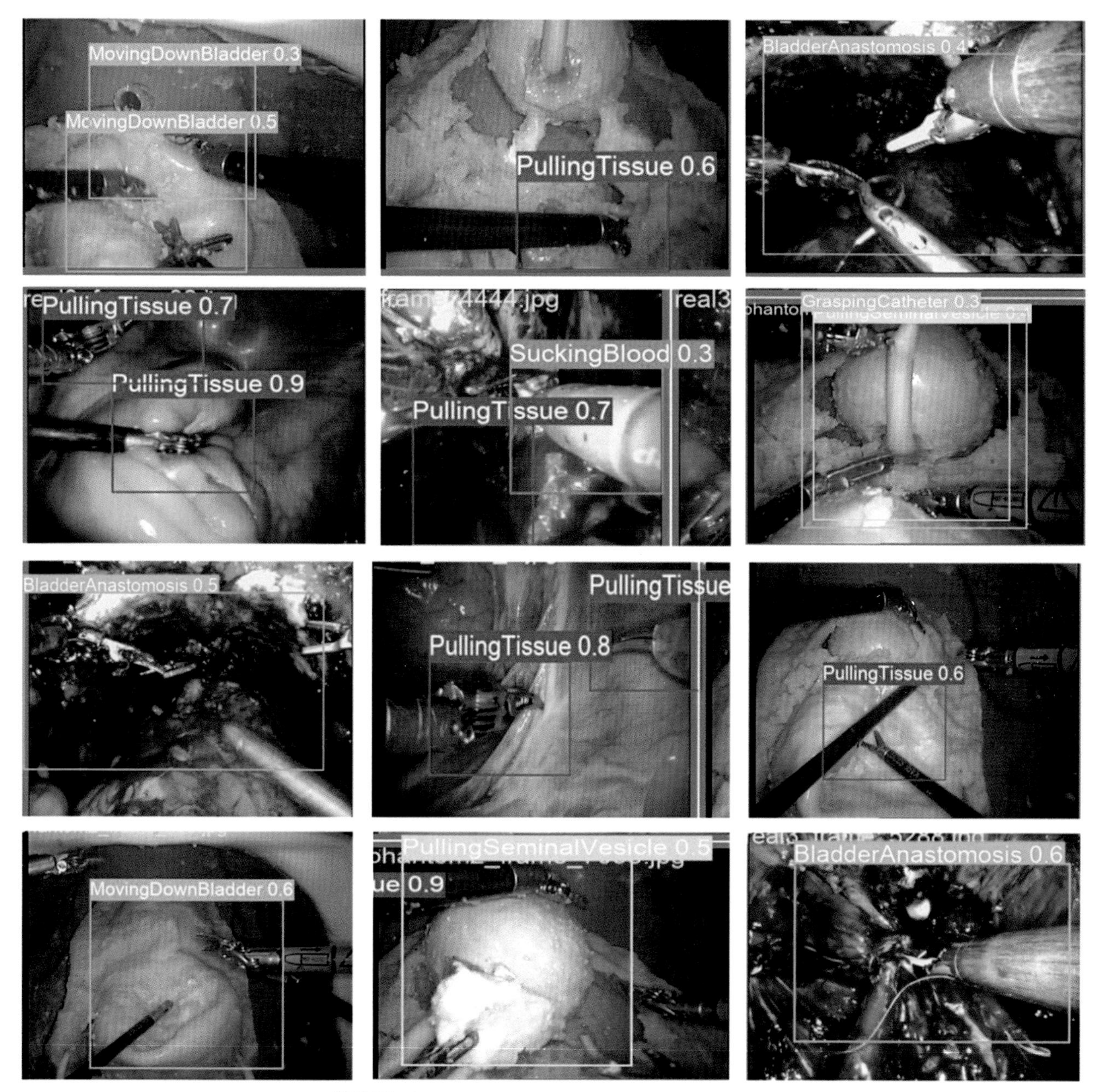

FIGURE 23.6 Results obtained by the proposed framework.

The methods of Li et al. [18] and Van Amsterdam et al. [19] are based on the self-attention and recurrent neural network (RNN) mechanism achieving a recognition accuracy of 83.6% and 86.5%. Li et al. [18] and Van Amsterdam et al. [19] can handle only a limited number of human actions. These methods failed to identify complex actions like PullingSeminalVesicle and PullingVasDeferens. To conclude, the proposed framework achieves superior performance on the ESAD dataset. The additional learner EfficientNet greatly supports the model to decrease false-positive rates. The training time of the framework is low compared to current state-of-the-art methods. Figs. 23.7 and 23.8 illustrate the confusion matrix and labels_correlogram generated by the proposed framework. Figs. 23.9–23.12 show the precision curve, recall curve, F1-score curve, and PR curve generated by the proposed framework at the 70th epoch.

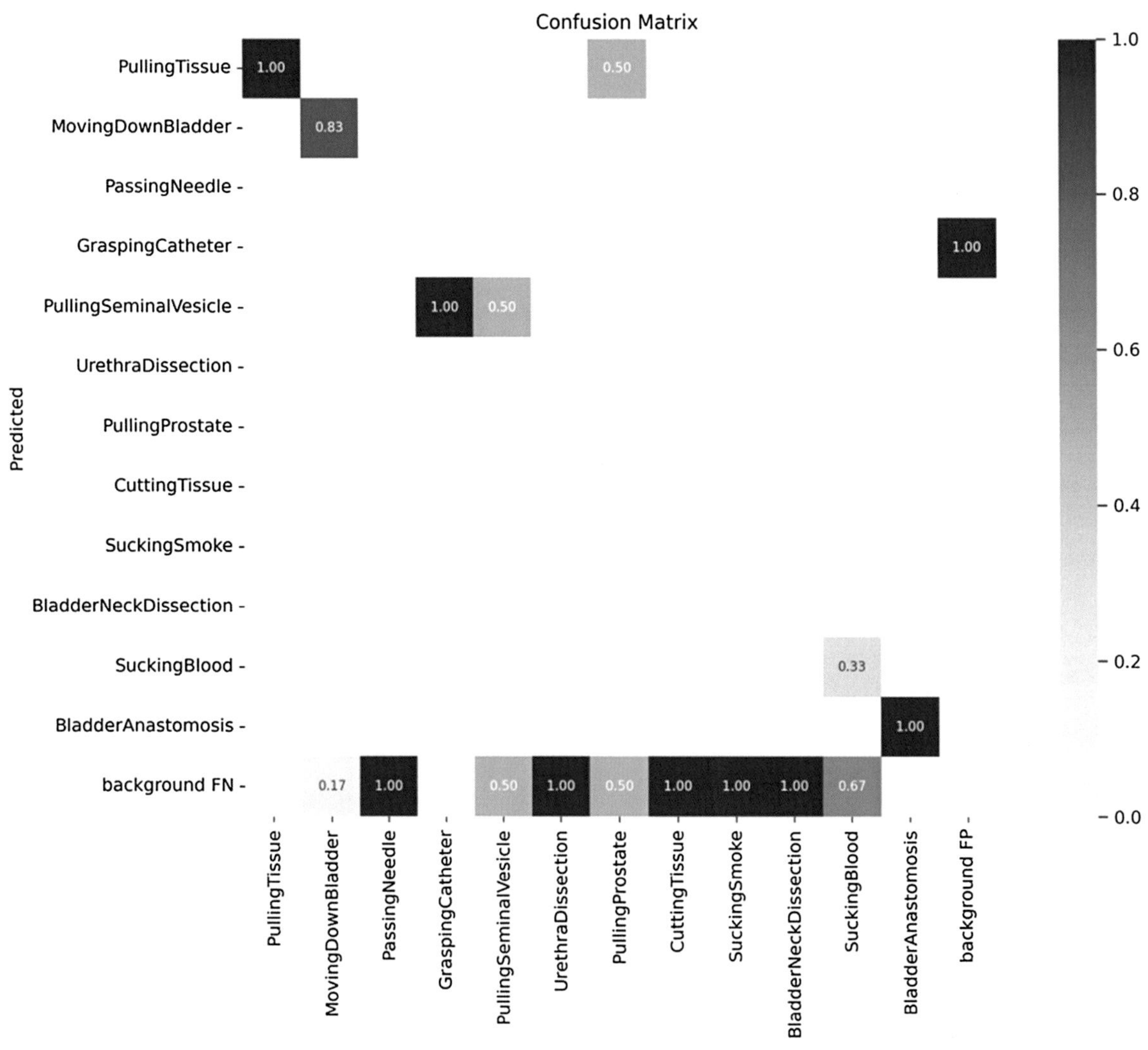

FIGURE 23.7 Confusion matrix.

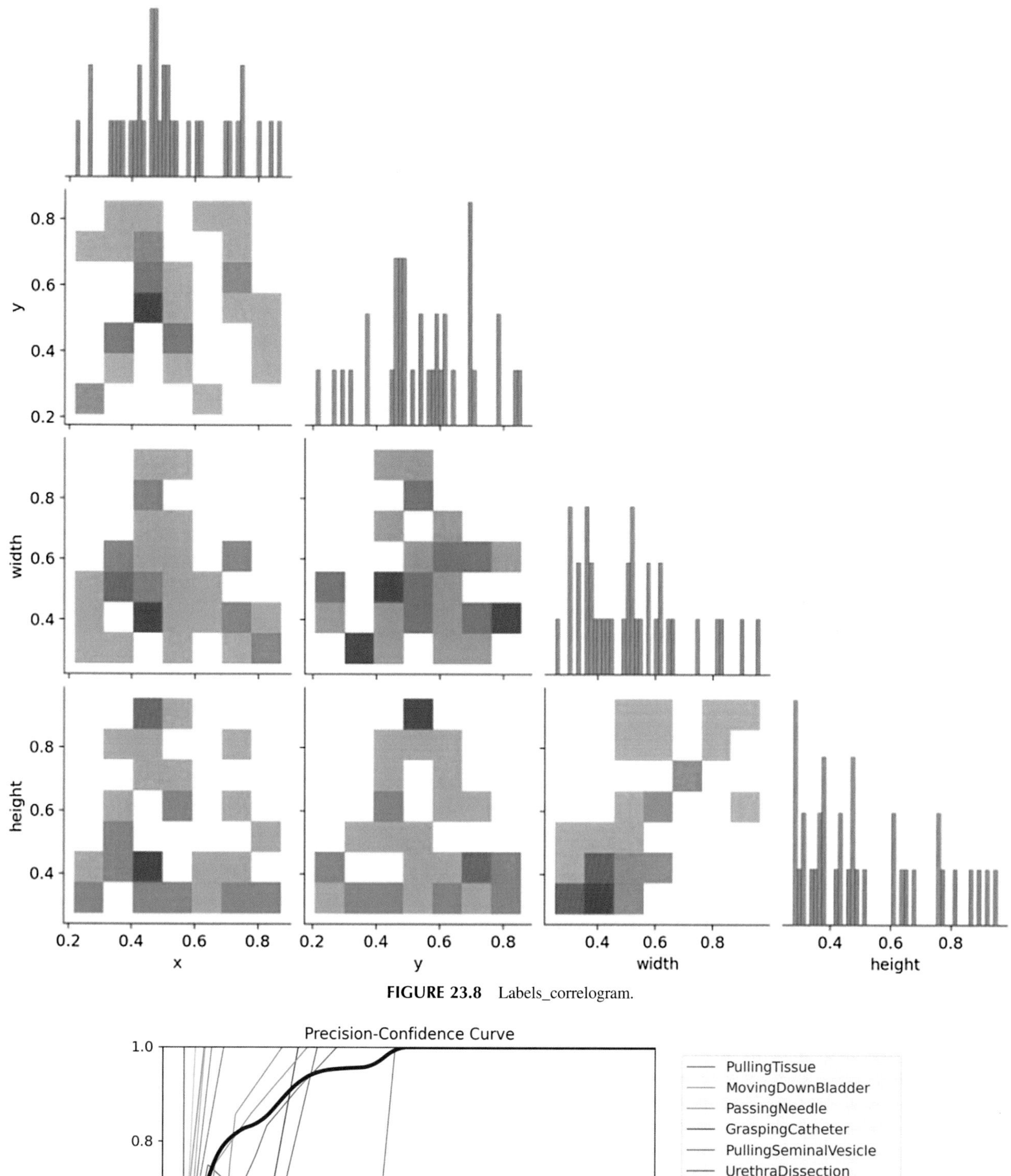

FIGURE 23.8 Labels_correlogram.

FIGURE 23.9 Precision curve generated at 70th epoch.

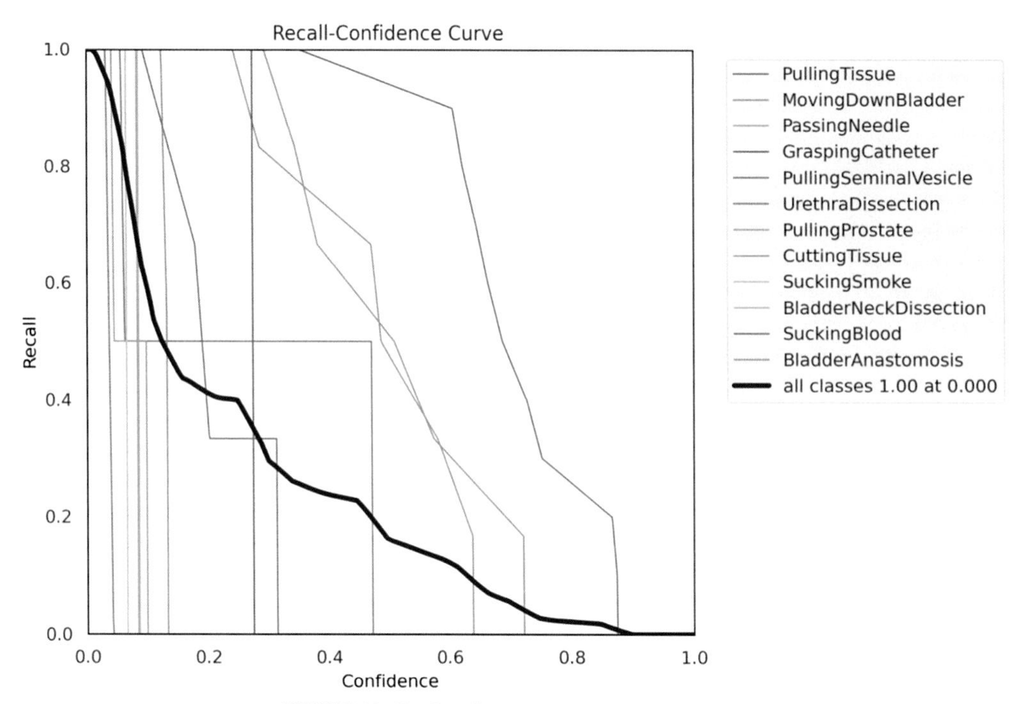

FIGURE 23.10 Recall curve generated at 70th epoch.

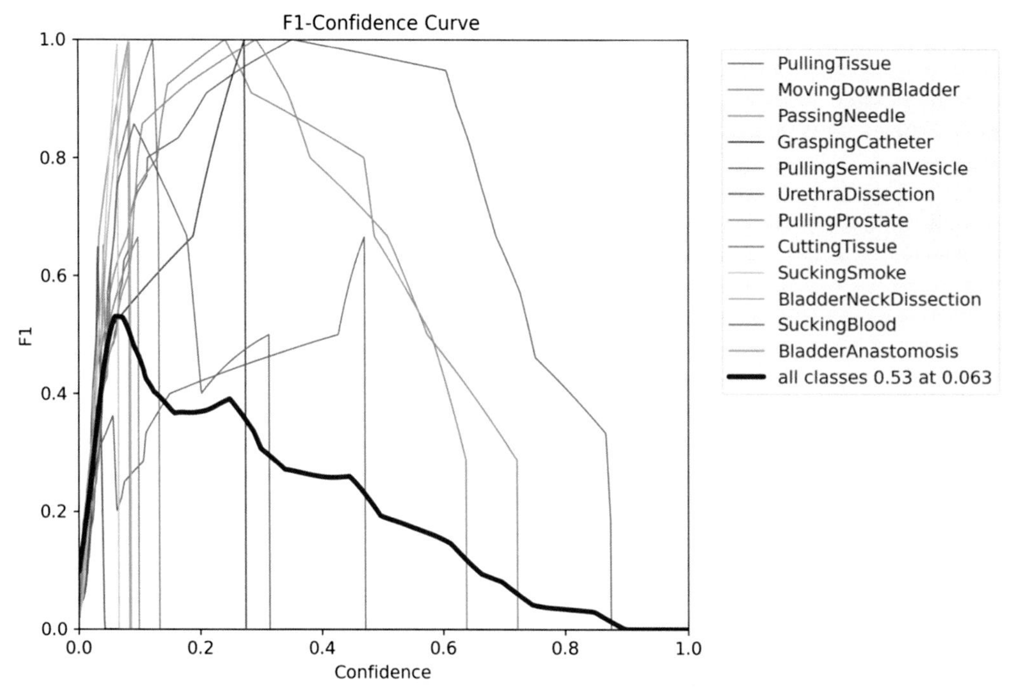

FIGURE 23.11 F1-score curve generated at 70th epoch.

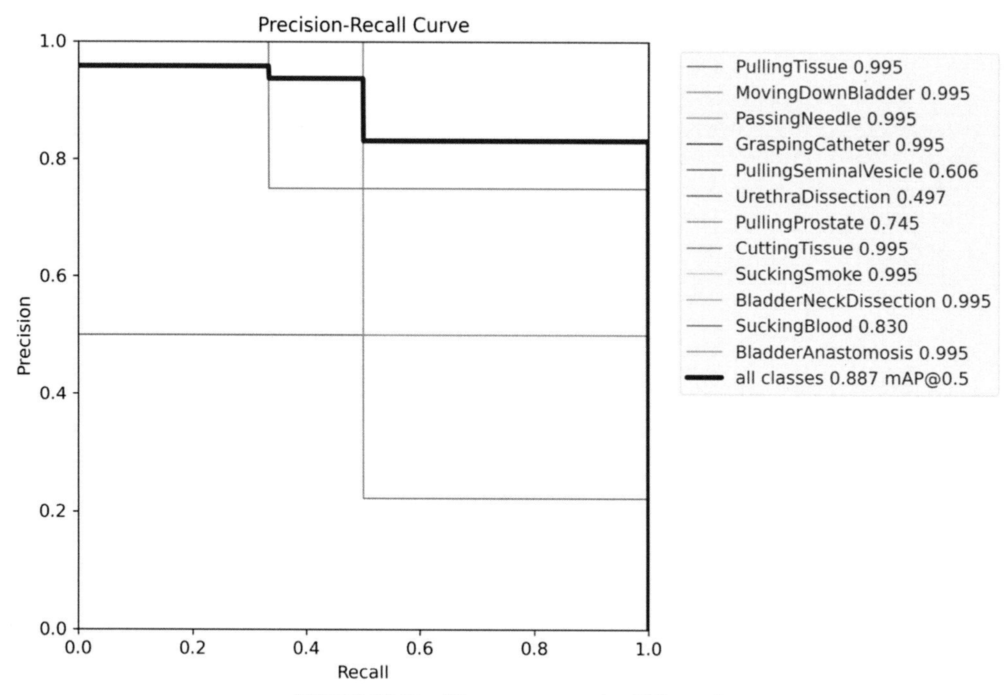

FIGURE 23.12 PR curve generated at 70th epoch.

10. Conclusion

The effective deployment of CNNs significantly enhances the performance of object detection algorithms. However, the individual object detector cannot handle surgeon actions because they are dynamic and have no fixed form. Due to their restricted field of view, object detectors are often fooled by medical components that resemble surgeon actions and produce false positives. This chapter introduces a new ensemble learning model for surgeon action detection and recognition to address these issues. YOLOv5 and EfficientDet, two potent object detectors with distinct specialities, are combined to increase the framework's robustness. Then, a learning network, EfficientNet, is integrated into the framework to reduce false positives. The proposed framework achieves superior surgeon action detection and recognition performance compared with poplar object detectors and state-of-the-art approaches. The significant improvements enable the model to function effectively in real-time action detection applications.

References

[1] S. Prabu, K. Joseph Abraham Sundar, A deep learning framework for grocery product detection and recognition, Food Analytical Methods (2022) 1–25, https://doi.org/10.1007/s12161-022-02384-2.

[2] E. Petlenkov, S. Nomm, J. Vain, F. Miyawaki, Application of self organizing Kohonen map to detection of surgeon motions during endoscopic surgery, in: Proceedings of IEEE International Joint Conference on Neural Networks, 2008, pp. 2806–2811, https://doi.org/10.1109/IJCNN.2008.4634193.

[3] S. Voros, G.D. Hager, Towards "real-time" tool-tissue interaction detection in robotically assisted laparoscopy, in: Proceedings of 2nd IEEE RAS and EMBS International Conference on Biomedical Robotics and Biomechatronics, 2008, pp. 562–567, https://doi.org/10.1109/BIOROB.2008.4762915.

[4] B. Kocev, F. Ritter, L. Linsen, Projector-based surgeon–computer interaction on deformable surfaces, International Journal of Computer Assisted Radiology and Surgery 9 (2) (2014) 301–312, https://doi.org/10.1007/s11548-013-0928-1.

[5] B. Van Amsterdam, H. Nakawala, E. De Momi, D. Stoyanov, Weakly supervised recognition of surgical gestures, in: Proceedings of 2019 International Conference on Robotics and Automation (ICRA), 2019, pp. 9565–9571, https://doi.org/10.1109/ICRA.2019.8793696.

[6] D.P. Azari, Y.H. Hu, B.L. Miller, B.V. Le, R.G. Radwin, Using surgeon hand motions to predict surgical maneuvers, Human Factors 61 (8) (2019) 1326–1339, https://doi.org/10.1177/0018720819838901.

[7] Y. Li, J. Ohya, T. Chiba, R. Xu, H. Yamashita, Subaction based early recognition of surgeons' hand actions from continuous surgery videos, IIEEJ Transactions on Image Electronics and Visual Computing 4 (2) (2016) 124–135.

[8] S. Ameur, A.B. Khalifa, M.S. Bouhlel, Hand-gesture-based touchless exploration of medical images with leap motion controller, in: Proceedings of 17th International Multi-Conference on Systems, Signals and Devices (SSD), 2020, pp. 6–11, https://doi.org/10.1109/SSD49366.2020.9364244.

[9] P. Sa-nguannarm, T. Charoenpong, C. Chianrabutra, K. Kiatsoontorn, A method of 3d hand movement recognition by a leap motion sensor for controlling medical image in an operating room, in: Proceedings of 2019 First International Symposium on Instrumentation, Control, Artificial Intelligence, and Robotics (ICA-SYMP), 2019, pp. 17–20, https://doi.org/10.1109/ICA-SYMP.2019.8645985.

[10] S. Prabu, A. Tripathi, K. Kaur, M.M. Krishna, A. Bora, M.F. Hasan, A comprehensive study of internet of things and digital business on the economic growth and its impact on human resource management, in: Proceedings of 2021 International Conference on Computing Sciences, ICCS), 2021, pp. 91–94, https://doi.org/10.1109/ICCS54944.2021.00026.

[11] Y. Cho, A. Lee, J. Park, B. Ko, N. Kim, Enhancement of gesture recognition for contactless interface using a personalized classifier in the operating room, Computer Methods and Programs in Biomedicine 161 (2018) 39–44, https://doi.org/10.1016/j.cmpb.2018.04.003.

[12] G. Forestier, F. Petitjean, P. Senin, F. Despinoy, A. Huaulmé, H.I. Fawaz, J. Weber, L. Idoumghar, P.A. Muller, P. Jannin, Surgical motion analysis using discriminative interpretable patterns, Artificial Intelligence in Medicine 91 (2018) 3–11, https://doi.org/10.1016/j.artmed.2018.08.002.

[13] S. Khalid, M. Goldenberg, T. Grantcharov, B. Taati, F. Rudzicz, Evaluation of deep learning models for identifying surgical actions and measuring performance, JAMA Network Open 3 (3) (2020) 1–10, https://doi.org/10.1001/jamanetworkopen.2020.1664.

[14] G. Huang, Surgical action recognition and prediction with transformers, in: Proceedings of 2022 IEEE 2nd International Conference on Software Engineering and Artificial Intelligence (SEAI), 2022, pp. 36–40, https://doi.org/10.1109/SEAI55746.2022.9832094.

[15] B. Zhang, J. Abbing, A. Ghanem, D. Fer, J. Barker, et al., Towards accurate surgical workflow recognition with convolutional networks and transformers, Computer Methods in Biomechanics and Biomedical Engineering: Imaging & Visualization 10 (4) (2022) 349–356, https://doi.org/10.1080/21681163.2021.2002191.

[16] C.I. Nwoye, C. Gonzalez, T. Yu, P. Mascagni, D. Mutter, J. Marescaux, N. Padoy, Recognition of instrument-tissue interactions in endoscopic videos via action triplets, in: International Conference on Medical Image Computing and Computer-Assisted Intervention, 2020, pp. 364–374, https://doi.org/10.1007/978-3-030-59716-0_35.

[17] W. Xu, R. Liu, W. Zhang, Z. Chao, F. Jia, Surgical action and instrument detection based on Multiscale information fusion, in: Proceedings of 2021 IEEE 13th International Conference on Computer Research and Development (ICCRD), 2021, pp. 11–15, https://doi.org/10.1109/ICCRD51685.2021.9386349.

[18] L. Li, X. Li, S. Ding, Z. Fang, M. Xu, H. Ren, S. Yang, SIRNet: fine-grained surgical interaction recognition, IEEE Robotics and Automation Letters 7 (2) (2022) 4212–4219, https://doi.org/10.1109/LRA.2022.3148454.

[19] B. Van Amsterdam, M.J. Clarkson, D. Stoyanov, Multi-task recurrent neural network for surgical gesture recognition and progress prediction, in: Proceedings of 2020 IEEE International Conference on Robotics and Automation (ICRA), 2022, pp. 1380–1386, https://doi.org/10.1109/ICRA40945.2020.9197301.

[20] H. Huynhnguyen, U.A. Buy, Toward gesture recognition in robot-assisted surgical procedures, in: Proceedings of 2020 2nd International Conference on Societal Automation (SA), 2021, pp. 1–4, https://doi.org/10.1109/SA51175.2021.9507175.

[21] R. Xu, H. Lin, K. Lu, L. Cao, Y. Liu, A forest fire detection system based on ensemble learning, Forests 12 (2) (2021) 1–16, https://doi.org/10.3390/f12020217.

[22] S. Prabu, K. Joseph Abraham Sundar, Enhanced attention-based encoder-decoder framework for text recognition, Intelligent Automation & Soft Computing 35 (2) (2023) 2071–2086, https://doi.org/10.32604/iasc.2023.029105.

[23] M. Sumathi, S. Prabu, Random forest based classification of user data and access protection, International Journal of Recent Technology and Engineering 8 (1) (2019) 1630–1635.

[24] S. Prabu, M. Sumathi, Evaluating phishing website detection on client side, International Journal of Recent Technology and Engineering 8 (6) (2019) 18–22.

[25] S. Prabu, K. Joseph Abraham Sundar, Object segmentation based on the integration of adaptive K-means and GrabCut algorithm, in: Proceedings of 2022 International Conference on Wireless Communications Signal Processing and Networking, 2022, pp. 213–216, https://doi.org/10.1109/WiSPNET54241.2022.9767099.

[26] V.S. Bawa, G. Singh, F. KapingA, I. Skarga-Bandurova, E. Oleari, A. Leporini, et al., The Saras Endoscopic Surgeon Action Detection (Esad) Dataset: Challenges and Methods, 2021, https://doi.org/10.48550/arXiv.2104.03178 arXiv preprint arXiv:2104.03178.

Further reading

[1] J.H. Kim, N. Kim, Y.W. Park, C.S. Won, Object detection and classification based on YOLO-V5 with improved maritime dataset, Journal of Marine Science and Engineering 10 (3) (2022) 1–14, https://doi.org/10.3390/jmse10030377.

Chapter 24

Understanding of healthcare problems and solutions using deep learning

Rajesh Kumar Shrivastava[1], Simar Preet Singh[1], Simranjit Singh[1] and Mohit Sajwan[2]
[1]*School of Computer Science Engineering and Technology (SCSET), Bennett University, Greater Noida, Uttar Pradesh, India;* [2]*Department of Information Technology, Netaji Subhash University of Technology, Dwarka, Delhi, India*

1. Introduction

Now a days, many repositories provide Electronic Health Record (EHR). But this EHR needs experts to understand, or we can say it is a black box for naïve users. These EHR need an expert to interpret them. Deep learning (DL) is the answer to this problem. Consequently, there is a growing demand for interpretable DL that enables end-users to assess the model. Before taking action, a user must decide whether to accept or reject predictions and recommendations. This chapter concentrates on the healthcare industry's capacity to comprehend DL models. To provide future researchers or clinical practitioners in this subject with a methodological reference, we begin by thoroughly outlining the approaches for interpretability. Along with the specifics of the strategies, we also analyze their benefits and drawbacks and the situations in which each is appropriate so that interested readers can compare them and decide which one to employ. Additionally, we go through how these approaches, which were first created to address problems in the public domain, have been modified and used to address issues in the healthcare industry and how they might aid physicians in comprehending these data-driven technologies. Overall, we believe this chapter will enable researchers and practitioners in the disciplines of clinical and artificial intelligence (AI) to understand better the techniques available for improving the interpretability of their DL models and select the most appropriate one.

The EHR data include both structured data (like patient demographics, diagnoses, and procedures) and unstructured data (like doctor notes and medical images) [1]. A class of machine learning (ML) models called DL, which is based on deep neural networks, has made a significant impact over the past 10 years on datasets for a variety of modalities, including pictures, natural language, and structured time series data [2]. A rise in research interest in creating a range of DL-based clinical decision support systems for diagnosis, prognosis, and therapy has been caused by the accessibility of vast amounts of data and previously unheard-of technological advancements [3].

Although DL is important to deal with voluminous and complex data, it also faced challenges in healthcare [4]. One major challenge is the opacity or "black box" aspect of DL algorithms, which makes it difficult to understand how the information about a case influences the model's decision. DL algorithms have numerous "high stakes" applications in healthcare, including predicting a patient's likelihood of hospital readmission [5], diagnosing a patient's disease [6], and recommending the best drug prescription and therapy plan [7]. There is considerable reluctance in the deployment of these models in these crucial use cases that involve clinical decision-making because the cost of model misclassification could be quite high [8]. DL models have also been shown to be fragile and susceptible to failure in the presence of both man-made and natural noise [9].

In this chapter, we concentrate on the interpretability of the DL models in healthcare. Although in some cases, DL systems may be biased [10]. One method for ensuring the freedom of such systems is to ensure interpretability from unfairness and bias when grading various racial and socioeconomic groups [10].

DL algorithms are applied in various nonhealthcare systems also to solve problems where a decision or recommendation is required, for example, agriculture, image processing, and statistical data analysis. But in this chapter, we discuss the utility and algorithms which are helpful in the healthcare system.

Deep Learning in Personalized Healthcare and Decision Support. **https://doi.org/10.1016/B978-0-443-19413-9.00016-3**

This chapter performs DL methods to the Annual Health Survey (AHS) 2012−13 [11] medical data. The goal of the AHS evaluation is to produce a thorough, accurate, and reliable dataset on the most critical health indicators, such as composite ones like infant mortality rate, maternal mortality ratio, and total fertility rate, as well as their covariates (process and outcome indicators) at the district level, and to map annual changes in those indicators. These standards would aid in a more thorough knowledge and prompt monitoring of the different factors that affect well-being, population health, in particular, reproductive, and pediatric health. The goal of the AHS is to produce a thorough, accurate, and reliable dataset on the most critical health indicators, such as composite ones like infant mortality rate, maternal mortality ratio, and total fertility rate, as well as their covariates (process and outcome indicators) at the district level, and to map annual changes in those indicators. These benchmarks will aid in a more thorough understanding of the many factors that affect population health and well-being, particularly reproductive and child health, as well as timely monitoring of those factors.

2. Related work

Chen et al. [12] proposed a hybrid deep-learning model to detect stroke risk protection (SRP). The problems of SRP with small and unbalanced stroke data have been solved in this work. Three essential elements make up our innovative Hybrid Deep Transfer Learning-based Stroke Risk Prediction (HDTL-SRP) framework. (3) Active Instance Transfer (AIT) for balancing the stroke data with the most insightful generated instances. (1) Network weight transfer (NWT) for using data from highly correlated diseases (such as hypertension or diabetes). (2) Generative Instance Transfer (GIT) for using external stroke data distribution among multiple hospitals while maintaining privacy. The suggested HDTL-SRP framework is found to perform better than the most advanced SRP models in both simulated and real-world circumstances.

Wickstrøm et al. [13] proposed a deep ensemble strategy for explainable CNNs. This is based on both simulated and actual data and the suggested strategy. Results show that deep ensembles can locate pertinent characteristics in clinical time series and that more comprehensible and reliable explanations can be offered by modeling the uncertainty in relevance scores. The idea and demonstration of a novel thresholding approach were made. Although just one thresholding was examined in this work, we think that other thresholding strategies may be relevant, which is an intriguing area of research for other works. The contributions of this research can help create decision support systems that are more reliable and accurate than earlier DL-based systems.

Kermany et al. [14] outline a generic AI platform for the diagnosis and referral of diabetic macular edema and choroidal neovascularization associated with neovascular AMD, two prevalent causes of significant vision loss. Our model demonstrated competitive OCT image analysis performed using a transfer learning approach without the need for a highly specialized DL machine or a database of millions of example photos (STAR Methods). Furthermore, the model's accuracy in identifying retinal disorders from OCT pictures was on par with that of retinal disease experts with extensive clinical training. The model maintained great performance in accuracy, sensitivity, specificity, and area under the ROC curve for getting the correct diagnosis and referral even after training with a substantially smaller number of images (about 1000 from each class).

Liu et al. [15] said that, in order to manage and enhance dental treatment accessibility and deliver home-based dental health care services more effectively, this article offers an iHome smart dental Health-IoT system based on intelligent hardware, DL, and a mobile terminal. The identification and classification of dental disorders were made possible by the trained model and client- and dentist-side application software (Apps) for mobile devices were developed. The service docking between the patient and the dentist's resources became a reality, thanks to the software platform's features, which include pre-examination of dental disease, consultation, appointment, and evaluation, among others. For seven dental illnesses, the AI algorithm had an identification rate of more than 90%. Authors [16,17] also showed their work in healthcare domain.

Harerimana et al. [18] had provided technical intuitions and insights into how to use DL techniques to benefit from the EHR data. They also dissected the technical aspects of numerous initiatives that have invested in using the enormous data sets from EHRs to apply DL models for clinical knowledge discovery. Even while DL has been clearly successful for other hospital operations like billing and patient administration, there is still more to be done in the application of DL to EHR data. The results that are now possible in this area still require the supervision of a medical domain expert. To bring AI and DL to the patient's bedside, more research must be done.

Waring et al. [19] presented a review on an automated machine learning (AutoML) for helping the healthcare professionals in efficient utilization of machine learning models "off-the-shelf" having little knowledge of data science. They reviewed 101 papers in the field of AutoML and identified numerous opportunities and barriers in using AutoML in the healthcare sector. They identify the main limitation of such a system as the ability of system to work efficiently at a large scale.

3. Methods

3.1 Image data processing

The standard approach to processing image data is given as follows:

The classification of an EHR using conventional methods is dependent on the use of statistical learning approaches and low-level features. Methods of segmentation in this area concentrate on the statistical information and estimate of the patient condition, which includes preprocessing techniques like data filtering, edge detection, image enhancement, image compression and sharpening, etc.

The classic EHR methods' fundamental process is as follows:

Capture, preparation, segmentation of the ROI, and feature extraction and classification, dimensionality reduction, and performance assessment, as visually summarized in Fig. 24.1.

The image data are mainly focused on 3Rs, that is, recognition, reconstruction, and reorganization. Medical image data may have a single object or multi objects in the image. As per the problem, the researcher must have to decide on the solution approaches. The problem associated with single object detection is classifying the object and localization of the problem or illness. In the case of multiobject detection very first, we have to segment the object and then localize the illness. In the image data, the processing is divided into the following steps:

1. Low-level processing: This part consists of various image processing techniques such as feature detection, matching, and early segmentation.
2. Mid-level processing: It will subgroup the image and try to understand the meaning of the attribute of the sub-image.
3. High-level processing: This process has the sense of the visual content and make identifies the medical illness up to the capabilities of humans understanding.

3.2 Object recognition

In medical images, it is a big challenge to detect or identify the area where a problem exists. Object recognition techniques are used to figure out the problem area.

For example, Fig. 24.2 shows an X-ray report. If we wish to search for an anomaly or illness in this image, we need an algorithm that helps to detect an object to target the illness in the image. Object detection refers to finding a small segment in a whole image. That is known as object detection which researchers do with a medical image. The research question is that is the detection of any object is tough. The answer is in Fig. 24.3.

In Fig. 24.3, we have various chest X-rays images. It is challenging to determine which image contains COVID-19 symptoms once we understand the purpose of object detection. Another research question is how to determine the objects from the images. Wang et al. [30] already discussed point to point correlation method. The author mentioned that information retrieval from medical images is complicated especially if multiple features are merged in the image. A medical image contains high-dimensional data with nonuseful information. Wang et al. [30] proposed a fine-grained image retrieval method that looks for local points and segment images in fine-grained samples.

This point-to-point correlation method is proper but not much effective with 3D images it also suffers in occlusion, changes in viewing angle, and articulation of parts. Following are a few research challenges to determining objects.

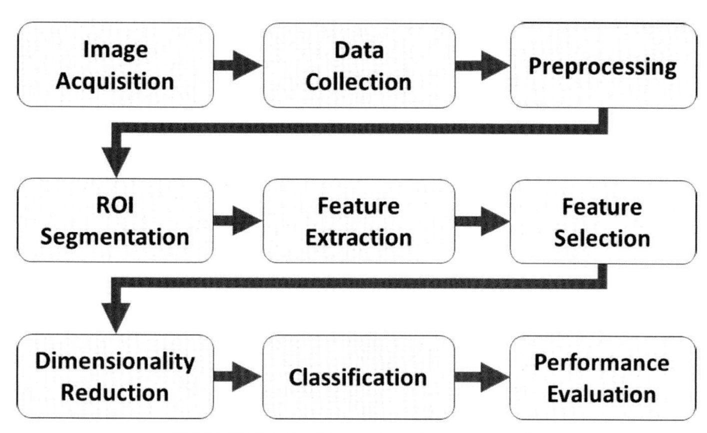

FIGURE 24.1 General analysis approach.

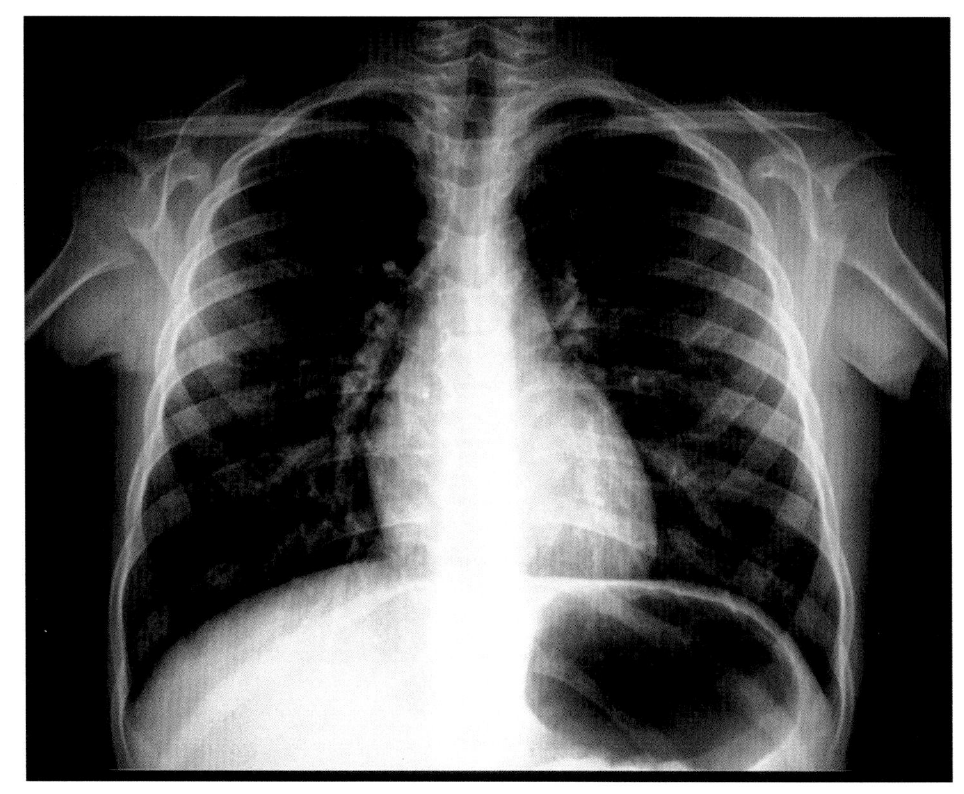

FIGURE 24.2 Chest X-ray.

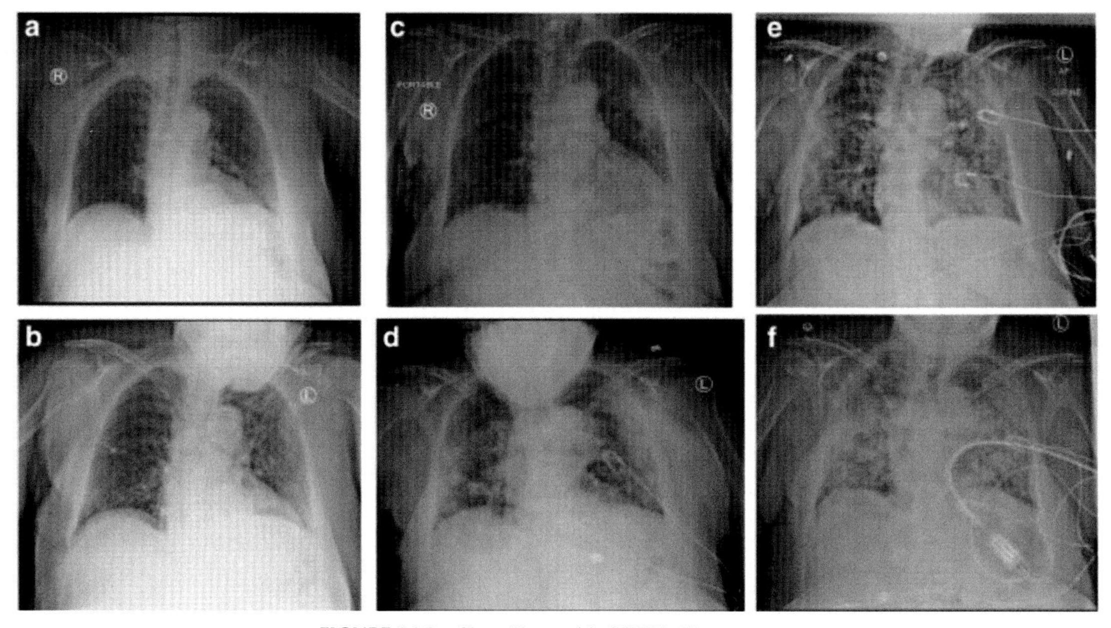

FIGURE 24.3 Chest X-ray with COVID-19 symptoms.

1. viewpoint variation
2. illumination
3. occlusion
4. scale
5. deformation
6. background clutter

Matching two different images and finding the similarity is the critical factor in identifying the correct solution. Alignment is the one way to do the same. Alignment is the process of fitting a model to a transformation between pairs of features (*matches*) in two images.

Another solution is the recognition by component. The fundamental assumption of recognition-by-components (RBC) [20] is that a modest set of generalized-cone components, called geons, can be derived from contrasts of five readily detectable properties of edges in a two-dimensional image: curvature, collinearity, symmetry, parallelism, and contamination. This method works on the hypothesis that there is a small number of geometric components that constitute the primitive elements of the object recognition system (like letters to form words). The properties of edges that are postulated to be relevant to the generation of the volumetric primitives have the desirable properties that they are invariant over changes in orientation and can be determined from just a few points on each edge. But this approach has limitations also, that is, the modeling has been limited to concrete entities with specified boundaries. This limitation is shared by many modern object detection algorithms.

Feature-based methods [21] are commonly used in medical images. This method combines *local* appearance, spatial constraints, invariants, and classification techniques from machine learning. But now in the present scenario, bag-of-feature (BOF) [22] method is used primarily (Fig. 24.4).

This model uses various approaches such as a combination of local and global methods, modeling context, integrating recognition, and segmentation. This model extract features from images and creates a visual vocabulary. That vocabulary acts like a dictionary and helps to determine the correct portion of illness.

4. Standard methods for learning

To understand the usage of DL in healthcare, we took the example of Pneumonia. Pneumonia is a lung inflammation that primarily affects the tiny air sacs known as alveoli. A productive or dry cough, chest pain, a fever, and breathing difficulties are typical symptoms. The disorder can range in severity. The most prevalent causes of pneumonia are infections with viruses or bacteria; other microbes, certain drugs, or illnesses, including autoimmune diseases, are less frequently to blame. Cystic fibrosis, chronic obstructive pulmonary disease (COPD), asthma, diabetes, heart failure, a history of smoking, a bad cough reflex, such as after a stroke, and having a weakened immune system are risk factors. The physical exam and symptoms are frequently used to make a diagnosis. Blood tests, sputum culture, and chest X-rays can all support the diagnosis.

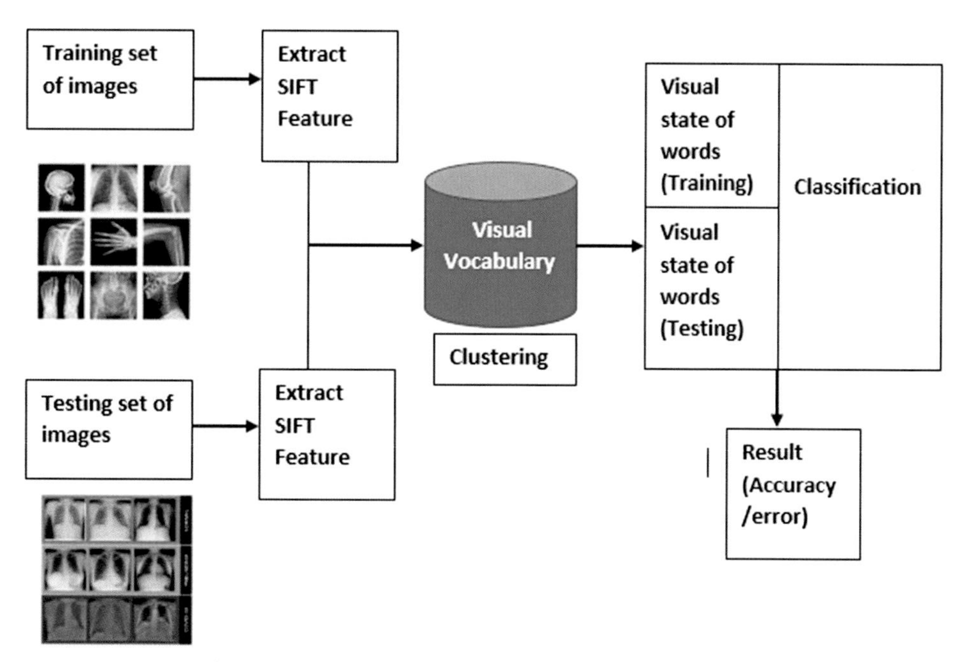

FIGURE 24.4 Bag-of-feature (BOW) model for medical image [23].

4.1 Data collection

Each image category (pneumonia/normal) has its subfolder within the dataset's three main folders such as training, testing, and validation. 5863 X-ray images are available [24], divided into two groups (pneumonia and normal). From retrospective cohorts of children patients aged one to five at the Guangzhou Women and Children's Medical Center in Guangzhou, chest X-ray images (anterior-posterior) were chosen. All chest X-ray imaging was done as part of the regular clinical treatment provided to patients. All chest radiographs were initially checked for quality control before being removed from the study of the chest X-ray pictures. Before the diagnosis for the photos could be used to train the AI system, they were graded by two experienced doctors. A third expert also reviewed the evaluation set to make sure there were no grading mistakes. The dataset is available on Kaggle. These data divide into two categories, images of normal chest X-rays and images of pneumonia chest X-rays. Fig. 24.5 shows the classification of data in the training set. In contrast, we divide data in an 80-20 ratio for training and testing. Further, we utilize 10% of training data for validation.

4.2 Experimental results

We applied several methods to verify DL in healthcare. The first approach that we discuss here is the CNN method. The standard CNN model has the following layers.

Fig. 24.6 explains the working of the CNN model. Very first, we have the input of medical images. That will process by the CNN layer. CNN layer divides images into small grids, that is, 3×3. They provide these small grids to the max pooling layer. The max pool layer reduces the size of an image by half. This layer multiplies the image by a 2×2 matrix, so the size of the image will become half. The Max pool layer will take the maximum value of the pixel while reducing it, so the image will not lose its features. Once we have reduced size, the flatten operation is performed on the image so the image will convert into a 1-dimensional array. Then, the CNN model applies a neural network (NN). In the NN, input data are processed with a hidden layer. The hidden layer uses the "RELU" activation function. The RELU function takes only supporting values, so the CNN model trains very fast. Our outcome is based on binary value, that is, either the patient is suffering from illness or not, so the CNN model uses the "Sigmoid" function for output. After completion of training, our CNN model gets 88% accuracy in the validation with a 32% of loss. Fig. 24.7 shows the experimental result of the CNN model. Where Fig. 24.7A is about the accuracy achieved in each model. And Fig. 24.7B shows the epoch-wise losses. We can easily observe that if epochs are increasing in the CNN model, then this model achieves better accuracy with minimum losses.

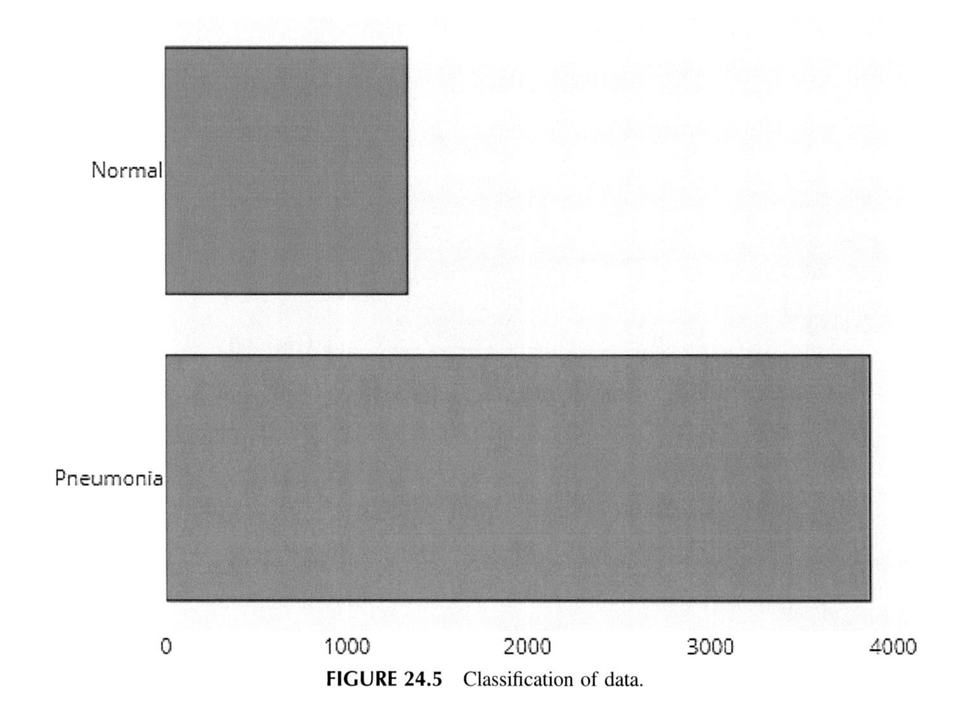

FIGURE 24.5 Classification of data.

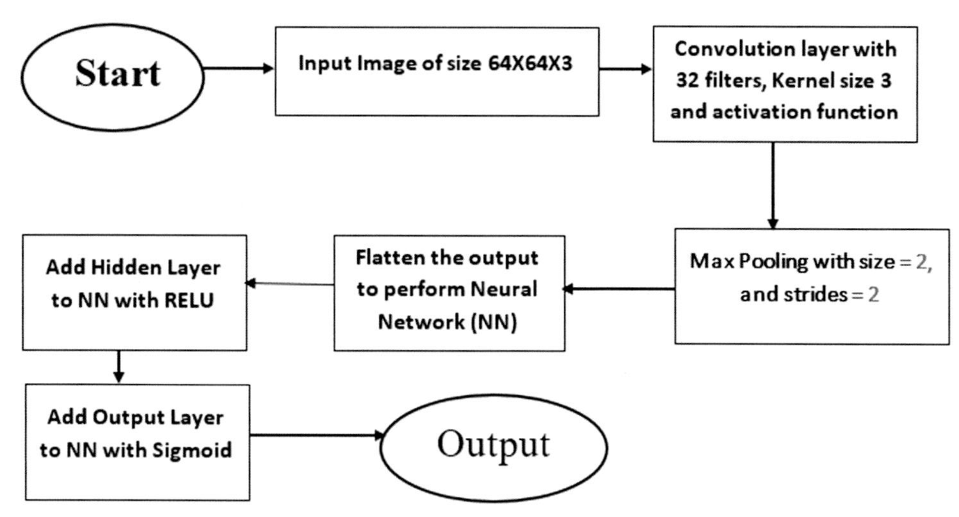

FIGURE 24.6 Convolutional neural network (CNN) model.

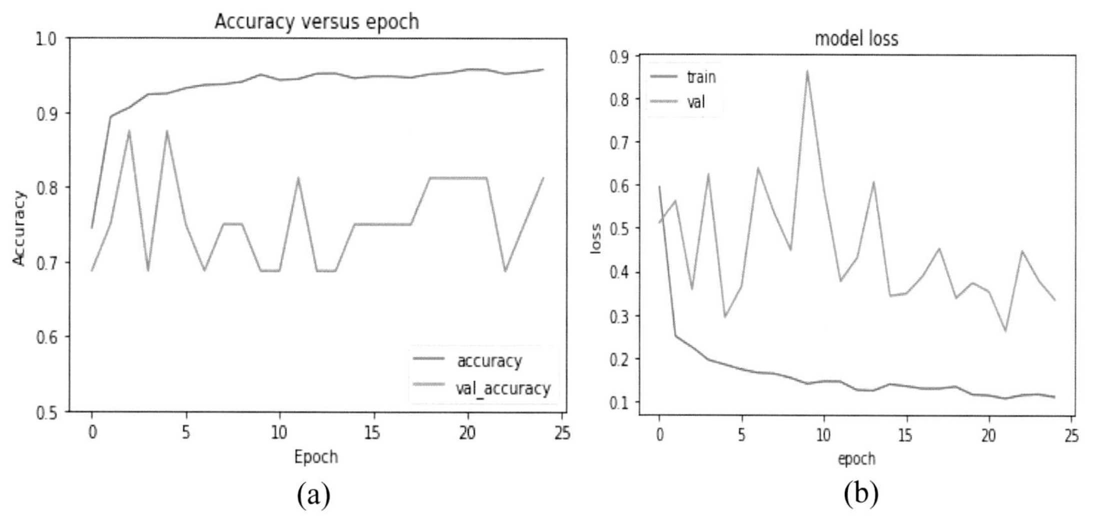

FIGURE 24.7 Experimental result of convolutional neural network (CNN). (A) accuracy versus epoch. (B) model loss versus epoch.

Fig. 24.8 shows the modified CNN model. CNN process depends on how many layers are involved in it. If we add more layers, it needs more computing power, but the result is better. We also evaluate the same principle and modify CNN to get better results.

Fig. 24.9 shows the total loss percentage. It is shown that when we run our model for a greater number of epochs, then the loss will reduce. Here, Fig. 24.9 also depicts that validation and training both losses are both symmetric. After a certain execution of epochs, the loss will remain constant. We achieve 13% of validation lose in this experiment.

Fig. 24.10 depicts the accuracy graph. We achieved 94% accuracy with our modified CNN model, as it is clear that when the algorithm grows with epochs, the accuracy increases. Accuracy also depends on data size and the batch size selected for the problem. But this model suffers in the time of testing. For testing purposes, we used a different set of data, and the model achieved 60% loss and 79% of accuracy. Hence, this model needs more strengths, and we have a scope for improvement.

To continue our discussion about the usage of ML/DL in healthcare, we try to inculcate another famous model Residual Networks (ResNet). This model is commonly used for benchmarking and comparison purposes.

Residual Networks (ResNet) [25–27]: A common neural network that serves as the foundation for many computer vision applications is called ResNet. This model was introduced in the 2015 ImageNet challenge. ResNet represented a significant advancement in that it successfully enabled us to train intense neural networks with more than 150 layers. This model is beneficial for training. It is simple to train networks with several layers, even thousands, without raising the training error percentage. By applying identity mapping, ResNets assist in solving the vanishing gradient problem. The

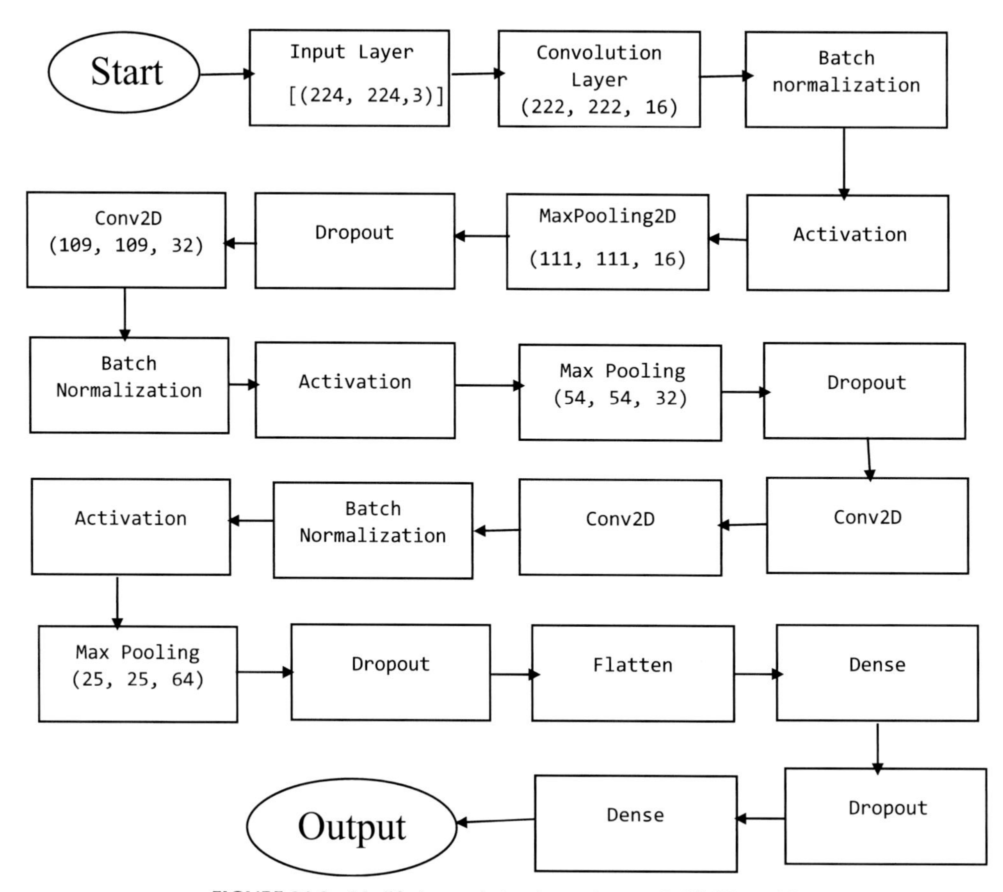

FIGURE 24.8 Modified convolutional neural network (CNN) model.

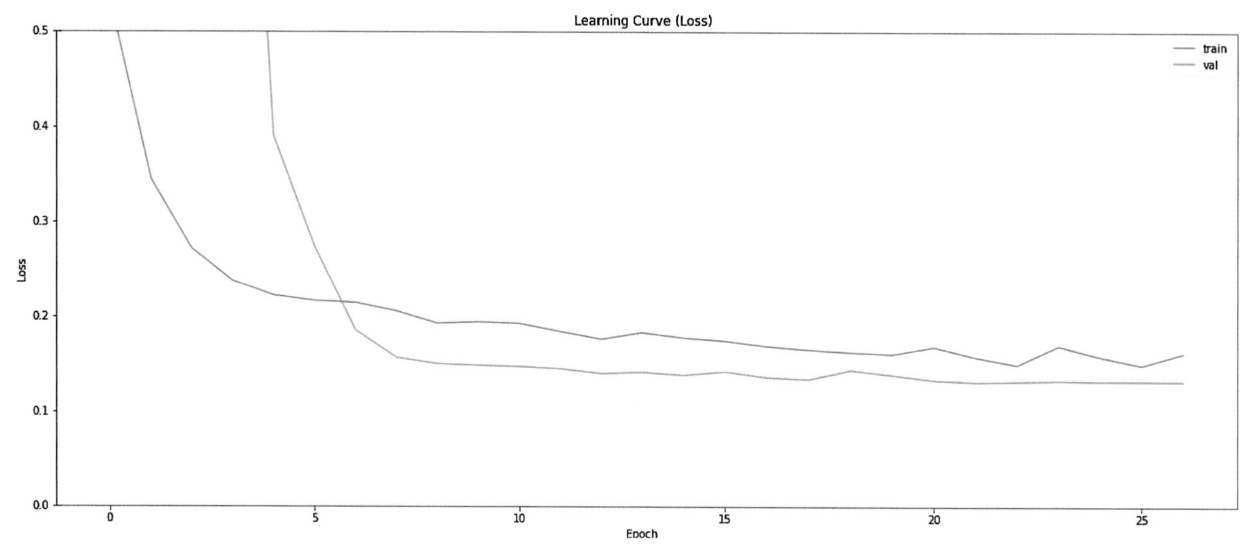

FIGURE 24.9 Loss curve of modified convolutional neural network (CNN) model.

most important thing is that ResNet is a CNN, but it overcomes the problem of vanishing gradient [28]. It is a gateless or open-gated variation of the HighwayNet [29], which was the first functionally complete, extremely deep feedforward neural network with hundreds of layers—much deeper than earlier neural networks. A ResNet is made up of multiple blocks, one for each layer. This is due to the fact that ResNets typically increases the number of operations within a block to go deeper, while the total number of layers four remains constant (Fig. 24.11).

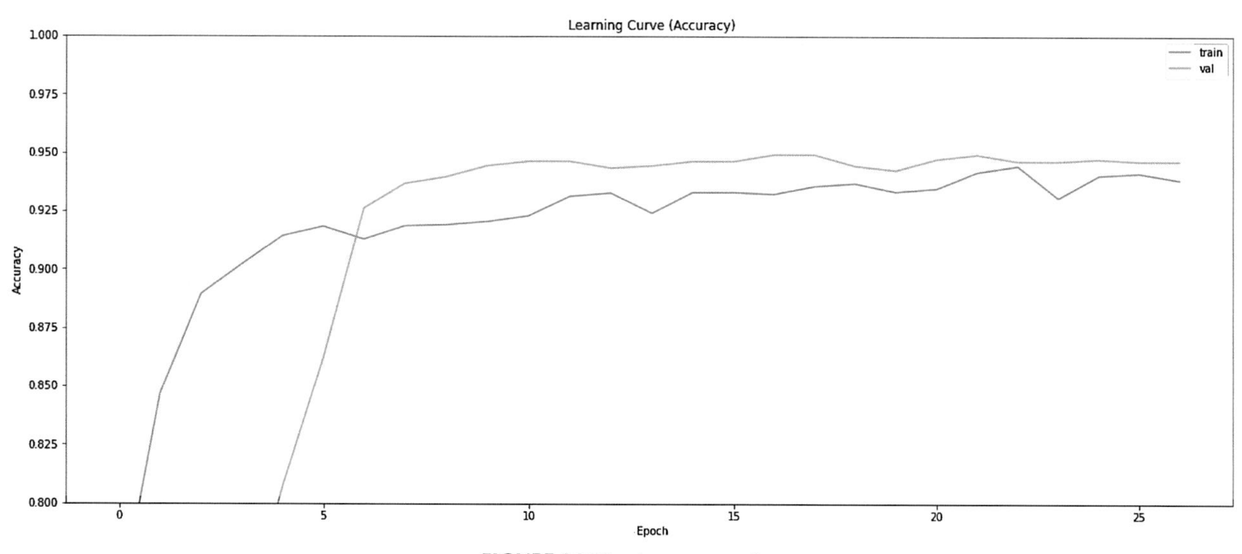

FIGURE 24.10 Accuracy graph.

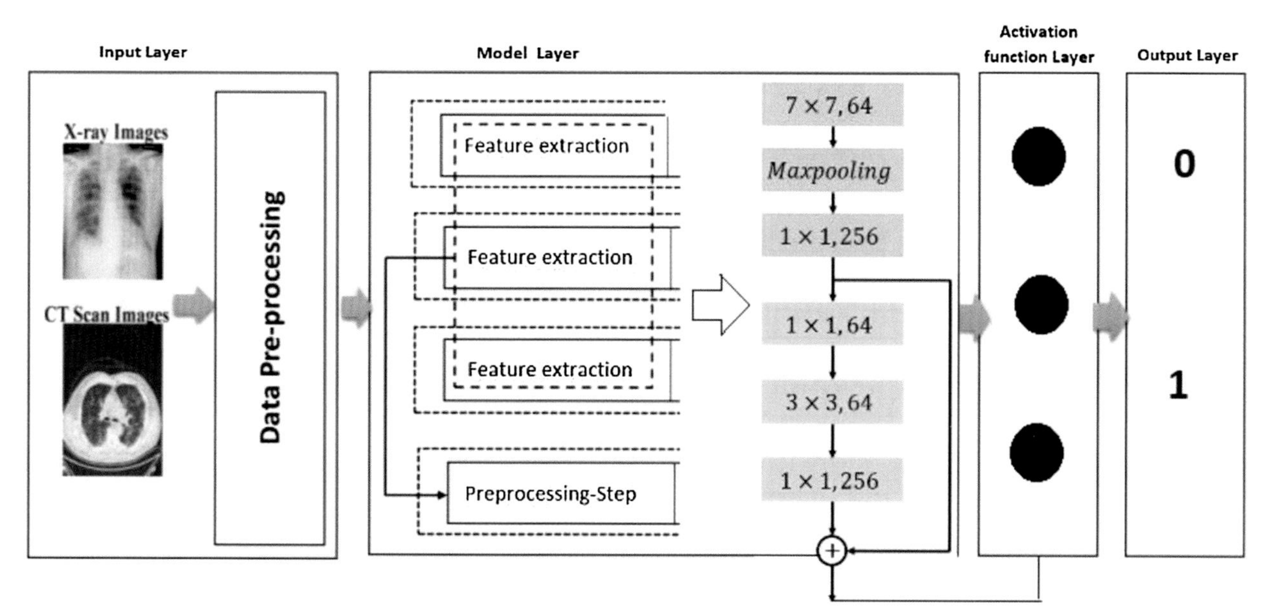

FIGURE 24.11 Example of ResNet architecture.

The Resnet model is mainly build-up with following layers

Layer	Output shape
input_1 (InputLayer)	[(None, 224, 224, 3)]
resnet152v2 (Functional)	(None, 7, 7, 2048)
global_average_pooling2d (Gl	(None, 2048)
dense (Dense)	(None, 128)
dropout (Dropout)	(None, 128)
dense_1 (Dense)	(None, 1)

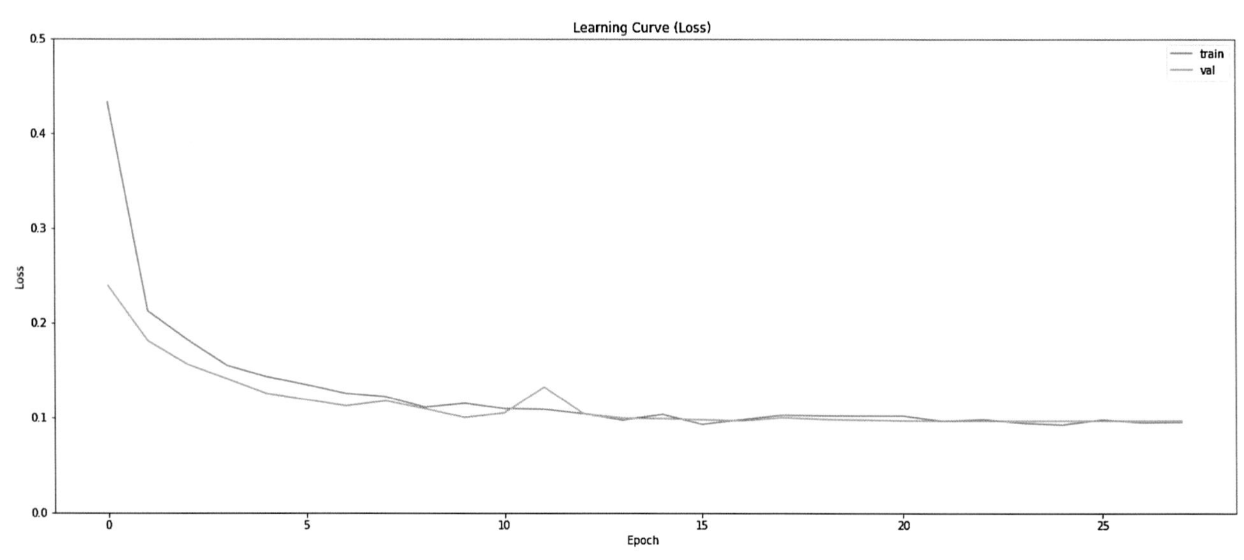

FIGURE 24.12 Loss in ResNet model.

The ResNet model shows the 09% of loss (Fig. 24.12) on training and validation which is better than our previous model. It also shows the 96% of accuracy (Fig. 24.13) in training time.

The ResNet model that we depict here reduced test loss (Fig. 24.14) by up to 41% and test accuracy by 84% (Fig. 24.15).

4.3 Confusion matrix

Another evaluation parameter that is used in DL is the confusion matrix. A confusion matrix is used to describe how well a classification system performs. The output of a classification algorithm is shown and summarized in a confusion matrix. Table 24.1 displays a confusion matrix where benign tissue is referred to be healthy and malignant tissue is regarded as cancerous (Fig. 24.16).

Basically, the confusion matrix is the evaluation of true positive (TP), false positive (FP), true negative (TN), and false negative (FN). Important predictive analytics like recall, specificity, accuracy, and precision are visualized using confusion matrices. Because they provide direct comparisons of variables like true positives, false positives, true negatives, and false negatives, confusion matrices are helpful.

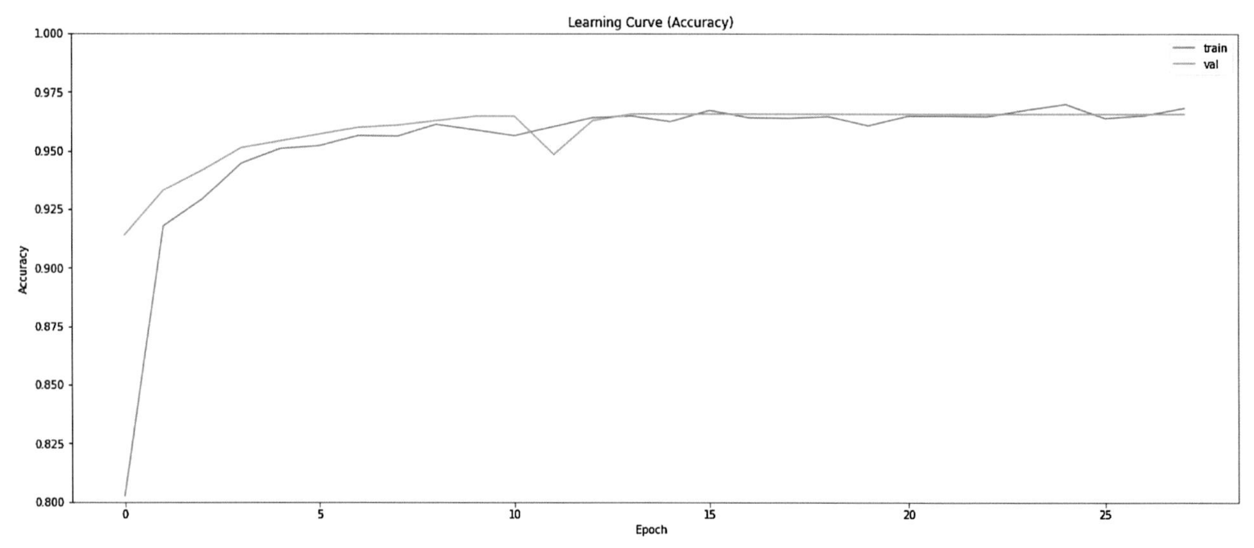

FIGURE 24.13 Accuracy achieved in ResNet model.

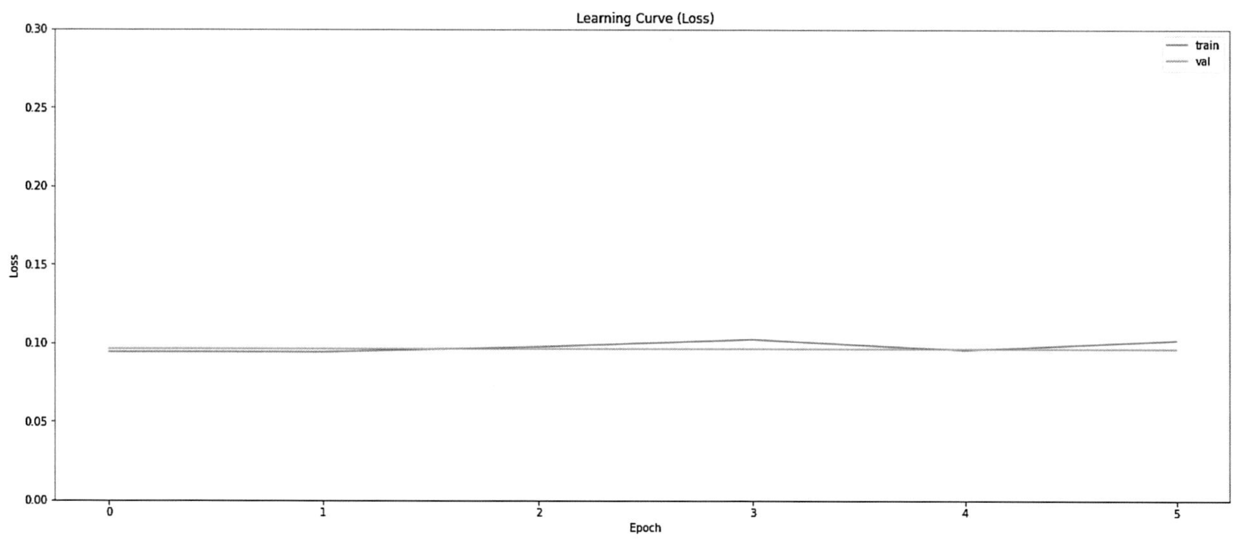

FIGURE 24.14 Loss in testing (RESNET).

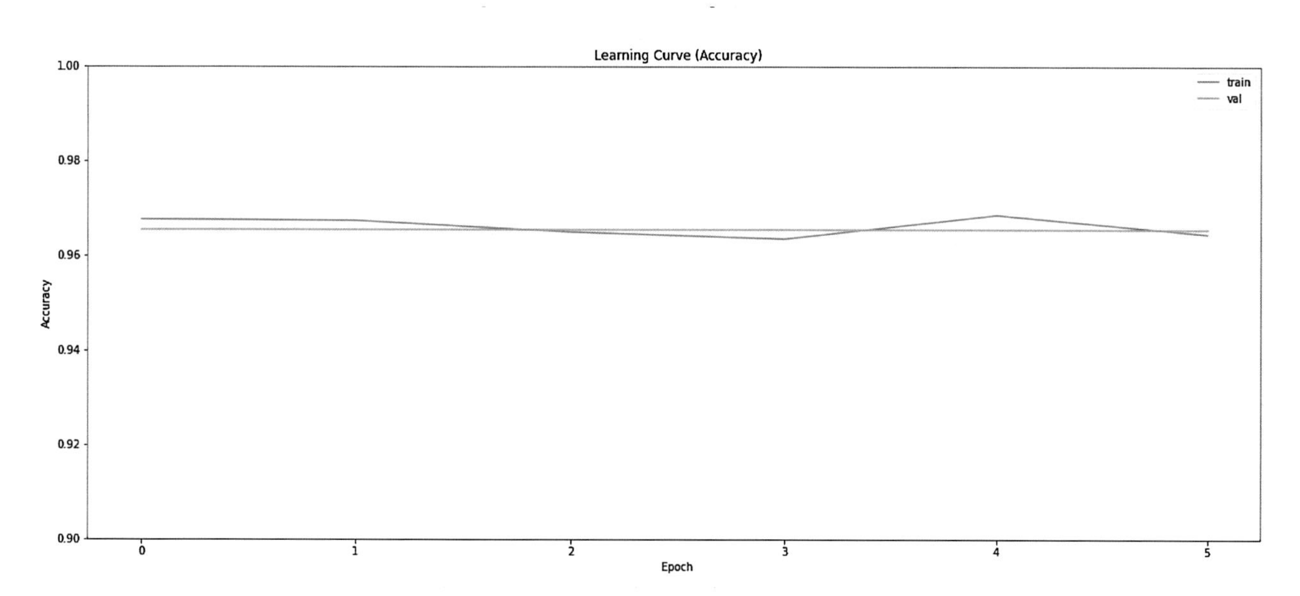

FIGURE 24.15 Accuracy in testing (RESNET).

TABLE 24.1 Confusion matrix.

	Precision	Recall	F1-score
Pneumonia (class 0)	0.95	0.63	0.76
Normal (class 1)	0.81	0.98	0.89
Accuracy	0.88	0.85	0.87
Weighted avg	0.86	0.85	0.84

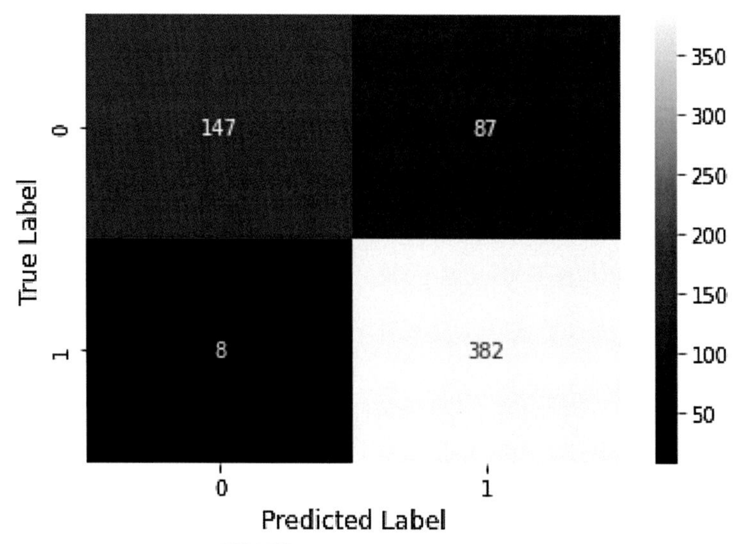

FIGURE 24.16 Confusion matrix.

5. Conclusion and future directions

In this book chapter, we give an insight into DL models. We also discussed how artificial intelligence helps us to make analysis easy. We also experimented with medical data that depict the usage of data learning in healthcare. Security is another concern in HER and healthcare. We will also look into this aspect in the future. In the end, we wish to explore the field of explainable AI (XAI). XAI lead us to get new heights in the healthcare industry.

References

[1] S. Santoshi, D. Sengupta, Artificial intelligence in precision medicine: a perspective in biomarker and drug discovery, in: Artificial Intelligence and Machine Learning in Healthcare, Springer, Singapore, 2021, pp. 71–88.

[2] Y.H. Wang, P.A. Nguyen, M.M. Islam, Y.C. Li, H.C. Yang, Development of deep learning algorithm for detection of colorectal cancer in EHR data, MedInfo 264 (2019) 438–441.

[3] S. Dash, S.K. Shakyawar, M. Sharma, S. Kaushik, Big data in healthcare: management, analysis and future prospects, Journal of Big Data 6 (1) (2019) 1–25.

[4] J. Petch, S. Di, W. Nelson, Opening the black box: the promise and limitations of explainable machine learning in cardiology, Canadian Journal of Cardiology (2021).

[5] K. Huang, J. Altosaar, R. Ranganath, Clinicalbert: modeling clinical notes and predicting hospital readmission, arXiv preprint arXiv:1904.05342 (2019).

[6] J. Davar, H.M. Connolly, M.E. Caplin, M. Pavel, J. Zacks, S. Bhattacharyya, et al., Diagnosing and managing carcinoid heart disease in patients with neuroendocrine tumors: an expert statement, Journal of the American College of Cardiology 69 (10) (2017) 1288–1304.

[7] K. Rough, A.M. Dai, K. Zhang, Y. Xue, L.M. Vardoulakis, C. Cui, et al., Predicting inpatient medication orders from electronic health record data, Clinical Pharmacology & Therapeutics 108 (1) (2020) 145–154.

[8] M. Jagielski, A. Oprea, B. Biggio, C. Liu, C. Nita-Rotaru, B. Li, Manipulating machine learning: poisoning attacks and countermeasures for regression learning, in: 2018 IEEE Symposium on Security and Privacy (SP), IEEE, May 2018, pp. 19–35.

[9] X. Yuan, P. He, Q. Zhu, X. Li, Adversarial examples: attacks and defenses for deep learning, IEEE Transactions on Neural Networks and Learning Systems 30 (9) (2019) 2805–2824.

[10] S. Murugesan, S. Malik, F. Du, E. Koh, T.M. Lai, Deepcompare: visual and interactive comparison of deep learning model performance, IEEE Computer Graphics and Applications 39 (5) (2019) 47–59.

[11] https://hfw.assam.gov.in/sites/default/files/swf_utility_folder/departments/hfw_lipl_in_oid_3/menu/information_and_services/FACTSHEET-Assam%20-%2012%20-13.pdf.

[12] J. Chen, Y. Chen, J. Li, J. Wang, Z. Lin, A.K. Nandi, Stroke risk prediction with hybrid deep transfer learning framework, IEEE Journal of Biomedical and Health Informatics 26 (1) (2021) 411–422.

[13] K. Wickstrøm, K.Ø. Mikalsen, M. Kampffmeyer, A. Revhaug, R. Jenssen, Uncertainty-aware deep ensembles for reliable and explainable predictions of clinical time series, IEEE Journal of Biomedical and Health Informatics 25 (7) (2020) 2435–2444.

[14] D.S. Kermany, M. Goldbaum, W. Cai, C.C. Valentim, H. Liang, S.L. Baxter, et al., Identifying medical diagnoses and treatable diseases by image-based deep learning, Cell 172 (5) (2018) 1122–1131.
[15] L. Liu, J. Xu, Y. Huan, Z. Zou, S.C. Yeh, L.R. Zheng, A smart dental health-IoT platform based on intelligent hardware, deep learning, and mobile terminal, IEEE Journal of Biomedical and Health Informatics 24 (3) (2019) 898–906.
[16] G. Deep, J. Kaur, S.P. Singh, S.R. Nayak, M. Kumar, S. Kautish, MeQryEP: a texture based descriptor for biomedical image retrieval, Journal of Healthcare Engineering 2022 (2022).
[17] A. Solanki, S. Kumar, C. Rohan, S.P. Singh, A. Tayal, Prediction of breast and lung cancer, comparative review and analysis using machine learning techniques, Smart Computing and Self-Adaptive Systems (2021) 251–271.
[18] G. Harerimana, J.W. Kim, H. Yoo, B. Jang, Deep learning for electronic health records analytics, IEEE Access 7 (2019) 101245–101259.
[19] J. Waring, C. Lindvall, R. Umeton, Automated machine learning: review of the state-of-the-art and opportunities for healthcare, Artificial Intelligence in Medicine 104 (2020) 101822.
[20] S. Saralajew, L. Holdijk, M. Rees, E. Asan, T. Villmann, Classification-by-components: probabilistic modeling of reasoning over a set of components, Advances in Neural Information Processing Systems 32 (2019).
[21] K. Kuppala, S. Banda, T.R. Barige, An overview of deep learning methods for image registration with focus on feature-based approaches, International Journal of Image and Data Fusion 11 (2) (2020) 113–135.
[22] L. Shuang, C. Deyun, C. Zhifeng, P. Ming, Multi-feature fusion method for medical image retrieval using wavelet and bag-of-features, Computer Assisted Surgery 24 (Suppl. 1) (2019) 72–80.
[23] C.H. Cao, H.L. Cao, The research on medical image classification algorithm based on PLSA-BOW model, Technology and Health Care 24 (s2) (2016) S665–S674.
[24] D. Kermany, K. Zhang, M. Goldbaum, Labeled optical coherence tomography (OCT) and chest X-ray images for classification, Mendeley Data V2 (2018), https://doi.org/10.17632/rscbjbr9sj.2.
[25] S. Targ, D. Almeida, K. Lyman, Resnet in resnet: generalizing residual architectures, arXiv preprint arXiv:1603.08029 (2016).
[26] G.N. Nguyen, N.H. Le Viet, M. Elhoseny, K. Shankar, B.B. Gupta, A.A. Abd El-Latif, Secure blockchain enabled Cyber–physical systems in healthcare using deep belief network with ResNet model, Journal of Parallel and Distributed Computing 153 (2021) 150–160.
[27] Z. Jiang, Z. Ma, Y. Wang, X. Shao, K. Yu, A. Jolfaei, Aggregated decentralized down-sampling-based ResNet for smart healthcare systems, Neural Computing & Applications (2021) 1–13.
[28] H.H. Tan, K.H. Lim, Vanishing gradient mitigation with deep learning neural network optimization, in: 2019 7th International Conference on Smart Computing & Communications (ICSCC), IEEE, June 2019, pp. 1–4.
[29] H. Zhijun, L. Weiming, H. Zengzhen, Research on methods of structural optimization based on Trimedia when developing province highway net-monitoring software system, in: 2013 Fourth International Conference on Intelligent Systems Design and Engineering Applications, IEEE, November 2013, pp. 494–497.
[30] X. Wang, R. Lan, H. Wang, Z. Liu, X. Luo, Fine-grained correlation analysis for medical image retrieval, Computers & Electrical Engineering 90 (2021) 106992, https://doi.org/10.1016/j.compeleceng.2021.106992.

Chapter 25

Deep convolution classification model-based COVID-19 chest CT image classification

R. Sujatha[1] and Jyotir Moy Chatterjee[2]
[1]*School of Information Technology and Engineering, Vellore Institute of Technology, Vellore, Tamil Nadu, India;* [2]*Department of Information Technology, Lord Buddha Education Foundation (Asia Pacific University), Kathmandu, Nepal*

1. Introduction

Coronavirus (COVID-19) is a global illness proclaimed as a pandemic by the WHO on March 11, 2020. At the time of writing, more than 60 million individuals have been affected by this viral disease, and more than 1.4 million died [1]. A swift conclusion is essential to stop the proliferation of the sickness and expand the viability of clinical therapy. Viral nucleic acid identification utilizing real-time polymerase chain reaction (RT-PCR) is the acknowledged approved indicative technique. In any case, numerous countries cannot provide adequate RT-PCR because of the infectiousness of the illness. Therefore, only individuals with apparent side effects are tested. in addition, it takes a few hours to obtain a result. In this way, quicker and more dependable screening methods that might be additionally confirmed by the PCR test (or replace it) are required.

COVID-19 is a breathing illness that is caused by a novel coronavirus. Typical manifestations in the affected individual are high temperature, cough, sore throat, and difficulty breathing [2]. Some patients can experience their sense of taste affected, tiredness, pain, and nasal blockage [3]. The time between contracting the illness and the main signs of side effects can reach 14 days [4]. Contamination of this infection is sent through the expiration of patients, for example, coughing and sneezing. On the off chance that an individual comes into contact with an infected person, they may become contaminated. The antibodies/medications for this disease were not accessible recently. Isolation and social distancing are the main answers to controlling the spread of this disease.

The most broadly utilized COVID-19 discovery method is RT-PCR. Notwithstanding, RT-PCR units are expensive and take 6–9 h to confirm infection [5]. Because of the reduced accuracy of RT-PCR, it gives false-negative results. To overcome this issue, radiological imaging strategies, for example, CXR and CT, are utilized to distinguish and analyze COVID-19 [6]. In this work, CXR is favored over CT. The reason for this is that X-ray machines are accessible in the most emergency clinics. X-ray machines also are less expensive than CT machines. In addition, X-rays have lower ionizing radiation than CT [7]. Coronavirus leaves some radiological marks that can be effectively identified through CXR. For this, radiologists are needed to investigate these marks. However, this is a time-consuming and error-prone task. Consequently, there is a need to mechanize the examination or CXR.

The onset of COVID-19 was witnessed across the globe at the end of 2019. The work carried out in this field is mainly about its classification and prediction. The involvement of different disciplines has been prominent. When searching for COVID-19 and image processing in the Scopus database a huge amount of research work carried out by researchers was revealed. VOSviewer is used to draft the bibliometric network from the cumulative results. Relevant terms are sketched using the text-mining functionality. Visualizing the same affords the inference that pandemics, pneumonia, X-ray, deep learning, and so on have a high influence in this field, which is demonstrated in Fig. 25.1 [8].

Deep Learning in Personalized Healthcare and Decision Support. https://doi.org/10.1016/B978-0-443-19413-9.00022-9

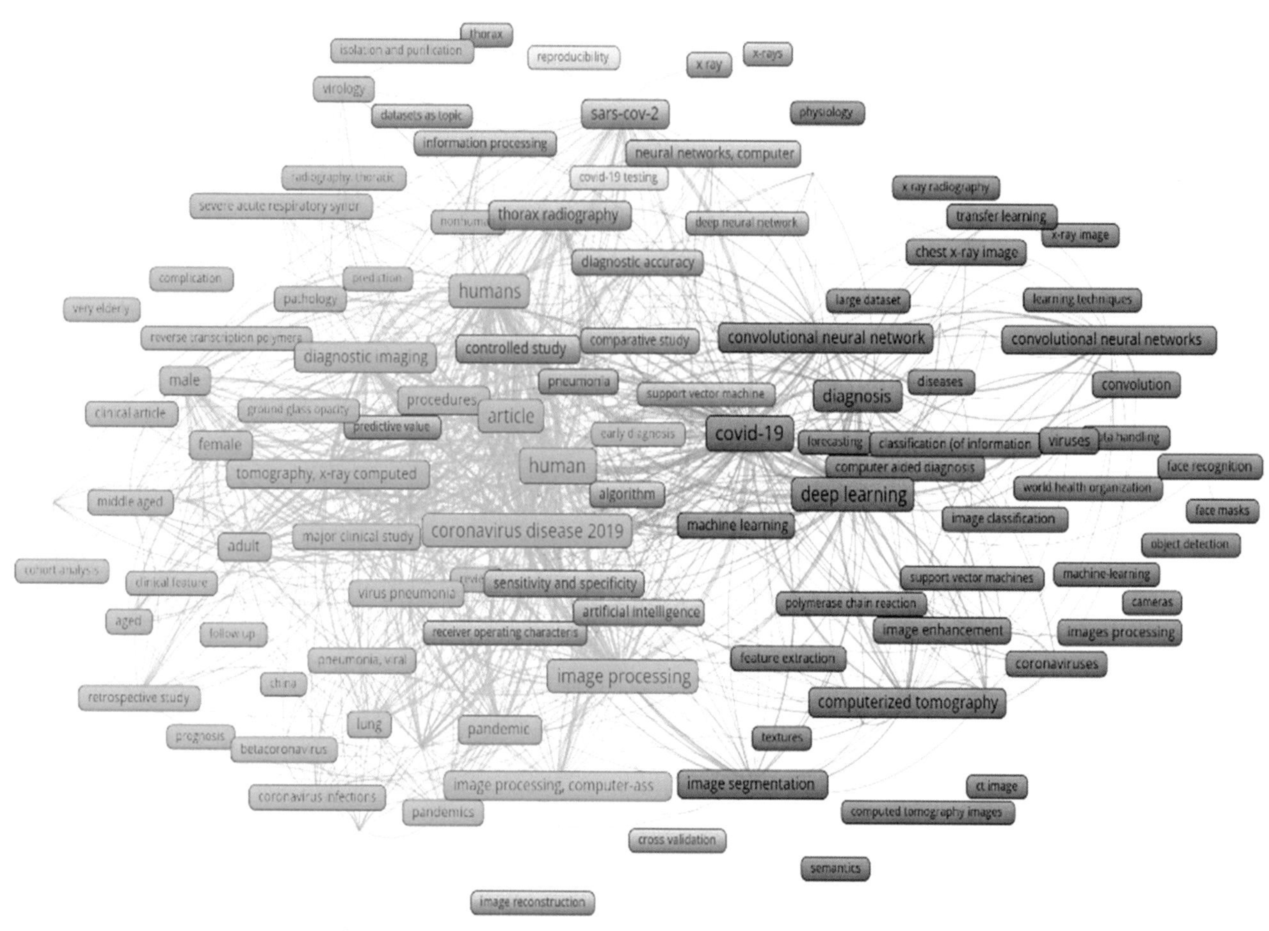

FIGURE 25.1 Bibliometric network for COVID-19 and image processing.

Computed tomography (CT) and X-ray imaging are by all accounts the best choices to identify COVID-19 [9] with a higher accuracy [10]. CT will probably turn out to be increasingly significant for the identification and treatments of COVID-19 pneumonia, especially bearing in mind the unending increases in worldwide cases.

Deep learning (DL) techniques have been broadly utilized in clinical imaging. Specifically, convolutional neural networks (CNNs) have been utilized for characterization and segmentation issues, additionally providing CT images [11]. Although CNNs have exhibited favorable results in this type of utilization, they need a great deal of information to be accurately prepared. Indeed, CT images of the lungs could be effectively misclassified, particularly in cases of pneumonia, because of various causes. There were different datasets for COVID-19 and some accessible ones have a large amount of CT images.

In the present work, our contribution is as follows:

- We aimed to obtain better classification results for the considered CT images for COVID-19 diagnosis from the existing works.
- We have proposed a deep 11-layered CNN architecture and named it DCCM (Deep Convolution Classification Model).
- The proposed approach has attained disease classification of at 93.68%, AUC of 89%, precision of 99.57%, and recall of 44.8%, respectively.

The rest of the chapter is structured as follows: Section 2 discusses the existing works in the current exploration area. The proposed approach with a detailed elaboration of utilized concepts has been elaborated in Section 3. Section 4 presents the investigational details. The experimental findings obtained and their detailed elaboration are presented in Section 5. Section 6 concludes the chapter in conjunction with potential forthcoming efforts.

2. Related works

The principal objective described in Ref. [12] is to suggest a medical decision support system utilizing the execution of a CNN. Ref. [13] presents a joint framework of CNN and RNN to analyze COVID-19 from CXR. Ref. [14] introduces a light CNN configuration dependent on the approach of the SqueezeNet, for the distinction between COVID-19 CT images from other CT images (community-acquired pneumonia and disease-free images) In Ref. [15]. a deep neural network (DNN) is proposed that utilizes faster regions with the CNN (Faster R−CNN) structure to discover COVID-19 sufferers from chest X-ray images utilizing an accessible open-access dataset. Ref. [16] suggests a DL technique dependent on combining CNN with long short-term memory (LSTM) to analyze COVID-19 mechanized from CXR. Ref. [17] suggested a progression of experiments utilizing supervised learning models to perform an accurate classification on datasets consisting of medical images from COVID-19 patients and medical images of some other related infections affecting the lungs. Ref. [18] presented a structure dependent on capsule networks, alluded to as COVID-CAPS, competent in taking care of modest datasets, which is noteworthy because of the abrupt and quick emergence of COVID-19. An advanced approach known as GSA-DenseNet121-COVID-19 relied on an advanced CNN architecture considering an optimization algorithm [19]. In Ref. [20] a mechanized deep transfer learning method was utilized to reveal COVID-19 problems in CXR by using an excessive rendition of the Inception (Xception) model. Ref. [21] introduced an efficient DL approach for the diagnosis of COVID-19 with a voting method. Ref. [22] proposed a novice dynamic method to improvise the fuzzy c-means (FCM) clustering approach for automated localization and segmentation of liver and hepatic injuries from CT scans. An advanced approach with two-dimensional (2D) curvelet change, chaotic salp swarm algorithm (CSSA), and DL approach was created to find out whether the sufferer had COVID-19 pneumonia from CXR in Ref. [23]. Two DL architectures have been suggested that automatically detect positive COVID-19 cases utilizing CT and CXR in Ref. [24]. To advance the automatic learning of latent features, a custom CNN design has been suggested in Ref. [25] which learns remarkable convolutional channel designs for each type of pneumonia Ref. [26] attempted to create a computer-aided diagnosis (CAD) model of CXR to identify COVID-19-contaminated pneumonia. Research in Ref. [27], with the help of multi-CNN, in conjunction with a few preprepared CNNs, enabled automated identification of COVID-19 from CXR. Ref. [28] suggested a calibrated random forest model helped by the AdaBoost calculation for COVID-19 patient health. Ref. [29] aimed to examine appropriate statistical neural network models and their advanced form for COVID-19 mortality expectation in Indian populations and to gauge the future COVID-19 cases for India. Ref. [30] introduced an approach that might be helpful in predicting the spread of COVID-2019 utilizing linear regression, multilayer perceptron, and a vector autoregression strategy on COVID-19 Kaggle information to understand the pandemic illustration of the disease and spread of COVID-19 in India.

For automatic COVID-19 detection utilizing unprocessed chest X-ray and CT scan images, a novel CNN model, dubbed CoroDet, has been proposed [31]. Ref. [32] introduced a machine learning (ML)-based COVID-19 cough classifier that can separate COVID-19-positive coughs from both COVID-19-negative and healthy coughs captured on a smart-phone. Ref. [33] used time series analysis, correlation analysis, the Granger test, and the GMDH method to forecast the impact of COVID-19 cases in India. Ref. [34] examined 30 well-known memes and the increase in the number of their captions during the COVID-19 lockdown period to study the Internet meme activity in social networks (September 2017 to August 2020). Ref. [35]. suggested creating a website with a healthcare chatbot to provide support and keep tabs on the COVID-19 scenario. Ref. [36] provides a broad survey of the continuous procedures, for example, conclusion, expectation, medication, and immunization improvement, and preventive measures utilized in fighting coronavirus alongside advancements utilized and their constraints.

Ref. [37] aimed to identify and order coronavirus utilizing ML. To recognize coronavirus in CT image and computer-aided design framework is projected to recognize and group the coronavirus by using the clinical examples obtained from coronavirus-contaminated patients with the assistance of some ML methods such as Decision Tree, Support Vector Machine, K-NN, etc. Ref. [38] proposed a framework where individuals are isolated from the foundation utilizing the Just go for it v4 object location method, and afterward the recognized individuals are followed by bouncing boxes utilizing the Deepsort procedure. Ref. [39] examined the essential AI innovation uses during the time spent recovering from COVID-19 according to three fundamental viewpoints: forecast, side effect discovery, and advancement, considering a broad literature review. Ref. [40] gauged the number of cases and deaths in Saudi Arabia utilizing time series and notable factual estimating procedures including exponential smoothing and linear regression.

3. Proposed architecture and workflow

In this section, we describe how the proposed DCCM will work. Fig. 25.2 illustrates the workflow of our work with an 11-layered deep convolution network for the classification of CT images. The image dataset must be processed, followed by

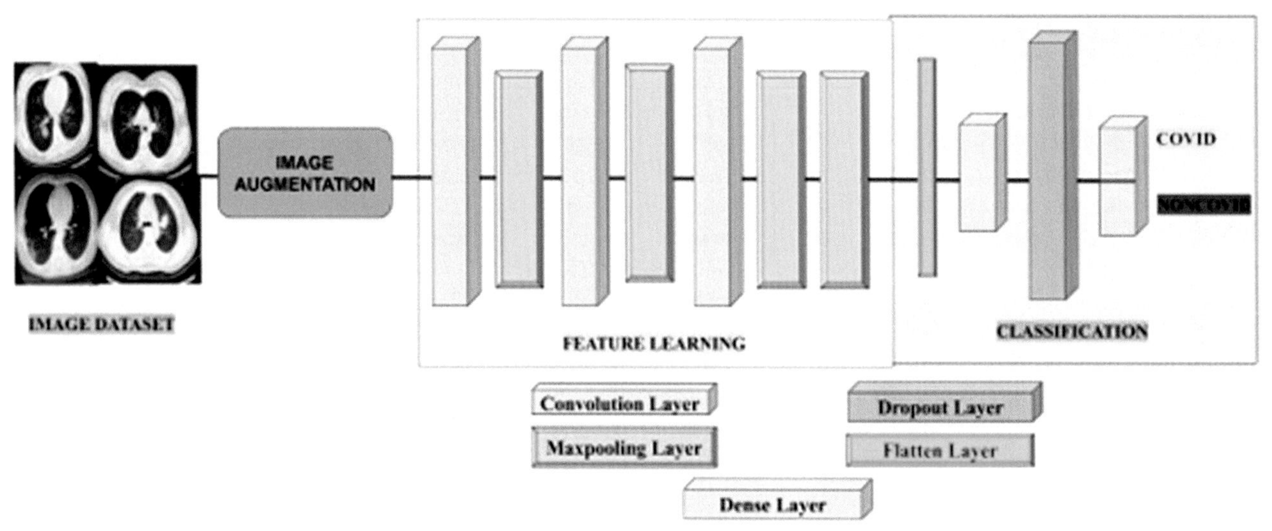

FIGURE 25.2 Proposed workflow.

the data augmentation module for optimization, and then followed by the feature learning and classification module. The proposed model is established on Google Colab employing the TensorFlow framework that Google gives for researchers to work in a GPU environment to deploy the DL algorithm [41].

3.1 Image augmentation

The considered CT images are subdivided into 70% for training and 30% for testing. To have better learning, it is advisable to proceed with the pedagogical manner of dividing [42]. Therefore, based on that training set and the testing set with two classes, namely COVID and non-COVID, it is made for further processing. To get better performance and avoid overfitting, it is mandatory to proceed with data augmentation in the DL vertical. To facilitate this, ImageDataGenerator API is available in Keras. Keras is an open-source programming library that gives a Python point of interaction to artificial neural networks, and it acts as a connection point for the TensorFlow library [43]. Data augmentation is an artificial method of populating the training images in different ways such as shifts, random, flip, and other methods. With the deployment of the image acquisition process, the model's performance is optimized and helps avoid overfitting problems [13,44,45]. As mentioned at the start, this is the strategy that will increase the diversity of the available training data and increase the model's accuracy and work as a regularizer. In our proposed model, the existing training set is expanded with the following parameters, as mentioned in Table 25.1.

3.2 Convolution layer (CL)

This is the primary and essential block in the case of deep neural network modeling. Filters are applied to get the feature map that helps in feature detection. The working of the convolution resembles the human brain that convolves the taken input and passes it to the CL. Various hyperparameters provided decide the quality of output and, based on the image dataset, considered fine-tuning of the parameters is mandated. Image classification results from the activities of the various layers from the input to the output layer [46]. Here the input is [$224 \times 224 \times 3$], which will carry the raw pixel

TABLE 25.1 ImageDataGenerator parameters.

Parameter	Value
rescale	1./255
shear_range	0.2
zoom_range	0.2
flip	horizontal

approximations of the image, for this situation, an image of breadth 32, height 32, and with three coloring channels R, G, and B. The equation for the CL is given as:

$$Z(a, b) = (G * H)(a, b) = \Sigma\Sigma \mathrm{I}\,(\mathrm{a}+\mathrm{e}, \mathrm{b}+\mathrm{f})K\,(p, q) \tag{25.1}$$

where G is the input matrix, H is a 2D filter of size $p \times q$, and Z is the outcome of a 2D feature map. The working of the convolutional layer is represented by $G * H$. The rectified linear unit (ReLU) layer is utilized in Ref. [47] to enhance the nonlinearity in feature maps. ReLU discovers activation by putting the threshold input at zero, which is represented as follows: f(x) = max (0, x) [25].

3.3 MaxPooling layer

Pooling is another building block in the neural network model that progressively decreases the spatial size of the depiction and normally operates independently over each feature map. The prefix "Max" in our proposed model indicates the max value is considered in each feature map. Most prominent is received as the output—precisely stated as a sample-based discretization approach. A typical CNN model architecture is to have several convolutions and pooling layers stacked one after the other. The pooling layer [48] executes the down-sampling of a given input measurement to reduce the number of parameters. Maxpooling is a widely recognized strategy, which delivers the most significant incentive in an input region [14]. The formula is as follows:

$$(m_h - g + 1)\ /\ r * (m_x - g + 1)\ /\ r * m_c \tag{25.2}$$

where,

m_h—feature map—height
m_x—feature map—width
m_c—feature map—number of channels
g—filter—size
r—stride length

3.4 Dropout layer

This layer helps avoid the overfitting of the model designed by Srivastava et al. [49]. It is the process of removing the units temporarily. Random selection of drop is the exciting phenomenon utilized in this process, and its analogy is derived from the sexual reproduction that is contributed using parts of the genes of each parent and a few mutations. An n-layer completely associated neural organization (discarding bias) can be characterized as [50]:

$$f(a; \{W_i\}_i \in \{1, \ldots, n\}) = \varnothing_n(W_n \varnothing_{n-1}(W_{n-1} \ldots (\varnothing_1(W_1 a)))) \tag{25.3}$$

where, ϕi is a nonlinearity (e.g., ReLU), Wi for $i \in \{1, \ldots, n\}$ are weight matrices, and a is the input.

3.5 Flatten layer

The process of converting the multi-dimensional matrix to a single one helps in the easy functioning of the successive dense layer. Flattening changes the information into a one-dimensional array to contribute to the next layer by flattening the yield of the CL to result in a solitary long feature vector. Also, it is associated with the last characterization model, which is a fully connected layer. As such, we put all the pixel information in one line and make associations with the last layer [51].

3.6 Dense layer

This layer maps the input to the correct class after obtaining the input from several layers. The dense layer = fully connected layer is the geography that depicts how the neurons are connected to the succeeding layer of neurons (every neuron is associated with each neuron in the succeeding layer) in a transitional layer (additionally called a hidden layer) [52].

3.7 Eleven-layered architecture

Sequential ordering of the various layers helps make the classification process perfect. Fig. 25.3 provides complete insight into the hyperparameters utilized in each phase of the workflow, and from the model, it is drawn with a neutron [53]. Additionally, the activation function used in the convolution layer is "relu," pool size is 2 in the maxpooling layer, and the dropout rate is 0.25 with "softmax" as activation for the dense layer. Here the batch size is 16, the filter is 32, the kernel size is 32, and the epoch is 75.

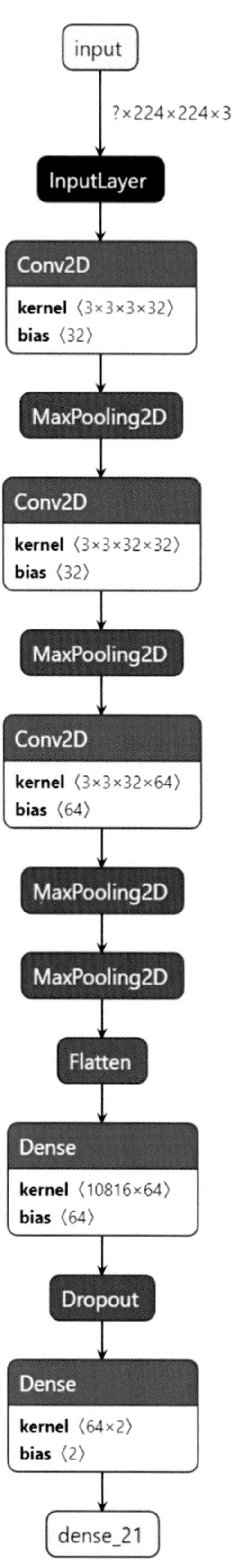

FIGURE 25.3 11-layered architecture.

4. Experiment

4.1 Dataset description

Recently, the search for COVID-19 images has been very high, and information provided by the dataset and the technology deployment will pave the way for early detection followed by treatment. We have utilized an openly accessible dataset [54]. The image dataset considered in this work was obtained from the Tongji Hospital, Wuhan, China, during the coronavirus outbreak between January and April. Discussion about this dataset mentioned that a CT slide provides extensive clinical evidence that helps in accurate decision-making. According to the source, the data are collected from various COVID-19-associated papers. The distribution of the dataset is 349 CTs of COVID-19-infected patients and 397 CTs of normal people.

The input shape for the images is taken as 224 × 224. A preview of the images is given in Fig. 25.4.

4.2 Experimental setup

The subtleties of the layers and their request in the intended model, the outcome state of each layer, the number of parameters (weights) in each layer, and the overall number of parameters (weights) are introduced in Fig. 25.5.

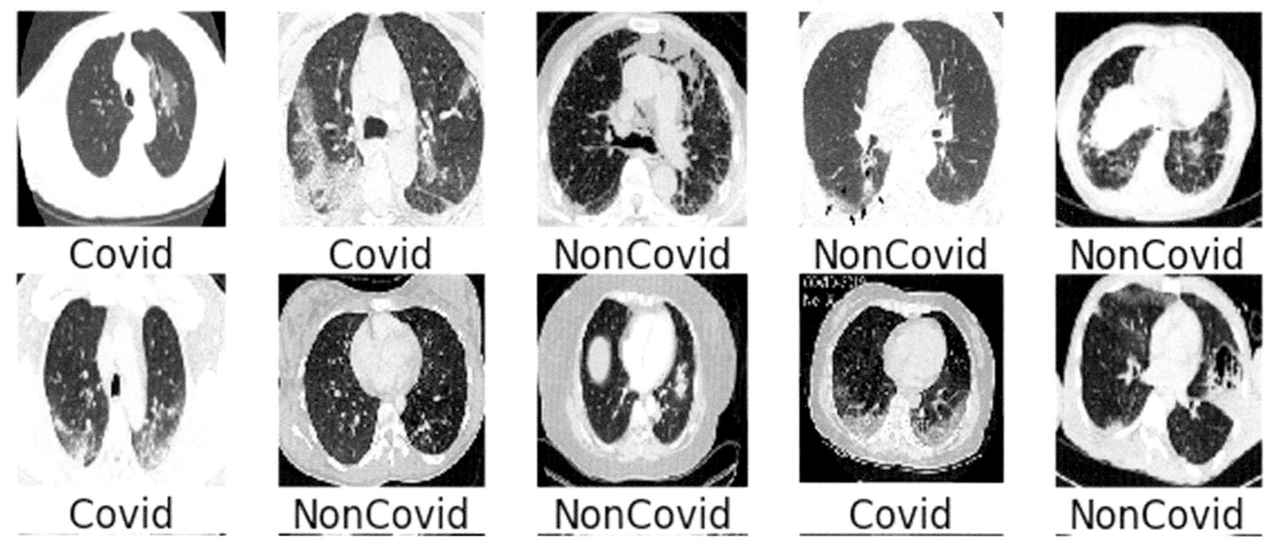

FIGURE 25.4 Dataset image [54].

Model: "*sequential_2*"		
Layer (Type)	**Output Shape**	**Param#**
conv2d_38 (Conv2D)	(None, 222, 222, 32)	896
max_pooling2d_40	(MaxPooling (None, 111, 111, 32)	0
conv2d_39 (Conv2D)	(None, 109, 109, 32)	9248
max_pooling2d_41	(MaxPooling (None, 54, 54, 32	0
conv2d_40 (Conv2D)	(None, 52, 52, 64)	18496
max_pooling2d_42	(MaxPooling (None, 26, 26, 64)	0
max_pooling2d_43	(MaxPooling (None, 13, 13, 64)	0
flatten_11 (Flatten)	(None, 10816)	0
dense_22 (Dense)	(None, 64)	692288
dropout_17 (Dropout)	(None, 64)	0
dense_23 (Dense)	(None, 2)	130
Total params: 721,058 Trainable params: 721,058 Non-trainable params: 0		

FIGURE 25.5 Layer types and parameters in DCCM.

TABLE 25.2 Comparison of parameter number.

Model	Parameter number
VGG-16	138,357,544
VGG-19	143,667,240
ResNet-50	25,557,032
DenseNet-169	14,149,480
EfficientNet-b1	7,794,184
Proposed model	**721,058**

All the software and libraries utilized in the described work are open source. Users must utilize Google Colab Notebook utilizing the GPU run-time type to imitate the outcomes. This software can be utilized without cost as Google provides it to investigate exercises utilizing a Tesla K80 GPU of 12 GB. Here the pre-prepared and scaled CNN models are utilized for transfer learning in image characterization issues. The model was created by Google AI in May 2019 and is accessible from Github stores. In outline, the software utilized here can be utilized without permit worries as it is free and openly accessible. The creators have utilized ImageDataGenerator as the virtual library in the proposed technique. We evaluate our approaches using five metrics, namely loss, accuracy, AUC, precision, and recall.

Table 25.2 presents the number of weight parameters in various DL models. The DCCM model for CT image classification for COVID-19 has used a much lower number of parameters in comparison with more sophisticated architectures such as VGG-19 but the performance is much better.

5. Experimental results and discussions

Table 25.3 presents the detailed experimental results received with the number of epochs, loss during classification, classification accuracy in disease prediction, area under the curve, precision, and recall, respectively.

In this experiment, we have considered a total of 75 epochs. We have presented a snapshot of the experiment conducted and not presented the complete values.

5.1 Classification performance metrics

The classification is performed using the proposed 11-layer architecture, and sequential flow maps of the images as COVID and non-COVID cases. The analysis of the results of DCCM shows better performance than the existing pretrained networks. The following section illustrates the results of the various performance metrics.

TABLE 25.3 Experimental results.

Epochs	Loss	Accuracy	AUC	Precision	Recall
1	0.302	0.8736	0.848	0.9121	0.418
14	0.253	0.8908	0.858	0.9144	0.389
23	0.243	0.9061	0.908	0.9347	0.521
32	0.241	0.9291	0.89	0.9764	0.475
43	0.211	0.908	0.897	0.9529	0.466
53	0.156	0.9464	0.886	0.9804	0.383
60	0.134	0.9464	0.882	0.9626	0.395
75	0.165	0.9368	0.89	0.9957	0.448

Fig. 25.6 presents the difference between the epoch and accuracy in classification. As the number of epochs is increasing the accuracy increases. In the present scenario, by using 75 epochs, we received a disease classification accuracy of 93.68%.

Fig. 25.7 presents the comparison between loss and epochs. As the number of epochs increases the training loss is minimized but, in the case of validation loss, it increases with an increase in the epoch.

$$Precision = \frac{True\ Positive\ (TP)}{True\ Positive\ (TP) + False\ Positive\ (FP)} \tag{25.4}$$

Fig. 25.8 shows that as the epochs increase, gradually the precision also increases, which means the disease classification by our proposed model is correct 93.68% of the time.

$$Recall = \frac{True\ Positive\ (TP)}{True\ Positive\ (TP) + False\ Negative\ (FN)} \tag{25.5}$$

As per Fig. 25.9, as the number of epochs increases the recall value goes down, i.e., out of all the clinical images present in the dataset, recall tells how many we have correctly identified as COVID and Non-COVID cases.

In Fig. 25.10, it is shown that as the number of epochs increases the AUC also gets higher.

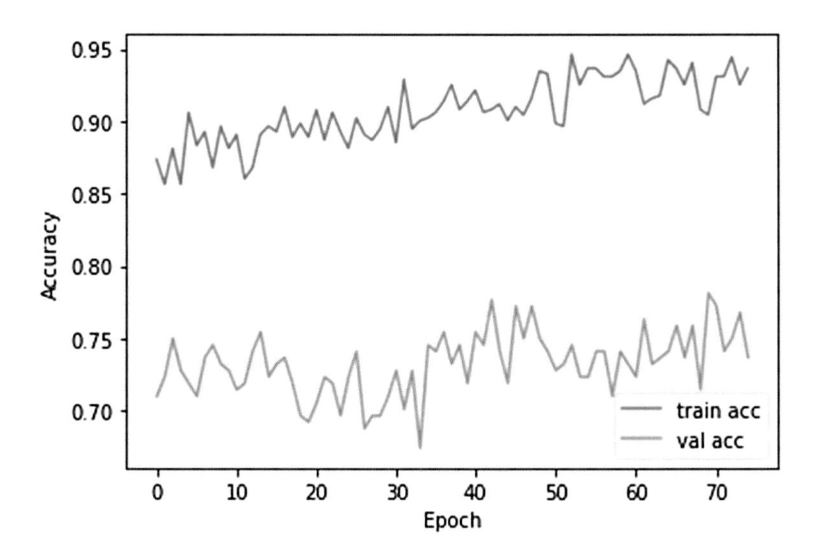

FIGURE 25.6 Accuracy versus epoch.

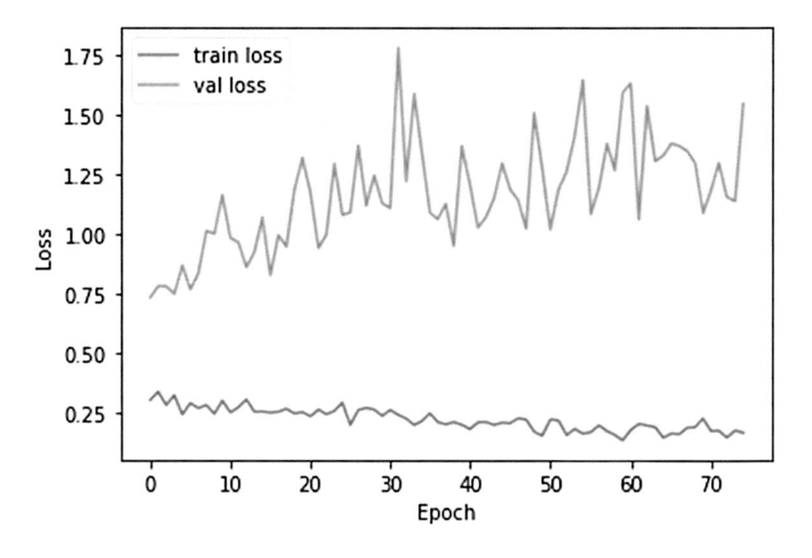

FIGURE 25.7 Loss versus epoch.

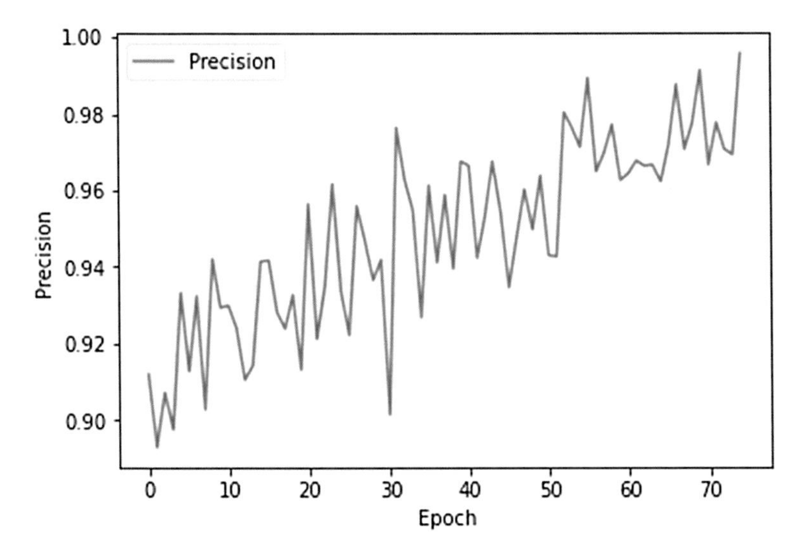

FIGURE 25.8 Precision versus epoch.

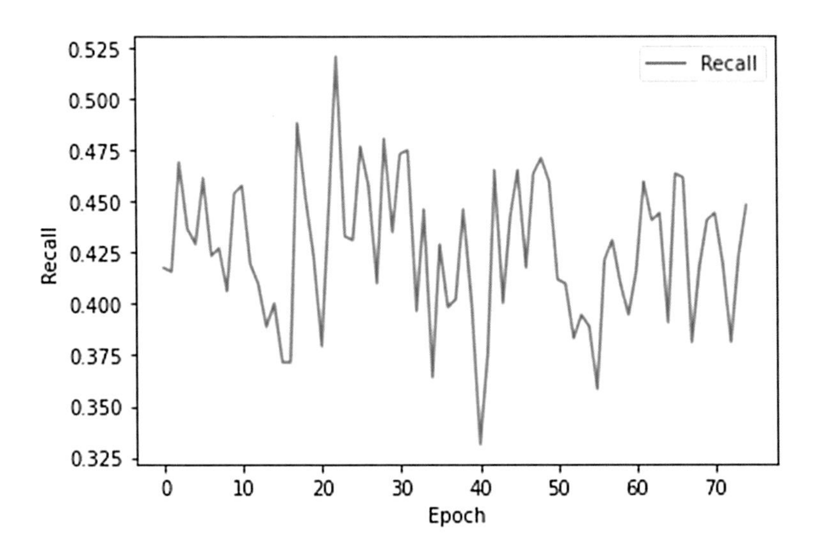

FIGURE 25.9 Recall versus epoch.

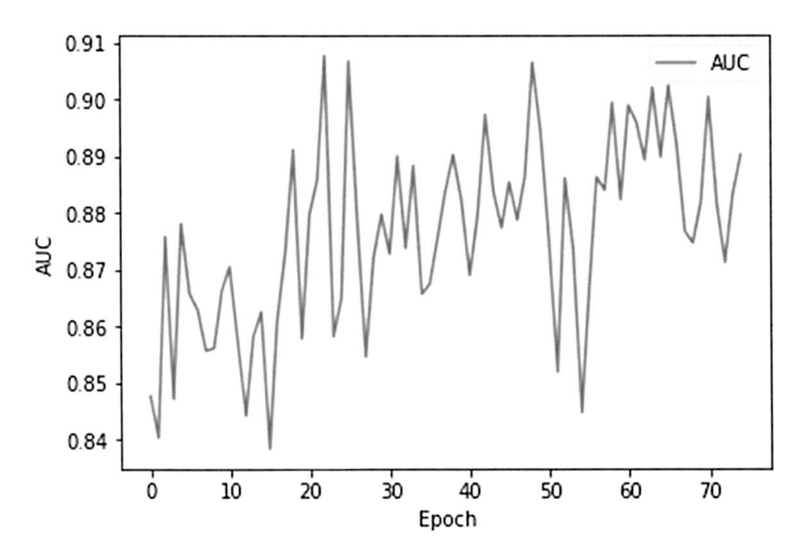

FIGURE 25.10 AUC versus epoch.

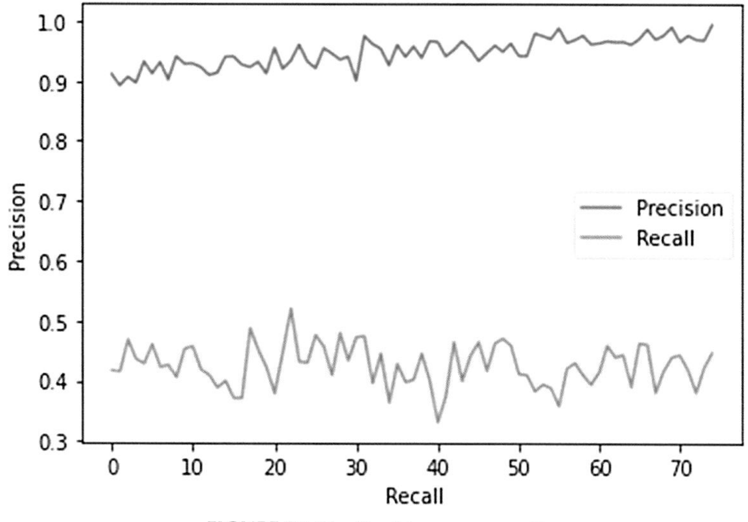

FIGURE 25.11 Precision versus recall.

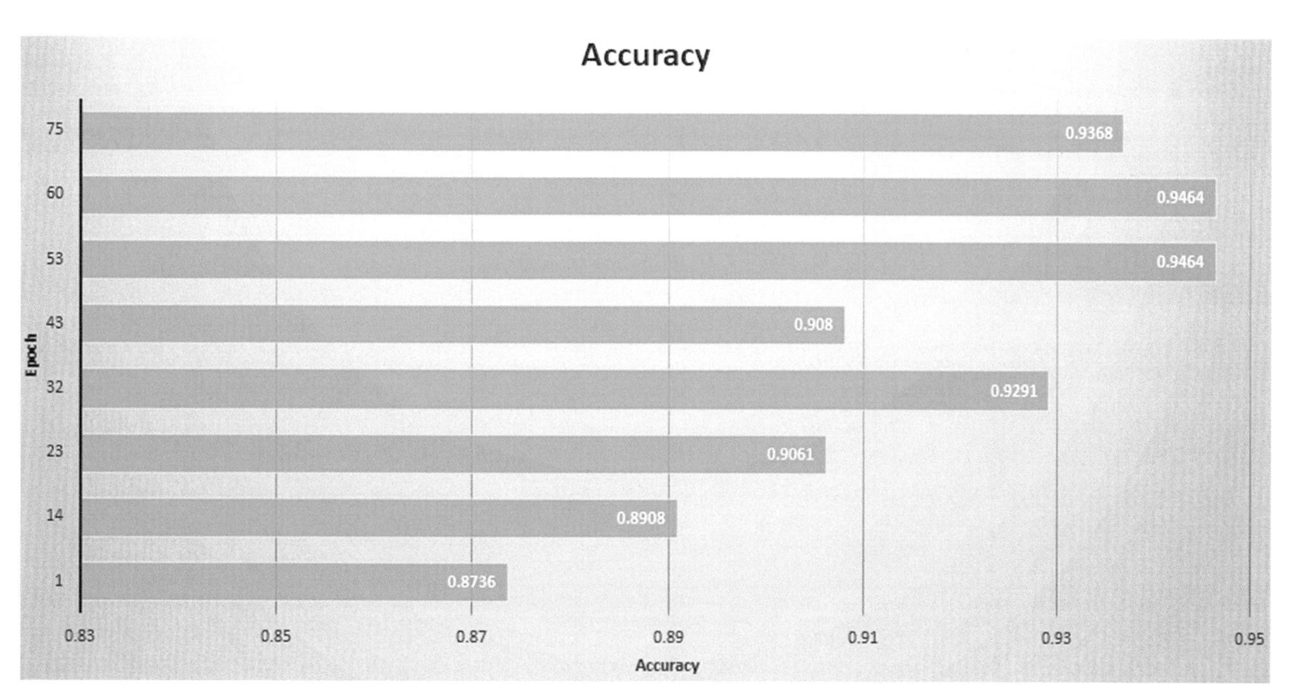

FIGURE 25.12 Classification accuracy is based on the number of epochs.

Fig. 25.11 presents the comparison between precision and recall.

Fig. 25.12 presents the disease classification accuracy received during the experiment based on the number of epochs. The highest accuracy of 94.64% is received during epochs 53 and 60, respectively.

Fig. 25.13 presents the evaluation metrics considered for evaluating the effectiveness of the intended CNN model. As the number of epochs grows, the loss minimizes—the precision, accuracy, and AUC increase as the number of epochs increases. The recall sometimes gets higher and sometimes goes lower as the number of epochs increases. From this, it can be said that the intended CNN model performs well in the case of disease classification.

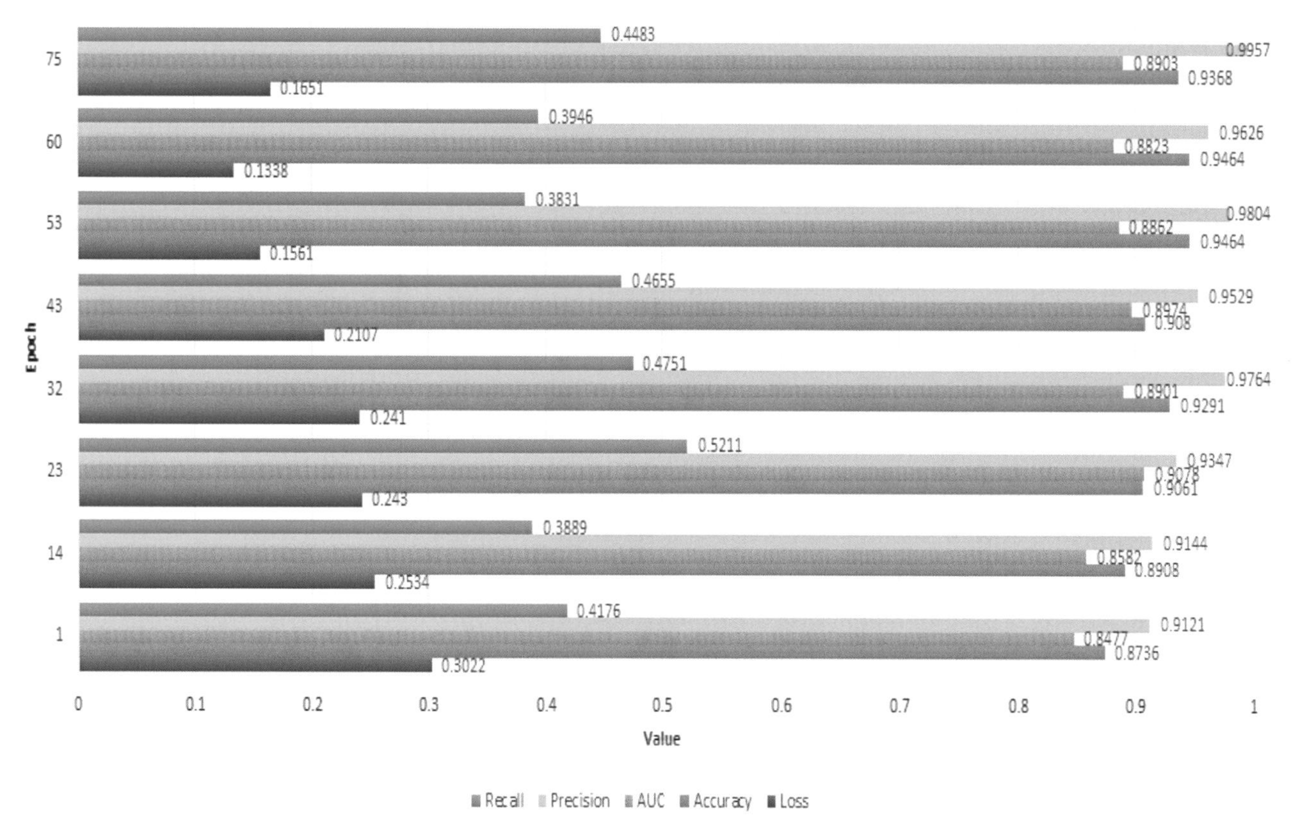

FIGURE 25.13 Evaluation parameters.

TABLE 25.4 Result comparison.

Reference no.	Approach		Accuracy
[55]	DenseNet-169	Random initialization	83%
		Transfer learning (TL)	87.10%
		TL + CSSL	89.10%
DCCM			**93.68%**

5.2 Comparison of DCCM with existing works

We compared our proposed model with the existing model (Table 25.4) to find the disease accuracy and found that the intended model operates better than the existing model. The disease classification accuracy we received is 93.68%, with AUC at 89%, precision at 99.57%, and recall at 44.8%, respectively.

6. Conclusions and future work

It is evident that information communication and technology dramatically influence the healthcare sector—the designed work was performed by applying data augmentation and stacking the deep layers. Finally, a fully connected layer helps with better classification. The disease classification accuracy obtained in DCCM is 93.68%, AUC 89%, precision 99.57%, and recall 44.8%, respectively. This work could be optimized by incorporating other related features that illustrate the patient's other symptoms. Severity can still be assessed better by incorporating other related features that illustrate the patient's condition. In our designed work, feature learning and classification are carried out with deep learning layers. The system's performance can be optimized by using machine learning classifiers like XGBoost, and Stochastic gradient descent in the final part. Therefore, hybridizing the system with other related attributes of the patient will provide an efficient system.

References

[1] Johns Hopkins Coronavirus Resource Center, COVID-19 Map - Johns Hopkins Coronavirus Resource Center, 2020. Available at: https://coronavirus.jhu.edu/map.html. (Accessed 14 August 2022).

[2] I. Kostrikov, D. Yarats, R. Fergus, Image augmentation is all you need: Regularizing deep reinforcement learning from pixels, arXiv (April 28, 2020) preprint arXiv:2004.13649.

[3] J. Wu, Introduction to convolutional neural networks, National Key Lab for Novel Software Technology. Nanjing University. China 5 (23) (May 1, 2017) 495.

[4] D. Singh, V. Kumar, M. Kaur, Classification of COVID-19 patients from chest CT images using multi-objective differential evolution–based convolutional neural networks, European Journal of Clinical Microbiology and Infectious Diseases 39 (7) (July 2020) 1379–1389.

[5] A.S. Lundervold, A. Lundervold, An overview of deep learning in medical imaging focusing on MRI, Zeitschrift für Medizinische Physik 29 (2) (May 1, 2019) 102–127.

[6] Vishakha. Pooling layer and its types explained! Data Science and Machine Learning. Kaggle. [cited 2022Aug14]. Available from: https://www.kaggle.com/general/175896.

[7] N. Srivastava, G. Hinton, A. Krizhevsky, I. Sutskever, R. Salakhutdinov, Dropout: a simple way to prevent neural networks from overfitting, Journal of Machine Learning Research 15 (1) (January 1, 2014) 1929–1958.

[8] Centre for Science and Technology Studies. (n.d.). VOSviewer - Visualizing scientific landscapes. VOSviewer. Retrieved October 8, 2022, from https://www.vosviewer.com/.

[9] M.D. Landry, L. Geddes L, A.P. Moseman, J.P. Lefler, S.R. Raman, J. van Wijchen, Early reflection on the global impact of COVID19, and implications for physiotherapy, Physiotherapy 107 (June 1, 2020) A1–A3.

[10] A. Sharifi-Razavi, N. Karimi, N. Rouhani, COVID-19 and intracerebral haemorrhage: causative or coincidental? New Microbes and New Infections 35 (May 1, 2020) 100669.

[11] L. Sarker, M.M. Islam, T. Hannan, Z. Ahmed, COVID-DenseNet: a deep learning architecture to detect COVID-19 from chest radiology images, Preprint (May 9, 2020) 2020050151.

[12] X. Xie, Z. Zhong, W. Zhao, C. Zheng, F. Wang, J. Liu, Chest CT for typical 2019-nCoV pneumonia: relationship to negative RT-PCR testing, Radiology 296 (February 12, 2020) E41–E45.

[13] M.Y. Ng MY, E.Y. Lee, J. Yang, F. Yang F, X. Li, H. Wang, M.M. Lui, C.S. Lo, B. Leung, P.L. Khong, C.K. Hui, Imaging profile of the COVID-19 infection: radiologic findings and literature review, Radiology: Cardiothoracic Imaging 2 (1) (February 2020).

[14] S. Latif, M. Usman, S. Manzoor, W. Iqbal, J. Qadir, G. Tyson, I. Castro, A. Razi, M.N. Boulos, A. Weller, J. Crowcroft, Leveraging data science to combat COVID-19: a comprehensive review, IEEE Transactions on Artificial Intelligence 1 (1) (August 2020) 85–103.

[15] F. Chua, D. Armstrong-James, S.R. Desai, J. Barnett, V. Kouranos, O.M. Kon, R. José, R. Vancheeswaran, M.R. Loebinger, J. Wong, M.T. Cutino-Moguel, The role of CT in case ascertainment and management of COVID-19 pneumonia in the UK: insights from high-incidence regions, The Lancet Respiratory Medicine 8 (5) (May 1, 2020) 438–440.

[16] Y. Fang, H. Zhang, J. Xie, M. Lin, L. Ying, P. Pang, W. Ji, Sensitivity of chest CT for COVID-19: comparison to RT-PCR, Radiology 296 (August 19, 2020) E115–E119.

[17] X. Xu, F. Zhou, B. Liu, D. Fu, X. Bai, Efficient multiple organ localization in CT image using 3D region proposal network, IEEE Transactions on Medical Imaging 38 (8) (January 24, 2019) 1885–1898.

[18] G. Marques, D. Agarwal, I. de la Torre Díez, Automated medical diagnosis of COVID-19 through EfficientNet convolutional neural network, Applied Soft Computing 96 (November 1, 2020) 106691.

[19] M.S. Al-Rakhami, M.M. Islam, M.Z. Islam, A. Asraf, A.H. Sodhro, W. Ding, Diagnosis of COVID-19 from X-rays using combined CNN-RNN architecture with transfer learning, medRxiv (January 1, 2021) 100088.

[20] M. Polsinelli, L. Cinque, G. Placidi, A light CNN for detecting COVID-19 from CT scans of the chest, Pattern Recognition Letters 140 (December 1, 2020) 95–100.

[21] K.H. Shibly, S.K. Dey, M.T. Islam, M.M. Rahman, COVID faster R–CNN: a novel framework to Diagnose Novel Coronavirus Disease (COVID-19) in X-Ray images, Informatics in Medicine Unlocked 20 (January 1, 2020) 100405.

[22] M.Z. Islam, M.M. Islam, A. Asraf, A combined deep CNN-LSTM network for the detection of novel coronavirus (COVID-19) using X-ray images, Informatics in Medicine Unlocked 20 (January 1, 2020) 100412.

[23] S. Varela-Santos, P. Melin, A new approach for classifying coronavirus COVID-19 based on its manifestation on chest X-rays using texture features and neural networks, Information Sciences 545 (February 4, 2021) 403–414.

[24] P. Afshar, S. Heidarian, F. Naderkhani, A. Oikonomou, K.N. Plataniotis, A. Mohammadi, Covid-caps: a capsule network-based framework for identification of covid-19 cases from x-ray images, Pattern Recognition Letters 138 (October 1, 2020) 638–643.

[25] D. Ezzat D, A.E. Hassanien, H.A. Ella, An optimized deep learning architecture for the diagnosis of COVID-19 disease based on gravitational search optimization, Applied Soft Computing 98 (January 1, 2021) 106742.

[26] N.N. Das, N. Kumar, M. Kaur, V. Kumar, D. Singh, Automated deep transfer learning-based approach for detection of COVID-19 infection in chest X-rays, Irbm 43 (July 3, 2020) 114−119.
[27] P. Silva, E. Luz, G. Silva, G. Moreira, R. Silva, D. Lucio, D. Menotti, COVID-19 detection in CT images with deep learning: a voting-based scheme and cross-datasets analysis, Informatics in Medicine Unlocked 20 (January 1, 2020) 100427.
[28] Machine Learning - Whats the difference between a dense layer and an output layer in a CNN?. Cross Validated. [cited 2022 Aug 14]. Available from: https://stats.stackexchange.com/questions/383727/whats-the-difference-between-a-dense-layer-and-an-output-layer-in-a-cnn.
[29] Netron | Apps | Electron. www.electronjs.org. [cited 2022 Aug 14]. Available from: https://www.electronjs.org/apps/netron.
[30] A.M. Anter, S. Bhattacharyya, Z. Zhang, Multi-stage fuzzy swarm intelligence for automatic hepatic lesion segmentation from CT scans, Applied Soft Computing 96 (November 1, 2020) 106677.
[31] E. Hussain, M. Hasan, M.A. Rahman, I. Lee, T. Tamanna, M.Z. Parvez, CoroDet: a deep learning based classification for COVID-19 detection using chest X-ray images, Chaos, Solitons and Fractals 142 (2021) 110495.
[32] M. Pahar, M. Klopper, R. Warren, T. Niesler, COVID-19 cough classification using machine learning and global smartphone recordings, Computers in Biology and Medicine 135 (2021) 104572.
[33] R. Sujatha, J.M. Chatterjee, Role of artificial intelligence in COVID-19 prediction based on statistical methods, in: Applications of Artificial Intelligence in COVID-19, Springer, Singapore, 2021, pp. 73−97.
[34] I. Priyadarshini, J.M. Chatterjee, R. Sujatha, N. Jhanjhi, A. Karime, M. Masud, Exploring Internet meme activity during COVID-19 Lockdown using artificial intelligence techniques, Applied Artificial Intelligence 36 (1) (2022) 2014218.
[35] A.K. Pandey, R.R. Janghel, R. Sujatha, S.S. Kumar, T.S. Kumar, J.M. Chatterjee, CoronaGo website integrated with chatbot for COVID-19 tracking, ISIC (January 2021) 521−527.
[36] W. Ding, J. Nayak, H. Swapnarekha, A. Abraham, B. Naik, D. Pelusi, Fusion of intelligent learning for COVID-19: a state-of-the-art review and analysis on real medical data, Neurocomputing 457 (2021) 40−66.
[37] O.R. Shahin, H.H. Alshammari, A.I. Taloba, R.M. Abd El-Aziz, Machine learning approach for autonomous detection and classification of COVID-19 virus, Computers & Electrical Engineering 101 (2022) 108055.
[38] S. Yadav, P. Gulia, N.S. Gill, J.M. Chatterjee, A real-time crowd monitoring and management system for social distance classification and healthcare using deep learning, Journal of Healthcare Engineering 2022 (2022).
[39] N. Patel, S. Trivedi, J.M. Chatterjee, Cmbatting COVID-19: artificial intelligence technologies and challenges, ScienceOpen Preprints (2022), https://doi.org/10.14293/S2199-1006.1.SOR-.PPVK63O.v1.
[40] S. Larabi-Marie-Sainte, S. Alhalawani, S. Shaheen, K.M. Almustafa, T. Saba, F.N. Khan, A. Rehman, Forecasting COVID19 parameters using time-series: KSA, USA, Spain, and Brazil comparative Case study, Heliyon (2022) e09578.
[41] A. Altan, S. Karasu, Recognition of COVID-19 disease from X-ray images by hybrid model consisting of 2D curvelet transform, chaotic salp swarm algorithm and deep learning technique, Chaos, Solitons and Fractals 140 (November 1, 2020) 110071.
[42] M.F. Aslan, M.F. Unlersen, K. Sabanci, A. Durdu, CNN-based transfer learning−BiLSTM network: a novel approach for COVID-19 infection detection, Applied Soft Computing 98 (January 1, 2021) 106912.
[43] Wikipedia contributors, Keras, September 8, 2022 from, https://en.wikipedia.org/wiki/Keras. (Accessed 23 October 2022).
[44] B. Abraham, M.S. Nair, Computer-aided detection of COVID-19 from X-ray images using multi-CNN and Bayesnet classifier, Biocybernetics and Biomedical Engineering 40 (4) (October 1, 2020) 1436−1445.
[45] R. Karthik, R. Menaka, M. Hariharan, Learning distinctive filters for COVID-19 detection from chest X-ray using shuffled residual CNN, Applied Soft Computing 99 (February 1, 2021) 106744.
[46] C. Iwendi, A.K. Bashir, A. Peshkar, R. Sujatha, J.M. Chatterjee, S. Pasupuleti, R. Mishra, S. Pillai, O. Jo, COVID-19 patient health prediction using boosted random forest algorithm, Frontiers in Public Health 8 (July 3, 2020) 357.
[47] E. Bisong, Building Machine Learning and Deep Learning Models on Google Cloud Platform, Apress, Berkeley, CA, 2019.
[48] A. Gholamy, V. Kreinovich, O. Kosheleva. Why 70/30 or 80/20 Relation between Training and Testing Sets: A Pedagogical Explanation.
[49] S. Dhamodharavadhani, R. Rathipriya, J.M. Chatterjee, COVID-19 mortality rate prediction for India using statistical neural network models, Frontiers in Public Health 8 (August 28, 2020) 441.
[50] M. D. Bloice, C. Stocker, A. Holzinger. Augmentor: an image augmentation library for machine learning. arXiv preprint arXiv:1708.04680. August 11, 2017.
[51] J. Reinhold, Dropout on convolutional layers is weird. Medium, Data Science (2019) [cited 2022Aug14]. Available from: https://towardsdatascience.com/dropout-on-convolutional-layers-is-weird-5c6ab14f19b2.
[52] J. Jeong, The most intuitive and easiest guide for CNN, The Medium (2019) [cited 2022Aug14]. Available from: https://towardsdatascience.com/the-most-intuitive-and-easiest-guide-for-convolutional-neural-network-3607be47480#:~:text=Flattening%20is%20converting%20the%20data.
[53] R. Sujath, J.M. Chatterjee, A.E. Hassanien, A machine learning forecasting model for COVID-19 pandemic in India, Stochastic Environmental Research and Risk Assessment 34 (7) (July 2020) 959−972.
[54] UCSD-AI4H. UCSD-AI4H/COVID-CT, GitHub (2020) [cited 2022Aug14]. Available from: https://github.com/UCSD-AI4H/COVID-CT.
[55] X. Yang, X. He, J. Zhao, Y. Zhang, S. Zhang, P. Xie. COVID-CT-dataset: a CT scan dataset about COVID-19. arXiv preprint arXiv:2003.13865. March 30, 2020.

Chapter 26

Internet of Medical Things in curbing pandemics

M.S. Sadiq[1], I.P. Singh[2] and M.M. Ahmad[3]
[1]*Department of Agricultural Economics and Extension, FUD, Dutse, Jigawa, Nigeria;* [2]*Department of Agricultural Economics, SKRAU, Bikaner, Rajasthan, India;* [3]*Department of Agricultural Economics and Extension, BUK, Kano, Nigeria*

1. Introduction

The Internet of Things (IoT) is a clearly defined network of connected mechanical, digital, and computer devices that can transmit data over the network without any input from humans at any level [1]. All of the gadgets that have been discussed are linked to specific, individual identifying numbers or codes. Today's well-established and tested IoT technology serves as a hub for numerous strategies, real-time analytics, machine learning theory, sensory products, etc. Additionally, IoT is acknowledged as having the ability to support human needs in real life through a variety of means, such as smart home security systems, smart lighting setups, and many other examples that are simple to handle through our regular use of smart speakers, smartphones, etc. [2].

All nations, including India, were battling COVID-19 in the current pandemic crisis, and they are all still looking for a workable, affordable solution to deal with the issues that are emerging in various ways [3]. Physical scientists and engineers are working to meet these challenges, develop new theories, characterize new research problems, produce user-centered explanations, and enlighten ourselves and the general public [4]. Sensors, machine learning, real-time analysis, and embedded systems are just a few of the technologies that have shaped and been shaped by the IoT concept. It relates to the idea of a "smart hospital" and other technology managed by wired or wireless internet [5]. In order to do the necessary work, smart gadgets can collect data and communicate it in daily life. IoT applications are being used in connected healthcare, smart cities, automobiles, devices, and entertainment systems. The deployment of IoT in the medical industry revolves around a variety of sensors, medical devices, artificial intelligence, diagnostic, and advanced imaging equipment. In both established and emerging sectors and communities, these gadgets increase productivity and quality of life.

IoT connects all digital, mechanical, and computational systems to transport data over the Internet without involving any human beings. During the COVID-19 pandemic, this technology saw a rise in the monitoring of healthcare [6]. Numerous deaths currently occur as a result of inaccurate and delayed health information. By using sensors, this technology can instantly alert users of potential health problems. The cloud was where all COVID-19 patient data were kept, making it easier to give patients the care they need. A person's everyday activities can be recorded by this technology, which can also send out notifications about potential health issues.

In the medical industry, having the right equipment is crucial for a successful surgery. IoT has a strong capacity for performing successful procedures and for analyzing postoperative improvement. The used of IoT during the COVID-19 pandemic aided in providing patients with better care [7]. IoT was successfully used for real-time monitoring, which prevented fatalities from conditions including diabetes, heart failure, asthma attacks, and high blood pressure. To efficiently communicate the necessary health data to the doctor, smart medical equipment are connected via a smartphone. Additionally, these gadgets record information on oxygen, blood pressure, weight, sugar levels, etc.

1.1 Ecosystem of medicine for highly infectious diseases

POC devices for infectious diseases have historically been developed primarily for low-resource environments [8]. The COVID-19 pandemic, for example, is a good example of how they might be used to contain extremely contagious disease

Deep Learning in Personalized Healthcare and Decision Support. https://doi.org/10.1016/B978-0-443-19413-9.00003-5

outbreaks. The creation, production, and implementation of a supportive ecosystem that includes home-screening, POC devices, and the IoMT for illness diagnosis and monitoring are encouraged by the necessary strategy of self-quarantining under such circumstances [9].

The present pandemic is pushing the healthcare sector to the brink of collapse; clinical centers are already overburdened with verified cases and suspected ones awaiting diagnostic confirmation [4]. The need for these services is growing, but there is a dearth of diagnostic tools and supplies, which restricts patient access and raises the risk for front-line healthcare professionals who are performing mission-critical tasks. The crisis might be mitigated, and lives could be saved with the help of a robust, encouraging medical ecosystem.

A more complete medical infrastructure is obviously required to combat highly contagious diseases. The development and use of POC, home-based diagnostic equipment that lessen testing and monitoring burdens and enforce stay-at-home protocols is prompted by a lack of clinical resources and the necessity of implementing self-quarantine methods. By merging such a strategy with the IoMT, information might be sent to healthcare facilities for the creation of appropriate patient care protocols and to governmental bodies for the proper allocation of supplies and equipment [10]. Numerous lives could be saved, thanks to this joint strategy, which would also preserve already-stressed economies and serve as a model for more effective future danger management.

1.2 IoT implementation in the medical sector

With the aid of cutting-edge technology, IoT has a great ability to produce solutions of the highest caliber. In the field of medicine, it becomes a new reality of an original concept that offers COVID-19 patients the greatest care and conducts accurate surgery [3]. During the ongoing pandemic, complicated situations are readily managed and controlled digitally. IoT uses new medical issues to develop top-notch support solutions for practitioners, patients, and patients. To implement the IoT successfully, many process processes are methodically defined (Fig. 26.1). Sensors are used to perceive, collect, and receive the necessary data regarding patient health and disease. Here, everything that is tangible is networked (linked to the Internet), and devices show real-time process monitoring. Specific doctors are given the necessary medical data in accordance with their needs.

2. Cognitive Internet of Things (CIoT)

A network of connected physical items, such as sensors, health monitoring equipment, smart meters, home appliances, autonomous vehicles, etc., is referred to as the IoT. These items can perceive, process, and communicate with one another, interact with people automatically, and intelligently serve users thanks to this vast connection [4]. Data traffic is anticipated to reach 4394 EB by 2030 due to the daily growth in applications and the enormous increase in wireless devices that are connected to each other.

The cognitive radio-based IoT known as cognitive IoT (CIoT) is a promising method for the effective exploitation of limited spectrum in order to fulfill this constantly growing bandwidth requirement [7]. The core concept of CIoT is the dynamic allocation of radio channels for the transmission of information between extremely densely connected items.

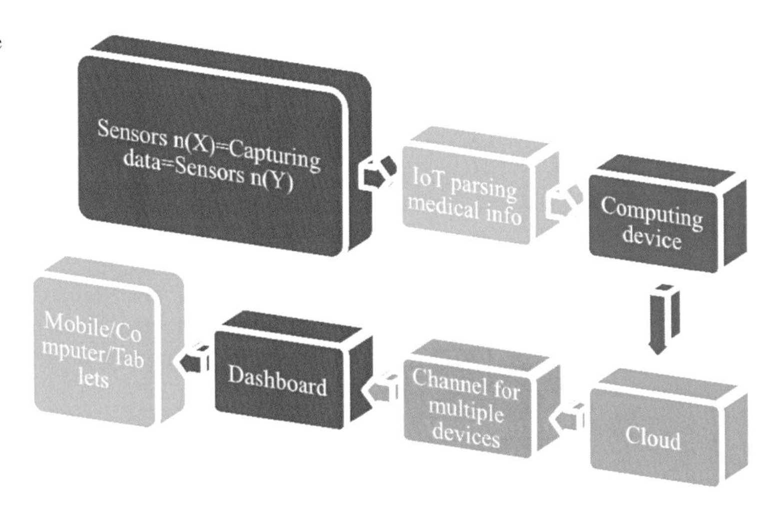

FIGURE 26.1 Process chart of IoT implementation in the medical field.

Given that every individual will be connected to and monitored by a vast network, the CIoT concept is most adapted to this pandemic. A large portion of activities, including e-commerce, e-learning, smart metering, e-surveillance, smart healthcare, and telemedicine services, takes place online as a result of the global lockdown and restrictions on movement and crowds. Through wireless networking and communication, which use bandwidth, these activities are made possible. By opportunistically looking for empty channels, the vast CIoT network sends brief packets, conserving bandwidth, and effectively using the spectrum resource [11].

2.1 Cognitive Internet of Medical Things (CIoMT)

The novel coronavirus (SARS-CoV-2) pandemic, also known as coronavirus disease 2019, has spread over the entire world since Wuhan, China, first reported it on December 31, 2019. Because it is the infectious virus that spreads fastest, it became a new threat to public health on a global scale. Presently, there are 5,555,749 COVID-19 infection cases globally, of which 348,220 deaths have been documented and 2,875,734 are active cases impacting 213 nations [12]. Every hour, these data dramatically increase. Medical professionals and academics are presently looking for innovative methods to detect for and stop the COVID-19 pandemic in this worldwide health catastrophe. In order to ensure appropriate patient isolation and prevent disease containment, rapid monitoring of viral infections is crucial for both healthcare personnel and the greater public health perspective [13]. The current digital technologies that can be used in this scenario to address significant clinical issues related to COVID-19 include IoT and Artificial Intelligence (AI) research.

In the current digital era of cutting-edge technology, recent advancements in IoT in 5G telecommunication network, AI that includes machine learning methods (Random Forest, Naive Bayes, Decision Tree, Extreme Learning Machine, Reinforcement learning, Long Short Term Memory Network, Convolutional Neural Network, etc.) and deep learning techniques, big-data analytics, cloud computing, Industry 4.0, and block-chain technology can provide long-term solutions to solve the COVID-1 [14]. These technological advancements can aid in the betterment of the disease's identification and treatment as well as its containment. As shown in Fig. 26.1, these interconnected technologies can help with real-time data gathering from people in remote regions using IoT; processing, analyzing, forecasting, and decision-making using AI and big-data analytics; backing-up the data using cloud computing; and this is strengthened by blockchain technology for secure data networks. One such technology is the cognitive IoT CIoT, which enables each and every physical entity in the globe to actively exchange information while guaranteeing guaranteed quality of service (QoS) criteria. The term "CIoT" refers to the cognitive radio (CR) enabled IoT, which facilitates machine-to-machine communication in a network of ever-growing wireless devices [15]. The technique for allocating dynamic spectrum that is CR-based can support a huge number of devices and a variety of applications. The class of CIoT known as Cognitive Internet of Medical Things (CIoMT) is specifically designed for the medical sector and is crucial to the development of intelligent healthcare [16]. Through IoMT, medical staff can remotely access the patient's real-time physiological data, such as body temperature, blood pressure, heart rate, glucose level, oxygen level, etc., as well as psychological data, such as speech and expression [17] (Fig. 26.2).

2.2 The uses of CIoMT

The use of CIoMT has increased throughout the years, as previously said, as a result of decades of study on smart healthcare [18]. By eschewing standard treatment approaches, this technology offers the chance to make important improvements in COVID-19 management. This crisis is being diagnosed, followed up on, tracked, and managed in real-time, which includes daily new cases of the illness [19]. It is difficult for a few people to manage the situation given that the entire population is impacted, unless there are live data updates available. CIoMT makes it possible to combine sensory data, autonomous processing, and network connectivity [20]. In order to combat COVID-19, CIoMT can be implicated in a number of key regions, as shown in Fig. 26.3. IoMT is extensively used in COVID-19 crisis management to provide patients with online medical services, appropriate healthcare, and test at home or a quarantine facility. In addition, as shown in Fig. 26.4, it can develop a medical platform for the administration of datasets valuable for governmental and healthcare services.

1. ***Real-time monitoring***
   Using this technology, it is possible to track the real-time daily update in COVID-19 cases around the world, including the number of cured patients, the number of fatalities, and the number of current cases in various regions. In order to help health authorities and policymakers make better decisions and be more prepared to control the disease, AI can be used to estimate the severity of the disease and predict its activity. Each person connected to the CIoMT network can access the government initiatives, healthcare precautions, and treatment process updates.

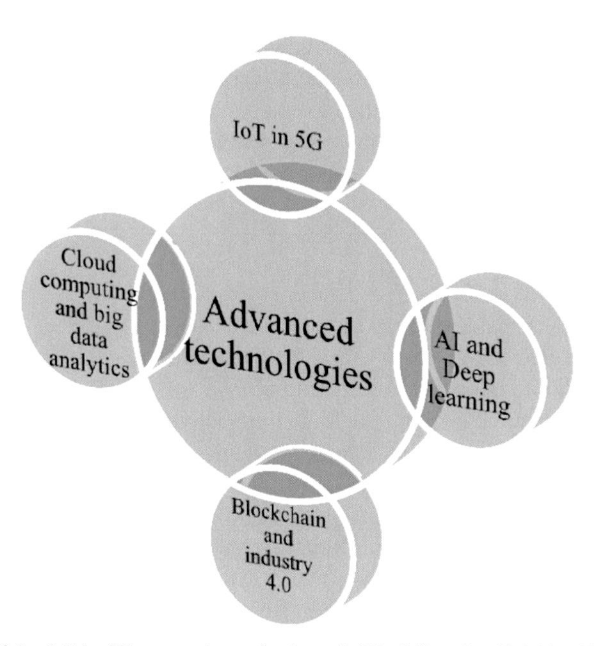

FIGURE 26.2 IoT in fifth generation technology (IoT in 5G) and artificial intelligence (AI).

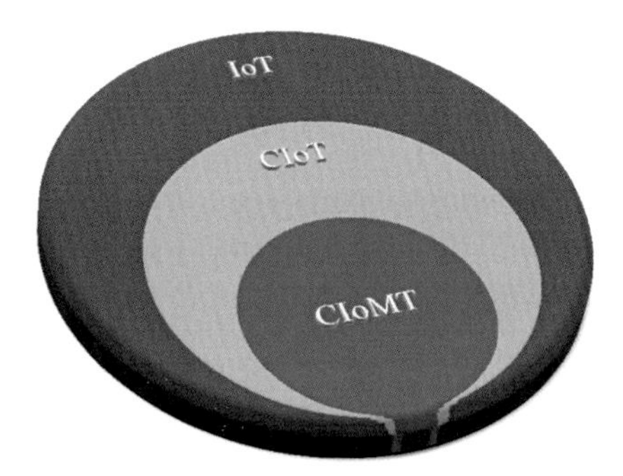

FIGURE 26.3 CIoT as a special case of IoT.

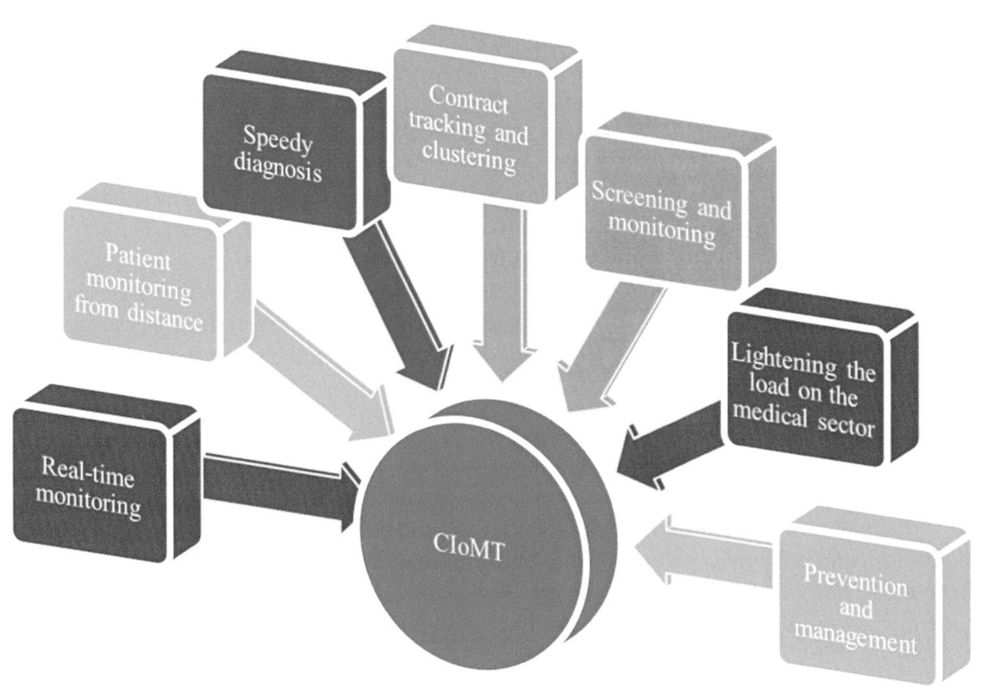

FIGURE 26.4 Major application areas of CIoMT for tackling COVID-19.

2. ***Patient monitoring from a distance***
   Doctors and other healthcare professionals were susceptible to COVID-19, while they were working because of how contagious it was. With the aid of fingertip medical data such as blood pressure, glucose levels, heart rate, electrocardiograms, electroencephalograms, electromyograms, pulse rates, temperature, and breathing rates, the CIoMT enables clinicians to remotely monitor patients' health conditions. Wearable IoT sensors can be used to collect data on this kind of clinical parameter. Real-time medical data sharing was available, thanks to the COVID-19 hospitals' broad internet connectivity, which saves time and effort. The use of CIoMT is beneficial for individuals with numerous illnesses or the elderly in particular.
3. ***Speedy diagnosis***
   Even if they do not exhibit any clinical signs, the suspected patients and migrants were placed in quarantine; thus, it is crucial that these cases are diagnosed quickly. Through specific network applications, CIoMT enables people with travel histories to connect to medical providers for quick diagnoses with little error. By live streaming video pictures that can be further analyzed by visual sensors with AI capabilities, the lab staff can remotely do an X-ray or computed tomography (CT) scan from the control room, speeding up the diagnosis and confirmation of the case. This permits contact-free and early viral identification as well.
4. ***Contact tracking and clustering***
   To restrict the spread of the pandemic, contact tracing of confirmed cases is crucial. This laborious task can be made easier if the location history of COVID-positive patients was easily accessible in a database that healthcare authorities can access. Depending on the quantity of confirmed instances via CIoMT, the area-wise grouping and categorization of the regions as containment zones, buffer zones, red zones, orange zones, green zones, etc. was swiftly updated. When the medical and healthcare facilities were connected via IoT, the location-specific number of positive instances can be gathered in real time. Through the use of AI framework, the government may quickly access these data and notify for health examinations for the impacted area. Various lockdown and social segregation regulations and orders can be put into effect, thanks to zone clustering.
5. ***Screening and monitoring***
   The general public and healthcare authorities can access the thermal imaging-based facial recognition data via CIoMT for screening and monitoring purposes at various entrance points of airports, railway stations, hostels, and other locations. The transmission of the virus may be slowed down by this automatic surveillance of suspected and positive cases.
6. ***Lightening the load on the medical sector***
   A limited number of healthcare personnel were helped by the CIoMT to handle a sizable portion of the population starting with diagnosis, monitoring, and therapy. Further reducing the workload, the CIoMT allowed for remote disease monitoring. For diagnosis, the lab technicians did not need to go to people's doors. Modeling and infection predictions were made possible by AI that was injected with IoT sensory data. By working with blockchain firms, hospitals made sure that telemedicine consultations and medicines deliveries arrived at the patient's door in a timely manner.
7. ***Prevention and management***
   By prompt medical and governmental action, as well as private vigilance, the virus can be contained from spreading. By using specific apps, the CIoMT enabled one to be aware of nearby positive cases and to be aware of potential threats (e.g., Arogya Setu used in India). Geographical clustering further curtailed the spread of illness.

3. Internet of Medical Things

As is well known, the IoT is a global collaboration of numerous appliances and devices that share data in real time [21]. The primary driving force behind IoT is the automation and intelligence of the task (in gadgets). By enhancing their capacity for intelligence processing and data analysis, the limitless amount of autonomous "smartness" makes it possible for them to interact with any manual procedure [22]. The Internet is therefore more active and effective on a global scale in the physical world. For instance, you can turn on the fan, AC, and light bulb using a smartphone app. Streetlights, traffic lights, thermostats, and motion sensors are other components that are connected to the Internet (IoT).

The Internet of Medical Things (IoMT) combines the IoT with medical equipment (IoT) [23]. Future healthcare systems will be based on IoMTs, where every medical gadget will be connected to the Internet and supervised by medical specialists. As it develops, this offers speedier and less expensive health treatment. Due to physical distance regulations that force healthcare practitioners to manage patients remotely using IoMT devices, telehealth practices have increased during the COVID-19 pandemic [24]. Additionally, the UN established the Sustainable Development Goals (SDG) by 2030 in 2015. One of the SDG's main objectives is to promote health and well-being. IoMT currently has the capacity to achieve the objective of health and wellbeing.

A significant portion of the population in India will be above the age of 60 by 2030, and as the population grows, so does the number of senior individuals. The cost and availability of healthcare gadgets are rising along with life expectancy in general. As is well known, a significant amount of medical personnel and healthcare service providers will also be needed [25]. Internet of Medical Things (IoMT) is a solution to such a problem. Cost control will be simpler using IoMT. The health metrics, such as blood pressure, heart rate, and others, that are monitored by medical equipment connected to the Internet or the web will also maintain tabs on patients' routines, such as how much they sleep and what they do each day.

These technologies will be used in the case of the pandemic illness such as COVID-19 to track the actions of sick people or patients. IoMT hardware will provide monitoring functions [26]. IoMT is able to send and record the status of patient activity and send reminders to both the patient and the healthcare service provider, such as a doctor or nurse, because patients and elderly patients sometimes forget to take medication or neglect to take it at a specific time. Wearable IoMT devices make it feasible to take measures, keep tabs on the patient's daily activities, and send out or inform an alarm if anything goes wrong or a diagnosis is made incorrectly [27]. Every aspect of life will result in an increase in real-time or statistical data, and it is a far more difficult process to separate and use that data for prediction. Similar to the COVID-19 example, the Arogya Setu app in India can offer real-time data. We should employ medical IoT (mIoT or IoMT) based on that data to take essential and preventative steps in the future.

As a result of the Internet of Medical Things, pharmaceutical corporations and healthcare providers will soon have access to large amounts of data on patients, including information on their blood pressure, ECG, respiratory rate, sleep patterns, pulse rates, diabetes, and cancer patients. Medical devices and sensors that are connected to the Internet and access data according to device or app configurations will retrieve or gather all of these data. Athletes and sportspeople will have already adopted these technologies to monitor athletes' physical habits. IoMT includes remote patient diagnosis and monitoring of individuals with long-term or chronic disease conditions, using patient mHealth devices (such as wristbands that calculate heart rate, number of steps, etc.) that communicate with the medical team or patients [28].

The applications of IoMT are increasing and enable devices to share information with other devices that are connected through the web or internet, just as in the case of IoT and as a result of more advancement in the IoT field by Near Field Communication (NFCs) or Radio Frequency Identifications (RFIDs) [29]. As hospitals or medical centers use RFIDs on their medical equipment or supplies, hospital employees and other relevant parties are kept informed of the stock levels and can take appropriate precautions as needed. Additionally, IoMT devices or equipment make it simple to do treatments in which patients must visit to hospitals and doctors' offices for minor medical exams or to get reports [30]. IoMT has a wide range of applications that can be employed in real-time situations in addition to patient health monitoring. We follow medical equipment like wheelchairs, oxygen pumps or cylinders, ECGs, the status of ventilator beds, nebulizers, defibrillators, and other monitoring devices in real time by tagging the items with sensors or IoT.

Real-time data will be gathered using the Arogya Setu app as the COVID-19 situation develops, allowing medical facilities or equipment to be assigned right away to the most affected or infected areas. Teams of doctors or medical staff can also be deployed at various locations by taking into account real-time data provided by mobile apps and IoMT devices are available at such locations IoT-enabled hygiene monitoring devices are very beneficial in these scenarios to prevent patients from contracting an infection since, as we all know, infection is a huge issue in this type of situation, and it is also a key concern factor in hospitals. The monitoring of environmental parameters, temperature control, humidity control, and inventory control in drugstores are just a few of the asset management tasks that IoT devices may assist with.

3.1 The idea behind the Internet of Medical Things

Cities were suddenly being placed under lockdown as COVID-19 spread more quickly, forcing nearly 10 billion people to self-quarantine at home. The demand for essential medical equipment and supplies was tremendous, and it was likely to exceed the ability to refill stocks quickly enough [31]. To access medical care, residents and potential patients had to leave their homes, which compromised isolation and quarantine efforts and endangered disease management. Also, the medical community was advising people with minor or suspected symptoms to stay at home due to the lack of isolation units and appropriate medical equipment. It was obvious that a different, at-home diagnostic test would provide an excellent solution to meet this critical and unmet demand.

The expanded, healthcare-focused version of the IoT is known as the IoMT [32]. If it were employed to address the current problem, it might be used to construct a medical platform that would assist patients in receiving proper treatment at home and a thorough illness management database for governmental and healthcare institutions. Equipment for diagnosis and healthcare (protective masks, thermometers, drugs, and POC (point-of-care) COVID-19 kits for diagnosing and monitoring infection) might be obtained by those with minor symptoms. Patients might routinely transmit their health status via the internet to the IoMT platform (clinical cloud storage), where it could be shared with surrounding hospitals, the Centers for Disease Control (CDC), and state and local health agencies [33]. The government (the CDC, and municipal and state health bureaus) might deploy equipment and designate quarantine stations (hotels or centralized quarantine facilities) if necessary. Hospitals could then provide online health consultations based on each patient's health condition.

People might dynamically track their disease condition and receive the necessary medical care via the IoMT platform without infecting others. In addition to providing a systematic database that would enable the government to adequately monitor illness transmission, properly distribute supplies, and implement emergency plans, this would lower national health costs and alleviate the pressures of a medical device scarcity.

3.2 Wireless technologies that enable IoMT

Sensors and gadgets make up IoT systems, which are connected by a network of cloud ecosystems with high-speed connectivity between each module [34]. The large storage space provided by cloud services receives the unprocessed data that have been acquired at these devices/sensors. To acquire further understanding from this data, it is further cleaned and then analyzed. In order to further aid in the visualization, analysis, processing, and administration of the data, additional software, tools, and applications are required. Smartphones, monitoring devices, sensors, smart wearables, and other medical devices are all connected via a variety of wireless technologies, including RFID, NFC, Bluetooth, long-term evolution, and 5G/6G (and beyond). Due to its enormous bandwidth and ultra-low latency advantages, 5G/6G and beyond are currently widely used in IoMT.

4. Smart e-healthcare

Hospitals that use intelligent automated and optimized modules (perhaps based on AI/ML) on the ICT infrastructure to enhance patient care processes and provide new capabilities are referred to as "smart hospitals" [35]. Telemedicine, telehealth, and remote robot surgery are just a few of the applications of smart hospitals. While telehealth focuses on providing nonclinical care remotely, telemedicine is used to deliver clinical treatment at a distance. In remote robot surgery, surgical procedures are carried out by medical robots under the direction of a surgeon who is located far away.

An example of a typical IoMT flow would be a smart healthcare system where the inbound data from multiple sources are first gathered (for example, by remote gathering or physical gathering) and transferred to the EHR. If the staff collects data as medical notes offline and on paper, it may be categorized as unstructured data. It is simpler to process data in subsequent systems, such the customer relationship management (CRM) system, if it is received from devices and sensors in an organized form utilizing predefined data fields for users to complete [36]. The CRM makes available the resources for data analysis and data assignment to the ecosystem's predetermined targets. The CRM system receives the necessary data and information from EHR systems and processes it. These processed data produce additional triggers for patients and medical staff throughout the ecosystem. In the form of customized health regimens, hospitals and medical professionals communicate with the patients directly. The same CRM software in the ecosystem notifies the doctors and other medical personnel about the reminders and other alerts.

4.1 IoMT architecture

Three levels make up the architecture for IoMTs: the items layer, the fog layer, and the cloud layer. This architecture is a modified version of that seen in Ref. [35]. In this design, healthcare professionals can also communicate directly over the router connecting the Thing and Fog layers and via the regional fog layer processing servers. Below is a description of each layer:

1. Patient monitoring equipment, sensors, actuators, medical records, pharmaceutical controls, diet plan generators, etc. are all included in the things layer. With the ecosystem's users, this layer is in close proximity. At this layer, information is gathered from components like wearables, patient monitoring, and remote care. To protect the integrity of the data collected, the devices utilized at this point should be securely installed. These devices' connections to the fog layer are made possible by the ecosystem's local routers. To provide useful information, the data are further processed at the cloud and fog layers. The medical professionals can also access the patient data over this router to shorten the wait.
2. Between the cloud and the thing layers, the fog layer functions. Local servers and gateway devices make up this layer of the sparsely distributed fog networking architecture. The bottom layer devices make use of the local computing power to provide consumers with real-time responses. These servers are also used to oversee and monitor the system's security and integrity. This layer's gateway components are in charge of routing the data from these servers to the cloud layer for additional processing. The medical professionals can also access the patient data over this router to shorten the wait.

3. The cloud layer is made up of computing and data storage capabilities that enable analysis of the data and the creation of decision-making systems based on it. Incorporating massive medical and healthcare systems into the cloud also provides a vast reach for handling their daily operations with simplicity. The data generated by the medical infrastructure will be kept in this layer's cloud resources, where analytical work may be done as needed in the future.

5. IoMTS emerging technologies

Following is a discussion of the many developing technologies and how they relate to IoMT, including blockchain, PUF, AI, and SDN:

a. **Blockchain technology:** Blockchain technology is a decentralized ledger that tracks the activities of the network's computing nodes. Blockchain provides a solution to many problems occurring in the security of participating entities in the healthcare system created around it, and the IoMT has increased growth in distributed computing industries [10]. The information shared between any two nodes in the network is stored and can be used for cross referencing in the blockchain, which is made up of blocks or nodes connected by a network. As a result of this technology, it is possible to pinpoint the precise location of network criminals. These blocks contain data from earlier blocks. As a result, unidentified blocks in the network are deleted, opening the door for the implementation of blockchain as a trusting method in IoMT and other information exchange systems.
Without the need for a centralized authority, entities can communicate with one another using blockchain technology. Blocks of data are used to store the data entries in a blockchain. As previously stated, these blocks have cryptographic protocols to bundle the data about their neighboring blocks in the chain so that it may be used safely. Other users can read these blocks and their data, but the data in these blocks cannot be altered. Additionally, smart contracts can be processed easily on the blockchain without the requirement for a central authority to do so. Since the design of these contracts allows for self-execution, oversight is not necessary. Ethereum is a well-known provider of "smart contracts," offering its services on blockchains. Infrastructure must be separated into smaller modules in order to embrace solutions based on blockchain in the healthcare industry.
The IoMT framework's relevant devices can then be integrated with these modules. The resulting system will have a dispersed structure and would enable network power decentralization. While the amount of data entering the healthcare ecosystem is constantly increasing, one advantage of using blockchain technology is that they can be trusted. The ever-increasing demand for data transfers via the healthcare infrastructure is expected to be satisfied by the blockchain. Hospital EHR systems are currently being evaluated with blockchain technology and this will be followed by several global clinical usage trials.
b. **Physically unclonable function (PUF) devices:** PUF devices produce a distinctive fingerprint for the weak points in the IoMT ecosystem. Due to the variations in how these gadgets are made, each one has its own fingerprint or signature. In the IoMT environment, where end devices (sensors) are vulnerable to assaults involving hardware tampering, these fingerprints can be used to generate secret keys (cryptography keys) that will safeguard the devices and the data they contain.
c. **AI in IoMT:** Precision medicine calls for speedy delivery of customized regimens and sophisticated diagnostics. AI provides real-time solutions in identifying novel pathways for treating particular ailments based on historical and current data, which offer a strong argument for this. Using AI-based solutions, the many components of the healthcare ecosystem can be changed. These will incorporate AI methods for building classifiers, such as automatic patient data collection, appointment scheduling, and selection of lab tests, treatment plans, drugs, and surgical procedures. Further training and help for decision-making processes could be provided by these classifiers. In order to extract data from such unstructured data points in the infrastructure for other classifiers that cannot be digitally recorded, NLP provides ways. These might take the form of lab results, notes from a physical examination, notes from an operation, and other patient discharge-related data. On the basis of past data, machine learning also makes predictions about the future. To forecast future circumstances, it uses supervised, unsupervised, or reinforced learning.
d. **SDN in IoMT:** There are two components to the network in IoMTs: the data plane and the control plane. While the control plane completes the essential duties to enable the data plane to make forwarding choices, the data plane forwards traffic in the direction of its destination. A standardized method of communication between the data plane and control plane is offered by software-defined networking (SDN). OpenFlow, OpenvSwitch Database Management protocol, and OpenFlow Configuration protocol (OF-CONFIG) are a few examples of common SDN protocols. Numerous different pieces of data from the data plane can be gathered from an external server (which may be situated in the cloud) using the standard OpenFlow protocol because the interface between the data plane and control plane can

be made standard using a standard SDN protocol. This makes it possible to create various e-healthcare applications because they can exist on the cloud layer.

6. POC biosensing for infectious illnesses with IoMT assistance

A development in information technology (IT) called the "Internet of Things" allows data to be sent wirelessly and operations to be automated without the need for human participation [37,38]. With the development of smart gadgets like the "smart house," "automotive sensors," "smart farming," and "smart meters" to conserve energy and labor, the IoT has now integrated itself into a number of organizations. In order to reach a brighter future, the idea of IoT technology has also infiltrated the health-tech sector and is known as the "Internet of Medical Things" [2]. The IoMT is a network of connected medical devices. Using an advanced health-tech environment, IoMT devices improve real-time diagnosis efficiency and quality [38]. Many healthcare organizations have begun making significant investments in this area in an effort to take use of the potential of IoMT to empower medical professionals. Through intelligent services for imaging, sensing, and diagnostics provided by the IoT and smartphones, this new era of digitalization generates smart environments. For better real-time results, optical microscopy, flow cytometry, and other imaging readouts can all benefit from using the resolution of a cell phone camera lens [8]. POCs with Bluetooth low energy (BLE) technology provide data communication up to a few feet, allowing devices like fitness trackers and noninvasive mouthguard biosensors for salivary glucose to continually monitor patients' vital signs and physical activity.

The goal of research is to create smart fabrics with sensors that can measure body temperature, ECG, blood pressure, and heart rate. Medical kiosks, fitness tracking devices (wristbands, smart watches, shorts, etc.), clinical grade wearables (smart belts, chest straps), remote patient monitoring technology, smart pills, and many other items are among the many Internet-connected devices that are readily available in the market in 2020. As the pathological level of these biomarkers has not yet been defined, a variety of noninvasive materials, including sweat, saliva, feces, tears, and breath, may be utilized in eHealth diagnostic devices (eDiagnostics) to find biomarkers of major diseases (viral infections, cancer, and HIV) [8]. Wearable eDiagnostic devices at eHealth systems could meet a variety of diagnostic demands. It is interesting to note that people with modest and low incomes use smartphones frequently, which encourages the adoption of POCT without incurring additional fees. Furthermore, it will expand medical services at lower costs in rural areas and isolated people. The demand for improved communication, information resources, and social networking at the POC is a major factor driving the fast adoption of smartphone-based computing devices in the health system. Other beneficial characteristics of IoMT devices include short messaging service, email, global positioning systems (GPS), and interactive voice recordings that can aid in automating POC diagnostics in data gathering and further continuous transmission into a main database via satellite networks. A sizable collection of databases that aid in the study of complicated diseases like cancer, asthma, coronary artery disease, diabetes, and Parkinson's disease can be created using patient health information. The envisioned smartphone-based POCs can conduct end point or real-time tests using a specific detection method, such as colorimetric, fluorescence emission, reflection, current, or turbidity [39]. Smart POCs, for instance, can detect the emergence of a color shift in an ELISA-based assay, and the images they collect can then be further analyzed with the help of tailored software to quantify the pathogens. Numerous optical add-ons must be applied to provide sensitive smartphone fluorescence detection. As with lateral flow, PCR, and loop-mediated isothermal nucleic acid amplification (LAMP), IoT-based POC devices have been described with simple read-out platforms for detection of many infectious diseases. These devices are used to identify new infections. Additionally, electrochemical IoMT uses built-in circuits to capture clinical data and enable real-time pathogen diagnosis. POC systems combined with ML algorithms, or artificial intelligence (AI), in recent years has led to a paradigm change in favor of decision-making based on data analytics. Medical professionals now have another choice when it comes to treating patients specifically and keeping track of their results thanks to AI. Health care professionals are able to evaluate patient risks for various ailments like heart attack, cancer, and trauma using these AI systems. The multistep test is first carried out by the AI sensors utilizing a portable analyzer, and the biomarker concentrations are then automatically digitalized.

With the aid of acquired, the range of diseases can be anticipated using a variety of algorithms, including as classification, cluster, pattern, and disease traits, to choose the best course of therapy. Various wireless devices, including mobile phones and personal digital assistants, complement M-Health, a WHO worldwide observatory used to show results to patients. In order to actively control and reduce the risk of developing diseases, health analytics offers patients individualized therapy. The healthcare industry has been changed by real-time POC diagnostics and machine learning. With the integration of AI into the linked ecosystem for the possibility of automated prescription, the IoMT market growth is anticipated to accelerate more [40].

7. IoT's primary benefits for the COVID-19 pandemic

IoT is a cutting-edge technology that makes sure that all those who have contracted this infection are quarantined. A good monitoring system is helpful when under quarantine. Through the use of an Internet-based network, all high-risk patients may be easily tracked. Biometric parameters like blood pressure, heart rate, and glucose level are taken using this technique [41]. We can expect to witness an increase in the productivity of medical staff and a decrease in their workload with the successful adoption of this technology. With less costs and errors, the same can be used to the COVID-19 pandemic.

8. Significant applications of IoT for COVID-19 pandemic

To build a smart network for the ideal health management system, IoT makes use of a huge number of interconnected devices. It monitors and tracks all diseases to increase the patient's safety. Without any human interaction, it digitally records the patient's data and information. For effective decision-making, this information is also useful [3]. In order to meet the crucial need of reducing the effects of the COVID-19 pandemic, IoT was deployed in a variety of applications. With the aid of the relevant gathered data, it has the potential to predict the upcoming circumstance. To effectively handle this pandemic, its applications were put to use. In order to receive individualized care, the patient can use IoT services to properly monitor their heart rate, blood pressure, glucose levels, and other activities. Monitoring older people's health conditions is helpful. The real-time tracking of medical equipment and gadgets for a seamless, delay-free course of treatment is one of this technology's key uses in healthcare. This technology can be used by healthcare insurance firms to spot fraud claims and increase system transparency. This enhances the patient's treatment workflow through effective performance and also aids in decision-making in challenging cases. IoT's primary uses for the COVID-19 pandemic included the following:

1. ***A hospital that is Internet-connected:*** The installation of IoT to support pandemics like COVID-19 requires a fully integrated network within hospital grounds.
2. ***Alert the concerned medical professionals in the event of an emergency:*** Thanks to this integrated network, both the patients and the staff will be able to act more promptly and efficiently as necessary.
3. ***Transparent COVID-19 treatment:*** Patients can receive the advantages provided without any favoritism or discrimination.
4. ***Automated treatment process:*** The choice of treatment modalities becomes effective and aids in the proper handling of the situations.
5. ***Telehealth consultation:*** This specifically uses well-connected teleservices to make therapy accessible to those in need in remote regions.
6. ***Using a wireless healthcare network to identify COVID-19 patients:*** Smartphones can be loaded with a number of reliable applications that can make the identification process easier and more successful.
7. ***Effective patient tracing:*** The effective patient tracing finally gave service providers the ability to manage cases more deftly.
8. ***Real-time information throughout the spread of this infection:*** Because the devices, locations, channels, etc. are informed and connected, proper case management is possible due to the timely information exchange.
9. ***Rapid COVID-19 screening:*** Using smart, connected treatment devices, the correct diagnosis will be tried as soon as the case is received or discovered. In the end, this improves the screening procedure as a whole.
10. ***Find a creative solution:*** The best supervision possible is the main objective. It can be done through bringing inventions to the masses with success.
11. ***Connect all medical equipment and devices to the internet:*** IoT connected all medical equipment and devices to the internet during the COVID-19 treatment, which provided real-time information during the procedure.
12. ***Accurate virus forecasting:*** Using a statistical method and the data report currently accessible, it is possible to estimate the situation in the future. It will also be helpful in planning for a better working environment among the government, medical professionals, academicians, etc.

9. Technologies of IoT for the healthcare during COVID-19 pandemic

In order to construct intelligent information systems that meet the specific COVID-19 patient criteria, the IoT connected medical instruments, machines, and gadgets. To enhance output, quality, and understanding of emerging diseases, an alternative interdisciplinary strategy is required. IoT technology track changes in crucial patient data to get pertinent data.

High-quality medical devices were greatly impacted by IoT technologies, which aid in providing a customized response during the COVID-19 pandemic. Digital data can be captured, saved, and analyzed using these technologies. All clinical records are kept digitally, and with the use of internet resources, patient information and data can be easily transferred in emergency situations, which help doctors work more effectively. We successfully monitored and regulated all the critical parameters such as temperature, blood sugar level, blood pressure, and information on COVID-19 patient health by employing smart sensors.

The optimum form of communication and monitoring relies heavily on software. For the finest care in the future, all records were kept in strict confidence. To improve treatment accuracy, efficiency, and dependability, AI improves the performance of physicians and surgeons. By using this technology, medical professionals may quickly identify bone abnormalities and lessen patient discomfort. The motion and control of the physical object are introduced by various actuators. The IoT's best technology for enhancing planning and real-time information is virtual reality. The many IoT technologies for use in healthcare during the COVID-19 pandemic were listed below.

1. **Big Data**
   - ➢ The best method available to collect, store, and analyze the COVID-19 patient's data appeared to be big data.
   - ➢ Medical data were historically stored in hard copy, incurring additional costs.
   - ➢ Big data efficiently maintained all patient data, the billing system, and the clinical record.
   - ➢ Data were saved systematically, which swiftly gave the best solution for healthcare.
   - ➢ Big data has outstanding capability to store data in digital form.
2. **Cloud Computing**
   - ➢ It saved on-demand data storage utilizing computer system resources and described data with the aid of the Internet.
   - ➢ In urgent situations, it immediately exchanged COVID-19 patient information.
   - ➢ Share confidential data sources that assisted medical professionals in performing their duties quickly and effectively.
   - ➢ It lowered the cost of data storage while also improving data quality.
3. **Smart Sensors**
   - ➢ A smart sensor's capacity to interface with a digital network and generate dependable results in the medical field is great.
   - ➢ It keeps track of and regulates all patient health-related parameters.
   - ➢ Easily monitored the COVID-19 patient's blood pressure, temperature, oxygen concentrator, sugar level, infusion, and fluid management system.
   - ➢ Useful for learning about one's health, damaged bones, and surrounding biological tissue.
4. **Software**
   - ➢ For the medical industry, there is specialized software available that enhances patient care, stores patient data, and provides treatment along with accompanying tests and diagnoses.
   - ➢ Aids in enhancing communication between patients and medical professionals.
   - ➢ The COVID-19 patient's medical history was stored, and software made it simple to identify and manage the patient's disease and other private information.
5. **Artificial Intelligence**
   - ➢ Provide exceptional ability to forecast and control viral infections.
   - ➢ Artificial intelligence helps to perform, assess, validate, predict, and analyze data in a predetermined context.
   - ➢ The efficiency, accuracy, and effectiveness of doctors and surgeons are increased with the aid of this technology.
   - ➢ It assessed the COVID-19 patient's discomfort while taking different medications.
6. **Actuators**
   - ➢ An actuator is a device that provides motion into a system and directs it to act in a specific environment.
   - ➢ The major uses of medical actuators are to uphold precision and regulate necessary parameters.
   - ➢ Aided in the development of hospital beds with adjustable heights to meet the needs of COVID-19 patients.
7. **Virtual Reality/Augmented Reality**
   - ➢ A better method of integrating people and technology to deliver real-time information
   - ➢ The effectiveness of COVID 19 patient therapy, patient safety, and planning all benefitted from the use of virtual reality.
   - ➢ Give COVID-19 patients and clinicians the information they need to increase the accuracy of surgery planning.
   - ➢ Additionally, augmented reality offers digital life information in the form of sound and image.

10. IoT-enabled healthcare helpful during COVID-19 pandemic

Millions of people's lives improved, thanks to IoT's good effects on healthcare. It extensively examines the healthcare system and discovers sickness. It gives each person individualized attention for their advantage. IoT technologies provided a variety of information during the COVID-19 pandemic, including appointment reminders, activity tracking, calorie counts, blood pressure readings, and disease status. IoT had a different application in the medical industry to spur innovation during the COVID-19 pandemic. It reduced waiting times because it was the greatest method for keeping track of both patients and workers. For the patient's comfort, it introduced many devices. The quality of inpatient care has improved because of smart gadgets including blood gas analyzers, thermometers, smart beds, glucose meters, ultrasound machines, and X-ray machines. The IoT can be used to improve or replace biological components. It had uses in connected imaging, clinical operations, drug distribution, patient monitoring, laboratory tests, and medication management during the COVID-19 pandemic, among other areas of medical equipment. IoT aids medical personnel in providing patients with the best care possible.

All hospital actions are digitally recorded in a centralized information system that is created by this technology. Data analytics can also be applied to COVID-19 pandemic problem-solving. With the use of this technology, patients' health can be simply tracked, and complex cases can be decided with accuracy. It warns about any impending ailments and offers a preventative measure by routinely monitoring the health state. In addition to serving as a timely reminder to take medication, it aids in the early detection of an asthma attack. Insights from IoT-enabled healthcare during the COVID-19 pandemic included the following:

i. **COVID-19 patient treatment:**
   - ➢ IoT with real-time location service employed for the COVID-19 patient's optimum care.
   - ➢ Nebulizers, scales, wheelchairs, pumps, and other medical equipment were utilized for monitoring in the framework of the IoT.
   - ➢ Additionally, verified, monitored, and maintained control over environmental factors including temperature and humidity.

ii. **Smart Medical Center:**
   - ➢ Using automation and a specialized network that is connected, IoT offered a smart hospital.
   - ➢ Software gave accurate information on the patient's ongoing problems.
   - ➢ The patient's wait time was cut in half, thanks to the total digitalization of the system in the smart hospital.
   - ➢ It offered a review of the patient's and process's records.
   - ➢ Data analysis supported everyday operations to enhance COVID-19 patient care.
   - ➢ IoT devices transmitted, effectively stored, and analyzed data of COVID-19 patients for future therapy.

iii. **Storage of COVID-19 patient data:**
   - ➢ Spread knowledge on the origin of this virus.
   - ➢ Checking and analyzing the patient's recovery is beneficial.

iv. **Use many sources and tools to create your content:**
   - ➢ IoT offers a variety of sources and devices that can be automatically analyzed in the healthcare industry.
   - ➢ The required data of the COVID-19 patients were automatically collected, reported, and stored by an IoT device.
   - ➢ It aids in the resolution of complex cases, hence lowering surgical and medical mistake.

v. **Making wise decisions:**
   - ➢ This technology records data and allows for the accurate and high-quality decision-making that is necessary for precise surgery.
   - ➢ Aided in the accurate recording of COVID-19 patient data, which was previously challenging for doctors to perform.

vi. **Track the COVID-19 patient's condition:**
   - ➢ It forecasted the patient's arrival and kept track of the state of the support systems.
   - ➢ Assisted in keeping an eye on infection and cleanliness in hospital and support facilities.
   - ➢ Accessed patient data for COVID-19 and maintained additional information.

vii. **Warning regarding the COVID-19 illness:**
   - ➢ This device alerted the human for COVID-19 disease in life-threatening conditions with real-time tracking.
   - ➢ It rapidly alerted users via connected devices.
   - ➢ Reported on the state of people's health and expressed an opinion accurately.
   - ➢ Offered more accurate monitoring, real-time alerts, and prompt treatment.

viii. **Information for the healthcare professional:**
  ➢ Appropriate lighting through personal management and intimate knowledge of the patient need
  ➢ Appropriate information to the healthcare professional, location, time identification, and database.
  ➢ The best way for doctors to communicate with patients' families.

ix. **The right medication:**
  ➢ It kept track of the COVID-19 patient's diet, protein consumption, and proper medicine administration to the body.
  ➢ Examined and tracked the patient's progress in day-to-day activities.

x. **Appropriate facilities:**
  ➢ During the COVID-19 pandemic, the patient process was simply automated with the use of IoT.
  ➢ Information wa exchanged to improve the efficacy of healthcare services.
  ➢ It appeared to be the most effective method for more effectively utilizing high-quality resources, which enhances the planning of difficult operation.

xi. **Check the glucose level:** Check the level and flow of glucose in accordance with the patient's needs.
  ➢ Automatically modify the insulin dosage to remain within a safe range.

xii. **Provide assistance in remote areas:**
  ➢ IoT can help remote patients locate doctors using smart mobile phone applications in areas far from clinics.
  ➢ Useful in examining a COVID-19 patient and determining the infection's origin.
  ➢ It will enhance patient care and boost hospital digitization.

xiii. **Discovering an asthma attack:**
  ➢ It immediately identifies asthma symptoms before an attack takes place.
  ➢ Informed about the attack and details relating to its prevention.

xiv. **A prompt to take your medication:**
  ➢ This technology's primary use is to remind patients when it's time to take their medications.
  ➢ When a dose is missed, it warns the patient to take it at the appropriate time.

xv. **Emergency situation:**
  ➢ Before approaching a facility or hospital, IoT analyzes the distance and accesses the patient profile in an emergency situation.
  ➢ This resulted in better emergency care and a decrease in related losses.

xvi. **Smart Bed:**
  ➢ Applications for the IoT are working to create a smart bed that can adjust in height to meet the needs of COVID-19 patients.
  ➢ The pressure and support for the patient can be automatically adjusted by this smart bed.

11. Conclusion

IoT's primary goal in medicine is to aid in the accurate treatment of various pandemic instances. By reducing risks and improving overall performance, it facilitates the surgeon's work. Doctors can quickly identify changes in a patient's vital parameters by employing this technique. As it advances toward the ideal approach for an information system to adapt world-class results and enables development of treatment systems in the hospital, this information-based service creates new healthcare opportunities. With improved illness detection training and guidance for the future, medical students may now detect diseases more accurately. The effective application of IoT can assist in effectively resolving several medical difficulties, such as complexity, speed, and cost. To track caloric intake and patient treatments for conditions like asthma, diabetes, and arthritis, it is simple to customize. The performance of healthcare generally during pandemic days can be improved by this digitally controlled health management system.

References

[1] S. Vishnu, S.J. Ramson, R. Jegan, Internet of medical things (IoMT)—an overview, in: 2020 5th International Conference on Devices, Circuits and Systems (ICDCS), IEEE, 2020, pp. 101–104.

[2] F. Al-Turjman, M.H. Nawaz, U.D. Ulusar, Intelligence in the internet of medical things era: a systematic review of current and future trends, Computer Communications 150 (2020) 644–660.

[3] R.P. Singh, M. Javaid, A. Haleem, R. Vaishya, S. Ali, Internet of medical things (IoMT) for orthopaedic in COVID-19 pandemic: roles, challenges, and applications, Journal of Clinical Orthopaedics and Trauma 11 (4) (2020) 713–717.

[4] S. Swayamsiddha, C. Mohanty, Application of cognitive internet of medical things for COVID-19 pandemic, Diabetes & Metabolic Syndrome: Clinical Research Reviews 14 (5) (2020) 911–915.
[5] A. Ghubaish, T. Salman, M. Zolanvari, D. Unal, A. Al-Ali, R. Jain, Recent advances in the internet-of-medical-things (IoMT) systems security, IEEE Internet of Things Journal 8 (11) (2020) 8707–8718.
[6] M.A. Rahman, M.S. Hossain, An internet-of-medical-things-enabled edge computing framework for tackling COVID-19, IEEE Internet of Things Journal 8 (21) (2021) 15847–15854.
[7] M. Morrison, G. Lăzăroiu, Cognitive internet of medical things, big healthcare data analytics, and artificial intelligence-based diagnostic algorithms during the COVID-19 pandemic, American Journal of Medical Research 8 (2) (2021) 23–36.
[8] S. Jain, M. Nehra, R. Kumar, N. Dilbaghi, T. Hu, S. Kumar, et al., Internet of medical things (IoMT)-integrated biosensors for point-of-care testing of infectious diseases, Biosensors and Bioelectronics 179 (2021) 113074.
[9] V. Malamas, F. Chantzis, T.K. Dasaklis, G. Stergiopoulos, P. Kotzanikolaou, C. Douligeris, Risk assessment methodologies for the internet of medical things: a survey and comparative appraisal, IEEE Access 9 (2021) 40049–40075.
[10] D. Połap, G. Srivastava, A. Jolfaei, R.M. Parizi, Blockchain technology and neural networks for the internet of medical things, in: IEEE INFOCOM 2020-IEEE Conference on Computer Communications Workshops (INFOCOM WKSHPS), IEEE, July 2020, pp. 508–513.
[11] A.E. Hassanien, A. Khamparia, D. Gupta, A. Slowik, Cognitive Internet of Medical Things for Smart Healthcare, Springer, New York, 2021.
[12] M.A. Jabbar, S.K. Shandilya, A. Kumar, S. Shandilya, Applications of cognitive internet of medical things in modern healthcare, Computers & Electrical Engineering 102 (2022) 108276.
[13] M.A. Scrugli, D. Loi, L. Raffo, P. Meloni, An adaptive cognitive sensor node for ECG monitoring in the Internet of Medical Things, IEEE Access 10 (2021) 1688–1705.
[14] G. Manogaran, M. Alazab, H. Song, N. Kumar, CDP-UA: cognitive data processing method wearable sensor data uncertainty analysis in the internet of things assisted smart medical healthcare systems, IEEE Journal of Biomedical and Health Informatics 25 (10) (2021) 3691–3699.
[15] B.D. Deebak, F.H. Memon, S.A. Khowaja, K. Dev, W. Wang, N.M.F. Qureshi, In the digital age of 5G networks: seamless privacy-preserving authentication for cognitive-inspired internet of medical things, IEEE Transactions on Industrial Informatics (2022).
[16] G. Hatzivasilis, O. Soultatos, S. Ioannidis, C. Verikoukis, G. Demetriou, C. Tsatsoulis, Review of security and privacy for the internet of medical things (IoMT), in: 2019 15th International Conference on Distributed Computing in Sensor Systems (DCOSS), IEEE, 2019, pp. 457–464.
[17] K. Tang, W. Tang, E. Luo, Z. Tan, W. Meng, L. Qi, Secure Information Transmissions in Wireless-Powered Cognitive Radio Networks for Internet of Medical Things, Security and Communication Networks, 2020, p. 2020.
[18] P. Thakur, G. Singh, Security and interference management in the cognitive-inspired internet of medical things, in: Intelligent Data Security Solutions for e-Health Applications, Academic Press, 2020, pp. 131–149.
[19] M.U. Alam, R. Rahmani, Cognitive internet of medical things architecture for decision support tool to detect early sepsis using deep learning, in: International Joint Conference on Biomedical Engineering Systems and Technologies, Springer, Cham, February 2020, pp. 366–384.
[20] H. Bedekar, G. Hossain, A. Goyal, Medical analytics based on artificial neural networks using cognitive internet of things, in: Fog Data Analytics for IoT Applications, Springer, Singapore, 2020, pp. 199–262.
[21] G. Thamilarasu, A. Odesile, A. Hoang, An intrusion detection system for internet of medical things, IEEE Access 8 (2020) 181560–181576.
[22] S. Kumar, A.K. Arora, P. Gupta, B.S. Saini, A review of applications, security and challenges of internet of medical things, in: Cognitive Internet of Medical Things for Smart Healthcare, 2021, pp. 1–23.
[23] Y. Sun, F.P.W. Lo, B. Lo, Security and privacy for the internet of medical things enabled healthcare systems: a survey, IEEE Access 7 (2019) 183339–183355.
[24] M. Papaioannou, M. Karageorgou, G. Mantas, V. Sucasas, I. Essop, J. Rodriguez, D. Lymberopoulos, A survey on security threats and countermeasures in internet of medical things (IoMT), Transactions on Emerging Telecommunications Technologies 33 (6) (2022) e4049.
[25] L. Sun, X. Jiang, H. Ren, Y. Guo, Edge-cloud computing and artificial intelligence in internet of medical things: architecture, technology and application, IEEE Access 8 (2020) 101079–101092.
[26] K. Wei, L. Zhang, Y. Guo, X. Jiang, Health monitoring based on internet of medical things: architecture, enabling technologies, and applications, IEEE Access 8 (2020) 27468–27478.
[27] N.M. Thomasian, E.Y. Adashi, Cybersecurity in the internet of medical things, Health Policy and Technology 10 (3) (2021) 100549.
[28] X. Li, H.N. Dai, Q. Wang, M. Imran, D. Li, M.A. Imran, Securing internet of medical things with friendly-jamming schemes, Computer Communications 160 (2020) 431–442.
[29] W.N. Ismail, M.M. Hassan, H.A. Alsalamah, G. Fortino, CNN-based health model for regular health factors analysis in internet-of-medical things environment, IEEE Access 8 (2020) 52541–52549.
[30] E.K. Wang, C.M. Chen, M.M. Hassan, A. Almogren, A deep learning based medical image segmentation technique in Internet-of-Medical-Things domain, Future Generation Computer Systems 108 (2020) 135–144.
[31] M.K. Kagita, N. Thilakarathne, T.R. Gadekallu, P.K.R. Maddikunta, A review on security and privacy of internet of medical things, in: Intelligent Internet of Things for Healthcare and Industry, 2022, pp. 171–187.
[32] T. Zhang, A.H. Sodhro, Z. Luo, N. Zahid, M.W. Nawaz, S. Pirbhulal, M. Muzammal, A joint deep learning and internet of medical things driven framework for elderly patients, IEEE Access 8 (2020) 75822–75832.
[33] T. Saba, K. Haseeb, I. Ahmed, A. Rehman, Secure and energy-efficient framework using internet of medical things for e-healthcare, Journal of Infection and Public Health 13 (10) (2020) 1567–1575.

[34] I.V. Pustokhina, D.A. Pustokhin, D. Gupta, A. Khanna, K. Shankar, G.N. Nguyen, An effective training scheme for deep neural network in edge computing enabled internet of medical things (IoMT) systems, IEEE Access 8 (2020) 107112–107123.
[35] S. Razdan, S. Sharma, Internet of medical things (IoMT): overview, emerging technologies, and case studies, IETE Technical Review (2021) 1–14.
[36] R.J.S. Raj, S.J. Shobana, I.V. Pustokhina, D.A. Pustokhin, D. Gupta, K.J.I.A. Shankar, Optimal feature selection-based medical image classification using deep learning model in internet of medical things, IEEE Access 8 (2020) 58006–58017.
[37] A.K. Kaushik, J.S. Dhau, H. Gohel, Y.K. Mishra, B. Kateb, N.Y. Kim, D.Y. Goswami, Electrochemical SARS-CoV-2 sensing at point-of-care and artificial intelligence for intelligent COVID-19 management, ACS Applied Bio Materials 3 (11) (2020) 7306–7325.
[38] M.A. Mujawar, H. Gohel, S.K. Bhardwaj, S. Srinivasan, N. Hickman, A. Kaushik, Nano-enabled biosensing systems for intelligent healthcare: towards COVID-19 management, Materials Today Chemistry 17 (2020) 100306.
[39] R. Gupta, A. Kumar, S. Kumar, A.K. Pinnaka, N.K. Singhal, Naked eye colorimetric detection of *Escherichia coli* using aptamer conjugated graphene oxide enclosed gold nanoparticles, Sensors and Actuators B: Chemical 329 (2021) 129100.
[40] S. Shrivastava, T.Q. Trung, N.E. Lee, Recent progress, challenges, and prospects of fully integrated mobile and wearable point-of-care testing systems for self-testing, Chemical Society Reviews 49 (6) (2020) 1812–1866.
[41] J.P.A. Yaacoub, M. Noura, H.N. Noura, O. Salman, E. Yaacoub, R. Couturier, A. Chehab, Securing internet of medical things systems: limitations, issues and recommendations, Future Generation Computer Systems 105 (2020) 581–606.

Index

'*Note:* Page numbers followed by "f " indicate figures and "t" indicate tables.'

A

Abdominal organs, 234
Accuracy (ACC), 126
  metrics, 126
  of RT-PCR, 343
Actinic keratosis, 76
Action detection (AD), 315
Action recognition (AR), 315
Active contours, 124–125
Active Instance Transfer (AIT), 330
Acute exacerbations of chronic obstructive pulmonary disease (AECOPD), 190
AdaBoost, 294–295
Adenocarcinoma, 131
Adult-onset diabetes. *See* Type 2 diabetes
Advanced metering infrastructure (AMI), 181
Adverse drug event (ADE), 267
Agility, 92
Alcohol-based sanitizer, 148
Algorithmic fairness and biases, 83
Alveoli, 333
Alzheimers disease (AD), 195, 229, 251
Ambient temperature parameter (ABC), 115
American College of Radiology, 85
American criminal justice system, 305–306
American Medical Association, 85
American trypanosomiasis, 271
Amino acids, 267
Annual Health Survey (AHS), 330
Antigen detection for COVID-19 vaccine, 150–151
Antimicrobials, 232–233
Antiviral drugs, 149
Anxiety, 289, 296
Application program interface (API), 182
Artificial intelligence (AI), 2, 40–41, 81, 95, 98, 115, 132–133, 154, 187, 203, 216, 225, 231, 245–246, 279, 315, 329, 357, 359, 365, 367
  AI-based building management and information system with multi-agent topology for energy-efficient building, 115
  algorithms, 2
  clinical decision support and predictive analytics, 85–89
    drug cycle in approval of pathway discovery, 85f
    faster drug screening in future, 87
    machine learning predictions rely on input data, 87
    natural language processing translate EHR jargon for patients, 86–87
  connection between quantifiable construction and function, 87–89
    healthcare eco-system in analytical decision point, 89f
    prescriptive modeling, 87–89, 90f
    support for clinical decision–making and predictive analytics, 87
  drug discovery and precision medicine with deep learning, 84–85
  ensuring transparency, explain ability, and intelligibility, 83–84
    algorithmic fairness and biases, 83
    concerns regarding privacy, 84
    data availability, 83–84
    concerning fair treatment and preconceptions in algorithms, 84f
    privacy concerns, 83
  healthcare using, 97–98
  natural language processing in drug, 89–91, 90f
  in neural network, 82f
  predictive analytics wide range of practical applications, 91–92
  privacy and security challenges, 82–83
  technologies, 185
  will challenge status quo in healthcare, 81–82
Artificial neural networks (ANNs), 41–42, 71, 133, 152, 189, 232, 249, 263–264, 278–280, 294–295, 346
Asthma, 89–90, 190
  attack, 369
Asthma management system (AMS), 190
AstraZeneca vaccine, 149
Atomic convolutional neural network (ACNN), 266
Atrial fibrillation (AF), 187
AUC, 350
Augmented reality (AR), 50–51
Autism spectrum disorder (ASD), 251–252
Autoencoders (AEs), 245–246, 249, 278
Automated machine learning (AutoML), 330
Automatic BCI system, 10–11
Automatic Ingestion Monitor (AIM), 50–51
Autonomous robotics, 184
Auxiliary classifier GAN (ACGAN), 75

B

Bacteria, 265
Bacteriocins, 265
Bag of Words (BoW), 219
Bag-of-Visual-Words (BoVW), 293
Bag–of–feature (BOF), 333
Bayesian networks (BNs), 133, 191–193
"Bennett recognition ratio" algorithm, 86–87
Bi-directional feature pyramid network (Bi-FPN), 318–319
Big data, 225, 367
  management, 226–227
Big social data, 286–289
Big-data clinical trials (BCT), 193–195
Binary trees, 205–206
Biobank, 237–238
  compiled from data on data from 10 largest biobanks in world, 238t
Biodiversity, 113
  action of, 113–114
  smart houses and smart buildings, 113–114
Biofuel, 112
Bioinformatics, 271
Biomedical text mining
  algorithms, 219–221
    case study, 221, 221f
    Naïve Bayes, 220–221
    random forest, 220
  natural language processing, 216–217
  system architecture of primary diagnosis of diseases, 217–221
    data collection, 217
    feature extraction, 218–219
    flow chart of primary diagnosis of diseases based on symptoms, 218f
    lemmatization and stemming, 218
    sample data subset of library, 219t
Biometric parameters, 177
Biosensors, 92
Black box, 329
  algorithms, 81
  systems, 81–82
Blockchain, 177
  technology, 364
Blood pressure (BP), 251–252
Blood–brain barrier (BBB), 266–267
Bluetooth, 245–246
Bluetooth low energy (BLE), 54, 365
Boltzmann machines (BMs), 40–41

Brain cancer, 250–251
Brain–computer interface (BCI), 10–11
Breast cancer, 251

C

Calorie estimation, 57–58
Cancer prognosis, 133
application in cancer prognosis and survival, 190
Cancerous diseases, 132
Carbon content, 180
Carbon dioxide emission (CO_2 emission), 110
Carbon power, 110–112
benefits, 112
rehabilitation strategies impact on environment, 110–112
European green deal website, 111t
Cardio vascular medicine, application in, 190
Cardiovascular diseases (CV diseases), 11, 187
Cardiovascular disorders, 3
Cardiovascular problems, 11
Cartesian coordinate system, 125
Categorization scale theory of Fitzpatrick skin type, 121
Cauchy-Schwarz inequality, 30
Center of Excellence (COE), 144
Centers for Disease Control and Prevention (CDC), 145–147, 362–363
Centralization, 177
Chagas disease (CD), 271
Chain codes, 120, 122–123
formulation of, 123
Chang Gung Memorial Hospital, 132
Chaotic salp swarm algorithm (CSSA), 345
Cheminformatics, 267
Chromatin immune precipitation (ChIP-Seq), 271
Chronic diseases, 85–86
application of DLM in patient care plan in chronic disease management, 189
Chronic obstructive pulmonary disease (COPD), 190, 226, 333
Classification of EHR, 331
Climate change disclosure laws, 112–113
Clinical assessment manuals, 291
Clinical decision support (CDS), 81–82
Clinical decision–making, 188
application of DLM in clinical decision-making in diagnosis of different diseases, 188–189
support for clinical decision–making and predictive analytics, 87
Clinical imaging, 234–236
schematic of convolutional neural network, 235f
Clinical outcome, 189
Clinical therapy, 343
Clinical trial management, 193–194
Cloud computing, 367
Clustering technique, 102
Coal crisis, 107–108
Cognitive Internet of Medical Things (CIOMT), 359
uses, 359–361
application areas of CIOMT for tackling COVID-19, 360f
CIOT as special case of IOT, 360f
Cognitive Internet of Things (CIOT), 358–361
cognitive internet of medical things, 359, 360f
uses of CIOMT, 359–361
Cognitive radio (CR), 359
Cognitive radio IOT (CRIOT), 184
Collaborative filtering (CF), 191–193
Colon Cancer Gene, 134
Colon data, 136–137
Colorectal cancer (CRC), 131
classification, 135–136
k-nearest neighbor, 135
random forest, 136
support vector machine, 135
methodology, 133–136
dimensionality reduction tool, 134
experimental dataset, 134
proposed system design, 134f
performance evaluation metrics, 136
related works, 132–133
research tool, 136
results, 136–140, 139f–140f
K-nearest neighbor principal component analysis, 138f
before PCA, 141t
after PCA, 141t
PCA with support vector machine, 138f
principal component analysis, 137f
ROC curve of K-nearest neighbor, 141f
Community well-being urgency in Kazakhstan, waste management technologies and technologies strengthen readiness for, 114
Composite, 330
Compound annual growth rate (CAGR), 168, 172
Computational deep learning frameworks for health monitoring, 3–4
details of deep learning frameworks data related to health informatics, models and intelligent applications, 4t–5t
Computational methods, 81
Computed tomography (CT), 234, 245–246, 249, 344, 361
images, 344
machines, 343
scan, 167
texture analysis, 132–133
Computer systems, 277
resources, 367
Computer vision, 185
algorithms, 119
Computer-aided diagnosis model (CAD model), 251, 345
Computer-assisted instruction (CAI), 231
Computer-assisted radiology (CAR), 90–91
Conditional random fields (CRF), 250–251
Conditional skin lesion synthesis, 75–76
Confidence interval (CI), 132–133
Confusion matrix, 140, 324, 338, 340f
Contact tracing, 230
Contrast Limited Adaptive Histogram Equalization (CLAHE), 119, 121–122
Contrastive divergence approach (CDA), 6
Convolution layer (CL), 346–347
Convolutional neural networks (CNN), 2, 53, 119, 147, 189, 232, 245–248, 266, 272, 293, 316, 344
Corona warriors, 165
Corona woes, 165
Coronavirus (COVID-19), 113–114, 143–144, 147–149, 155, 208, 228, 245–246, 305–306, 343, 362
acceptable approved cases of integrating biodiversity in COVID-19 and rehabilitation programs response to, 114
case detection, 179
Drugs Management Cell, 170
global technology developments to quickly treat, 179
high-risk groups, 149
IOT applications to fight, 179–181
IOT challenges in aftermath of, 181–182
IOT processes for combatting, 178
step-up flow for using IOT to curb COVID-19, 178f
IOT impacts in relation to COVID-19 concerns, 178
literature, 144–147
symptoms of COVID-19, 145–147
materials and methods, 152–154, 152f
artificial neural network, 152
model of neuron, 153–154
pandemic, 95, 143–144, 163–164, 170, 179, 285, 357, 367
IOT during, 177
IOT for curbing spread of COVID-19, 178f
IOT-enabled healthcare helpful during, 368–369
IOT's primary benefits for, 366
machine learning forecasting model for COVID–19 pandemic in India, 115
significant applications of IOT for, 366
technologies of IOT for healthcare during, 366–367
patients, 179–180
pneumonia, 344
problems of COVID-19 battle, 183–185
renewable volumes auction continue to break records, with minor delays caused by, 108
research gaps and motivation, 144
results, 155–158, 155f–156f
top10 states, 158, 160f
testing kits, 170
treatment, 164
vaccination, 228–229
vaccine against COVID-19, 149–151

Vaccine Intelligence Network, 149
Correlation coefficient operators
application example of medical diagnosis, 32–34
Fermatean fuzzy data of five diseases, 32t
Fermatean fuzzy medical information of patients, 33t
Fermatean fuzzy sets and, 24–27
new Fermatean fuzzy correlation operators, 27–32
Covaxin, 151, 170
COVID-19 mortality risk prediction model (COVID-19 MRPMC), 251–252
Covishield vaccine, 151, 167, 170
Criminal database, 305
Criminal investigation problem
case study of, 310–311
criminal persons provided proposed system, 311f
Criminal patient database, 307
Cross stagepartial network (CSPNet), 317
Cross validation (CV), 281
Cross-auto encoder technique, 293
Cryptographic algorithms, 182
Cryptographic protocols, 364
Customer relationship management (CRM), 363
Cutting-edge model sizing technique, 318–319
Cutting-edge technology, 358, 366
Cyclostationary detection, 184

D

Data acquisition spectacles, 54–55
Data calculation process, 279
Data concentrator unit (DCU), 181
Data mining techniques, 132
Data preprocessing, 99, 281
total number of times supported for given inputs, 281t
Data quirkiness, 46
Data science, 286, 296–297
Data size, 46
Data source, 45
Data storage, 176
Data-driven based optimization models for energy-efficient, comparative analysis of, 115
Data-driven systems, 203
daVinciTM system, 315
De-novo drug design (DNDD), 269
Decision process, 206
Decision systems, 208, 212
Decision tree (DT), 99, 132, 205–206, 251, 294–295
algorithm, 279
classifier, 219, 282
Decision-making, 203, 349
systems, 208–209, 364, 368
Decision-making module (DMM), 319
Deep belief networks (DBN), 40–41, 245–246, 272
Deep brain network, 46
Deep CNN, 119
Deep Convolution Classification Model (DCCM), 344
deep convolution classification model-based COVID-19 chest CT image, 343–344, 344f
experiment, 349–350
dataset description, 349
experimental setup, 349–350, 350t
experimental results, 350–354, 350t
classification performance metrics, 350–353, 351f
comparison of DCCM with existing works, 354, 354t
proposed architecture and workflow, 345–347, 346f
convolution layer, 346–347
dense layer, 347
dropout layer, 347
eleven-layered architecture, 347
flatten layer, 347
image augmentation, 346, 346t
maxpooling layer, 347
related works, 345
Deep convolutional GAN (DCGAN), 75–76
Deep feed forward neural network, 245–246
Deep generative models, 86–87
Deep learning (DL), 1, 40–41, 70, 81, 90–91, 95, 97, 132, 188, 225–226, 245–252, 248f, 250f, 263–264, 279, 296–297, 316, 329, 343–344. *See also* Machine learning (ML)
action of biodiversity, 113–114
advanced architectures and core concepts of deep learning in smart health, 5–6
advantages of deep learning in smart medical healthcare analytics, 9–10
intelligent healthcare system, 10f
algorithms, 189, 329
applications for disease prediction, 10–12
approaches of deep learning for drug target, 267–269
architecture, 272
for arrhythmia detection and phenol typing, 190
biobank, 237–238
carbon power, 110–112, 110f
case study, 114
acceptable approved cases of integrating biodiversity in response to COVID-19 and rehabilitation programs, 114
securing Georgia's forest by space, 114
waste management technologies and technologies strengthen readiness for community well-being urgency in Kazakhstan, 114
climate change disclosure laws, 112–113
clinical imaging, 234–236, 234f
coal crisis creates need for alternatives, 107–108
comparative analysis of deep learning frameworks for different disease detection, 6
deep learning models efficiency values through applying on different diseases, 7t–9t
computational deep learning frameworks for health monitoring, 3–4
contact tracing and pandemic management, 230
data analysis and statistical modules, 227–228
deep learning and advancement, 227–228
deep learning and immunization management, 228
drug discovery and precision medicine, 84–85, 236–237, 237f
discrimination and unequal treatment, 84–85
production of data and availability, 85
supervision of quality, 85
EfficientDet, 318–319
EfficientNet, 319
electronic health records, 2f, 232–234
employment creation part of sustainable recovery, 109–110
organic agriculture ability to create jobs, 109–110
experiment, 320
dataset, 320, 321f, 322t
training environment, 320, 322t
findings and motivation, 107
future challenges of deep learning in smart health diagnosis and treatment, 15
future of deep learning in healthcare, 92
future pandemic preparedness, 114–115
concerned legal wildlife trade in order to prevent next pandemic, 115
genomics, 236
green infrastructure measures in legislature, 108–109
hospital, clinical research, and healthcare facilities, 231–232
architecture of main deep learning models, 233f
illustration of deep neural network, 232f
pillars of machine learning for healthcare sector, 233f
IoT in medical and healthcare, 247–252, 248f, 250f
deep autoencoder architecture, 251f
recurrent neural network architecture, 249f
restricted Boltzmann machine architecture, 250f
review of literature, 246–247
subfields of artificial intelligence, 248f
limitations of deep learning frameworks, 15
literature review, 315–317
machine learning and deep learning, 225–226
with minor delays caused by COVID-19, renewable volumes auction continue to break records, 108
models, 190–191
neural networks, 230
online healthcare education, 231

Deep learning (DL) (*Continued*)
and opportunities in healthcare, 226–227
big data management, 226–227
deep learning opportunities in healthcare system, 227f
proposed approach, 317
surgical action detection, 317f
proposed ensemble network, 319
public health misinformation, 229–230
recent literature, 115
AI-based building management and information system with multi-agent topology for energy-efficient building, 115
comparative analysis of data-driven based optimization models for energy-efficient, 115
machine learning forecasting model for COVID-19 pandemic in India, 115
remote healthcare, 230
research and development, 14
result, 320–324
ablation study, 320, 322t
comparison with state-of-art approaches, 320–324, 323f, 327f
role in COVID-19 vaccination, 228–229
role in future disease prediction, 229
smart buildings, 112
telemedicine, 230–231
transfer learning, 296–297
virtual health care platforms, 231
YOLOv5, 317
Deep learning machine (DLM), 187
application in
burden of disease, 189f
cardio vascular medicine, 190
clinical decision-making in diagnosis of different diseases, 188–189
patient care plan in chronic disease management and rehabilitation, 189
effectiveness of DLM in screening and referral of clinical cases in remote and poor access areas, 195
facilitated diagnosis and treatment in healthcare, 193
facilitated research and development in healthcare, 193
role in mental health support and personalized care, 191–193
role in population level future disease prediction, 194–195
Deep neural networks (DNNs), 3, 41, 228, 232, 245–246, 249, 266, 345
Deep reinforcement learning, 269
Deep tensor imaging (DTI), 266
Deep–Q learning, 39, 43
experimental results, 44–45
daily rewards for treatment policies TP1 to TP4 after testing phase, 45f
daily rewards for treatment policies TP1 to TP4 after training phase, 45f
performance comparisons with current popular methods, 46f
future directions, 45–46
proposed mechanism, 42–44
architecture, 43f
input values with linked feature set, 44t
old Q-values replaced with new Q-values, 44f
output Q-values with action for agent, 44t
related work, 40–42
Dengue virus, 269
Denial of service (DOS), 176
Denoising Auto encoder (DAE), 249
Dense layer, 347
Density-based spatial clustering of applications (DBSCAN), 295
Depression, 289, 296
Detect routine activities (DCNN), 11–12
Detection, 181
Dia-Glass, 49–50
diabetes, 51–52
related works, 50–51
result and analysis, 59–64
dataset, 59–60
performance analysis, 60–64
system methodology, 52–58
calorie estimation, 57–58
data acquisition through spectacles, 54–55
food recognition using faster R-CNN, 55–57
information rendering and user notification, 58
schematic framework of Dia-Glass, 54f
user information insertion module, 53
Diabetes, 3, 51–52, 330
categories, 51
concerns, 51–52, 52t
prevalence, 52
prediction of diabetes prevalence over several age groups, 53f
Diabetic patients, 39
Diagnosis, 203, 225
Diagonal matrix, 125
Dice similarity coefficient (DSC), 125
Digital footprints, 286
Digital newsletters, 228
Digital society data sets, 305–306
Digital telehealth, 179–180
Digital transition, 95
Dimensionality reduction tool, 134
Discovery process, 263–264
Diseases, 215
application of DLM in clinical decision-making in diagnosis of different, 188–189
deep leering applications for disease prediction, 10–12
categories of recent research conducted in medical diagnosis and differentiation applications, 13t
classification of newly conducted researches in home-based and personal health care applications, 14t
classification of recent studies in disease prediction applications, 12t
deep learning in health care domains and major objectives, 11t
system architecture of primary diagnosis of, 217–221
Disjunctive normal form (DNF), 203
Dropout layer, 347
Drug development process, 263–264
Drug discovery, 2, 236–237
approaches of machine and deep learning for drug target, 267–269
accelerating de-novo drug design against novel proteins using deep learning, 270f
evolutionary chemical binding similarity method, 268f
linear discriminant analysis and novel descriptors for automatic change detection, 270f
schematic view of DD pipeline, 268f
with deep learning, 84–85
literature review, 265–267
practical applications of QSAR-based virtual screening, 270–272
omics data types and disease research methodologies, 271f
process, 271
toolset and methods based on machine learning, 264t
Drug molecules, 265
Drug screening in future, 87, 89f
Drug target, approaches of machine and deep learning for, 267–269

E

E-commerce, 92, 278
E-health records frameworks, 46
E-healthcare systems, 40–42, 207, 230
E-mental health, 296, 299
Economic growth, 108
Edge computing–based smart health care system
efficient health care system with improved performance, 99–102
dataset, 99–102
predictive analysis and machine learning workflow, 100f
suggested approach in smart healthcare, 99f
healthcare using artificial intelligence, 97–98, 98f
literature, 96–97
research carried out using machine learning, 96f
EfficientDet, 318–319
architecture, 318f–319f
EfficientNet, 319
architecture, 319
Ejection fraction (EF), 187
Elaboration Likelihood Model (ELM), 230
Electric power, 113
Electricity service providers (ESPs), 181
Electro dermal activity (EDA), 3
Electrocardiogram (ECG), 252–253

Electrocardiography, 179–180
Electromyography (EMG), 252–253
Electronic actuators, 208
Electronic health records (EHRs), 40–41, 227, 232–234, 245–246, 329
Electronic health systems, 95
Electronic medical record (EMR), 188, 226, 252–253, 263–264
Eleven-layered architecture, 347, 348f
Emotional well-being, 191–193
Employment creation part of sustainable recovery, 109–110
Endoscopic Surgeon Action Detection dataset (ESAD), 316
  dataset, 320
Endoscopic surgery, 315
Energy detection, 184
Energy efficiency of buildings, 113
Energy transition, 109
Energy-efficient building
  AI-based building management and information system with multi-agent topology for, 115
  comparative analysis of data-driven based optimization models for, 115
Environmental sensors, 92
Eosin (E), 132
  eosin-stained sections, 133
Epi-informatics, 267
Epidermal growth factor receptor (EGFR), 132
Epilepsy, diagnosis of, 249–250
Eradication, 179
Ethereum, 364
Euclidean distance method, 102, 124–125
European Economic Area (EEA), 285
Evolutionary chemical binding similarity (ECBS), 267–268
Evolutionary clustering, 267–268
Evolutionary computation (EC), 267–268
Extra Trees classifier (ET classifier), 281
Eye disease, 251

F

18F-fluorodeoxyglucose (18F-FDG), 132
Facebook, 285–288
False negative (FN), 62, 76
False positive (FP), 62, 76, 338
Familial hypercholesterolemia, 206
Faster regions with the CNN (Faster R –CNN), 345
Fatty liver disease, 194
Feature engineering, 293, 294f
Feature extraction, 218–219
  representation of index values, 219t
  with labels, 220t
Feature pyramid network (FPN), 316
Fermatean fuzzy correlation coefficient (FFCC), 24
Fermatean fuzzy correlation operators, 27–32
  computational example, 32
Fermatean fuzzy data, 32
Fermatean fuzzy distance measure (FFDM), 23
Fermatean fuzzy environment, 26
Fermatean fuzzy linguistic set (FFLS), 23
Fermatean fuzzy number (FFN), 25
Fermatean fuzzy operators, 26
Fermatean fuzzy sets (FFS), 23–27
  computational values, 27t
  existing correlation operators for, 26–27
  graphical representation of, 25f
Filter-based approaches, 280
Financial domain, 278
Financial fraud, 280
Flatten layer, 347
Fluorescence emission, 365
Fog computing–based smart health care system
  efficient health care system with improved performance, 99–102
    dataset, 99–102
    predictive analysis and machine learning workflow, 100f
    suggested approach in smart healthcare, 99f
  healthcare using artificial intelligence, 97–98, 98f
  literature review, 96–97
  research carried out using machine learning, 96f
Fog-based technology, 95
Follow the Leader (FTL), 209
Food and Drug Administration (FDA), 85, 149
Food recognition using faster R-CNN, 55–57
Foreign Direct Investment (FDI), 167
Forensic science approach, 305–306
Forum membership, 290
Fossil energy, 108
Fully convolutional neural networks (FCNN), 250–251
Fuzzy C-means clustering (FCM clustering), 132, 345
Fuzzy knowledge graph, 305–306
Fuzzy set theory (FST), 305
Fuzzy sets, 24, 305

G

Gain ratio, 293
Gated recurrent unit (GRU), 248–249
Gaussian function, 125
Gaussian Mixture Models (GMM), 316
Generative adversarial networks (GANs), 73, 90–91, 269
  for skin lesion synthesis, 73–75, 74f–75f
Generative instance transfer distribution, 330
Genetic algorithm (GA), 115
Genome-wide association studies (GWASs), 237–238
Genomics, 236
Genomics technologies, 188–189
Geolocation, 183–184
Georgia's forest by space, 114
Gestational diabetes, 51
Global coronavirus crisis, 182–183
Global environmental challenges, 107
Global GHG emissions, 110
Global health, 228–229
Global Health Security Index (GHS Index), 169
Global Positioning System (GPS), 252–253, 365
Global vector representations (GloVe), 191–193
Google, 318–319
Government hospitals, 164
Gradient boosting methods (GBM), 265
Granger test, 345
Green economy
  account of power situation in Iceland, 108
  approaches within direction of, 108–109
  NEOM, 108–109
  Saudi Arabia, Line, 109
  Saudi Arabia's Vision 2030, 108
Green energy, 110
Green infrastructure measures in legislature, 108–109
  approaches within direction of green economy, 108–109
Greenhouse gas (GHG), 109
  emissions, 108
Gross domestic product (GDP), 163–164
Gross national income (GNI), 164
GSA-DenseNet121-COVID-19, 345

H

Health, 245–246
  catastrophe, 359
  systems, 208, 231
    decision-making, 190–191
    process, 85
Health Information System (HIS), 311
Health information technology (HIT), 90
Healthcare, 39, 81–82
  AI will challenge status Quo in, 81–82
  using artificial intelligence, 97–98
  authorities, 361
  background study, 280
  component
    diagnostics, 168
    hospitals, 166–167, 166f
    medical devices and equipment, 167
    medical insurance, 168–169
    medical tourism, 169
    pharmaceutical industry, 167
    telemedicine, 168
  data, 91
    analytic sand modeling, 190–191
  data preprocessing, 281
  deep learning and opportunities in, 226–227
  efforts made to lessen potential dangers to healthcare organizations' security, 91–92, 92f
  facilities, 231–232
  future of deep learning in, 92
  industry, 188, 226
  industry, 227, 277, 329
  information and management system society, 172
  of machine learning models, 330

Healthcare (*Continued*)
management systems, 12
materials and methods, 280–282
working model, 280
working model of proposed work, 281f
methods, 331–333
image data processing, 331
object recognition, 331–333, 332f
partitioning of datasets, 281
related work, 330
remote, 230
sampling of collected data, 281
sectors, 225, 358
standard machine learning models, 282
standard methods for learning, 333–338
confusion matrix, 338
data collection, 334
experimental results, 334–338, 335f–336f, 338f
survey of frequently used algorithms for machine learning, 279–280
artificial neural network algorithm, 279–280
decision tree algorithm, 279
random forest algorithm, 279
SVM algorithm, 280
systems, 3, 11, 83, 163, 175, 187, 215–216, 226, 252–253
efforts for healthcare system development, 164
IOT in, 176–177
technologies of IOT for healthcare during COVID-19 pandemic, 366–367
Healthcare professionals (HCPs), 232–233
Healthy tissue, 338
Heart disease, 251
Heating–cooling systems, 115
Hematoxylin (H), 132
Hepatic diseases, application in, 190
Hidden Markov model (HMM), 191–193, 316
Hierarchical RNN architecture, 296
High-pass filter (HPF), 122
High-throughput screens (HTS), 236
Hospitalizations, 179–180
Hospitals, 163–164
clinical research, 231–232
Human identification, 305
Human languages, 216–217
Human–robot interaction (HRI), 3
Hybrid automated segmentation techniques, 120
Hybrid Deep Transfer Learning framework, 330
Hybrid methods, 125, 133
Hybrid–Chain code Euclidean Active Contour (h-CEAC), 119
EDRS and active contours, 124–125
feature extraction and chain codes, 122–123, 123f
formulation of chain code, 123
method, 122
pixel interdependence, 124
pixel neighborhoods, 124f
segmentation, 122–125
Hypercholesterolemia, 206
Hypertension (HTN), 187, 330

I

Iceland, account of power situation in, 108
Iesintra retinal cystoid fluid (IRC), 190
iHome smart dental Health-IoT system, 330
Image analysis, 293
Image augmentation, 346
Image contouring process, 124
Image data, 331
general analysis approach, 331f
processing, 331
Image enhancement, 121–122
Image processing, 343
Imbalanced credit card dataset, 280
Immobile stroke patients suffer infections, 194
Immune disease, 89–90
Immune system, 280, 333
Immune-checkpoint inhibitors (ICIs), 234
Immunization management, deep learning and, 228
Immunology detection for COVID-19 vaccine, 150–151
In silico technique, 131
Inception C-net (IC-Net), 235
Inception score (IS), 90–91
Indeterminacy degree
for IFS, 24
for PFS, 25
India
economic stability, 108
machine learning forecasting model for COVID-19 pandemic in, 115
Indian Council of Medical Research (ICMR), 150
Indian healthcare system, 164, 172
Indian pharmaceutical industry, 167
Industrial, Scientific, and Medical (ISM), 183
Industrial Internet of Things (IIOT), 180–181
Infectious diseases, 252–253
Inference matrix relation, 308
Infinite queuing system, 216
Information system, 83
with multi-agent topology for energy-efficient building, 115
Information technology (IT), 215, 365
Informational energy, 26
Inorganic compounds, 87
Instagram, 287–288
Instance-based learning, 135
Insulin-dependent diabetes. *See* Type 1 diabetes
Intelligent health monitoring system, 11
Intelligent healthcare, 99
Intelligent telemedicine diagnosis systems, 230
Intensive Care Unit (ICU), 163–164, 187
Intensive computational processes, 269
International community, 109
International Diabetes Federation (IDF), 49
International Renewable Energy Agency, 109
Internet of Healthcare Things (IOHT), 175–176, 180–181, 247
global technology developments to quickly treat COVID-19 cases, 179
IOT
applications to fight COVID-19, 179–181
challenges in aftermath of COVID-19, 181–182
during COVID-19 pandemic, 177
in healthcare systems, 176–177
impacts in relation to COVID-19 concerns, 178
processes for combating COVID-19, 178
problems of COVID-19 battle, 183–185
SWOT, 182–183
Internet of Medical Things (IOMT), 176, 357–358, 361–363
cognitive internet of things, 358–361
ecosystem of medicine for highly infectious diseases, 357–358
emerging technologies, 364–365
idea behind IOMT, 362–363
IOT implementation in medical sector, 358
IOT-enabled healthcare helpful during COVID-19 pandemic, 368–369
IOT's primary benefits for COVID-19 pandemic, 366
POC biosensing for infectious illnesses with IOMT assistance, 365
significant applications of IOT for COVID-19 pandemic, 366
smart e-healthcare, 363–364
technologies of IOT for healthcare during COVID-19 pandemic, 366–367
wireless technologies IOMT, 363
Internet of things (IoT), 95, 99, 172, 175, 208, 226, 230, 245–246, 357. *See also* Cognitive Internet of Things (CIOT)
applications, 175, 252–253
deep learning applications for Internet of things, 257f
to fight COVID-19, 179–181
illustrates IoT-linked devices used in COVID-19 and in healthcare system, 254t–256t
challenges in aftermath of COVID-19, 181–182
during COVID-19 pandemic, 177
explosion, 96
in healthcare systems, 176–177
implementation in medical sector, 358
process chart of IOT implementation in medical field, 358f
IOT-enabled healthcare helpful during COVID-19 pandemic, 368–369
IoT-enabled hygiene monitoring devices, 362
IoT-enabled wearable systems, 252–253
IoT-oriented s-fitness programs, 96

overall impacts in relation to COVID-19 concerns, 178
primary benefits for COVID-19 pandemic, 366
processes for combating COVID-19, 178
significant applications of IOT for COVID-19 pandemic, 366
system, 42
Intrusion detection, 183–184
Intuitionistic fuzzy number (IFN), 24
Intuitionistic fuzzy set (IFS), 23, 305
Ischemic and structural heart disease imaging, 190
Isolation beds, 170

J

Johns Hopkins University, 315
Just-in-time adaptive interventions (JITAI), 191–193

K

K-means clustering algorithm, 216
K-nearest neighbor (kNN), 135–136, 188–189, 191–193
Kaggle (public websites), 280
Kazakhstan, waste management technologies and technologies strengthen readiness for community well-being urgency in, 114
Keratinocyte carcinomas, 234
Kirstenrat sarcoma viral oncogene homolog (V-Ki-ras2), 133
Knot tying, 316, 319
Kohonen map-based segmentation method, 315
Kristen rat sarcoma (KRAS), 132–133

L

Laboratory information management systems, 252–253
Lancet Commission, 285
Lazy learning, 135
Learning, standard methods for, 333–338
Ledger system, 177
Left atriums enlargement (LAE), 187
Left ventricles (LV), 245–246
Left ventricular dysfunction (LVD), 187
Left ventricular hypertrophy (LVH), 188–189
Legal wildlife trade in order to prevent next pandemic, 115
Legislature, green infrastructure measures in, 108–109
Lemmatization, 218
Lesion indexing network (LIN), 70
Life First Emergency Traffic Control (LIFE), 180
Ligand pharmacophore modeling, 269
Ligand-based virtual screening, 267
Light-weight index generation, 97
Linear binary pattern histogram (LBPH), 247
Linear discriminant analysis (LDA), 269, 291, 316
Linear model, 132
Linear regression, 291–292
Linguistic indicators, 285–286
Logistic regression, 282
Long short-term memory (LSTM), 6, 41–42, 155, 248–249, 269, 345
Low-income and middle-income countries (LMICs), 226
Lymph node metastasis (LNM), 133

M

Machine algorithms, 131
Machine learning (ML), 2, 40–41, 81, 131–132, 144, 155, 187, 225–226, 245–246, 263–264, 277, 329, 345, 357. *See also* Deep learning (DL)
algorithms, 95, 203, 220, 225–226, 293
approaches of machine learning for drug target, 267–269
forecasting model for COVID-19 pandemic in India, 115
predictions rely on input data, 87
survey of frequently used algorithms for, 279–280
techniques, 286
Machine learning clinical decision support systems (ML CDSS), 208
Machine vision systems, 120
Machine-to-machine (M2M), 180–181
Magnetic resonance imaging (MRI), 167, 229, 234, 245–246
Malaria, 270
Malignant melanomas, 234
futurescope, 126–127
materials and methods, 120–122
datasets, 120–121, 121f
image enhancement, 121–122
motivation and contribution, 119–120
paper organization, 120
proposed methodology, 122–125
h-CEAC segmentation, 122–125
preprocessing phase, 122
workflow of our proposed methodology, 122f
related works, 120
results, 125–126
automated segmentation results in malignant melanoma, 127f
comparison with existing algorithms, 126, 128t
segmented image results dice coefficient, 126t
Malignant skin lesions, 119, 122
Malignant tissue, 338
Mammalian cells, 271
Managed print service (MPS), 265
Mann–Whitney U tests, 293
Matching filter (MF), 184
Maternal mortality ratio, 330
Maxpooling layer, 347
Mean absolute error (MAE), 90–91
Mean square error (MSE), 156
Medical image analysis, 226
Medical imaging and deep learning (MIDL), 320
Medical insurance of healthcare component, 168–169
Medical IOT (mIOT), 362
Medical oxygen, 163–164
Medical practice management software (MPM), 227
Medical Subject Headings (MeSH), 216
Medical tourism, 169
Medical tourism of healthcare component, 169
Medical waste, 114
Medicine for highly infectious diseases, ecosystem of, 357–358
Medicines, 170
Melanocytes, 69–70, 119
Melanocytic tumor lesion, 119
Melanoma, 74, 92, 119
diagnosis, 69–70
skin cancer, 71
skin lesion, 126
Mental disorder prediction, 291
Mental disorders, 286–289
common mental disorders definition, 287t
research on e-mental health social media, 288f
social media overtime, 288f
Mental Disorders Fact Sheet, 285
Mental health (MH), 191–193, 285
diagnostics and classifications, 298
therapy, 97
Mental state detection, 298
Merkle Root, 177
Meta-analysis, 132–133
Meter Data Management System (MDMS), 181
Metric tons per day (MTPD), 163–164
MicroRNAs (miRNA), 131
Microsatellite instability (MSI), 132
Microsoft COCO datasets, 317–318
Mild cognitive impairment (MCI), 251
Millimeter wave (MM wave), 183–184
Minimally Invasive Surgery (MIS), 315
MinMaxScaler class, 137
Mitral regurgitation (MR), 187
Mixed reality (MR), 50–51
Mobile communication protocols, 245–246
Molecular docking using ML, 267
Molecular dynamics simulations, 269
Monitor human health (MHMS), 247
Mortality rates, 147
Multi-agent topology for energy-efficient building, AI-based building management and information system with, 115
Multi-head cross-attention (MHCA), 316–317
Multi-task recurrent neural network (MTRNN), 316–317

Multilayer perceptron (MLP), 152, 235, 251, 316
Multimodal fusion, 234
Multiomics methodologies, 271
Multiple sclerosis (MS), 250
Multiscale fusion feature pyramid network (MSF-FPN), 316

N

N-95 masks, 170
N-gram language models, 291–292
Naive Bayes (NB), 191–193, 221, 265–266
  algorithm, 204–205
  approach, 220–221
  classifier, 282
Nanotechnology, 112
Narrow-band imaging (NBI), 132–133
Narrowband IoT (NB-IoT), 97
National Board for Financial Development, 108
National Center for Biotechnology Information (NCBI), 236
National Dairy Development Board (NDDB), 150
National Health Service (NHS), 145–147
National Institute for Health and Care Excellence (NICE), 192
National Institutes of Health (NIH), 87, 236
Nations healthcare system, 229
Natural forests, 114
Natural language processing (NLP), 89–90, 176, 216–217, 228, 245–246, 266, 320–323
  in drug, 89–91, 90f
  EHR jargon for patients, 86–87, 88f
  prediction process in, 217f
  schematic diagram of, 217f
Near Field Communication (NFCs), 362
Needle passing, 316, 319
Neural network algorithm (NNA), 115
Neural networks (NNs), 12, 40–41, 81, 132, 152, 225, 251–252, 269, 334
  models, 345
  techniques, 135
Neural organization, 347
Neurodegenerative diseases, 195
Neurons, 153, 265
  model of, 153–154
  recurrent neural network, 153–154
    artificial intelligence's vital role in COVID-19 prediction, 154f
    flow chart of long-short-term memory, 153f
Neuroscience, 97
Neutron, 347
New Enterprise Operating Model (NEOM), 108–109
New Future, 108–109
Nitric oxide, 150–151
No Regret Deep Learning (NRDL), 211
Non healthcare systems, 329
Non maximum suppression technique (NMS technique), 319
Non-insulin-dependent diabetes. *See* Type 2 diabetes
Nonalcoholic fatty liver disease (NAFLD), 190
Nonalcoholic steato-hepatitis (NASH), 190
Nongovernmental organizations (NGOs), 165
Nonlinear optimization technique, 115
Nonparametric learning, 135
No–regret deep learning framework
  aspects of disease detector for clinical decision-making, 205f
  experimental results, 211–212
    comparative analysis of popular algorithms used for clinical decision-making, 211f
    plot of learning rate *vs.* size of dataset, 211f
  findings, 206
  future directions, 212
  mechanism of reinforcement learning, 204f
  motivation, 207
  operational logic of decision tree, 203f
  proposed mechanism, 208–210
  related works, 207–208, 207f
  research gaps, 206
Novel coronavirus, 114
Novel hybrid segmentation approach, 120
Nucleic acid, 343, 365
  nucleic acids-based vaccines, 149

O

Object detection algorithms, 318
Object recognition, 331–333, 332f
Obsessive-compulsive personality disorder (OCPD), 193–194
Oil-based fuel, 108
Omicron, 172
Omics, 271
  information, 272
Online healthcare education, 231
Openflow Configuration protocol (OF-CONFIG), 364–365
Ophthalmology, application in, 190
Optical character recognition (OCR), 228
Optical coherence tomography (OCT), 251
Organic agriculture ability to create jobs, 109–110
  brief explanation, 110
Outpatient door (OPD), 164
Overfitting, 46
Oxygen pumps, 208

P

Pandemic
  concerned legal wildlife trade in order to prevent next, 115
  emergencies, 165
  management, 230
Paper organization, 120
Parkinson's disease (PD), 83–84, 249
Particle emission tomography (PET), 132
Particle swarm optimization (PSO), 115
Pascal VOC datasets, 317–318
Path aggregation network (PANet), 317
Path-based fingerprint fingerprint (PF2), 266
Pathogens, 144
Patient-facing systems, 9
Peak signal-to-noise ratio (PSNR), 90–91
Pearson correlation coefficients, 183–184, 293
Pediatric health, 330
Performance evaluation metrics, 136
Personal health records (PHR), 227
PH2 data sets, 120–121
Pharmaceutical industry of healthcare component, 167
Pharmaceutical stakeholders, 10
Physically unclonable function (PUF), 364
Picture Dataset (Picture DTs), 133, 137
Picture fuzzy database, 305–307
  for applications, 307–308
Picture fuzzy equivalence, 307
Picture fuzzy set (PFS), 305–307
  experimental results, 310–311
    case study of criminal investigation problem, 310–311
    evaluation, 311, 312f
  proposed model, 308–310
    and application, 309–310, 309t–310t
    process steps in summary of, 308
  research background, 306–308
    picture fuzzy database for applications, 307–308
    terms and definitions, 306–307
Picture fuzzy tolerance relation, 308
Picture fuzzy tuple, 308
Pixel cell, 124
Pneumonia, 333–334
Point to point correlation method, 331
Point-of-care (POC), 362–363
  biosensing for infectious illnesses with IOMT assistance, 365
Population level future disease prediction, role of DLM in, 194–195
Post-COVID-19 Indian healthcare system
  complacency, 169–170
  corona warriors and woes, 165
  creation of robust healthcare system, 163
  efforts for healthcare system development, 164
  future of Indian healthcare system, 172
  healthcare component
    diagnostics, 168
    hospitals, 166–167, 166f
    medical devices and equipment, 167
    medical insurance, 168–169
    medicaltourism, 169
    pharmaceutical industry, 167
    telemedicine, 168
  pandemonium scenes, 163–164
  providing treatment to all amidst difficulties, 165
  SWOT analysis of Indian healthcare industry in post-COVID-19 era, 171
  transformation of Indian healthcare sector post COVID-19, 165–166
    growth trend of India's healthcare sector, 167f
    imports and exports in India'smedical devices sector, 168t
    India's telemedicine market, 169f
Posts annotation, 291
Power efficiency, 113–114
Precision, 350
Precision medicine, 236–237
  with deep learning, 84–85

Prediction
algorithms, 293–297
deep learning, 296–297
reinforcement learning, 296
semi-supervised learning, 295–296
supervised learning, 293–295
unsupervised learning, 295
models, 147
Predictive analytics, 87, 91–92
efforts made to lessen potential dangers to healthcare organizations' security, 91–92, 92f
future of deep learning in healthcare, 92
predictive modeling in healthcare, 91f
Predictive models, 290
Prescriptive modeling, 87–89, 90f
Pressure swing absorption (PSA), 170
Primary diagnosis of diseases, system architecture of, 217–221
Primary user (PU), 184
Principal Component Analysis (PCA), 137
Privacy concerns, 83–84
Probabilistic and dynamic neural inference (PRIME), 265
Projector-Kinect calibration data, 315
Proteins, 266
Proteochemometric modeling (PCM), 266–267
Psychological well-being, 191–193
Psychometric test, 291
Public data, 289
Public health misinformation, 229–230
Public healthcare centers (PHCs), 164
Pythagorean fuzzy correlation coefficient, 24
PYTHON, 136

Q

Quality of service (QOS), 181, 359
Quantitative structure property relationship (QSPR), 265
Quantitative structure–activity relationship (QSAR), 265, 269
QSAR-based virtual screening, practical applications of, 270–272
Quantum chemistry methods, 87

R

Radial basis function (Rbf), 294–295
Radio frequency (RF), 266
Radio frequency identification (RFID), 179, 252–253, 362
Radiographic scans, 229
Radiology Images, 245–246
Random forest (RF), 99, 132, 136–137, 191–193, 220, 250, 294–295
algorithm, 279
classifier, 282
Random tree functions, 282
Randomized clinical/control trials (RCTs), 193–194
Real-time polymerase chain reaction (RT-PCR), 343
Recall, 350
Received signal strength indicators (RSSIs), 183–184
Recognition-by-components (RBC), 333
Recombinant proteins, 149
Recommender systems (RS), 191–193
Rectified linear unit (RELU), 346–347
Recurrent neural networks (RNNs), 2, 40–41, 143–144, 153–154, 232, 248–249, 272, 324
Reddit, 290
Region of interest (ROI), 56
pooling layer, 57
Region Proposal Network (RPN), 55–57
Regression, 99
and classification techniques, 99
Rehabilitation
acceptable approved cases of integrating biodiversity in response to COVID-19 and, 114
application of DLM in patient care plan in, 189
strategies, 110–112
Reinforcement learning (RL), 232, 266, 296
Reinforcement Learning for Structural Evolution (ReLeaSE), 266
Remote robot surgery, 363
Renal diseases, application in, 190
Renewable Energy and Energy Efficiency Partnerships (REEEP), 110
Reproducibility, 46
Residual Networks (ResNet), 335
loss in testing, 339f
ResNet50, 71
Resource distribution, 179
Resting-state functional magnetic resonance imaging (RfMRI), 251–252
Restricted Boltzmann machine (RBM), 5, 190–191, 245–246, 249
Retinal artificial intelligence diagnosis system (RAIDS), 189
Reverse-Phase Protein Array, 236
RNA sequencing (RNA-Seq), 271
Robotic-assisted laparoscopy, 315
Robust healthcare system, creation of, 163
ROC curve, 138
Rule-based methods, 305–306
Rule-based system, 218

S

Sanitizers, 170
SARS-CoV-2, 143–144, 151
coronaviruses, 144
pandemic, 359
Saudi Arabia, Line, 109
Saudi Arabia's Vision (2030), 108
Scanned conventional hematoxylin, 133
Schistosoma genus, 270–271
Schistosoma mansoni thioredoxin glutathione reductase (SmTGR), 270–271
Schistosomiasis, 270–271
Schizophrenia, 191–193
Scopus database, 343
Screening and referral of clinical cases in remote and poor access areas, effectiveness of DLM in, 195
Seborrheic keratoses, 234
Segmentation, 120, 123
Self-organized neural networks, 279–280
Semi-supervised learning, 279, 295–296
Sensors, 358, 363
Sentiment analysis, 293
Serum Institute of India (SII), 151, 167
Sexual reproduction, 347
Sharpening process, 122
Short-term memory networks (LSTM), 245–246
Sigmoid function, 334
Simplified molecular-input line-entry system (SMILES), 266
Single health system, 107
Single nucleotide, 188–189
Single nucleotide polymorphism (SNP), 188–189, 236
Skin cancers, 69
Skin cells, 69
Skin lesion classification methodology, 71–72
CNN architecture with modified VGG16, 71–72, 72f
dataset, 71, 71f
Smart Bed, 369
Smart buildings, 112–114
benefits, 113
concise description, 113–114
measures, 113
Smart e-healthcare, 363–364
IOMT architecture, 363–364
Smart fitness (s-fitness), 96
Smart health
advanced architectures and core concepts of deep learning in, 5–6
future challenges of deep learning in smart health diagnosis and treatment, 15
Smart healthcare systems, 246–247
Smart houses, 113–114
benefits, 113
concise description, 113–114
measures, 113
Smart medical healthcare analytics, advantages of deep learning in, 9–10
intelligent healthcare system, 10f
Smart meters (SMs), 181
Smart phones, 366
Smart sensors, 367
Smart watches, 3
Smart wireless network communication environment, 95
Social big data, 288
Social data, 290
analysis, 288–289
configuration, 291–293, 292f
feature engineering, 293, 294f
preprocessing, 292
Social interactions, 296
analysis, 293
Social media, 285–287
analysis, 298–299
assessment strategies, 289–291, 289f
forum membership, 290
posts annotation, 291
questionnaires and surveys, 289–290
self-declared mental health status, 290
evaluation, 297–298
confusion matrix, 298t
mental disorders and big social data, 286–289

Social media (*Continued*)
platforms, 293
prediction algorithms, 293–297
social data configuration, 291–293, 292f
social media-based e-mental health assessment, 286
status, challenges, and future direction, 298–299
Social networks, 285, 293
platforms, 291, 293
Software, 367
Software-defined networking (SDN), 364–365
Solar collection systems, 113–114
Solar electricity, 108
Solar photovoltaic (PV), 108
Spearman rank correlation coefficients, 293
Specificity metrics (SPEC metrics), 126
Spectrum resource, 358–359
Sputnik V, 151
Stacked Auto encoder (SAE), 249
Standard clustering algorithms, 295
Standard deviation (std), 137
Standard machine learning models, 282
cost of failure with given methods, 282f
Standard methods for learning, 333–338
StandardScaler class, 137
State-of-art-method, 126
comparison with, 320–324
Status quo in healthcare, AI will challenge, 81–82
Stemming method, 218
Sternal seism cardiogram (SCG), 3–4
Stochastic Gradient Descent (SGD), 247, 320
Strength weakness threat opportunity (SWOT), 171, 182–183
analysis of Indian healthcare industry in post-COVID-19 era, 171
analysis of IOT from perspective of global pandemic, 182t
Stress, 289
Stroke, 333
Stroke risk protection (SRP), 330
Structural similarity index (SSIM), 90–91
Structure-based virtual screening (SBVS), 267
Subretinal fluid (SRF), 190
Suicidality, 289
Supervised learning (SL), 226, 234, 293–295
distribution of supervised learning techniques in selected articles, 295f
techniques, 99
Supervised ML approach for data collection, 100
Support vector machine (SVM), 132, 135, 137, 188–189, 251, 265, 294–295, 316
algorithm, 280
polynomial kernel, 135
Support vector regression (SVR), 147
Surgeon actions, 316, 319
detection, 315
Surgical activity detection (SAD), 315
Surgical interaction recognition network (SIRNet), 316–317
Surgical maneuvers prediction system, 316
Surgical masks, 170
Surgical robots, 277
Surgical workflow, 316
Sustainable Development Goals (SDG), 361
Sustainable recovery, employment creation part of, 109–110
Suturing, 316, 319
Symptoms of patients, 217
Syndrome coronavirus, 175–176
Synthetic medical image augmentation, 69–71
background and motivation, 69–70
contributions, 70
experimental results, 76–77
dataset evaluation and performance metrics, 76
implementation specifications, 76–77
training and test data, 77
generation of synthetic skin lesions, 73–76
performance comparison, 77, 78t
related works, 70–71
skin lesion classification methodology, 71–72
Synthetic skin lesions
conditional skin lesion synthesis, 75–76
generation of, 73–76
generative adversarial networks for skin lesion synthesis, 73–75
traditional data augmentation, 73, 73f
Systematized Nomenclature of Medicine-Clinical Terms (SNOMED-CT), 226

T

Telehealth, 363
Telemedicine, 97, 168, 172, 230–231, 363
Term frequency (TF), 220
Text mining, 216
Text similarity feature (TSF), 228
The Cancer Genome Atlas (TCGA), 236
Therapeutic techniques, 132
Thermometers, 362–363
Thorax diseases, 235
Three-dimension (3D), 269
accelerometer, 96
spatiotemporal signals, 272
structure of receptors, 269
Three-dimensional printed pharmaceutical (3DP pharmaceutical), 265
Threshold-based algorithm, 320
Thyroid, 206
Traditional data augmentation, 73, 73f
Transfer learning, 2, 296–297
Transformer network, 316
True negative (TN), 62, 76, 338
True positive (TP), 62, 76, 338
Tuebinger Mole Analyzer system, 121
Twitter, 285–288, 290
API, 290
data analysis, 292
Two-dimensional curvelet (2D curvelet), 345
Type 1 diabetes, 51
Type 2 diabetes, 51

U

UCI Machine Learning Repository, 99
UDP-glucuronosyl transferase (UGT), 265
Undirected Graph (UGRNN), 266
Unified Medical Language System (UMLS), 226
Universe set, 306–307
Unsupervised learning (UL), 99, 226, 279, 295

V

Vaccines, 171
approved in India, 151
covaxin, 151
covishield, 151
sputnik V, 151
against COVID-19, 149–151
in COVID-19, 149–150
immunology and antigen detection for COVID-19 vaccine, 150–151
potential vaccine-related threats, 151
Variational Auto encoder (VAE), 249, 269
Vector analysis, 123
Vector transformation technique, 96
VGG-19, 350
Virtual health care platforms, 231
Virtual screening (VS), 266, 270–271
Virus, 147, 361
Voxel image analysis, 120

W

Waste disposal managing system, 114
Waste management technologies and technologies strengthen readiness for community well-being urgency in Kazakhstan, 114
Wearables, 97
BCG sensors, 3
health monitoring devices, 3
networks, 184
Weather, 113–114
Web-based technology, 168
Wi-Fi, 245–246
Wireless local area network (WLAN), 179
Wireless sensor network (WSN), 252–253
Wireless technologies IOMT, 363
World Health Organization (WHO), 145–147, 166, 171, 285

X

X-ray machines, 343

Y

YOLOv5, 317, 319
algorithm, 317
architecture, 318f